TOPOLOGY

VOLUME I

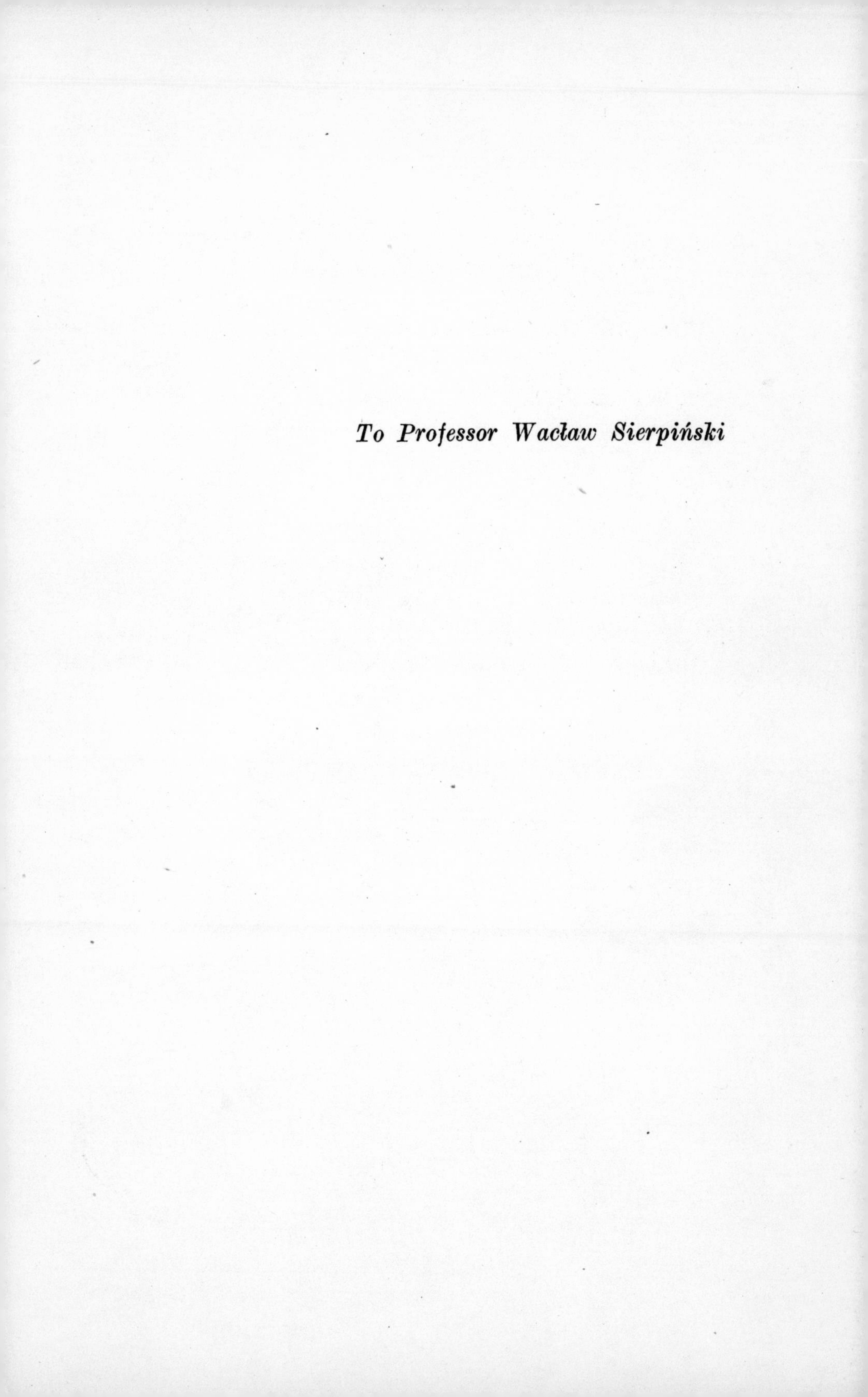

To Professor Wacław Sierpiński

TOPOLOGY

VOLUME I

New edition, revised and augmented

K. KURATOWSKI

Professor of Mathematics, University of Warsaw

Translated from French by
J. Jaworowski

1966
ACADEMIC PRESS
NEW YORK AND LONDON

PWN — POLISH SCIENTIFIC PUBLISHERS
WARSZAWA

This book is a translation of the original French
Topologie, volume I
published by PWN—Polish Scientific Publishers, 1958
in the series "Monografie Matematyczne" edited by
the Polish Academy of Sciences

The English edition of this book has been published by PWN jointly with

ACADEMIC PRESS INC.
111 Fifth Avenue
New York, New York 10003

ACADEMIC PRESS INC. (LONDON) LTD.
Berkeley Square House
Berkeley Square
London W. 1, England

Printed in Poland

PREFACE TO THE FIRST VOLUME

This volume is divided into three chapters.

The first chapter is devoted essentially to general topological spaces. However, more specialized aspects of topological spaces are also examined, such as $\mathscr{T}_1$-spaces, regular, completely regular, and normal spaces. We are also concerned in this chapter with fundamental notions such as base, subbase, cover, and continuous mapping, and with operations such as the cartesian product $\mathscr{X} \times \mathscr{X}$ or $2^{\mathscr{X}}$ (exponential topology) and $\mathscr{X}/\varrho$ (quotient topology).

Great use is made in this chapter (and in the following ones) of closure algebra. For instance, we try—whenever it is possible and useful to do—to express our definitions, theorems, and proofs in terms of Boolean algebra, augmented by the closure operation. The axioms of a topological space are also expressed in the same way.

The second chapter is devoted to the study of metric spaces, starting with more general spaces, having the limit as its primitive notion. In Sections B and C the space is assumed to be metric separable, and this includes problems of cardinality and of dimension. A part of the section which is devoted to dimension theory has the combinatorial aspect (with notions like simplex, complex, polyhedron, etc.). However, algebraic methods (homology and cohomology groups, etc.) are not used, as they would require a special treatment far exceeding the scope of this book.

The last section of Chapter 2 is devoted to the theory of Borel sets, Baire functions, and related topics. Their treatment is motivated by the needs of the general theory of mappings which are not assumed to be continuous; so this section differs from the preceding ones, which have a more geometrical aspect. The reader who is not interested in general function theory may omit this section.

In Chapter 3 we are concerned with complete spaces. The major part of the chapter is devoted to problems of general function theory which can be expressed in topological terms. Here we consider analytic sets, projective sets, and related problems. It is worthwhile noting that these ideas have recently acquired interesting applications in mathematical logics.

Due to the kindness of Professor Mostowski, there is an appendix on applications of topology to mathematical logics at the end of the volume. Another appendix—written by Professor Sikorski—is on the applications of topology to functional analysis.

Let us add that Volume II (as in the case of the French edition) will be devoted to the concepts of compactness, connectedness, and local connectedness, to some problems on retraction, on homotopy, and cohomotopy, and to problems on disconnecting the n-dimensional euclidean spaces with special emphasis on the complex number plane.

A reader familiar with the French edition of this book will certainly notice that a large part of the material concerning metric spaces (contained in both volumes) has been extended to topological spaces, and consequently this has been inserted in the first chapter (thus the first chapter has increased from 63 pages to 150 pages). The author is aware of the fact that this process of extending theorems on metric spaces to topological spaces has not been exhausted. However, in order to reach this limit, one would have to postpone the publication of this treatise for a rather long period of time; this did not seem appropriate to the author or to the editors.

May I add that a large part of the material of Chapter 1 was not contained in the French edition of this volume. In fact, many theorems of §§ 17 and 18 have been shown quite recently (some of them by the author).

The following list contains the best known treatises and textbooks on topology and the author will refer to many of them quite often.

P. Alexandrov, *Combinatorial topology* (in Russian), Moscow 1947.

— *Einfachste Grundbegriffe der Topologie*, Springer, 1932.

—, H. Hopf, *Topologie I*, Berlin 1935.

M. Antonowskii, W. Boltianskii, T. Sarymsakov, *Topological Boolean algebras* (in Russian), Tashkent 1963.

C. Berge, *Espaces topologiques*, Dunod 1959.

N. Bourbaki, *Topologie générale*, Paris, Hermann No. 1045, 1084, 1142, 1143, 1235 and foll., 1949–1962.

S. S. Cairns, *Introductory topology*, New York 1961.

E. Čech, *Topologicke prostory*, Praha 1959.

G. Choquet, *Cours d'analyse*. Tome II. *Topologie*, Paris 1964.

A. Császár, *Fondements de la topologie générale*, Budapest 1960, English edition, Pergamon Press 1963.

S. Eilenberg, N. Steenrod, *Foundations of algebraic topology*, Princeton 1952.

M. Fréchet, *Les espaces abstraits*, Monogr. Borel 1928.

W. Franz, *Allgemeine Topologie I*, Göschen 1960.

H. Hahn, *Reelle Funktionen I*, Leipzig 1932.

—, A. Rosenthal, *Set functions*, Albuquerque 1948.

D. W. Hall, G. L. Spencer, *Elementary topology*, New York 1955.

F. Hausdorff, *Mengenlehre*, Gruyter 1927.

J. G. Hocking, G. S. Young, *Topology*, Addison-Wesley 1961.

W. Hurewicz, H. Wallman, *Dimension theory*, Princeton 1941.

J. L. Kelley, *General topology*, Van Nostrand 1955.

H. J. Kowalsky, *Topologische Räume*, Basel 1961.

K. Kuratowski, *Introduction to set theory and topology*, Warszawa and Oxford 1961.

S. Lefschetz, *Algebraic topology*, Coll. Publ. 1942.

— *Introduction to topology*, Princeton 1949.

Z. P. Mamuzič, *Introduction to general topology*, Nordhoff 1963.

R. L. Moore, *Foundations of point set theory*, Coll. Publ. 1932.

M. H. A. Newman, *Elements of the topology of plane sets of points*, Cambridge 1939.

G. Nöbeling, *Grundlagen der analytischen Topologie*, Springer 1954.

E. M. Patterson, *Topology*, Interscience 1956.

A. Rosenthal, L. Zoretti, *Encyklopädie d. Math. Wiss.* II CGA, Leipzig 1924.

H. Seifert, W. Threlfall, *Lehrbuch der Topologie*, Teubner 1934.

W. Sierpiński, *General topology*, Toronto 1952.

G. F. Simmons, *Introduction to topology and modern analysis*, McGraw-Hill 1963.

H. Tietze, L. Vietoris, *Encyklopädie d. Math. Wiss.* III AB 13, Leipzig 1931.

R. Vaidyanathaswamy, *Treatise on set topology I*, Madras 1947.

G. T. Whyburn, *Analytic topology*, Coll. Publ. 1942.

It is a pleasure to acknowledge my deep indebtedness to Dr. Engelking for his many valuable remarks and discussions.

I wish to thank also Mrs Karłowicz for her help in preparing the manuscript of this volume and I wish to mention the numerous colleagues who have helped me in preparing the four previous French editions of this book, starting with the first in 1933. These are: Čech, Hurewicz, Knaster, Otto, Posament, Marczewski, Zygmund, Sikorski, Rasiowa, Császár, Katětov, Mazur, and Mrówka.

Finally, my thanks go to the Academic Press and to Polish Scientific Publishers and personally to Mr Muszyński for their assistance in the publication of this book.

K. Kuratowski

December 1963
WARSZAWA

CONTENTS

CHAPTER TWO

METRIC SPACES

A. *Relations to topological spaces. $\mathscr{L}^*$-spaces*

B. *Cardinality problems*

C. *Problems of dimension*

CHAPTER THREE

Complete spaces

INTRODUCTION

We shall recall here some notations and elementary theorems of general set theory and the algebra of logic([1]). The concepts of set theory will be used throughout (except the $(\mathscr{A})$-operation which is more specialized). The notation of the algebra of logic will be used whenever it leads to simplifying the arguments (e.g. in §§ 31, 37–40).

§ 1. Operations in logic and set theory

I. Algebra of logic. Let α and β be two *propositions* (*sentences*). Then by $\neg\alpha$ (or α') we denote the *negation* of α (i.e. "non-α") by $\alpha\vee\beta$ the *disjunction* ("α or β") and by $\alpha\wedge\beta$ the *conjunction* ("α and β" written also $\alpha\beta$). Also $\alpha \Rightarrow \beta$ represents α *implies* β, $\alpha \equiv \beta$ means that α is *equivalent to* β.

Let us mention the following theorems:

$\neg\neg\alpha \equiv \alpha$ (the law of double negation);

$(\alpha \Rightarrow \beta) \equiv [(\neg\beta) \Rightarrow (\neg\alpha)]$ (the law of contraposition);

$\neg(\alpha\wedge\beta) \equiv (\neg\alpha)\vee(\neg\beta)$ and $\neg(\alpha\vee\beta) \equiv (\neg\alpha)\wedge(\neg\beta)$ (the de Morgan laws);

$(\alpha \Rightarrow \beta) \equiv [(\neg\alpha)\vee\beta]$;

$[\alpha\wedge(\beta\vee\gamma)] \equiv [(\alpha\wedge\beta)\vee(\alpha\wedge\gamma)]$.

Each proposition has either the value 1 ("true") or the value 0 ("false"). We have

$\alpha\wedge(\neg\alpha) \equiv 0$ (the law of contradiction);

$\alpha\vee(\neg\alpha) \equiv 1$ (the law of excluded middle).

II. Algebra of sets. Let us denote by 1 a given set (considered as the *space* in the sequel). Its elements will be denoted by small letters $a, b, x, y, \ldots$, its subsets by capitals $A, B, X, \ldots$, and families of sets by bold face $\boldsymbol{A}, \boldsymbol{B}, \boldsymbol{X}, \ldots$; $x \in A$ means that x is an element of A (belongs to A). $A \cup B$ denotes the *union* of A and B, i.e. the set composed of elements which belong either to A or to B

([1]) Compare my textbook *Introduction to set theory and topology* (English translation), Pergamon Press—PWN, 1961.

(or to both). $A \cap B$ denotes the *intersection* of A and B, i.e. the set composed of elements which belong simultaneously to A and B. $A-B$ denotes the *difference* of A minus B, i.e. the set composed of elements which belong to A but which are not in B. The *void* set is denoted by 0. $A \subset B$ means that A is a *subset* of B (is *contained* in B). The *complement* of A is the set $-A = 1-A$ (denoted also by A^c). The *quotient* $A:B$ equals $A \cup (-B)$. Finally, the *symmetric difference* $A \dot{-} B = (A-B) \cup (B-A)$.

We have the following equivalences, which show duality between the algebra of logics and the algebra of sets:

$$x \in (-A) \equiv \neg(x \in A), \qquad x \in (A \cup B) \equiv (x \in A) \vee (x \in B),$$

$$x \in (A \cap B) \equiv (x \in A) \wedge (x \in B), \qquad (x \in 1) \equiv 1, \qquad (x \in 0) \equiv 0,$$

$$(A \subset B) \equiv [(x \in A) \Rightarrow (x \in B)], \qquad (A = B) \equiv [(x \in A) \equiv (x \in B)].$$

Let us mention the following formulas:

$$A \cup (-A) = 1, \qquad A \cap (-A) = 0, \qquad -(-A) = A,$$

$$A = (A \cap B) \cup (A-B), \qquad (A \cap B) \subset A \subset (A \cup B),$$

$$\left.\begin{aligned} -(A \cup B) &= (-A) \cap (-B) \\ -(A \cap B) &= (-A) \cup (-B) \end{aligned}\right\} \quad \text{(the de Morgan laws)},$$

$$-(A:B) = B-A, \qquad A:1 = A, \qquad A:0 = 1,$$

$$(A-B) \cup B = A \cup B, \qquad (A:B) \cap B = A \cap B,$$

$$A \dot{-} (B \dot{-} C) = (A \dot{-} B) \dot{-} C, \qquad A \cap (B \dot{-} C) = (A \cap B) \dot{-} (A \cap C),$$

$$(A \subset B) \equiv [(A \cup B) = B] \equiv [(A \cap B) = A] \equiv (A-B = 0),$$

$$(A \subset C) \wedge (B \subset D) \Rightarrow [(A \cup B) \subset (C \cup D)] \wedge [(A \cap B) \subset (C \cap D)],$$

$$(A \subset C) \wedge (B \subset C) \equiv (A \cup B) \subset C,$$

$$(C \subset A) \wedge (C \subset B) \equiv C \subset (A \cap B).$$

The sets A and B are called *disjoint* if $A \cap B = 0$. The set composed of one element a is denoted by (a).

III. Propositional functions. Let $\varphi(x)$ be a propositional function whose variable x ranges over the space 1 (non-void). $\varphi(x)$ means a condition and if x satisfies this condition, $\varphi(x)$ becomes a true proposition; in the opposite case, it becomes a false proposition. So, for instance, $x > 0$ is a propositional function (in the space of real numbers).

$\bigvee_x \varphi(x)$ means: there is an x such that $\varphi(x)$ (i.e. an x satisfying the considered condition).

$\bigwedge_x \varphi(x)$ means: for each x we have $\varphi(x)$ (i.e. each x satisfies the considered condition)(1).

For example:

$$\bigwedge_x (x+1 > x), \qquad \bigvee_x (x^2 = 1).$$

The following formulas can easily be shown:

$$\left.\begin{array}{l} \neg[\bigvee_x \varphi(x)] \equiv \bigwedge_x [\neg\varphi(x)] \\ \neg[\bigwedge_x \varphi(x)] \equiv \bigvee_x [\neg\varphi(x)] \end{array}\right\} \quad \text{(the generalized de Morgan laws)},$$

$$[\bigwedge_x \varphi(x)] \Rightarrow [\bigvee_x \varphi(x)],$$

$$[\bigvee_x \varphi(x)] \vee [\bigvee_x \psi(x)] \equiv \bigvee_x [\varphi(x) \vee \psi(x)],$$

$$[\bigwedge_x \varphi(x)] \wedge [\bigwedge_x \psi(x)] \equiv \bigwedge_x [\varphi(x) \wedge \psi(x)],$$

$$\bigvee_x [\varphi(x) \wedge \psi(x)] \Rightarrow [\bigvee_x \varphi(x)] \wedge [\bigvee_x \psi(x)],$$

$$[\bigwedge_x \varphi(x) \vee \bigwedge_x \psi(x)] \Rightarrow \bigwedge_x [\varphi(x) \vee \psi(x)].$$

The quantifiers $\bigvee$ and $\bigwedge$ are generalizations of $\vee$ and $\wedge$. Namely, if the space 1 is finite, say $1 = (a_0, a_1, \ldots, a_n)$, then

$$\bigvee_x \varphi(x) = \varphi(a_0) \vee \varphi(a_1) \vee \ldots \vee \varphi(a_n),$$

$$\bigwedge_x \varphi(x) = \varphi(a_0) \wedge \varphi(a_1) \wedge \ldots \wedge \varphi(a_n).$$

Obviously, any proposition a may be considered as a propositional function which is identically true or identically false. Thus we have

$$\bigvee_x a \equiv a \equiv \bigwedge_x a, \qquad \bigvee_x [a \wedge \varphi(x)] \equiv a \wedge \bigvee_x \varphi(x),$$

$$a \vee \bigwedge_x \varphi(x) \equiv \bigwedge_x [a \vee \varphi(x)].$$

---

(1) The symbols $\exists^x$ and $\forall^x$ are used also instead of $\bigvee_x$ and $\bigwedge_x$.

IV. The operation $\underset{x}{E}$. We denote by $\underset{x}{E}\varphi(x)$ the set of all x's such that $\varphi(x)$ (¹).

For example, $\underset{x}{E}(x>0)$ is the set of positive numbers. $\underset{x}{E}(x^2=x)$ is composed of two numbers 0 and 1.

The following formulas can easily be shown:

$$t \in \underset{x}{E}\varphi(x) \equiv \varphi(t), \tag{1}$$

$$\underset{x}{E}[\neg\varphi(x)] = -[\underset{x}{E}\varphi(x)], \tag{2}$$

$$\underset{x}{E}[\varphi(x)\vee\psi(x)] = \underset{x}{E}\varphi(x) \cup \underset{x}{E}\psi(x), \tag{3}$$

$$\underset{x}{E}[\varphi(x)\wedge\psi(x)] = \underset{x}{E}\varphi(x) \cap \underset{x}{E}\psi(x). \tag{4}$$

V. Infinite operations on sets. Let T be a given set. Let us assign to each t a set A_t (subset of the given space 1). The sets $\bigcup_t A_t$ and $\bigcap_t A_t$ are defined as follows:

$$(x\in\bigcup_t A_t) \equiv \bigvee_t (x\in A_t), \qquad (x\in\bigcap_t A_t) \equiv \bigwedge_t (x\in A_t). \tag{1}$$

Obviously, if T is finite, say $T=(1,2,\ldots,n)$, we have

$$\bigcup_t A_t = A_1\cup A_2\cup\ldots\cup A_n, \qquad \bigcap_t A_t = A_1\cap A_2\cap\ldots\cap A_n.$$

In the case where T is the set of positive integers, the variable n can be used instead of t and one writes

$$\bigcup_{n=1}^{\infty} A_n \text{ for } \bigcup_t A_t, \quad \text{and} \quad \bigcap_{n=1}^{\infty} A_n \text{ for } \bigcap_t A_t.$$

There are two important countable operations:

$$\operatorname*{Lim\,inf}_{n=\infty} A_n = \bigcup_{n=0}^{\infty}\bigcap_{k=0}^{\infty} A_{n+k} \quad\text{and}\quad \operatorname*{Lim\,sup}_{n=\infty} A_n = \bigcap_{n=0}^{\infty}\bigcup_{k=0}^{\infty} A_{n+k}.$$

If $\operatorname*{Lim\,inf}_{n=\infty} A_n = \operatorname*{Lim\,sup}_{n=\infty} A_n$, the sequence $A_0, A_1, \ldots$ is said to be *convergent*, and one writes

$$\operatorname*{Limes}_{n=\infty} A_n = \operatorname*{Lim\,inf}_{n=\infty} A_n = \operatorname*{Lim\,sup}_{n=\infty} A_n.$$

(¹) The symbol $\{x: \varphi(x)\}$ is frequently used instead of $\underset{x}{E}\varphi(x)$.

By using formula (1) one obtains equivalent formulas to those of Section III, replacing $\varphi(x)$ by A_t, $\bigvee_x$ by $\bigcup_t$ and so on. For example, the generalized de Morgan laws read as follows:

$$-\bigcup_t A_t = \bigcap_t (-A_t), \tag{2}$$

$$-\bigcap_t A_t = \bigcup_t (-A_t). \tag{3}$$

To these the following formulas may be added:

$$\bigwedge_t (A_t \subset B) \equiv [(\bigcup_t A_t) \subset B], \tag{4}$$

$$\bigwedge_t (B \subset A_t) \equiv [B \subset \bigcap_t A_t]. \tag{5}$$

Besides the operations $\bigcup_t A_t$ and $\bigcap_t A_t$ we consider the operations $\mathrm{S}(\boldsymbol{R})$ and $\mathrm{P}(\boldsymbol{R})$ on families of subsets of a given set 1. Namely, $\mathrm{S}(\boldsymbol{R})$ is the union and $\mathrm{P}(\boldsymbol{R})$ is the intersection of all sets belonging to the family $\boldsymbol{R}$, that is

$$[x \in \mathrm{S}(\boldsymbol{R})] \equiv \bigvee_X (x \in X \in \boldsymbol{R}),$$

$$[x \in \mathrm{P}(\boldsymbol{R})] \equiv [\bigwedge_X (X \in \boldsymbol{R}) \Rightarrow (x \in X)].$$

Thus, if $\boldsymbol{R}$ is finite, say $\boldsymbol{R} = (A_1, \ldots, A_n)$, then

$$\mathrm{S}(\boldsymbol{R}) = A_1 \cup \ldots \cup A_n \quad \text{and} \quad \mathrm{P}(\boldsymbol{R}) = A_1 \cap \ldots \cap A_n.$$

If $\boldsymbol{R}$ is void,

$$\mathrm{S}(\boldsymbol{R}) = 0 \quad \text{and} \quad \mathrm{P}(\boldsymbol{R}) = 1.$$

Are the members of the family $\boldsymbol{R}$ represented as A_t where $t \in T$, then $\mathrm{S}(\boldsymbol{R}) = \bigcup_t A_t$ and $\mathrm{P}(\boldsymbol{R}) = \bigcap_t A_t$.

VI. The family of all subsets of a given set. This family will be denoted by $(2^X)_{\mathrm{set}}$, or briefly 2^X, where no confusion can occur (the index is used to distinguish this family from the space of all closed subsets of X which will be considered later). Thus

$$A \in 2^X \equiv A \subset X. \tag{1}$$

It follows by V(4) that

$$(\bigcup_t A_t) \in 2^B \equiv \bigwedge_t (A_t \in 2^B). \tag{2}$$

By V(5):

$$2^{A \cap B} = 2^A \cap 2^B \quad \text{and generally} \quad 2^{\cap A_t} = \bigcap 2^{A_t}, \text{ where } t \in T. \tag{3}$$

Obviously $S(2^X) = X$.

VII. Ideals and filters. A (non-void) family $\boldsymbol{R}$ of subsets of a given set X is called an *ideal* if it is *hereditary* and *additive*, that means if:

$$A \in \boldsymbol{R} \text{ and } B \subset A \text{ imply } B \in \boldsymbol{R}, \tag{1}$$

$$A \in \boldsymbol{R} \text{ and } B \in \boldsymbol{R} \text{ imply } (A \cup B) \in \boldsymbol{R}. \tag{2}$$

An ideal is called a *proper ideal* if $X \notin \boldsymbol{R}$. A proper ideal is called *maximal* if it is contained in no other proper ideal; this is equivalent to saying

$$\text{either } A \in \boldsymbol{R} \text{ or } (-A) \in \boldsymbol{R} \text{ whenever } A \subset X. \tag{3}$$

One can show (with the help of the axiom of choice) that every proper ideal is contained in a maximal ideal.

The notion of a *filter* is dual to that of ideal. Namely a (non-void) family $\boldsymbol{S}$ of sets is called a filter(¹) if

$$A \in \boldsymbol{S} \text{ and } A \subset B \text{ imply } B \in \boldsymbol{S}, \tag{4}$$

$$A \in \boldsymbol{S} \text{ and } B \in \boldsymbol{S} \text{ imply } (A \cap B) \in \boldsymbol{S}. \tag{5}$$

A filter is *proper* if $0 \notin \boldsymbol{S}$. A proper filter is *maximal* (an *ultrafilter*) if it is contained in no other proper filter; this is equivalent to saying

$$\text{either } A \in \boldsymbol{S} \text{ or } (-A) \in \boldsymbol{S} \text{ whenever } A \subset X. \tag{6}$$

Obviously, $\boldsymbol{S}$ is a filter if and only if the family $\boldsymbol{R}$ of sets $(-A)$, where $A \in \boldsymbol{S}$, is an ideal.

A family $\boldsymbol{B}$ of sets is called *centered* if

$$\boldsymbol{B}_1 \subset \boldsymbol{B} \Rightarrow P(\boldsymbol{B}_1) \neq 0 \text{ for finite } \boldsymbol{B}_1.$$

Dual concept can be defined for ideals.

(¹) This concept is due to H. Cartan. See N. Bourbaki, *Topologie Générale*, Chap. I, § 6, No. 1.

§ 2. Cartesian products

I. Definition. The *cartesian product* of the sets X and Y is the set of all ordered pairs $\langle x, y\rangle$ where $x \epsilon X$ and $y \epsilon Y$. This set is denoted by $X \times Y$. Thus we have

$$[\langle x, y\rangle \epsilon (X \times Y)] \equiv (x \epsilon X) \wedge (y \epsilon Y). \tag{1}$$

More generally, given a finite system of sets $X_1, X_2, \ldots, X_n$, their cartesian product $X_1 \times X_2 \times \ldots \times X_n$ is the set of all sequences $\langle x_1, x_2, \ldots, x_n\rangle$ where $x_k \epsilon X_k$ for $k = 1, 2, \ldots, n$.

If all sets $X_1, \ldots, X_n$ are identical, so that $X_k = X$, we denote their cartesian product by X^n.

EXAMPLES. $\mathscr{I}^2$ is the square, $\mathscr{I}^3$ the cube. $\mathscr{E}^2$ is the plane, $\mathscr{E}^n$ the n-dimensional euclidean space.

A cylinder can be considered as the cartesian product of the circumference of the circle (base) and a closed interval (height). The surface of a torus is the cartesian product of the circumferences of two circles.

II. Rules of cartesian multiplication. The following formulas are easily shown with the help of I(1).

$$(X_1 \cup X_2) \times (Y_1 \cup Y_2)$$
$$= (X_1 \times Y_1) \cup (X_1 \times Y_2) \cup (X_2 \times Y_1) \cup (X_2 \times Y_2), \tag{1}$$

$$(X_1 \cap X_2) \times (Y_1 \cap Y_2) = (X_1 \times Y_1) \cap (X_2 \times Y_2), \tag{2}$$

$$(X_1 - X_2) \times Y = (X_1 \times Y) - (X_2 \times Y), \tag{3}$$

$$(X_1 \subset X_2 \text{ and } Y_1 \subset Y_2) \equiv (X_1 \times Y_1 \subset X_2 \times Y_2)$$
$$(\text{if } X_1 \neq 0 \neq Y_1), \tag{4}$$

$$A \times B = (A \times Y) \cap (X \times B) \quad (\text{where } A \subset X \text{ and } B \subset Y), \tag{5}$$

$$-(A \times B) = \big((-A) \times Y\big) \cup \big(X \times (-B)\big), \tag{6}$$

$$[(X_1 \times Y_1) = (X_2 \times Y_2)] \Rightarrow [X_1 = X_2 \text{ and } Y_1 = Y_2], \tag{7}$$

if all the factors are nonvoid.

Let us note the following rules concerning infinite unions and intersections:

$$\bigcup_s A_s \times \bigcup_t B_t = \bigcup_{st} A_s \times B_t, \tag{8}$$

$$\bigcap_s A_s \times \bigcap_t B_t = \bigcap_{st} A_s \times B_t. \tag{9}$$

III. Axes, coordinates, and projections. Given two sets X and Y, we call them—like in analytical geometry—*axes* of the product $X \times Y$. Each element $\mathfrak{z}$ of $X \times Y$ being of the form $\mathfrak{z} = \langle x, y \rangle$, we call x and y the *coordinates* of $\mathfrak{z}$ (its abscissa and ordinate). x and y are *projections* of $\mathfrak{z}$ into the axes. More generally, the projection of $\mathfrak{A} \subset X \times Y$ on the X-axis is the set of abscissas of $\mathfrak{A}$; hence it is the set

$$\underset{x}{E} \bigvee_{y} [\langle x, y \rangle \in \mathfrak{A}].$$

IV. Propositional functions of many variables. A propositional function $\varphi(x, y)$ of two variables $x \in X$ and $y \in Y$ can be considered as a propositional function of one variable $\mathfrak{z} = \langle x, y \rangle \in (X \times Y)$. The formulas of § 1, III hold true for functions of two (or more) variables if x is replaced by $\langle x, y \rangle$ (or $\langle x, y, z \rangle$, etc.). Furthermore:

$$\bigvee_{xy} \varphi(x, y) \equiv \bigvee_{x} \bigvee_{y} \varphi(x, y) \equiv \bigvee_{y} \bigvee_{x} \varphi(x, y),$$

$$\bigwedge_{xy} \varphi(x, y) \equiv \bigwedge_{x} \bigwedge_{y} \varphi(x, y) \equiv \bigwedge_{y} \bigwedge_{x} \varphi(x, y),$$

$$\bigvee_{x} \varphi(x) \wedge \bigvee_{x} \psi(x) \equiv \bigvee_{xx^*} [\varphi(x) \wedge \psi(x^*)],$$

$$\bigwedge_{x} \varphi(x) \vee \bigwedge_{x} \psi(x) \equiv \bigwedge_{xx^*} [\varphi(x) \vee \psi(x^*)],$$

$$\bigvee_{x} \varphi(x) \vee \bigwedge_{x} \psi(x) \equiv \bigvee_{x} \bigwedge_{x^*} [\varphi(x) \vee \psi(x^*)] \equiv \bigwedge_{x^*} \bigvee_{x} [\varphi(x) \vee \psi(x^*)],$$

$$\bigvee_{x} \varphi(x) \wedge \bigwedge_{x} \psi(x) \equiv \bigvee_{x} \bigwedge_{x^*} [\varphi(x) \wedge \psi(x^*)] \equiv \bigwedge_{x^*} \bigvee_{x} [\varphi(x) \wedge \psi(x^*)],$$

$$\bigvee_{x} \bigwedge_{y} \varphi(x, y) \Rightarrow \bigwedge_{y} \bigvee_{x} \varphi(x, y).$$

If x and y range over the same space $X = Y$, we have (replacing $\Rightarrow$ by $\rightarrow$):

$$\begin{array}{ccccccc}
 & \nearrow & \bigvee_{x} \bigwedge_{y} \varphi(x, y) & \rightarrow & \bigwedge_{y} \bigvee_{x} \varphi(x, y) & \searrow & \\
\bigwedge_{xy} \varphi(x, y) & \rightarrow & \bigwedge_{x} \varphi(x, x) & \rightarrow \!\!\!\!\times & \bigvee_{x} \varphi(x, x) & \rightarrow & \bigvee_{xy} \varphi(x, y). \\
 & \searrow & \bigvee_{y} \bigwedge_{x} \varphi(x, y) & \rightarrow & \bigwedge_{x} \bigvee_{y} \varphi(x, y) & \nearrow &
\end{array}$$

EXAMPLE. The continuity of a function at each point x is expressed by the following condition

$$\bigwedge_{\varepsilon}\bigwedge_{x}\bigvee_{\delta}\bigwedge_{h}(|h|<\delta)\Rightarrow(|f(x+h)-f(x)|<\varepsilon)$$

where the domain of variability of ε and δ is the set of positive numbers and that of x and h is $\mathscr{E}$.

If we interchange the order of the quantifiers $\bigwedge_{x}$ and $\bigvee_{\delta}$, we obtain the condition for *uniform* continuity of f.

Remark. It can be noticed that the sign of implication in the formula

$$\bigvee_{y}\bigwedge_{x}\varphi(x,y)\Rightarrow\bigwedge_{x}\bigvee_{y}\varphi(x,y)$$

cannot be replaced by the equivalence sign.

However there is the following formula

$$\bigwedge_{x}\bigvee_{y}\varphi(x,y)\equiv\bigvee_{f}\bigwedge_{x}\varphi[x,f(x)]$$

where f is a mapping of X into Y (see § 3, I).

V. Connections between the operators E and $\bigvee$.

$$\mathop{E}_{x}\bigvee_{y}\varphi(x,y)=\bigcup_{y}\mathop{E}_{x}\varphi(x,y). \tag{1}$$

Proof. By § 1, IV(1), we have $t\in\mathop{E}_{x}\bigvee_{y}\varphi(x,y)\equiv\bigvee_{y}\varphi(t,y)$. Put $A_y=\mathop{E}_{x}\varphi(x,y)$. Hence $\varphi(t,y)\equiv t\in\mathop{E}_{x}\varphi(x,y)\equiv t\in A_y$. It follows by § 1, V(1) that

$$\bigvee_{y}\varphi(t,y)\equiv\bigvee_{y}(t\in A_y)\equiv t\in\bigcup_{y}A_y\equiv t\in\bigcup_{y}\mathop{E}_{x}\varphi(x,y).$$

Similarly:

$$\mathop{E}_{x}\bigwedge_{y}\varphi(x,y)=\bigcap_{y}\mathop{E}_{x}\varphi(x,y). \tag{2}$$

THEOREM [1]. *The set $\mathop{E}_{x}\bigvee_{y}\varphi(x,y)$ is the projection of the set $\mathop{E}_{xy}\varphi(x,y)$ on the X-axis.*

---

[1] Cf. E. Schröder, *Vorlesungen über die Algebra der Logik*, vol. III, 1890–1905, p. 52. See also the paper of A. Tarski and myself *Les opérations logiques et les ensembles projectifs*, Fund. Math. 17 (1931), p. 243.

Proof. Let $A = \underset{xy}{E}\varphi(x, y)$. Hence

$$\varphi(x, y) \equiv (\langle x, y\rangle \in A) \quad \text{and} \quad \underset{x}{E}\underset{y}{\bigvee}\varphi(x, y) = \underset{x}{E}\underset{y}{\bigvee}(\langle x, y\rangle \in A).$$

This completes the proof (see III).

VI. Multiplication by an axis. Let us note the following formulas:

$$\underset{xy}{E}\varphi(y) = X\times\underset{y}{E}\varphi(y), \tag{1}$$

$$\underset{xy}{E}[\varphi(x)\vee\psi(y)] = [\underset{x}{E}\varphi(x)\times Y]\cup[X\times\underset{y}{E}\psi(y)] \tag{2}$$

(compare § 1, IV(3)),

$$\underset{xy}{E}[\varphi(x)\wedge\psi(y)] = [\underset{x}{E}\varphi(x)\times Y]\cap[X\times\underset{y}{E}\psi(y)] = \underset{x}{E}\varphi(x)\times\underset{y}{E}\psi(y) \tag{3}$$

(compare II(5)).

EXAMPLE 1. Let $\mathscr{G}$ be the set of integers and $\mathscr{R}$ the set of rational numbers. Then we have

$$\mathscr{R} = \underset{x}{E}\underset{y}{\bigvee}\underset{z}{\bigvee}[(y\in\mathscr{G})\wedge(z\in\mathscr{G})\wedge(xz = y)\wedge(z\neq 0)].$$

Put

$$\mathfrak{A} = \underset{xyz}{E}[(y\in\mathscr{G})\wedge(z\in\mathscr{G})\wedge(xz = y)\wedge(z\neq 0)].$$

By virtue of § 1, IV(4), $\mathfrak{A}$ is the intersection of four sets:

(i) the set $\underset{xyz}{E}(y\in\mathscr{G}) = X\times\underset{y}{E}(y\in\mathscr{G})\times Z$, i.e. the union of planes parallel to $X\times Z$ crossing the Y-axis at integer points,

(ii) the set $\underset{xyz}{E}(z\in\mathscr{G})$, equally composed of planes,

(iii) the hyperbolic paraboloid $y = xz$,

(iv) the space $X\times Y\times Z$ minus the plane $z = 0$.

According to the theorem in Section V, we obtain the set $\mathscr{R}$ by projecting first $\mathfrak{A}$ on the plane $X\times Y$ and then projecting this projected set on the X-axis.

EXAMPLE 2. Let $f_1, f_2, \ldots$ be a sequence of continuous functions of the real variable x. The set C of points of convergence of this sequence is by definition

$$C = \underset{x}{E} \underset{k}{\bigwedge} \underset{n}{\bigvee} \underset{m}{\bigwedge} [|f_{n+m}(x) - f_n(x)| \leqslant 1/k].$$

Put

$$A_{n,m,k} = \underset{x}{E}[|f_{n+m}(x) - f_n(x)| \leqslant 1/k],$$

hence

$$C = \bigcap_{k=1}^{\infty} \bigcup_{n=1}^{\infty} \bigcap_{m=1}^{\infty} A_{n,m,k}.$$

From the last formula follows the important property of the set C of being $\boldsymbol{F}_{\sigma\delta}$ (compare § 30, XII).

VII. Relations. The quotient-family. A propositional function of two variables $\varphi(x, y)$ is also called a *relation*; it is often written in the form $x \varrho y$.

If the range of variability of x is the set X, and Y is the range of y, the relation ϱ can be *identified* with the subset $\underset{xy}{E}\, x\varrho y$ of $X \times Y$.

The relation ϱ is called *reflexive* if $x \varrho x$ for each $x \epsilon X$; ϱ is *symmetric* if $x\varrho y$ implies $y \varrho x$; ϱ is *transitive* if $x\varrho y$ and $y\varrho z$ imply $x \varrho z$. A relation which is reflexive, symmetric and transitive is called an *equivalence relation*.

Let ϱ be an equivalence relation between variables x and y having the same range $X = Y$ of variability. It is easily seen that ϱ leads to a decomposition of X ("*induced by* ϱ") into disjoint sets (called *equivalence-sets*) such that two elements x_1 and x_2 belong to the same set if and only if $x_1 \varrho x_2$ (this is called the *principle of abstraction*). The family of all equivalence-sets is called the *quotient-family* X/ϱ.

Conversely, every decomposition $\boldsymbol{D}$ of X (i.e. every family $\boldsymbol{D}$ of disjoint sets such that $S(\boldsymbol{D}) = X$) leads to an equivalence relation ϱ (called *induced by* $\boldsymbol{D}$). Namely

$$x\varrho y \equiv \bigvee_{D} (D \epsilon \boldsymbol{D}) \wedge (x \epsilon D) \wedge (y \epsilon D).$$

VIII. Congruence modulo an ideal. Let $\boldsymbol{I}$ be an ideal. We say that A *is congruent to* B *modulo* $\boldsymbol{I}$ and we write $A \sim B \bmod \boldsymbol{I}$, if $(A \dot{-} B) \epsilon \boldsymbol{I}$; or equivalently if $(A - B) \epsilon \boldsymbol{I}$ and $(B - A) \epsilon \boldsymbol{I}$.

One shows easily the following two statements(1):

(1) *the relation* $A \sim B$ mod $\boldsymbol{I}$ *is an equivalence relation*,

(2) *if* $A_1 \sim B_1$ *and* $A_2 \sim B_2$, *then* $(A_1 \cup A_2) \sim (B_1 \cup B_2)$, $(A_1 \cap A_2) \sim (B_1 \cap B_2)$ *and* $A_1 - A_2 \sim B_1 - B_2$,

(3) $A \sim B$ mod $\boldsymbol{I}$ *if and only if* A *is of the form*

$$A = (B-P) \cup Q \quad \text{where} \quad P \in \boldsymbol{I} \text{ and } Q \in \boldsymbol{I}. \tag{α}$$

In order to prove (3), let us suppose first that (α) is fulfilled. Hence $A - B \subset Q$ and $B - A \subset P$. It follows that $A \sim B$ mod $\boldsymbol{I}$.

Next, assume that $(A-B) \in \boldsymbol{I}$ and $(B-A) \in \boldsymbol{I}$. Put $P = B-A$ and $Q = A-B$. Formula (α) follows at once.

(4) *if* $\boldsymbol{I}$ *is an* σ*-ideal, i.e. if the union of any countable family of sets belonging to* $\boldsymbol{I}$ *belongs to* $\boldsymbol{I}$, *then*:

$$\text{if } A_n \sim B_n \text{ mod } \boldsymbol{I}, \text{ then } (\bigcup_n A_n) \sim (\bigcup_n B_n) \text{ mod } \boldsymbol{I}. \tag{β}$$

This follows from the obvious formulas

$$\bigcup_n A_n - \bigcup_n B_n \subset [\bigcup_n (A_n - B_n)] \in \boldsymbol{I}, \quad \bigcup_n B_n - \bigcup_n A_n \subset [\bigcup_n (B_n - A_n)] \in \boldsymbol{I}.$$

§ 3. Mappings. Orderings. Cardinal and ordinal numbers

I. Terminology and notation. Let f be a *mapping* (a *function*) defined by the relation $y = f(x)$ where x ranges over X (the *domain* of f) and $y \in Y$. Thus

$$f = \underset{xy}{E}\, [y = f(x)] \subset X \times Y, \tag{1}$$

and we say that f is a mapping of X *into* Y. This is written:

$$f\colon X \to Y \quad \text{or} \quad X \xrightarrow{f} Y. \tag{2}$$

The set of all mappings of X into Y is denoted by

$$(Y^X)_{\text{set}} \quad \text{or more concisely} \quad Y^X$$

where no confusion can occur (the index is used to distinguish between this set and the set of continuous mappings which will be considered later).

(1) See e.g. R. Sikorski, *Boolean algebras*, Berlin–Göttingen–Heidelberg 1960, p. 27.

Obviously

$$Y^X \subset 2^{X \times Y}.$$

If each element of Y is a value of f, then f is said to be a mapping of X *onto* Y.

If for a given $f: X \to Y$, the range of variability of x is restricted to a set $E \subset X$, we obtain a *restricted* (a *partial*) function $g = f|E$. Thus

$$g(x) = f(x) \text{ for } x \in E, \quad \text{and} \quad g \subset f. \tag{3}$$

Owing to the last inclusion we call f an *extension* of g.

Let X, Y, and Z be three sets and f and g two mappings such that $f: X \to Y$ and $g: Y \to Z$. Then the *composed* mapping h: $X \to Z$ denoted by $h = g \circ f$ (or briefly $h = gf$) is defined by the formula

$$h(x) = g[f(x)] \quad \text{for} \quad x \in X. \tag{4}$$

The composition of mappings is *associative*, that is

$$(f_2 \circ f_1) \circ f_0 = f_2 \circ (f_1 \circ f_0). \tag{5}$$

Thus $f_2 f_1 f_0$ is well defined, provided $f_n: X_n \to X_{n+1}$ for $n = 0, 1, 2$.

A mapping $f: X \to Y$ is called *one-to-one* if

$$[f(x_1) = f(x_2)] \Rightarrow (x_1 = x_2);$$

in other terms, if

$$(x_1 = x_2) \equiv [f(x_1) = f(x_2)].$$

Obviously, if $f: X \to Y$ and $g: Y \to Z$ are one-to-one, so is $h = g \circ f$.

II. Images and counterimages. Let $f: X \to Y$. We call the *image* of $A \subset X$ the set

$$f^1(A) = \mathop{E}_{y} \bigvee_{x} [(x \in A) \wedge (y = f(x))] \tag{1}$$

(if no confusion can occur, we write $f(A)$ instead of $f^1(A)$).

Conversely, if $B \subset Y$, we call the *counterimage* of B the set

$$f^{-1}(B) = \mathop{E}_{x} [f(x) \in B]. \tag{2}$$

Thus (compare § 2, III), $f(A)$ is the projection on the Y-axis of the set

$$\underset{xy}{E}\big[(x \in A) \wedge \big(y = f(x)\big)\big]$$

and $f^{-1}(B)$ is the projection on the X-axis of the set

$$\underset{xy}{E}[(f(x) \in B) \wedge (y = f(x))] = \underset{xy}{E}[(y \in B) \wedge (y = f(x))].$$

One sees easily that

$$f(A) = \underset{y}{E}[A \cap f^{-1}(y) \neq 0]. \tag{3}$$

By definition,

$$f^1\colon 2^X \to 2^Y \quad \text{and} \quad f^{-1}\colon 2^Y \to 2^X. \tag{4}$$

In particular, f^{-1} is defined for every one-element-set (y) where $y \in Y$. We shall agree to write

$$\overset{-1}{f}(y) = f^{-1}\big((y)\big).$$

Hence

$$\overset{-1}{f}\colon Y \to 2^X, \quad \text{and} \quad [x \in \overset{-1}{f}(y)] \equiv [y = f(x)]. \tag{5}$$

Obviously, if $y \in Y - f(X)$, then $\overset{-1}{f}(y) = 0$.

Consider now the *decomposition* $\boldsymbol{D}_f$ (or briefly $\boldsymbol{D}$) *induced by the mapping* f, i.e. $\boldsymbol{D}_f$ is the family of all sets $\overset{-1}{f}(y)$ where y ranges over $f(X)$. Thus

$$X = \bigcup_y \overset{-1}{f}(y) \quad \text{and} \quad \overset{-1}{f}(y) \cap \overset{-1}{f}(y') = 0 \quad \text{for} \quad y \neq y'. \tag{6}$$

We shall agree to define $\overset{1}{f}$ for $D \in \boldsymbol{D}$ by the formula

$$\big(\overset{1}{f}(D)\big) = f^1(D), \quad \text{hence} \quad \overset{1}{f}\colon \boldsymbol{D} \to Y. \tag{7}$$

Obviously $\overset{1}{f}$ is one-to-one and, if $Y = f(X)$, $\overset{-1}{f}$ is its inverse:

$$\overset{1}{f}\overset{-1}{f} = \text{identity} = \overset{-1}{f}\overset{1}{f} \quad \text{and} \quad \overset{-1}{f}\colon Y \to \boldsymbol{D}. \tag{8}$$

III. Operations on images and counterimages. Let $f: X \to Y$, $A_t \subset X$ and $B_t \subset Y$, where $t \in T$. The following 12 formulas can easily be established:

$$f(A_1 \cup A_2) = f(A_1) \cup f(A_2); \quad (1) \qquad f(\bigcup_t A_t) = \bigcup_t f(A_t); \qquad (1a)$$

$$f(A_1 \cap A_2) \subset f(A_1) \cap f(A_2); \quad (2) \qquad f(\bigcap_t A_t) \subset \bigcap_t f(A_t); \qquad (2a)$$

$$f(A_1) - f(A_2) \subset f(A_1 - A_2); \qquad (3)$$

$$(A_1 \subset A_2) \Rightarrow \big(f(A_1) \subset f(A_2)\big); \qquad (4)$$

$$\big(f(A) = 0\big) \equiv (A = 0); \qquad (5)$$

$$f^{-1}(B_1 \cup B_2) = f^{-1}(B_1) \cup f^{-1}(B_2), \qquad (6)$$

$$f^{-1}(\bigcup_t B_t) = \bigcup_t f^{-1}(B_t); \qquad (6a)$$

$$f^{-1}(B_1 \cap B_2) = f^{-1}(B_1) \cap f^{-1}(B_2), \qquad (7$$

$$f^{-1}(\bigcap_t B_t) = \bigcap_t f^{-1}(B_t); \qquad (7a)$$

$$f^{-1}(B_1 - B_2) = f^{-1}(B_1) - f^{-1}(B_2); \qquad (8)$$

$$(B_1 \subset B_2) \Rightarrow \big(f^{-1}(B_1) \subset f^{-1}(B_2)\big); \qquad (9)$$

$$\big(f^{-1}(B) = 0\big) \equiv \big(B \cap f(X) = 0\big); \qquad (10)$$

$$A \subset f^{-1}f(A); \qquad (11)$$

$$ff^{-1}(B) = B \cap f(X). \qquad (12)$$

$$f(A) \cap B = f[A \cap f^{-1}(B)]; \qquad (13)$$

hence

$$[f(A) \cap B = 0] \equiv [A \cap f^{-1}(B) = 0], \qquad (13')$$

and equivalently

$$[f(A) \subset B] \equiv [A \subset f^{-1}(B)]; \qquad (13'')$$

consequently

$$(f^{-1})^{-1}(2^A) = 2^{-f(-A)}. \qquad (13''')$$

In order to prove (13), let us assume that $y \in f(A) \cap B$. Then there exists an $x \in A$ such that $y = f(x) \in A$, whence $x \in f^{-1}(B)$ and $y \in f[A \cap f^{-1}(B)]$. Conversely, by (2) and (12),

$$f[A \cap f^{-1}(B)] \subset f(A) \cap ff^{-1}(B) = f(A) \cap B.$$

The formulas (14) and (15) concern partial functions:

$$\textit{If } g = f|A, \textit{ then } g^{-1}(B) = A \cap f^{-1}(B). \tag{14}$$

$$\textit{If } X = \bigcup_t A_t \textit{ and } g_t = f|A_t, \textit{ then } f^{-1}(B) = \bigcup_t g_t^{-1}(B). \tag{15}$$

Formulas (14) and (15) can be established as follows:

$$[x \in g^{-1}(B)] \equiv [g(x) \in B] \equiv [x \in A \text{ and } f(x) \in B] \equiv [x \in A \cap f^{-1}(B)],$$

$$f^{-1}(B) = \bigcup_t f^{-1}(B) \cap A_t = \bigcup_t g_t^{-1}(B).$$

$$\mathrm{S}[\overset{-1}{f}(B)] = f^{-1}(B). \tag{16}$$

Since

$$X \in \overset{-1}{f}(B) \equiv \bigvee_{y \in B} [X = \overset{-1}{f}(y)] \equiv \bigvee_{y \in B} \left[X = f^{-1}\big((y)\big)\right],$$

it follows

$$\mathrm{S}[\overset{-1}{f}(B)] = \bigcup_{y \in B} f^{-1}\big((y)\big) = f^{-1}[\bigcup_{y \in B}(y)] = f^{-1}(B).$$

$$\overset{-1}{f}(B) = \boldsymbol{D}_f \cap \mathop{E}_{A}[A \subset f^{-1}(B)]. \tag{17}$$

For

$$\left[(A \in \boldsymbol{D}_f) \wedge \big(A \subset f^{-1}(B)\big)\right] \equiv \bigvee_y [(y \in B) \wedge (A = \overset{-1}{f}(y))] \equiv A \in \overset{-1}{f}(B).$$

It is to be noted that *if f is one-to-one, formulas* (2), (2a), *and* (3) *can be expressed as identities.* Namely we have in this case

$$f(A_1 \cap A_2) = f(A_1) \cap f(A_2); \tag{2'}$$

$$f(\bigcap_t A_t) = \bigcap_t f(A_t); \tag{2'a}$$

$$f(A_1) - f(A_2) = f(A_1 - A_2). \tag{3'}$$

Furthermore, if f is one-to-one, then

$$(x \in A) \equiv [f(x) \in f(A)]. \tag{18}$$

The following rules of the *algebra of functions* can easily be established:

$$(f = g) \Rightarrow (f^1 = g^1), \tag{18'}$$

i.e. if $f(x) = g(x)$ for each $x \in X$, then $f(A) = g(A)$ for each $A \subset X$.

Denoting by *id* the identity function (for a suitable domain), we have (compare (11) and Section V):

$$id \subset f^{-1}f^{1}, \qquad (f^{-1}f^{1} = id) \equiv (f \text{ is one-to-one}). \tag{19}$$

By (12) we have

$$f^{1}f^{-1} \subset id, \tag{20}$$

$$(f^{1}f^{-1} = id) \equiv (f \text{ is onto}). \tag{21}$$

Now, let us consider the *complex-mapping* $f = (f_0, f_1)$, where $f_0\colon X \to Y_0$ and $f_1\colon X \to Y_1$; f is defined by the condition:

$$f(x) = \big(f_0(x), f_1(x)\big), \qquad \text{hence} \qquad f\colon X \to Y_0 \times Y_1.$$

One has the following formulas

$$f(A) \subset f_0(A) \times f_1(A) \qquad \text{where} \qquad A \subset X, \tag{22}$$

$$f^{-1}(B_0 \times B_1) = f_0^{-1}(B_0) \cap f_1^{-1}(B_1) \qquad \text{where} \qquad B_j \subset Y_j. \tag{23}$$

For

$$\begin{aligned}
&(y_0, y_1) \epsilon f(A) \\
&\quad \equiv \bigvee_x (x \epsilon A) \big(y_0 = f_0(x)\big) \big(y_1 = f_1(x)\big) \Rightarrow [y_0 \epsilon f_0(A)][y_1 \epsilon f_1(A)],
\end{aligned}$$

$$\begin{aligned}
[x \epsilon f^{-1}(B_0 \times B_1)] &\equiv [f(x) \epsilon (B_0 \times B_1)] \equiv [f_0(x) \epsilon B_0][f_1(x) \epsilon B_1] \\
&\equiv [x \epsilon f_0^{-1}(B_0) \cap f_1^{-1}(B_1)].
\end{aligned}$$

Finally consider the *product-mapping* $g = f_0 \times f_1$, where $f_0\colon X_0 \to Y_0$ and $f_1\colon X_1 \to Y_1$; g is defined as follows:

$$g(x_0, x_1) = \big(f_0(x_0), f_1(x_1)\big); \qquad \text{hence} \qquad g\colon (X_0 \times X_1) \to (Y_0 \times Y_1).$$

It is easy to see that

$$g(A_0 \times A_1) = f_0(A_0) \times f_1(A_1) \qquad \text{where} \qquad A_j \subset X_j, \tag{24}$$

$$g^{-1}(B_0 \times B_1) = f_0^{-1}(B_0) \times f_1^{-1}(B_1) \qquad \text{where} \qquad B_j \subset Y_j. \tag{25}$$

IV. Commutative diagrams. Let $f\colon X \to Y$, $g\colon Y \to Z$ and $h\colon X \to Z$. If $h = g \circ f$ we say that the following (triangular) diagram is *commutative*

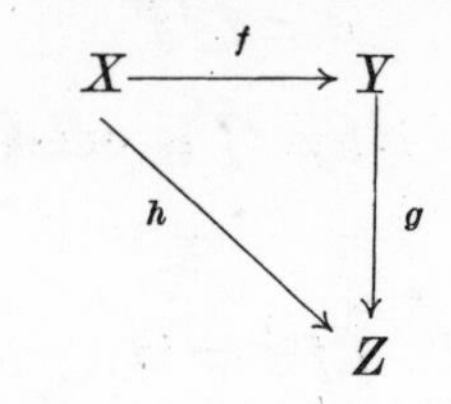

Obviously

$$h^{-1} = f^{-1} \circ g^{-1}, \quad \text{i.e.} \quad h^{-1}(C) = f^{-1}[g^{-1}(C)] \quad \text{for} \quad C \subset Z. \tag{1}$$

Here $g^{-1}\colon 2^Z \to 2^Y$ and $f^{-1}\colon 2^Y \to 2^X$.

More generally, the (rectangular) diagram

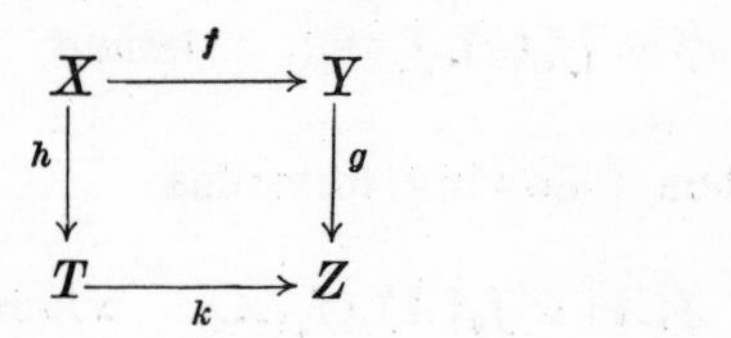

is said to be *commutative* if

$$g \circ f = k \circ h, \quad \text{i.e.} \quad g[f(x)] = k[h(x)]. \tag{2}$$

A triangular diagram is, of course, a particular case of the rectangular one; namely when $T = Z$ and k is the identity.

It follows by (1) that

$$f^{-1}g^{-1} = h^{-1}k^{-1}. \tag{3}$$

THEOREM 1. *Formula* (2) *is equivalent to*

$$fh^{-1} \subset g^{-1}k, \tag{4}$$

and to

$$hf^{-1} \subset k^{-1}g. \tag{5}$$

Proof. (2) $\Rightarrow$ (4). For (compare I(5), III(19)–(20)):

$$fh^{-1} \subset g^{-1}gfh^{-1} = g^{-1}khh^{-1} \subset g^{-1}k.$$

(4) $\Rightarrow$ (2). For

$$gf \subset gfh^{-1}h \subset gg^{-1}kh \subset kh,$$

whence $gf(x) \subset kh(x)$, which means that $gf(x) = kh(x)$.

Formulas (4) and (5) being symmetrical (relative to the diagonal XZ), the proof of the equivalence $(2) \equiv (5)$ is analogous.

DEFINITION. A diagram is called *bi-commutative*[1] if

$$fh^{-1} = g^{-1}k, \tag{6}$$

or symmetrically, if

$$hf^{-1} = k^{-1}g. \tag{7}$$

In other words (according to Theorem 1), if it is commutative and satisfies the inclusion inverse to (4) (or to (5)), i.e.

$$g^{-1}k \subset fh^{-1}, \quad \text{or symmetrically,} \quad hf^{-1} \subset k^{-1}g. \tag{8}$$

THEOREM 2. *A commutative diagram is bi-commutative if and only if the following statement holds:*

If $g(y) = k(t)$, *there is an* x *such that* $y = f(x)$ *and* $t = h(x)$. (9)

Proof. It suffices to show that $(8) \equiv (9)$.

$(8) \Rightarrow (9)$. Suppose $g(y) = k(t)$. Hence $y \in g^{-1}k(t)$ and by (8), $y \in fh^{-1}(t)$. Therefore, there is an x such that $y = f(x)$ and $h(x) = t$.

$(9) \Rightarrow (8)$. Let $y \in g^{-1}k(t)$. Hence $g(y) = k(t)$. It follows by (9) that $x \in h^{-1}(t)$ and $y = f(x) \in fh^{-1}(t)$, which completes the proof.

Remark 1. A commutative diagram does not need to be bi-commutative. This is shown in the following example: $X = Y = T = (a, b)$, $Z = (a)$, $f = h =$ identity, $g = k =$ const.

Remark 2. If $f = id$, the bi-commutativity of the (triangular) diagram means that

$$h^{-1} = g^{-1}k \tag{10}$$

and hence—in the case where h is onto—that k is one-to-one. By (10), $hh^{-1} = hg^{-1}k$, but $hh^{-1} = id$ and $g^{-1} = h^{-1}k^{-1}$. Hence $k^{-1}k = id$, which means that k is one-to-one (by III(19)).

V. Set-valued mappings. A function is called *set-valued* whose values are sets (subsets of a given set). So, for instance, the mappings f^1: $2^X \to 2^Y$ and f^{-1}: $2^Y \to 2^X$, considered in Section II, are set-valued.

Let F_t: $Y \to 2^X$ be set-valued mappings where $t \in T$; thus $F_t(y) \subset X$.

---

[1] or *exact*, see P. J. Hilton, *Homotopy and duality* (mimeographed), Cornell University 1959 (July), § 6, p. 32.

We agree to write

$$F_0 \subset^* F_1 \quad \text{whenever} \quad F_0(y) \subset F_1(y) \quad \text{for each} \quad y \in Y$$

(or briefly $F_0 \subset F_1$ when no confusion can occur with the notation $g \subset f$ as used in I(3)),

$$F = F_0 \cup F_1 \quad \text{whenever} \quad F(y) = F_0(y) \cup F_1(y) \quad \text{for} \quad y \in Y,$$

$$F = F_0 \cap F_1 \quad \text{whenever} \quad F(y) = F_0(y) \cap F_1(y) \quad \text{for} \quad y \in Y,$$

and so on. Thus the set $(2^X)^Y$ can be considered as the space of a Boolean algebra.

THEOREM 1. *Let $A \subset X$. The following formulas hold:*

$$F_0 \subset F_1 \Rightarrow F_1^{-1}(2^A) \subset F_0^{-1}(2^A); \tag{1}$$

$$\textit{if } F = F_0 \cup F_1, \textit{ then } F^{-1}(2^A) = F_0^{-1}(2^A) \cap F_1^{-1}(2^A), \tag{2}$$

and generally

$$(\bigcup_t F_t)^{-1}(2^A) = \bigcap_t F_t^{-1}(2^A); \tag{2a}$$

$$\textit{if } F = F_0 \cap F_1, \textit{ then } [F_0^{-1}(2^A) \cup F_1^{-1}(2^A)] \subset F^{-1}(2^A) \tag{3}$$

and generally

$$\bigcup_t F_t^{-1}(2^A) \subset (\bigcap_t F_t)^{-1}(2^A). \tag{3a}$$

Proof. In order to show (1), assume that $y \in F_1^{-1}(2^A)$. Hence $F_1(y) \in 2^A$, i.e. $F_1(y) \subset A$. As $F_0 \subset F_1$, it follows that $F_0(y) \subset F_1(y) \subset A$ and therefore $y \in F_0^{-1}(2^A)$.

(2a) follows from the equivalences (compare § 1, VI(2)):

$$y \in (\bigcup_t F_t)^{-1}(2^A) \equiv (\bigcup_t F_t(y)) \in 2^A \equiv \bigwedge_t [F_t(y) \in 2^A]$$
$$\equiv \bigwedge_t [y \in F_t^{-1}(2^A)] \equiv y \in \bigcap_t F_t^{-1}(2^A).$$

(3a) follows from (1) by virtue of the inclusion

$$(\bigcap_t F_t) \subset F_t,$$

and of § 1, V(4).

THEOREM 2. *Let $f: X \to Y$. Put $F = \overset{-1}{f}$. Then we have*

$$F^{-1}(2^{X-A}) = Y - f(A) \quad \textit{for each} \quad A \subset X.$$

For by II(3) we have: $[y \in Y - f(A)] \equiv [A \cap F(y) = 0]$, and $[A \cap F(y) = 0] \equiv [F(y) \subset X - A] \equiv [F(y) \in 2^{X-A}] \equiv [y \in F^{-1}(2^{X-A})]$.

VI. Sets of equal power. Cardinal numbers. Two sets X and Y are said to be of *equal power* (or *equivalent* in the sense of set theory) if there exists a one-to-one mapping of X onto Y. We write in this case $X \sim Y$.

It is easily seen that the relation $X \sim Y$ is an equivalence relation. This leads to the introduction of cardinal numbers. Namely we assign to each set X an object called a *cardinal number*, denoted by X, in such a way that the same cardinal number is assigned to two sets X and Y if and only if $X \sim Y$. Note that the equivalence class corresponding to this relation is not a set in the sense of set theory; however, if the variable sets X and Y are supposed to be subsets of a given set A—as is the case of most problems of Topology—then the equivalence classes are sets and they can be considered, by definition, to be cardinal numbers (relative to A).

With the usual notation, $\aleph_0$ denotes the cardinal number of the set of integers, and $\mathfrak{c}$ the cardinal number of the set of real numbers.

The following formulas, analogous to well-know arithmetic rules, hold:

$$Y^{X \cup T} \sim Y^X \times Y^T \quad \text{provided} \quad X \cap T = 0, \tag{1}$$

$$(Y \times Z)^X \sim (Y^X) \times (Z^X), \tag{2}$$

$$Y^{X \times T} \sim (Y^X)^T. \tag{3}$$

Proof of (1). Put $\alpha(f) = \langle f|X, f|T \rangle$ for $f \in Y^{X \cup T}$. One sees easily that α is an one-to-one mapping of the first member of (1) onto the second.

Proof of (2). Let $f(x) = \langle f_1(x), f_2(x) \rangle$ for $f \in (Y \times Z)^X$. Put

$$\beta(f) = \langle f_1, f_2 \rangle.$$

Proof of (3). Let $f \in Y^{X \times T}$. Put

$$g_t(x) = f(x, t). \tag{4}$$

Hence $g_t \in Y^X$ and therefore $g \in (Y^X)^T$. Finally put $\gamma(f) = g$, then γ is the required mapping.

VII. Characteristic functions. We call the *characteristic function of a set A* (contained in the given space X) the function f_A defined as follows

$$f_A(x) = \begin{cases} 1 & \text{for} \quad x \in A, \\ 0 & \text{for} \quad x \in X - A. \end{cases}$$

Thus f: $2^X \to (0, 1)^X$. It is easily seen that this mapping is onto and one-to-one. Thus

$$2^X \sim (0, 1)^X$$

(which shows that the notations Y^X and 2^X are in agreement).

The following formulas hold true:

$$f_X \equiv 1; \tag{1}$$

$$f_0 \equiv 0; \tag{2}$$

$$f_{-A}(x) = 1 - f_A(x); \tag{3}$$

$$f_{A \cap B} = f_A \cdot f_B; \tag{4}$$

$$f_{A-B} = f_A - f_{A \cap B}; \tag{5}$$

$$\textit{if } A = \bigcup_t A_t, \textit{ then } f_A(x) = \max_t f_{A_t}(x); \tag{6}$$

$$\textit{if } A = \bigcap_t A_t, \textit{ then } f_A(x) = \min_t f_{A_t}(x); \tag{7}$$

$$(A = \operatorname*{Limes}_{n=\infty} A_n) \equiv \big(f_A(x) = \lim_{n=\infty} f_{A_n}(x)\big). \tag{8}$$

The concept of a characteristic function of a set can be extended to a sequence of sets, or more generally, to a multivalued function. Namely, let F: $T \to 2^X$, i.e. $F_t \subset X$ for $t \in T$. Then the characteristic function f_F of F attaches to each $x \in X$ a function $f_F(x) \in (0, 1)^T$ defined as follows:

$$f_F^t(x) = \begin{cases} 1 & \text{if} \quad x \in F_t, \\ 0 & \text{if} \quad x \in X - F_t. \end{cases} \tag{9}$$

Thus the characteristic function of a sequence $F_1, F_2, \ldots, F_n, \ldots$ assumes as values sequences of numbers $x^{(1)}, x^{(2)}, \ldots$ such that $x^{(n)} = 1$ if $x \in F_n$ and $x^{(n)} = 0$ if $x \in (-F_n)$.

VIII. Generalized cartesian products. The notion of a set-valued function leads to a generalization of cartesian multiplication. Namely, let $F\colon T \to 2^X$, i. e. $F(t) \subset X$. Then the cartesian product

$$\mathop{P}_{t\in T} F(t)$$

is, by definition, the set of all functions $\mathfrak{z}\in X^T$ such that $\mathfrak{z}(t)\in F(t)$ for $t\in T$. Thus we have

$$\big(\mathfrak{z}\in \mathop{P}_{t} F(t)\big) \equiv \bigwedge_t \mathfrak{z}(t)\in F(t). \tag{0}$$

Note that, if $F(t) = X$ for each $t\in T$, then $\mathop{P}_{t} F(t) = X^T$.

The above definition of generalized cartesian product coincides with the definition given in § 2, I in the case where T is finite.

We shall frequently write X_t instead of $F(t)$ and $\mathfrak{z}^t$ instead of $\mathfrak{z}(t)$; and as in the finite case, we shall call the sets X_t the *axes* of the product $\mathop{P}_{t} X_t$; $\mathfrak{z}^t$ is called the tth *coordinate* of the point $\mathfrak{z}$ (or its *projection* on X_t; we write also $\mathrm{pr}_t\mathfrak{z}$ instead of $\mathfrak{z}^t$).

Let $A_t \subset X_t$. The following formulas are generalizations of § 2, II(5), § 2, II(6), § 2, II(2), VI(1), and VI(2)([1]):

$$\mathop{P}_{t} A_t = \bigcap_t \mathop{P}_{t'} A_{t,t'} \text{ where } A_{t,t} = A_t \text{ and } A_{t,t'} = X_{t'} \text{ for } t' \neq t; \tag{1}$$

$$-\mathop{P}_{t} A_t = \bigcup_t \mathop{P}_{t'} B_{t,t'} \text{ where } B_{t,t} = -A_t \text{ and } B_{t,t'} = X_{t'} \text{ for } t' \neq t; \tag{2}$$

$$\bigcap_s \mathop{P}_{t} A_{s,t} = \mathop{P}_{t} \bigcap_s A_{s,t}; \tag{2'}$$

$$Y^{\cup X_t} \sim \mathop{P}_{t} Y^{X_t} \text{ provided } X_t \cap X_{t'} = 0 \text{ for } t' \neq t; \tag{3}$$

$$\big(\mathop{P}_{t} Y_t\big)^X \sim \mathop{P}_{t} (Y_t^X). \tag{4}$$

The following formula concerns the associativity of cartesian multiplication:

if $T = T' \cup T''$ *and* $T' \cap T'' = 0$,

$$\text{then} \quad \mathop{P}_{t\in T} Y_t \sim \big[\mathop{P}_{t'\in T'} Y_{t'} \times \mathop{P}_{t''\in T''} Y_{t''}\big]. \tag{5}$$

Let us note the following generalization of the formula given in the remark of § 2, IV.

([1]) Compare N. Bourbaki, *Théorie des ensembles*, Paris 1958, p. 33.

Let $\varphi_t(x)$ be a propositional function where $t \in T$ and $x \in X_t$. Then

$$\bigwedge_t \bigvee_{x \in X_t} \varphi_t(x) \equiv \bigvee_{\mathfrak{z}} \bigwedge_t \varphi_t(\mathfrak{z}^t) \text{ where } \mathfrak{z} \text{ ranges over } \underset{t}{P} X_t. \tag{6}$$

Consider now a *complex mapping* $\mathfrak{f}: X \to \underset{t}{P} Y_t$, where $\mathfrak{f} = \{f_t\}$ and $f_t: X \to Y_t$. Thus

$$[\mathfrak{y} = \mathfrak{f}(x)] \equiv \bigwedge_t [\mathfrak{y}^t = f_t(x)]. \tag{7}$$

Formulas (22) and (23) of Section III can be easily generalized as follows:

$$\mathfrak{f}(A) \subset \underset{t}{P} f_t(A) \quad \text{where} \quad A \subset X, \tag{8}$$

$$\mathfrak{f}^{-1}(\underset{t}{P} B_t) = \bigcap_t f_t^{-1}(B_t) \quad \text{where} \quad B_t \subset Y_t, \tag{9}$$

in particular, if $B_t = Y_t$ for all t except t_0, then

$$\mathfrak{f}^{-1}(\underset{t}{P} B_t) = f_{t_0}^{-1}(B_{t_0}). \tag{10}$$

Finally consider a *product mapping* $\mathfrak{g} = \underset{t}{P} f_t$, where $f_t: X_t \to Y_t$. That means:

$$[\mathfrak{y} = \mathfrak{g}(\mathfrak{z})] \equiv \bigwedge_t [\mathfrak{y}^t = f_t(\mathfrak{z}^t)]. \tag{11}$$

One has the following generalizations of III(24) and III(25):

$$\mathfrak{g}(\underset{t}{P} A_t) = \underset{t}{P} f_t(A_t) \quad \text{where} \quad A_t \subset X_t, \tag{12}$$

$$\mathfrak{g}^{-1}(\underset{t}{P} B_t) = \underset{t}{P} f_t^{-1}(B_t) \quad \text{where} \quad B_t \subset Y_t, \tag{13}$$

and in particular

$$\mathfrak{g}^{-1}(\mathfrak{y}) = \underset{t}{P} f_t^{-1}(\mathfrak{y}^t) \quad \text{where} \quad \mathfrak{y} \in \underset{t}{P} Y_t. \tag{14}$$

Proof. Applying (11), (0), and (6), gives

$$\mathfrak{y} \in \mathfrak{g}(\underset{t}{P} A_t) \equiv \bigvee_{\mathfrak{z}} [\mathfrak{y} = \mathfrak{g}(\mathfrak{z})](\mathfrak{z} \in \underset{t}{P} A_t) \equiv \bigvee_{\mathfrak{z}} \bigwedge_t [\mathfrak{y}^t = f_t(\mathfrak{z}^t)](\mathfrak{z}^t \in A_t)$$

$$\equiv \bigwedge_t \bigvee_x [\mathfrak{y}^t = f_t(x)](x \in A_t) \equiv \bigwedge_t \mathfrak{y}^t \in f_t(A_t) \equiv \mathfrak{y} \in \underset{t}{P} f_t(A_t);$$

$$\mathfrak{z} \in \mathfrak{g}^{-1}(\underset{t}{P} B_t) \equiv \mathfrak{g}(\mathfrak{z}) \in \underset{t}{P} B_t \equiv \bigwedge_t f_t(\mathfrak{z}^t) \in B_t \equiv \bigwedge_t \mathfrak{z}^t \in f_t^{-1}(B_t)$$

$$\equiv \mathfrak{z} \in \underset{t}{P} f_t^{-1}(B_t).$$

IX. Examples of countable products. Denoting by $\mathscr{D}$ the set of all positive integers, the product $X^{\mathscr{D}}$ is the set of all infinite sequences

$$\mathfrak{z} = [\mathfrak{z}^1, \mathfrak{z}^2, \ldots, \mathfrak{z}^n, \ldots] \quad \text{where} \quad \mathfrak{z}^n \in X \text{ for } n = 1, 2, \ldots. \tag{1}$$

It seems appropriate to write in this case $X^{\aleph_0}$ instead of $X^{\mathscr{D}}$ (similarly we wrote X^n instead of $X^{(0,1,\ldots,n-1)}$; this notation will be extended in Section XII to $X^{\aleph_\alpha}$ for arbitrary α).

The following examples have become classical.

1. $\mathscr{E}^{\aleph_0}$—the *Fréchet space*—has as elements infinite sequences composed of arbitrary reals. If these reals are restricted to the interval $\mathscr{I}$, we obtain the *Hilbert cube* $\mathscr{I}^{\aleph_0}$.

2. The set $\mathscr{D}^{\aleph_0}$ can be identified with the set $\mathscr{N}$ of all irrational numbers between 0 and 1. Namely, each element $\mathfrak{z} \in \mathscr{D}^{\aleph_0}$ determines in a unique way the irrational number (developed in a continuous fraction)

$$\frac{1|}{|\mathfrak{z}^1} + \frac{1|}{|\mathfrak{z}^2} + \ldots + \frac{1|}{|\mathfrak{z}^n} + \ldots. \tag{2}$$

3. If A is composed of two elements $A^{\aleph_0}$ can be identified with the *Cantor discontinuum* $\mathscr{C}$, defined as follows (putting $A = (0, 2)$)(¹).

According to (1), $\mathfrak{z}^n$ is 0 or 2, and we put

$$\mathfrak{z} = \frac{\mathfrak{z}^1}{3} + \frac{\mathfrak{z}^2}{9} + \ldots + \frac{\mathfrak{z}^n}{3^n} + \ldots. \tag{3}$$

Thus $\mathscr{C}$ is the set of all numbers in the interval $\mathscr{I}$ which can be written in the ternary system of calculation without using the digit 1.

Another (geometrical) approach to $\mathscr{C}$ is the following. Let us divide the interval $\mathscr{I}$ into three equal intervals and remove the middle open interval. We divide the remaining two intervals $(0, 1/3)$ and $(2/3, 1)$ into three equal intervals and remove their (open) middle parts. Continuing in this way we obtain an infinite sequence of deleted open intervals. Deleting from the interval $\mathscr{I}$ their union we obtain the set $\mathscr{C}$ (which was defined previously arithmetically).

(¹) G. Cantor, Math. Ann. 21 (1883), p. 590.

The Cantor discontinuum has numerous applications. Here we call attention to the fact that, given an infinite sequence of subsets $A_1, A_2, \ldots, A_n, \ldots$ of X and denoting its characteristic function by f (see Section VII), $2f$ is a mapping of X into $\mathscr{C}$ and $2f^n(x) = 2$ if $x \in A_n$ and $= 0$ if $x \in X - A_n$(1).

X. Orderings. Let there be given a set X and a relation among its elements written as $x \leqslant y$. Consider the following four conditions (of reflexivity, antisymmetry, transitivity and connectedness):

1. for all x, $x \leqslant x$;
2. if $x \leqslant y$ and $y \leqslant x$, then $x = y$;
3. if $x \leqslant y$ and $y \leqslant z$, then $x \leqslant z$;
4. for each pair x, y, either $x \leqslant y$ or $y \leqslant x$.

If conditions 1–3 are satisfied, we say that the relation $x \leqslant y$ is an *ordering* of X (or that the set X is *ordered*); the relation $x \leqslant y$ is a *quasi-ordering* if it satisfies conditions 1 and 3 only; it is a *linear ordering* if it satisfies conditions 1–4.

For example, the family 2^X is ordered by the relation of inclusion $A \subset B$.

If a family $\boldsymbol{R} \subset 2^X$ is linearly ordered by the above relation, we say that $\boldsymbol{R}$ is *monotone*.

A quasi-ordered set is called a *directed set* if to each pair x, y there exists z such that $x \leqslant z$ and $y \leqslant z$. Again, such is the family 2^X (since $A \subset A \cup B$ and $B \subset A \cup B$).

An ordered set X is said to be *cofinal* with the set $Y \subset X$, if to each $x \in X$ corresponds an $y \in Y$ such that $x \leqslant y$.

For example, the set of all real numbers is cofinal with the set of positive integers.

Obviously, if X contains the last element a, it is cofinal with (a).

We say that the relation $\leqslant$ which orders the set X and the relation $\leqslant^*$ which orders the set Y establish *similar orderings* if there exists a one-to-one mapping f (called a *similarity mapping*)

(1) See E. Szpilrajn-Marczewski, Fund. Math. 26 (1936), p. 302, and 31 (1938), p. 207.

of X onto Y such that

$$(x_1 \leqslant x_2) \equiv \big(f(x_1) \leqslant^* f(x_2)\big).$$

The relation between two ordered sets making them similar is an equivalence relation. Consequently, one can assign *order types* to ordered sets so that the same order type is assigned to two ordered sets if and only if they are similar (this procedure is completely analogous to the procedure of assigning to sets their cardinal numbers).

XI. Well ordering. Ordinal numbers. A linear ordering of a set X is called a *well ordering* if every non-empty subset of X has a first element. The order types of well-ordered sets are called *ordinal numbers* (such as, in particular, $0, 1, \ldots, n, \ldots$; ω is the ordinal number of the set of all non-negative integers given in their natural order).

By the *Zermelo theorem*, every set can be well ordered. This theorem is a consequence of the *axiom of choice* which can be formulated as follows:

For each X there exists a mapping $f\colon 2^X \to X$ such that $f(A) \in A$ whenever $A \neq 0$.

For further use, where applicable, we shall mention the following statement (called *Zorn's lemma*):

Let X be an ordered set such that for each linearly ordered set $A \subset X$ there is in X an element following all the elements of A (or equal to the last element of A, if it exists); then there exists in X a maximal element(1).

Ordinal numbers have various applications in the *definitions by transfinite induction*. These definitions refer to the following theorem.

Denote by $\Gamma(a)$ the set of all $\xi < a$. Let X be a given set, a an ordinal and $h\colon 2^X \to X$. Then there exists an $f\colon \Gamma(a+1) \to X$ such that

$$f(\xi) = h[f(\Gamma(\xi))] \quad \textit{for every} \quad \xi \leqslant a.$$

Let us finally recall the following notation. An order type a of a well-ordered set Z is called *initial* if it is the least ordinal among all the ordinal numbers corresponding to all possible well-orderings

(1) For other similar statements, see J. Kelley, *op. cit.*, p. 33; see also my paper *Une méthode d'élimination des nombres transfinis des raisonnements mathématiques*, Fund. Math. 3 (1922), p. 89.

of Z. Thus ω (also denoted by ω_0) is an initial ordinal number. The next initial number is Ω $(= \omega_1)$, the least ordinal corresponding to uncountable sets. Generally, one considers ω_α for arbitrary α.

By definition $\aleph_\alpha = \overline{\overline{\Gamma(\omega_\alpha)}}$. Thus $\aleph_1$ is the smallest cardinal of uncountable sets; it is the next greater cardinal to $\aleph_0$. The hypothesis that $\aleph_1 = \mathfrak{c}$ is called the *continuum-hypothesis*; it is independent of the usual system of axioms of set theory.

XII. The set $X^{\aleph_\alpha}$. We agree to write (as we did in Section IX for $\alpha = 0$)

$$X^{\aleph_\alpha} = X^{\Gamma(\omega_\alpha)}. \tag{1}$$

The well-known formula

$$2\mathfrak{m} = \mathfrak{m} = \mathfrak{m}^2 \quad \text{(for } \mathfrak{m} \text{ infinite)}, \tag{2}$$

which can be expressed in the form

$$\Gamma(\omega_\alpha)\times(0,1) \sim \Gamma(\omega_\alpha) \sim \Gamma(\omega_\alpha)\times\Gamma(\omega_\alpha), \tag{3}$$

leads to the conclusion that

$$X^{\aleph_\alpha}\times X^{\aleph_\alpha} \sim X^{\aleph_\alpha} \sim (X^{\aleph_\alpha})^{\aleph_\alpha}. \tag{4}$$

Moreover if φ: $\Gamma(\omega_\alpha)\times(0,1) \to \Gamma(\omega_\alpha)$ is one-to-one and onto, then the mapping χ defined by the condition

$$\chi(f) = f\circ\varphi \quad \text{for each} \quad f \in X^{\aleph_\alpha}$$

is the required one-to-one mapping of $X^{\aleph_\alpha}$ onto $X^{\aleph_\alpha}\times X^{\aleph_\alpha}$.

In a similar way, denoting by ψ a one-to-one mapping of $\Gamma(\omega_\alpha)\times\Gamma(\omega_\alpha)$ onto $\Gamma(\omega_\alpha)$, $f\circ\psi$ defines a one-to-one mapping of $X^{\aleph_\alpha}$ onto $(X^{\aleph_\alpha})^{\aleph_\alpha}$.

The proof follows easily from the formulas (compare VI (1) and (3)):

$$X^{\aleph_\alpha}\times X^{\aleph_\alpha} \sim X^{\Gamma(\omega_\alpha)\times(0,1)} \quad \text{and} \quad (X^{\aleph_\alpha})^{\aleph_\alpha} \sim X^{\Gamma(\omega_\alpha)\times\Gamma(\omega_\alpha)}. \tag{5}$$

XIII. Inverse systems, inverse limits. Let T be a directed set. Let X be a set-valued function, X: $T \to 2^A$; thus $X_t \subset A$ for each $t \in T$. Let f be a function defined on $T\times T$ for pairs t_0, t_1 where $t_0 \leqslant t_1$, and such that

$$f_{t_0 t_1}\colon X_{t_1} \to X_{t_0}. \tag{1}$$

We assume further that

$$f_{t_0t_1} \circ f_{t_1t_2} = f_{t_0t_2} \quad \text{for} \quad t_0 \leqslant t_1 \leqslant t_2 \quad \text{(transitivity)} \tag{2}$$

and

$$f_{tt} = \text{identity}. \tag{3}$$

Then we call the triple (T, X, f) an *inverse system*(1).

The *inverse limit* of the system (T, X, f), denoted by

$$X_\infty \quad \text{or} \quad \underleftarrow{\mathrm{Lim}}(T, X, f) \quad \text{or} \quad \underset{t, t_0 \leqslant t_1}{\mathrm{Lim}} \{X_t, f_{t_0t_1}\},$$

is the subset of the cartesian product $\underset{t \in T}{P} X_t$ composed of elements $\mathfrak{z} = \{\mathfrak{z}^t\}$ such that

$$f_{t_0t_1}(\mathfrak{z}^{t_1}) = \mathfrak{z}^{t_0}. \tag{4}$$

In other words, we have for $\mathfrak{z} \in X_\infty$

$$f_{t_0t_1} \circ \mathrm{pr}_{t_1} = \mathrm{pr}_{t_0}. \tag{5}$$

We shall agree to write

$$f_t = \mathrm{pr}_t | X_\infty, \quad \text{i.e.} \quad f_t(\mathfrak{z}) = \mathfrak{z}^t. \tag{6}$$

Consequently

$$f_{t_0t_1} \circ f_{t_1} = f_{t_0}, \quad \text{hence} \quad f_{t_0}^{-1} = f_{t_1}^{-1} \circ f_{t_0t_1}^{-1}. \tag{7}$$

Consider two inverse systems (T, X, f) and (T, Y, g). Suppose that h attaches to each t a map

$$h_t \colon X_t \to Y_t$$

such that commutativity holds in the diagram (for $t_0 \leqslant t_1$)

$$\begin{array}{ccc} X_{t_0} & \xleftarrow{f_{t_0t_1}} & X_{t_1} \\ {\scriptstyle h_{t_0}}\downarrow & & \downarrow{\scriptstyle h_{t_1}} \\ Y_{t_0} & \xleftarrow[g_{t_0t_1}]{} & Y_{t_1} \end{array}, \tag{8}$$

(1) See S. Eilenberg and N. Steenrod, *Foundations of Algebraic Topology*, Princeton 1952, Chap. VIII. Compare P. S. Alexandrov, Ann. of Math. 30 (1928).

i.e.

$$h_{t_0} \circ f_{t_0 t_1} = g_{t_0 t_1} \circ h_{t_1}. \tag{9}$$

Then we may define a map

$$h_\infty: X_\infty \to Y_\infty$$

such that the following diagram is commutative for each $t \in T$:

$$\begin{array}{ccc} X_t & \xleftarrow{f_t} & X_\infty \\ {\scriptstyle h_t}\downarrow & & \downarrow{\scriptstyle h_\infty} \\ Y_t & \xleftarrow[g_t]{} & Y_\infty \end{array} \tag{10}$$

We put $\mathfrak{y} = h_\infty(\mathfrak{z})$ for $\mathfrak{z} \in X_\infty$, where

$$h_t(\mathfrak{z}^t) = \mathfrak{y}^t, \quad \text{i.e.} \quad h^t_\infty(\mathfrak{z}) = h_t(\mathfrak{z}^t). \tag{11}$$

It is easily seen that

if each h_t is a one-to-one mapping onto, so is h_∞. (12)

***XIV. The $(\mathscr{A})$-operation**(1). Let $\{A_{k_1 \dots k_n}\}$ be a system of sets defined for each finite sequence $k_1, \dots, k_n$ of positive integers. The set

$$R = \bigcup_{k_1 \dots k_n \dots} \bigcap_{n=1}^{\infty} A_{k_1 \dots k_n}$$

is called the *result of the $(\mathscr{A})$-operation* applied to the system $\{A_{k_1 \dots k_n}\}$.

In particular, if $A_{k_1 \dots k_n} = B_{k_1}$ or $A_{k_1 \dots k_n} = B_n$, we have

$$R = \bigcup_{k=1}^{\infty} B_k \quad \text{or} \quad R = \bigcap_{n=1}^{\infty} B_n, \quad \text{respectively}.$$

Denoting by $\mathfrak{z} = [\mathfrak{z}^1, \mathfrak{z}^2, \dots, \mathfrak{z}^n, \dots]$ an arbitrary irrational number between 0 and 1 ($\mathfrak{z}^n$ is a positive integer, see IX, 2), we have

$$R = \bigcup_{\mathfrak{z} \in \mathscr{N}} \bigcap_{n=1}^{\infty} A_{\mathfrak{z}^1 \dots \mathfrak{z}^n} \tag{0}$$

(1) See M. Souslin and N. Lusin, C. R. Paris 164 (1917), p. 88. Comp. F. Hausdorff, *Mengenlehre*, § 19.

There are important properties, such as measurability or Baire property, which are invariant of the $(\mathscr{A})$-operation (see § 11).

where $\mathcal{N}$ denotes the set of all irrational numbers between 0 and 1.

The system $\{A_{\mathfrak{z}^1\ldots\mathfrak{z}^n}\}$ is called *regular* if

$$A_{\mathfrak{z}^1\ldots\mathfrak{z}^n\mathfrak{z}^{n+1}} \subset A_{\mathfrak{z}^1\ldots\mathfrak{z}^n}.$$

Each system can be *regularized* with no effect on the result of the $(\mathscr{A})$-operation. Namely, we put

$$\overset{*}{A}_{\mathfrak{z}^1\ldots\mathfrak{z}^n} = A_{\mathfrak{z}^1} \cap A_{\mathfrak{z}^1\mathfrak{z}^2} \cap \ldots \cap A_{\mathfrak{z}^1\mathfrak{z}^2\ldots\mathfrak{z}^n}. \tag{1}$$

Let us mention the following formulas concerning regular systems[1]:

$$\bigcup_m \bigcup_{\mathfrak{z}} \bigcap_k A_{m\mathfrak{z}^1\ldots\mathfrak{z}^k} = \bigcup_{\mathfrak{z}} \bigcap_k A_{\mathfrak{z}^1\ldots\mathfrak{z}^k}, \tag{2}$$

and more generally,

$$\bigcup_m \bigcup_{\mathfrak{z}} \bigcap_k A_{\mathfrak{y}^1\ldots\mathfrak{y}^i m\mathfrak{z}^1\ldots\mathfrak{z}^k} = \bigcup_{\mathfrak{z}} \bigcap_k A_{\mathfrak{y}^1\ldots\mathfrak{y}^i\mathfrak{z}^1\ldots\mathfrak{z}^k}; \tag{3}$$

$$A_{\mathfrak{z}^1\ldots\mathfrak{z}^k} \subset B_{\mathfrak{z}^1\ldots\mathfrak{z}^k} \quad \text{implies} \quad \bigcup_{\mathfrak{z}} \bigcap_k A_{\mathfrak{z}^1\ldots\mathfrak{z}^k} \subset \bigcup_{\mathfrak{z}} \bigcap_k B_{\mathfrak{z}^1\ldots\mathfrak{z}^k}; \tag{4}$$

$$\text{the set of terms in } \bigcup_{\mathfrak{z}} \bigcup_k A_{\mathfrak{z}^1\ldots\mathfrak{z}^k} \text{ is countable.} \tag{5}$$

THEOREM 1. *If the system is regular, then*

$$A - \bigcup_{\mathfrak{z}} \bigcap_k A_{\mathfrak{z}^1\ldots\mathfrak{z}^k} \subset \bigcup_{\mathfrak{z}} \bigcup_k \Big(A_{\mathfrak{z}^1\ldots\mathfrak{z}^k} - \bigcup_m A_{\mathfrak{z}^1\ldots\mathfrak{z}^k m}\Big) \tag{6}$$

where we put $A_{\mathfrak{z}^1\ldots\mathfrak{z}^k} = A$ *if* $k = 0$.

Proof. Let $p \in A$. Suppose that p does not belong to the right-hand side of (6); hence

$$\bigwedge_{\mathfrak{z}} \bigwedge_k [(p \in A_{\mathfrak{z}^1\ldots\mathfrak{z}^k}) \Rightarrow \bigvee_m (p \in A_{\mathfrak{z}^1\ldots\mathfrak{z}^k m})],$$

which means that for each system $m_1 \ldots m_k$ $(k \geqslant 0)$ of indices such that $p \in A_{m_1\ldots m_k}$, there is an index m such that $p \in A_{m_1\ldots m_k m}$. As $p \in A$, it follows that m_1 exists such that $p \in A_{m_1}$, whence similarly $p \in A_{m_1 m_2}$, etc. Consequently there is an infinite sequence m_1,

---

[1] See N. Lusin and W. Sierpiński, *Sur quelques propriétés des ensembles* (A), Bull. Acad. Sc. Cracovie 1918, p. 35.

$m_2, \ldots$ such that

$$p \in \bigcap_k A_{m_1 \ldots m_k} \quad \text{and} \quad p \in \bigcup_{\mathfrak{z}} \bigcap_k A_{\mathfrak{z}^1 \ldots \mathfrak{z}^k},$$

which proves that p does not belong to the left-hand side of (6).

THEOREM 2. *If the system is regular and*

$$[(\mathfrak{z}^1 \ldots \mathfrak{z}^n) \neq (\mathfrak{y}^1 \ldots \mathfrak{y}^n)] \Rightarrow (A_{\mathfrak{z}^1 \ldots \mathfrak{z}^n} \cap A_{\mathfrak{y}^1 \ldots \mathfrak{y}^n} = 0), \tag{7}$$

then

$$\bigcup_{\mathfrak{z}} \bigcap_n A_{\mathfrak{z}^1 \ldots \mathfrak{z}^n} = \bigcap_n \bigcup_{\mathfrak{z}} A_{\mathfrak{z}^1 \ldots \mathfrak{z}^n}. \tag{7'}$$

Proof. The inclusion obtained from (7′) replacing $=$ by $\subset$ is always true (for non-regular systems also; comp. § 2, IV).

Now suppose that p belongs to the right-hand side of (7). Hence there is one, and only one, index m_1 such that $p \in A_{m_1}$. Similarly, there exists a pair q_1, m_2 such that $p \in A_{q_1 m_2}$. As $A_{q_1 m_2} \subset A_{q_1}$, it follows that $p \in A_{q_1}$ and consequently $q_1 = m_1$. We prove in an analogous way that m_3 exists such that $p \in A_{m_1 m_2 m_3}$, etc. Put $\mathfrak{z} = (m_1, m_2, m_3, \ldots)$. Thus $p \in A_{\mathfrak{z}^1 \ldots \mathfrak{z}^n}$ for each n. This means that p belongs to the left-hand side of (7′).

THEOREM 3(1). *Let R satisfy* (0) *and let us put for $\alpha < \Omega$:*

$$A^0_{\mathfrak{z}^1 \ldots \mathfrak{z}^n} = A_{\mathfrak{z}^1 \ldots \mathfrak{z}^n}, \tag{8}$$

$$A^{\alpha+1}_{\mathfrak{z}^1 \ldots \mathfrak{z}^n} = A^{\alpha}_{\mathfrak{z}^1 \ldots \mathfrak{z}^n} \cap \bigcup_k A^{\alpha}_{\mathfrak{z}^1 \ldots \mathfrak{z}^n k}, \tag{9}$$

$$A^{\lambda}_{\mathfrak{z}^1 \ldots \mathfrak{z}^n} = \bigcap_{\xi < \lambda} A^{\xi}_{\mathfrak{z}^1 \ldots \mathfrak{z}^n} \quad \textit{for } \lambda \textit{ limit}. \tag{10}$$

Moreover put

$$E_\alpha = \bigcup_k A^\alpha_k, \tag{11}$$

$$T_\alpha = \bigcup_{\mathfrak{z}} \bigcup_n (A^{\alpha}_{\mathfrak{z}^1 \ldots \mathfrak{z}^n} - A^{\alpha+1}_{\mathfrak{z}^1 \ldots \mathfrak{z}^n}), \tag{12}$$

$$K_\alpha = E_\alpha - T_\alpha. \tag{13}$$

Then

$$\bigcup_{\alpha < \Omega} K_\alpha = R = \bigcap_{\alpha < \Omega} E_\alpha. \tag{14}$$

---

(1) W. Sierpiński, *Sur une propriété des ensembles* (A), Fund. Math. 8 (1926), p. 362. Further citations are found in this paper.

Proof. Let $x \in K_\alpha$. Hence $x \in E_\alpha$ and $x \notin T_\alpha$. It follows from the first formula, by (11), that there is a k_1 such that $x \in A^\alpha_{k_1}$. By the second formula, we have $x \notin (A^\alpha_{k_1} - A^{\alpha+1}_{k_1})$; this follows from (12) substituting $\mathfrak{z} = [k_1, 1, 1, \ldots]$ and $n = 1$. From this it is seen that $x \in A^{\alpha+1}_{k_1}$. We infer by (9) that there is a k_2 such that $x \in A^\alpha_{k_1 k_2}$. Therefore—substituting in (12) $\mathfrak{z} = [k_1, k_2, 1, 1, \ldots]$ and $n = 2$—we get $x \notin (A^\alpha_{k_1 k_2} - A^{\alpha+1}_{k_1 k_2})$, whence $x \in A^{\alpha+1}_{k_1 k_2}$.

One is led in this way to an infinite sequence $k_1, k_2, \ldots$ such that $x \in A^\alpha_{k_1 \ldots k_n}$ for each n. In other words,

$$x \in \bigcap_n A^\alpha_{\mathfrak{z}^1 \ldots \mathfrak{z}^n} \tag{15}$$

where $\mathfrak{z} = [k_1, k_2, \ldots]$.

On the other hand it is easy to show, applying transfinite induction, that

$$A^\alpha_{\mathfrak{z}^1 \ldots \mathfrak{z}^n} \supset A^\beta_{\mathfrak{z}^1 \ldots \mathfrak{z}^n} \quad \text{for} \quad \alpha < \beta. \tag{16}$$

Thus

$$x \in \bigcap_n A^0_{\mathfrak{z}^1 \ldots \mathfrak{z}^n} = \bigcap_n A_{\mathfrak{z}^1 \ldots \mathfrak{z}^n}, \quad \text{whence} \quad x \in R.$$

Therefore

$$\bigcup_{\alpha < \Omega} K_\alpha \subset R. \tag{17}$$

In order to infer (14) we shall show that

$$R \subset \bigcap_{\alpha < \Omega} E_\alpha \tag{18}$$

and

$$\bigcap_{\alpha < \Omega} T_\alpha = 0. \tag{19}$$

We have, for each α and m,

$$\bigcap_n A_{\mathfrak{z}^1 \ldots \mathfrak{z}^n} \subset A^\alpha_{\mathfrak{z}^1 \ldots \mathfrak{z}^m}. \tag{20}$$

This can be shown by transfinite induction in view of (8), (10), and the formula

$$A^\alpha_{\mathfrak{z}^1 \ldots \mathfrak{z}^n} \cap A^\alpha_{\mathfrak{z}^1 \ldots \mathfrak{z}^n \mathfrak{z}^{n+1}} \subset A^\alpha_{\mathfrak{z}^1 \ldots \mathfrak{z}^n} \cap \bigcup_k A^\alpha_{\mathfrak{z}^1 \ldots \mathfrak{z}^n k} = A^{\alpha+1}_{\mathfrak{z}^1 \ldots \mathfrak{z}^n},$$

which proves that if inclusion (20) is true for α (and for each m), then it is true for $\alpha+1$.

Substituting $m = 1$ in (20), we get

$$\bigcap_n A_{\delta^1 \ldots \delta^n} \subset A^{\alpha}_{\delta^1} \subset \bigcup_k A^{\alpha}_k = E_\alpha, \tag{21}$$

which leads to inclusion (18).

In order to prove (19), let us suppose that $x \in T_\alpha$ for each $\alpha < \Omega$. Consequently, to each α there corresponds a system of indices $k_1, \ldots, k_n$ such that

$$x \in (A^{\alpha}_{k_1 \ldots k_n} - A^{\alpha+1}_{k_1 \ldots k_n}). \tag{22}$$

It follows at once that there are two different α and β to which the same system of indices corresponds. Thus, besides (22) we have

$$x \in (A^{\beta}_{k_1 \ldots k_n} - A^{\beta+1}_{k_1 \ldots k_n}). \tag{23}$$

Let $\alpha < \beta$. It follows by (23) and (16), $x \in A^{\alpha+1}_{k_1 \ldots k_n}$ which is a contradiction to (22).

Thus (18) and (19) have been established. It now only remains to show (14). To this end, in view of (18) and (17), we have to prove that

$$\bigcap_\alpha E_\alpha \subset \bigcup_\alpha K_\alpha.$$

But this follows from (19), because if $x \in E_\alpha$ for each α, then there is an α such that

$$x \in (E_\alpha - T_\alpha) = K_\alpha \subset \bigcup_{\alpha < \Omega} K_\alpha.$$

***XV. Lusin sieve**(1). Let $\mathscr{R}_0$ be the set of binary fractions

$$r = \frac{1}{2^{m_1}} + \ldots + \frac{1}{2^{m_n}} \quad \text{where} \quad 1 \leqslant m_1 < \ldots < m_n. \tag{1}$$

A mapping W which assigns to each $r \in \mathscr{R}_0$ a set $W_r \subset X$ (where X is a fixed set) is called a *sieve*. The set A composed of all elements x such that there is an infinite sequence $r_1, r_2, \ldots$ satisfying conditions

$$r_1 < r_2 < \ldots \quad \text{and} \quad x \in W_{r_1} \cap W_{r_2} \cap \ldots, \tag{2}$$

is said to be *sieved* by the sieve W.

---

(1) Compare N. Lusin, Fund. Math. 10 (1927), p. 9.

In other words, if

$$M_x = \underset{r}{E}(x \in W_r), \tag{3}$$

or equivalently, if

$$W_r = \underset{x}{E}(r \in M_x), \tag{4}$$

then

$$A = \underset{x}{E}(M_x \text{ is not well ordered by the relation } r \geqslant s). \tag{5}$$

Remark. If $X = \mathscr{E}$, W_r can be imagined to be a subset of the horizontal line $y = r$. Then M_{x_0} becomes the intersection of $\bigcup_r W_r$ with the vertical line $x = x_0$.

Let us call *constituents* of $-A$ (relative to the sieve W) the sets A_α defined for $\alpha < \Omega$ as follows:

$$A_\alpha = \underset{x}{E}(M_x \text{ has order-type } \alpha); \tag{6}$$

hence

$$(-A) = \bigcup_\alpha A_\alpha.$$

In order to establish some useful relations between the $(\mathscr{A})$-operation and the sieve-operation, let us put

$$I(r) = [m_1, m_2 - m_1, \ldots, m_n - m_{n-1}] \tag{7}$$

where r is given by formula (1).

I is a one-to-one correspondence between $\mathscr{R}_0$ and the set of all finite systems of positive integers, because each system $k_1, \ldots, k_n$ corresponds to the number

$$r = \frac{1}{2^{k_1}} + \frac{1}{2^{k_1+k_2}} + \ldots + \frac{1}{2^{k_1+\ldots+k_n}}. \tag{8}$$

THEOREM[1]. *Let $\{A_{k_1\ldots k_n}\}$ be regular, and let $W_r = A_{I(r)}$. Let R be given by* XIV(0), *and A by* XV(5). *Then $R = A$, which means that the result of the $(\mathscr{A})$-operation is sieved by the sieve W.*

---

[1] See N. Lusin and W. Sierpiński, Journ. de Math. II (1923), p. 65–68. Cf. W. Sierpiński, *Le crible de M. Lusin et l'opération $(\mathscr{A})$ dans les espaces abstraits*, Fund. Math. 11 (1928), p. 16, N. Lusin, Fund. Math. 10 (1927), p. 20 et E. Selivanowski, C. R. Paris 184 (1927), p. 1311.

Proof. Let $x \in R$. Then there is a sequence $k_1, k_2, \ldots$ such that $x \in A_{k_1} \cap A_{k_1 k_2} \cap \ldots$ Define r_n according to (8). Hence (2) is fulfilled, i.e. $x \in A$.

Conversely, assume that $x \in A$, i.e. that (2) is fulfilled. Put

$$\lim_{n=\infty} r_n = \sum_{n=1}^{\infty} \frac{1}{2^{m_n}} \quad \text{where} \quad 1 \leqslant m_1 < m_2 < \ldots.$$

Put $k_1 = m_1$ and $k_n = m_n - m_{n-1}$ for $n > 1$. For n fixed, denote by j_n an index such that

$$\sum_{i=1}^{n} \frac{1}{2^{m_i}} < r_{j_n} < \sum_{i=1}^{\infty} \frac{1}{2^{m_i}}.$$

Put

$$r_{j_n} = \frac{1}{2^{q_1}} + \ldots + \frac{1}{2^{q_s}} \quad \text{where} \quad 1 \leqslant q_1 < \ldots < q_s.$$

Hence $q_1 = m_1, \ldots, q_n = m_n$. It follows that the first n terms of the system $I(r_{j_n})$ are identical to $k_1, \ldots, k_n$, respectively. The system $\{A_{k_1 \ldots k_n}\}$ being regular, we infer that $A_{I(r_{j_n})} \subset A_{k_1 \ldots k_n}$, whence $x \in A_{k_1 \ldots k_n}$ for $n = 1, 2, \ldots$ Therefore $x \in R$.

***XVI. Application to the Cantor discontinuum $\mathscr{C}$.** Let $\mathscr{R}_0 = [r_1, r_2, \ldots]$. The element $\mathfrak{z}$ of $\mathscr{C}$ being represented in the form (compare IX(3))

$$\mathfrak{z} = \frac{\mathfrak{z}^1}{3} + \frac{\mathfrak{z}^2}{9} + \ldots \quad \text{where} \quad \mathfrak{z}^n = 0 \text{ or } 2,$$

$R_{\mathfrak{z}}$ is defined by the formula

$$(r_n \in R_{\mathfrak{z}}) \equiv (\mathfrak{z}^n = 2). \tag{1}$$

Denote by $\bar{\mathfrak{z}}$ the order-type of $R_{\mathfrak{z}}$ (in respect to the relation $r \geqslant s$).

All the order-types τ of countable sets are assumed by $\bar{\mathfrak{z}}$ when $\mathfrak{z}$ ranges over $\mathscr{C}$.

This follows from the fact that $\mathscr{R}_0$ has a dense order type and consequently contains subsets similar to any countable ordered set given in advance.

Thus if we put

$$L_\tau = \underset{\mathfrak{z}}{E}(\bar{\mathfrak{z}} = \tau), \quad \text{then} \quad L_\tau \neq 0. \tag{2}$$

In particular, if α is an ordinal and

$$L = \underset{\mathfrak{z}}{E}(\bar{\mathfrak{z}} < \Omega) \quad \text{then} \quad L = \bigcup_{\alpha < \Omega} L_\alpha, \tag{3}$$

which is a decomposition of L in non-empty sets.

The sets L_α are the constituents of L relative to the sieve C defined by the identity

$$C_r = \underset{\mathfrak{z}}{E}(r \in R_{\mathfrak{z}}).$$

For, $\mathscr{C} - L$ is sieved by C according to the formula

$$(\mathfrak{z} \in C_{r_n}) \equiv (r_n \in R_{\mathfrak{z}}) \equiv (\mathfrak{z}^n = 2). \tag{4}$$

CHAPTER ONE

TOPOLOGICAL SPACES

§ 4. Definitions. Closure operation

I. Definitions. A *topological space* is a set 1 (whose elements are called *points*) and a function (called *closure*) assigning to each set $X \subset 1$ a set $\overline{X} \subset 1$ satisfying the following four axioms:

AXIOM 1. $\overline{X \cup Y} = \overline{X} \cup \overline{Y}$.

AXIOM 2. $X \subset \overline{X}$.

AXIOM 3. $\overline{0} = 0$.

AXIOM 4. $\overline{\overline{X}} = \overline{X}$.

If, moreover, the following axiom is satisfied:

AXIOM 5. $\overline{(p)} = (p)$ where $p \in 1$,

the space is called a *$\mathcal{T}_1$-space*(1) (or *topological in the strict sense* in contrast to general topological space satisfying Axioms 1–4).

If not otherwise stated, "space" will always mean a $\mathcal{T}_1$-space

(1) Analogous axioms were introduced by F. Riesz, *Stetigkeitsbegriff und abstrakte Mengenlehre*, Atti del IV Congr. Int. d. Mat., vol. II, Roma 1909. See also my paper *Sur l'opération $\overline{A}$ de l'Analysis Situs*, Fund. Math. 3 (1922), pp. 182–199.

$\mathcal{T}_1$-spaces are called *accessibles* by M. Fréchet. See his book *Espaces abstraits*, Paris 1928, p. 185. Compare also E. H. Moore, *On a form of general analysis*, Yale Coll. 1910.

A considerable part of the topology can be developed in terms of the Boolean algebra completed by the operation $\overline{A}$ and without involving the notion of point, i.e. in terms of closure algebra; see R. Sikorski, *Closure algebras*, Fund. Math. 36 (1949), pp. 165–206; H. Rasiowa and R. Sikorski, *The mathematics of metamathematics*, Monografie Matematyczne 41, Warszawa 1963, Chapter III (*Topological Boolean algebras*); G. Nöbeling, *Grundlagen der analytischen Topologie*, Berlin (Springer) 1954, J. C. C. McKinsey and A. Tarski, *The algebra of topology*, Annals of Mathematics 45 (1944), pp. 141–191.

For a further analysis of the system of axioms, see A. Monteiro, *Caractérisation de l'opération de fermeture par un seul axiome*, Portugaliae Math. 4 (1945), pp. 158–160; K. Iseki, *On definition of topological space*, Journ. Osaka Inst. 1 (1949), pp. 97–98. See also (for some generalizations) P. C. Hammer, *Extended topology*, Nieuw Archief 9 (1961), 10 (1962), and Proc. Acad. Amsterdam 1963.

II. Geometrical interpretation[1]. In the case where 1 denotes the euclidean space (of dimension n), $\overline{X}$ is the set X augmented by its accumulation points. We shall prove that the axioms are then satisfied.

First suppose that $p \in \overline{X \cup Y}$; hence $p = \lim p_n$, where $p_n \in \overline{X \cup Y}$. There exists therefore an infinite number of p_n's which all belong either to X or to Y; in the first case $p \in \overline{X}$, and in the second case $p \in \overline{Y}$. Hence, in each case, $p \in \overline{X} \cup \overline{Y}$. Therefore $\overline{X \cup Y} \subset \overline{X} \cup \overline{Y}$.

On the other hand, it is evident that if p belongs to the closure of X, then it must belong to the closure of every set containing X, and hence to $\overline{X \cup Y}$. Consequently, $\overline{X} \cup \overline{Y} \subset \overline{X \cup Y}$ and we see that Axiom 1 is satisfied.

Axioms 2 and 3 are manifestly satisfied. It therefore remains only to prove Axiom 4. By the definition, $\overline{X} \subset \overline{\overline{X}}$. To prove that $\overline{\overline{X}} \subset \overline{X}$, suppose that $p \in \overline{\overline{X}}$ and that S is a ball (of dimension n) which contains p in its interior. Since the point p belongs to the closure of the set $\overline{X}$, there exists in the interior of S a point $r \in \overline{X}$; the last condition implies the existence of a point s belonging to $S \cap X$. Therefore each ball containing p in its interior contains a point of X. This implies the formula $p \in \overline{X}$.

III. Rules of topological calculus:

1. $X \subset Y$ *implies* $\overline{X} \subset \overline{Y}$;
2. $\overline{X \cap Y} \subset \overline{X} \cap \overline{Y}$;
3. $\overline{X} - \overline{Y} \subset \overline{X - Y}$;

3a. $\overline{X} : \overline{Y} \subset \overline{X : Y}$;

4. $\overline{\bigcap_\iota X_\iota} \subset \bigcap_\iota \overline{X_\iota}$;
5. $\bigcup_\iota \overline{X_\iota} \subset \overline{\bigcup_\iota X_\iota}$;
6. *if* X *is finite, then* $\overline{X} = X$;
7. $\overline{1} = 1$.

The first five rules follow from Axiom 1.

To prove rule 1 we remark that (according to § 1, II) the inclusion $X \subset Y$ is equivalent to the condition $Y = X \cup Y$ which implies $\overline{Y} = \overline{X \cup Y}$. Hence, by Axiom 1, $\overline{Y} = \overline{X} \cup \overline{Y}$ which is equivalent to the inclusion $\overline{X} \subset \overline{Y}$.

[1] For an interpretation in the domain of mathematical logic, see A. Tarski, *Der Aussagenkalkül und die Topologie*, Fund. Math. 31 (1938), p. 103.

Rule 1 implies 2, since the inclusions $X \cap Y \subset X$ and $X \cap Y \subset Y$ imply $\overline{X \cap Y} \subset \overline{X}$ and $\overline{X \cap Y} \subset \overline{Y}$, whence $\overline{X \cap Y} \subset \overline{X} \cap \overline{Y}$.

The identity $X \cup Y = (X - Y) \cup Y$ implies by virtue of Axiom 1 that $\overline{X} \cup \overline{Y} = \overline{X - Y} \cup \overline{Y}$ and, by taking the intersection of both sides of this identity with $1 - \overline{Y}$, we obtain $\overline{X} - \overline{Y} = \overline{X - Y} - \overline{Y} \subset \subset \overline{X - Y}$ and rule 3 follows.

Rule 3a follows easily from rule 3.

Rule 4 is a generalization of 2 (for an arbitrary, countable or uncountable set of factors) and it can be proved in an analogous way. Thus, for each index $\varkappa$, we have $\bigcap_\iota X_\iota \subset X_\varkappa$, whence $\overline{\bigcap_\iota X_\iota} \subset \overline{X}_\varkappa$ and $\overline{\bigcap_\iota X_\iota} \subset \bigcap_\varkappa \overline{X}_\varkappa$.

Similarly, the inclusion $X_\varkappa \subset \bigcup_\iota X_\iota$ implies $\overline{X}_\varkappa \subset \overline{\bigcup_\iota X_\iota}$, whence $\bigcup_\varkappa \overline{X}_\varkappa \subset \overline{\bigcup_\iota X_\iota}$. This proves rule 5.

Rule 6 is a consequence of Axioms 1 and 5.

Rule 7 follows from Axiom 2.

Formula 5 can be replaced by the more precise formula

8. $\overline{\bigcup_\iota X_\iota} = \bigcup_\iota \overline{X}_\iota \cup \bigcap_{\iota_1 \dots \iota_k} \overline{\bigcup_\varkappa X_\varkappa}$ where $\varkappa \neq \iota_j$ for $j \leqslant k$

(the operator $\bigcap$ being extended to all finite systems of values of ι).

According to formulas 5 and 1 the right-hand side of 8 is a subset of its left-hand side. In order to prove the inclusion in the opposite direction, let us consider a system $\iota_1 \dots \iota_k$. We have, by Axiom 1,

$$\overline{X}_{\iota_1} \cup \dots \cup \overline{X}_{\iota_k} \cup \overline{\bigcup_\varkappa X_\varkappa} = \overline{\bigcup_\iota X_\iota}.$$

Hence

$$\overline{\bigcup_\iota X_\iota} \subset \bigcup_\iota \overline{X}_\iota \cup \overline{\bigcup_\varkappa X_\varkappa}.$$

This formula being true for every system $\iota_1 \dots \iota_k$, the inclusion from left to right in 8 follows.

A particular case of 8 is

9. $$\overline{\bigcup_{n=1}^{\infty} X_n} = \bigcup_{n=1}^{\infty} \overline{X}_n \cup \bigcap_{n=1}^{\infty} \overline{X_n \cup X_{n+1} \cup \dots}$$

Remark. It is easily seen that all the above formulas with the exception of formula 6 are valid in all topological spaces (not necessarily $\mathscr{T}_1$-spaces).

It is also seen that in the definition of $\mathcal{T}_1$-spaces Axiom 2 can be omitted (this follows from Axioms 1 and 5, X being the union of one-element sets); Axiom 3 can be omitted too if the space is supposed to contain more than one point.

IV. Relativization. If E is a fixed set of points and X is an arbitrary subset of E, we call $E \cap \overline{X}$ the closure of X *relative to* E. The relative closure satisfies Axioms 1–5 relativized to E; that is to say, if X and Y are two arbitrary subsets of E, then

(1_E) $E \cap \overline{X \cup Y} = (E \cap \overline{X}) \cup (E \cap \overline{Y})$;

(4_E) $E \cap \overline{E \cap \overline{X}} = E \cap \overline{X}$;

(5_E) *if* X *is empty or contains at the most one point, then*

$$E \cap \overline{X} = X.$$

Proof. Propositions (1_E) and (5_E) are direct consequences of Axioms 1, 3, and 5. As to proposition (4_E) we have, by rule 2 and Axiom 4,

$$\overline{E \cap \overline{X}} \subset \overline{E} \cap \overline{\overline{X}} \subset \overline{\overline{X}} = \overline{X}, \quad \text{whence} \quad E \cap \overline{E \cap \overline{X}} \subset E \cap \overline{X}.$$

Since the reverse inclusion is a consequence of Axiom 2, the identity (4_E) follows.

It has been proved, therefore, that Axioms 1–5 can be relativized to an arbitrary set E. Consequently the same is true for the theorems which follow from Axioms 1–5: they remain valid if one considers as a space an arbitrary subset E of 1 (and if the closure is relativized).

We saw in Section II that Axioms 1–5 are satisfied in the euclidean space. It follows that these axioms are also satisfied when one takes as a space an arbitrary subset of the euclidean space.

V. Logical analysis of the system of axioms. The Axioms 1, 4, and 5 are independent. In fact, if we take as the space a set composed of two elements a and b, and if we put $\overline{0} = 0$, $\overline{(a)} = (a)$, $\overline{(b)} = (b)$ and $\overline{(a, b)} = 0$, then Axioms 4 and 5 are satisfied, but Axiom 1 is not. If, for a non-empty space, we put $\overline{X} = 0$ for every X, then Axioms 1 and 4 are satisfied, but 5 is not. Finally, to show the independence of Axiom 4, consider the following very instructive example[1]: The space is composed of all real functions of real

[1] Due to M. Fréchet; see his Thesis, Rend. del Circolo Matem. di Palermo 22 (1906), p. 15.

For an extensive study of spaces for which Axiom 4 is not assumed, see the treatise of E. Čech, *Topologické prostory*.

variable. X being a subset of this space, let every limit function of a sequence of functions of X belong to $\overline{X}$. This function space satisfies Axioms 1 and 5 but does not satisfy 4. In fact, if A denotes the set of continuous functions, then $\overline{\overline{A}} \neq \overline{A}$, for the Dirichlet function equal to 1 for rational points and to 0 for irrational points belongs to $\overline{\overline{A}}$ but does not belong to $\overline{A}$.

Each of the four Axioms 1–4 can be expressed in the form $F(X_1, \ldots, X_k) = 0$, where the function F involves only Boolean algebra operations and the operation $\overline{X}$. It is remarkable that no other axiom of this form exists which would be independent of the considered system and which would be satisfied in the n-dimensional euclidean space([1]).

A partial solution of this problem is furnished by the following table([2]).

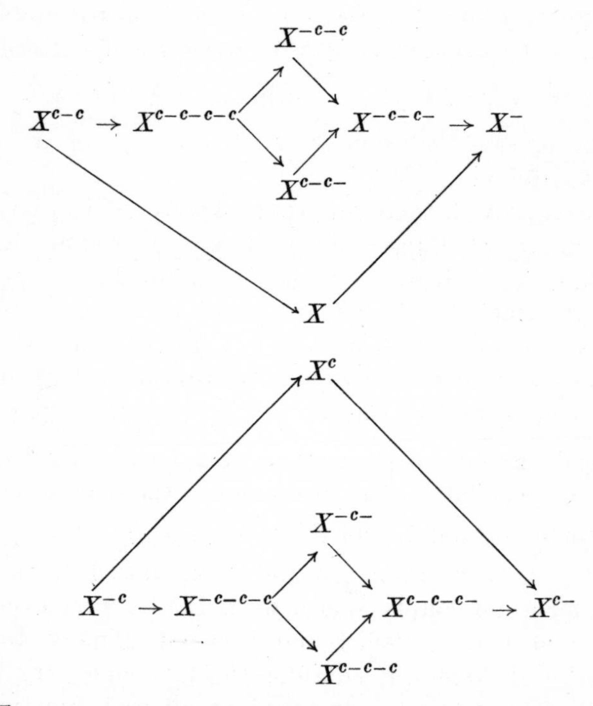

([1]) Theorem of J. C. C. McKinsey and A. Tarski, *The Algebra of Topology*, Ann. of Math. 45 (1944), pp. 141–191.

([2]) For typographic reasons we write X^{-} instead of $\overline{X}$; here the arrow replaces the inclusion sign $\subset$.

Suppose that we apply, to a set X, two operations, $\overline{X}$ and X^c $(= 1-X)$. What is the number of sets that we obtain?

One can prove that *this number is* 14 (1).

The 14 sets in question are given in the table above. The same table contains all the inclusions generally valid.

Remark. The above theorem is a particular case of the following theorem on ordered sets:

Let A *be an ordered set* (see § 3, X) *and let* $f\colon A \to A$ *be an increasing* (*isotonic*), *expanding and idempotent mapping, and let* $g\colon A \to A$ *be a decreasing involution*; in other terms

$$x \leqslant y \text{ implies } f(x) \leqslant f(y) \text{ and } g(y) \leqslant g(x),$$

$$x \leqslant f(x), \quad ff(x) = f(x), \quad gg(x) = x.$$

Then, the semigroup generated by f *and* g *consists of* 14 *elements* (*at most*). *Namely* (*we put* $gfg = i$): *identity*, g, f, fg, gf, i, fgf, fi, if, gfi, fif, $fgfi$, $gfif$, *and* ifi.

A table analogous to the table given above exhibits all the order relations valid for elements of the semi-group under consideration.

The proof is quite elementary (see P. C. Hammer, *op. cit.*). One proves first that the mapping i (which corresponds to Int, see § 6) is increasing, shrinking and idempotent. Next, one shows that the mappings fi and if are idempotent. This completes the proof.

§ 5. Closed sets, open sets

I. Definitions. The set X is *closed* if $\overline{X} = X$ (2). The set X is *open*, if the complement of X is closed, i.e. if $\overline{1-X} = 1-X$, or else, if $X = 1-\overline{1-X}$.

EXAMPLES. In the space of real numbers, the set of integers, as well as any interval $a \leqslant x \leqslant b$ are closed; the interval $a < x < b$ is open (it is not closed); in the plane, the latter set is not open.

(1) For a proof, see my paper of Fund. Math. 3, p. 196, quoted above. For some connected problems, see P. C. Hammer, *Kuratowski's closure theorem*, Nieuw Archief 8 (1960), pp. 74–80, and T. A. Chapman, *An extension of the Kuratowski closure and complementation problem*, Math. Magaz. 35 (1962), pp. 31–35.

(2) Notion due to G. Cantor, Math. Ann. 21 (1883), p. 51.

In the space of integers, every set is simultaneously closed and open.

If f is a bounded function defined in the interval $a \leqslant x \leqslant b$, then its graph, i.e. the set $\underset{xy}{E}\big(y = f(x)\big)$ is closed (in the plane) if and only if the function f is continuous (see § 20, V, Theorem 8).

II. Operations.

THEOREM 1. *The union of two closed sets is a closed set.*

This follows from Axiom 1 if we put $\overline{X} = X$, $\overline{Y} = Y$.

THEOREM 2. *The intersection* (of a finite or infinite number) *of closed sets is closed.*

Proof. If we put $\overline{X}_\iota = X_\iota$ in rule 4, we have $\overline{\bigcap_\iota X_\iota} \subset \bigcap_\iota X_\iota$ and since, by Axiom 2, $\bigcap_\iota X_\iota \subset \overline{\bigcap_\iota X_\iota}$, hence $\overline{\bigcap_\iota X_\iota} = \bigcap_\iota X_\iota$.

Using the de Morgan formula (see § 1, II and V), according to which $1-(X \cap Y) = (1-X) \cup (1-Y)$, and, in general, $1-\bigcap_\iota X_\iota = \bigcup_\iota (1-X_\iota)$, we infer from the preceding proposition that *the intersection of two open sets is open* and that *the union of an arbitrary family of open sets is open.*

According to Axiom 3 and rule 7, the sets 0 and 1 are simultaneously closed and open.

By Axiom 4, $\overline{X}$ is closed. Moreover, $\overline{X}$ *is the smallest closed set containing* X, i.e. $\overline{X}$ is the intersection of all closed sets containing X. This is because the condition $X \subset F$, where $F = \overline{F}$, implies $\overline{X} \subset F$ (cf. § 4, III, 1).

Remark 1. The definition of a *topological space* (see § 4, I) can be equivalently formulated assuming *closed set* as a primitive term (instead of closure) and supposing that the following axioms are satisfied:

AXIOM 1′. *The union of a finite family of closed sets is a closed set.*

AXIOM 2′. *The intersection of an arbitrary family of closed sets is a closed set.*

Since the case of a void family is also included, it follows that (see § 1, V): the void set as well as the whole space are both closed.

Defining in this space the closure $\overline{X}$ of X as the intersection of all closed sets containing X, it is easily seen that Axioms 1–4 of Section I are satisfied.

Conversely, as we have shown, Axioms 1′ and 2′ are satisfied in every topological space.

R e m a r k 2. Another equivalent form of defining topological space is the following (dual to the preceding) [1]. We assume the notion of the *open* set as a primitive term satisfying the following axioms:

AXIOM 1″. *The intersection of a finite family of open sets is an open set.*

AXIOM 2″. *The union of an arbitrary family of open sets is an open set.*

Since a closed set is defined as the complement of open set, an argument similar to the preceding one proves the equivalence of the above definition of topological space and of that given in Section I based on Axioms 1–4.

III. Properties. One can define the closed sets to be sets of the form $\bar{X}$. Similarly, the open sets coincide with the sets of the form $(1-\bar{X})$.

THEOREM. *If G is open, then $G \cap \bar{X} \subset \overline{G \cap X}$ for every X.*

For, by definition of an open set, $G = 1-\overline{1-G}$, whence $G \cap \bar{X} = \bar{X}-\overline{1-G} \subset \overline{X-(1-G)} = \overline{G \cap X}$, by rule 3.

The inclusion $G \cap \bar{X} \subset \overline{G \cap X}$ implies $\overline{G \cap \bar{X}} \subset \overline{G \cap X}$ and, by the formula $X \subset \bar{X}$, we have $\overline{G \cap X} \subset \overline{G \cap \bar{X}}$; this yields an important identity

$$\overline{G \cap \bar{X}} = \overline{G \cap X}. \tag{1}$$

IV. Relativization. According to the terminology adopted in § 4, IV, a set X is *closed relative to* E, if $X = E \cap \bar{X}$. The set X is relatively open, if $X \subset E$ and $(E-X)$ is relatively closed (with respect to E); in other words, if $X = E-\overline{E-X}$.

THEOREM. *A necessary and sufficient condition for a set to be closed (open) relative to E is that it is to be the intersection of E and of a closed (open) set.*

(1) See P. Alexandrov, Math. Ann. 94 (1925), p. 208, and W. Sierpiński, Math. Ann. 97 (1926), p. 335, and *Topologia* (1928). See also § 7, II below. Compare A. Monteiro, *Les ensembles fermés et les fondements de la Topologie*, Portug. Math. 2 (1941), pp. 56–66.

Proof. The condition is sufficient, for if X is relatively closed, then $X = E \cap \overline{X}$, and if X is relatively open, then $X = E-\overline{E-X}$ $= E \cap (1-\overline{E-X})$.

To show that the condition is sufficient, suppose first that $X = E \cap F$ and $F = \overline{F}$. We have to prove that X is closed relative to E, i.e. that $X = E \cap \overline{X}$ or that $E \cap F = E \cap \overline{E \cap F}$. Thus, by rule 1, $\overline{E \cap F} \subset \overline{F}$ and since $\overline{F} = F$, then $E \cap \overline{E \cap F} \subset E \cap F$. The reverse inclusion is a direct result of Axiom 2.

Finally, if $X = E \cap G$ and G is open, the set $(E-X)$, equal to $E-E \cap G = E \cap (1-G)$, is an intersection of E with a closed set; therefore it is closed relative to E. Consequently, X is relatively open, q.e.d.

In particular, if E is closed (open), the property of being closed (open) relative to E implies the same property in the absolute sense. This results from the preceding theorem by the fact that the intersection of two closed sets is closed and that the intersection of two open sets is open (see Section II).

The same theorem implies that the property of being relatively closed is transitive, i.e. that *if X is closed in Y and Y is closed in E, then X is closed in E.*

Since, by the hypothesis, $X = Y \cap \overline{X}$ and $Y = E \cap \overline{Y}$, so $X = E \cap \overline{X} \cap \overline{Y}$ and X is relatively closed in E as the intersection of E with a closed set.

The same holds for the property of being relatively open.

V. F_σ-sets, G_δ-sets.

The union of a countable(1) *family of closed sets is said to be an* F_σ; *the intersection of a countable family of open sets is said to be a* G_δ(2).

One sees readily that the complement of an F_σ-set is a G_δ-set and the complement of a G_δ-set is an F_σ-set. The union of an

(1) Note that by a *countable* set we mean a set whose elements can be arranged in a finite or infinite sequence in this sense, every finite set is countable.

(2) These notions which are generalizations of the notion of a closed set and of an open set were studied specifically for purposes of the theory of functions. However, they have proved to be very useful in geometrical problems of the topology of complete metric spaces (see Chapter III). They are due to W. H. Young, Ber. Ges. Wiss. Leipzig 55 (1903),p. 287.

infinite countable family of $\boldsymbol{F}_\sigma$-sets is evidently an $\boldsymbol{F}_\sigma$-set. The intersection of two $\boldsymbol{F}_\sigma$ is an $\boldsymbol{F}_\sigma$; for, if $A = \bigcup_{n=1}^{\infty} A_n$ and $B = \bigcup_{n=1}^{\infty} B_n$, then $A \cap B = \bigcup_{n,m=1}^{\infty} A_n \cap B_m$ and, the sets A_n and B_m being closed, their intersection $A_n \cap B_m$ is also closed. Hence $A \cap B$ is an $\boldsymbol{F}_\sigma$-set. By symmetry, the intersection of an infinite countable family of $\boldsymbol{G}_\delta$-sets is a $\boldsymbol{G}_\delta$-set; and the union of two $\boldsymbol{G}_\delta$-sets is a $\boldsymbol{G}_\delta$-set.

Every $\boldsymbol{F}_\sigma$-set is the union of an *increasing* sequence of closed sets. For

$$F_1 \cup F_2 \cup F_3 \ldots = F_1 \cup (F_1 \cup F_2) \cup (F_1 \cup F_2 \cup F_3) \cup \ldots$$

and the sets in the parentheses are closed.

Similarly, every $\boldsymbol{G}_\delta$-set is the intersection of a *decreasing* sequence of open sets.

In order that X should be an $\boldsymbol{F}_\sigma$-set (a $\boldsymbol{G}_\delta$-set) *relative to E*, it is necessary and sufficient that X should be the intersection of E with an $\boldsymbol{F}_\sigma$-set (a $\boldsymbol{G}_\delta$-set).

This is a consequence of the identities:

$$(F_1 \cap E) \cup (F_2 \cap E) \cup \ldots = (F_1 \cup F_2 \cup \ldots) \cap E$$

and

$$(G_1 \cap E) \cap (G_2 \cap E) \cap \ldots = (G_1 \cap G_2 \cap \ldots) \cap E.$$

In particular, if E is an $\boldsymbol{F}_\sigma$-set (a $\boldsymbol{G}_\delta$-set), then so is X.

By Axiom 5, every countable set is an $\boldsymbol{F}_\sigma$.

EXAMPLES. The set of rational numbers in the space of real numbers is an $\boldsymbol{F}_\sigma$-set. The set of irrational numbers is a $\boldsymbol{G}_\delta$-set. We shall see later that the set of irrational numbers is not an $\boldsymbol{F}_\sigma$-set.

VI. Borel sets. By generalizing the notions of closed and open set with the aid of set theory operations (as has been done in the preceding section) we are led to consider the sets which can be derived from the closed (or open) sets by the operations of countable union and intersection, and by the operation of subtraction. We make the following definition.

DEFINITION (1). The family $\boldsymbol{F}$ of *Borel sets* is the smallest family satisfying the following conditions:

(1) every closed set belongs to $\boldsymbol{F}$;

(2) if X belongs to $\boldsymbol{F}$, then $1-X$ belongs to $\boldsymbol{F}$;

(3) if X_n $(n = 1, 2, \ldots)$ belong to $\boldsymbol{F}$, then the set $\bigcap_{n=1}^{\infty} X_n$ belongs to $\boldsymbol{F}$.

It is seen that condition (1) can be replaced by

(1′) every open set belongs to $\boldsymbol{F}$,

and that condition (3) an be replaced by (3′) obtained by substituting $\bigcup$ for $\bigcap$ in (3).

A more detailed study of Borel sets is found in Chapters II and III.

The sets encountered mostly in applications of topology are Borel sets. On the other hand, as we shall see, examples of not Borel sets are known.

For relative Borel sets, we have the following theorem.

THEOREM. *Given a set E, the Borel sets relative to E are identical with sets of the form $B \cap E$, where B is a variable Borel set.*

Proof. The family of sets X, such that $X \cap E$ is a Borel set relative to E, satisfies conditions (1)–(3), for

(i) if X is closed, $X \cap E$ is closed in E, hence a Borel set relative to E;

(ii) if $X \cap E$ is a Borel set in E, then so is $(1-X) \cap E$, since $(1-X) \cap E = E-(X \cap E)$;

(iii) if each $E \cap X_n$ is a Borel set in E, then so is $E \cap \bigcap_{n=1}^{\infty} X_n$, since $E \cap \bigcap_{n=1}^{\infty} X_n = \bigcap_{n=1}^{\infty} (E \cap X_n)$.

Thus, the family of Borel sets being the smallest family satisfying conditions (1)–(3), is contained in the family of sets X in question. That is to say, if X is a Borel set, then $X \cap E$ is a Borel set relative to E.

Conversely, the family of sets $X = B \cap E$, where B is a Borel set, satisfies conditions (1)–(3) relativized to E, since

(1) See E. Borel, *Leçons sur la théorie des fonctions*, Paris 1898, p. 46, and F. Hausdorff, *Mengenlehre*, § 18 "Borelsche Systeme", and *Grundzüge*, p. 304.

(i) every closed set in E belongs to this family;

(ii) if $X = B \cap E$, the set $E - X = E \cap (1 - B)$, as an intersection of E with a Borel set, belongs to the considered family;

(iii) if $X_n = E \cap B_n$, then $\bigcap_{n=1}^{\infty} X_n = E \cap \bigcap_{n=1}^{\infty} B_n$, hence $\bigcap_{n=1}^{\infty} X_n$ also belongs to the considered family.

Consequently, this family contains the family of Borel sets relativé to E. In other words, if X is a Borel set relative to E, then X is an intersection of E with a Borel set.

VII. Cover of a space. Refinement. A family $\boldsymbol{A}$ of open sets is a *cover* (open) of the space $\mathscr{X}$ if each point of $\mathscr{X}$ belongs to some member of $\boldsymbol{A}$: $\mathscr{X} = \mathrm{S}(\boldsymbol{A})$.

$\boldsymbol{B}$ is a *subcover* of $\boldsymbol{A}$ if $\boldsymbol{B}$ is a cover of $\mathscr{X}$ and $\boldsymbol{B} \subset \boldsymbol{A}$.

$\boldsymbol{B}$ is a *refinement* of $\boldsymbol{A}$ if $\boldsymbol{B}$ is a cover of $\mathscr{X}$ and each element of $\boldsymbol{B}$ is contained in some element of $\boldsymbol{A}$; in this case we write $\boldsymbol{A} \leqslant \boldsymbol{B}$ ($\boldsymbol{A}$ precedes or equals $\boldsymbol{B}$).

Obviously, the relation $\boldsymbol{A} \leqslant \boldsymbol{B}$ establishes a quasi-order in the set of all covers of $\mathscr{X}$. Moreover, this set is a *directed* set (see § 3, X). For $\boldsymbol{A}$ and $\boldsymbol{B}$ being two covers, denote by $\boldsymbol{C}$ the family of all sets of the form $U \cap V$, where $U \in \boldsymbol{A}$ and $V \in \boldsymbol{B}$. It follows that $\boldsymbol{C}$ is a cover and that $\boldsymbol{A} \leqslant \boldsymbol{C}$ and $\boldsymbol{B} \leqslant \boldsymbol{C}$.

A family $\boldsymbol{A}$ is said to be *locally finite*, if each point p is contained in an open set G which intersects only a finite number of the elements of $\boldsymbol{A}$.

Locally finite families have the following interesting property:

THEOREM. *If* $\{X_t\}$, $t \in T$ *(T arbitrary), is a locally finite family, then*

$$\overline{\bigcup_t X_t} = \bigcup_t \overline{X}_t. \tag{1}$$

Proof. In view of § 4, III, 5, it is only necessary to show that

$$\overline{\bigcup_t X_t} \subset \bigcup_t \overline{X}_t. \tag{2}$$

Let $p \in \overline{\bigcup_t X_t} \cap G$, where G is open such that

$$G \cap X_{t_i} \neq 0 \text{ for } 1 \leqslant i \leqslant n \quad \text{and} \quad G \cap X_t = 0 \text{ for } X_t \neq X_{t_i}. \tag{3}$$

We shall prove that there is an i such that $p \in \overline{X}_{t_i}$.

Suppose, on the contrary, that

$$p \notin (\overline{X}_{t_1} \cup \ldots \cup \overline{X}_{t_n}).$$

Put $H = G - (\overline{X}_{t_1} \cup \ldots \cup \overline{X}_{t_n})$. It follows that $p \in H$ and (by (3)) that $H \cap \bigcup_t X_t = 0$. But this contradicts our assumption that $p \in \overline{\bigcup_t X_t}$.

Let us add that a family is said to be *σ-locally finite* if it is the union of a countable number of locally finite families.

The notions defined above give rise to a number of important classes of spaces which will be studied later (also in Volume II).

$\mathcal{X}$ is said to be *compact* (or *bicompact*) if every cover of $\mathcal{X}$ contains a finite subcover. $\mathcal{X}$ is *countably compact* if every countable cover contains a finite subcover. $\mathcal{X}$ is called a *Lindelöf space* if every cover contains a countable subcover.

$\mathcal{X}$ is *paracompact* if each cover has a locally finite refinement (¹).

VIII. Hausdorff spaces. A topological space is called a *Hausdorff* space (or a $\mathcal{T}_2$-space), if for each pair of points $p \neq q$ there exist two open sets G and H such that

$$p \in G, \quad q \in H, \quad G \cap H = 0 \text{(}^2\text{)}. \tag{1}$$

Obviously a euclidean space is Hausdorff.

THEOREM 1. *Each Hausdorff space is a $\mathcal{T}_1$-space.*

Let p be a given point. By assumption, each $x \neq p$ belongs to an open G_x such that p is not in G_x. Consequently, $1 - (p) = \bigcup_{x \neq p} G_x$. Thus $1 - (p)$ is open, and (p) is closed.

THEOREM 2. *Each subset of a Hausdorff space is Hausdorff.*

Let $p \neq q$ be in E. The space being Hausdorff, let G and H be open and satisfying (1). Then $G \cap E$ and $H \cap E$ are open relative to E and contain p and q, respectively.

(¹) For other notions related to the concept of a cover, see R. Arens and J. Dugundji, *Remark on the concept of compactness*, Portug. Math. 9 (1950), pp. 141–143.

(²) For other equivalent definitions of Hausdorff spaces, see Bourbaki, *l. c.*, p. 85 (called there *separated spaces*), and J. Sebastiao e Silva, *Sur l'axiomatique des espaces de Hausdorff*, Portug. Math. 2 (1941), pp. 93–109.

Remark 1. The definition of a Hausdorff space can be expressed in terms of *closure algebra* as follows: if $A \neq 0 \neq B$ and $A \cap B = 0$, there are two open sets G and H such that $A \cap G \neq 0 \neq B \cap H$ and $G \cap H = 0$(¹).

Remark 2. *There exist $\mathcal{T}_1$-spaces which are not Hausdorff.* Such is the space composed of the points $1/n$ for $n = 1, 2, \ldots$ and the point 0 with the following topology: Open sets which do not contain the point 1 are identical with open sets in the usual topology of real numbers; sets containing 1 are open if and only if they are complements of finite sets. Obviously, each open set containing 0 is infinite, and consequently the points 0 and 1 cannot be separated by means of disjoint open sets. (Note that 0 and 1 are the only two accumulation points of the space (see § 9). Of course, if one of the points p or q is isolated, condition (1) is satisfied.)

IX. $\mathcal{T}_0$-spaces. A topological space is called a *$\mathcal{T}_0$-space* if for each pair of different points there exists an open set which contains one of the points and does not contain the other one(²).

An equivalent statement to this is: if no two different points have the same closure.

Obviously, *each $\mathcal{T}_1$-space is a $\mathcal{T}_0$-space.*

Remark 1. The inverse is not true. An interesting example of a *$\mathcal{T}_0$-space which is not a $\mathcal{T}_1$-space*, is the following(³). Let us consider a simplex $p_0 \ldots p_n$ (see Chapter II, § 31). Our space has as points all the edges of this simplex (including the simplex itself). The closure of a point of this space, i.e. of a simplex S, is the set having as its elements all the edges of S; the closure of a set $\boldsymbol{E}$ is the union of closures of elements of $\boldsymbol{E}$. Thus the considered (finite) space is $\mathcal{T}_0$ but is not $\mathcal{T}_1$.

Remark 2. On the other hand, the space composed of two elements such that the closure of each of them is the whole space, is a *topological space but not a $\mathcal{T}_0$-space.*

(¹) See G. Nöbeling, *Grundlagen der analytischen Topologie*, p. 79.

(²) The notion of a $\mathcal{T}_0$-space was introduced by Kolmogorov. See Alexandrov-Hopf, p. 58.

(³) *Ibidem*, pp. 26, 40. For a more special study, see C. E. Aull and W. J. Thron, *Separation axioms between T_0 and T_1*, Scandin. Math. 24 (1962), p. 26.

X. Regular spaces. A notion less general than that of a Hausdorff space is the notion of a regular space(1).

DEFINITION. A topological space is called *regular* if every point p and every closed set F which does not contain p can be separated by open sets; i.e. if there are two open sets G_0 and G_1 such that

$$p \in G_0, \ F \subset G_1 \quad \text{and} \quad G_0 \cap G_1 = 0. \tag{1}$$

This can be stated equivalently: if there exists an open set G such that

$$p \in G \quad \text{and} \quad \bar{G} \cap F = 0. \tag{2}$$

A regular $\mathcal{T}_1$-space is called a $\mathcal{T}_3$-space.

THEOREM. *A subset of a regular space is regular.*

Proof. Let E be a subset of a regular space, and let $p \in (E-F)$ where F is closed relative to E, i.e. $F = \bar{F} \cap E$ (see Section IV). As p does not belong to $\bar{F}$, there are two open sets G_0 and G_1 such that $p \in G_0$, $\bar{F} \subset G_1$, and $G_0 \cap G_1 = 0$. Set $H_0 = G_0 \cap E$ and $H_1 = G_1 \cap E$. Hence H_0 and H_1 are open in E and $p \in H_0$, $F \subset H_1$ and $H_0 \cap H_1 = 0$. This means that E is regular.

Remark 1. The definition of a regular space can be expressed in terms of *closure algebra* as follows: if $A - \bar{B} \neq 0$ there is an open set G such that $\bar{B} \subset G$ and $A - \bar{G} \neq 0$(2).

Remark 2. *A Hausdorff space may fail to be regular.* An example of this is the following space. Let $A = (1, 1/2, \ldots, 1/n, \ldots)$. We define topology in the interval $\mathcal{I}$ as follows. Closed are all sets closed in the usual sense; in addition A is considered as being closed.

XI. Base and subbase. A family $\boldsymbol{B}$ of open sets is called a (open) *base* of the space if each open set can be represented as the union of elements of a subfamily of $\boldsymbol{B}$.

A family $\boldsymbol{S}$ of open sets is called a (open) *subbase* if the family of all finite intersections of elements of $\boldsymbol{S}$ is a base of the space(3).

(1) See L. Vietoris, Mon. f. Math. u. Ph. 31 (1921), p. 173, and H. Tietze, *Beiträge zur allgemeinen Topologie*, Math. Ann. 88 (1923), p. 301.

For more details, see N. Bourbaki *loc. cit.*, p. 91.

(2) See G. Nöbeling, *loc. cit.*

(3) Cf. Kelley, *loc. cit.* p. 48.

EXAMPLE 1. The family of all open intervals $r < x < s$, with r and s rational, is a base of the space $\mathscr{E}$ of all real numbers.

The family of all open circles in $\mathscr{E}^2$ with rational radius and with centre having rational coordinates is a base for $\mathscr{E}^2$.

EXAMPLE 2. The family of rays of the form $x > r$ and of the form $x < r$, with r rational, is a subbase of $\mathscr{E}$.

EXAMPLE 3. Consider the Cantor discontinuum $\mathscr{C}$ (see § 3, IX), i.e. the set of all numbers t of the form $t = (0, t^{(1)}, t^{(2)}, \ldots)_3$ where $t^{(n)} = 0$ or 2, or, what is equivalent, the set $(0, 2)^{\mathscr{D}}$. The sets

$$C_{n,a} = \underset{t}{E}(t^{(n)} = a), \quad \text{where } a \text{ is } 0 \text{ or } 2,$$

form a subbase for $\mathscr{C}$ (with its usual topology).

EXAMPLE 4. For the space $\mathscr{N}$ of irrational numbers, i.e. for the space $\mathscr{D}^{\mathscr{D}}$ of sequences of positive integers $\mathfrak{z} = (\mathfrak{z}^{(1)}, \mathfrak{z}^{(2)}, \ldots)$ (see § 3, IX), the sets

$$N_{n,m} = \underset{\mathfrak{z}}{E}(\mathfrak{z}^{(n)} = m) \quad \text{where} \quad n = 1, 2, \ldots, \ m = 1, 2, \ldots$$

form a subbase.

Remark 1. Let X be a set of arbitrary elements and $\boldsymbol{S}$ a family of subsets of X. As soon as we assume that $\boldsymbol{S}$ is a subbase of X, X becomes a topological space.

Remark 2. The spaces $\mathscr{E}$, $\mathscr{E}^2$, $\mathscr{C}$ and $\mathscr{N}$, considered in examples 1–4, have countable bases. The same is true of $\mathscr{E}^n$, $\mathscr{E}^{\aleph_0}$ and of their subspaces.

We shall show later (Chapter II) that in any regular $\mathscr{T}_1$-space having a countable base one can define a distance (in other words, the space is metrizable).

Remark 3. A space having a countable base (but having no finite base) is said to be of *weight* $\aleph_0$.

Generally, the *weight of a space* is the least cardinal $\mathfrak{m}$ such that the space has a base of cardinality $\mathfrak{m}$.

Remark 4. Besides the notions of open bases and subbases one may consider the dual notions of closed bases and closed subbases. Namely, a family $\boldsymbol{A}$ of closed sets is called a *closed base* of the space if each closed set can be represented as the intersection of some elements of $\boldsymbol{A}$. A family $\boldsymbol{R}$ of closed sets is called a *closed subbase* if the family of all finite unions of elements of $\boldsymbol{R}$ is a closed base of the space.

Obviously, $F \in \boldsymbol{A} \equiv (-F) \in \boldsymbol{B}$ and $F \in \boldsymbol{R} \equiv (-F) \in \boldsymbol{S}$.

Finally let us consider the notion of a *local* (*open*) *base at a point* p. Every family of open sets containing p and such that each open set containing p contains a member of that family is so called.

Topological spaces with a *countable local base at each point* are of great importance (they are said to satisfy the *first countability axiom*) (1). Such are, for example, metric spaces.

This leads in a natural way to the notions of *local subbase* and of *local weight*.

The spaces with a countable open base (or with a countable local base) will be studied in more detail in Chapter II. Here, let us show the following important relation of these spaces to the Lindelöf spaces (see Section VII):

LINDELÖF THEOREM (2). *Every topological space with a countable open base is a Lindelöf space.*

In other words, every (uncountable) family of open sets $\{G_t\}$ contains a countable sequence $G_{t_1}, G_{t_2}, \ldots$ such that

$$\bigcup_{n=1}^{\infty} G_{t_n} = \bigcup_{t} G_t.$$

Proof. Let $R_1, R_2, \ldots$ be the base of the space and let $R_{k_1}, R_{k_2}, \ldots$ be the sequence of all the sets contained in the sets of the family $\{G_t\}$. By the axiom of choice, for each index k_n there exists an index t_n such that $R_{k_n} \subset G_{t_n}$. Therefore

$$\bigcup_{n=1}^{\infty} R_{k_n} \subset \bigcup_{n=1}^{\infty} G_{t_n} \subset \bigcup_{t} G_t.$$

On the other hand, if $p \in G_t$, there exists an index j such that $p \in R_j \subset G_t$. Since the index j belongs to the sequence $k_1, k_2, \ldots$, it follows that

$$p \in \bigcup_{n=1}^{\infty} R_{k_n}.$$

---

(1) Axiom introduced by F. Hausdorff, *Grundzüge der Mengenlehre*, p. 263.

(2) C. R. Paris, vol. 137 (1903), p. 697. Compare also W. H. Young, Proc. London Math. Soc. (1) 35 (1903), p. 384.

Hence

$$\bigcup_t G_t \subset \bigcup_{n=1}^{\infty} R_{k_n},$$

and the proof is completed.

Remark. The Lindelöf theorem can be restated as follows: Every (uncountable) family of closed sets $\{F_t\}$ contains a (countable) sequence $F_{t_1}, F_{t_2}, \ldots$ such that

$$\bigcap_{n=1}^{\infty} F_{t_n} = \bigcap_t F_t.$$

It suffices to let G_t denote the complement of F_t and to apply the de Morgan law (§ 1, V) to the Lindelöf theorem.

§ 6. Boundary and interior of a set

I. Definitions(1). The *boundary* of a set X is the set

$$\mathrm{Fr}(X) = \overline{X} \cap \overline{1-X}.$$

The *interior* of X is the set

$$\mathrm{Int}(X) = 1-\overline{1-X}.$$

Examples. In the euclidean plane, if X denotes the circle $x^2+y^2 \leqslant 1$, $\mathrm{Fr}(X)$ is the circumference of the circle, $\mathrm{Int}(X)$ is the open disc. Nothing will change, if we replace the sign $\leqslant$ by $<$.

In the space of natural numbers, the boundary of every set is empty.

In the space of real numbers, the boundary of the set of rational numbers is the entire space.

If X is an arbitrary subset of the space of real numbers, and if f is the function defined by the conditions: $f(x) = 1$, for $x \epsilon X$, and $f(x) = 0$ otherwise, then the boundary of X constitutes the set of points of discontinuity of f (see § 13).

---

(1) See G. Cantor, Göttinger Nachr. 1879, p. 128, and C. Jordan, Journ. de Math. (4) 8 (1892), p. 72.

II. Formulas. We shall use in the sequel the following formulas [1]:

$$\mathrm{Int}(X \cap Y) = \mathrm{Int}(X) \cap \mathrm{Int}(Y), \tag{1}$$

$$X \subset Y \quad \textit{implies} \quad \mathrm{Int}(X) \subset \mathrm{Int}(Y), \tag{1'}$$

$$\bigcup_\iota \mathrm{Int}(X_\iota) \subset \mathrm{Int}(\bigcup_\iota X_\iota), \tag{2}$$

$$\mathrm{Int}(X) = X - \overline{1-X} = X - \mathrm{Fr}(X) \subset X, \tag{3}$$

$$\mathrm{Fr}(1-X) = \mathrm{Fr}(X), \tag{4}$$

$$\mathrm{Fr}(\overline{X}) \subset \mathrm{Fr}(X), \tag{5}$$

$$\mathrm{Fr}(X) = X \cap \overline{1-X} \cup (\overline{X} - X), \tag{6}$$

$$X \cup \mathrm{Fr}(X) = \overline{X}, \tag{7}$$

$$\mathrm{Fr}(X) \cup \mathrm{Fr}(Y) = \\ = \mathrm{Fr}(X \cup Y) \cup \mathrm{Fr}(X \cap Y) \cup \big(\mathrm{Fr}(X) \cap \mathrm{Fr}(Y)\big)\,[2], \tag{8}$$

$$\mathrm{Int}[\mathrm{Int}(X)] = \mathrm{Int}(X), \tag{9}$$

$$\mathrm{Int}(X) \cap \mathrm{Fr}(X) = 0, \tag{10}$$

$$\mathrm{Fr}\big(\mathrm{Int}(X)\big) \subset \mathrm{Fr}(X), \tag{11}$$

$$\overline{\mathrm{Int}\big(\mathrm{Fr}(X)\big)} = \overline{X \cap \mathrm{Int}\big(\mathrm{Fr}(X)\big)} = \overline{\mathrm{Int}\big(\mathrm{Fr}(X)\big) - X}, \tag{12}$$

$$\mathrm{Fr}\{\mathrm{Fr}[\mathrm{Fr}(X)]\} = \mathrm{Fr}[\mathrm{Fr}(X)], \tag{13}$$

$$\mathrm{Int}(X - Y) \subset \mathrm{Int}(X) - \mathrm{Int}(Y), \tag{14}$$

$$\overline{\mathrm{Int}(\overline{X \cup Y})} = \overline{\mathrm{Int}(\overline{X})} \cup \overline{\mathrm{Int}(\overline{Y})}. \tag{15}$$

Formulas (1)–(3) can be established as follows:

$$1 - \overline{1-(X \cap Y)} = 1 - \overline{(1-X) \cup (1-Y)}$$
$$= 1 - (\overline{1-X} \cup \overline{1-Y}) = (1 - \overline{1-X}) \cap (1 - \overline{1-Y}).$$

---

[1] Some of them can be found in a paper by M. Zarycki, *Quelques notions fondamentales de l'Analysis Situs au point de vue de l'Algèbre de la Logique*, Fund. Math. 9 (1927), pp. 3–15. The author considers also other set-to-set functions, such as the "border" $= X \cap \overline{1-X}$, and the "exterior" $= 1 - \overline{X}$. See also in that direction P. C. Hammer, *Extended topology* (Structure of isotonic functions), Journ. f.r.u.a. Math. 213(1964), pp. 174–186, and of the same author, Nieuw Archief 9(1961) and 10(1962) and Indag. Math. 1963.

[2] See A. H. Stone, Trans. Amer. Math. Soc. 65 (1949), p. 428.

$$\bigcup_\iota \operatorname{Int}(X_\iota) = 1-\bigcap_\iota \overline{1-X_\iota} \subset 1-\bigcap_\iota (1-X_\iota) = \operatorname{Int}(\bigcup_\iota X_\iota).$$

$$1-\overline{1-X} \subset 1-(1-X) = X;$$

hence

$$\operatorname{Int}(X) = X \cap (1-\overline{1-X}) = X-\overline{1-X} = X-(X \cap \overline{1-X})$$
$$= X-(X \cap \overline{X} \cap \overline{1-X}) = X-(\overline{X} \cap \overline{1-X}) = X-\operatorname{Fr}(X) \subset X.$$

Formulas (4) and (5) are evident.

$$\overline{X} \cap \overline{1-X} = \big((\overline{X} \cap \overline{1-X}) \cap X\big) \cup \big(\overline{X} \cap \overline{1-X} \cap (1-X)\big),$$

whence we obtain (6) by the inclusions $X \subset \overline{X}$ and $1-X \subset \overline{1-X}$.

$$X \cup \operatorname{Fr}(X) = X \cup (X \cap \overline{1-X}) \cup (\overline{X}-X) = X \cup (\overline{X}-X) = \overline{X},$$

whence formula (7) is obtained.

$$\operatorname{Fr}(X \cup Y) = \overline{X \cup Y} \cap \overline{1-(X \cup Y)} = \overline{X \cup Y} \cap \overline{(1-X) \cap (1-Y)}$$
$$\subset (\overline{X} \cap \overline{1-X} \cap \overline{1-Y}) \cup (\overline{Y} \cap \overline{1-X} \cap \overline{1-Y})$$
$$\subset (\overline{X} \cap \overline{1-X}) \cup (\overline{Y} \cap \overline{1-Y}) = \operatorname{Fr}(X) \cup \operatorname{Fr}(Y).$$

$$\operatorname{Fr}(X \cap Y) = \operatorname{Fr}\big(1-(X \cap Y)\big) = \operatorname{Fr}\big((1-X) \cup (1-Y)\big)$$
$$\subset \operatorname{Fr}(1-X) \cup \operatorname{Fr}(1-Y) = \operatorname{Fr}(X) \cup \operatorname{Fr}(Y).$$

On the other hand

$$X = (X \cap Y) \cup (X-Y) \quad \text{and} \quad 1-X = (Y-X) \cup \big(1-(X \cup Y)\big),$$

hence

$$\operatorname{Fr}(X) \subset (\overline{X \cap Y} \cap \overline{1-X}) \cup (\overline{X-Y} \cap \overline{Y-X}) \cup$$
$$\cup \big(\overline{X-Y} \cap \overline{1-(X \cup Y)}\big)$$
$$\subset \operatorname{Fr}(X \cap Y) \cup \big(\operatorname{Fr}(X) \cap \operatorname{Fr}(Y)\big) \cup \operatorname{Fr}(X \cup Y).$$

This completes the proof of (8).

Formulas (9) and (10) are evident and formula (11) follows from (4) and (5).

By (1′) and (3), the inclusion $\operatorname{Fr}(X) \subset \overline{X}$ implies that $\operatorname{Int}\big(\operatorname{Fr}(X)\big) \subset \overline{X}$. Denoting by G the (open) set $\operatorname{Int}\big(\operatorname{Fr}(X)\big)$, we have $G = G \cap \overline{X}$, whence, by § 5, III, $\bar{G} = \overline{G \cap \bar{X}} = \overline{G \cap X}$ which proves the first part of (12). The second part follows by (4) after substituting $1-X$ for X.

The proof of (13) and (14) we leave to the reader. Formula (15) will be established in § 8.

III. Relations to closed and to open sets. Evidently, *every boundary is a closed set* and *every interior is an open set*(1). Moreover, *the interior is the largest open subset of* X; for, if G is an open subset of X, then $1-X \subset 1-G = \overline{1-G}$, whence $\overline{1-X} \subset 1-G$ and therefore $G \subset 1-\overline{1-X} = \mathrm{Int}(X)$.

If X *is closed, then* $\mathrm{Fr}(X) = X \cap \overline{1-X}$; *if* X *is open, then* $\mathrm{Fr}(X) = \bar{X}-X$.

Each of these identities characterizes the closed and open sets, respectively, because it follows from (6) that, in the first case, $\bar{X}-X = 0$, whence $\bar{X} = X$, and, in the second case, $X \cap \overline{1-X} = 0$, hence $\overline{1-X} = 1-X$.

In particular, *the condition* $\mathrm{Fr}(X) = 0$ *is equivalent to the statement that* X *is simultaneously closed and open.*

THEOREM. *A necessary and sufficient condition that* X *is the difference of two closed sets is that* $\bar{X}-X$ *should be closed*(2).

Proof. Let $X = E-F$, where E and F are closed. Then

$$X = \bar{X} \cap (E-F) = (\bar{X} \cap E)-(\bar{X} \cap F),$$

and since $X \subset E$, then $\bar{X} \subset E$, whence $\bar{X} \cap E = \bar{X}$. Consequently $X = \bar{X}-(\bar{X} \cap F)$, whence $\bar{X}-X = \bar{X} \cap F$; therefore $\bar{X}-X$ is closed.

Conversely, if $\bar{X}-X$ is closed, the set X, being equal to $\bar{X}-(\bar{X}-X)$, is a difference of two closed sets.

IV. Addition theorem. The following theorem is related to formula (2) (it will be used in § 8):

---

(1) This last statement is equivalent to Axiom 4. It was considered by certain authors as an axiom under the name "Hedrick's condition". See E. R. Hedrick, Trans. Amer. Math. Soc. 12 (1911), p. 285 and M. Fréchet, *Espaces abstraits*, p. 201.

(2) See a paper by W. Sierpiński and myself, *Sur les différences de devx ensembles fermés*, Tôhoku Math. Journ. 20 (1921), p. 22.

THEOREM. *If* $\{X_\iota\}$ *is a family (of an arbitrary power) of sets open relative to the union* $\bigcup_\iota X_\iota$, *then*

$$\mathrm{Int}(\bigcup_\iota X_\iota) = \bigcup_\iota \mathrm{Int}(X_\iota), \tag{i}$$

$$\overline{\mathrm{Int}(\overline{\bigcup_\iota X_\iota})} = \overline{\bigcup_\iota \mathrm{Int}(\overline{X}_\iota)}. \tag{ii}$$

Proof. Let $S = \bigcup_\iota X_\iota$; we have $X_\iota = S - \overline{S-X_\iota} \subset 1-\overline{S-X_\iota}$, whence $S \subset \bigcup_\iota (1-\overline{S-X_\iota})$. Also

$$\begin{aligned}\mathrm{Int}(S) &= S \cap \mathrm{Int}(S) \subset \bigcup_\iota (1-\overline{S-X_\iota}) \cap (1-\overline{1-S}) \\ &= \bigcup_\iota \big(1-\overline{(S-X_\iota) \cup (1-S)}\big) = \bigcup_\iota (1-\overline{1-S \cap X_\iota}) \\ &= \bigcup_\iota (1-\overline{1-X_\iota}) = \bigcup_\iota \mathrm{Int}(X_\iota),\end{aligned}$$

and, by (2), we obtain (i).

By § 5, Section III, we have $\mathrm{Int}(\overline{S}) = \mathrm{Int}(\overline{S}) \cap \overline{S} \subset \overline{\mathrm{Int}(\overline{S}) \cap S}$. We have proved that $S \subset \bigcup_\iota (1-\overline{S-X_\iota})$; therefore $\mathrm{Int}(\overline{S}) \subset \overline{\mathrm{Int}(\overline{S}) \cap \bigcup_\iota (1-\overline{S-X_\iota})} = \overline{\bigcup_\iota \big(\mathrm{Int}(\overline{S})-\overline{S-X_\iota}\big)}$. Now $\mathrm{Int}(\overline{S})-\overline{S-X_\iota} \subset \overline{S}-\overline{S-X_\iota} \subset \overline{S-(S-X_\iota)} = \overline{X}_\iota$; the first member of this inclusion being open is contained in the interior of $\overline{X}_\iota$. Thus $\mathrm{Int}(\overline{S})-\overline{S-X_\iota} \subset \mathrm{Int}(\overline{X}_\iota)$. By the preceding formula we obtain $\mathrm{Int}(\overline{S}) \subset \overline{\bigcup_\iota \mathrm{Int}(\overline{X}_\iota)}$, hence $\overline{\mathrm{Int}(\overline{S})} \subset \overline{\bigcup_\iota \mathrm{Int}(\overline{X}_\iota)}$.

It remains to prove the reverse inclusion. By (1′), $\mathrm{Int}(\overline{X}_\iota) \subset \mathrm{Int}(\overline{S})$. Hence $\bigcup_\iota \mathrm{Int}(\overline{X}_\iota) \subset \mathrm{Int}(\overline{S})$ and, finally, $\overline{\bigcup_\iota \mathrm{Int}(\overline{X}_\iota)} \subset \overline{\mathrm{Int}(\overline{S})}$.

V. Separated sets.

DEFINITION. X and Y are said to be *separated sets*[1] if

$$\overline{X} \cap Y = 0 = X \cap \overline{Y}. \tag{1}$$

Their complement, $1-(X \cup Y)$, is said *to separate* X *from* Y (as well as each $V \subset X$ from each $W \subset Y$).

Two separated sets are *disjoint*, the converse, however, may fail to be true, as seen on simple examples.

[1] According to S. Mazurkiewicz, Fund. Math. 1 (1920), p. 66. See also a paper by B. Knaster and myself in Fund. Math. 2 (1921), p. 206.

THEOREM 1. *If X and Y are separated and $V \subset X$ and $W \subset Y$, then V and W are separated.*

This follows directly from (1).

THEOREM 2. *If A and B are closed (or open), the sets $A-B$ and $B-A$ are separated.*

If $\bar{A} = A$, then $\overline{A-B} \subset \bar{A} = A$, whence $\overline{A-B} \cap (B-A) = 0$.

The case of open sets reduces to the case of closed sets by virtue of the identity $A-B = (1-B)-(1-A)$.

THEOREM 3. *If A and B are both closed (or both open) and $A \cap B = 0$, then A and B are separated.*

This follows from Theorem 2 since $A-B = A$ and $B-A = B$.

THEOREM 4. *If X is separated from Y and from Z, then X is separated from $Y \cup Z$.*

This is because $\bar{X} \cap (Y \cup Z) = (\bar{X} \cap Y) \cup (\bar{X} \cap Z) = 0$ and on the other hand $X \cap \overline{Y \cup Z} = (X \cap \bar{Y}) \cup (X \cap \bar{Z}) = 0$.

THEOREM 5. *X and Y are separated if and only if they are disjoint and closed in their union.*

Proof. If X and Y are separated, we have by (1)

$$X = X \cup (\bar{X} \cap Y) = \bar{X} \cap (X \cup Y),$$

which means that X (as well as Y) is closed in $X \cup Y$.

On the other hand, if $X \cap Y = 0$ and $X = \bar{X} \cap (X \cup Y)$, it follows that $\bar{X} \cap Y \subset X \cap Y = 0$.

THEOREM 6. $\mathrm{Fr}(X)$ *separates the interior of X from the interior of its complement.*

This is because

$$1-\mathrm{Fr}(X) = 1-(\bar{X} \cap \overline{1-X}) = (1-\bar{X}) \cup (1-\overline{1-X}).$$

VI. Duality between the operations $\bar{A}$ and $A^\circ = \mathrm{Int}(A)$. The formula $-\bar{X} = \mathrm{Int}(-X)$, i.e. $-(X)^- = (-X)^\circ$, permits the duality of Boolean algebra (based on the de Morgan laws) to be extended to statements expressed with the help of the operations A^- and A°. In other words, these operations being dual, one can obtain from any true statement of that kind a true statement by replacing all the operations by its duals (recall the dual operations: $A \cup B$ and $A \cap B$, $A-B$ and $A:B$, $A \subset B$ and $A \supset B$, 0 and 1, comp. § 1, II).

For example, one obtains in this way II(1) from § 4, Axiom 1; II(2) from § 4, III, 4; II(14) from § 4, III(3a) and so on.

It should be noticed that, instead of $\overline{X}$, $\mathrm{Int}(X)$ *can be taken as a primitive term* of the definition of topological space([1]). The respective axioms are

α. $\mathrm{Int}(X \cap Y) = \mathrm{Int}(X) \cap \mathrm{Int}(Y)$;

β. $\mathrm{Int}(X) \subset X$;

γ. $\mathrm{Int}(1) = 1$;

δ. $\mathrm{Int}[\mathrm{Int}(X)] = \mathrm{Int}(X)$.

§ 7. Neighbourhood of a point. Localization of properties

I. Definitions. The set X is said to be a *neighbourhood of the point* p, if $p \in \mathrm{Int}(X)$, i.e. if p is an interior point of X; in other words, if p does not belong to $\overline{1-X}$.

An open set is a neighbourhood of each of its points. A neighbourhood of p contains an *open* neighbourhood of p; its interior is such an open neighbourhood.

Every set containing a neighbourhood of p is itself a neighbourhood of p. The intersection of two neighbourhoods of p is a neighbourhood of p, since (compare § 6, II(1)): $\mathrm{Int}(X \cap Y) = \mathrm{Int}(X) \cap \mathrm{Int}(Y)$. Thus *the family of neighbourhoods of p is a filter.*

A subset X of E is said to be a *relative neighbourhood of p with respect to E,* if p belongs to the interior of X relative to E, i.e. $p \in E - \overline{E-X}$.

THEOREM. *If X_j, for $j = 0$ and 1, is a neighbourhood of p relative to E_j, then $X_0 \cup X_1$ is a neighbourhood of p relative to $E_0 \cup E_1$.*

Proof. We have $p \in E_0 \cap E_1 - (\overline{E_0 - X_0} \cup \overline{E_1 - X_1})$. Since $(E_0 - X_0) \cup (E_1 - X_1) \supset \big(E_0 - (X_0 \cup X_1)\big) \cup \big(E_1 - (X_0 \cup X_1)\big) = (E_0 \cup E_1) - (X_0 \cup X_1)$, we obtain $p \in (E_0 \cup E_1) - \overline{(E_0 \cup E_1) - (X_0 \cup X_1)}$.

II. Equivalences.

THEOREM 1. *In order that $p \in \overline{X}$ it is necessary and sufficient that each neighbourhood E of p satisfies the inequality $X \cap E \neq 0$.*

([1]) The notion of boundary can be taken too as primitive term for a topological space. See M. Zarycki, *op. cit.* et J. Albuquerque, *La notion de "frontière" en topologie*, Portug. Math. 2 (1941), pp. 280–289.

Proof. Suppose that $p \in \overline{X}$ and let E be an arbitrary neighbourhood of p. This means that p does not belong to $\overline{1-E}$, and hence $p \in \overline{X} - \overline{1-E} \subset \overline{X-(1-E)} = \overline{X \cap E}$. Consequently $X \cap E \neq 0$.

On the other hand, suppose that each neighbourhood E of p satisfies $X \cap E \neq 0$. It then follows that the set $1-X$ is not a neighbourhood of p, whence $p \in 1 - \overline{(1-X)} = \overline{X}$.

This theorem implies directly the following corollary.

COROLLARY 1a. *In order that* $p \in \mathrm{Fr}(X)$ *it is a necessary and sufficient condition that each neighbourhood* E *of* p *should satisfy the double inequality* $E \cap X \neq 0 \neq E - X$.

In the two preceding statements, the term neighbourhood of p can be replaced by open set containing p.

Remark. The definition of a topological space (§ 4, I) can be equivalently formulated assuming *open neighbourhood* (*o.n.*) *of a point* as a primitive term and supposing the following axioms to be satisfied(1).

A. To every point p there corresponds at least one o.n. Each o.n. of p contains p.

B. If U and V are two o.n. of p, there exists an o.n. of p contained in $U \cap V$.

C. If U is an o.n. of p, and $q \in U$, there exists an o.n. V of q contained in U.

It is easily seen that, in a topological space the propositions A–C are fulfilled. Conversely, if the closure in a space satisfying A–C is defined by the condition expressed in Theorem 1, the space becomes topological (in the sense of § 4, I)(2).

---

(1) See F. Hausdorff, *Grundzüge der Mengenlehre*, p. 213. For the first approach of that kind, see D. Hilbert, Göttinger Nachrichten 1902 (reproduced in *Grundlagen der Geometrie*, Leipzig 1913, p. 165).

Also compare the (V)-spaces ("neighbourhood" spaces) of M. Fréchet, *Espaces abstraits*, p. 172.

(2) Problems of equivalence of closure topology and neighbourhood topology under weaker assumptions (not all conditions 1–4 on closure are assumed) were considered by M. M. Day, *Convergence, closure and neighbourhoods*, Duke Math. Journ. 11 (1944), pp. 181–199. See also H. Ribeiro, *Une extension de la notion de convergence*, Portug. Mathematica 2 (1941), pp. 153–161.

III. Converging filters. We say that a (proper) filter $\boldsymbol{F}$ (see § 1, VII) *converges* to a point p if each neighbourhood of p is a member of $\boldsymbol{F}$ (i.e. if $\boldsymbol{F}$ is finer than the filter of all neighbourhoods of p)(¹).

If the space is Hausdorff, any (proper) filter $\boldsymbol{F}$ *can converge to one point only, called the limit of* $\boldsymbol{F}$.

For suppose that $a \neq b$ are limit points of $\boldsymbol{F}$, and let A and B be two disjoint neighbourhoods of a and b. Then $A \in \boldsymbol{F}$ and $B \in \boldsymbol{F}$, and consequently $(A \cap B) \in \boldsymbol{F}$. But $A \cap B = 0$, whence 0 would be a member of $\boldsymbol{F}$, which is impossible.

IV. Localization. If $\mathbf{P}$ is a property of sets, we denote by $\boldsymbol{P}$ the family of sets which have this property. In many cases this leads to the following definition of localization:

DEFINITION. The set X *has the property* $\mathbf{P}$ *at the point* p if there exists a neighbourhood E of p such that $(X \cap E) \in \boldsymbol{P}$. The symbol X^* denotes *the set of points* p *(which may or may not belong to* X*) at which* X *does not have the property* $\mathbf{P}$.

Thus, for example, if $\boldsymbol{P}$ is the family composed of the empty set, then, by the theorem of Section II, $X^* = \overline{X}$. By localization of the properties of being a finite set and being a countable set we shall obtain later the notions of the derived set and of the set of condensation points of X (§ 9 and § 23).

We are going to study the operation X^* by requiring the family $\boldsymbol{P}$ to be an *ideal* (see § 1, VII), i.e.

(i) *the conditions* $X \in \boldsymbol{P}$ *and* $Y \subset X$ *imply* $Y \in \boldsymbol{P}$,

(ii) *the conditions* $X \in \boldsymbol{P}$ *and* $Y \in \boldsymbol{P}$ *imply* $(X \cup Y) \in \boldsymbol{P}$.

Consequences of (i). If we assume condition (i), we can require the neighbourhood E in the above definition to be open. The reason for this is that each neighbourhood E of p contains an open neighbourhood G of p (Section I). Hence, if $X \cap E \in \boldsymbol{P}$, then the inclusion $X \cap G \subset X \cap E$ implies $X \cap G \in \boldsymbol{P}$.

It follows that the set $1 - X^*$, as a union of open sets, is open. Consequently, X^* *is closed.*

Since $\boldsymbol{P} \neq 0$, then $0 \in \boldsymbol{P}$.

(¹) For further definitions and theorems, see N. Bourbaki, *Topologie Générale*, Chapter I, §§ 6 and 7.

We shall prove that

$$X \subset Y \textit{ implies } X^* \subset Y^*; \tag{1}$$

$$X^{**} \subset X^* = \overline{X^*} \subset \overline{X}; \tag{2}$$

$$\textit{if } G \textit{ is open, then } G \cap X^* = G \cap (G \cap X)^*. \tag{3}$$

Let $X \subset Y$, $p \in X^*$ and let G be an open set containing p. Then $X \cap G \notin \boldsymbol{P}$ and, since $X \cap G \subset Y \cap G$, it follows from (i) that $Y \cap G \notin \boldsymbol{P}$. Hence $p \in Y^*$.

Since $(1-\overline{X}) \cap X = 0 \in \boldsymbol{P}$, we have $1-\overline{X} \subset 1-X^*$, i.e. $X^* \subset \overline{X}$. Therefore $(X^*)^* \subset \overline{X^*} = X^*$.

Finally, suppose that $p \in G \cap X^*$. If H is a neighbourhood of p, then $H \cap G \cap X \notin \boldsymbol{P}$ (since $H \cap G$ is a neighbourhood of p). It follows that $p \in (G \cap X)^*$. Hence $G \cap X^* \subset (G \cap X)^*$. On the other hand, by (1), the inclusion $G \cap X \subset X$ implies $(G \cap X)^* \subset X^*$, whence $G \cap (G \cap X)^* \subset G \cap X^*$ which implies (3).

Condition (3) means that the condition that X possesses or does not possess a property at a point p depends on a neighbourhood of p; it is, therefore, a "local" condition.

Finally, we remark that statement (1) implies (see § 4, III) the following formulas:

$$(X \cap Y)^* \subset X^* \cap Y^*; \tag{4}$$

$$(\bigcap_\iota X_\iota)^* \subset \bigcap_\iota X_\iota^*; \tag{5}$$

$$\bigcup_\iota X_\iota^* \subset (\bigcup_\iota X_\iota)^*. \tag{6}$$

Consequences of (i) and (ii):

$$(X \cup Y)^* = X^* \cup Y^*; \tag{7}$$

$$X^* - Y^* \subset (X-Y)^*. \tag{8}$$

Proof. If $p \notin X^*$, there exists a neighbourhood G of p such that $X \cap G \in \boldsymbol{P}$. Similarly, if $p \notin Y^*$, there exists a neighbourhood H of p such that $Y \cap H \in \boldsymbol{P}$. By (i), $X \cap G \cap H \in \boldsymbol{P}$ and $Y \cap G \cap H \in \boldsymbol{P}$. Hence, by (ii), $(X \cup Y) \cap G \cap H \in \boldsymbol{P}$ which proves that $p \notin (X \cup Y)^*$. Hence $(X \cup Y)^* \subset X^* \cup Y^*$. The reverse inclusion is a particular case of (6).

Formula (8) follows from (7) (compare § 4, III, rule 3).

V. Locally closed sets

DEFINITION. The set X ($\subset 1$) is said to be *locally closed at the point* $p \in X$ if there exists an open set G containing p such that $G \cap X$ is closed in G(¹).

THEOREM 1. *The set of all points p of X such that X is not locally closed at p is identical to $\overline{\overline{X}-X} \cap X$ (called the residue of X*(²)*).*

Proof. First, let us assume that $p \notin (\overline{\overline{X}-X} \cap X)$. Put $G = 1-\overline{\overline{X}-X}$. Thus G is open and contains p. Moreover

$$G \cap \overline{\overline{X}-X} = 0, \qquad \text{hence} \qquad G \cap \overline{X}-X = 0,$$

and therefore $(G \cap \overline{X}) \subset X$. It follows that $(G \cap \overline{X}) \subset (G \cap X) \subset (G \cap \overline{X})$, and finally $G \cap X = G \cap \overline{X}$. Thus $G \cap X$ is closed in G.

Next, let us assume that X is closed at p. Let G be an open set containing p and closed in G, i.e. $G \cap X = \overline{G \cap X} \cap G$. We have to show that

$$G \subset 1-\overline{\overline{X}-X}, \qquad \text{i.e.} \qquad \overline{\overline{X}-X} \subset 1-G, \text{ or } \overline{X}-X \subset 1-G,$$

G being open. Now

$$\overline{X}-X \subset \overline{X}-(G \cap X) = \overline{X}-(\overline{G \cap X} \cap G) = (\overline{X}-\overline{G \cap X}) \cup (\overline{X}-G).$$

By § 4, III, 3,

$$\overline{X}-\overline{G \cap X} \subset \overline{X-G \cap X} = \overline{X-G} \subset \overline{1-G} = 1-G,$$

thus $\overline{X}-X \subset 1-G$.

COROLLARY. *The set X is locally closed at each of its points if and only if X is the difference of two closed sets or* (equivalently stated, see § 6, III) *if $\overline{X}-X$ is closed.*

According to the theorem, the condition to be locally closed at each point means that $X \cap \overline{\overline{X}-X} = 0$. This identity is equivalent to the inclusion $\overline{\overline{X}-X} \subset \overline{X}-X$, which means that $\overline{X}-X$ is closed.

Remark 1. For *regular* spaces the above definition of the property of being locally closed can be expressed in terms of the general definition of local properties stated in Section IV. Namely:

(¹) See N. Bourbaki, *loc. cit.*, p. 41.

(²) For more details about the residues, see § 12, VII and VIII.

THEOREM 2. *In order that the subset X of a regular space should be locally closed at the point $p \in X$, it is necessary and sufficient that there should exist a neighbourhood E of p (relative to the space) such that $E \cap X$ is closed.*

For, suppose that X is locally closed at $p \in X$. Let G be an open neighbourhood of p such that $G \cap X$ is closed in G, i.e. $G \cap X = \overline{G \cap X} \cap G$. The space being regular there exists an open set H such that $p \in H$ and $\overline{H} \subset G$. Hence

$$\overline{H} \cap X = \overline{H} \cap G \cap X = \overline{H} \cap \overline{G \cap X} \cap G = \overline{H} \cap \overline{G \cap X}.$$

Thus the set $E = \overline{H}$ is a neighbourhood of p having a closed intersection with X. Therefore the condition is necessary.

In order to show that the condition is sufficient, we put $G = \mathrm{Int}(E)$. As $E \cap X$ is closed, the set $G \cap X = G \cap (E \cap X)$ is closed in G.

Remark 2([1]). The Theorem 2 does not hold true in spaces which are not regular.

In other words, if the space is not regular, it contains a point p and an open set G containing p such that no closed set $F \subset G$ is a neighbourhood of p. Therefore the set $X = G$ is locally closed at p (compare Corollary), although the condition stated in Theorem 2 is not fulfilled.

§ 8. Dense sets, boundary sets and nowhere dense sets

I. Definitions. 1. X is a *dense set,* if $\overline{X} = 1$ ([2]).

2. X is a *boundary set* if its complement is dense, i.e. if $\overline{1-X} = 1$.

3. X is a *nowhere dense set,* if its closure is a boundary set, i.e. if $\overline{1-\overline{X}} = 1$([3]).

4. A space containing a countable dense subset is called *separable*([4]).

([1]) This remark is due to Mrs. M. Całczyńska-Karłowicz.

([2]) G. Cantor, Math. Ann. 15 (1879), p. 2.

([3]) This notion is due to P. du Bois-Reymond, *Die allgemeine Funktionentheorie I*, Tübingen 1882.

([4]) After M. Fréchet, Rend. Circ. Mat. di Palermo 22 (1906), p. 23. For locally separable spaces (a notion due to P. Urysohn, Fund. Math. 9 (1927), p. 119), see P. Alexandrov, Math. Ann. 92 (1924), pp. 294–301, and W. Sierpiński, Fund. Math. 21 (1933), pp. 107–113.

EXAMPLES. In the space of real numbers, the set of rational numbers is dense and boundary. However, this set is not nowhere dense.

In the plane, the circumference of a circle is nowhere dense.

In the interval 01, the Cantor set $\mathscr{C}$ is nowhere dense.

It is easily seen that any set containing a dense set is dense; any subset of a boundary set is a boundary set; any subset of a nowhere dense set is nowhere dense. The closure of a nowhere dense set is nowhere dense.

Each nowhere dense set is a boundary set. Conversely, each closed boundary set is nowhere dense.

A boundary set cannot be open unless it is empty.

II. Necessary and sufficient conditions(1). The condition $\overline{1-X} = 1$ is equivalent to the formula $\mathrm{Int}(X) = 0$. Therefore the boundary sets can be defined by the condition that they *contain no interior point* or that they contain no open non-empty set. Since the condition $\mathrm{Int}(X) = 0$ is equivalent to the inclusion $X \subset \mathrm{Fr}(X)$ (see § 6, II(3)), a boundary set can be defined as a set *contained in its boundary* or as a set satisfying the inclusion $X \subset \overline{1-X}$. It follows that $\overline{X}$ is a boundary set if and only if $\overline{X} \subset \overline{1-\overline{X}}$ or, if and only if

$$X \subset \overline{1-\overline{X}}.$$

The latter property therefore characterizes the nowhere dense sets.

Another necessary and sufficient condition for X to be nowhere dense is the following: *Every non-empty open set contains a non-empty open subset disjoint to X.*

For, if X is nowhere dense, the set $\overline{X}$, as a boundary set, contains no open non-empty set. Therefore, if G is an arbitrary open set $(\neq 0)$, the set $G-\overline{X}$ is non-empty, open and disjoint to X.

On the other hand, if X is not nowhere dense, $\overline{X}$ is not a boundary set, and hence $\mathrm{Int}(\overline{X}) \neq 0$. Thus, if G is a non-empty open subset of $\mathrm{Int}(\overline{X})$, then $G \subset \mathrm{Int}(\overline{X}) \subset \overline{X}$, whence $G \cap \overline{X} \neq 0$ and $G \cap X \neq 0$ (see § 5, III) which proves that the non-empty open set $\mathrm{Int}(\overline{X})$ contains no open non-empty subset which would be disjoint to X.

---

(1) For numerous properties of dense, nowhere dense, and boundary sets see A. D. Wallace, *On non-boundary sets*, Bull. Amer. Math. Soc. 45 (1939), p. 420.

III. Operations.

THEOREM 1. *The union of a boundary set and a nowhere dense set is a boundary set*[1].

Proof. Let $\overline{1-X} = 1 = \overline{1-\overline{Y}}$. It follows by § 4, III, 3 that

$$1-\overline{Y} = \overline{1-X}-\overline{Y} \subset \overline{(1-X)-Y} = \overline{1-(X \cup Y)};$$

hence $1 = \overline{1-\overline{Y}} \subset \overline{1-(X \cup Y)}$ and finally $\overline{1-(X \cup Y)} = 1$.

THEOREM 2. *The union of two nowhere dense sets is nowhere dense*[2]. *Hence the family of all nowhere dense sets is an ideal.*

Proof. Let X and Y be nowhere dense. Hence $\overline{X}$ and $\overline{Y}$ are also nowhere dense and, by Theorem 1, $\overline{X} \cup \overline{Y}$ is a boundary set and—being closed—is nowhere dense.

This completes the proof.

THEOREM 3. *Let $\{X_\iota\}$ be a family of sets open relative to $\bigcup_\iota X_\iota$. Then, if each X_ι is a boundary set (respectively a nowhere set) so is $\bigcup_\iota X_\iota$.*

This follows from the theorem of § 6, IV, by assuming $\mathrm{Int}(X_\iota) = 0$, respectively $\mathrm{Int}(\overline{X}_\iota) = 0$.

IV. Decomposition of the boundary.

THEOREM 1. *The boundary of a set X is composed of two boundary sets, namely the set $X \cap \overline{1-X} = X-\mathrm{Int}(X)$ and $\overline{X}-X$.*

Proof. The formula $\mathrm{Fr}(X) = (X \cap \overline{1-X}) \cup (\overline{X}-X)$ was proved in § 6 (formula 6). It remains to show that $X \cap \overline{1-X}$ is a boundary set, as the set $\overline{X}-X$ is obtained from $X \cap \overline{1-X}$ by substituting $1-X$ in place of X. Now we have

$$1 = \overline{1-X} \cup (1-\overline{1-X}) \subset \overline{1-X} \cup \overline{1-\overline{1-X}} = \overline{(1-X) \cup (1-\overline{1-X})}$$
$$= \overline{1-X \cap \overline{1-X}}.$$

The proof is completed.

---

[1] However, the union of two boundary sets is not necessarily a boundary set; this is the case of the set of rational numbers and of the set of irrational numbers in the space $\mathscr{E}$.

[2] Cf. S. Janiszewski, *Thèse* (1911), p. 26.

Remark. At the same time we have proved that the boundary sets can be defined as sets of the form $X \cap \overline{1-X}$ (or as sets of the form $\overline{X}-X$).

THEOREM 2. *If X and Y are separated, then $\overline{X} \cap \overline{Y}$ is nowhere dense.*

As $\overline{X} \cap Y = 0$, we have:

$$\overline{X} \cap \overline{Y} = (\overline{X} \cap \overline{Y} \cap Y) \cup (\overline{X} \cap \overline{Y}-Y) = (\overline{X} \cap \overline{Y})-Y \subset \overline{Y}-Y,$$

and by Theorem 1, $\overline{Y}-Y$ is nowhere dense.

THEOREM 3. *If X is closed or open, then* $\mathrm{Fr}(X)$ *is nowhere dense.*

For, if $X = \overline{X}$, then $\mathrm{Fr}(X) = X \cap \overline{1-X}$ which is, by Theorem 1, a boundary set and hence (being closed) a nowhere dense set. The case where X is open is similar.

V. Open sets modulo nowhere dense sets. Let $\boldsymbol{I}$ denote the ideal of all nowhere dense sets.

A is said *open modulo* $\boldsymbol{I}$ if there is an open G such that A is congruent to G mod $\boldsymbol{I}$ (written $A \sim G$, compare § 2, VIII); in other terms if $A-G$ and $G-A$ are nowhere dense.

Each closed set F is open modulo $\boldsymbol{I}$. For $F = \mathrm{Int}(F) \cup \mathrm{Fr}(F)$ and $\mathrm{Fr}(F)$ is nowhere dense (IV, Theorem 3).

Similarly, *each open set is closed modulo* $\boldsymbol{I}$. Thus the families of open sets modulo $\boldsymbol{I}$ and of closed sets modulo $\boldsymbol{I}$ are identical.

THEOREM 1. *The family of all open sets modulo nowhere dense sets is a field, i.e. if A_1 and A_2 are open sets modulo $\boldsymbol{I}$, then so are $A_1 \cup A_2$, $A_1 \cap A_2$ and $1-A_1$.*

Proof. Suppose $A_1 \sim G_1$ and $A_2 \sim G_2$, where G_1 and G_2 are open. Then (see § 2, VIII), $A_1 \cup A_2 \sim G_1 \cup G_2$ and $A_1 \cap A_2 \sim G_1 \cap G_2$.

It remains to show that, if $A \sim G$, then there is an open H such that $1-A \sim H$.

Now $A \sim G$ yields $1-A \sim 1-G$ and $1-G$ being closed, hence open modulo $\boldsymbol{I}$, there is an open H such that $1-G \sim H$. Therefore $1-A \sim H$.

THEOREM 2. *A is open* mod $\boldsymbol{I}$ *if and only if there is an open H such that $H \subset A$ and $A-H$ is nowhere dense.*

Proof. The condition is obviously sufficient. Now suppose that $A-G$ and $G-A$ are nowhere dense. Put $H = G \cap \mathrm{Int}(A)$. Hence

$$A-H = (A-G) \cup (A \cap \overline{1-A}) = (A-G) \cup (G \cap A \cap \overline{1-A})$$

and as $G \cap \overline{1-A} \subset \overline{G-A}$ (since G is open), it follows that

$$A-H \subset (A-G) \cup \overline{G-A}.$$

This completes the proof (according to Theorem 2 of Section III).

THEOREM 3. *A is open* mod $\boldsymbol{I}$ *if and only if* $\mathrm{Fr}(A)$ *is nowhere dense.*

Proof. 1. Suppose that A is open mod $\boldsymbol{I}$ and put according to Theorem 2, $A = H \cup N$ where H is open and N nowhere dense. Now $\mathrm{Fr}(A) \subset \mathrm{Fr}(H) \cup \mathrm{Fr}(N)$ (by § 6, II(8)). As each of the sets $\mathrm{Fr}(H)$ and $\mathrm{Fr}(N)$ is nowhere dense, so is their union (by Theorem 2 of Section III).

2. Suppose that $\mathrm{Fr}(A)$ is nowhere dense. As $A = \mathrm{Int}(A) \cup \cup [A \cap \mathrm{Fr}(A)]$, it follows that A is open mod $\boldsymbol{I}$.

The preceding theorems yield the following corollaries:

COROLLARY 4. *X is open* mod $\boldsymbol{I}$ *if and only if X is a difference of a closed set and a nowhere dense set.*

For, $1-(G \cup N) = (1-G)-N$.

COROLLARY 5. *In the family of sets open* mod $\boldsymbol{I}$, *the notions of boundary set and nowhere dense set coincide.*

For if Y is a boundary set of the form $X = G \cup N$, the open set G is empty, hence X is equal to the nowhere dense set N.

Remark 1. As we shall see in § 9, VI, *each scattered set is open* mod *nowhere dense sets*. Another interesting family of open sets mod nowhere dense sets will be examined in § 12. In § 13, VI we shall consider some relations between pointwise discontinuous functions and open sets mod nowhere dense sets.

Remark 2. It is worthy noticing that the sets *closed-open* mod $\boldsymbol{I}$, i.e. sets of the form $A = (H-N) \cup (N-H)$ where H is closed-open and N is nowhere dense, can be characterized by the condition[1]

$$\mathrm{Int}(\bar{A}) = \overline{\mathrm{Int}(A)}.$$

---

[1] Theorem of Norman Levine, *On the commutivity of the closure and interior operators in topological spaces*, Amer. Math. Monthly 68 (1961), pp. 474–477. This theorem has interesting connections with the "*metastonean*" spaces in the sense of J. Dixmier (*Sur certains espaces considérés par M. H. Stone*, Summa Brasil. Math. 2 (1951), 151–181). Compare H. Choda and K. Matoba, *On a theorem of Levine*, Proc. Japan Acad. 37 (1961), pp. 462–463.

A more general problem in this direction is considred by T. Chapman[1]. He considers namely the semigroup of all operators generated by the closure and interior operators (they are 7 in number, compare § 4, V) and gives conditions on A for which $O_1(A) = O_2(A)$ where O_1 and O_2 are elements of that semigroup.

VI. Relativization. If $X \subset E$, then, by definition, X is a dense set, a boundary set, or a nowhere dense set *relative to* E if

$$\bar{X} \cap E = E, \quad \overline{E-X} \cap E = E, \quad \overline{E-\bar{X}} \cap E = E,$$

i.e. if $E \subset \bar{X}$, $E \subset \overline{E-X}$, $E \subset \overline{E-\bar{X}}$, respectively.

By II, the condition for X to be a boundary set (a nowhere dense set, respectively) relative to E can also be expressed by the inclusion

$$X \subset \overline{E-X} \qquad (X \subset \overline{E-\bar{X}}, \text{ respectively}).$$

It is easily seen that:

THEOREM 1. *X is dense in $\bar{X}$; $\bar{X}-X$ is a boundary set in $\bar{X}$; $X \cap \mathrm{Int}\big(\mathrm{Fr}(X)\big)$ is simultaneously a dense set and a boundary set in* $\mathrm{Int}\big(\mathrm{Fr}(X)\big)$ (by § 6, II(12)).

THEOREM 2. *If X is dense in E, X is dense in every subset of E (containing X); if X is a boundary (or nowhere dense) set in E, then so is X relative to every set containing E.*

THEOREM 3. *If X is dense in Y, and Y is dense in Z, then X is dense in Z. In particular, if X is dense in Y, then X is dense in $\bar{Y}$.*

THEOREM 4. *If X is nowhere dense in $\bar{E}$, then $X \cap E$ is nowhere dense in E.*

Proof. We have, by the hypothesis, $X \subset \overline{\bar{E}-\bar{X}}$. Hence, by the formula $\bar{E}-\bar{X} = \bar{E}-\bar{\bar{X}} \subset \overline{E-\bar{X}}$, we obtain $X \subset \overline{E-\bar{X}}$, whence

$$X \cap E \subset \overline{E-\bar{X}} \subset \overline{E-\overline{X \cap E}}.$$

THEOREM 5. *If G is open and X is a boundary (nowhere dense) set, then so is $X \cap G$ relative to G, and $X \cap \bar{G}$ relative to $\bar{G}$.*

---

[1] *A further note on closure and interior operators*, Amer. Math. Monthly 69 (1962), pp. 524–529.

For, if X is a boundary set, then $X \subset \overline{1-X}$, whence $X \cap G \subset \overline{1-X} \cap G$ and, since, by § 5, III, $\overline{1-X} \cap G \subset \overline{G-X}$, we obtain $X \cap G \subset \overline{G-X} = \overline{G-(X \cap G)}$.

Similarly, if X is nowhere dense, then $X \subset \overline{1-\overline{X}}$, whence

$$X \cap G \subset \overline{1-\overline{X}} \cap G \subset \overline{G-\overline{X}} \subset \overline{G-\overline{X \cap G}}.$$

Moreover, $X \cap \overline{G} = (X \cap G) \cup \big(X \cap (\overline{G}-G)\big)$ and since the set $\overline{G}-G$, and hence also $X \cap (\overline{G}-G)$, are by Theorem 1 nowhere dense in $\overline{G}$, the rest of Theorem 5 follows from III.

VII. Localization. X is said to be a *boundary* (*nowhere dense*) *set at a point* p if there exists a neighbourhood G of p such that $X \cap G$ is a boundary (nowhere dense) set.

Thus, for example in the plane, the set composed of a circle (with its interior) and a segment having a single point p in common with the circle is locally nowhere dense at each point of the segment except the point p, but this set is not nowhere dense at any point of the circle.

Since every subset of a boundary (nowhere dense) set is a boundary (nowhere dense) set, we can replace "a neighbourhood" by "an open neighbourhood" in the preceding definition (compare § 7, IV).

THEOREM 1. *X is nowhere dense at a point p if and only if $\overline{X}$ is a boundary set at p.*

For, if X is not nowhere dense at p, and if G is an open set containing p, $G \cap X$ is not nowhere dense. Hence there exists an open set H with $0 \neq H \subset \overline{G \cap X}$. Therefore (compare § 5, III) $H = H \cap \overline{G \cap X} \subset \overline{H \cap G \cap X}$, whence $H \cap G \neq 0$, and, since $H \cap G \subset G \cap \overline{G \cap X} \subset G \cap \overline{X}$, it follows that $G \cap \overline{X}$ is not a boundary set, whence $\overline{X}$ is not a boundary set at p.

Conversely, if $\overline{X}$ is not a boundary set at $p \epsilon G$, there exists an open H such that $0 \neq H \subset G \cap \overline{X}$. Hence $H \subset G \cap \overline{X} \subset \overline{G \cap X}$, which shows that X is not a nowhere dense set at p.

THEOREM 2. *$\overline{\mathrm{Int}(X)}$ is the set of points at which X is not locally a boundary set; $\overline{\mathrm{Int}(\overline{X})}$ is the set where X is not a locally nowhere dense set.*

For, let $p \in \overline{\mathrm{Int}(X)}$ and let G be an open neighbourhood of p. Then $G \cap \overline{\mathrm{Int}(X)} \neq 0$ implies $0 \neq G \cap \mathrm{Int}(X) \subset G \cap X$ which proves that X is not a boundary set at p.

Conversely, if $p \in 1 - \overline{\mathrm{Int}(X)}$, the set $G = 1 - \overline{\mathrm{Int}(X)}$ is an open neighbourhood of p such that $G \cap X$ is a boundary set since

$$\mathrm{Int}(G \cap X) = \mathrm{Int}(G) \cap \mathrm{Int}(X) = \left(1 - \overline{\mathrm{Int}(X)}\right) \cap \mathrm{Int}(X) = 0.$$

The second part of the theorem follows from the first one by Theorem 1.

THEOREM 3. *The set of points where X or $\bar{X}$ is a locally boundary (locally nowhere dense) set is a boundary (nowhere dense) set.*

In particular, if X is a locally boundary (locally nowhere dense) set at each of its points, then X is a boundary (nowhere dense) set.

Proof. $X - \overline{\mathrm{Int}(X)} \subset X - \mathrm{Int}(X)$ and, since the latter set is a boundary set by Section IV, then so is $X - \overline{\mathrm{Int}(X)}$.

Similarly, $X - \overline{\mathrm{Int}(\bar{X})} \subset \bar{X} - \mathrm{Int}(\bar{X})$ and, since the latter is a closed boundary set, and hence a nowhere dense set, then so is $X - \overline{\mathrm{Int}(\bar{X})}$.

Remark. By Theorem 2 of Section III, the family of nowhere dense sets is an ideal. Therefore, in formulas (1)-(7) of § 7, IV, we can substitute $\overline{\mathrm{Int}(\bar{X})}$ for X^*. In particular, we obtain

$$\text{(i)} \quad \overline{\mathrm{Int}(\bar{X})} \cup \overline{\mathrm{Int}(\bar{Y})} = \overline{\mathrm{Int}(\overline{X \cup Y})}, \quad \bigcup_\iota \overline{\mathrm{Int}(\bar{X}_\iota)} \subset \overline{\mathrm{Int}(\overline{\bigcup_\iota X_\iota})}.$$

Theorem 5 of Section VI implies that

(ii) *If G is open and X is a boundary (nowhere dense) set at a point p, then so is $X \cap G$ at p relative to G (if $p \in G$) and so is $X \cap \bar{G}$ relative to $\bar{G}$ (if $p \in \bar{G}$).*

For, by this assumption, there exists a neighbourhood H of p such that $H \cap X$ is a boundary (nowhere dense) set. By substituting $H \cap X$ for X in VI, Theorem 5, it follows that $H \cap G \cap X$ is a boundary (nowhere dense) set relative to G and so is $H \cap \bar{G} \cap X$ relative to $\bar{G}$. Also $H \cap G$ and $H \cap \bar{G}$ are neighbourhoods of p relative to G and $\bar{G}$, respectively.

VIII. Closed domains(¹). A set X is said to be a *closed domain* provided that it is closed and is not locally nowhere dense at any of its points; in other words (see Section VII) if $X = \overline{\mathrm{Int}(X)}$, i.e. if $X = \overline{1-\overline{1-X}}$.

The closed domains can also be defined to be the *closures of open sets*(²). For, the preceding formula shows that every closed domain is the closure of an open set. Conversely, if G is open and $X = \bar{G}$, then G is an open subset of X. Consequently $G \subset \mathrm{Int}(X) \subset X$, whence $\bar{G} \subset \overline{\mathrm{Int}(X)} \subset \bar{X} = \bar{G}$ and, therefore, $X = \overline{\mathrm{Int}(X)}$.

The inclusion $\mathrm{Fr}(X) \subset \overline{\mathrm{Int}(X)}$ *characterizes the closed domains among the closed sets*. It is satisfied if X is a closed domain since $\mathrm{Fr}(X) \subset X$. Conversely, if $\mathrm{Fr}(X) \subset \overline{\mathrm{Int}(X)}$, then

$$\bar{X} = \mathrm{Fr}(X) \cup \mathrm{Int}(X) = \mathrm{Fr}(X) \cup \overline{\mathrm{Int}(X)} = \overline{\mathrm{Int}(X)}.$$

If X is a closed domain, then $\mathrm{Fr}(X) = \mathrm{Fr}(\overline{1-X}) = \mathrm{Fr}\big(\mathrm{Int}(X)\big)$.

For $\mathrm{Fr}(X) = X \cap \overline{1-X} = \overline{1-\overline{1-X}} \cap \overline{1-X} = \mathrm{Fr}(\overline{1-X})$.

The union of two closed domains is a closed domain.

More generally: *If $\{D_\iota\}$ is a family of closed domains, then $\overline{\bigcup_\iota D_\iota}$ is a closed domain.*

This is a consequence of VII(i).

If X is a subset of a closed domain D and p is a point of D, then a necessary and sufficient condition that X is a boundary (nowhere dense) set at p is that X is a boundary (nowhere dense) set relatively to D at p.

The condition is necessary, since in VII(ii) $\bar{G}$ can be replaced by D. Conversely, if G is open and $G \cap X$ is a boundary (nowhere dense) set relative to D, then it is also relative to the entire space; this shows that the condition is sufficient.

Thus we see that, under the same hypotheses concerning X and D, the property of X to be a *boundary set, a nowhere dense set, or a closed*

(¹) For this term, compare H. Lebesgue, Fund. Math. 2 (1921), p. 273. For theorems, see my papers of Fund. Math. 3 (1922), pp. 192–195, and Fund. Math. 5 (1924), p. 117.

(²) Since $\overline{\mathrm{Int}(X)}$ is a closed domain, it follows that

$$\overline{\mathrm{Int}\,\overline{\mathrm{Int}(X)}} = \overline{\mathrm{Int}(X)}, \quad \text{whence} \quad \overline{1-\overline{1-\overline{1-\overline{X}}}} = \overline{1-\overline{X}}.$$

Compare the table given in § 4, V.

domain is equivalent, respectively, to the same property relativized with respect to D since the first (the second) property means that X is a locally boundary (locally nowhere dense) set at each of its points, and the third property means that X is not nowhere dense at any of its points.

It also follows that *the property to be a relative closed domain is transitive*, i.e. if X is a closed domain with respect to Y and Y is a closed domain with respect to Z, then X is a closed domain with respect to Z.

IX. Open domains. An *open domain* is the complement of a closed domain(¹).

The open domains can be characterized by the identity $X = \mathrm{Int}(\overline{X})$. For, if $X = \mathrm{Int}(\overline{X}) = 1-\overline{1-\overline{X}}$, the set X, as the complement of the closed domain $\overline{1-\overline{X}}$, is an open domain. Conversely, if X is an open domain and if we put $D = 1-X$, then $1-\overline{1-\overline{X}} = 1-1-\overline{1-D} = 1-D = X$ since D is a closed domain.

The open domains can also be defined to be open sets satisfying the inclusion $\mathrm{Fr}(X) \subset \mathrm{Int}(\overline{1-X})$. For $\mathrm{Fr}(X) = \mathrm{Fr}(1-X)$ and $1-X$ is a closed domain if and only if $\mathrm{Fr}(1-X) \subset \overline{\mathrm{Int}(1-X)}$.

If X is an open domain, then $\mathrm{Fr}(X) = \mathrm{Fr}(\overline{X}) = \mathrm{Fr}(1-\overline{X})$.

For $\mathrm{Fr}(X) = \overline{X}-X = \overline{X}-\mathrm{Int}(\overline{X}) = \overline{X} \cap \overline{1-\overline{X}} = \mathrm{Fr}(\overline{X})$.

Since the union of two closed domains is a closed domain, *the intersection of two open domains is an open domain.*

§ 9. Accumulation points

I. Definitions. p is an *accumulation point* of the set X, if $p \in \overline{X-p}$. The set X^d of accumulation points of X is called the *derived set* of X.

p is an *isolated* point, if $p \in X - X^d$(²).

We shall assume that the space under consideration is $\mathscr{T}_1$.

(¹) For an interesting application of this notion to the problem of topological interpretation of the propositional calculus see A. Tarski, Fund. Math. 31 (1938), p. 127. For an application to the Boolean algebra, see G. Birkhoff, *Lattice Theory*, Coll. Publ. 25, New York 1940, p. 103; M. H. Stone, Trans. Amer. Math. Soc. 41 (1937), p. 375 and Fund. Math. 29 (1937), p. 263; P. R. Halmos, *Lectures on Boolean algebras*, Princeton 1963, § 4.

(²) Notions due to G. Cantor, Math. Ann. 5 (1872), p. 129. Cantor used also the terms "coherence" and "adherence" to mean the sets $X \cap X^d$ and $X - X^d$, respectively.

EXAMPLES. Every real number is an accumulation point of the space of real numbers. Every natural number is an isolated point of the space of all natural numbers. The set A of the numbers $1/n+1/m$ $(n = 1, 2, \ldots, m = 1, 2, \ldots)$ has as its derived set the set composed of the numbers $1/n$ and of the number 0; its second derived set (i.e. the derived set of the derived set) is composed only of the number 0; its third derived set is empty.

In Chapter II, § 24, Section IV, we shall study derived sets of a transfinite order.

II. Equivalences. *$p \in X^d$ if and only if each neighbourhood E of p satisfies $X \cap E - p \neq 0$.*

p is an isolated point if and only if there exists a neighbourhood E of p such that $X \cap E = p$.

For, by § 7, II, the condition $p \in \overline{X-p}$ can be expressed by the inequality $(X-p) \cap E \neq 0$.

By the same proposition, the term neighbourhood can be replaced by *open neighbourhood.*

The statement "$X \cap E - p \neq 0$" can be replaced by the condition "$X \cap E$ is infinite".

For, if E is a neighbourhood of p such that the set $X \cap E$ is finite, then $X \cap E - p$ is closed and the set $A = E - (X \cap E - p)$ is a neighbourhood of p such that $X \cap A - p = 0$.

Thus it follows that the condition $p \in X^d$ can be expressed by saying that X *is not locally finite at the point* p.

III. Formulas(¹). Since the family of all finite sets is an ideal, formulas of § 7 concerning the localization can be applied to the operation X^d. In particular

(1) $(X \cup Y)^d = X^d \cup Y^d$;

(2) $X^d - Y^d \subset (X-Y)^d$;

(3) $X^{dd} \subset X^d$;

(4) $(\bigcap_\iota X_\iota)^d \subset \bigcap_\iota X_\iota^d$;

(¹) Analogous formulas concerning the operation $X \cap X^d$ (the coherence of X) were proved by M. Zarycki, *Allgemeine Eigenschaften der Cantorschen Kohärenzen*, Trans. Amer. Math. Soc. 30, p. 498.

(5) $\bigcup_\iota X_\iota^d \subset (\bigcup_\iota X_\iota)^d$;

(6) $\overline{X^d} = X^d$ (1);

(7) $X \subset Y$ *implies* $X^d \subset Y^d$.

Moreover, we have

(8) $\overline{X} = X \cup X^d$.

For, if $p \in \overline{X}$ and $p \notin X$, then $X - p = X$ and $p \in \overline{X - p}$, whence $p \in X^d$. Conversely, if $p \in X^d$, then $p \in \overline{X - p} \subset \overline{X}$.

Since $p^d = 0$, it follows from (1) that the derived set of every finite set is empty and that

(9) $(X - p)^d = X^d = (X \cup p)^d$,

i.e. that the derived set is not changed by adding to or by removing from the given set a finite number of points.

IV. Discrete sets. A set composed exclusively of isolated points is said to be *discrete*.

Every finite set is discrete. Every subset of a discrete set is discrete.

The set $X - X^d$ is discrete since each point of $X - X^d$, as an isolated point of X, is an isolated point of $X - X^d$.

The condition for p to be an isolated point of the space means that $p \notin \overline{1 - p}$, i.e. that *the point p constitutes an open set.* The space is discrete if and only if $1^d = 0$; that is, if and only if every subset of the space is closed.

V. Sets dense in themselves. A set X is said to be *dense in itself*, if X contains no isolated points, i.e. if $X \subset X^d$ (2).

If X is closed and dense in itself, it is said to be a *perfect* set. This condition can be expressed by the formula $X = X^d$ (since the condition $X = \overline{X}$ can be expressed, according to III(8) by the inclusion $X^d \subset X$).

THEOREM 1. *If X is dense in itself, then $\overline{X}$ is a perfect set.*

By the hypothesis, $X \subset X^d$, hence, by III(8), $X^d = X \cup X^d = \overline{X}$. Applying III(1) and (3), we obtain $(\overline{X})^d = X^d \cup X^{dd} = X^d = \overline{X}$.

(1) For an analysis of this statement done by C. G. Yang, see J. Kelley, *loc. cit.*, p. 56.

(2) G. Cantor, Math. Ann. 23 (1884), p. 471.

THEOREM 2. *The union of an arbitrary number of sets dense in themselves is dense in itself* (by formula III(5)).

THEOREM 3. *If the space is dense in itself, then every open set and every dense set is dense in itself.*

We have $1 \subset 1^d$; if G is open, then $G = 1-F$ and $F^d \subset F$. It follows that $G = 1-F \subset 1-F^d \subset 1^d-F^d \subset (1-F)^d = G^d$.

Now let $\overline{X} = 1$. It follows that $X \cup X^d = 1$, whence $X^d \cup X^{dd} = 1^d$. Since $X^{dd} \subset X^d$ and $1 \subset 1^d$, it follows that $1 \subset X^d$ and $X \subset X^d$.

THEOREM 4. *If X is a dense and boundary set, then the space is dense in itself.*

By the hypothesis, $\overline{X} = \overline{1-X} = 1$. Let $p \in X$; then $1-X \subset 1-p$, $1 = \overline{1-X} \subset \overline{1-p}$, and therefore $p \in \overline{1-p}$. Hence $p \in 1^d$ and it follows that $X \subset 1^d$. By symmetry, $1-X \subset 1^d$, and hence $1 \subset 1^d$.

THEOREM 5. *The sets* $\mathrm{Int}(\mathrm{Fr}(X))$ *and* $X \cap \mathrm{Int}(\mathrm{Fr}(X))$ *are dense in themselves, for every* X.

By § 8, VI, Theorem 1, the set $X \cap \mathrm{Int}(\mathrm{Fr}(X))$ is a dense and boundary set in $\mathrm{Int}(\mathrm{Fr}(X))$. It follows from Theorem 4 that the latter is dense. As a dense subset of a dense in itself set, $X \cap \mathrm{Int}(\mathrm{Fr}(X))$ is dense in itself by Theorem 3.

EXAMPLE. The Cantor discontinuum (see § 3, IX) is perfect (and, at the same time, is nowhere dense) in the interval 01.

VI. Scattered sets. X is said to be a *scattered* set[1] if it contains no dense in itself non-empty subset.

Every isolated set is scattered. A subset of a scattered set is scattered.

Remark. The closure of a scattered (or even discrete) set need not be scattered. For example, let us write rational numbers of the interval 01 as irreducible fractions p/q. Then the points $(p/q, 1/q)$ of the plane form an discrete set, but its closure contains the entire interval 01.

THEOREM 1. *In a space dense in itself every scattered set is nowhere dense. Its complement is therefore dense in itself.*

---

[1] "Separierte Menge" of G. Cantor, *ibid.* For a detailed study of scattered sets, see Z. Semadeni, *Sur les ensembles clairsemés*, Rozpr. Mat. 19 (1954).

Proof. If X is not nowhere dense, then $G = \mathrm{Int}(\overline{X})$ is a non-empty open set. It follows that $G = G \cap \overline{X} \subset \overline{G \cap X}$ which implies that $G \cap X$ is dense in G. But G is open and, by V, Theorem 3, it is dense in itself. By the second part of the same theorem, $G \cap X$ is dense in itself. It follows that X is not scattered. By V, Theorem 3, the complement of a boundary set is dense in itself.

THEOREM 2. *The union of two scattered sets is scattered.*

Proof. Let X and Y be scattered and let Z be dense in itself such that $0 \neq Z \subset X \cup Y$. It follows that $Z-(Z \cap X) \subset Y$ and, as Z is dense in itself and $Z \cap X$ scattered, we have $Z-(Z \cap X) \neq 0$. Furthermore, by Theorem 1 (for $1 = Z$), the set $Z-(Z \cap X)$ is dense in itself.

Therefore Y cannot be scattered.

THEOREM 3. *Every space is the union of two disjoint sets, one*[1] *perfect and the other scattered (one of them may, of course, be empty).*

Proof. Let P be the union of all sets dense in themselves. By V, Theorems 2 and 1, P and $\overline{P}$ are dense in themselves. Since every dense in itself set is a subset of P, it follows that $\overline{P} \subset P$ which means that P is closed; being also dense in itself, P is perfect. Finally, $1-P$ does not contain any dense in itself subset ($\neq 0$).

THEOREM 4. *The boundary of a scattered set is nowhere dense.*

Proof. The set $X \cap \mathrm{Int}[\mathrm{Fr}(X)]$ (which is dense in itself by V, Theorem 5) is a subset of X, hence it is empty, if X is scattered. It follows that $\mathrm{Int}[\mathrm{Fr}(X)] = 0$ since $X \cap \mathrm{Int}[\mathrm{Fr}(X)]$ is dense in $\mathrm{Int}[\mathrm{Fr}(X)]$ (by § 8, VI, Theorem 1). Therefore $\mathrm{Fr}(X)$ is a boundary set; being closed, it is nowhere dense.

Remark. It follows by § 8, V that *a scattered set is the union of an open set and a nowhere dense set* (hence the difference of a closed set and a nowhere dense set); *if a scattered set is a boundary set, it is nowhere dense.*

---

[1] It is called the *kernel* of the space. Formulas concerning the kernel, analogous to those concerning the closure, derived set and so on, were established by Zarycki, *Über den Kern einer Menge*, Jahresber. d. D. Math. Ver. 39 (1930), p. 154. For an extension of Theorem 3 to Boolean algebras, see A. Tarski, *A lattice-theoretical fixpoint theorem and its application*, Pacific Journ. Math. 5 (1955), p. 306.

THEOREM 5. *X is a scattered set if and only if for every perfect set P, $X \cap P$ is nowhere dense in P*(1).

Proof. Let X be a scattered set. By Theorem 1, if P is perfect (or, more generally, a set dense in itself) and if P is regarded as the initial space, the set $X \cap P$ is nowhere dense. Therefore the condition is necessary.

To show that it is sufficient, suppose that X is not scattered. Let D be a set, dense in itself and non-empty, contained in X. Let $P = \bar{D}$. If the condition of the theorem is satisfied, $X \cap \bar{D}$ is nowhere dense in $\bar{D}$. Hence (§ 8, VI, Theorem 4), $X \cap D$ is nowhere dense in D which is impossible since $X \cap D = D \neq 0$.

Remark. The term *perfect* in Theorem 5 can be replaced by *dense in itself.*

§ 10. Sets of the first category (meager sets)

I. Definition. A set is said to be of the *first category*, if it is the union of a countable sequence of nowhere dense sets(2).

EXAMPLES AND REMARKS. In the space $\mathscr{E}$ of real numbers, the set of rational numbers is evidently of the first category. However, the set of irrational numbers is not: This follows from the fact that $\mathscr{E}$ is not of the first category (with respect to itself).

The latter statement(3) can be established as follows: Let $Q = \bigcup_{n=1}^{\infty} N_n$ be a set of the first category (N_n are nowhere dense). Since the set N_1 is nowhere dense, there exists a closed interval I_1 such that $I_1 \cap N_1 = 0$.

We procede by induction; given a finite sequence of intervals each of which is contained in the preceding one: $I_1 \supset I_2 \supset \ldots \supset I_{n-1}$, let I_n be an interval such that $I_n \subset I_{n-1}$ and $I_n \cap N_n = 0$ (such an interval exists by § 8, II, since N_n is nowhere dense).

(1) Compare M. Fréchet, *Quelques propriétés des ensembles abstraits,* Fund. Math. 10 (1927), p. 330. See also A. Denjoy, Journ. de Math. 1916, where this condition is admitted for a definition of scattered sets.

(2) Notion introduced by R. Baire, Ann. di Mat. (3), 3 (1899), p. 65. A. Denjoy employed the term "*gerbé*" for the sets of the first category and the term "*résiduel*" for their complementary sets. See Journ. de Math. (7), 1 (1915), pp. 123–125. Sets of the first category are also called *meager*; see J. Kelley, *loc. cit.*, p. 201.

(3) This is a particular case of the Baire theorem (see Chapter III, § 34, IV).

By the classical Ascoli theorem, there exists a point common to all intervals I_n, $n = 1, 2, \ldots$ This point does not belong to Q and hence $\mathcal{E} \neq Q$.

The notion of first category sets is frequently used in the theory of functions. We quote, for example, the following theorem (see § 31, X): Given a convergent sequence of continuous functions $\{f_n\}$, the set of points of discontinuity of the function $f(x) = \lim_{n=\infty} f(x)$ is of the first category.

In numerous problems of topology, the notion of set of the first category is analogous to that of set of measure 0 in the measure theory ("negligible" sets).

It is remarkable that the family of subsets of the first category of the interval is—if we assume the continuum hypothesis—equivalent, in the sense of set theory, to the family of sets of measure 0; i.e. there exists a one-to-one transformation of the interval onto itself which establishes a one-to-one correspondence between the elements of the two families([1]).

II. Properties. The family of sets of the first category is a *σ-ideal*, i.e. is *hereditary* and *countably additive* (every subset of a set of the first category is of the first category and the union of a countable family of sets of the first category is again of the first category).

Every boundary $\boldsymbol{F}_\sigma$*-set is of the first category.*

For, if $\bigcup_{n=1}^{\infty} F_n$ is a boundary set, then each F_n is a boundary set; as a closed boundary set, F_n is nowhere dense.

Every set of the first category is contained in an $\boldsymbol{F}_\sigma$*-set of the first category.*

For $X = \bigcup_{n=1}^{\infty} N_n \subset \bigcup_{n=1}^{\infty} \overline{N}_n$ and, since the set N_n is nowhere dense, $\overline{N}_n$ is also nowhere dense (see § 8, I).

([1]) W. Sierpiński, *Sur la dualité entre la première catégorie et la mesure nulle*, Fund. Math. 22 (1934), p. 276, and *Hypothèse du continu*, Monografie Matematyczne v. 4 (1934), p. 77.

See also E. Szpilrajn-Marczewski, *On the equivalence of some classes of sets*, Fund. Math. 30 (1938), p. 325, and J. C. Oxtoby and S. M. Ulam, *On the equivalence of any set of first category to a set of measure zero*, Fund. Math. 31 (1938), p. 201.

III. Union theorem. *If $\{X_\iota\}$ is a family (of an arbitrary power) of sets open relative to the union $S = \bigcup_\iota X_\iota$ and if each X_ι is of the first category, then S is also of the first category*(1).

Proof. Let $G_1, G_2, \ldots, G_a, \ldots$ be a (transfinite) well-ordered sequence of non-empty disjoint open sets satisfying two conditions: (i) $S \cap G_a$ is of the first category, (ii) the sequence is saturated, i.e. there exists no open non-empty set G disjoint with all terms of the given sequence and such that $S \cap G$ is of the first category.

Evidently we have $S = (\bigcup_a S \cap G_a) \cup (S - \bigcup_a G_a)$.

To prove the theorem it is sufficient to show that:

(1) $\bigcup_a S \cap G_a$ is of the first category,

(2) $S - \bigcup_a G_a$ is nowhere dense.

1. Since, by the hypothesis, $S \cap G_a$ is of the first category, $S \cap G_a = N_1^a \cup N_2^a \cup \ldots \cup N_n^a \cup \ldots$, where the sets N_n^a $(n = 1, 2, \ldots)$ are nowhere dense. Putting $N_n = N_n^1 \cup N_n^2 \cup \ldots \cup N_n^a \cup \ldots$, we have $\bigcup_a S \cap G_a = N_1 \cup N_2 \cup \ldots \cup N_n \cup \ldots$.

It remains to prove that N_n is nowhere dense. Since the sets G_a are disjoint, the inclusion $N_n^a \subset G_a$ implies $N_n^a \cap G_\beta = 0$ for every $\beta \neq a$. Hence $N_n^\beta = N_n^\beta \cap G_\beta = \bigcup_a N_n^a \cap G_\beta = N_n \cap G_\beta$ which shows that N_n^β is open in N_n. Applying Theorem 3 of § 8, III, we conclude that N_n is nowhere dense.

2. To prove that $S - \bigcup_a G_a$ is nowhere dense it is sufficient to show that $1 - \bigcup_a G_a$ is nowhere dense. Since the latter set is closed, we have to show that it is a boundary set.

Suppose that H is an open set such that $0 \neq H \subset 1 - \bigcup_a G_a$. By the definition of the sequence $\{G_a\}$, the set $S \cap H$ is not of the first category. Let X_ι be a set such that $H \cap X_\iota \neq 0$. Consider the open set $G = H - \overline{S - X_\iota}$.

(1) S. Banach, *Théorème sur les ensembles de première catégorie*, Fund. Math. 16 (1930), p. 395.

Since X_ι is open in S, it follows that $X_\iota = S - \overline{S - X_\iota}$, whence $S \cap G = S \cap H - \overline{S - X_\iota} = H \cap X_\iota \subset X_\iota$. Therefore $S \cap G$ is of the first category. On the other hand $G \neq 0$ since, as we have just proved, $0 \neq H \cap X_\iota \subset G$. Finally $G \cap G_a = 0$ for every a, since $G \subset H \subset 1 - \bigcup_a G_a$.

But this contradicts the definition of the sequence $\{G_a\}$.

IV. Relativization.

THEOREM 1. *If X is of the first category relative to E, then so is X relative to any set containing E.*

THEOREM 2. *If X is of the first category relative to $\bar{E}$, then $X \cap E$ is of the first category relative to E.*

THEOREM 3. *If G is open and X is of the first category, then so is $X \cap G$ relative to G and $X \cap \bar{G}$ relative to $\bar{G}$.*

These three theorems are immediate consequences of Theorems 2, 4, and 5 of § 8, VI.

V. Localization. X is said to be of the *first category at a point* p, if there exists a neighbourhood G of p such that the set $X \cap G$ is of the first category.

The set of points where X is not of the first category (the points where X is of the "second" category) will be denoted by $D(X)$.

Since the family of sets of the first category is an ideal (comp. Section II), we can replace X^* by $D(X)$ in § 7, IV. Thus we conclude that, in the preceding definition, we can replace the term *neighbourhood* by *open neighbourhood*. Moreover, we have the following relations:

$$D(X \cup Y) = D(X) \cup D(Y); \tag{1}$$

$$D(X) - D(Y) \subset D(X - Y); \tag{2}$$

$$D(\bigcap_\iota X_\iota) \subset \bigcap_\iota D(X_\iota); \tag{3}$$

$$\bigcup_\iota D(X_\iota) \subset D(\bigcup_\iota X_\iota);\text{[1]} \tag{4}$$

$$X \subset Y \textit{ implies } D(X) \subset D(Y); \tag{5}$$

$$\textit{If } G \textit{ is open, then } G \cap D(X) = G \cap D(G \cap X). \tag{6}$$

---

[1] The identity sign does not necessarily hold here; for example, let X_n be the interval $1/n \leqslant x \leqslant 1$ in the interval $0 \leqslant x \leqslant 1$. Compare however (13).

By the theorem in Section III, if X is of the first category at each of its points, then X is of the first category. For, by the hypothesis, each point p of X belongs to an open set G_p such that $X \cap G_p$ is of the first category. Therefore X is a union of sets open in X which are of the first category and, by the mentioned theorem, X itself is of the first category[1]. Thus

$$(X \textit{ is of the first category}) \equiv \big(X \cap D(X) = 0\big) \equiv \big(D(X) = 0\big). \quad (7)$$

For, the condition $D(X) = 0$ implies $X \cap D(X) = 0$ and this implies, as we have seen, that X is of the first category. Conversely, if X is of the first category, then $D(X) = 0$.

We conclude that

$$D\big(X - D(X)\big) = 0, \quad (8)$$

i.e. the set of points of X where X is of the first category is of the first category. For, by (5),

$$D\big(X - D(X)\big) \subset D(X),$$

whence

$$\big(X - D(X)\big) \cap D\big(X - D(X)\big) \subset \big(X - D(X)\big) \cap D(X) = 0$$

which implies formula (8), by virtue of (7).

Formula (8), together with (2), implies that $D(X) - D\big(D(X)\big) = 0$, and hence $D(X) \subset D\big(D(X)\big)$. Since the reverse inclusion holds by § 7, IV(2), it follows that

$$D\big(D(X)\big) = D(X). \quad (9)$$

By § 8, VII, the set of points where X is not nowhere dense is equal to $\overline{\mathrm{Int}(\overline{X})}$. It follows that

$$D(X) \subset \overline{\mathrm{Int}(\overline{X})} \subset \overline{X}. \quad (10)$$

Since $D(X)$ is closed (§ 7, IV), it follows from (10) that $D\big(D(X)\big) \subset \overline{\mathrm{Int}\big(D(X)\big)} \subset D(X)$. By (9) we get

$$D(X) = \overline{\mathrm{Int}\big(D(X)\big)}, \quad (11)$$

---

[1] This proposition can be proved in a more direct way (without referring to the theorem in Section III), if we assume that the space has a countable base composed of open sets $R_1, R_2, \ldots$. For, under this hypothesis, the open set G_p can be replaced by an $R_{n(p)}$ of the first category and, since $X \cap R_{n(p)}$ is of the first category, then so is the countable union $\bigcup_{p \in X} X \cap R_{n(p)} = X$.

which shows that $D(X)$ is a closed domain (§ 8, VIII). Therefore, $D(X) \neq 0$ implies that X is not of the first category at any point of the open non-empty set $\operatorname{Int} D(X)$.

$$[D(Y) = 0] \Rightarrow [D(X \cup Y) = D(X) = D(X-Y)], \tag{12}$$

i.e. X remains to be of the first category at p after adjoining to or removing from it a set of the first category. This proposition is an immediate consequence of formula (1).

$$\textit{The set } D(\bigcup_{n=1}^{\infty} X_n) - \bigcup_{n=1}^{\infty} D(X_n) \textit{ is nowhere dense.} \tag{13}$$

It suffices to show that if G is a non-empty open set, then there exists a non-empty open set H such that

$$H \cap D(\bigcup_{n=1}^{\infty} X_n) - \bigcup_{n=1}^{\infty} D(X_n) = 0 \quad \text{and} \quad H \subset G.$$

If, for every n, $G \cap D(X_n) = 0$, then $G \cap X_n \subset X_n - D(X_n)$. Hence

$$G \cap \bigcup_{n=1}^{\infty} X_n \subset \bigcup_{n=1}^{\infty} \big(X_n - D(X_n)\big)$$

and, by (8), the set $G \cap \bigcup_{n=1}^{\infty} X_n$ is of the first category. Consequently $D(G \cap \bigcup_{n=1}^{\infty} X_n) = 0$ which implies by (6) that $G \cap D(\bigcup_{n=1}^{\infty} X_n) = 0$. Setting $H = G$ we obtain the required conclusion.

Now suppose that there exists an n such that $G \cap D(X_n) \neq 0$. Let $H = G \cap \operatorname{Int}\big(D(X_n)\big)$. By (11), $H \neq 0$ and since $H \subset D(X_n)$, it follows that H satisfies the required identity.

VI. Decomposition formulas. Obviously

$$X = \big(X - D(X)\big) \cup \big(X \cap D(X)\big); \tag{14}$$

$$\begin{aligned} X &= X \cap \overline{X - D(X)} \cup \big(X - \overline{X - D(X)}\big) \\ &= X - \operatorname{Int}\big(D(X)\big) \cup \big(X \cap \operatorname{Int}\big(D(X)\big)\big). \end{aligned} \tag{15}$$

These formulas give a decomposition of X into two disjoint parts such that the first one is of the first category and the second one is not of the first category at any of its points. The first member of the union is open relative to X in (14) and is closed in (15).

Proof. By formula (8), the set $X-D(X)$ is of the first category; hence, by (12),

$$D(X \cap D(X)) = D(X), \quad \text{and} \quad X \cap D(X) \subset D(X) = D(X \cap D(X)),$$

which shows that the set $X \cap D(X)$ is not of the first category at any of its points.

On the other hand, the set $X \cap \overline{X-D(X)}$ is of the first category, as a union of two sets $X \cap D(X) \cap \overline{X-D(X)}$ and $(X-D(X)) \cap \cap \overline{X-D(X)}$, the first one being nowhere dense as a subset of a nowhere dense set $D(X) \cap \overline{1-D(X)} = \mathrm{Fr}(D(X))$, and the second one being of the first category as a subset of the set $X-D(X)$ (compare § 8, V and (8)).

Since the set $X \cap \overline{X-D(X)}$ is of the first category, it follows from (12) that $D(X-\overline{X-D(X)}) = D(X)$ and we have

$$X-\overline{X-D(X)} \subset X-(X-D(X)) = X \cap D(X) \subset D(X)$$
$$= D(X-\overline{X-D(X)})$$

which proves that the set $X-\overline{X-D(X)}$ is not of the first category at any of its points.

*ULAM THEOREM. *In a space dense in itself, every set of the power $\aleph_1$ which is not of the first category can be decomposed into a countable family of disjoint sets none of which is of the first category* (¹).

Proof. By a theorem of the general set theory (²), if Z is a set of the power $\aleph_1$ and N is a family of subsets of Z such that, for every sequence $A_1, A_2, \ldots, A_n, \ldots$ of sets belonging to N, the difference $Z-\bigcup_{n=1}^{\infty} A_n$ is uncountable, then there exists an uncountable family of disjoint subsets of Z which do not belong to N.

Let us denote by N the family of subsets of Z that are of the first category. Since each individual point of a set is of the first category (the space is dense in itself), it follows from the above theorem that there exists in Z an uncountable family of disjoint subsets

(¹) S. Ulam, *Über gewisse Zerlegungen von Mengen*, Fund. Math. 20 (1933), p. 222.

(²) Theorem of S. Ulam, Fund. Math. 16 (1930), p. 145. This theorem was later generalized by W. Sierpiński to all less than the first inaccessible aleph. Consequently one can generalize in an analogous manner the theorem of the text. See W. Sierpiński, Fund. Math. 20 (1933), p. 214.

none of which is of the first category. By augmenting one of these subsets with all points of Z that do not belong to the others, we obtain the required decomposition.

We mention without proof the following theorem:

*LUSIN THEOREM. *Every subset Z of the interval $\mathcal{I}$ that is not of the first category at any point of $\mathcal{I}$ contains two disjoint subsets that have the same property*(1).

Moreover, one proves that it contains an uncountable family of such sets, if one assumes the continuum hypothesis(2).

*§ 11. Open sets modulo first category sets. Baire property

I. Definition. Let $\boldsymbol{J}$ denote the σ-ideal of all first category sets.

A is said to be *open modulo* $\boldsymbol{J}$ if there is an open G such that $A \sim G \bmod \boldsymbol{J}$ (compare § 2, VIII); in other words: if $A-G$ and $G-A$ are first category sets.

The property of a set to be open mod $\boldsymbol{J}$ is also called the *Baire property* (in the large sense)(3). We denote by $\boldsymbol{B}$ the family of these sets.

The statement (3) of § 2, VIII yields the following theorem.

THEOREM. *A is open* mod $\boldsymbol{J}$ *if and only if A is of the form*

$$A = (G-P) \cup R \tag{1}$$

where G is open and P and R are first category sets.

II. General remarks. The sets encountered "in practice" have the Baire property. On the other hand, there exist sets which do not have this property (see Section IVa). The role played by the Baire property in topology is analogous to that of measurability (of sets or functions) in analysis. We shall return to these questions in Chapter III, § 40.

Obviously, *each set open* (*or closed*) *modulo nowhere dense sets* (see § 8, V) *is a member of* $\boldsymbol{B}$. (Such are in particular closed sets and scattered sets.)

---

(1) See W. Sierpiński, *Hypothèse du continu*, p. 172.

(2) W. Sierpiński, *ibid.*, p. 115.

(3) This notion is related to the thesis of R. Baire, Ann. di Mat. (3) 3 (1899). Lebesgue used the term "ensembles Z" for sets of this kind (see Journ. de math. S. 6, vol. 1, p. 186) and H. Hahn called them "offene Mengen bis auf eine Menge erster Kategorie" (see *Reelle Funktionen*, p. 137).

They are also obviously *closed* modulo first category sets. Thus, in the definition of $\boldsymbol{B}$ the term "closed" can be substituted to "open".

III. Operations. Like in the case of sets open modulo nowhere dense sets, one shows:

THEOREM 1. *$\boldsymbol{B}$ is a field. Moreover, if $B_n \epsilon \boldsymbol{B}$ for $n = 1, 2, \ldots$, then $(\bigcup_n B_n) \epsilon \boldsymbol{B}$ and $(\bigcap_n B_n) \epsilon \boldsymbol{B}$.*

Proof. As $\boldsymbol{J}$ is a σ-ideal, the formula $(\bigcup_n B_n) \epsilon \boldsymbol{B}$ follows directly from § 2, VIII (4). Applying the de Morgan rule we obtain $(\bigcap_n B_n) \epsilon \boldsymbol{B}$.

COROLLARY. *Every Borel set belongs to $\boldsymbol{B}$*(1).

Proof. The family $\boldsymbol{B}$ satisfies the following three conditions: (i) it contains all closed sets, (ii) it contains the complements of sets which belong to it, and (iii) it contains countable intersections of sets which belong to it. Since the family of Borel sets is the smallest family satisfying these three conditions (§ 5, VI), it must be a part of the family $\boldsymbol{B}$.

IV. Equivalences(2). *Each of the following conditions is necessary and sufficient in order that X have the Baire property:*

1. *There exists a set P of the first category such that $X-P$ is closed and open relative to $1-P$;*

2. *X is a union of a $\boldsymbol{G}_\delta$-set and a set of the first category;*

3. *X is a difference of an $\boldsymbol{F}_\sigma$-set and a set of the first category;*

4. *The set $D(X) \cap D(1-X)$ is nowhere dense*; in other words, *every open non-empty set contains a point where either X or $1-X$ is of the first category*(3).

5. *The set $D(X)-X$ is of the first category.*

---

(1) Theorem of Lebesgue, *loc. cit.*, p. 187. The converse theorem is not true (see § 39).

(2) Compare W. Sierpiński, *Sur l'invariance topologique de la propriété de Baire*, Fund. Math. 4 (1923), p. 319, and *La propriété de Baire des fonctions et leurs images*, *ibid.*, 11 (1928), p. 305, E. Szpilrajn-Marczewski, *O mierzalności i warunku Baire'a*, C. R. du I Congrès des Math. des Pays Slaves, Varsovie 1929, p. 209, and my paper *Sur la propriété de Baire dans les espaces métriques*, Fund. Math. 16 (1930), p. 390.

(3) This is the condition which was originally taken as a definition of the Baire property.

Proof. Let $X \in \boldsymbol{B}$. Then, by Section I,

$$X = (G - P_1) \cup P_2 = (F - P_3) \cup P_4,$$

where G is open, F is closed, and P_n are of the first category. If we put $P = P_1 \cup P_2 \cup P_3 \cup P_4$, then $X - P = G - P = F - P$ which proves that $X - P$ is simultaneously open and closed in $1 - P$.

Having proved this, let us put $X - P = G \cap (1 - P)$. According to § 10, II, let R be an $\boldsymbol{F}_\sigma$-set of the first category containing P. We have

$$X = (X - R) \cup (X \cap R) = (X - P - R) \cup (X \cap R)$$
$$= (G - P - R) \cup (X \cap R) = (G - R) \cup (X \cap R).$$

Since $G - R$ is a $\boldsymbol{G}_\delta$-set and $X \cap R$ is of the first category, we have decomposed X into a $\boldsymbol{G}_\delta$-set and a set of the first category.

If $X \in \boldsymbol{B}$, then $1 - X$ is also a set of the family $\boldsymbol{B}$ (by Section III, 1). Therefore $1 - X = M \cup N$, where M is a $\boldsymbol{G}_\delta$-set and N is of the first category. It follows that $X = (1 - M) - N$ which shows that X is a difference of an $\boldsymbol{F}_\sigma$-set and a set of the first category.

Conversely, since every $\boldsymbol{F}_\sigma$-set, every $\boldsymbol{G}_\delta$-set and every set of the first category belong to the family $\boldsymbol{B}$, it follows (by Section III) that conditions 2 and 3 are satisfied.

Therefore we have proved that each of the conditions 1, 2, and 3 is necessary and sufficient. To prove that also the remaining conditions are necessary and sufficient, suppose that $X \in \boldsymbol{B}$ and let, according to the definition, $X = (G - P) \cup R$. It follows that $1 - X = (1 - G) - R \cup (P - R)$.

It was shown that the set $D(E)$ does not change by adding to E or removing from it a set of the first category (§ 10, V (12)). Consequently $D(X) = D(G)$ and $D(1 - X) = D(1 - G)$. Since $D(G) \subset \overline{G}$ and $D(1 - G) \subset \overline{1 - G}$ (§ 10, V (10)), we have $D(X) \cap D(1 - X) \subset \overline{G} \cap \overline{1 - G} = \overline{G} - G$ and, since the set $\overline{G} - G$ is nowhere dense, it follows that $D(X) \cap D(1 - X)$ is nowhere dense.

Therefore the set $D(X) - X$ is of the first category, since

$$D(X) - X = \big(D(X) - X\big) \cap D(1 - X) \cup \big(D(X) - X - D(1 - X)\big)$$
$$\subset D(X) \cap D(1 - X) \cup \big((1 - X) - D(1 - X)\big),$$

where $D(X) \cap D(1 - X)$ is nowhere dense by the hypothesis and $(1 - X) - D(1 - X)$ is of the first category by § 10, V (8).

Finally, if $D(X)-X$ is of the first category, then the identity

$$X = D(X)-\big(D(X)-X\big) \cup \big(X-D(X)\big),$$

where $D(X)$ is closed and the sets $\big(D(X)-X\big)$ and $\big(X-D(X)\big)$ are of the first category, yields $X \in \boldsymbol{B}$.

This completes the proof.

COROLLARY 1. *Every set X is contained in an $\boldsymbol{F}_\sigma$-set Z such that the condition $X \subset B \in \boldsymbol{B}$ implies that $(Z-B)$ is of the first category* (1).

Proof. Since the set $X-D(X)$ is of the first category (by § 10, V (8)), there exists an $\boldsymbol{F}_\sigma$-set W of the first category such that $X-D(X) \subset W$. Consequently $Z = W \cup D(X)$ is an $\boldsymbol{F}_\sigma$-set containing X. Moreover, if $X \subset B$, then $D(X) \subset D(B)$ (§ 10, V (5)), whence

$$Z-B = W-B \cup \big(D(X)-B\big) \subset W \cup \big(D(B)-B\big).$$

By the preceding theorem the set $D(B)-B$ is of the first category. Therefore $(Z-B)$ is of the first category.

COROLLARY 2. *If a set X having the Baire property is not of the first category at any point of the space, then the set $(1-X)$ is of the first category; if it is not of the first category at any of its points, then it contains a point where $(1-X)$ is of the first category (provided that $X \neq 0$).*

Proof. The hypothesis $1 = D(X)$ implies, by condition 4, that $D(1-X)$ is nowhere dense. On the other hand, by § 10, V(11), $D(1-X)$ is a closed domain. These two properties imply that $D(1-X) = 0$, hence (§ 10, V(7)) that $(1-X)$ is of the first category.

On the other hand, if $X \subset D(X)$, then the inclusion $X \subset D(1-X)$ does not hold, since the set X would then be nowhere dense (by condition 4), which contradicts the hypothesis.

COROLLARY 3. *If X has the Baire property, then the condition $X \cap Y = 0$ implies* $\operatorname{Int}\big(D(X)\big) \cap \operatorname{Int}\big(D(Y)\big) = 0$.

Proof. The condition $X \cap Y = 0$ implies (§ 10, V(5)) that $D(Y) \subset D(1-X)$, whence $[D(X) \cap D(Y)] \subset [D(X) \cap D(1-X)]$. Therefore, by condition 4, the set $D(X) \cap D(Y)$ is nowhere dense. It follows that $\operatorname{Int}\big(D(X) \cap D(Y)\big) = 0$ which implies the required identity (by § 6, II(1)).

---

(1) Theorem of E. Szpilrajn-Marczewski, *loc. cit.*, p. 299.

IVa. Existence theorems. By using Theorem IV, 4, we shall establish *the existence of sets which do not have the Baire property in the space $\mathscr{E}$ of real numbers.*

To do this, let us decompose the set $\mathscr{E}$ into disjoint subsets by letting two numbers belong to the same subset if and only if their difference is rational. By the axiom of choice, there exists a set V_0 containing a single element from each of these subsets. We shall prove that V_0 does not have the Baire property(¹).

Let $r_1, r_2, \ldots, r_n, \ldots$ be the sequence of rational numbers ($\neq 0$). Denote by V_n the set obtained from V_0 by the translation $y = x+r_n$. One readily sees that $\mathscr{E} = \bigcup_{n=0}^{\infty} V_n$. On the other hand, $V_0 \cap V_n = 0$ (for $n \neq 0$), for otherwise V_0 would contain a number y of the form $x+r_n$, where $x \in V_0$. One would then have $y-x = r_n$ while, by the definition, V_0 contains no pair of points whose difference is rational.

Since the space $\mathscr{E}$ is not of the first category in itself (§ 10, I), it follows that one of the sets V_n is not of the first category either. Therefore V_0 is not of the first category, since V_n is obtained from V_0 by a translation, so that they both have the same category. Consequently, there exists an interval ab such that V_0 is not of the first category at any point of this interval (§ 10, V (11)).

Suppose that V_0 has the Baire property. By condition 4, the interval ab contains a subinterval cd ($a < c < d < b$) such that the set $(cd - V_0)$ is of the first category. Let r_n be a rational number such that $0 < r_n < c-a$.

The condition $V_0 \cap V_n = 0$ implies that $V_n \cap cd \subset cd - V_0$. Therefore the set $V_n \cap cd$ is of the first category and the same is true for the part of V_0 contained in the interval $c-r_n$, $d-r_n$ (since it is obtained by a translation of $V_n \cap cd$). But this contradicts the hypothesis that V_0 is not of the first category at any point of ab.

Remarks. (i) Every set of power $\aleph_1$ which does not have the Baire property contains an uncountable family of disjoint subsets which do not have the Baire property.

(¹) This is the construction used by G. Vitali (*Sul problema della misura dei gruppi di punti di una retta*, Bologna 1905) to prove the existence of sets non-measurable in the sense of Lebesgue.

A proof of the existence of sets which do not have the Baire property was also given by H. Lebesgue, *Contributions à l'étude des correspondances de M. Zermelo*, Bull. Soc. Math. de France 35 (1907), pp. 202–212.

To prove this we take $\boldsymbol{N}$ in the theorem of Ulam (§ 10, VI) to be the family of subsets of Z which have the Baire property.

(ii) The proof of the existence of sets which do not have the Baire property in the space $\mathscr{E}$ is *ineffective*, i.e. no way of *pointing out an individual set of this kind* is given(1).

Similar is the problem of finding an effective proof of the existence of sets non-measurable in the sense of Lebesgue.

V. Relativization.

THEOREM 1. *The Baire property is transitive, i.e. if X has the Baire property relative to E, then the condition $E \in \boldsymbol{B}$ implies $X \in \boldsymbol{B}$.*

Proof. Let $X = U \cup P$, where U is a $\boldsymbol{G}_\delta$-set relative to E, and P is of the first category relative to E. Consequently $U = V \cap E$, where V is a $\boldsymbol{G}_\delta$-set (see § 5, V). The set U belongs to $\boldsymbol{B}$ as the intersection of two sets belonging to $\boldsymbol{B}$. Finally, P is of the first category, therefore the set $X = U \cup P$ belongs to $\boldsymbol{B}$.

THEOREM 2. *If X has the Baire property relative to $\bar{E}$, then $X \cap E$ has the Baire property relative to E.*

Proof. We have $X = G - P \cup R$, where G is open in $\bar{E}$ and P and R are of the first category in $\bar{E}$. By taking the common part with E we have $X \cap E = (G \cap E) - (P \cap E) \cup (R \cap E)$. The set $G \cap E$ is open in E and the sets $P \cap E$ and $R \cap E$ are (by § 10, IV, Theorem 2) of the first category in E. Thus our proposition has been proved.

VI. Baire property in the restricted sense.

DEFINITION. X has the *Baire property in the restricted sense*, if for every E, the set $X \cap E$ has the Baire property relative to E; we then write $X \in \boldsymbol{B}_r$.

We shall prove that the range of E in this definition can be restricted to that of *perfect* sets.

Let E be an arbitrary set and let $\bar{E} = A \cup C$ be the decomposition of $\bar{E}$ into a perfect set and a scattered set (§ 9, VI, Theorem 3). Hence $X \cap \bar{E} = (X \cap A) \cup (X \cap C)$. By the hypothesis, $X \cap A$ has

(1) This problem was posed by R. Baire. See a remark on this subject by H. Lebesgue in Journ. de Math. 6, 1, p. 186.

the Baire property relative to A, whence, by Section V, 1, it has this property relatively to $\bar{E}$. Since $X \cap C$ is scattered, it has the Baire property relative to $\bar{E}$ (see Section II). It follows that the union $(X \cap A) \cup (X \cap C) = X \cap \bar{E}$ has the Baire property relative to $\bar{E}$. Therefore, by Section V, Theorem 2, $X \cap E$ has the Baire property relative to E, q.e.d.

It follows that the range of E can also be restricted to that of *closed* sets.

By relativizing the theorems of the preceding sections we obtain the corresponding propositions concerning the family $\boldsymbol{B}_r$. In particular, if X is a *Borel set*, then $X \cap E$ is a Borel set relative to E (see § 5, VI) and therefore it has the Baire property relative to E, so that $X \in \boldsymbol{B}_r$. Similarly, every scattered set belongs to $\boldsymbol{B}_r$.

The theorem of Section IV implies that $X \in \boldsymbol{B}_r$ *if and only if every set Z closed in X is a union of a Borel set and a set of the first category in Z*[1].

(In perfectly normal spaces (see § 14, VI) the term *Borel set* can be replaced by $\boldsymbol{G}_\delta$-set.)

Proof. If $X \cap \bar{Z}$ has the Baire property relative to $\bar{Z}$, then $Z = X \cap \bar{Z} = M \cup P$, where M is a $\boldsymbol{G}_\delta$-set relative to $\bar{Z}$ and P is a set of the first category in $\bar{Z}$. The set M, as an intersection of a $\boldsymbol{G}_\delta$-set and a closed set $\bar{Z}$, is a Borel set; the set P, as a set of the first category in $\bar{Z}$, is also of the first category in Z (§ 10, IV, Theorem 2).

Suppose now that the condition of the theorem is satisfied. It suffices to show that $X \in \boldsymbol{B}_r$, i.e. that if F is an arbitrary closed set, then $X \cap F$ has the Baire property relative to F. By the hypothesis, $X \cap F = M \cup P$, where M is a Borel set (relative to the space) and P is of the first category in $X \cap F$. It follows that M is a Borel set relative to F and P is of the first category relative to F (§ 10, IV, Theorem 1). The set $X \cap F$ has, therefore, the Baire property relative to F.

Remark. In spaces where the notion of a Borel set is a topological invariant, as for example in complete metric spaces, the latter condition implies directly the *topological invariance of the Baire property in the restricted sense* (see § 35).

---

[1] Theorem of Sierpiński, *Sur l'invariance topologique de la propriété de Baire*, Fund. Math. 4 (1923), p. 319.

VII. ($\mathscr{A}$)-operation.

THEOREM. *The Baire property is invariant under* ($\mathscr{A}$)*-operation* [1].

Proof. Let

$$X = \bigcup_{\mathfrak{z}} \bigcap_{n=1}^{\infty} X_{\mathfrak{z}^1 \ldots \mathfrak{z}^n}, \tag{1}$$

and let the sets $X_{\mathfrak{z}^1 \ldots \mathfrak{z}^n}$ have the Baire property (for the notations see § 3, XIV).

Since the intersection of a finite number of sets with the Baire property is also of the same kind, we can replace $X_{\mathfrak{z}^1 \ldots \mathfrak{z}^n}$ by $X_{\mathfrak{z}^1} \cap \cap X_{\mathfrak{z}^1 \mathfrak{z}^2} \cap \ldots \cap X_{\mathfrak{z}^1 \ldots \mathfrak{z}^n}$. Therefore we can assume that the system $X_{\mathfrak{z}^1 \ldots \mathfrak{z}^n}$ is regular.

By IV, Corollary 1, there exists an $\boldsymbol{F}_\sigma$-set Z such that

$$X \subset Z, \tag{2}$$

if $B \in \boldsymbol{B}$ *and* $X \subset B$, *then* $Z - B$ *is of the first category.* (3)

In general, there exists a $Z_{\mathfrak{y}^1 \ldots \mathfrak{y}^i}$ with the Baire property such that

$$\bigcup_{\mathfrak{z}} \bigcap_{n=1}^{\infty} X_{\mathfrak{y}^1 \ldots \mathfrak{y}^i \mathfrak{z}^1 \ldots \mathfrak{z}^n} \subset Z_{\mathfrak{y}^1 \ldots \mathfrak{y}^i}, \tag{2a}$$

if $B \in \boldsymbol{B}$ *and* $\bigcup_{\mathfrak{z}} \bigcap_{n=1}^{\infty} X_{\mathfrak{y}^1 \ldots \mathfrak{y}^i \mathfrak{z}^1 \ldots \mathfrak{z}^n} \subset B$, *then* $Z_{\mathfrak{y}^1 \ldots \mathfrak{y}^i} - B$ *is of the first category.* (3a)

Moreover, we can suppose that

$$Z_{\mathfrak{y}^1 \ldots \mathfrak{y}^i} \subset X_{\mathfrak{y}^1 \ldots \mathfrak{y}^i}, \tag{4}$$

since the set $Z_{\mathfrak{y}^1 \ldots \mathfrak{y}^i} \cap X_{\mathfrak{y}^1 \ldots \mathfrak{y}^i}$ obviously satisfies the conditions imposed on $Z_{\mathfrak{y}^1 \ldots \mathfrak{y}^i}$.

By virtue of the identity $X = Z - (Z - X)$ and since Z is an $\boldsymbol{F}_\sigma$-set, it suffices to prove that $(Z - X)$ is of the first category. By the successive application of propositions (1), (4) and § 3, XIV (6), we obtain

$$Z - X =$$

$$= Z - \bigcup_{\mathfrak{y}} \bigcap_{i=1}^{\infty} X_{\mathfrak{y}^1 \ldots \mathfrak{y}^i} \subset Z - \bigcup_{\mathfrak{y}} \bigcap_{i=1}^{\infty} Z_{\mathfrak{y}^1 \ldots \mathfrak{y}^i} \subset \bigcup_{\mathfrak{y}} \bigcup_{i=0}^{\infty} (Z_{\mathfrak{y}^1 \ldots \mathfrak{y}^i} - \bigcup_{m=1}^{\infty} Z_{\mathfrak{y}^1 \ldots \mathfrak{y}^i m}).$$

---

[1] See O. Nikodym, *Sur une propriété de l'opération* ($\mathscr{A}$), Fund. Math. 7 (1925), p. 149, and C. R. Soc. Sc. de Varsovie 19 (1926), p. 294; N. Lusin and W. Sierpiński, *Sur quelques propriétés des ensembles* (A), Bull. Acad. Cracovie 1918, p. 35; E. Szpilrajn-Marczewski, *loc. cit.*; N. Lusin, C. R. Paris 164 (1917).

Since the union $\bigcup_{\mathfrak{y}}\bigcup_{i=0}^{\infty}$ is countable (compare § 3, XIV (5)), it remains to prove that the set $(Z_{\mathfrak{y}^1\ldots\mathfrak{y}^n} - \bigcup_{m=1}^{\infty} Z_{\mathfrak{y}^1\ldots\mathfrak{y}^{i_m}})$ is of first category. But this follows from (3a), where we can put $B = \bigcup_{m=1}^{\infty} Z_{\mathfrak{y}^1\ldots\mathfrak{y}^{i_m}}$ by virtue of the formula

$$\bigcup_{\mathfrak{z}}\bigcap_{n=1}^{\infty} X_{\mathfrak{y}^1\ldots\mathfrak{y}^i\mathfrak{z}^1\ldots\mathfrak{z}^n} = \bigcup_{m=1}^{\infty}\bigcup_{\mathfrak{z}}\bigcap_{n=1}^{\infty} X_{\mathfrak{y}^1\ldots\mathfrak{y}^{i_m}\mathfrak{z}^1\ldots\mathfrak{z}^n} \subset \bigcup_{m=1}^{\infty} Z_{\mathfrak{y}^1\ldots\mathfrak{y}^{i_m}},$$

which is a consequence of propositions § 3, XIV(2) and (2a).

COROLLARY. *The Baire property in the restricted sense is invariant under the* $(\mathscr{A})$*-operation.*

For, if E is a fixed set, then by (1),

$$E \cap X = \bigcup_{\mathfrak{z}}\bigcap_{n=1}^{\infty}(E \cap X_{\mathfrak{z}^1\ldots\mathfrak{z}^n}).$$

Therefore, if we suppose that $E \cap X_{\mathfrak{z}^1\ldots\mathfrak{z}^n}$ has the Baire property relative to E, then, by the preceding theorem, $E \cap X$ has the same property.

Remarks. The invariance of the Baire property is a particular case of the following theorem of the general set theory.

Let $\boldsymbol{S}$ be a family of subsets of a given space satisfying the following conditions: Firstly the union of an infinite countable number of elements of $\boldsymbol{S}$ belongs to $\boldsymbol{S}$; secondly the complement of any element of $\boldsymbol{S}$ belongs to $\boldsymbol{S}$, and thirdly to every set X (of the given space) there corresponds a set $Z \supset X$ of the family $\boldsymbol{S}$ such that the conditions $X \subset S \in \boldsymbol{S}$ and $Y \subset Z - S$ imply $Y \in \boldsymbol{S}$.

Under these hypotheses the property of being an element of $\boldsymbol{S}$ *is invariant under the* $(\mathscr{A})$*-operation* [1].

The family of sets with the Baire property is such an $\boldsymbol{S}$-family (by III and IV, Corollary 1). The sets measurable in the sense of Lebesgue constitute another important example of an $\boldsymbol{S}$-family. For, the first two conditions are evidently fulfilled and so is the third one; here, Z may be taken to be a $\boldsymbol{G}_\delta$-set with the measure equal to the exterior measure of X and then the set $Z - S$ has the measure zero.

Therefore, the measurability in the sense of Lebesgue is invariant under the $(\mathscr{A})$*-operation.*

---

[1] Theorem of Szpilrajn-Marczewski, *loc. cit.*, p. 300.

*§ 12. Alternated series of closed sets

I. Formulas of the general set theory(1). Let

$$X_0, X_1, \ldots, X_\xi, \ldots, X_\alpha \tag{1}$$

be a decreasing transfinite sequence of sets, i.e. such that $\xi > \zeta$ implies $X_\xi \subset X_\zeta$. Moreover, suppose that

$$X_0 = 1. \tag{2}$$

$$X_\lambda = \bigcap_{\xi<\lambda} X_\xi \text{ if } \lambda \text{ is a limit ordinal or if } \lambda = \alpha. \tag{3}$$

We easily prove that

$$\begin{aligned} 1 &= X_0 - X_1 \cup X_1 - X_2 \cup \ldots \cup X_\xi - X_{\xi+1} \cup \ldots \cup X_\alpha \\ &= \bigcup_{\xi<\alpha} (X_\xi - X_{\xi+1}) \cup X_\alpha . \end{aligned} \tag{4}$$

The sets $(X_0 - X_1 \cup X_2 - X_3 \cup \ldots)$ and $(X_1 - X_2 \cup X_3 - X_4 \cup \cup \ldots \cup X_\alpha)$ are disjoint, therefore

$$1 - (X_0 - X_1 \cup X_2 - X_3 \cup \ldots) = X_1 - X_2 \cup X_3 - X_4 \cup \ldots \cup X_\alpha . \tag{4a}$$

II. Definition. *A set of the form*

$$E = F_1 - F_2 \cup F_3 - F_4 \cup \ldots \cup F_\xi - F_{\xi+1} \cup \ldots ,$$

where $\{F_\xi\}$ forms a decreasing sequence of closed sets, is said to be resolvable into an alternated series of decreasing closed sets or, simply, resolvable.

Remark. In complete spaces, the resolvable sets coincide with sets which are simultaneously $\boldsymbol{F}_\sigma$ and $\boldsymbol{G}_\delta$ (Chapter III, § 34, VI). Many important properties of $\boldsymbol{F}_\sigma$ and $\boldsymbol{G}_\delta$-sets are consequences of their resolvability. This is one of the reasons why the resolvable sets deserve study. See also § 13, VI.

III. Separation theorems. Resolution into alternating series. Let E and H be two arbitrary sets and let us assume that sequence (1) satisfies conditions (2) and (3) as well as the condition

$$X_{\xi+1} = \overline{X_\xi \cap E} \cap \overline{X_\xi \cap H}.$$

(1) See F. Hausdorff, *Mengenlehre*, p. 80.

Obviously sequence (1) is composed of closed sets. Moreover, it is decreasing, since $X_{\xi+1} \subset \bar{X}_\xi = X_\xi$. Therefore, from a certain index onwards, say $(\alpha-1)$, its terms are identical. Hence

$$X_\alpha = \overline{X_\alpha \cap E} \cap \overline{X_\alpha \cap H}.$$

Let us put $P_\xi = X_\xi - \overline{X_\xi \cap E}$, $R_\xi = X_\xi - \overline{X_\xi \cap H}$. Therefore $X_\xi - X_{\xi+1} = P_\xi \cup R_\xi$, whence, by (4),

$$1 = \bigcup_{\xi<\alpha} P_\xi \cup \bigcup_{\xi<\alpha} R_\xi \cup X_\alpha; \quad \text{hence} \quad 1 - \bigcup_{\xi<\alpha} P_\xi \subset \bigcup_{\xi<\alpha} R_\xi \cup X_\alpha.$$

On the other hand,

$$P_\xi = X_\xi - \overline{X_\xi \cap E} \subset X_\xi - (X_\xi \cap E) = X_\xi - E \subset 1 - E,$$

and hence

$$\bigcup_{\xi<\alpha} P_\xi \subset 1-E \quad \text{and therefore} \quad E \subset 1 - \bigcup_{\xi<\alpha} P_\xi \subset \bigcup_{\xi<\alpha} R_\xi \cup X_\alpha.$$

Similarly, $\bigcup_{\xi<\alpha} R_\xi \subset 1-H$. Thus we obtain

$$E - X_\alpha \subset \bigcup_{\xi<\alpha} R_\xi, \quad H \cap \bigcup_{\xi<\alpha} R_\xi = 0.$$

Moreover, *the set*

$$\bigcup_{\xi<\alpha} R_\xi = \bigcup_{\xi<\alpha} (X_\xi - \overline{X_\xi \cap H})$$
$$= (1-\bar{H}) \cup (\bar{E} \cap \bar{H}) - \overline{\bar{E} \cap H} \cup \overline{E \cap \bar{H}} \cap \overline{\bar{E} \cap H} - \ldots$$

is the union of an alternating series of decreasing closed sets since $X_{\xi+1} = \overline{X_\xi \cap E} \cap \overline{X_\xi \cap H} \subset \overline{X_\xi \cap H}$.

In particular, we conclude that

1° *if the only root of the equation* $X = \overline{X \cap E} \cap \overline{X \cap H}$ *is* $X = 0$, *then there exists a resolvable set* D (namely *the set* $D = \bigcup_{\xi<\alpha} R_\xi$) *such that* $E \subset D$ *and* $H \cap D = 0$ [1].

2° *putting* $H = 1-E$ *we obtain*

$$X_{\xi+1} = \overline{X_\xi \cap E} \cap \overline{X_\xi - E} = \text{the boundary of } X_\xi \cap E \text{ relative to } X_\xi, \tag{5}$$

$$X_\alpha = \overline{X_\alpha \cap E} \cap \overline{X_\alpha - E} \quad \text{and} \quad E - X_\alpha = \bigcup_{\xi<\alpha} R_\xi, \tag{6}$$

---

[1] We make use of this proposition in the proof of an important theorem of Baire on functions of the first class (see Chapter II, § 31, X).

since $\bigcup\limits_{\xi<\alpha} R_\xi \subset 1-H = E$ and $R_\xi \subset X_\xi - X_{\xi+1}$ so that the set $\bigcup\limits_{\xi<\alpha} R_\xi$, as a subset of $\bigcup\limits_{\xi<\alpha}(X_\xi - X_{\xi+1})$, is disjoint to X_α. Therefore $\bigcup\limits_{\xi<\alpha} R_\xi \subset E - X_\alpha$.

Thus by taking away from E the "*remainder*" $X_\alpha \cap E$ we obtain a resolvable set. Consequently

3° *if the remainder* $X_\alpha \cap E$ *is zero, then*

$$E = \bigcup_{\xi<\alpha} R_\xi$$

$$= (1-\overline{1-E}) \cup (\bar{E} \cap \overline{1-E} - \overline{\bar{E}-E}) \cup (\overline{E \cap \overline{1-E}} \cap \overline{\bar{E}-E}) - \ldots \quad \text{(i)}$$

$$E = 1-H = 1-\bigcup_{\xi<\alpha} P_\xi = 1-\bigcup_{\xi<\alpha} (X_\xi - \overline{X_\xi \cap E})$$

$$= \bigcup_{\xi<\alpha} (\overline{X_\xi \cap E} - X_{\xi+1})$$

$$= \bar{E} - (\bar{E} \cap \overline{1-E}) \cup \overline{E \cap \overline{1-E}} - (\overline{E \cap \overline{1-E}} \cap \overline{\bar{E}-E}) \cup \ldots \quad \text{(ii)}$$

(by I (4a)).

IV. Properties of the remainder. The set X_α in formula (6) is *the largest set* satisfying the equation

$$X = \overline{X \cap E} \cap \overline{X-E}. \tag{7}$$

Proof. If X satisfies (7), then $X \subset X_0 = 1$. Now, if $X \subset X_\xi$, then $X \cap E \subset X_\xi \cap E$ and $X-E \subset X_\xi - E$, whence $\overline{X \cap E} \cap \overline{X-E} \subset \overline{X_\xi \cap E} \cap \overline{X_\xi - E}$ and, by (5) and (7), $X \subset X_{\xi+1}$. Finally, if for every $\xi < \lambda$ (λ is a limit ordinal), $X \subset X_\xi$, then $X \subset \bigcap\limits_{\xi<\lambda} X_\xi = X_\lambda$.

By the transfinite induction principle, X is a subset of each X_ξ, hence of X_α.

Observe that (7) is equivalent to

$$\overline{X \cap E} = X = \overline{X-E}. \tag{8}$$

Proof. (7) implies that $X \subset \overline{X \cap E}$ and $X \subset \overline{X-E}$ and, on the other hand, $\overline{X \cap E} \subset \bar{X}$ and $\overline{X-E} \subset \bar{X}$. The required equivalence follows from the fact that, by (7), $X = \bar{X}$.

V. Necessary and sufficient conditions. *Each of the following conditions is necessary and sufficient in order that the set E be resolvable:*

(i) *the formula* (8) *implies that* $X = 0$; in other words, *for every closed set* $F \neq 0$, *the boundary of* $F \cap E$ *relative to* F, i.e. *the set* $\overline{F \cap E} \cap \overline{F-E}$, *is* $\neq F$;

(ii) *for every closed set* F, *the boundary of* $F \cap E$ *relative to* F *is nowhere dense in* F;

(iii) *the remainder vanishes, i.e.* $X_a \cap E = 0$.

Proof. 1. Condition (i) is necessary. Suppose that E is resolvable into an alternating series of decreasing closed sets:

$$E = F_1 - F_2 \cup F_3 - F_4 \cup \ldots \cup F_\xi - F_{\xi+1} \cup \ldots \quad (\xi+1 < a). \tag{9}$$

Obviously we can suppose that the limit indices in this resolution are omitted. Assuming that $F_0 = 1$, $F_\lambda = \bigcap_{\xi<\lambda} F_\xi$ (for a limit ordinal λ) and $F_a = \bigcap_{\xi<a} F_\xi$, we infer from (4) that

$$1 - E = F_0 - F_1 \cup F_2 - F_3 \cup \ldots \cup F_a. \tag{10}$$

We shall prove that condition (8) implies $X \subset F_\xi$ for every ξ. Firstly, $X \subset F_0 = 1$. Assume that $X \subset F_\xi$. If ξ is even, (9) implies that $X \cap E \subset F_{\xi+1}$ (since all the differences preceding $F_{\xi+1}$ are disjoint to F_ξ, hence also to X, while all the terms following $F_{\xi+1}$ are contained in $F_{\xi+1}$). Therefore $X = \overline{X \cap E} \subset F_{\xi+1}$. Similarly, if ξ is odd, (10) implies that $X - E \subset F_{\xi+1}$, and hence $X = \overline{X-E} \subset F_{\xi+1}$. Finally, if $X \subset F_\xi$ for every $\xi < \lambda$, then

$$X \subset \bigcap_{\xi<\lambda} F_\xi = F_\lambda.$$

Thus we have proved that $X \subset F_\xi$, for every ξ. Therefore $X \subset F_a$. This and (10) imply that $X \subset 1-E$, whence $X \cap E = 0$ and $X = \overline{X \cap E} = 0$.

2. Condition (i) implies (ii). By § 6, II (12), $\overline{\mathrm{Int}(\mathrm{Fr}(E))} = \overline{\mathrm{Int}(\mathrm{Fr}(E)) \cap E} = \overline{\mathrm{Int}(\mathrm{Fr}(E)) - E}$. Therefore we can substitute $X = \overline{\mathrm{Int}(\mathrm{Fr}(E))}$ in (8) since, obviously, the identity $\overline{X} = \overline{X \cap E}$ implies $\overline{X} = \overline{\overline{X} \cap E}$ (for $\overline{X \cap E} \subset \overline{\overline{X} \cap E} \subset \overline{X}$). Now, if we suppose that (8) implies $X = 0$, then $\overline{\mathrm{Int}(\mathrm{Fr}(E))} = 0$. It follows that the boundary of E does not contain interior points, i.e. it is nowhere dense.

Moreover, if we consider $F \cap E$ in place of E, and the boundary of $F \cap E$ relative to F, instead of the boundary of E, we infer that this relative boundary is nowhere dense in F, since condition (i) implies the same conditions relativized to F (and the latter means that condition (i) is fulfilled for every $X \subset F$).

3. Condition (ii) implies (iii). Putting $F = X_a$ in (ii) we infer that the boundary of $X_a \cap E$ relative to X_a is nowhere dense in X_a. Hence it cannot coincide with X_a unless $X_a = 0$. Therefore, by (6), $X_a = 0$, whence $X_a \cap E = 0$.

4. Condition (iii) is sufficient by III, 3°.

Thus the theorem has been proved.

We have shown in § 8, V, that the boundary of a set is nowhere dense if and only if the set is a union of an open set and a nowhere dense set (equivalently, if it is the difference of a closed set and a nowhere dense set). It follows by the preceding theorem that

A necessary and sufficient condition in order that the set E be resolvable is that, relative to every closed set F, the set $E \cap F$ should be a union of an open set and of a nowhere dense set (or, equivalently, *that $E \cap F$ should be a difference of a closed set and a nowhere dense set*).

It follows from condition (i) of the preceding theorem that:

Every non-empty closed set F contains a point where either $F \cap E$ or $(F-E)$ is "locally empty" relative to F(1), *i.e. such that there exists a neighbourhood G of this point with either $G \cap F \cap E = 0$ or $G \cap F-E = 0$.*

Proof. If $\overline{F \cap E} = F = \overline{F-E}$, then $G \cap F \neq 0$ implies $G \cap F \cap$ $\cap E \neq 0 \neq G \cap F-E$ (see § 5, III). Thus the condition in question implies (i). Conversely, if $p \in F-\overline{F \cap E}$, we put $G = 1-\overline{F \cap E}$; hence $G \cap F \cap E = 0$. If $p \in F-\overline{F-E}$, we put $G = 1-\overline{F-E}$.

VI. Properties of resolvable sets.

THEOREM 1. *Resolvable sets constitute a field, i.e. the union and the intersection of two resolvable sets and the complement of a resolvable set are resolvable.*

This is a consequence of condition (ii) of the theorem of Section V and of the fact that the sets with nowhere dense boundaries constitute a field (§ 8, V).

---

(1) Compare § 7, IV and § 11, IV, 4.

THEOREM 2. *Every resolvable set has the Baire property in the restricted sense.*

For, if E is resolvable and F is closed, then $E \cap F$ is the union of an open set and of a set nowhere dense in F (compare Section V).

THEOREM 3. *If a boundary set is resolvable, it is nowhere dense.*

For, in the domain of sets with nowhere dense boundaries, the notions of a boundary set and of a nowhere dense set coincide (§ 8, V).

THEOREM 4. *Every scattered set is resolvable.*

This is because the boundary of a scattered set is nowhere dense (§ 9, VI, 4).

VII. Residues[1]. The set $X \cap \overline{\overline{X}-X}$ is called the *residue* of X (in the sense of Hausdorff).

The set $X_a \cap E$ (the remainder of E) coincides with its residue.

For, if E and X are two sets satisfying formula (8), then $X \cap E$ is its residue. Because $X = \overline{X-E} = \overline{X-X \cap E} = \overline{\overline{X \cap E}-X \cap E}$ since $X = \overline{X \cap E}$. It therefore follows that $X \cap E = X \cap E \cap$ $\cap \overline{\overline{X \cap E}-X \cap E}$.

THEOREM 1. *E is a resolvable set if and only if every non-empty set Y closed in E does not coincide with its residue; in other terms* (see § 7, V), *if Y contains a point at which Y is closed.*

Proof. Suppose that $Y = \overline{Y} \cap E$ and $Y = Y \cap \overline{\overline{Y}-Y}$. Then $\overline{Y} = \overline{\overline{Y} \cap E}$ and, on the other hand, $\overline{Y} \subset \overline{\overline{Y}-Y} = \overline{\overline{Y}-\overline{Y} \cap E}$ $= \overline{\overline{Y}-E} \subset \overline{Y}$. Hence $\overline{\overline{Y} \cap E} = \overline{Y} = \overline{\overline{Y}-E}$ and, by replacing X with $\overline{Y}$ in V, (i), we have $Y = 0$. Thus the condition is necessary. It is also sufficient, for it implies, by virtue of the preceding proposition, that $X_a \cap E = 0$, and hence, by V, (iii), that the set E is resolvable.

THEOREM 2. *A necessary and sufficient condition for a set to be the difference of two closed sets is that its residue should be empty.*

This follows from the corollary of § 7, V.

VIII. Residues of transfinite order. Given a set E, we form a sequence of residues of all orders in the following manner: R_1 is the residue of E, $R_{\xi+1}$ is the residue of R_ξ, and $R_\lambda = \bigcap_{\xi<\lambda} R_\xi$ if λ

[1] F. Hausdorff, *Grundzüge der Mengenlehre*, p. 280.

is a limit ordinal. Since the sequence of residues so defined is decreasing, there exists a certain ordinal β such that $R_\beta = R_{\beta+1} = \dots$. By I (4), we have $E - R_\beta = E - R_1 \cup R_1 - R_2 \cup \dots$.

The terms of this alternated series are not closed, but by virtue of the identity $X - \overline{\bar{X} - X} = \bar{X} - \overline{\bar{X} - X}$ (since $\bar{X} - X - \overline{\bar{X} - X} = 0$), we have $R_\xi - R_{\xi+1} = \bar{R}_\xi - \overline{\bar{R}_\xi - R_\xi}$, hence

$$E - R_\beta = \bar{E} - \overline{\bar{E} - E} \cup \overline{E \cap \overline{\bar{E} - E}} - \dots \cup \bar{R}_\xi - \overline{\bar{R}_\xi - R_\xi} \cup \dots \quad (1)$$

In particular, *if E is resolvable, then the "last" residue R_β of E is empty* (see Section VII), and formula (1) presents a resolution of E into a series of decreasing closed sets.

The last residue R_β is a set which is not locally closed at any of its points, since it coincides with its own residue. We shall prove that

Among the sets closed in E the set R_β is the largest one that is not locally closed at any of its points.

Proof. Suppose that the set Y is closed in E and $Y - R_\beta \neq 0$. We have to show that Y contains a point where it is locally closed. By using decomposition (1) we conclude that there exists an index ξ such that $Y \cap R_\xi - R_{\xi+1} \neq 0$. Let ξ be the smallest index with this property. Then $Y \subset R_\xi$ and hence Y is closed in R_ξ. Since $R_{\xi+1}$ is the residue of R_ξ, the set R_ξ is locally closed at each point of $R_\xi - R_{\xi+1}$; hence it is locally closed at each point of the non-empty set $Y \cap R_\xi - R_{\xi+1}$. This implies that the set Y is also locally closed at each point of the latter set.

§ 13. Continuity. Homeomorphism

I. Definition. Let $\mathscr{X}$ and $\mathscr{Y}$ be topological spaces and let $f: \mathscr{X} \to \mathscr{Y}$. f is said to be *continuous at the point x* if[1]

$$(x \in \bar{A}) \Rightarrow [f(x) \in \overline{f(A)}] \quad \text{for each} \quad A \subset \mathscr{X}; \quad (0)$$

equivalently (see § 3, II (2)) if

$$(x \in \bar{A}) \Rightarrow [x \in f^{-1}\overline{f(A)}] \quad \text{for each} \quad A \subset \mathscr{X}. \quad (0')$$

---

[1] Compare F. Hausdorff, *Grundzüge der Mengenlehre*, Chap. 9, § 1.

In the case where the spaces $\mathscr{X}$ and $\mathscr{Y}$ have common points, it is desirable to distinguish between the closure in $\mathscr{X}$ and in $\mathscr{Y}$. We omit this distinction in order to simplify the notation.

If f is continuous at every point, it is said briefly to be *continuous*. The family of all continuous mappings $f\colon \mathscr{X} \to \mathscr{Y}$ into, i.e. such that $f(\mathscr{X}) \subset \mathscr{Y}$, will be denoted by $\mathscr{Y}^{\mathscr{X}}$.

In the case where $\mathscr{X}$ and $\mathscr{Y}$ are sets of real numbers, the definition of continuity considered here coincides with that of the classical analysis (called "*Heine definition*"). Thus, for example, $\mathscr{E}^{\mathscr{I}}$ denotes the set of real-valued continuous functions defined in the interval $0 \leqslant x \leqslant 1$.

II. Necessary and sufficient conditions.

THEOREM. *A necessary and sufficient condition for f to be continuous at the point x is that for every neighbourhood B of $f(x)$, the set $f^{-1}(B)$ should be a neighbourhood of x; in other words, that*

$$[f(x) \in \mathrm{Int}(B)] \Rightarrow [x \in \mathrm{Int}(f^{-1}(B))] \quad \textit{for each} \quad B \subset \mathscr{Y}, \tag{1}$$

or equivalently:

$$[x \in f^{-1}(\mathrm{Int}(B))] \Rightarrow [x \in \mathrm{Int}(f^{-1}(B))] \quad \textit{for each} \quad B \subset \mathscr{Y}, \tag{1'}$$

or (by replacing B with $\mathscr{Y}-B$) that

$$[x \in \overline{f^{-1}(B)}] \Rightarrow [x \in f^{-1}(\bar{B})] \quad \textit{for each} \quad B \subset \mathscr{Y}. \tag{2}$$

Proof. We deduce (2) from (0) by putting $A = f^{-1}(B)$ (compare § 3, III, (12)). Conversely by putting $B = f(A)$ in (2) (and by § 3, III, (11)) we obtain (0).

By using the fact that every neighbourhood of a point contains an open neighbourhood of this point, we obtain the following condition ("*Cauchy definition*").

COROLLARY. *f is continuous at the point x if and only if for every open set H containing $f(x)$ there exists an open set G containing x such that $f(G) \subset H$.*

III. The set $D(f)$ of points of discontinuity. Let $f\colon \mathscr{X} \to \mathscr{Y}$. Write briefly $D = D(f)$. By (0′), (1′), and (2) we have

$$D = \bigcup_A [\bar{A} - f^{-1}(\overline{f(A)})] = \bigcup_B \big(f^{-1}(\mathrm{Int}(B)) - \mathrm{Int}(f^{-1}(B))\big)$$
$$= \bigcup_B \big(\overline{f^{-1}(B)} - f^{-1}(\bar{B})\big), \tag{1}$$

where A ranges over all subsets of $\mathscr{X}$, and B over subsets of $\mathscr{Y}$.

It follows that

$$f(D) = \bigcup_A [f(\bar{A}) - \overline{f(A)}]. \tag{2}$$

For

$$f(D) = \bigcup_A f[\bar{A} - f^{-1}\overline{(f(A))}] = \bigcup_A f[\bar{A} \cap f^{-1}(\mathscr{Y} - \overline{f(A)})]$$
$$= \bigcup_A [f(\bar{A}) \cap (\mathscr{Y} - \overline{f(A)})] = \bigcup_A [f(\bar{A}) - \overline{f(A)}],$$

by § 3, III, (13).

Let us note that one can suppose that the variable B ranges over the family of open sets in the third term of (1) and over that of closed sets in the fourth term.

For, by putting $\operatorname{Int}(B) = G$ we have

$$f^{-1}(\operatorname{Int}(B)) - \operatorname{Int}(f^{-1}(B)) \subset f^{-1}(G) - \operatorname{Int}(f^{-1}(G)),$$

and by putting $\bar{B} = F$ we have

$$\overline{f^{-1}(B)} - f^{-1}(\bar{B}) \subset \overline{f^{-1}(F)} - f^{-1}(F).$$

Thus

$$D = \bigcup_G [f^{-1}(G) - \operatorname{Int}(f^{-1}(G))] = \bigcup_F (\overline{f^{-1}(F)} - f^{-1}(F)). \tag{3}$$

Moreover, if $\boldsymbol{S}$ is a *subbase* of $\mathscr{Y}$, the range or variability of G may be restricted to $\boldsymbol{S}$, i.e.

$$D = \bigcup_{G \in \boldsymbol{S}} \{f^{-1}(G) - \operatorname{Int}[f^{-1}(G)]\}. \tag{4}$$

In other words, if for a given $x \in \mathscr{X}$, the implication

$$x \in f^{-1}(G) \Rightarrow x \in \operatorname{Int}[f^{-1}(G)] \tag{5}$$

holds true for each $G \in \boldsymbol{S}$, then it is true of any open $G \subset \mathscr{Y}$ (i.e. f is continuous at x).

To show this, we have to prove that

(i) whenever (5) is true of G_1 and G_2, it is true of $G_1 \cap G_2$;

(ii) whenever (5) is true of G_ι for each $\iota \in I$, then it is true of $\bigcup_\iota G_\iota$.

Now these two statements follow easily from the formulas (see § 6, II (1) and (2)):

$$\operatorname{Int}(A \cap B) = \operatorname{Int}(A) \cap \operatorname{Int}(B) \quad \text{and} \quad \bigcup_\iota \operatorname{Int}(A_\iota) \subset \operatorname{Int}(\bigcup_\iota A_\iota).$$

Similarly, if $\boldsymbol{T}$ is a *closed subbase* of $\mathscr{Y}$, the range of variability of F in (3) can be restricted to $\boldsymbol{T}$.

The first part of (3) implies that $D \subset X^d$ (the derived set of $\mathscr{X}$), for every isolated point of the space, as an interior point of every set containing it, does not belong to $f^{-1}(G) - \mathrm{Int}\big(f^{-1}(G)\big)$.

The following identities define two important classes of functions: (i) $D = 0$ the *continuous* functions; (ii) $\overline{\mathscr{X} - D} = \mathscr{X}$ the functions *pointwise discontinuous.*

IV. Continuous mappings. Formulas I (0′), II (2), and II (1′) (as well as III (1) to (3)) yield directly the following theorem.

THEOREM 1. *The following conditions are necessary and sufficient for f to be continuous:*

(1) $f(\bar{A}) \subset \overline{f(A)}$, *for every* $A \subset \mathscr{X}$, *i.e.* $\bar{A} \subset f^{-1}\overline{f(A)}$;

(2) $\overline{f^{-1}(B)} \subset f^{-1}(\bar{B})$, *for every* $B \subset \mathscr{Y}$, *i.e.* $f\overline{f^{-1}(B)} \subset \bar{B}$;

(3) $f^{-1}(G)$ *is open, for every open set* $G \subset \mathscr{Y}$;

(4) $f^{-1}(F)$ *is closed, for every closed set* $F \subset \mathscr{Y}$.

The operation f^{-1} being additive and multiplicative (§ 3, III, (6a) and (7a)), we infer from (3) and (4) that:

(5) *if* B *is an* $\boldsymbol{F}_\sigma$*-set or a* $\boldsymbol{G}_\delta$*-set, respectively, then so is* $f^{-1}(B)$.

THEOREM 2. *The composition of two continuous mappings is a continuous mapping.*

More precisely, *if f is continuous at the point x and g is continuous at the point* $f(x)$, *then the composition* $h = g \circ f$ *is continuous at the point x.*

Proof. By I(0), $x \in \bar{A}$ implies $f(x) \in \overline{f(A)}$ and this, by the continuity of g, implies that $g\big(f(x)\big) \in \overline{g\big(f(A)\big)}$.

THEOREM 3. *Let* $f\colon \mathscr{X} \to \mathscr{Y}$ *be continuous onto. If A is dense in* $\mathscr{X}$, $f(A)$ *is dense in* $\mathscr{Y}$.

Consequently, separability is invariant under continuous mappings.

The proof follows directly from (1).

V. Relativization. Restriction. Retraction. We recall that $f \mid A$ denotes the function obtained from $f\colon \mathscr{X} \to \mathscr{Y}$ by restricting the set of its arguments to A (§ 3, I). The partial function $f \mid A$ is said to be *continuous at the point x relative to A* if $x \in A$ and if the condition $x \in \overline{X \cap A}$ implies $f(x) \in \overline{f(X \cap A)}$ for each $X \subset \mathscr{X}$.

THEOREM 1. *The continuity of a function at a point p is a local property, i.e. if A is a neighbourhood of p, then the continuity of the partial function $f \mid A$ at p implies the continuity of f at p.*

Proof. Let $p \in \overline{X}$. We have $X = X \cap A \cup (X-A)$, whence $\overline{X} = \overline{X \cap A} \cup \overline{X-A} \subset \overline{X \cap A} \cup \overline{\mathcal{X}-A}$. By the hypothesis, p does not belong to $\overline{\mathcal{X}-A}$, whence $p \in \overline{X \cap A}$ and, by the continuity of the partial function, $f(p) \in \overline{f(X \cap A)} \subset \overline{f(X)}$.

THEOREM 2. *If $\mathcal{X} = A \cup B$, $p \in A \cap B$ and the functions $f \mid A$ and $f \mid B$ are continuous at the point p, then so is the function f.*

Proof. If $p \in \overline{X} = \overline{X \cap A} \cup \overline{X \cap B}$, then $p \in \overline{X \cap A}$ or $p \in \overline{X \cap B}$. Therefore $f(p) \in \overline{f(X \cap A)} \subset \overline{f(X)}$ or $f(p) \in \overline{f(X \cap B)} \subset \overline{f(X)}$.

THEOREM 3. *If $\mathcal{X} = A \cup B$ is a decomposition of $\mathcal{X}$ into two closed or two open sets and if the functions $f \mid A$ and $f \mid B$ are continuous, then f is continuous over $\mathcal{X}$.*

Proof. If $p \in A \cap B$, the function f is continuous at p by Theorem 2; if $p \notin A$, then p is an interior point of B and hence f is continuous by Theorem 1.

DEFINITION. A subset A of $\mathcal{X}$ is a *retract of* $\mathcal{X}$ ([1]), if there exists a continuous mapping f (called a *retraction*) of $\mathcal{X}$ onto A such that $f(x) = x$ for $x \in A$. In other words, if the identity restricted to A admits a continuous extension f over the whole space $\mathcal{X}$ such that $f(\mathcal{X}) = A$.

Retraction can be regarded as a generalization of projection; given the euclidean plane $\mathcal{X} \times \mathcal{Y}$, the formula $f(x, y) = x$ defines a retraction of the product $\mathcal{X} \times \mathcal{Y}$ to the $\mathcal{X}$-axis.

THEOREM 4. *If A is a retract of $\mathcal{X}$, then every continuous function defined on A (with values in a space $\mathcal{Y}$) admits a continuous extension f^* over $\mathcal{X}$; i.e. $f = f^* \mid A$.*

For, if g is a retraction of $\mathcal{X}$ to A, we can let $f^*(x) = f(g(x))$ for $x \in \mathcal{X}$.

THEOREM 5. *Let $\mathcal{X}$ be Hausdorff and let $f\colon \mathcal{X} \to \mathcal{X}$ continuous. Then the set of fixed points under f, i.e. the set*

$$\underset{x}{E}[f(x) = x]$$

is closed.

([1]) See K. Borsuk, *Sur les rétractes,* Fund. Math. 17 (1931), p. 153.

We have to show that $\underset{x}{E}[f(x) \neq x]$ is open. Now $\mathscr{X}$ being Hausdorff, we have

$$\big(f(x) \neq x\big) \equiv \bigvee_{G,H} \big(f(x) \in G\big)(x \in H)(G \cap H = 0) \qquad (G \text{ and } H \text{ open}).$$

Equivalently stated (the operator $\bigcup$ being extended to all open G, H such that $G \cap H = 0$):

$$\underset{x}{E}[f(x) \neq x] = \bigcup_{G,H} \underset{x}{E}\big(f(x) \in G\big)(x \in H) = \bigcup_{G,H} \underset{x}{E}\big(x \in f^{-1}(G)\big)(x \in H)$$
$$= \bigcup_{G,H} \big(f^{-1}(G) \cap H\big).$$

f being continuous, $f^{-1}(G)$ is open. This completes the proof.

THEOREM 6. *Each retract of a Hausdorff space is closed.*

This follows from the preceding theorem, for f being a retraction of $\mathscr{X}$ onto R, we have $R = \underset{x}{E}[f(x) = x]$.

VI. Real-valued functions. Characteristic functions. Consider the case $\mathscr{Y} = \mathscr{E}$ (= space of real numbers), $\mathscr{X}$ being an arbitrary topological space.

THEOREM 1. *Let f: $\mathscr{X} \to \mathscr{E}$. f is continuous at x_0 if and only if there is for each $\varepsilon > 0$ an open G containing x_0 and such that*

$$|f(x) - f(x_0)| < \varepsilon \quad \textit{for each} \quad x \in G. \tag{1}$$

To show this, we substitute to H in II (Corollary) the open interval $\underset{y}{E}[|y - f(x_0)| < \varepsilon]$.

THEOREM 2. *If f: $X \to \mathscr{E}$ is continuous, then the sets*

$$\underset{x}{E}[f(x) \leqslant a], \quad \underset{x}{E}[f(x) \geqslant a], \quad \underset{x}{E}[a \leqslant f(x) \leqslant b]$$

are closed and the following sets are open:

$$\underset{x}{E}[f(x) < a], \quad \underset{x}{E}[f(x) > a], \quad \underset{x}{E}[a < f(x) < b].$$

This is true in virtue of IV (3) and (4), because these sets are inverse images of closed, respectively open, sets:

$$\underset{y}{E}(y \leqslant a), \quad \underset{y}{E}(y \geqslant a), \quad \underset{y}{E}(a \leqslant y \leqslant b),$$
$$\underset{y}{E}(y < a), \quad \underset{y}{E}(y > a), \quad \underset{y}{E}(a < y < b).$$

The concept of *uniform convergence* can be introduced in the same way as in elementary analysis.

Let $f_n: \mathscr{X} \to \mathscr{E}$ for $n = 1, 2, \ldots$ and let $f: \mathscr{X} \to \mathscr{E}$. We say that the sequence $f_1, f_2, \ldots$ converges uniformly to f if

$$\bigwedge_{\varepsilon>0} \bigvee_{k} \bigwedge_{n\geqslant k} \bigwedge_{x} |f_n(x) - f(x)| < \varepsilon. \tag{2}$$

The following classical theorem can be extended to the case where $\mathscr{X}$ is an arbitrary topological space([1]).

THEOREM 3. *The limit of a uniformly convergent sequence of continuous functions is a continuous function.*

Proof. Put $f(x) = \lim\limits_{n=\infty} f_n(x)$. Let $\varepsilon > 0$ and let x_0 be a given point of $\mathscr{X}$. In virtue of (2), k exists such that

$$|f_k(x) - f(x)| < \varepsilon/3 \quad \text{for each} \quad x \in \mathscr{X}. \tag{3}$$

Therefore, substituting $x = x_0$, we have

$$|f_k(x_0) - f(x_0)| < \varepsilon/3. \tag{4}$$

Since f_k is continuous at the point x_0, there is an open set G containing x_0 such that (compare (1))

$$|f_k(x) - f_k(x_0)| < \varepsilon/3 \quad \text{for each} \quad x \in G. \tag{5}$$

Inequalities (3) to (5) yield condition (1), which means that x_0 is a point of continuity of f.

Recall that the *characteristic function* f_A of the set A is the function which assumes the value 1 for points of A and the value 0 for points of the complement of A.

THEOREM 4. *The following identity holds*

$$\mathrm{Fr}(A) = D(f_A), \tag{6}$$

i.e. the boundary of A coincides with the set of points of discontinuity of the characteristic function of A.

Proof. Let $f(x) = 1$ (the case $f(x) = 0$ is similar).

If $x \in \mathrm{Fr}(A)$, the set $A = f^{-1}(1)$ is not a neighbourhood of x although the set composed of the point 1 alone is a neighbourhood of $f(x)$ in the space $(0, 1)$. Hence by the theorem of Section II, $x \in D$.

([1]) A further extension obtained by replacing $\mathscr{E}$ by an arbitrary metric space will be considered in Chapter II. The proof will be essentially the same.

Conversely, if $x \in A - \mathrm{Fr}(A)$, then $x \in \mathrm{Int}(A)$ and, by the same theorem, x is a point of continuity of f.

THEOREM 5. *The characteristic function of a set A is, respectively, continuous or pointwise discontinuous if and only if this set is simultaneously closed and open or, respectively, has a nowhere dense boundary.*

For by (6), the condition $D = 0$ is equivalent to $\mathrm{Fr}(A) = 0$ which means that A is closed and open.

From § 12, V, (ii) the following corollary is obtained.

COROLLARY. *The characteristic function of a set A is pointwise discontinuous on each closed set if and only if the set A is resolvable into an alternated series of decreasing closed sets.*

VII. One-to-one continuous mappings. Comparison of topologies. Suppose now that the mapping $f: \mathscr{X} \to \mathscr{Y}$ is one-to-one. The following properties are invariant under one-to-one continuous mappings.

1. The property of being an *accumulation point* of the space; for, $x \in \overline{\mathscr{X} - x}$ implies $f(x) \in \overline{f(\mathscr{X} - x)} = \overline{\mathscr{Y} - f(x)}$, since $f(\mathscr{X} - x) = f(\mathscr{X}) - f(x)$ (see § 3, III, (3′));

2. the property of being *dense in itself* (this is an immediate consequence of the preceding proposition);

3. the property of being a *boundary set* in the space; for the identity $\mathscr{X} = \overline{\mathscr{X} - X}$ implies $f(\mathscr{X}) = f(\overline{\mathscr{X} - X}) \subset \overline{f(\mathscr{X} - X)} = \overline{f(\mathscr{X}) - f(X)}$.

The totality of all topological spaces which have the same cardinality can be *ordered* assuming that $\mathscr{Y} \leqslant \mathscr{X}$ if there exists a one-to-one continuous mapping f, of $\mathscr{X}$ onto $\mathscr{Y}$.

The sets $\mathscr{X}$ and $\mathscr{Y}$ being of the same cardinality, we can express this definition in the following terms ("identifying" x with $f(x)$).

A topological space being a pair, $(1, F)$, where 1 is the set of points and F a function (the closure) satisfying Axioms § 4, I, 1–4, we have

$$(1, F_2) \leqslant (1, F_1) \quad \text{if} \quad F_1(X) \subset F_2(X) \quad \text{for each} \quad X \subset 1,$$

and we say that the topology of 1 given by F_1 is *finer* than that given by F_2.

In other words, *each open set in the coarser topology is open in the finer one.* Thus, the finer topology is reached in open sets (hence in closed sets).

The discrete topology is the finest topology (since every set is open). If we do not require the space to be $\mathcal{T}_1$, there exists the coarsest topology, namely the topology where the closure of each non void set is the whole space; that is, where the only closed sets are the void set and the whole space.

The coarsest $\mathcal{T}_1$-topology is the topology where the only closed sets are: the void set, the whole space, and the finite sets.

VIII. Homeomorphism. If the mapping f of the space $\mathcal{X}$ onto (the whole) space $\mathcal{Y}$ is continuous and one-to-one, and its inverse f^{-1} is also continuous, then we say that f is a *homeomorphism*[1] and the spaces $\mathcal{X}$ and $\mathcal{Y}$ are said to be *homeomorphic* (or of the same topological type). We then write

$$\mathcal{X} \underset{\text{top}}{=} \mathcal{Y} \quad \text{(topological equivalence)}.$$

If $\mathcal{X} = \mathcal{Y}$, a homeomorphism is called a *topological automorphism.*

The homeomorphism relation is clearly *reflexive, symmetric,* and *transitive.*

THEOREM 1. *Each of the following conditions is necessary and sufficient for a one-to-one mapping f to be a homeomorphism:*

$$f(\bar{A}) = \overline{f(A)} \quad \textit{for every} \quad A \subset \mathcal{X}; \tag{1}$$

$$f^{-1}(\bar{B}) = \overline{f^{-1}(B)} \quad \textit{for every} \quad B \subset \mathcal{Y}. \tag{2}$$

Proof. By IV, (1), the inclusion $f(\bar{A}) \subset \overline{f(A)}$ is equivalent to the continuity of f, while the inclusion $\overline{f(A)} \subset f(\bar{A})$, by IV (2), is equivalent to the continuity of f^{-1}. The second part of the theorem can be proved in a similar way.

THEOREM 2. *A necessary and sufficient condition for f to be a homeomorphism is the following:*

$$\bar{A} = f^{-1}\left(\overline{f(A)}\right) \quad \textit{for every} \quad A \subset \mathcal{X}; \tag{3}$$

or equivalently

$$(x \in \bar{A}) \equiv \left(f(x) \in \overline{f(A)}\right). \tag{3a}$$

---

[1] This term was introduced by H. Poincaré, Journ. Ec. Polyt. (2), 1 (1895), p. 9.

Proof. If f is one-to-one, our statement is true, for then (3) is equivalent to (1). It remains to show that a function satisfying (3) is one-to-one. Let $f(p) = f(q)$. Then $\bar{p} = f^{-1}(\overline{f(p)}) = f^{-1}(\overline{f(q)}) = \bar{q}$, whence $\bar{p} = \bar{q}$.

IX. Topological properties. Each property of the space which is invariant under homeomorphisms is called a *topological property*. It follows from VIII (3a) and § 3, III (18) that every property expressed in terms of the operation $\bar{A}$ (and of operations of set theory and of logics) is topological.

More generally, if a point a (or a set A, or a family $\boldsymbol{A}$ of sets, and so on) has a given property with respect to the space $\mathscr{X}$ and if f is a homeomorphism which maps $\mathscr{X}$ onto a space $\mathscr{Y}$, then the point $f(a)$ has the same property with respect to $\mathscr{Y}$ (provided that the property is expressed as above).

Thus *it is impossible to distinguish between two homeomorphic spaces by any topological means.* Similarly, if A and B are two sets situated in the spaces $\mathscr{X}$ and $\mathscr{Y}$, respectively, and if there exists a homeomorphism of $\mathscr{X}$ onto $\mathscr{Y}$ which maps A onto B, then the sets A and B are indistinguishable in their spaces from the topological point of view (with respect to the spaces $\mathscr{X}$ and $\mathscr{Y}$).

It should be remarked that two sets may be homeomorphic and, at the same time, they may be situated in the space in a different manner so that there is no topological equivalence between them. For example, in the space of real numbers, the set composed of a point, a segment, and a second point (in that order) is not topologically equivalent (though it is homeomorphic) to the set composed of two points and a segment following them. However, the same sets (regarded as subsets of the plane) are equivalent with respect to the plane.

A property of a set A situated in a space $\mathscr{X}$ which is invariant with respect to homeomorphisms of A onto subsets of $\mathscr{X}$—or, more generally, onto subsets of spaces of a family $\mathfrak{F}$—is said to be an *intrinsic invariant* with respect to $\mathscr{X}$ or with respect to the family $\mathfrak{F}$, respectively.

Thus, for example (as we shall see later), the property of being an open subset A of the euclidean space $\mathscr{E}^n$ is an intrinsic topological invariant with respect to $\mathscr{E}^n$, so also is the number of regions into which a closed and bounded set A cuts $\mathscr{E}^n$. These properties being "exterior" properties of A are, at the same time, equivalent (in the domain of subsets of $\mathscr{E}^n$) to "interior" properties of A (i.e. those which can be formulated in topological terms while considering A as a space). This fact implies their topological invariance.

The role of the space is essential here. By taking the interval $\mathscr{I}$, instead of $\mathscr{E}^n$, as the space, we can see that these two properties are not intrinsic invariants.

As we shall see in § 35, IV, the property of being a $\boldsymbol{G}_\delta$-set is an intrinsic invariant with respect to the family of complete spaces.

All the properties of the space, of subsets, of points, an so on, considered so far are topological properties, i.e. invariants of homeomorphisms.

X. Topological rank. The space $\mathscr{X}$ is said (after P. Alexandrov) to be *topologically contained* in the space $\mathscr{Y}$ if it is homeomorphic to a subset of $\mathscr{Y}$. We write

$$\mathscr{X} \underset{\text{top}}{\subset} \mathscr{Y} \qquad \text{(topological inclusion)}.$$

If $\mathscr{X}$ is topologically contained in $\mathscr{Y}$ and $\mathscr{Y}$ in $\mathscr{X}$, then we say that $\mathscr{X}$ and $\mathscr{Y}$ are of the same *topological rank*(1). If this relation holds only in one direction, then we say that the topological rank of one space is *higher* than that of the other.

Evidently, two spaces may have the same topological rank without being homeomorphic. Such are, for example, the unlimited straight line and a closed interval. However, by a theorem of general set theory(2), the following proposition holds.

If $\mathscr{X}$ and $\mathscr{Y}$ have the same topological rank, then there exists a set $A \subset \mathscr{X}$ and a set $B \subset \mathscr{Y}$ such that A is homeomorphic to B and $\mathscr{X}-A$ is homeomorphic to $\mathscr{Y}-B$.

The topological ranks of two spaces may be *uncomparable*; as, for example, in the case of a circumference and a set composed of three segments having one end-point in common.

XI. Homogeneous spaces. We say that the set A is *homogeneous* in the space, if, for every pair of points a, b of A, there exists a homeomorphism f of the space onto itself (an automorphism) such

(1) Called "type de dimensions" by Fréchet (see Math. Ann. 68 (1910), pp. 145–168); or "Homoie" by Mahlo.

(2) S. Banach, Fund. Math. 6 (1924), pp. 236–239.

that $f(a) = b$. In particular, the space itself is homogeneous, if all its points are topologically equivalent (in the sense of Section IX).

The circumference and the euclidean n-dimensional space are homogeneous. Moreover, these spaces are *bihomogeneous*, i.e. there exists a homeomorphism carrying b to a and a to b simultaneously. A space, however, can be homogeneous without being bihomogeneous(1).

One can show that(2), if A is simultaneously closed and open and if a and b are two equivalent points such that $a \in A$ and $b \in 1-A$, then the points a and b are bi-equivalent, i.e. there exists a homeomorphism f of the space onto itself such that $f(a) = b$ and $f(b) = a$(3).

A set A may be not homogeneous in a space containing it, but, at the same time, it may be homogeneous when considered as a space for itself.

It follows easily from the definition that *if A is a homogeneous set, then a topological property which holds at one point of A also holds at all other points.* In particular, by taking the property of being an interior point, an accumulation point, a point where the given set is of the first category, we conclude that *a homogeneous set is*

(i) *either an open set or a boundary set*;

(ii) *either dense in itself or discrete*;

(iii) *either a set of the first category or a set which is not of the first category at any of its points.*

Also the following related theorem holds(4).

---

(1) See my paper *Un problème sur les ensembles homogènes*, Fund. Math. 3 (1922), pp. 14–19 (a problem of Knaster).

(2) ibidem, p. 16.

(3) In a paper *Über topologisch homogene Kontinua*, Fund. Math. 15 (1930), p. 102, van Dantzig discusses other types of homogenity, in particular, an involution homogenity. These notions can also be localized.

(4) This theorem is an extension of a certain theorem of S. Banach concerning topological groups to topological spaces (see the following section). The theorem of Banach has applications in functional analysis. See S. Banach, *Théorie des opérations linéaires*, Monografie Matematyczne Vol. I (1932), p. 20; see also my note, *Sur la propriété de Baire dans les groupes métriques*, Studia Math. 4 (1933).

* THEOREM. *Let $\mathfrak{H}$ be a family of automorphisms of a space such that to each pair of points x, y there exists an automorphism h of the family $\mathfrak{H}$ with $y = h(x)$. Let Z be a set such that, for every element h of $\mathfrak{H}$,*

$$\text{either} \quad Z = h(Z) \quad \text{or} \quad Z \cap h(Z) = 0. \tag{4}$$

Under these assumptions, if Z has the Baire property, then Z is either of the first category or it is simultaneously closed and open.

Proof. First, we observe (see Section IX) that if Z has a (topological) property at a point $z \in Z$, then $h(Z)$ has the same property at the point $h(z)$. Moreover, Z is homogeneous, for, to each $z_1 \in Z$ there exists an automorphism h such that $z_1 = h(z)$ and, since $z_1 \in Z \cap h(Z)$, it follows from (4) that $Z = h(Z)$.

By (iii), if Z is not of the first category, then it is not of the first category at any of its points. It follows by the Baire property (§ 11, IV, cor. 2) that there exists a point $z \in Z$ at which $(1-Z)$ is of the first category. Let G be an open set such that $z \in G$ and $(G-Z)$ is of the first category. We shall prove that $G-Z = 0$.

Suppose that $p \in G-Z$. Let h be an automorphism belonging to $\mathfrak{H}$ such that $p = h(z)$. Since $p \in h(Z)-Z$, we have by (4) that $Z \cap$ $\cap\, h(Z) = 0$; hence $h(Z) \subset 1-Z$ and $G \cap h(Z) \subset G-Z$ which proves that $G \cap h(Z)$ is of the first category. Consequently $h(Z)$ is of the first category at the point $p = h(z)$. But this contradicts the remark made at the beginning of the proof, since Z is not of the first category at the point z.

Now, since $G \subset Z$, it follows that z is an interior point of Z, and since Z is homogeneous, it follows that Z is open (by (i)).

It remains to prove that Z is closed.

Let $p \in \bar{Z}$ and $z \in Z$. Let, as before, $p = h(z)$. Since Z is open and h is an automorphism of the space, then $h(Z)$ is open. It follows that the formulas $p \in \bar{Z}$ and $p \in h(Z)$ imply $Z \cap h(Z) \neq 0$. Hence $Z = h(Z)$ and $p \in Z$.

***XII. Applications to topological groups.** A topological space (denote it by G) is called a *topological group* if for every pair of points x, y of there corresponds their "sum" $z = x+y$ in such a way that (i) $(x+y)+z = x+(y+z)$; (ii) there exists a zero element θ such that $x+\theta = x = \theta+x$; (iii) there exists an element

$(-x)$ such that $x+(-x) = \theta$; (iv) the operations $(x+y)$ and $-x$ are continuous[1].

Let $h_{-a}(x) = a+x$. We readily see that the condition $y = h_a(x)$ implies $x = h_{-a}(y)$. It follows that $h_a(x)$ (for a fixed a) is an automorphism of the space and that the family $\mathfrak{H}$ of all functions h_a satisfies the hypothesis of the theorem of Section XI.

A *subgroup* is a set which, for every x, contains $(-x)$ and, for every pair x, y, contains $x+y$. We easily verify that if Z is a subgroup, then condition XI (4) is fulfilled. By the previous result it follows that *every subgroup is homogeneous* (in particular, every topological group is homogeneous). By virtue of the theorem of Section XI, *every subgroup having the Baire property is either a set of the first category or a set which is simultaneously closed and open.*

XIII. Open mappings. Closed mappings. By IV (3) and (4), if f is continuous, the inverse image of an open set is open, and the inverse image of a closed set is closed. However, this is not true of images, as is seen in simple examples.

Thus we are led to the following definition.

DEFINITION. A continuous mapping f: $\mathscr{X} \to \mathscr{Y}$ is *open* if the image $f(G)$ of each open $G \subset \mathscr{X}$ is open in $\mathscr{Y}$. Similarly, *closed* mappings are defined, replacing the term "open" by "closed".

For example, the projection of the plane $\mathscr{E}^2$ onto one of the axes is an open mapping. It is not closed, since the projection of the hyperbola $y = 1/x$ on the x-axis is not closed.

Each continuous mapping of the interval $\mathscr{I}$ is closed (this is true, as we shall see later, of every compact space).

Obviously, *each one-to-one continuous open* (*or closed*) *mapping is a homeomorphism*[2].

[1] See, for example, L. Pontriagin, *Topological groups,* Princeton 1946, or D. Montgomery and L. Zippin, *Topological transformation groups,* New York 1955.

[2] The open mappings are also called *interior* mappings. See S. Stoïlow, Ann. Ec. Norm. Sup. III, 45 (1928).

See also G. T. Whyburn, *Open and closed mappings,* Duke 17 (1950), pp. 69–74, and *Open mappings on locally compact spaces,* Memoirs Amer. Math. Soc.; A. Archangielskii, *On open and almost-open mappings of topological spaces,* Dokl. Akad. Nauk URSS, 58 (1962), p. 999; A. D. Wallace, *Some characterizations of interior transformations,* Amer. Journ. Math. 61 (1939), pp. 757–763.

THEOREM 1. *Let* $f\colon \mathscr{X} \to \mathscr{Y}$ *be open (closed) and let* $B \subset \mathscr{Y}$. *Put* $g = f \mid f^{-1}(B)$. *Then* $g\colon f^{-1}(B) \to B$ *is open (closed).*

Proof. Let H be open in $f^{-1}(B)$; i.e. $H = G \cap f^{-1}(B)$ where G is open in $\mathscr{X}$. Now

$$g(H) = g\big(G \cap f^{-1}(B)\big) = f\big(G \cap f^{-1}(B)\big) = f(G) \cap B$$

by formula (13) of § 3, III. Thus $g(H)$ is open relative to B.

The proof in the case where f is closed is similar.

THEOREM 2. *Let* $f\colon \mathscr{X} \to \mathscr{Y}$ *be continuous onto and* $g\colon \mathscr{Y} \to \mathscr{Z}$ *continuous. If* $h = gf$ *is open (closed), so is* g.

Proof. Let h be open and let $G \subset \mathscr{Y}$ be open. Since f is onto, we have $G = ff^{-1}(G)$. Hence $g(G) = gff^{-1}(G) = hf^{-1}(G)$

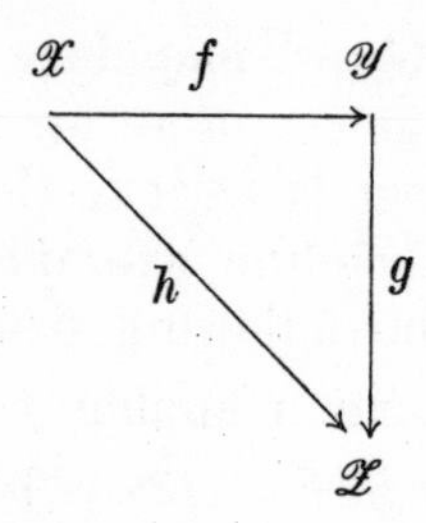

and as f is continuous and h open, $hf^{-1}(G)$ is open.

The case of h closed is similar.

XIV. Open and closed mappings at a given point.

DEFINITION. A continuous mapping $f\colon \mathscr{X} \to \mathscr{Y}$ is *open at the point* $y_0 \in \mathscr{Y}$, if for each G open in $\mathscr{X}$,

$$y_0 \in f(G) \Rightarrow y_0 \in \operatorname{Int}\big(f(G)\big). \tag{1}$$

Equivalently stated, if for each $A \subset \mathscr{X}$,

$$y_0 \in f\big(\operatorname{Int}(A)\big) \Rightarrow y_0 \in \operatorname{Int}\big(f(A)\big). \tag{2}$$

f is closed at y_0, if for each F closed in $\mathscr{X}$,

$$y_0 \in \overline{f(F)} \Rightarrow y_0 \in f(F). \tag{3}$$

Equivalently stated, if for each $A \subset \mathscr{X}$,

$$y_0 \in \overline{f(A)} \Rightarrow y_0 \in f(\bar{A}). \tag{4}$$

Obviously, *if f is open (closed) at each $y \in \mathscr{Y}$, then f is open (closed).*

THEOREM 1. *f is open at y_0 if and only if for each $B \subset \mathscr{Y}$,*

$$y_0 \in \overline{B} \Rightarrow f^{-1}(y_0) \subset \overline{f^{-1}(B)}. \tag{5}$$

Proof 1. Suppose that f is open at y_0 and that the inclusion in (5) is not true. Put $G = \mathscr{X} - \overline{f^{-1}(B)}$. Hence $f^{-1}(y_0) \cap G \neq 0$ and therefore

$$y_0 \in f(G) = f\big(\mathscr{X} - \overline{f^{-1}(B)}\big) \subset ff^{-1}(\mathscr{Y} - B) \subset \mathscr{Y} - B.$$

On the other hand, as $y_0 \in f(G)$, we have by (1)

$$y_0 \in \operatorname{Int}\big(f(G)\big) \subset \operatorname{Int}(\mathscr{Y} - B) = \mathscr{Y} - \overline{B},$$

i.e. $y_0 \notin \overline{B}$. Thus the condition is necessary.

2. Suppose now that G is open and $y_0 \notin \operatorname{Int}\big(f(G)\big)$, i.e. $y_0 \in \overline{\mathscr{Y} - f(G)}$. Suppose that (5) is satisfied for $B = \mathscr{Y} - f(G)$. By assumption, $y_0 \in \overline{B}$. Hence

$$f^{-1}(y_0) \subset \overline{f^{-1}(B)} = \overline{f^{-1}(\mathscr{Y}) - f^{-1}f(G)} \subset \overline{\mathscr{X} - G} = \mathscr{X} - G,$$

since $f^{-1}f(G) \supset G$. Thus $f^{-1}(y_0) \cap G = 0$ and finally $y_0 \notin f(G)$.

It follows that the condition is sufficient.

COROLLARY 2. *f is open if and only if for each $B \subset \mathscr{Y}$,*

$$f^{-1}(\overline{B}) \subset \overline{f^{-1}(B)};\text{[1]} \tag{6}$$

equivalently (see IV (2)), *if*

$$f^{-1}(\overline{B}) = \overline{f^{-1}(B)}. \tag{7}$$

Obviously, *f is closed if and only if for each $A \subset \mathscr{X}$,*

$$\overline{f(A)} \subset f(\overline{A}); \tag{8}$$

equivalently (see IV (1)), *if*

$$\overline{f(A)} = f(\overline{A}). \tag{9}$$

THEOREM 3. *f is closed at y_0 if and only if for each open $G \subset \mathscr{X}$ such that $f^{-1}(y_0) \subset G$ there exists an open $H \subset \mathscr{Y}$ such that*

$$y_0 \in H, \tag{10}$$

$$f^{-1}(H) \subset G. \tag{11}$$

---

[1] See R. Sikorski, *Closure homeomorphisms and interior mappings*, Fund. Math. 41 (1955), p. 13, and A. D. Wallace, *loc. cit.*

Proof. 1. Suppose that f is closed at y_0 and that G is open and $f^{-1}(y_0) \subset G$. Put $H = \mathscr{Y} - \overline{f(\mathscr{X}-G)}$. Suppose that $y_0 \notin H$, i.e. $y_0 \in \overline{f(\mathscr{X}-G)}$. Hence $y_0 \in f(\mathscr{X}-G)$ by (3), but this contradicts the inclusion $f^{-1}(y_0) \subset G$.

Thus, (10) is true. (11) is also true, for

$$f^{-1}(H) = \mathscr{X} - f^{-1}\overline{f(\mathscr{X}-G)} \subset \mathscr{X} - f^{-1}f(\mathscr{X}-G) \subset G.$$

Thus our condition is necessary.

2. Suppose now that our condition is satisfied and that F is closed and $y_0 \notin f(F)$. We have to show that $y_0 \notin \overline{f(F)}$.

Put $G = \mathscr{X} - F$. As $y_0 \notin f(F)$, we have $f^{-1}(y_0) \cap F = 0$, i.e. $f^{-1}(y_0) \subset G$. Hence, by assumption, there is an open $H \subset \mathscr{Y}$ satisfying (10) and (11).

By (11), $f^{-1}(H) \cap F = 0$, whence $H \cap f(F) = 0$. It follows that $H \cap \overline{f(F)} = 0$ (since H is open); hence by (10), $y_0 \notin \overline{f(F)}$.

This means that our condition is sufficient.

Remark 1. If $y_0 \in \mathscr{Y} - f(\mathscr{X})$, f is open at y_0; if $y_0 \in \mathscr{Y} - \overline{f(\mathscr{X})}$, f is closed at y_0.

This follows directly from (1) and (3).

Remark 2. The range of variability of G in (1) can be restricted to an *open base* of $\mathscr{X}$.

This follows easily from § 3, III (1a) and § 6, II (1').

Remark 3. The definition given in this section can be generalized as follows:

f is open at the set $B \subset \mathscr{Y}$ if for each G open in $\mathscr{X}$,

$$[B \cap \operatorname{Int}(f(G)) = 0] \Rightarrow [B \cap f(G) = 0]; \tag{12}$$

f is closed at the set $B \subset \mathscr{Y}$ if for each F closed in $\mathscr{X}$,

$$[B \cap f(F) = 0] \Rightarrow [B \cap \overline{f(F)} = 0]. \tag{13}$$

Without loss of generality, we may assume in (12) that $B \subset f(\mathscr{X})$ and in (13) that $B \subset \overline{f(\mathscr{X})}$.

XV. Bicontinuous mappings. Let $f\colon \mathscr{X} \to \mathscr{Y}$ be continuous and onto. If $G \subset \mathscr{Y}$ is open, then $f^{-1}(G)$ is open. The converse is not generally true. This leads to the following definition.

DEFINITION. f is called *bicontinuous*[1] if it is onto and if

$$(A \text{ is open}) \equiv (f^{-1}(A) \text{ is open}). \tag{1}$$

Equivalently: if

$$(A \text{ is closed}) \equiv (f^{-1}(A) \text{ is closed}). \tag{2}$$

THEOREM 1. *Let $f: \mathscr{X} \to \mathscr{Y}$ be open or closed onto. Then f is bicontinuous.*

Proof. Let f and $f^{-1}(A)$ be open. f being onto, we have $A = ff^{-1}(A)$, and f being open, $ff^{-1}(A)$ is open.

The case of f closed is similar.

THEOREM 2. *If f is bicontinuous and one-to-one, then f is a homeomorphism.*

Proof. Let $G \subset \mathscr{X}$ be open. f being one-to-one, $G = f^{-1}f(G)$. Hence $f(G)$ is open. Thus f is open. It follows that f is a homeomorphism.

THEOREM 3. *Let $f: \mathscr{X} \to \mathscr{Y}$ be bicontinuous and $g: \mathscr{Y} \to \mathscr{Z}$. If $h = gf$ is continuous, then so is g. If h is bicontinuous, then so is g.*

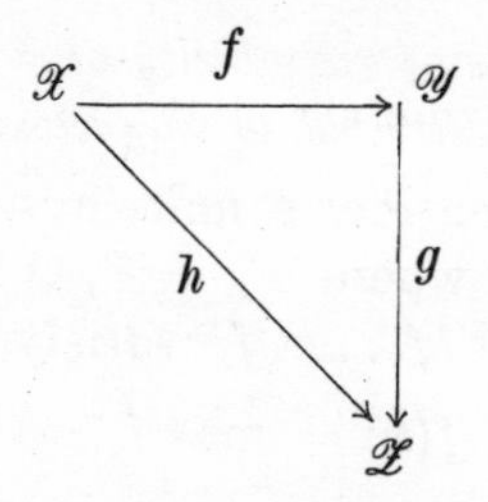

Proof. Let $G \subset \mathscr{Z}$ be open. Then $h^{-1}(G)$ is open.

But $h^{-1}(G) = f^{-1}[g^{-1}(G)]$. As f is bicontinuous, it follows that $g^{-1}(G)$ is open.

Next, let us assume that h is bicontinuous, and that $g^{-1}(A)$ is open. Then $f^{-1}[g^{-1}(A)]$ is open. But $f^{-1}[g^{-1}(A)] = h^{-1}(A)$. Consequently, A is open since h is bicontinuous.

Remarks. There are bicontinuous mappings which are neither open nor closed. This follows from the existence of closed not open mappings and of open not closed mappings.

---

[1] Also called *quasi-compact* (and continuous), see G. T. Whyburn, *Open and closed mappings*, Duke Math. J. 17 (1950), p. 69; or *strongly continuous*, see Alexandrov-Hopf, *Topologie* I, § 5, 4, p. 65.

On the other hand, each mapping of a discrete space into a space which is not discrete is continuous without being bicontinuous.

The notion of bicontinuity is of great importance for the quotient topology (see § 19).

§ 14. Completely regular spaces. Normal spaces

I. Completely regular spaces.

DEFINITION. A topological space $\mathscr{X}$ is called *completely regular* if for each point p and each closed set F which does not contain p there is a continuous mapping $f\colon \mathscr{X} \to \mathscr{I}$ such that

$$f(p) = 0 \quad \text{and} \quad f(x) = 1 \quad \text{for} \quad x \in F\,(^1). \tag{1}$$

THEOREM 1. *Each completely regular space is regular.*

Proof. Let p, F, and f be as defined previously. Put $G = \underset{x}{E}[f(x) < 1/2]$. Then G is open, $p \in G$, and $\bar{G} \cap F = 0$.

THEOREM 2. *A subset of a completely regular space is completely regular.*

The proof is obvious.

THEOREM 3. *The range of variability of F in the definition of complete regularity can be restricted to any closed subbase of $\mathscr{X}$.*

Proof. Let us first consider a finite system $F_1, \ldots, F_n$ of closed sets, a point $p \in \mathscr{X} - F_0$ where $F_0 = F_1 \cup \ldots \cup F_n$, and a system of continuous functions $f_1, \ldots, f_n$ satisfying conditions (1). Put

$$f(x) = \max_{1 \leqslant i \leqslant n} f_i(x). \tag{2}$$

Obviously f satisfies conditions (1) for F_0.

Moreover f is continuous. This follows (compare § 13, VI, Theorem 2) from the identity:

$$\underset{x}{E}[u < f(x) < v] = \bigcup_{i=1}^{n} \underset{x}{E}[u < f_i(x)] \cap \bigcap_{j=1}^{n} \underset{x}{E}[f_j(x) < v],$$

according to which the set $\underset{x}{E}[u < f(x) < v]$ is open for each pair of real numbers $u < v$.

Now let us consider an arbitrary closed F such that $p \in \mathscr{X} - F$. We shall show that there is a continuous f satisfying (1).

---

(1) See A. Tychonov, *Ueber topologische Erweiterung von Räumen*, Math. Ann. 102 (1930), p. 545. Compare P. Urysohn, Math. Ann. 94 (1925), p. 292. A completely regular $\mathscr{T}_1$-space is also called a *Tychonov space*.

Let $\boldsymbol{B}$ be a closed base of $\mathscr{X}$. Then there is a system of members $F_1, \ldots, F_n$ of $\boldsymbol{B}$ such that

$$p \in \mathscr{X} - F_0 \quad \text{and} \quad F \subset F_0 \quad \text{where} \quad F_0 = F_1 \cup \ldots \cup F_n.$$

The function f defined by formula (2) obviously satisfies (1).

Remark 1. *Each topological group is completely regular*([1]).

Remark 2. *A regular space may fail to be completely regular.* Moreover, there are regular $\mathscr{T}_1$-spaces on which every real continuous function is constant([2]).

II. Normal spaces.

DEFINITION([3]). A topological space is called *normal* if, for each pair of closed disjoint sets F_0, F_1, there are two open sets G_0 and G_1 such that

$$F_0 \subset G_0, \quad F_1 \subset G_1, \quad \text{and} \quad G_0 \cap G_1 = 0. \tag{3}$$

This can be stated equivalently: if, for each pair of open sets G_0 and G_1 such that $\mathscr{X} = G_0 \cup G_1$, there are two closed sets F_0 and F_1 such that

$$F_0 \subset G_0, \quad F_1 \subset G_1, \quad \text{and} \quad F_0 \cup F_1 = \mathscr{X}. \tag{4}$$

It should be noticed that the sets F_0 and F_1 in (4) may be supposed to be of the form

$$F_0 = \overline{H}_0 \quad \text{and} \quad F_1 = \overline{H}_1 \tag{5}$$

where H_0 and H_1 are open.

For, let $\mathscr{X}$ be normal and suppose that (4) is satisfied. Then there are two open sets H_0 and K_0 such that

$$F_0 \subset H_0, \quad \mathscr{X} - G_0 \subset K_0, \quad \text{and} \quad H_0 \cap K_0 = 0.$$

Consequently, $H_0 \subset \mathscr{X} - K_0$, whence $\overline{H}_0 \subset \mathscr{X} - K_0 \subset G_0$. Similarly, there is an open H_1 such that $F_1 \subset H_1$ and $\overline{H}_1 \subset G_1$.

([1]) Pontrjagin theorem. Compare D. Montgomery and L. Zippin, *Topological transformation groups*, Intersc. Tracts, New York 1955, p. 29.

([2]) See E. Hewitt, *On two problems of Urysohn*, Ann. of Math. 47 (1946), p. 503, and J. Novák, *Regular space on which every continuous function is constant*, Časopis Pěstov. Mat. Fys. 73 (1948), p. 58.

([3]) Compare H. Tietze, Math. Ann. 88 (1923), p. 301.

A similar argument shows that in (3) the identity $G_0 \cap G_1 = 0$ may be replaced by

$$\bar{G}_0 \cap \bar{G}_1 = 0. \tag{6}$$

Finally, it may be noted that condition (3) can be replaced by the following (analogous to condition (2) from § 5, X): *there exists an open G such that*

$$F_0 \subset G \quad \textit{and} \quad \bar{G} \cap F_1 = 0. \tag{7}$$

Remark. *Every normal space is completely regular.* This follows directly from the Urysohn lemma given in Section IV.

On the other hand, *there exist completely regular spaces which are not normal*(¹). However, the following is true:

THEOREM 1. *A regular Lindelöf space* (see § 5, VII) *is normal*(²).

Proof. Let $A = \bar{A}$, $B = \bar{B}$ and $A \cap B = 0$. We have to define two open sets G and H such that $A \subset G$, $B \subset H$, and $G \cap H = 0$.

By virtue of the regularity of the space, there is for each $a \in A$ an open set G_a such that $a \in G_a$ and $\bar{G}_a \cap B = 0$. Thus the family composed of the sets G_a and of the set $-A$ is a cover. Consequently it contains a countable subcover $-A, G_{a_1}, G_{a_2}, \ldots$. Hence

$$A \subset G_{a_1} \cup G_{a_2} \cup \ldots .$$

Similarly $B \subset H_{b_1} \cup H_{b_2} \cup \ldots$ where H_{b_n} is open and $\bar{H}_{b_n} \cap A = 0$.

Put $U_1 = G_{a_1}$, $V_1 = H_{b_1} - \bar{U}_1$, for $n > 1$:

$$U_n = G_{a_n} - (\bar{V}_1 \cup \ldots \cup \bar{V}_{n-1}), \qquad V_n = H_{b_n} - (\bar{U}_1 \cup \ldots \cup \bar{U}_n),$$

and

$$G = U_1 \cup U_2 \cup \ldots \quad \text{and} \quad H = V_1 \cup V_2 \cup \ldots .$$

As $A \cap \bar{V}_n \subset A \cap \bar{H}_{b_n} = 0$, we have $A \subset G$. Similarly $B \subset H$. Finally $G \cap H = 0$, i.e. $U_n \cap V_m = 0$. For, if $m \leqslant n-1$, we have (by the definition of U_n) that $U_n \cap \bar{V}_m = 0$, and if $n \leqslant m$, then $\bar{U}_n \cap V_m = 0$.

COROLLARY 1a. *A regular $\mathscr{T}_1$-space with a countable base is normal.*

(¹) See W. T. van Est and H. Freudenthal, *Trennung durch stetige Funktionen in topologischen Räumen*, Indag. Math. 13 (1951), p. 365.

See also an example of Niemytzki quoted in Alexandrov-Hopf, p. 31.

(²) Theorem of Tychonov. See *Über einen Metrisationssatz von P. Urysohn*, Math. Ann. 95 (1925), pp. 139–142.

For, by the theorem of § 5, Section XI, each topological space with a countable base is Lindelöf.

THEOREM 2. *The property of being normal is invariant under closed mappings.*

Proof. Let $f\colon \mathcal{X} \to \mathcal{Y}$ be a closed mapping onto.

Let $G_j \subset \mathcal{Y}$, $j = 0, 1$, be open and $\mathcal{Y} = G_0 \cup G_1$. Then $f^{-1}(G_j)$ is open and

$$\mathcal{X} = f^{-1}(G_0) \cup f^{-1}(G_1).$$

$\mathcal{X}$ being normal, there are two closed sets F_0 and F_1 such that (compare (4)):

$$F_0 \subset f^{-1}(G_0), \quad F_1 \subset f^{-1}(G_1), \quad \text{and} \quad F_0 \cup F_1 = \mathcal{X}.$$

It follows that

$$f(F_0) \subset G_0,\ f(F_1) \subset G_1 \quad \text{and} \quad f(F_0) \cup f(F_1) = \mathcal{Y}.$$

Since f is closed, $f(F_j)$ is closed, and consequently $\mathcal{Y}$ is normal.

Remark. Normality is not invariant under *open* mappings. See § 17.

We shall show that, in the definition of normal spaces, the *disjoint closed sets* can be replaced by *separated* F_σ*-sets.*

THEOREM 3 (1). *Let $\mathcal{X}$ be normal and let A and B be two separated F_σ-sets. Then there exist two open sets G and H such that*

$$A \subset G, \quad B \subset H, \quad \textit{and} \quad G \cap H = 0. \tag{8}$$

Proof. Put $A = A_1 \cup A_2 \cup \ldots$ and $B = B_1 \cup B_2 \cup \ldots$ where A_i and B_j are closed. We shall define two sequences of open sets $G_1, G_2, \ldots$ and $H_1, H_2, \ldots$ satisfying conditions

$$A_i \subset G_i, \quad B_j \subset H_j, \quad \overline{G}_i \cap \overline{H}_j = 0 \quad \text{for} \quad i, j = 1, 2, \ldots, \tag{9}$$

and then we shall put $G = G_1 \cup G_2 \cup \ldots$ and $H = H_1 \cup H_2 \cup \ldots$; thus, conditions (8) will be fulfilled.

Proceed by induction. As $A_1 \cap \overline{B} = 0 = B_1 \cap \overline{A}$, there are open sets (compare (6)): P_1, Q_1, U_1, and V_1 such that

$$A_1 \subset P_1, \quad \overline{B} \subset Q_1, \quad B_1 \subset U_1, \quad \overline{A} \subset V_1, \quad \overline{P}_1 \cap \overline{Q}_1 = 0 = \overline{U}_1 \cap \overline{V}_1.$$

---

(1) See E. Čech, *Topologické prostory*, p. 107.

Put $G_1 = P_1 \cap V_1$ and $H_1 = U_1 \cap Q_1$.

Obviously conditions (9) are satisfied for $i = 1 = j$. Moreover

$$\bar{G}_1 \cap \bar{B} = 0 = \bar{H}_1 \cap \bar{A}. \tag{10}$$

Now, let us assume that conditions (9) and (10) are satisfied for indices $\leqslant n$. Like in the case $n = 1$, there exist open sets P_{n+1}, Q_{n+1}, U_{n+1}, and V_{n+1} such that

$$A_{n+1} \subset P_{n+1}, \quad \bar{H}_1 \cup \ldots \cup \bar{H}_n \cup \bar{B} \subset Q_{n+1}, \quad \bar{P}_{n+1} \cap \bar{Q}_{n+1} = 0,$$
$$B_{n+1} \subset U_{n+1}, \quad \bar{G}_1 \cup \ldots \cup \bar{G}_n \cup \bar{A} \subset V_{n+1}, \quad \bar{U}_{n+1} \cap \bar{V}_{n+1} = 0.$$

We put $G_{n+1} = P_{n+1} \cap V_{n+1}$ and $H_{n+1} = U_{n+1} \cap Q_{n+1}$.

It is easy to verify that (9) and (10) are satisfied for $i, j \leqslant n+1$. This completes the proof.

III. Combinatorially similar systems of sets in normal spaces. Two finite systems of sets $A_1, \ldots, A_n$ and $B_1, \ldots, B_n$ are said to be *combinatorially similar*(1), if the equivalence

$$\{A_{i_1} \cap \ldots \cap A_{i_k} = 0\} \equiv \{B_{i_1} \cap \ldots \cap B_{i_k} = 0\} \tag{1}$$

holds for every system of indices ($\leqslant n$).

The same term is also applicable to infinite sequences satisfying condition (1) for every k.

THEOREM(2). *If $F_1, \ldots, F_n$ is a finite system of closed sets in a normal space, then each F_i $(1 \leqslant i \leqslant n)$ is contained in an open set G_i such that the system $\bar{G}_1, \ldots, \bar{G}_n$ is combinatorially similar to the given system.*

Proof. Consider all the intersections of the form $F_{i_1} \cap \ldots \cap F_{i_k}$ where $F_1 \cap F_{i_1} \cap \ldots \cap F_{i_k} = 0$. Let S be their union. The set S being closed and disjoint to F_1, there exists, by II (7), an open set G_1 such that $F_1 \subset G_1$ and $\bar{G}_1 \cap S = 0$. We shall prove that the system $\bar{G}_1, F_2, \ldots, F_n$ is similar to the system $F_1, F_2, \ldots, F_n$.

(1) After P. Alexandrov, Annals of Math. 30 (1928), p. 16.

(2) Compare W. Hurewicz, Math. Ann. 100 (1928). This theorem is a generalization of statement II (3). For an extension to the case of an infinite system of sets in a metric space, see my note in Fund. Math. 24 (1935), p. 259. Compare also K. Morita, Journ. Math. Soc. Japan 2 (1950), p. 22, and M. Katětov, Journ. Math. Tchecoslov. 2 (1952), p. 336.

In order to prove this, consider an empty intersection whose members belong to the second system. We must show that the intersection of the corresponding members of the first system is also empty. Obviously, we can confine ourselves to the case where this intersection involves F_1. Consequently, this intersection is of the form $F_1 \cap F_{i_1} \cap \ldots \cap F_{i_k} = 0$. It follows that $F_{i_1} \cap \ldots \cap F_{i_k} \subset S$, and since $\bar{G}_1 \cap S = 0$, therefore $\bar{G}_1 \cap F_{i_1} \cap \ldots \cap F_{i_k} = 0$.

Now we proceed by induction. Suppose that the system $F_1, \ldots, F_n$ is similar to the system $\bar{G}_1, \ldots, \bar{G}_{k-1}, F_k, \ldots, F_n$, where $F_1 \subset G_1, \ldots, F_{k-1} \subset G_{k-1}$. It follows that there exists an open set G_k such that $F_k \subset G_k$ and that the system

$$\bar{G}_1, \ldots, \bar{G}_{k-1}, F_k, F_{k+1}, \ldots, F_n$$

is similar to

$$\bar{G}_1, \ldots, \bar{G}_{k-1}, \bar{G}_k, F_{k+1}, \ldots, F_n.$$

Therefore the latter system is similar to $F_1, \ldots, F_n$, since the property of being similar is transitive.

The proof of the theorem is completed.

COROLLARY. *Given a system of open sets $G_1, \ldots, G_n$ such that $\mathscr{X} = G_1 \cup \ldots \cup G_n$, there exists a system of open sets $H_1, \ldots, H_n$ such that $\mathscr{X} = H_1 \cup \ldots \cup H_n$ and $\bar{H}_i \subset G$.*

Applying the preceding theorem to the sets $F_i = \mathscr{X} - G_i$, we deduce the existence of open sets V_i such that $F_i \subset V_i$ and that $\bar{V}_1 \cap \ldots \cap \bar{V}_n = 0$. Let $H_i = \mathscr{X} - \bar{V}_i$. Then

$$\bar{H}_i = \overline{\mathscr{X} - \bar{V}_i} \subset \overline{\mathscr{X} - V_i} = \mathscr{X} - V_i \subset \mathscr{X} - F_i = G_i$$

and

$$H_1 \cup \ldots \cup H_n = \mathscr{X} - (\bar{V}_1 \cap \ldots \cap \bar{V}_n) = \mathscr{X}.$$

Remark. In metric spaces the preceding statement can be extended to arbitrary families of open sets (see § 21, XII).

IV. Real-valued functions defined on normal spaces. The following property of normal spaces, stated in the Urysohn Lemma, is a strengthening of the property I (1) of completely regular spaces. It is, as can easily be seen, a characteristic property of normal spaces.

URYSOHN LEMMA [1]. *Given two disjoint closed sets A and B in a normal space $\mathscr{X}$, there exists a continuous function $f\colon \mathscr{X} \to \mathscr{I}$ such that*

$$f(x) = 0 \text{ for } x \in A \quad \text{and} \quad f(x) = 1 \text{ for } x \in B. \tag{1}$$

Proof. First we shall assign to every fraction of the form $r = k/2^n$ $(k = 0, 1, \ldots, 2^n)$, an open set $G(r)$ so that

(i) $A \subset G(0)$, $\mathscr{X} - B = G(1)$,

(ii) the condition $r < r'$ implies $\overline{G(r)} \subset G(r')$.

We proceed by induction with respect to the exponent n.

For $n = 0$ the conditions (i) and (ii) are fulfilled by II (7). Suppose that they are fulfilled for $n-1$. We must define $G(k/2^n)$ for an odd k. By hypothesis

$$\overline{G[(k-1)/2^n]} \subset G[(k+1)/2^n].$$

By II (7), there exists an open set, which we denote by $G(k/2^n)$, such that

$$\overline{G[(k-1)/2^n]} \subset G(k/2^n) \quad \text{and} \quad \overline{G(k/2^n)} \subset G[(k+1)/2^n].$$

Thus the function $G(r)$ is defined for every r.

Let $f(x) = 0$ for $x \in G(0)$ and $f(x) =$ least upper bound of the r's such that $x \in \mathscr{X} - G(r)$ for $x \notin G(0)$. By (i), $f(x) = 0$ for $x \in A$ and $f(x) = 1$ for $x \in B$.

It remains to prove that the function f is continuous, i.e. (§ 13, VI, Theorem 1) that for every x_0 and every natural number n there exists an open set H containing x_0 such that the condition $x \in H$ implies $|f(x_0) - f(x)| < 1/2^n$.

Let r be a (finite dyadic) fraction such that

$$f(x_0) < r < f(x_0) + 1/2^{n+1}. \tag{2}$$

Let $H = G(r) - \overline{G(r-1/2^n)}$ with the convention $G(s) = 0$ for $s < 0$ and $G(s) = \mathscr{X}$ for $s > 1$. It follows that $x_0 \in H$. For, the inequality $f(x_0) < r$ implies $x_0 \in G(r)$, while the inequality $r - 1/2^{n+1} < f(x_0)$ implies

$$x_0 \in \mathscr{X} - G(r - 1/2^{n+1}) \subset \mathscr{X} - \overline{G(r-1/2^n)}.$$

---

[1] P. Urysohn, *Über die Mächtigkeit der zusammenhängenden Mengen*, Math. Ann. 94 (1925), p. 290.

Moreover, the hypothesis $x \in H$ implies $x \in G(r)$, hence $f(x) \leqslant r$. It also implies

$$x \in \mathscr{X} - \overline{G(r-1/2^n)} \subset \mathscr{X} - G(r-1/2^n);$$

hence $r-1/2^n \leqslant f(x)$. Therefore

$$f(x_0)-1/2^n < f(x) < f(x_0)+1/2^n.$$

The Urysohn Lemma is a particular case of a fundamental Tietze theorem which will be proved using the following auxiliary theorem.

AUXILIARY THEOREM. *Given a continuous function f defined in a closed subset F of a normal space $\mathscr{X}$ and satisfying $|f(x)| \leqslant c$, where $c > 0$, there exists a continuous function g defined in the whole space $\mathscr{X}$ and satisfying the conditions*

$$|g(x)| \leqslant \frac{1}{3}c \quad \text{for} \quad x \in \mathscr{X}, \tag{3}$$

$$|f(x)-g(x)| \leqslant \frac{2}{3}c \quad \text{for} \quad x \in F. \tag{4}$$

Proof. Let

$$A = \mathop{E}_{x}\left[f(x) \leqslant -\frac{1}{3}c\right], \quad B = \mathop{E}_{x}\left[f(x) \geqslant \frac{1}{3}c\right]$$

$$\text{and} \quad J = \mathop{E}_{y}\left[|y| \leqslant \frac{1}{3}c\right]. \tag{5}$$

The sets A and B being closed and disjoint (compare § 13, VI), there is, by the Urysohn Lemma, a continuous function $g\colon \mathscr{X} \to J$ such that $g(x) = -\frac{1}{3}c$ for $x \in A$ and $g(x) = \frac{1}{3}c$ for $x \in B$. Obviously g satisfies the conditions of the lemma.

TIETZE EXTENSION THEOREM[1]. *Given a continuous real valued*

---

[1] H. Tietze, *Über Funktionen, die auf einer abgeschlossenen Menge stetig sind*, Journ. f. Math. 145 (1915), pp. 9–14. For the proof compare P. Urysohn, *loc. cit.*, p. 193.

For various extensions of the Tietze theorem, see J. Dugundji, *An extension of Tietze's theorem*, Pacific Journ. Math. 1 (1951), pp. 353–367; O. Hanner, *Retraction and extension of mappings of metric and non-metric spaces*, Arkiv Math. 2 (1952), pp. 315–360; C. H. Dowker, *On a theorem of Hanner*, *ibid.* pp. 307–313.

Compare also K. Borsuk, Bull. Acad. Pol. 1933 where an extension of a function is defined so as the "extension operation" to be linear with respect to the given functions.

function f defined in a closed set F of a normal space $\mathscr{X}$, there exists a continuous real valued function f^* defined in the entire space $\mathscr{X}$ such that

$$f^*(x) = f(x) \quad \textit{for} \quad x \in F. \tag{6}$$

Moreover, if the function f is bounded:

$$|f(x)| \leqslant c, \quad \textit{where} \quad c > 0, \tag{7}$$

then

$$|f^*(x)| \leqslant c \quad \textit{for} \quad x \in \mathscr{X}. \tag{8}$$

Proof. First assume that the function f is bounded, i.e. that (7) holds. Let

$$f^*(x) = \sum_{n=0}^{\infty} g_n(x), \tag{9}$$

where the functions g_n are defined by induction as follows.

Let $g_0(x) \equiv 0$. Given $n \geqslant 0$, assume that

$$\left| f(x) - \sum_{i=0}^{n} g_i(x) \right| \leqslant \left(\frac{2}{3}\right)^n c \quad \text{for} \quad x \in F \tag{10}$$

(this is the case for $n = 0$, by (7)).

Replace in the auxiliary theorem $f(x)$ by $f(x) - \sum_{i=0}^{n} g_i(x)$ and c by $\left(\frac{2}{3}\right)^n c$. It follows that there exists a continuous function $g_{n+1}(x)$ defined on $\mathscr{X}$ such that

$$|g_{n+1}(x)| \leqslant \frac{2^n}{3^{n+1}} c \quad \text{for} \quad x \in \mathscr{X}, \tag{11}$$

$$\left| f(x) - \sum_{i=0}^{n+1} g_i(x) \right| \leqslant \left(\frac{2}{3}\right)^{n+1} c \quad \text{for} \quad x \in F. \tag{12}$$

Thus a sequence of functions $g_1, g_2, \ldots$ is defined and formulas (10)–(12) are fulfilled for $n = 0, 1, \ldots$.

Since the functions g_n are continuous and series (9) is uniformly convergent on $\mathscr{X}$ (by (11)), the function f^* is continuous on $\mathscr{X}$ (by Theorem 3 of § 13, VI).

Formula (6) follows directly from inequality (10).

Finally, (8) follows from (11): if $x \in \mathcal{X} - F$, then

$$|f^*(x)| = \left|\sum_{n=1}^{\infty} g_n(x)\right| \leqslant \sum_{n=0}^{\infty} |g_{n+1}(x)| \leqslant c \sum_{n=0}^{\infty} \frac{2^n}{3^{n+1}} = c.$$

Thus the theorem has been proved for f bounded. The case of f unbounded can be reduced to the former case as follows(1). $\mathcal{E}$ being homeomorphic to the open interval $I_0 = (-1 < y < 1)$, we may assume that $f\colon F \to I_0$. As shown, there is a continuous extension $f^*\colon \mathcal{X} \to \bar{I}_0$ of f. Let B denote the two-element set $(-1, 1)$ and let $H = f^{*-1}(B)$. Then H is closed and $H \cap F = 0$. By the Urysohn Lemma, there is a continuous $h\colon \mathcal{X} \to \mathcal{I}$ such that $h(x) = 0$ for $x \in H$ and $h(x) = 1$ for $x \in F$. Put $g(x) = f^*(x) \cdot h(x)$. Then $g\colon \mathcal{X} \to I_0$ and g is the required continuous extension of f.

V. Hereditary normal spaces.

DEFINITION. A space is called *hereditary* (or completely) *normal* if each of its subsets is normal; equivalently: if each open subset is normal (as is easily seen using condition II(4)).

Remark 1. *A normal space may fail to be hereditary normal.*

Let $X = \underset{\alpha}{E}(\alpha \leqslant \Omega)$ and $Y = \underset{\beta}{E}(\beta \leqslant \omega)$ with the natural (order) topology. The required space is the cartesian product $X \times Y$ (compare § 15, I)(2). Let $A = (X \times Y) - (\Omega, \omega)$. Then $X \times Y$ is normal, while A is not. Namely the sets $\Omega \times Y$ and $X \times \omega$, with the point (Ω, ω) removed, have no disjoint neighbourhoods.

Let us add that *each metric space is hereditary normal* (see § 21).

Remark 2. *Each F_σ-subset of a normal space is normal.* This follows from Theorem 3 of Section II (since two disjoint sets, closed relatively to an F_σ-set, are separated F_σ-sets).

In the Theorems 1–4, which follow, the space 1 is supposed to be hereditary normal(3).

---

(1) See H.-J. Kowalsky, *Topologische Räume*, p. 135.

(2) Called also *Tychonov plank*. See J. L. Kelley, *General Topology*, p. 132.

(3) For the special case of metric spaces, the proofs given in this section can be essentially simplified. See § 21.

Formula II(7), valid for arbitrary normal spaces, can be strengthened as follows for hereditary normal spaces.

THEOREM 1 [1]. *If A and B are separated, there is an open G such that*

$$A \subset G \quad \textit{and} \quad \bar{G} \cap B = 0. \tag{1}$$

Consequently (compare § 6, V, Theorem 6), $\mathrm{Fr}(G)$ *separates A from B.*

Proof. Let $E = 1 - \bar{A} \cap \bar{B}$. Hence the sets $\bar{A} - \bar{B}$ and $\bar{B} - \bar{A}$ are disjoint and closed in E. As the space is hereditary normal, there is a G open in E (hence in 1) such that

$$\bar{A} - \bar{B} \subset G \quad \text{and} \quad \bar{G} \cap E \cap (\bar{B} - \bar{A}) = 0.$$

Now $A \cap \bar{B} = 0$, hence $A \subset \bar{A} - \bar{B} \subset G$. As $B \cap \bar{A} = 0$, we have $\bar{G} \cap B \subset \bar{G} \cap \bar{B} - \bar{A} \subset \bar{G} \cap E \cap \bar{B} - \bar{A} = 0$.

This completes the proof.

THEOREM 2. *For each pair A, B of closed sets there is a pair A^*, B^* of closed sets such that*

$$A^* \cup B^* = 1, \quad A^* \cap (A \cup B) = A, \quad B^* \cap (A \cup B) = B. \tag{2}$$

Proof. By Theorem 2 of § 6, V, $A - B$ and $B - A$ are separated. Hence, by virtue of Theorem 1, there is an open G such that

$$A - B \subset G \quad \text{and} \quad \bar{G} \cap B - A = 0. \tag{3}$$

Put

$$A^* = \bar{G} \cup (A \cap B) \quad \text{and} \quad B^* = (-G) \cup (A \cap B). \tag{4}$$

It follows that $A^* \cup B^* \supset \bar{G} \cup (-G) = 1$.

Next $A \cup B = (A - B) \cup (A \cap B) \cup (B - A)$. Hence by (3) and (4):

$$\begin{aligned} A^* \cap (A \cup B) &= [\bar{G} \cap (A - B)] \cup (A \cap B) \cup [\bar{G} \cap (B - A)] \\ &= (A - B) \cup (A \cap B) \cup 0 = A, \\ B^* \cap (A \cup B) &= [(-G) \cap (A - B)] \cup (A \cap B) \cup [(-G) \cap (B - A)] \\ &= 0 \cup (A \cap B) \cup (B - A) = B. \end{aligned}$$

---

(1) H. Tietze, Math. Ann. 88 (1923), p. 301; P. Alexandrov and P. Urysohn, Math. Ann. 92 (1924), pp. 258–266; P. Urysohn, Math. Ann. 94 (1925), pp. 262–295.

Remark 3. We shall show that the condition stated in Theorem 2 is characteristic of hereditary normal spaces([1]).

Suppose that Theorem 2 is valid in a topological space 1. Let $E \subset 1$. We have to show that E is normal.

In other words, given two disjoint sets M_0 and M_1, closed in E, i.e.

$$M_0 \cap M_1 = 0, \qquad M_j = \overline{M}_j \cap E \quad (j = 0, 1), \tag{5}$$

we have to define (compare II(4)) two sets F_0 and F_1 closed in E and such that

$$F_0 \cup F_1 = E, \qquad F_j \cap M_j = 0 \quad (j = 0, 1). \tag{6}$$

By hypothesis there are two closed sets C_0 and C_1 such that

$$C_0 \cup C_1 = 1, \qquad C_j \cap (\overline{M}_0 \cup \overline{M}_1) = \overline{M}_j. \tag{7}$$

Put $F_j = C_{1-j} \cap E$. By (7), $F_0 \cup F_1 = (C_0 \cup C_1) \cap E = E$. Next,

$$F_j \cap M_j = C_{1-j} \cap M_j = C_{1-j} \cap (\overline{M}_0 \cup \overline{M}_1) \cap M_j = \overline{M}_{1-j} \cap M_j.$$

But the sets M_{1-j} and M_j are separated, since (compare § 6, V, Theorem 5) they are disjoint and closed in E and hence in $M_0 \cup M_1$. Therefore $\overline{M}_{1-j} \cap M_j = 0$, which completes the proof.

Theorem 1 implies the following slightly more general statement.

THEOREM 3. *If $A \cap \overline{B} = 0$, then there exists an open set G such that*

$$A \subset G, \qquad G \cap \overline{B} = 0 \quad \text{and} \quad \overline{G} \cap B \subset \overline{A}.$$

Proof. The sets A and $B-\overline{A}$ are separated, since

$$A \cap \overline{B-\overline{A}} \subset A \cap \overline{B} = 0 = \overline{A} \cap (B-\overline{A}).$$

Hence by Theorem 1 there is an open set G_0 such that

$$A \subset G_0 \quad \text{and} \quad \overline{G}_0 \cap (B-\overline{A}) = 0, \quad \text{i.e.} \quad \overline{G}_0 \cap B \subset \overline{A}.$$

We put $G = G_0 - \overline{B}$.

THEOREM 4. *Suppose that each pair A_i, A_j $(i \neq j)$ belonging to the system $A_1, \ldots, A_n$ is separated. Then there exists a system of open sets $G_1, \ldots, G_n$ such that*

$$A_i \subset G_i \quad \text{and} \quad \overline{A}_i \cap \overline{A}_j = \overline{G}_i \cap \overline{G}_j \quad \text{for} \quad i \neq j. \tag{8}$$

([1]) Compare in that direction H. Ribeiro, *Caractérisations des espaces réguliers normaux et complètement normaux au moyen de l'opération de dérivation*, Portug. Math. 2 (1944), pp. 13–19.

Proof. Consider first the case $n = 2$ (1). Put $A = \bar{A}_1 - \bar{A}_2$ and $B = \bar{A}_2 - \bar{A}_1$. The sets A and B being separated (by Theorem 2 of § 6, V), there exists by Theorem 1 an open G such that

$$\bar{A}_1 - \bar{A}_2 \subset G \tag{9}$$

and

$$(\bar{G} \cap \bar{A}_2) \subset \bar{A}_1. \tag{10}$$

By virtue of the same theorem applied to the pair of sets $\bar{A}_2 - \bar{G}$ and $\bar{G} - \bar{A}_2$, there is an open H such that

$$\bar{A}_2 - \bar{G} \subset H \tag{11}$$

and

$$(\bar{H} \cap \bar{G}) \subset \bar{A}_2. \tag{12}$$

Now, because the sets A_1 and A_2 are separated, we have $A_1 = A_1 - \bar{A}_2$ and $A_2 = A_2 - \bar{A}_1$. Therefore by (9) $A_1 \subset G$, and by (10) and (11):

$$A_2 = A_2 - \bar{A}_1 \subset A_2 - (\bar{G} \cap \bar{A}_2) = A_2 - \bar{G} \subset H,$$

and hence $(\bar{A}_1 \cap \bar{A}_2) \subset (\bar{G} \cap \bar{H})$.

The converse inclusion follows from (12) and (10):

$$(\bar{H} \cap \bar{G}) \subset (\bar{G} \cap \bar{A}_2) \subset (\bar{A}_1 \cap \bar{A}_2).$$

This completes the proof for the case $n = 2$.

Now, let us procede by induction. Suppose that the theorem is true for $n-1$. Therefore, there exists a system of open sets $G_1, \ldots, G_{n-2}, H$ such that:

$$A_1 \subset G_1, \ \ldots, \ A_{n-2} \subset G_{n-2}, \qquad (A_{n-1} \cup A_n) \subset H$$

and

$$\bar{A}_i \cap \bar{A}_j = \bar{G}_i \cap \bar{G}_j, \qquad \bar{A}_i \cap \overline{A_{n-1} \cup A_n} = \bar{G}_i \cap \bar{H}$$
$$\text{for} \quad j < i \leqslant n-2.$$

As shown before, there are open sets H_1 and H_2 such that

$$A_{n-1} \subset H_1, \qquad A_n \subset H_2, \qquad \bar{H}_1 \cap \bar{H}_2 = \bar{A}_{n-1} \cap \bar{A}_n.$$

Now, put $G_{n-1} = H \cap H_1$ and $G_n = H \cap H_2$. Then the system $G_1, \ldots, G_n$ satisfies the conditions (8).

---

(1) See K. Menger, Ergebnisse Math. Koll. 1, Wien 1931, p. 16.

For, if $i \leqslant n-2$, we have

$$(\bar{G}_i \cap \bar{G}_{n-1}) \subset (\bar{G}_i \cap \bar{H} \cap \bar{H}_1) = (\bar{A}_i \cap \overline{A_{n-1} \cup A_n} \cap \bar{H}_1)$$

$$\subset (\bar{A}_i \cap \bar{A}_{n-1}) \cup (\bar{A}_i \cap \bar{H}_2 \cap \bar{H}_1) = (\bar{A}_i \cap \bar{A}_{n-1}) \cup (\bar{A}_i \cap \bar{A}_{n-1} \cap \bar{A}_n)$$

$$= (\bar{A}_i \cap \bar{A}_{n-1}) \subset (\bar{G}_i \cap \bar{G}_{n-1}),$$

and hence $\bar{A}_i \cap \bar{A}_{n-1} = \bar{G}_i \cap \bar{G}_{n-1}$.

On the other hand

$$(\bar{G}_{n-1} \cap \bar{G}_n) \subset (\bar{H}_1 \cap \bar{H}_2) = (\bar{A}_{n-1} \cap \bar{A}_n) \subset (\bar{G}_{n-1} \cap \bar{G}_n),$$

and hence $\bar{A}_{n-1} \cap \bar{A}_n = \bar{G}_{n-1} \cap \bar{G}_n$.

Remark 4. The sets $G_1, \ldots, G_n$ are *disjoint*.

This is because, by Theorem 2 of § 8, IV, $\bar{A}_i \cap \bar{A}_j$ is nowhere dense.

Remark 5. In metric spaces some extensions of Theorem 4 to infinite systems are valid (see § 21, XI). To another extension leads the notion of *collectionwise normal* spaces(1).

VI. Perfectly normal spaces.

DEFINITION. A topological space $\mathscr{X}$ is called *perfectly normal*(2) if for each pair of disjoint closed sets A and B, there is a continuous function $f\colon \mathscr{X} \to \mathscr{I}$ such that

$$[f(x) = 0] \equiv (x \in A) \quad \text{and} \quad [f(x) = 1] \equiv (x \in B). \tag{1}$$

As seen this condition is stronger than the condition IV (1) in the Urysohn Lemma. Thus, every perfectly normal space is normal.

Moreover, *every perfectly normal space is hereditary normal*(3).

For each open subset is F_σ (see Theorem 2 below) and each F_σ-subset is normal (see V, Remark 2).

On the other hand, *a hereditary normal space may fail to be perfectly normal.* Consider the space $\mathscr{X}$ defined as follows(4). $\mathscr{X}$ is uncountable and contains one "singular" point p such that each set

(1) See R. H. Bing, *Metrization of topological spaces*, Canadian Journ. of Math. 3 (1951), p. 176. Compare M. Katětov, *On extending locally finite covers* (Russian), Coll. Math. 6 (1958), pp. 145–151.

(2) Compare "F-property" by P. Urysohn, Math. Ann. 94 (1925), p. 286. See also E. Čech, *Sur la dimension des espaces parfaitement normaux*, Bull. Acad. Bohême 33 (1932), pp. 38–55, and *Topologické Prostory*, p. 110.

(3) This theorem is due to P. Urysohn, Math. Ann. 94 (1925), p. 286.

(4) This example is due to E. Čech.

containing p is closed. Besides this the only closed sets are finite sets.

It is easily seen that the one-element set (p) is not a G_δ-set, and hence $\mathscr{X}$ is not perfectly normal (by Theorem 2 which follows).

As we shall see later, *every metric space is perfectly normal.*

THEOREM 1. *For each A closed in a perfectly normal space $\mathscr{X}$, there is a continuous function $f\colon \mathscr{X} \to \mathscr{I}$ such that*

$$[f(x) = 0] \equiv [x \in A]. \tag{2}$$

This follows from the definition by putting $B = 0$.

In order to establish an important characterization of perfectly normal spaces, we shall first prove the following lemma.

LEMMA[1]. *Let A be a closed G_δ-set in a normal space. Let B be closed and $A \cap B = 0$. Then there is a continuous function $f\colon \mathscr{X} \to \mathscr{I}$ satisfying formula (2) and the implication*

$$(x \in B) \Rightarrow \big(f(x) = 1\big). \tag{3}$$

Proof. Put

$$A = G_1 \cap G_2 \cap \ldots, \qquad \text{where} \quad G_n \text{ is open.} \tag{4}$$

We may assume that $G_n \cap B = 0$, since otherwise we could replace G_n by $G_n - B$.

By the Urysohn Lemma (see Section IV) there is for each n a continuous function $f_n : \mathscr{X} \to \mathscr{I}$ such that

$$f_n(x) = \begin{cases} 0 & \text{for} \quad x \in A, \\ 1 & \text{for} \quad x \in \mathscr{X} - G_n. \end{cases}$$

Put

$$f(x) = \sum_{n=1}^{\infty} \frac{1}{2^n} f_n(x).$$

As this series is uniformly convergent, it follows that f is continuous (compare § 13, VI, Theorem 3) and $f\colon \mathscr{X} \to \mathscr{I}$.

The implication (3) is fulfilled since, for each n, $B \subset \mathscr{X} - G_n$ and, consequently, $f_n(x) = 1$ for $x \in B$.

---

[1] For the lemma and Theorem 2, see N. Vedenisov, *Sur les fonctions continues dans des espaces topologiques*, Fund. Math. 27 (1936), p. 235; see also Compos. Math. 7 (1940), pp. 194–200, where the converse of Theorem 1 is shown.

As $A \subset G_n$, we have for $x \in A$, $f_n(x) = 0$, and hence $f(x) = 0$.

It remains to be shown that if $x \notin A$, $f(x) \neq 0$. Now if $x \notin A$, there is by (4) an index n such that $x \notin G_n$, hence $f_n(x) = 1$. It follows that $f(x) > 0$.

THEOREM 2 (of Vedenisov). *$\mathscr{X}$ is perfectly normal if and only if $\mathscr{X}$ is normal and each closed set is a G_δ-set.*

Proof 1. Let A be a closed subset of a perfectly normal space. Let f be the function considered in Theorem 1. We put

$$G_n = \underset{x}{E}[f(x) < 1/n].$$

Hence

$$\bigcap_{n=1}^{\infty} G_n = \underset{x\ n}{E\bigwedge}[f(x) < 1/n] = \underset{x}{E}[f(x) = 0] = A.$$

2. Now let $\mathscr{X}$ be normal and such that each closed set is a G_δ-set. Let A and B be two closed and disjoint sets. We have to define a continuous function $f\colon \mathscr{X} \to \mathscr{I}$ satisfying (1).

By virtue of the lemma, there is a continuous function $g\colon \mathscr{X} \to \mathscr{I}$ such that

$$(x \in A) \equiv [g(x) = 0] \quad \text{and} \quad (x \in B) \Rightarrow [g(x) = 1].$$

Put

$$G = \underset{x}{E}[g(x) < 1/2], \quad F = \underset{x}{E}[g(x) = 1/2], \quad H = \underset{x}{E}[g(x) > 1/2].$$

Therefore $G \cup F$ and $H \cup F$ are closed and $(G \cup F) \cap B = 0$. Hence the lemma can be applied to the sets $G \cup F$ and B (taking the interval $1/2, 1$ instead of $0, 1$); thus there is a continuous function h defined on $\mathscr{X}$ and such that

$$1/2 \leqslant h(x) \leqslant 1, \quad [x \in (G \cup F)] \Rightarrow [h(x) = 1/2],$$
$$(x \in B) \equiv [h(x) = 1].$$

Put

$$f(x) = \begin{cases} g(x) & \text{for} \quad x \in (G \cup F), \\ h(x) & \text{for} \quad x \in (H \cup F). \end{cases}$$

f is defined and continuous because $(G \cup F) \cap (H \cup F) = F$ and for $x \in F$ we have $g(x) = 1/2 = h(x)$. Moreover $(G \cup F) \cup (H \cup F) = \mathscr{X}$, hence $f\colon \mathscr{X} \to \mathscr{I}$ (compare § 13, V, Theorem 3).

Finally, it can easily be seen that (1) is satisfied.

THEOREM 3. *Each regular $\mathcal{T}_1$-space with a σ-locally finite base* $\boldsymbol{B}$ (compare § 5, VII) *is perfectly normal* (*even metrizable*, see § 21, XVII).

Proof. Let

$$\boldsymbol{B} = \boldsymbol{B}_1 \cup \boldsymbol{B}_2 \cup \ldots \quad \text{and} \quad \boldsymbol{B}_n = \{G_{n,t}\} \quad \text{where} \quad t \in T_n, \tag{5}$$

$\boldsymbol{B}_n$ being locally finite.

Let us prove first that the space $\mathcal{X}$ is normal.

Let A and B be closed and disjoint. As $\mathcal{X}$ is regular, one can attach to each $x \in A$ a pair of indices $n(x)$ and $t(x)$ such that

$$x \in G_{n(x),t(x)} \quad \text{and} \quad B \cap \overline{G}_{n(x),t(x)} = 0. \tag{6}$$

Put

$$M_k = \mathop{E}_{x}(n(x) = k) \quad \text{and} \quad G_k = \bigcup_{x \in M_k} G_{n(x),t(x)}. \tag{7}$$

It follows that $A = M_1 \cup M_2 \cup \ldots$, and hence

$$A \subset G_1 \cup G_2 \cup \ldots \tag{8}$$

As the family $\boldsymbol{B}_k$ is locally finite, it follows from (7) by the theorem of § 5, VII, that

$$\overline{G}_k = \overline{\bigcup_{x \in M_k} G_{n(x),t(x)}} = \bigcup_{x \in M_k} \overline{G}_{n(x),t(x)}.$$

Hence by (6)

$$B \cap \overline{G}_k = 0. \tag{9}$$

Similarly, there exists a sequence of open sets $H_1, H_2, \ldots$ such that

$$B \subset H_1 \cup H_2 \cup \ldots \quad \text{and} \quad A \cap \overline{H}_k = 0. \tag{10}$$

Put

$$U = \bigcup_{k=1}^{\infty}[G_k - (\overline{H}_1 \cup \ldots \cup \overline{H}_k)], \quad V = \bigcup_{k=1}^{\infty}[H_k - (\overline{G}_1 \cup \ldots \cup \overline{G}_k)].$$

Consequently $U \cap V = 0$, and by formulas (8)–(10), $A \subset U$ and $B \subset V$. As the sets U and V are open, this means that the space is normal.

We shall now show that $\mathcal{X}$ is perfectly normal. According to Theorem 2, it remains to show that each open A is a $\boldsymbol{F}_\sigma$-set.

Now denoting by x a point of A, and by B the set $\mathscr{X}-A$, formulas (6) to (9) hold. By (9) we have $\bar{G}_k \subset A$, whence by (8), $A = \bar{G}_1 \cup \cup \bar{G}_2 \cup \dots$. This completes the proof.

Remark. The Corollary 1a of Section II stating that each regular $\mathscr{T}_1$-space with a countable base is normal, is obviously also a consequence of Theorem 3.

§ 15. Cartesian product $\mathscr{X}\times\mathscr{Y}$ of topological spaces

I. Definition. $\mathscr{X}$ and $\mathscr{Y}$ being two topological spaces, the topology in $\mathscr{X}\times\mathscr{Y}$ (see § 2, I) is introduced in the following way:

DEFINITION. A set $\mathfrak{G} \subset \mathscr{X}\times\mathscr{Y}$ is called *open in* $\mathscr{X}\times\mathscr{Y}$ if and only if it is the union of cartesian products $G\times H$ where G and H are open subsets of $\mathscr{X}$ and $\mathscr{Y}$ respectively.

In other words, the family of all sets $G\times H$ is a *base* of $\mathscr{X}\times\mathscr{Y}$.

By § 2, II, (2), $(G_1\times H_1) \cap (G_2\times H_2) = (G_1 \cap G_2)\times (H_1 \cap H_2)$.

Therefore, assuming that G_1, G_2, H_1, and H_2 are open sets (in $\mathscr{X}$ and $\mathscr{Y}$ respectively), the intersection $(G_1\times H_1) \cap (G_2\times H_2)$ is open (in $\mathscr{X}\times\mathscr{Y}$). It follows that the intersection of any two open sets in $\mathscr{X}\times\mathscr{Y}$ is open.

As the union of an arbitrary family of open sets in $\mathscr{X}\times\mathscr{Y}$ is open, the following theorem is deduced (see § 5, II, Remark 2):

THEOREM 1. *The cartesian product of two topological spaces is a topological space.*

As $(\mathscr{X}\times\mathscr{Y})-(x_0, y_0) = \big((\mathscr{X}-x_0)\times\mathscr{Y}\big) \cup \big(\mathscr{X}\times(\mathscr{Y}-y_0)\big)$ (by § 2, II (6), the next theorem is true.

THEOREM 2. *The cartesian product of two $\mathscr{T}_1$-spaces is a $\mathscr{T}_1$-space.*

This theorem follows also from Theorem 5 below.

EXAMPLE. In the case of the plane $\mathscr{E}^2$, the usual topology agrees with the above definition. For every open set in $\mathscr{E}^2$ can be represented as the union of squares with sides parallel to the $\mathscr{X}$ and $\mathscr{Y}$ axes.

Another theorem can also be easily shown (compare § 2, II (8)).

THEOREM 3. *If $\{B_\iota\}$ is a base of $\mathscr{X}$, and $\{C_\varkappa\}$ a base of $\mathscr{Y}$, then $\{B_\iota\times C_\varkappa\}$ is a base of $\mathscr{X}\times\mathscr{Y}$.*

The same remains true of subbases.

THEOREM 4. *$\mathscr{X}$ and $\mathscr{Y}$ being topological spaces, the family of sets $G\times\mathscr{Y}$ and of sets $\mathscr{X}\times H$, where G is open in $\mathscr{X}$ and H in $\mathscr{Y}$, is a subbase of $\mathscr{X}\times\mathscr{Y}$.*

Because $G\times H = (G\times\mathscr{Y}) \cap (\mathscr{X}\times H)$.

Also the following theorem is true (see § 5, XI, Remark 4).

THEOREM 5. *The sets $F\times\mathscr{Y}$, as well as the sets $\mathscr{X}\times K$, where F is closed in $\mathscr{X}$ and K in $\mathscr{Y}$, are closed, and their family is a closed subbase of $\mathscr{X}\times\mathscr{Y}$.*

Remark. Recall (see § 2, VII) that a relation is a subset of $\mathscr{X}\times\mathscr{Y}$, namely the set $\underset{xy}{E}\, x\varrho y$. Consequently *a relation* will be called *closed* if this set is closed (in $\mathscr{X}\times\mathscr{Y}$).

II. Projections and continuous mappings. Given $z = (x, y) \in \mathscr{X}\times\mathscr{Y}$ consider x as function of z. Put $x = f(z)$ and similarly $y = g(z)$. Thus

$$f\colon \mathscr{X}\times\mathscr{Y} \to \mathscr{X} \quad \text{and} \quad g\colon \mathscr{X}\times\mathscr{Y} \to \mathscr{Y},$$

where f and g are called *projections* of $\mathscr{X}\times\mathscr{Y}$ on the axes $\mathscr{X}$ and $\mathscr{Y}$; $f(z)$ is the abscissa of z and $g(z)$ its ordinate.

THEOREM 1. *The projections are open mappings.*

Proof. If G is open in $\mathscr{X}$, we have $f^{-1}(G) = G\times\mathscr{Y}$, which is open by definition. Thus f is continuous.

It is also open. For suppose that G and H are open (in $\mathscr{X}$ and $\mathscr{Y}$ respectively); then $f(G\times H) = G$ is open.

THEOREM 2. *Let $h\colon \mathscr{T} \to \mathscr{X}\times\mathscr{Y}$. Put $h(t) = \big(h_1(t), h_2(t)\big)$ where $h_1(t) \in \mathscr{X}$ and $h_2(t) \in \mathscr{Y}$. Then h is continuous if and only if h_1 and h_2 are continuous.*

More precisely, *h is continuous at t_0 if and only if h_1 and h_2 are continuous at t_0.*

Proof. Suppose that h is continuous at t_0. As $h_1(t) = fh(t)$, where f is the projection of $\mathscr{X}\times\mathscr{Y}$ on the $\mathscr{X}$-axis, it follows by Theorem 1 that h_1 is continuous at t_0.

Suppose that h_1 and h_2 are continuous at t_0. Let $\mathfrak{G} \subset \mathscr{X}\times\mathscr{Y}$ be open and let $t_0 \in h^{-1}(\mathfrak{G})$. We have to show that $t_0 \in \operatorname{Int} h^{-1}(\mathfrak{G})$. According to § 13, III (5), we can assume that $\mathfrak{G}$ belongs to a subbase of $\mathscr{X}\times\mathscr{Y}$. Put (see I, Theorem 4) $\mathfrak{G} = G\times\mathscr{Y}$. Hence (see § 3, II (23)), $h^{-1}(\mathfrak{G}) = h_1^{-1}(G)$ and therefore $t_0 \in h_1^{-1}(G)$. As h_1 is continuous at t_0, it follows that $t_0 \in \operatorname{Int} h_1^{-1}(G) = \operatorname{Int} h^{-1}(\mathfrak{G})$.

THEOREM 3. *$\varphi(x, y)$ being a propositional function such that the set $\underset{xy}{E}\,\varphi(x, y)$ is open in $\mathscr{X}\times\mathscr{Y}$, the set $\underset{x}{E}\,\underset{y}{\bigvee}\,\varphi(x, y)$ is open in $\mathscr{X}$.*

If $\underset{xy}{E}\varphi(x,y)$ *is supposed to be closed (instead of being open), the set* $\underset{x}{E}\underset{y}{\bigwedge}\varphi(x,y)$ *is closed.*

This follows directly from Theorem 1.

III. Operations on cartesian products. Let $\mathscr{X}$ and $\mathscr{Y}$ be two topological spaces. Let $A\subset\mathscr{X}$ and $B\subset\mathscr{Y}$.

The following formulas hold true:

$$\overline{A\times B}=\bar{A}\times\bar{B}, \tag{1}$$

$$\operatorname{Int}(A\times B)=\operatorname{Int}(A)\times\operatorname{Int}(B), \tag{2}$$

$$\operatorname{Fr}(A\times B)=\big(\operatorname{Fr}(A)\cap\bar{B}\big)\cup\big(\bar{A}\times\operatorname{Fr}(B)\big), \tag{3}$$

$$(A\times B)^d=(A^d\times\bar{B})\cup(\bar{A}\times B^d). \tag{4}$$

Proof. 1. Let $z=(x,y)\in\overline{A\times B}$ and (as in II) put $f(z)=x$. Since f is continuous (by II, 1), $z\in\overline{A\times B}$ yields $f(z)\in\overline{f(A\times B)}$, that is $x\in\bar{A}$. Similarly $y\in\bar{B}$. Hence $(x,y)\in\bar{A}\times\bar{B}$.

Next, suppose that $x\in\bar{A}$ and $y\in\bar{B}$. Let $(x,y)\in\mathfrak{G}$ where $\mathfrak{G}$ is open. Hence there are G and H open and such that $(x,y)\in G\times H\subset\mathfrak{G}$. It follows that $x\in G$ and $y\in H$ and hence by hypothesis $G\cap A\neq 0\neq H\cap B$. Consequently $(G\times H)\cap(A\times B)\neq 0$, therefore $\mathfrak{G}\cap(A\times B)\neq 0$ which yields $(x,y)\in\overline{A\times B}$.

2. Let us recall that (§ 2, II (6):

$$(\mathscr{X}\times\mathscr{Y})-(A\times B)=\big((\mathscr{X}-A)\times\mathscr{Y}\big)\cup\big(\mathscr{X}\times(\mathscr{Y}-B)\big).$$

Therefore

$$\begin{aligned}\operatorname{Int}(A\times B)&=\mathscr{X}\times\mathscr{Y}-\overline{\mathscr{X}\times\mathscr{Y}-A\times B}\\&=\big(\mathscr{X}\times\mathscr{Y}-\overline{(\mathscr{X}-A)\times\mathscr{Y}}\big)\cap\big(\mathscr{X}\times\mathscr{Y}-\overline{\mathscr{X}\times(\mathscr{Y}-B)}\big)\\&=\big((\mathscr{X}\times\mathscr{Y}-\overline{\mathscr{X}-A})\times\bar{\mathscr{Y}}\big)\cap(\mathscr{X}\times\mathscr{Y}-\bar{\mathscr{X}}\times\overline{\mathscr{Y}-B})\\&=\big((\mathscr{X}-\overline{\mathscr{X}-A})\times\mathscr{Y}\big)\cap\big(\mathscr{X}\times(\mathscr{Y}-\overline{\mathscr{Y}-B})\big)\\&=\big(\operatorname{Int}(A)\times\mathscr{Y}\big)\cap\big(\mathscr{X}\times\operatorname{Int}(B)\big)\\&=\operatorname{Int}(A)\times\operatorname{Int}(B).\end{aligned}$$

3. $\operatorname{Fr}(A\times B) = \overline{A\times B} \cap \overline{\mathscr{X}\times\mathscr{Y}-A\times B} = (\bar{A}\times\bar{B}) \cap (\overline{\mathscr{X}-A}\times\overline{\mathscr{Y}} \cup \overline{\mathscr{X}}\times\overline{\mathscr{Y}-B}) = \operatorname{Fr}(A)\times\bar{B} \cup \bar{A}\times\operatorname{Fr}(B)$.

4. $(a, b)\in(A\times B)^d \equiv (a, b)\in\overline{A\times B-(a, b)} \equiv (a,b)\in(\overline{(A-a)\times B} \cup \overline{A\times(B-b)}) \equiv (a, b)\in(\overline{A-a}\times\bar{B} \cup \bar{A}\times\overline{B-b}) \equiv ((a\in\overline{A-a}$ and $b\in\bar{B})$ or $(a\in\bar{A}$ and $b\in\overline{B-b})) \equiv ((a, b)\in A^d\times\bar{B}$ or $(a, b)\in\bar{A}\times B^d) \equiv ((a, b) \in (A^d\times\bar{B}) \cup (\bar{A}\times B^d))$.

IV. Diagonal. We call *diagonal* of $\mathscr{X}^2 = \mathscr{X}\times\mathscr{X}$ the set

$$\Delta = \mathop{E}_{xy}(x = y). \tag{1}$$

THEOREM 1. *We have* $\Delta \underset{\text{top}}{=} \mathscr{X}$.

The required homeomorphism is the projection $(x, y) \to x$.

THEOREM 2. *If* $\mathscr{X}$ *is Hausdorff, the diagonal is closed (in* $\mathscr{X}^2$).

Proof. Put $\nabla = \mathscr{X}^2-\Delta$. We have to show that ∇ is open, i.e. that given a point (x, y) of ∇, there are two open sets G and H such that $x\in G$, $y\in H$ and $G\times H \subset \nabla$. Now, as $x \neq y$ there are ($\mathscr{X}$ being Hausdorff) two open sets G and H such that $x\in G$, $y\in H$, and $G \cap H = 0$. Consequently $(G\times H) \cap \Delta = 0$, i.e. $(G\times H) \subset \nabla$.

Theorem 2 can be generalized as follows.

THEOREM 3. *Let* f_j: $\mathscr{X}_j \to \mathscr{Y}$ *be continuous for* $j = 0, 1$. If $\mathscr{Y}$ *is Hausdorff, the set*

$$\Gamma = \mathop{E}_{x_0x_1}(f_0(x_0) = f_1(x_1))$$

is closed (in $\mathscr{X}_0\times\mathscr{X}_1$).

In other words (see Section I, Remark), *the relation*

$$(x_0 \varrho x_1) \equiv (f_0(x_0) = f_1(x_1))$$

is closed.

Proof. Put $g(x_0, x_1) = (f_0(x_0), f_1(x_1))$. Thus g: $\mathscr{X}_0\times\mathscr{X}_1 \to \mathscr{Y}^2$, and by II, 2 g is continuous. Hence, denoting by Δ the diagonal of $\mathscr{Y}^2$, $g^{-1}(\Delta)$ is closed. But $g^{-1}(\Delta) = \Gamma$, because

$$[(x_0, x_1)\in g^{-1}(\Delta)] \equiv [g(x_1, x_2)\in\Delta] \equiv [(f_0(x_0), f_1(x_1))\in\Delta]$$
$$\equiv [f_0(x_0) = f_1(x_1)].$$

In the particular case where $\mathscr{X}_0 = \mathscr{X}_1$, there is the following corollary.

COROLLARY 3a. *If $f\colon \mathscr{X}\to\mathscr{Y}$ is continuous and $\mathscr{Y}$ is Hausdorff, the induced equivalence relation*

$$(x_0 \sim x_1) \equiv \big(f(x_0) = f(x_1)\big)$$

is closed.

For open mappings the converse of Theorem 3 is true.

THEOREM 4. *Let $f_j\colon \mathscr{X}_j\to\mathscr{Y}$ be open onto. If the set Γ is closed, $\mathscr{Y}$ is Hausdorff.*

In particular, if Δ is closed, $\mathscr{X}$ is Hausdorff.

Proof. Let $y_0\neq y_1$ and put $y_j = f_j(x_j)$. It follows that $(x_0, x_1) \in \mathscr{X}_0\times\mathscr{X}_1-\Gamma$. The set $\mathscr{X}_0\times\mathscr{X}_1-\Gamma$ being open, there is G_j $(j = 0, 1)$ open in $\mathscr{X}_j$ such that $x_j\in G_j$ and $(G_0\times G_1)\cap\Gamma = 0$. It follows that $f_0(G_0)\cap f_1(G_1) = 0$. For, otherwise there would exist $x_j'\in G_j$ such that $f_0(x_0') = f_1(x_1')$ and therefore $(x_0', x_1')\in\Gamma$.

Since f_j is open, $f_0(G_0)$ and $f_1(G_1)$ are two open disjoint sets containing y_0 and y_1, respectively.

Remark 1. The notion of diagonal can be considered for the more general case where X and Y are *subsets* of a given space. This does not affect the definition (1) of the diagonal.

Then instead of Theorem 1 one has the more general theorem:

$$\Delta \underset{\text{top}}{=} X\cap Y.$$

If $X\cup Y$ is Hausdorff, Δ is closed in $X\times Y$.

Remark 2. Let us consider a propositional function $\varphi(x, y)$, where $x, y\in\mathscr{X}$, and a topological property of the set $\underset{xy}{E}\varphi(x, y)\wedge(x = y)$ relative to the diagonal $\Delta = \underset{xy}{E}(x = y)$; then the set $\underset{x}{E}\varphi(x, x)$ has the same property relative to the $\mathscr{X}$-axis.

This follows at once from Theorem 1.

V. Properties of f considered as subset of $\mathscr{X}\times\mathscr{Y}$. Let $f\colon \mathscr{X}\to\mathscr{Y}$. As is commonly known, f is identical with the set (called the *graph* of f);

$$\mathfrak{I} = \underset{xy}{E}\big(y = f(x)\big).$$

THEOREM 1. *If f is continuous, we have $\mathfrak{I} \underset{\text{top}}{=} \mathscr{X}$.*

Put $\mathfrak{z}(x) = \big(x, f(x)\big)$. Then $\mathfrak{z}$ is a homeomorphism of $\mathscr{X}$ onto $\mathfrak{I}$ (by Theorems 1 and 2 of Section II).

THEOREM 2. *If $\mathscr{Y}$ is Hausdorff and f continuous, $\mathfrak{J}$ is closed (in $\mathscr{X}\times\mathscr{Y}$).*

Put $\mathfrak{y}(x, y) = (f(x), y)$. Then $\mathfrak{y}^{-1}(\Delta) = \mathfrak{J}$ where Δ is the diagonal of $\mathscr{Y}\times\mathscr{Y}$. Δ being closed (by IV, Theorem 2) and $\mathfrak{y}$ continuous, $\mathfrak{J}$ is closed.

THEOREM 3. *Put $\mathfrak{r}(x, y) = (x, f(x))$. Then if f is continuous, $\mathfrak{r}$: $\mathscr{X}\times\mathscr{Y} \to \mathfrak{J}$ is a retraction.*

Theorem 2 can be generalized as follows.

THEOREM 4. *If $\mathscr{Y}$ is Hausdorff and f is continuous at x_0, then*

$$(x_0, y) \in \overline{\mathfrak{J}} \Rightarrow (x_0, y) \in \mathfrak{J}, \tag{1}$$

which means that $y = f(x_0)$.

Proof. Put $y_0 = f(x_0)$. Let $y_1 \neq y_0$. Then there are in $\mathscr{Y}$ two open sets H_0 and H_1 such that

$$y_0 \in H_0, \quad y_1 \in H_1, \quad \text{and} \quad H_0 \cap H_1 = 0. \tag{2}$$

Because f is continuous at x_0, there is a G open in $\mathscr{X}$ such that

$$x_0 \in G \quad \text{and} \quad x \in G \Rightarrow f(x) \in H_0, \quad \text{i.e.} \quad (\mathfrak{J} \cap (G\times\mathscr{Y})) \subset G\times H_0. \tag{3}$$

By (2), $H_0 \cap H_1 = 0$, and consequently

$$(\mathfrak{J} \cap (G\times H_1)) \subset (G\times H_0) \cap (G\times H_1) = 0. \tag{4}$$

As $G\times H_1$ is a neighbourhood of (x_0, y_1), it follows from (4) that (x_0, y_1) does not belong to $\overline{\mathfrak{J}}$.

THEOREM 5. *Suppose that $\mathscr{Y}$ is Hausdorff and dense in itself and that the set D_f of points of discontinuity of f is a boundary set. Then $\mathfrak{J}$ is nowhere dense.*

Proof. Suppose that $\mathfrak{J}$ is not nowhere dense, that is $\overline{\mathfrak{J}}$ contains an open set ($\neq 0$). Then there is a G open in $\mathscr{X}$ and an H open in $\mathscr{Y}$ (both non-void) such that $G\times H \subset \overline{\mathfrak{J}}$. Let $x \in G$. Now $x\times H$ is open in $x\times\mathscr{Y}$ and hence it is dense in itself and consequently contains more than one point. On the other hand

$$x\times H \subset G\times H \subset \overline{\mathfrak{J}}, \quad \text{whence} \quad x\times H \subset \overline{\mathfrak{J}} \cap (x\times\mathscr{Y}),$$

and thus $\overline{\mathfrak{J}} \cap (x\times\mathscr{Y})$ does not reduce to a single point.

By Theorem 4, x is a point of discontinuity of f. Thus $G \subset D_f$ and consequently D_f is not a boundary set.

VI. Horizontal and vertical sections. Cylinder on $A\subset\mathscr{X}$. We call *horizontal section* of $\mathscr{X}\times\mathscr{Y}$ the set

$$\mathscr{X}\times(y_0)=\underset{xy}{E}(y=y_0). \tag{1}$$

The set $(x_0)\times\mathscr{Y}$ is called *vertical section* of $\mathscr{X}\times\mathscr{Y}$.

More generally, given $A\subset\mathscr{X}$, we call the *cylinder on A* the set

$$A\times\mathscr{Y}=\underset{xy}{E}(x\in A). \tag{2}$$

THEOREM 1. *Each horizontal section is homeomorphic to $\mathscr{X}$, i.e.*

$$\mathscr{X}\times(y_0)\underset{\text{top}}{=}\mathscr{X},\qquad hence\qquad \mathscr{X}\underset{\text{top}}{\subset}\mathscr{X}\times\mathscr{Y}. \tag{3}$$

Namely the projection $(x,y_0)\to x$ is the required homeomorphism.

The proof is immediate (the theorem is, of course, a particular case of Theorem V, 1, namely the case where f is constant).

The next theorem follows easily.

THEOREM 2. *Given a propositional function $\varphi(x,y)$, the projection on the $\mathscr{X}$-axis transforms the set $\underset{xy}{E}\varphi(x,y)\cap\big(\mathscr{X}\times(y_0)\big)$ onto $\underset{x}{E}\varphi(x,y_0)$.*

VII. Invariants of cartesian multiplication.

THEOREM 1. *Let $A\subset\mathscr{X}$, $B\subset\mathscr{Y}$, $A\neq 0\neq B$. Then $A\times B$ is closed, open, and dense, if and only if both sets A and B have the respective properties.*

Proof. This follows from III (1) and (2) (compare also § 2, II (7)):

$$(\overline{A\times B}=A\times B)\equiv(\bar{A}\times\bar{B}=A\times B)\equiv(\bar{A}=A \text{ and } \bar{B}=B),$$

$$\{\mathrm{Int}(A\times B)=A\times B\}\equiv\{\mathrm{Int}(A)\times\mathrm{Int}(B)=A\times B\}$$
$$\equiv\{\mathrm{Int}(A)=A \text{ and } \mathrm{Int}(B)=B\},$$

$$(\overline{A\times B}=\mathscr{X}\times\mathscr{Y})\equiv(\bar{A}\times\bar{B}=\mathscr{X}\times\mathscr{Y})\equiv(\bar{A}=\mathscr{X} \text{ and } \bar{B}=\mathscr{Y}).$$

COROLLARY 1a. *(a,b) is an isolated point of $\mathscr{X}\times\mathscr{Y}$ if and only if a is isolated in $\mathscr{X}$ and b in $\mathscr{Y}$.*

Proof. (a,b) is isolated in $\mathscr{X}\times\mathscr{Y}$ if and only if the set composed of this point is open in $\mathscr{X}\times\mathscr{Y}$. But this is equivalent by Theorem 1, to assume that (a) is open in $\mathscr{X}$ and (b)—in $\mathscr{Y}$.

THEOREM 2. *$A\times B$ is a boundary set, a nowhere dense set, a dense in itself set, if and only if one of the sets A or B has the respective property.*

Proof. This follows from the equivalences:

$$\{\operatorname{Int}(A\times B)=0\}\equiv\{\operatorname{Int}(A)\times\operatorname{Int}(B)=0\}$$
$$\equiv\{\operatorname{Int}(A)=0 \text{ or } \operatorname{Int}(B)=0\}$$
$$\{\operatorname{Int}(\overline{A\times B})=0\}=\{\operatorname{Int}(\bar{A}\times\bar{B})=0\}$$
$$\equiv\{\operatorname{Int}(\bar{A})=0 \text{ or } \operatorname{Int}(\bar{B})=0\}.$$

The last part follows directly from Corollary 1a.

COROLLARY 2a[1]. *Let $A\subset\mathcal{X}$. If A has one of the following properties: of being closed, open, F_σ, G_δ, a boundary set, nowhere dense, set of first category, set with Baire property, dense in itself—then the set $A\times\mathcal{Y}$ has the respective property.*

In other words, *if, for a given propositional function $\varphi(x)$, the set $\underset{x}{E}\,\varphi(x)$ has one of the mentioned properties, so does the set $\underset{xy}{E}\,\varphi(x)$.*

THEOREM 3. *The following properties are invariant of cartesian multiplication:*

(a) *of being Hausdorff,*

(b) *of being regular,*

(c) *of being completely regular.*

Proof. (a) Suppose that $\mathcal{X}$ and $\mathcal{Y}$ are Hausdorff. Let $\mathfrak{z}_1=(x_1,y_1)$, $\mathfrak{z}_2=(x_2,y_2)$, and $\mathfrak{z}_1\neq\mathfrak{z}_2$. Then, either $x_1\neq x_2$ or $y_1\neq y_2$. We may suppose that $x_1\neq x_2$. Since $\mathcal{X}$ is Hausdorff, there are two open sets $G_1\subset\mathcal{X}$ and $G_2\subset\mathcal{X}$ such that $x_1\in G_1$, $x_2\in G_2$, and $G_1\cap G_2=0$. It follows that $\mathfrak{z}_1\in(G_1\times\mathcal{Y})$, $\mathfrak{z}_2\in(G_2\times\mathcal{Y})$, and that $G_1\times\mathcal{Y}$ and $G_2\times\mathcal{Y}$ are open and disjoint.

(b) Let $\mathfrak{z}=(x,y)\in\mathfrak{G}$ where $\mathfrak{G}$ is open in $\mathcal{X}\times\mathcal{Y}$. We have to show that there is an open $\mathfrak{H}$ such that $\mathfrak{z}\in\mathfrak{H}$ and $\overline{\mathfrak{H}}\subset\mathfrak{G}$.

We may obviously restrict ourselves to the case where $\mathfrak{G}$ belongs to the base of the space $\mathcal{X}\times\mathcal{Y}$. Thus we may put $\mathfrak{G}=A\times B$, where A and B are open in $\mathcal{X}$ and $\mathcal{Y}$, respectively. Since $\mathcal{X}$ and $\mathcal{Y}$ are regular, there are open sets C and D such that

$$a\in C,\quad \bar{C}\subset A,\quad \text{and}\quad b\in D,\quad \bar{D}\subset B.$$

Put $\mathfrak{H}=C\times D$. Hence $\mathfrak{z}\in\mathfrak{H}$ and

$$\overline{\mathfrak{H}}=\overline{C\times D}=\bar{C}\times\bar{D}\subset A\times B=\mathfrak{G}.$$

---

[1] See also § 22, IV. Compare in that direction J. C. Oxtoby, *Cartesian products of Baire spaces*, Fund. Math. 49 (1961), pp. 157–166.

(c) Let $\mathfrak{z}_0 = (x_0, y_0) \in \big((\mathcal{X}\times\mathcal{Y})-\mathfrak{F}\big)$ where $\mathfrak{F}$ is closed. In order to show that $\mathcal{X}\times\mathcal{Y}$ is completely regular, we may restrict ourselves to the case where $\mathfrak{F}$ belongs to a closed subbase of $\mathcal{X}\times\mathcal{Y}$, i.e. to the case $\mathfrak{F} = A\times\mathcal{Y}$, where $A = \bar{A} \subset \mathcal{X}$ (see Theorem 3 of § 14, I). It follows that $x_0 \in \mathcal{X}-A$. Now, as $\mathcal{X}$ is supposed to be completely regular, there exists a continuous mapping $g\colon \mathcal{X}\to\mathcal{I}$ such that $g(x_0) = 0$ and $g(x) = 1$ for $x \in A$. Put $f(x, y) = g(x)$. Thus, $f(\mathfrak{z}_0) = 0$ and $f(\mathfrak{z}) = 1$ for $\mathfrak{z}\in\mathfrak{F}$.

Remark 1. The converse theorem to Theorem 3 is also true; that means that *if $\mathcal{X}\times\mathcal{Y}$ has one of the properties* (a)–(c), *then both $\mathcal{X}$ and $\mathcal{Y}$ have the respective property.*

This follows from the fact that these properties are hereditary and that $\mathcal{X} \underset{\text{top}}{\subset} \mathcal{X}\times\mathcal{Y}$ (see VI (3)).

Remark 2. *The property of being normal is not invariant under cartesian multiplication*(¹).

Moreover, a product of a normal space and a metric space need not to be normal(²).

Remark 3. Corollary 2a is not true of the Baire property in the restricted sense (see § 40, VIII).

LEMMA. *Let $\mathfrak{Z} \subset \mathcal{X}\times\mathcal{Y}$ be dense in itself and let a be an isolated point of the projection A of $\mathfrak{Z}$ onto the $\mathcal{X}$-axis. Then the set $\big((a)\times\mathcal{Y}\big)\cap\mathfrak{Z}$ is dense in itself, and consequently the projection of $\mathfrak{Z}$ on the $\mathcal{Y}$-axis contains a dense in itself set* ($\neq 0$).

Proof. The set (a) being open in A, there exists a set G open in $\mathcal{X}$ such that $(a) = G\cap A$. A being the projection of $\mathfrak{Z}$ into $\mathcal{X}$, we have $\mathfrak{Z}\subset A\times\mathcal{Y}$. Therefore (see also § 2, II (2)):

$$\big((a)\times\mathcal{Y}\big)\cap\mathfrak{Z} = \big((G\cap A)\times\mathcal{Y}\big)\cap\mathfrak{Z} = (G\times\mathcal{Y})\cap(A\times\mathcal{Y})\cap\mathfrak{Z}$$
$$= (G\times\mathcal{Y})\cap\mathfrak{Z}.$$

Thus $\big((a)\times\mathcal{Y}\big)\cap\mathfrak{Z}$ is open in $\mathfrak{Z}$ (since $G\times\mathcal{Y}$ is open in $\mathcal{X}\times\mathcal{Y}$ by Theorem 1), and is consequently dense in itself (by § 9, V, 3).

(¹) See R. H. Sorgenfrey, *On the topological product of paracompact spaces*, Bull. Amer. Math. Soc. 53 (1947), p. 631. Compare J. L. Kelley, *loc. cit.* p. 134.

Concerning products of *hereditary* normal spaces, see M. Katětov, *Complete normality of cartesian products*, Fund. Math. 36 (1948), pp. 271–274.

(²) See E. Michael, Bull. Math. Soc. 69 (1963), p. 375, and K. Morita, Proc. Japan Academy 39 (1963), p. 148. Compare in that direction J. Dieudonné, *Un critère de normalité pour les produits*, Coll. Math. 6 (1958), pp. 29–32.

Finally, as $(a)\times\mathcal{Y} \underset{\text{top}}{=} \mathcal{Y}$, $\mathcal{Y}$ contains a set homeomorphic to $\big((a)\times\mathcal{Y}\big)\cap\mathfrak{Z}$, hence a dense in itself $(\neq 0)$ set.

THEOREM 4 [1]. *The space $\mathcal{X}\times\mathcal{Y}$ is scattered if and only if $\mathcal{X}$ and $\mathcal{Y}$ are scattered* ($\mathcal{X}\neq 0\neq\mathcal{Y}$).

Proof. Suppose that one of the sets $A\subset\mathcal{X}$ or $B\subset\mathcal{Y}$ is dense in itself $(\neq 0)$. Then by Theorem 2, $A\times B$ is dense in itself and thus $\mathcal{X}\times\mathcal{Y}$ is not scattered.

Conversely, if $\mathfrak{Z}\subset\mathcal{X}\times\mathcal{Y}$ is dense in itself $(\neq 0)$, then by the lemma, either the projection of $\mathfrak{Z}$ in $\mathcal{X}$ or the projection of $\mathfrak{Z}$ in $\mathcal{Y}$ contains a dense in itself $(\neq 0)$ set.

THEOREM 5. *Let $f_j\colon \mathcal{X}_j\to\mathcal{Y}_j$ for $j=0,1$. Then the product mapping $f=f_0\times f_1$ of $\mathcal{X}_0\times\mathcal{X}_1$ into $\mathcal{Y}_0\times\mathcal{Y}_1$* (see § 3, III) *is open if and only if f_0 and f_1 are open.*

More precisely, *f is open at the point $(y_0,y_1)\in f(\mathcal{X}_0\times\mathcal{X}_1)$ if and only if f_j is open at y_j for $j=0,1$.*

Proof. 1. Suppose that f is open at (y_0,y_1). Let $y_0\in\overline{B}\subset\mathcal{Y}_0$. We have to show (compare § 13, XIV (5)) that $f_0^{-1}(y_0)\subset\overline{f_0^{-1}(B)}$.

Let $\mathfrak{B}=B\times\mathcal{Y}_1$. By III (1), $\overline{\mathfrak{B}}=\overline{B}\times\mathcal{Y}_1$; hence $(y_0,y_1)\in\overline{\mathfrak{B}}$, and as f is open at (y_0,y_1) it follows by § 13, XIV (5) (see also § 3, III (25)) that

$$f^{-1}(y_0,y_1)\subset\overline{f^{-1}(\mathfrak{B})}=\overline{f^{-1}(B\times\mathcal{Y}_1)}=\overline{f_0^{-1}(B)\times f_1^{-1}(\mathcal{Y}_1)}$$
$$=\overline{f_0^{-1}(B)}\times\overline{f_1^{-1}(\mathcal{Y}_1)}=\overline{f_0^{-1}(B)}\times\mathcal{X}_1.$$

But by § 3, III (25),

$$f^{-1}(y_0,y_1)=f_0^{-1}(y_0)\times f_1^{-1}(y_1),$$

whence

$$f_0^{-1}(y_0)\times f_1^{-1}(y_1)\subset\overline{f_0^{-1}(B)}\times\mathcal{X}_1.$$

As $f^{-1}(y_0,y_1)\neq 0$, and consequently $f_1^{-1}(y_1)\neq 0$, it follows finally (see § 2, II (4)) that $f_0^{-1}(y_0)\subset\overline{f_0^{-1}(B)}$.

2. Let us assume that f_j is open at y_j for $j=0,1$. Let $\mathfrak{G}$ be open in $\mathcal{X}_0\times\mathcal{X}_1$ and $(y_0,y_1)\in f(\mathfrak{G})$. Assume (compare § 13, XIV, Remark 2) that $\mathfrak{G}$ belongs to the base of $\mathcal{X}_0\times\mathcal{X}_1$. Thus $\mathfrak{G}=G_0\times G_1$ where G_j is open in $\mathcal{X}_j$. As (see § 3, III (24)) $f(G_0\times G_1)=f_0(G_0)\times$

---

[1] See the paper of S. Ulam and of myself *Quelques propriétés topologiques du produit combinatoire*, Fund. Math. 19 (1932), p. 248.

$\times f_1(G_1)$, we have $y_j \epsilon f_j(G_j)$ and—f_j being open at y_j—it follows that $y_j \epsilon \operatorname{Int} f_j(G_j)$. Consequently (compare III(2)):

$$(y_0, y_1) \epsilon [\operatorname{Int} f_0(G_0)] \times [\operatorname{Int} f_1(G_1)] = \operatorname{Int}[f_0(G_0) \times f_1(G_1)] = \operatorname{Int} f(G_0 \times G_1).$$

Therefore, f is open at (y_0, y_1).

§ 16. Generalized cartesian products

I. Definition. Recall that by the definition of $\underset{t}{P} \mathscr{X}_t$, where $t \epsilon T$, we have (compare § 3, VIII(0))

$$\mathfrak{z} \epsilon \underset{t}{P} \mathscr{X}_t \equiv \bigwedge_t \mathfrak{z}^t \epsilon \mathscr{X}_t, \tag{0}$$

where $\mathfrak{z}^t$ is the t-th coordinate of the point $\mathfrak{z}$.

Now, let us suppose that $\mathscr{X}_t$ is a topological space for each $t \epsilon T$. We introduce the topology (called *Tychonov topology*) in $\underset{t}{P} \mathscr{X}_t$ by means of the following definition.

DEFINITION[1]. The family of sets of the form $\underset{t}{P} X_t^*$, where $X_{t_0}^* = G$ for a given t_0 and for a given open G in $\mathscr{X}_{t_0}$, and $X_t^* = \mathscr{X}_t$ for $t \neq t_0$ is a *subbase* of $\underset{t}{P} \mathscr{X}_t$.

Otherwise stated, the subbase is composed of sets $\mathfrak{G}$ of the form

$$\mathfrak{G}_{t,G} = \underset{\mathfrak{z}}{E}(\mathfrak{z}^t \epsilon G) \tag{1}$$

where G is open in $\mathscr{X}_t$.

It follows that the *base* of $\underset{t}{P} \mathscr{X}_t$ has as elements sets of the form $\underset{t}{P} G_t$, where G_t is open in $\mathscr{X}_t$ and, except for a finite set of elements of T, is identical to $\mathscr{X}_t$. Moreover, G_t may be assumed to belong to the base of $\mathscr{X}_t$. It follows also that if T is composed of two elements, the above definition is in agreement with the definition given in § 15, I.

In the particular case of the Cantor discontinuum $\mathscr{C} = (0, 2)^{\aleph_0}$ and of the space of irrational numbers $\mathscr{N} = \mathscr{D}^{\aleph_0}$, the topology given by our definition is identical with the usual topology of $\mathscr{C}$ and $\mathscr{N}$ considered as subsets of $\mathscr{E}$ (see Examples 2 and 3 of § 3, IX).

---

(1) Tychonov, *Ueber einen Funktionenraum*, Math. Ann. 111 (1935), p. 763, and *Ueber die topologische Erweiterung von Räumen*, *ibidem* 102 (1930), p. 544.

Many theorems stated in § 15 for the case of a product of two spaces can easily be extended to the general case of an arbitrary class of spaces as shown by the following theorems.

THEOREM 1. *The family of sets of the form* $\underset{t}{\mathbf{P}} F_t$, *where* F_t *is closed in* $\mathscr{X}_t$ *and, for all* t *with one exception,* $F_t = \mathscr{X}_t$, *is a closed subbase of* $\underset{t}{\mathbf{P}} \mathscr{X}_t$.

THEOREM 2. *If for each* $t \in T$, $\mathscr{X}_t$ *is a* $\mathscr{T}_1$-*space,* $\underset{t}{\mathbf{P}} \mathscr{X}_t$ *is a* $\mathscr{T}_1$-*space.*

Proof. $\{a_t\}_{t \in T}$ being a point of $\underset{t}{\mathbf{P}} \mathscr{X}_t$, put $F_{t_0} = \underset{t}{\mathbf{P}} X_t^*$ where $X_{t_0}^* = (a_{t_0})$ and $X_t^* = \mathscr{X}_t$ for $t \neq t_0$. Then by Theorem 1, F_{t_0} is closed and so is $\underset{t}{\mathbf{P}}(a_t) = \bigcap_t F_t$.

II. Projections and continuous mappings. Consider $\mathfrak{z}^t$ as function of $\mathfrak{z}$ (called *projection on the* $\mathscr{X}_t$-*axis*):

$$\mathrm{pr}_t(\mathfrak{z}) = \mathfrak{z}^t \quad \text{and} \quad \mathrm{pr}_{t_0}\colon (\underset{t}{\mathbf{P}} \mathscr{X}_t) \to \mathscr{X}_{t_0}. \tag{1}$$

Obviously, if $A \subset \mathscr{X}_{t_0}$, then

$$\mathrm{pr}_{t_0}^{-1}(A) = \underset{\mathfrak{z}}{E}(\mathfrak{z}^{t_0} \in A) = \underset{t}{\mathbf{P}} X_t^*, \tag{2}$$

where $X_{t_0}^* = A$ and $X_t^* = \mathscr{X}_t$ for $t \neq t_0$.

THEOREM 1. *The projections are open mappings of the product space onto the axes.*

For, A being supposed open in $\mathscr{X}_{t_0}$, the set $\mathrm{pr}_{t_0}^{-1}(A)$ is open in the product space by (2) and I (1).

On the other hand, if $\mathfrak{G}$ is open in $\underset{t}{\mathbf{P}} \mathscr{X}_t$, so is $\mathrm{pr}_t(\mathfrak{G})$. One can obviously assume that $\mathfrak{G}$ is a member of the base; put $\mathfrak{G} = G_{t_1} \times \ldots \times G_{t_n} \times \underset{t}{\mathbf{P}} \mathscr{X}_t$ where $t \neq t_k$ $(k = 1, 2, \ldots, n)$. Then $\mathrm{pr}_{t_k}(\mathfrak{G}) = G_{t_k}$ and $\mathrm{pr}_t(\mathfrak{G}) = \mathscr{X}_t$ for $t \neq t_k$.

In the case where $\mathscr{X}_t = \mathscr{Y}$ for each $t \in T$, we deduce from Theorem 1:

THEOREM 2. *Let* $g \in (\mathscr{Y}^T)_{\mathrm{set}}$; *i.e.* $g\colon T \to \mathscr{Y}$. *Then, for* g *variable the function* $f_t\colon (\mathscr{Y}^T)_{\mathrm{set}} \to \mathscr{Y}$, *defined by means of the identity*

$$f_t(g) = g(t),$$

is continuous.

THEOREM 3. *Let* $\varphi\colon \mathscr{Z} \to \underset{t}{\mathbf{P}} \mathscr{X}_t$; i.e. $\varphi^t\colon \mathscr{Z} \to \mathscr{X}_t$. *Then we have*

$$(\varphi \textit{ is continuous at } z_0) \equiv \bigwedge_t (\varphi^t \textit{ is continuous at } z_0).$$

The proof is completely similar to that of Theorem 2, § 15, II. In the second part of the argument, we assume that $\mathfrak{G} = \underset{t}{P} X_t^*$ where $X_{t_0}^* = G$ open in $\mathscr{X}_{t_0}$ and $X_{t_0}^* = \mathscr{X}_t$ for $t \neq t_0$. Then we use the formula (see § 3, VIII (9)) $\varphi^{-1}(\mathfrak{G}) = (\varphi^{t_0})^{-1}(G)$.

THEOREM 4. (Associativity of cartesian multiplication.) *Let $t_0 \in T$ and $\mathscr{Y} = \underset{t \neq t_0}{P} \mathscr{X}_t$. Then*

$$\underset{t}{P} \mathscr{X}_t \underset{\text{top}}{=} \mathscr{X}_{t_0} \times \mathscr{Y}.$$

Namely, let us put for $\mathfrak{z} \in \underset{t}{P} \mathscr{X}_t$:

$$\mathfrak{x}(\mathfrak{z}) = \mathfrak{z}^{t_0}, \quad \mathfrak{y}(\mathfrak{z}) = \{\mathfrak{z}^t\}_{t \neq t_0}, \quad \textit{and} \quad \mathfrak{w}(\mathfrak{z}) = \big(\mathfrak{x}(\mathfrak{z}), \mathfrak{y}(\mathfrak{z})\big).$$

Then $\mathfrak{w}$ is the required homeomorphism.

Proof. Obviously, $\mathfrak{w}$ is a one-to-one mapping onto. By Theorems 1 and 3 the mappings $\mathfrak{x}$ and $\mathfrak{y}$ are continuous, hence (by Theorem 3) $\mathfrak{w}$ is continuous.

It remains to show that $\mathfrak{w}$ transforms open sets into open sets. It suffices to show it for the case of open sets belonging to a subbase of $\underset{t}{P} \mathscr{X}_t$. Thus, let us suppose that G is open in $\mathscr{X}_t$ and put $\mathfrak{G} = \underset{\mathfrak{z}}{E}(\mathfrak{z}^t \in G)$. We have to show that $\mathfrak{w}(G)$ is open.

Now, if $t = t_0$, we have $\mathfrak{w}(\mathfrak{G}) = G \times \mathscr{Y}$. If $t = t_1 \neq t_0$,

$$\mathfrak{w}(\mathfrak{G}) = \mathscr{X}_{t_0} \times \underset{t \neq t_0}{P} X_t^* \quad \text{where} \quad X_{t_1}^* = G \text{ and } X_t^* = \mathscr{X}_t \text{ if } t_0 \neq t \neq t_1.$$

Hence $\underset{t \neq t_0}{P} X_t^*$ is open in $\mathscr{Y}$ and consequently $\mathfrak{w}(\mathfrak{G})$ is open in $\mathscr{X}_{t_0} \times \mathscr{Y}$.

Theorem 1 of § 15, VI can be extended to the case of T arbitrary as follows.

THEOREM 5. *Let $t_0 \in T$ and $\mathfrak{z}_0 \in \underset{t}{P} \mathscr{X}_t$. Then*

$$\underset{\mathfrak{z}}{E} \bigwedge_{t \neq t_0} (\mathfrak{z}^t = \mathfrak{z}_0^t) \underset{\text{top}}{=} \mathscr{X}_{t_0}, \quad \textit{thus} \quad \mathscr{X}_{t_0} \underset{\text{top}}{\subset} \underset{t}{P} \mathscr{X}_t.$$

Proof. Put $\mathscr{Y} = \underset{t \neq t_0}{P} \mathscr{X}_t$ and $\mathfrak{y}_0 = \{\mathfrak{z}_0^t\}_{t \neq t_0}$. Then $\underset{\mathfrak{z}}{E} \bigwedge_{t \neq t_0} (\mathfrak{z}^t = \mathfrak{z}_0^t)$ is homeomorphic to $\mathscr{X}_{t_0} \times \mathfrak{y}_0$, and this completes the proof according to § 15, VI (3).

Let T and U be arbitrary sets and $f\colon U \to T$ be one-to-one and onto. Then we obviously have

$$\underset{t\in T}{\mathsf{P}}\, \mathscr{X}_t \underset{\text{top}}{=} \underset{u\in U}{\mathsf{P}}\, \mathscr{X}_{f(u)} \tag{3}$$

and the required homeomorphism $\mathfrak{h}$ can be defined by the condition

$$\mathfrak{h}^u(\mathfrak{z}) = \mathfrak{z}^{\mathfrak{h}(u)} \qquad \text{for each} \qquad \mathfrak{z} \in \underset{t\in T}{\mathsf{P}}\, \mathscr{X}_t \quad \text{and} \quad u \in U.$$

In the particular case where $\mathscr{X}_t = \mathscr{X}$ for each $t \in T$, we have

$$\mathscr{X}^T \underset{\text{top}}{=} \mathscr{X}^U. \tag{4}$$

This leads to the following theorem:

THEOREM 6. $\mathscr{X}^{\aleph_a} \times \mathscr{X}^{\aleph_a} \underset{\text{top}}{=} \mathscr{X}^{\aleph_a} \underset{\text{top}}{=} (\mathscr{X}^{\aleph_a})^{\aleph_a}$.

The proof follows directly from formulas (1), (3), and (5) of § 3, XII (compare also § 3, XII (4)).

COROLLARY 6a. *$\mathscr{C}$ denoting the Cantor discontinuum, we have*

$$\mathscr{C}^{\aleph_0} \underset{\text{top}}{=} \mathscr{C}, \tag{5}$$

$$\mathscr{I}^{\aleph_0} \textit{ (as well as } \mathscr{I}^n\textit{) is a continuous image of } \mathscr{C}. \tag{6}$$

Formula (5) follows from Theorem 6 since $\mathscr{C}$ can be represented as $(0, 1)^{\aleph_0}$.

In order to prove (6), let us note first that the interval $\mathscr{I}$ is a continuous image of $\mathscr{C}$. Namely given an $x \in \mathscr{C}$, we have

$$x = \frac{c_1}{3^1} + \frac{c_2}{3^2} + \ldots + \frac{c_m}{3^m} + \ldots \qquad (c_m = 0 \text{ or } 2), \tag{7}$$

and we put[1]

$$f(x) = \frac{c_1}{2^2} + \frac{c_2}{2^3} + \ldots + \frac{c_m}{2^{m+1}} + \ldots. \tag{8}$$

Obviously f is a continuous mapping of $\mathscr{C}$ onto $\mathscr{I}$.

f induces a continuous mapping g of $\mathscr{C}^n$ onto $\mathscr{I}^n$ $(n \leqslant \aleph_0)$. Namely we put

$$g^i(\mathfrak{z}) = f(\mathfrak{z}^i) \qquad \text{for} \qquad \mathfrak{z} \in \mathscr{C}^n. \tag{9}$$

---

[1] This is the "step-function" (which can easily be extended to the whole interval $\mathscr{I}$). Compare G. Cantor, Acta Math. 7 (1885) and H. Lebesgue, Journ. de Math. (6) 1 (1905), p. 210.

This completes the proof since $\mathscr{C}^n$ is a continuous image of $\mathscr{C}$ (by formula (5)).

COROLLARY 6b. (Generalized Peano theorem.)(1) *The cubes $\mathscr{I}^n$ ($n \leqslant \aleph_0$) are continuous images of the interval $\mathscr{I}$.*

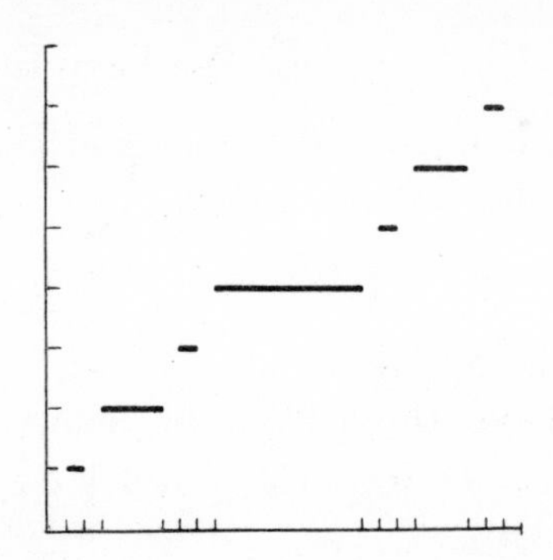

According to Corollary 6a, let $h: \mathscr{C} \to \mathscr{I}^n$ be onto. On each continuous interval to $\mathscr{C}$ extend h linearly. The extended function transforms $\mathscr{I}$ onto $\mathscr{I}^n$.

COROLLARY 6c. (Generalized Tietze extension theorem.) *Let $\mathscr{X}$ be normal, $F = \overline{F} \subset \mathscr{X}$ and $f: F \to \mathscr{I}^n$ (resp. $\mathscr{E}^n$) continuous ($n \leqslant \aleph_0$). Then $f \subset g: \mathscr{X} \to \mathscr{I}^n$ (resp. $\mathscr{E}^n$), where g is continuous.*

Proof. Since $f: F \to \mathscr{I}^n$ is continuous, so is $f^i: F \to \mathscr{I}$. Hence by Tietze theorem (§ 14, IV) there is $g: \mathscr{X} \to \mathscr{I}^n$ such that $g^i: \mathscr{X} \to \mathscr{I}$ is a continuous extension of f^i. By Theorem 3, g is continuous.

COROLLARY 6d. *Let $\{A_t\}$, $t \in T$, be an indexed family of subsets of the space $\mathscr{X}$. Then its characteristic function $f: \mathscr{X} \to (0,1)^T$ is continuous if and only if each A_t is closed-open.*

By definition (see § 3, VII (9)), f^t is the characteristic function of A_t. Hence f^t is continuous if and only if A_t is closed-open (by Theorem 5 of § 13, VI).

III. Operations on cartesian products. Formula III (1) of § 15 can be extended to the general case, where $A_t \subset \mathscr{X}_t$ and $t \in T$. Namely:

$$\overline{\underset{t}{P}\, A_t} = \underset{t}{P}\, \overline{A}_t. \tag{1}$$

(1) See G. Peano, Math. Ann. 36 (1890), p. 157, and H. Lebesgue, *loc. cit.*

Proof. Let $\mathfrak{z} \in \overline{\underset{t}{P} A_t}$. The projection pr_t being continuous, it follows that

$$\mathfrak{z}^t = \mathrm{pr}_t(\mathfrak{z}) \in \overline{\mathrm{pr}_t(\underset{s}{P} A_s)} = \bar{A}_t.$$

Hence $\mathfrak{z} \in \underset{t}{P} \bar{A}_t$.

On the other hand, let $\mathfrak{a} \in \underset{t}{P} \bar{A}_t$. We have to show that $\mathfrak{G}$ being open and containing $\mathfrak{a}$, $\mathfrak{G}$ contains a point of $\underset{t}{P} A_t$. We may suppose that $\mathfrak{G}$ belongs to the base, i.e. that

$$\mathfrak{a} \in \mathfrak{G} = \underset{\mathfrak{z}}{E} (\mathfrak{z}^{t_1} \in G_{t_1}) \dots (\mathfrak{z}^{t_n} \in G_{t_n})$$

where G_{t_k} is open in $\mathscr{X}_{t_k}$ $(k = 1, 2, \dots, n)$.

As $\mathfrak{a}^{t_k} \in G_{t_k}$, we have $G_{t_k} \cap \bar{A}_{t_k} \neq 0$, hence $G_{t_k} \cap A_{t_k} \neq 0$.

Let $\mathfrak{b} = \{\mathfrak{b}^t\}$, where $\mathfrak{b}^{t_k} \in (G_{t_k} \cap A_{t_k})$ and $\mathfrak{b}^t \in A_t$ for $t \neq t_k$. It follows that $\mathfrak{b} \in (\mathfrak{G} \cap \underset{t}{P} A_t)$.

The following formula yields a generalization of § 15, III (2):

$$\mathrm{Int}(\underset{t}{P} A_t) \subset \underset{t}{P}\, \mathrm{Int}(A_t), \tag{2}$$

and if $A_t = \mathscr{X}_t$ except for a finite number of t's, then

$$\mathrm{Int}(\underset{t}{P} A_t) = \underset{t}{P}\, \mathrm{Int}(A_t). \tag{3}$$

Moreover, if $\mathrm{Int}(\underset{t}{P} A_t) \neq 0$, there exists a finite system $t_1, \dots, t_n$ such that $A_t = \mathscr{X}_t$ for $t \neq t_k$ $(k = 1, 2, \dots, n)$.

To prove the second part of the above statement put $\mathfrak{z} \in \mathrm{Int}(\underset{t}{P} A_t)$ and denote by $\mathfrak{G}$ a member of the base (considered in Section I) containing $\mathfrak{z}$ and contained in $\mathrm{Int}(\underset{t}{P} A_t)$. We may obviously suppose that $\mathfrak{G}$ is of the form

$$\mathfrak{G} = G_{t_1} \times G_{t_2} \times \dots \times G_{t_n} \times \underset{t \neq t_k}{P} \mathscr{X}_t.$$

As $\mathfrak{z} \in \mathfrak{G} \subset \mathrm{Int}(\underset{t}{P} A_t)$, it follows that $G_{t_k} \subset A_{t_k}$ and $\mathscr{X}_t = A_t$ for $t \neq t_k$. Hence $\mathfrak{z}^{t_k} \in \mathrm{Int}(A_{t_k})$ and finally $\mathfrak{z} \in \underset{t}{P}\, \mathrm{Int}(A_t)$.

IV. Diagonal. Let $\mathscr{X}_t = \mathscr{X}$ for each $t \in T$. The diagonal of $\underset{t}{P} \mathscr{X}_t = (\mathscr{X}^T)_{\mathrm{set}}$ is the set

$$\Delta = \underset{\mathfrak{z}}{E} \bigwedge_{tt'} (\mathfrak{z}^t = \mathfrak{z}^{t'}).$$

THEOREM 1. $\Delta \underset{\mathrm{top}}{=} \mathscr{X}$.

The required homeomorphism is pr_{t_0} (where t_0 is arbitrary). (See II, Theorems 1 and 3.)

THEOREM 2. *If $\mathscr{X}$ is Hausdorff, the diagonal Δ of $(\mathscr{X}^T)_{\mathrm{set}}$ is closed.*

The proof is analogous to the proof of Theorem 2 of § 15, IV. We consider the complement of Δ, i.e. the set

$$\nabla = \mathop{E}_{\mathfrak{z}} \bigvee_{tt'} (\mathfrak{z}^t \neq \mathfrak{z}^{t'}) = \bigcup_{tt'} \mathop{E}_{\mathfrak{z}} (\mathfrak{z}^t \neq \mathfrak{z}^{t'}).$$

Let $\mathfrak{z}_0 \in \nabla$ be given, as well as t and t' such that $\mathfrak{z}_0^t \neq \mathfrak{z}_0^{t'}$. $\mathscr{X}$ being Hausdorff, there exist G and H open and such that $\mathfrak{z}_0^t \in G$, $\mathfrak{z}_0^{t'} \in H$, and $G \cap H = 0$. Put

$$\mathfrak{A} = \mathop{E}_{\mathfrak{z}} (\mathfrak{z}^t \in G)(\mathfrak{z}^{t'} \in H).$$

Obviously $\mathfrak{A}$ is open and $\mathfrak{z}_0 \in \mathfrak{A}$. Moreover $\mathfrak{A} \subset \nabla$, for otherwise there would exist $\mathfrak{z} \in \mathfrak{A}$ such that $\mathfrak{z}^t = \mathfrak{z}^{t'}$, but this is impossible since $G \cap H = 0$. Thus ∇ is open and Δ closed.

Now let us consider

$$f_t\colon \mathscr{X}_t \to \mathscr{Y} \quad \text{continuous}.$$

Put

$$\mathfrak{Z} = \mathop{E}_{\mathfrak{z}} \bigwedge_{t't''} \left(f_{t'}(\mathfrak{z}^{t'}) = f_{t''}(\mathfrak{z}^{t''})\right) \tag{1}$$

and consider, for $\mathfrak{z} \in \mathop{P}_{t} \mathscr{X}_t$, the point $\mathfrak{f}(\mathfrak{z}) \in (\mathscr{Y}^T)_{\mathrm{set}}$ defined as follows:

$$\mathfrak{f}^t(\mathfrak{z}) = \{f_t(\mathfrak{z}^t)\}. \tag{2}$$

THEOREM 3. *For each $t_0 \in T$ we have*

$$f_{t_0} \circ \mathrm{pr}_{t_0}(\mathfrak{Z}) = \bigcap_t f_t(\mathscr{X}_t). \tag{3}$$

Proof. $y = f_{t_0} \circ \mathrm{pr}_{t_0}(\mathfrak{z}) = f_{t_0}(\mathfrak{z}^{t_0}) = f_t(\mathfrak{z}^t)$ for each $t \in T$ and $\mathfrak{z} \in \mathfrak{Z}$. Thus, $y \in f_t(\mathscr{X}_t)$ for each $t \in T$.

On the other hand, if $y \in \bigcap_t f_t(\mathscr{X}_t)$, there is a $\mathfrak{z} \in \mathop{P}_{t} \mathscr{X}_t$ such that $y = f_t(\mathfrak{z}^t)$ for each t. Therefore $\mathfrak{z} \in \mathfrak{Z}$ and $y \in f_t \circ \mathrm{pr}_t(\mathfrak{Z})$.

THEOREM 4. *If the mappings f_t are one-to-one (respectively homeomorphisms), the same is true of $f_t \circ \mathrm{pr}_t \mid \mathfrak{Z}$.*

Proof. Suppose that the mappings f_t are one-to-one. Then by (2)

$$[\mathfrak{f}(\mathfrak{z}) = \mathfrak{f}(\mathfrak{y})] \equiv \bigwedge_t [f_t(\mathfrak{z}^t) = f_t(\mathfrak{y}^t)] \Rightarrow \bigwedge_t (\mathfrak{z}^t = \mathfrak{y}^t) \equiv (\mathfrak{z} = \mathfrak{y}).$$

On the other hand, $\mathfrak{f}^{-1}$ is defined on $\mathfrak{f}(\mathfrak{Z})$ and $\mathfrak{f}^{-1}(\mathfrak{y})$ has as coordinates $f_t^{-1}(\mathfrak{y})$, $t \in T$. If we assume that the mappings f_t^{-1} are continuous, $\mathfrak{f}^{-1}$ is continuous (by II, 3).

THEOREM 5. *If $\mathscr{Y}$ is Hausdorff, $\mathfrak{Z}$ is closed.*

This follows from the fact that $f\colon \underset{t}{P}\mathscr{X}_t \to (\mathscr{Y}^T)_{\text{set}}$ is continuous and that $\mathfrak{Z} = f^{-1}(\Delta)$, where Δ is the diagonal of $(\mathscr{Y}^T)_{\text{set}}$. As Δ is closed (by Theorem 2), so is $\mathfrak{Z}$.

Remark. As in the case of two factors, one can consider the diagonal of $\underset{t}{P}\mathscr{X}_t$ when $\mathscr{X}_t \subset \mathscr{X}$ for each t. The definition (1) remains valid. Theorem 1 has to be replaced by

$$\Delta \underset{\text{top}}{=} \bigcap_t \mathscr{X}_t.$$

If $\mathscr{X}$ is Hausdorff, Δ is closed in $\underset{t}{P}\mathscr{X}_t$.

V. Invariants of cartesian multiplications.

THEOREM 1. *Let $0 \neq A_t \subset \mathscr{X}_t$. Then $\underset{t}{P}A_t$ is closed or dense in $\underset{t}{P}\mathscr{X}_t$ if and only if each A_t has the respective property.*

This follows from III (1) (compare § 2, II (7):

$$(\overline{\underset{t}{P}A_t} = \underset{t}{P}A_t) \equiv (\underset{t}{P}\bar{A}_t = \underset{t}{P}A_t) \equiv \bigwedge_t(\bar{A}_t = A_t),$$

$$(\overline{\underset{t}{P}A_t} = \underset{t}{P}\mathscr{X}_t) \equiv (\underset{t}{P}\bar{A}_t = \underset{t}{P}\mathscr{X}_t) = \bigwedge_t(\bar{A}_t = \mathscr{X}_t).$$

THEOREM 2. *If for an infinity of values of t the space $\mathscr{X}_t$ contains more than one point, then $\underset{t}{P}\mathscr{X}_t$ is dense in itself.*

Proof. Let $\mathfrak{G}$ be a member of the base containing a given $\mathfrak{z} \in \underset{t}{P}\mathscr{X}_t$. We may assume that $\mathfrak{G}$ is of the form $G_{t_1} \times \ldots \times G_{t_n} \times \underset{t}{P}\mathscr{X}_t$ where $t \neq t_k$ $(k = 1, 2, \ldots, n)$. Let $\mathfrak{y}$ be a point such that $\mathfrak{y}^{(t_k)} = \mathfrak{z}^{(t_k)}$ for $k = 1, 2, \ldots, n$ and $\mathfrak{y}^t \neq \mathfrak{z}^t$ for an infinity of values of t. Then $\mathfrak{y} \in \mathfrak{G}$ and $\mathfrak{y} \neq \mathfrak{z}$.

Thus every neighbourhood of $\mathfrak{z}$ contains a point different of $\mathfrak{z}$. Hence $\mathfrak{z}$ is an accumulation point of $\underset{t}{P}\mathscr{X}_t$.

THEOREM 3. *$\underset{t}{P}A_t$ is a boundary set if and only if either one of the factors is a boundary set or $A_t \neq \mathscr{X}_t$ for an infinity of values of t.*

$\underset{t}{P}A_t$ is nowhere dense if and only if either one of the factors is nowhere dense or $\bar{A}_t \neq \mathscr{X}_t$ for an infinity of values of t.

The proof follows easily from III (2).

THEOREM 3'. *If one of the sets A_t is of the first category, so is $\underset{t}{P} A_t$.*

Proof. Suppose that A_{t_0} is of the first category. Then by § 15, VII, Corollary 2a, $A_{t_0} \times \underset{t \neq t_0}{P} \mathscr{X}_t$ is of the first category in $\underset{t}{P} \mathscr{X}_t$. This completes the proof since $\underset{t}{P} A_t \subset (A_{t_0} \times \underset{t \neq t_0}{P} \mathscr{X}_t)$.

THEOREM 4. *The following properties are invariant of general cartesian multiplication: of being Hausdorff, regular, completely regular.*

The proof is essentially the same as in the case of two factors (Theorem 3 of § 15, VII).

COROLLARY 4a. *Let a be any ordinal number. Then $\mathscr{I}^{\aleph_a}$, as well as every subset of $\mathscr{I}^{\aleph_a}$, is completely regular.*

This follows from the facts that the interval $\mathscr{I}$ is completely regular and that the property of being completely regular is hereditary.

The converse to Corollary 4a is also true.

THEOREM 5. *Let $\mathscr{X}$ be a completely regular $\mathscr{T}_1$-space. Then $\mathscr{X} \underset{\text{top}}{\subset} I^{\aleph_a}$ for sufficiently large a.*

More precisely, $\aleph_a$ is the power of $\Phi = \mathscr{I}^{\mathscr{X}}$ and

$$\mathscr{X} \underset{\text{top}}{\subset} (\mathscr{I}^{\Phi})_{\text{set}}. \tag{1}$$

Proof. Define for each $x \in \mathscr{X}$ the function $\mathfrak{z}(x)$ (of the variable $f \in \Phi$) by the condition

$$\mathfrak{z}^f(x) = f(x). \tag{2}$$

Thus $\mathfrak{z}(x) \in (\mathscr{I}^{\Phi})_{\text{set}}$. We shall show that

$$\mathfrak{z} \textit{ is the required homeomorphism stated in } (1). \tag{3}$$

For a given $f \in \Phi$, $\mathfrak{z}^f$ is obviously continuous (by virtue of (2)). Therefore $\mathfrak{z}$ is continuous (by Theorem 3 of Section II). $\mathfrak{z}$ is one-to-one. For, let us assume that $x_0 \neq x_1$. As $\mathscr{X}$ is a completely regular $\mathscr{T}_1$-space, there is an $f \in \Phi$ such that $f(x_0) = 0$ and $f(x_1) = 1$. Therefore by (2) $\mathfrak{z}^f(x_0) \neq \mathfrak{z}^f(x_1)$ and consequently $\mathfrak{z}(x_0) \neq \mathfrak{z}(x_1)$.

Now, let G be open in $\mathscr{X}$; we have to show that $\mathfrak{z}(G)$ is open in $\mathfrak{z}(\mathscr{X})$. It suffices to show that, for $x_0 \in G$, there is an open $\mathfrak{G}$ such that $\mathfrak{z}(x_0) \in \mathfrak{G}$ and

$$\mathfrak{G} \cap \mathfrak{z}(\mathscr{X}) \subset \mathfrak{z}(G). \tag{4}$$

Let $f \in \Phi$ be such that

$$f(x_0) = 0 \quad \text{and} \quad f(x) = 1 \quad \text{for} \quad x \in \mathscr{X} - G. \tag{5}$$

Put $\mathfrak{G} = \underset{\mathfrak{y}}{E}(\mathfrak{y}^f \neq 1)$ where $\mathfrak{y} \in (\mathscr{I}^{\Phi})_{\text{set}}$.

The projection of $\mathfrak{y}$ being a continuous function of $\mathfrak{y}$ (by II, Theorem 1), $\mathfrak{G}$ is open in $(\mathscr{I}^{\Phi})_{\text{set}}$.

Moreover $\mathfrak{z}(x_0) \in \mathfrak{G}$ since $\mathfrak{z}^f(x_0) = f(x_0) \neq 1$ by (2) and (5).

Finally, to show (4) put $\mathfrak{y} \in \mathfrak{G} \cap \mathfrak{z}(\mathscr{X})$. Hence $\mathfrak{y}^f \neq 1$ and there is an x such that $\mathfrak{y} = \mathfrak{z}(x)$. It follows that $\mathfrak{y}^f = \mathfrak{z}^f(x)$, whence by (2) $\mathfrak{y}^f = f(x)$ and therefore $f(x) \neq 1$. Consequently $x \in G$ by (5). As $x \in G$ and $\mathfrak{y} = \mathfrak{z}(x)$ we have $\mathfrak{y} \in \mathfrak{z}(G)$.

Remarks. Thus a completely regular $\mathscr{T}_1$-space is—from the topological point of view—nothing else but a subset of the generalized cube $\mathscr{I}^{\aleph_\alpha}$ ($\aleph_\alpha$ may be assumed to be the cardinality of a base of the space). As we shall see later the subsets of $\mathscr{I}^{\aleph_0}$ are topologically equivalent with metric separable spaces. We shall see also in Volume II (Chapter 4) that the closed subsets of $\mathscr{I}^{\aleph_\alpha}$ are topologically equivalent with compact Hausdorff spaces.

The closed subsets of $\mathscr{E}^{\aleph_\alpha}$ are topologically equivalent with the Hewitt *Q-spaces* (1).

THEOREM 6 (2). *If $\mathscr{X}_n$ is separable for $n = 1, 2, \ldots$, then the infinite product $\mathscr{X}_1 \times \mathscr{X}_2 \times \ldots$ is separable.*

In particular, *$\mathscr{I}^{\aleph_0}$ and $\mathscr{E}^{\aleph_0}$ are separable spaces.*

---

(1) See E. Hewitt, *Rings of real-valued continuous functions, I*, Trans. Amer. Math. Soc. 64 (1948), pp. 54–99, M. Katětov, *On real-valued functions in topological spaces*, Fund. Math. 38 (1951), pp. 85–91, T. Shirota, *A class of topological spaces*, Osaka Math. Journ. 4 (1952), pp. 23–40, and L. Gillman and M. Jerison, *Rings of continuous functions*, New York 1960.

Compare also S. Mrówka, *Some properties of Q-spaces*, Bull. Acad. Pol. Sci. 5 (1957), pp. 947–950, and R. Engelking and S. Mrówka, *On E-compact spaces*, Bull. Acad. Pol. Sci. 6 (1958), pp. 429–436, where analogies with compact spaces are emphasized.

(2) This theorem remains true for the cartesian product of $\mathfrak{c}$ separable spaces, see E. Pondiczery, *Power problems in abstract spaces*, Duke Math. Journ. 11 (1944), p. 835; E. Hewitt, *A remark on density characters*, Bull. Amer. Math. Soc. 52 (1946), pp. 641–643, and E. Marczewski, *Séparabilité et multiplication cartésienne des espaces topologiques*, Fund. Math. 34 (1947), pp. 127–143.

Proof. Let R_n be countable and dense in $\mathscr{X}_n$. Let r_n be a fixed point of R_n. Denote by $\mathfrak{R}$ the set of all points $\mathfrak{r} = (\mathfrak{r}^1, \mathfrak{r}^2, \ldots)$ such that

(i) $\mathfrak{r}^n \in R_n$ for each n,

(ii) for sufficiently large m (depending on $\mathfrak{r}$) we have

$$\mathfrak{r}^m = r_m, \quad \mathfrak{r}^{m+1} = r_{m+1}, \quad \ldots.$$

Obviously, $\mathfrak{R}$ is countable and has points in common with each set of the form $G_1 \times \ldots \times G_k \times \mathscr{X}_{k+1} \times \mathscr{X}_{k+2} \times \ldots$, where G_i is non-void and open in $\mathscr{X}_i$ for $i \leqslant k$. Hence $\mathfrak{R} \cap \mathfrak{G} \neq 0$ for each open $\mathfrak{G} (\neq 0)$.

Remark. It is worth noticing that if $\mathscr{X}_n = \mathscr{E}$ (or $\mathscr{I}$) for $n = 1, 2, \ldots$, one can assume that $\mathfrak{R}$ is formed by sequences all of whose terms are rational, only a finite number of them being different from zero.

THEOREM 7. *If A_n has the Baire property for $n = 1, 2, \ldots$, then $\underset{n}{P} A_n$ has also this property.*

Proof. By § 3, VIII (1):

$$\underset{n}{P} A_n = \bigcap_n (\mathscr{X}_1 \times \ldots \times \mathscr{X}_{n-1} \times A_n \times \mathscr{X}_{n+1} \times \ldots)$$

and by § 15, VII, Corollary 2a, the set in brackets has the Baire property. This completes the proof, since the Baire property is countably multiplicative (see § 11, III, Theorem 1).

Remark. $\underset{n}{P} A_n$ can have the Baire property (can even be nowhere dense) although none of the sets A_n has this property. Such is the case where $\mathscr{X}_i = \mathscr{I}$ and A_i is a set without the Baire property contained in the interval $0, 1/2$.

THEOREM 8. *Let f_t: $\mathscr{X}_t \to \mathscr{Y}_t$ for $t \in T$, and let $\mathfrak{f}$ denote the product mapping: $\underset{t}{P} \mathscr{X}_t \to \underset{t}{P} \mathscr{Y}_t$, i.e.* (see § 3, VII (11))

$$[\mathfrak{y} = \mathfrak{f}(\mathfrak{x})] \equiv \bigwedge_t [\mathfrak{y}^t = f_t(\mathfrak{x}^t)].$$

If $\mathfrak{f}$ is open at the point $\mathfrak{y}_0 \in f(\underset{t}{P} \mathscr{X}_t)$, f_t is open at $\mathfrak{y}_0^t$ for each $t \in T$.

Conversely, *if for each t, f_t is open at $\mathfrak{y}_0^t$ and for each t, except a finite number of t's, f_t is onto, then $\mathfrak{f}$ is open at $\mathfrak{y}_0$.*

The proof is completely similar to that of Theorem 5 of § 15, VII, except that in the first part instead of § 3, III (25) one has to use § 3, VIII (13). In the second part one assumes that the set $\mathfrak{G}$ (belonging to the base of $\underset{t}{P}\mathcal{X}_t$) is of the form

$$\mathfrak{G} = G_{t_1}\times\ldots\times G_{t_n}\times\underset{t'}{P}\mathcal{X}_{t'} \quad \text{where} \quad t' \neq t_k \quad \text{for} \quad k = 1, \ldots, n,$$

and instead of III (2) and § 3, III (24), one has to use III (3) and § 3, VIII (12).

VI. Inverse limits. Let us consider an inverse system $(T, \mathcal{X}, f)$, where (see § 3, XIII) T is a directed set, $\mathcal{X}_t$ is a topological space defined for each $t \in T$, and $f_{t_0 t_1}$ is continuous and defined for pairs t_0, t_1 such that $t_0 \leqslant t_1$, and $f_{t_0 t_1}\colon \mathcal{X}_{t_1} \to \mathcal{X}_{t_0}$. As in § 3 (XIII (2) and (3)) f has the transitivity and identity properties. Recall that the inverse limit of the system $(T, \mathcal{X}, f)$, denoted by $\mathcal{X}_\infty$, is defined by the conditions:

$$\mathcal{X}_\infty \subset \underset{t}{P}\mathcal{X}_t \quad \text{and} \quad \mathfrak{z} \in X_\infty \equiv \bigwedge_{t_0 < t_1} [f_{t_0 t_1}(\mathfrak{z}^{t_1}) = \mathfrak{z}^{t_0}]. \tag{1}$$

THEOREM 1. *If each $\mathcal{X}_t$ for $t \in T$ is Hausdorff, respectively completely regular, then so is $\mathcal{X}_\infty$.*

Moreover $\mathcal{X}_\infty$ is closed in $\underset{t}{P}\mathcal{X}_t$.

Proof. By Theorem 4 of Section V, $\underset{t}{P}\mathcal{X}_t$ is Hausdorff, respectively completely regular. Hence, so is $\mathcal{X}_\infty$.

In order to show that $\mathcal{X}_\infty$ is closed, put $\mathfrak{z} \in (\underset{t}{P}\mathcal{X}_t) - \mathcal{X}_\infty$. We have to define an open $\mathfrak{G}$ such that

$$\mathfrak{z} \in \mathfrak{G} \quad \text{and} \quad \mathfrak{G} \cap \mathcal{X}_\infty = 0. \tag{2}$$

As $\mathfrak{z} \notin \mathcal{X}_\infty$, there is by (1) a pair $t_0 < t_1$ such that $f_{t_0 t_1}(\mathfrak{z}^{t_1}) \neq \mathfrak{z}^{t_0}$. The space $\mathcal{X}_{t_0}$ being Hausdorff, there are two sets U and V open in $\mathcal{X}_{t_0}$ such that

$$\mathfrak{z}^{t_0} \in U, \quad f_{t_0 t_1}(\mathfrak{z}^{t_1}) \in V, \quad \text{and} \quad U \cap V = 0. \tag{3}$$

Put

$$\mathfrak{G} = \underset{\mathfrak{y}}{E}(\mathfrak{y}^{t_0} \in U)[f_{t_0 t_1}(\mathfrak{y}^{t_1}) \in V]. \tag{4}$$

$\mathfrak{G}$ is open since $f_{t_0 t_1}$, as well as the projection, are continuous.

Obviously $\mathfrak{z} \in \mathfrak{G}$. Finally $\mathfrak{G} \cap \mathcal{X}_\infty = 0$. For, if $\mathfrak{y} \in X_\infty$ and $f_{t_0 t_1}(\mathfrak{y}^{t_1}) \in V$, it follows by (1) that $\mathfrak{y}^{t_0} \in V$. Therefore $\mathfrak{y}^{t_0} \notin U$ since $U \cap V = 0$. Thus $\mathfrak{y} \notin \mathfrak{G}$.

THEOREM 2. *The family of sets* $f_t^{-1}(G_t) = \mathcal{X}_\infty \cap \underset{\mathfrak{z}}{E}(\mathfrak{z}^t \in G_t)$, *where* $t \in T$ *and* G_t *ranges over all open subsets of* $\mathcal{X}_t$, *is a base of* $\mathcal{X}_\infty$.

Proof. As the family of sets $\mathrm{pr}_t^{-1}(G_t)$ is a subbase of $\underset{t}{P}\mathcal{X}_t$ (see I (1) and II (2)), the sets $\mathcal{X}_\infty \cap \mathrm{pr}_t^{-1}(G_t)$ form a subbase of $\mathcal{X}_\infty$. In fact this subbase is a base of $\mathcal{X}_\infty$, what means that each finite intersection of sets of that kind is of that kind too. For, let $G_{t_1}, \ldots, G_{t_k}$ be a finite system of sets open in $\mathcal{X}_{t_1}, \ldots, \mathcal{X}_{t_k}$, and let $t \geqslant t_i$ for each $i \leqslant k$. Put

$$G_t = f_{t_1 t}^{-1}(G_{t_1}) \cap \ldots \cap f_{t_k t}^{-1}(G_{t_k}). \tag{5}$$

It follows (compare § 3, XIII (7)) that

$$f_t^{-1}(G_t) = f_t^{-1} f_{t_1 t}^{-1}(G_{t_1}) \cap \ldots \cap f_t^{-1} f_{t_k t}^{-1}(G_{t_k}) = f_{t_1}^{-1}(G_{t_1}) \cap \ldots \cap f_{t_k}^{-1}(G_{t_k}).$$

THEOREM 3. *For each* $\mathfrak{Z} \subset \mathcal{X}_\infty$, *we have*

$$\mathfrak{z} \in \overline{\mathfrak{Z}} \equiv \bigwedge_t (\mathfrak{z}^t \in \overline{\mathfrak{Z}^t}) \quad \textit{where} \quad \mathfrak{z} \in \mathcal{X}_\infty. \tag{6}$$

Equivalently stated $\overline{\mathfrak{Z}} = \bigcap_t \underset{\mathfrak{z}}{E}(\mathfrak{z}^t \in \overline{\mathfrak{Z}^t}) = \bigcap_t \mathrm{pr}_t^{-1}(\overline{\mathfrak{Z}^t})$, *where* $\mathfrak{Z}^t$ *denotes the sets of points* $\mathfrak{z}^t$ *such that* $\mathfrak{z} \in \mathfrak{Z}$.

Proof. The inclusion $\mathfrak{z} \in \overline{\mathfrak{Z}} \Rightarrow \mathfrak{z}^t \in \overline{\mathfrak{Z}^t}$ follows from the continuity of the projection.

Conversely, let $\mathfrak{z} \in \mathcal{X}_\infty - \overline{\mathfrak{Z}}$. Then, by Theorem 2, there is a $t \in T$ and an open G_t in $\mathcal{X}_t$ such that $\mathfrak{z}^t \in G_t$ and that $\mathfrak{Z} \cap \underset{\mathfrak{y}}{E}(\mathfrak{y}^t \in G_t) = 0$, which means that $\mathfrak{Z}^t \cap G_t = 0$, whence $\overline{\mathfrak{Z}^t} \cap G_t = 0$ (G_t being open). It follows that $\mathfrak{z}^t \notin \overline{\mathfrak{Z}^t}$.

THEOREM 4. *Let* $(T, \mathcal{X}, f)$ *and* $(T, \mathcal{Y}, g)$ *be two inverse system and let* h_t: $\mathcal{X}_t \to \mathcal{Y}_t$ *be continuous and* h_∞: $\mathcal{X}_\infty \to \mathcal{Y}_\infty$ *with the commutativity assumptions like in* § 3, XIII. *Then* h_∞ *is continuous.*

This follows at once from the formula (see § 3, XIII (11)):

$$h_\infty^t(\mathfrak{z}) = h_t(\mathfrak{z}^t),$$

since the projection and the mapping h_t are continuous.

COROLLARY 4a. *Under the same assumptions as in Theorem* 4, *suppose that* h_t *is a homeomorphism onto. Then* h_∞ *is a homeomorphism of* $\mathcal{X}_\infty$ *onto* $\mathcal{Y}_\infty$.

This follows from Theorem 4 and from the remark (12) made in § 3, XIII, stating that if each h_t is one-to-one and onto, so is h_∞.

§ 17. The space $2^{\mathscr{X}}$. Exponential topology

I. Definition. Let $\mathscr{X}$ be a topological space, $2^{\mathscr{X}}$ denotes the set of all closed subsets of $\mathscr{X}$:

$$(F \in 2^{\mathscr{X}}) \equiv (\bar{F} = F). \tag{1}$$

Now let $A \subset \mathscr{X}$. We agree to denote by 2^A (rel. to $\mathscr{X}$) the set of all $F = \bar{F} \subset A$. Thus, we have for $F \in 2^{\mathscr{X}}$:

$$(F \in 2^A) \equiv (F \subset A), \quad \text{i.e.} \quad 2^A = \mathop{E}_{F}(F \subset A). \tag{2}$$

Consequently for $F \in 2^{\mathscr{X}}$:

$$\mathop{E}_{F}(F \cap A \neq 0) = 2^{\mathscr{X}} - 2^{\mathscr{X}-A}. \tag{3}$$

The topology in $2^{\mathscr{X}}$ (called the *exponential topology*) is the coarsest topology in which the sets 2^A for A open are open (in $2^{\mathscr{X}}$) and for A closed are closed (in $2^{\mathscr{X}}$). In other words, we assume that the family of all sets 2^G and of all sets $2^{\mathscr{X}} - 2^{\mathscr{X}-G}$, where G is open in $\mathscr{X}$, is an open subbase of $2^{\mathscr{X}}$.

We shall use the following notation: $A_0, A_1, \ldots, A_n$ being any given finite system of arbitrary sets ($n \geqslant 0$), let

$$\boldsymbol{B}(A_0, A_1, \ldots, A_n) = 2^{A_0} \cap (2^{\mathscr{X}} - 2^{\mathscr{X}-A_1}) \cap \ldots \cap (2^{\mathscr{X}} - 2^{\mathscr{X}-A_n}), \tag{4}$$

i.e.

$$F \in \boldsymbol{B}(A_0, A_1, \ldots, A_n) \equiv (F \subset A_0)(F \cap A_1 \neq 0)\ldots(F \cap A_n \neq 0). \tag{4'}$$

Thus, the sets $\boldsymbol{B}(G_0, G_1, \ldots, G_n)$, where $G_0, \ldots, G_n$ are open, form an open base of $2^{\mathscr{X}}$.

We may suppose, of course, that $A_i \subset A_0$ for $i = 1, \ldots, n$, since $\boldsymbol{B}(A_0, A_1, \ldots, A_n) = \boldsymbol{B}(A_0, A_1 \cap A_0, \ldots, A_n \cap A_0)$.

The topology being thus defined, $2^{\mathscr{X}}$ *is a topological space*[1].

The void set is an isolated element of this space.

II. Fundamental properties. The following formulas hold in an arbitrary $\mathscr{T}_1$-space $\mathscr{X}$:

$$2^{A_0 \cap A_1} = 2^{A_0} \cap 2^{A_1} \quad \text{and generally} \quad 2^{\bigcap_t A_t} = \bigcap_t 2^{A_t} \tag{1}$$

---

[1] Compare E. Michael, *Topologies on spaces of subsets*, Trans. Amer. Math. Soc. 71 (1951), pp. 152–183, L. Vietoris, Monatsh. f. Math. u. Phys. 31 (1921), pp. 173–204, G. Choquet, *Convergences*, Ann. Univ. Grenoble 23 (1947), pp. 55–112, O. Frink, *Topology in lattices*, Trans. Amer. Math. Soc. 51 (1942), pp. 569–582.

$$(A \subset B) \equiv (2^A \subset 2^B), \quad \text{hence} \quad (A = B) \equiv (2^A = 2^B), \tag{2}$$

$$\overline{2^A} = 2^{\bar{A}}, \tag{3}$$

$$\operatorname{Int}(2^A) = 2^{\operatorname{Int}(A)}. \tag{4}$$

Formulas (1) and (2) are obvious. Moreover (2) implies

$$2^A \subset 2^{\bar{A}} \quad \text{and} \quad 2^{\operatorname{Int}(A)} \subset 2^A,$$

whence $\overline{2^A} \subset 2^{\bar{A}}$ and $2^{\operatorname{Int}(A)} \subset \operatorname{Int}(2^A)$ since $2^{\bar{A}}$ is closed and $2^{\operatorname{Int}(A)}$ is open (in $2^{\mathscr{X}}$).

It remains to show that

$$2^{\bar{A}} \subset \overline{2^A}, \tag{3'}$$

$$\operatorname{Int}(2^A) \subset 2^{\operatorname{Int}(A)}. \tag{4'}$$

Proof of (3′). Let $F_0 \in 2^{\bar{A}}$, i.e. $F_0 \subset \bar{A}$. Let $\boldsymbol{B}(G_0, \ldots, G_n)$ be an arbitrary member of the base containing F_0. Hence $F_0 \subset G_0$ and $F_0 \cap G_i \neq 0$ for $i = 1, \ldots, n$. As $F_0 \subset \bar{A}$, there is $p_i \in A \cap G_0 \cap G_i$. Put $F = (p_1, \ldots, p_n)$. Hence $F \subset A \cap G_0$ and $F \cap G_i \neq 0$ for $i = 1, \ldots, n$. Therefore $F \in 2^A \cap \boldsymbol{B}(G_0, \ldots, G_n)$. It follows that $F_0 \in \overline{2^A}$.

Proof of (4′). Let $F_0 \notin 2^{\operatorname{Int}(A)}$, i.e. $F_0 \not\subset \mathscr{X} - \overline{\mathscr{X} - A}$; hence $F_0 \cap \overline{\mathscr{X} - A} \neq 0$. Let $\boldsymbol{B}(G_0, \ldots, G_n)$ be an arbitrary member of the base containing F_0. Hence $F_0 \subset G_0$ and $F_0 \cap G_i \neq 0$. As $F_0 \cap \overline{\mathscr{X} - A} \neq 0$, there is $p \in G_0 - A$. Put $F = F_0 \cup (p)$. Consequently $F \not\subset A$. Moreover $F \subset G_0$ and $F \cap G_i \neq 0$ for $i = 1, \ldots, n$, i.e. $F \in \boldsymbol{B}(G_0, \ldots, G_n)$. Thus each neighbourhood of F_0 (in $2^{\mathscr{X}}$) contains an F such that $F \not\subset A$, i.e. such that $F \in 2^{\mathscr{X}} - 2^A$. It follows that $F_0 \in \overline{2^{\mathscr{X}} - 2^A}$, whence $F_0 \notin \operatorname{Int}(2^A)$.

Formulas (3) and (4) are particular cases of the following formula (which will be used later)(¹): *If* $A_i \subset A_0$ *for* $i = 1, \ldots, n$, *then*

$$\overline{\boldsymbol{B}(A_0, A_1, \ldots, A_n)} = \boldsymbol{B}(\bar{A}_0, \bar{A}_1, \ldots, \bar{A}_n). \tag{5}$$

Proof. By I (4) and II (3)–(4), we have

$$\overline{\boldsymbol{B}(A_0, A_1, \ldots, A_n)} \subset \overline{2^{A_0}} \cap \overline{2^{\mathscr{X}} - 2^{\mathscr{X} - A_1}} \cap \ldots \cap \overline{2^{\mathscr{X}} - 2^{\mathscr{X} - A_n}}$$

$$\subset 2^{\bar{A}_0} \cap (2^{\mathscr{X}} - 2^{\mathscr{X} - \bar{A}_1}) \cap \ldots \cap (2^{\mathscr{X}} - 2^{\mathscr{X} - \bar{A}_n}) = \boldsymbol{B}(\bar{A}_0, \bar{A}_1, \ldots, \bar{A}_n).$$

(¹) See Michael, *loc. cit.*, p. 156.

In order to prove the converse inclusion, let

$$F \in \boldsymbol{B}(\bar{A}_0, \bar{A}_1, \ldots, \bar{A}_n) \quad \text{and} \quad F \in \boldsymbol{G}, \tag{6}$$

where $\boldsymbol{G}$ is open. We have to show that

$$\boldsymbol{G} \cap \boldsymbol{B}(A_0, A_1, \ldots, A_n) \neq 0. \tag{7}$$

We may assume, of course, that $\boldsymbol{G}$ belongs to the base of $2^{\mathscr{X}}$, i.e. that $\boldsymbol{G} = \boldsymbol{B}(G_0, G_1, \ldots, G_m)$ where G_j is open (for $j \geqslant 0$) and, in addition, $G_j \subset G_0$.

Thus we have, by (6) and I (4′),

$$F \subset \bar{A}_0, F \cap \bar{A}_i \neq 0 \text{ for } i \geqslant 1, \quad F \subset G_0, F \cap G_j \neq 0 \text{ for } j \geqslant 1. \tag{8}$$

Consequently

$$G_0 \cap \bar{A}_i \neq 0 \neq G_j \cap \bar{A}_0 \quad \text{hence} \quad G_0 \cap A_i \neq 0 \neq G_j \cap A_0.$$

Put $p_i \in G_0 \cap A_i$ and $q_j \in G_j \cap A_0$. Let $K = (p_1, \ldots, p_n, q_1, \ldots, q_m)$. One can easily verify that K belongs to the left member of (7). This completes the proof.

THEOREM 1. *The sets* $\underset{F}{E}(F \subset A)$ *and* $\underset{F}{E}(F \cap A \neq 0)$ *are closed (open) in* $2^{\mathscr{X}}$ *if and only if* A *is closed (open) in* $\mathscr{X}$.

The sufficiency of this condition follows directly from the definition of the exponential topology. Its necessity is an easy consequence of (3) and (4).

THEOREM 2. *Let* $\mathscr{X}$ *be a* $\mathscr{T}_1$*-space and* $A \subset \mathscr{X}$ *(arbitrary). Then the set* $\underset{F}{E}(A \subset F)$ *is closed.*

Proof. Obviously $(A \not\subset F) \equiv \bigvee_{x \in A} \big(F \subset \mathscr{X} - (x)\big)$. Therefore

$$\underset{F}{E}(A \not\subset F) = \bigcup_{x \in A} \underset{F}{E}\big(F \subset \mathscr{X} - (x)\big).$$

The space being $\mathscr{T}_1$, $\mathscr{X} - (x)$ is open and hence $\underset{F}{E}\big(F \subset \mathscr{X} - (x)\big)$ is open, and so is $\underset{F}{E}(A \not\subset F)$.

THEOREM 3. *If* $\mathscr{X}$ *is* $\mathscr{T}_1$, $2^{\mathscr{X}}$ *is* $\mathscr{T}_1$.

Proof. Let $K \in 2^{\mathscr{X}}$. We have to show that $\underset{F}{E}(F = K)$ is closed. Now

$$(F = K) \equiv [(F \subset K) \text{ and } (K \subset F)].$$

Therefore

$$\underset{F}{E}(F = K) = \underset{F}{E}(F \subset K) \cap \underset{F}{E}(K \subset F).$$

The last two sets being closed (by Theorems 1 and 2), so is their intersection.

THEOREM 4. *The family of all finite subsets of a $\mathscr{T}_1$-space $\mathscr{X}$ is dense in $2^{\mathscr{X}}$.*

Proof. Let $\boldsymbol{B}$ be a member of the base of $2^{\mathscr{X}}$ (compare I (4)). Then there is a system $G_0, \ldots, G_n$ of open sets in $\mathscr{X}$ such that $\boldsymbol{B}$ is the family of all sets F satisfying I (4)). We have to show that among them there is a finite set.

If $G_0 = 0$ (and consequently $n = 0$), the void set satisfies I (4). Suppose $G_0 \neq 0$. Let F be any set satisfying I (4) and let

$$p_0 \in F \quad \text{and} \quad p_i \in F \cap G_i \quad \text{for} \quad i = 1, 2, \ldots, n.$$

The set $F^* = (p_0, p_1, \ldots, p_n)$ is the required set.

THEOREM 5. *Let $f: \mathscr{X} \to \mathscr{Y}$ be continuous onto. Let $\boldsymbol{D}$ be the family of all sets $\overset{-1}{f}(y)$ (i.e. $\boldsymbol{D}$ is the decomposition of $\mathscr{X}$ induced by f) and let $\overset{1}{f}$ denote the inverse mapping to $\overset{-1}{f}$ (i.e. $\big(\overset{1}{f}(D)\big) = f(D)$ for $D \in \boldsymbol{D}$, (see* § 3, II (7)). *Then $\overset{1}{f}$ is a continuous mapping: $\boldsymbol{D} \to \mathscr{Y}$ ($\boldsymbol{D}$ having the exponential topology and $\mathscr{Y}$ being a $\mathscr{T}_1$-space).*

Proof. Let $G \subset \mathscr{Y}$ be open. We have to show that $\overset{-1}{f}(G)$ is open in $\boldsymbol{D}$, i.e. that it is an intersection of $\boldsymbol{D}$ with an open subset of $2^{\mathscr{X}}$. Now, we have (cf. § 3, III (17))

$$\overset{-1}{f}(G) = \boldsymbol{D} \cap \underset{F}{E}[F \subset f^{-1}(G)] \quad \text{where} \quad F \in 2^{\mathscr{X}}.$$

f being continuous, $f^{-1}(G)$ is open and so is $\underset{F}{E}[F \subset f^{-1}(G)]$.

III. Continuous set-valued functions. Let $\mathscr{X}$ and $\mathscr{Y}$ be two topological spaces. Let $F_t: \mathscr{Y} \to 2^{\mathscr{X}}$. We have the following formulas whose proofs are completely similar to the proofs of analogous formulas of set theory (see § 3, V):

$$(F_0 \subset F_1) \Rightarrow [F_1^{-1}(2^A) \subset F_0^{-1}(2^A)], \tag{i}$$

$$(F_0 \cup F_1)^{-1}(2^A) = F_0^{-1}(2^A) \cap F_1^{-1}(2^A). \tag{ii}$$

If A is closed we have the more general formula:

$$\overline{\big(\bigcup_t F_t\big)}^{-1}(2^A) = \bigcap_t F_t^{-1}(2^A). \tag{iii}$$

Namely

$$\big(\overline{\bigcup_t F_t(y)} \subset A\big) \equiv \bigwedge_t \big(F_t(y) \subset A\big),$$

and hence

$$y \in \overline{\big(\bigcup_t F_t\big)}^{-1}(2^A) \equiv \big(\overline{\bigcup_t F_t(y)} \subset A\big) \equiv \bigwedge_t \big(F_t(y) \subset A\big) \equiv y \in \bigcap_t F_t^{-1}(A).$$

THEOREM 1. *F is continuous if and only if the set*

$$F^{-1}(2^A) = \underset{y}{E}[F(y) \in 2^A] = \underset{y}{E}[F(y) \subset A] \tag{1}$$

is open in $\mathscr{Y}$ whenever A is open in $\mathscr{X}$, and is closed in $\mathscr{Y}$ whenever A is closed in $\mathscr{X}$.

Equivalently, if for each closed (resp. open) $A \subset \mathscr{X}$ the set

$$\mathscr{Y} - F^{-1}(2^{\mathscr{X}-A}) = \underset{y}{E}[F(y) \cap A \neq 0] \tag{2}$$

is closed (resp. open) in $\mathscr{Y}$.

More precisely, F is continuous at y_0 if and only if both implications hold true:

$$y_0 \in F^{-1}(2^G) \Rightarrow y_0 \in \mathrm{Int}\big(F^{-1}(2^G)\big) \textit{ whenever } G \textit{ is open in } \mathscr{X}, \tag{3}$$

and

$$y_0 \in \overline{F^{-1}(2^K)} \Rightarrow y_0 \in F^{-1}(2^K) \textit{ whenever } K \textit{ is closed in } \mathscr{X}. \tag{4}$$

Proof. By § 13, III (3), if F is continuous at y_0, each of the following implications is satisfied:

$$y_0 \in F^{-1}(G) \Rightarrow y_0 \in \mathrm{Int}\, F^{-1}(G) \text{ whenever } G \text{ is open in } 2^{\mathscr{X}}, \tag{5}$$

and

$$y_0 \in \overline{F^{-1}(F)} \Rightarrow y_0 \in f^{-1}(F) \text{ whenever } F \text{ is closed in } 2^{\mathscr{X}}. \tag{6}$$

Then, replacing G by 2^G for G open, and F by 2^K for K closed, we obtain (3) and (4).

Conversely, if the implication (5) holds true, F is continuous at y_0 (by § 13, III (3)). Moreover, the range of variability of G can be restricted to a subbase of $2^{\mathscr{X}}$ (see § 13, III), so that we may assume that either $G = 2^A$ or $G = 2^{\mathscr{X}} - 2^{\mathscr{X}-A}$ with A open. In the first case (5) follows directly from (3); in the second case, it can be easily deduced from (4).

COROLLARY 1a. *Let $D(F)$ denote the set of points of discontinuity of F (compare § 13, III (3)). Then*

$$D(F) = \bigcup_G \{F^{-1}(2^G) - \mathrm{Int}[F^{-1}(2^G)]\} \cup \bigcup_K \{\overline{F^{-1}(2^K)} - F^{-1}(2^K)\} \quad (7)$$

where G ranges over the open subsets of $\mathscr{X}$ and K over its closed subsets.

THEOREM 2. *Let $f: \mathscr{X} \to \mathscr{Y}$ be continuous. Then the inverse mapping $f^{-1}: 2^{\mathscr{Y}} \to 2^{\mathscr{X}}$ is continuous if and only if f is simultaneously closed and open.*

In particular (compare also Corollary 3a),

$$\{\overset{-1}{f}: \mathscr{Y} \to 2^{\mathscr{X}} \text{ is continuous}\} \equiv \{f \text{ is closed-open}\}.$$

If the mapping $f: \mathscr{X} \to \mathscr{Y}$ is closed, then the mapping $f^1: 2^{\mathscr{X}} \to 2^{\mathscr{Y}}$ is continuous.

Proof. The first part of the theorem follows directly from Theorem 1 (1), Theorem 1, II, and from § 3, III (13‴) (compare also § 3, II (3)).

Now, let us assume that f is closed. Let $B = \bar{B} \subset \mathscr{Y}$. We have to show that $\underset{A}{E}[f^1(A) \subset B]$ is closed and $\underset{A}{E}[f^1(A) \cap B = 0]$ is open, whenever $A \in 2^{\mathscr{X}}$. Now, by § 3, III (13′) and (13″), we have

$$\underset{A}{E}[f^1(A) \subset B] = \underset{A}{E}[A \subset f^{-1}(B)],$$

$$\underset{A}{E}[f^1(A) \cap B = 0] = \underset{A}{E}[A \cap f^{-1}(B) = 0]$$

and as $f^{-1}(B) \in 2^{\mathscr{X}}$, the set $\underset{A}{E}[A \subset f^{-1}(B)]$ is closed and the set $\underset{A}{E}[A \cap f^{-1}(B) = 0]$ is open according to the definition of topology in $2^{\mathscr{X}}$ (A denotes a variable member of $2^{\mathscr{X}}$).

Remark 1. Theorem 2 can be stated in the following more general form:

Let f be continuous. Then f^{-1} is continuous at B if and only if f is simultaneously closed and open at B.

For the proof, see § 18, III, Theorem 5a.

THEOREM 3. *Let* $f\colon \mathscr{Y} \to \mathscr{X}$ *be continuous. Put* $F(y) = (f(y))$. *Then the mapping* $F\colon \mathscr{Y} \to 2^{\mathscr{X}}$ *is continuous* ($\mathscr{X}$ *is supposed to be* $\mathscr{T}_1$).

For $\underset{y}{E}[F(y) \subset A] = \underset{y}{E}[f(y) \in A] = f^{-1}(A)$.

COROLLARY 3a. *Let* $F(x) = (x)$. *Then the mapping* $F\colon \mathscr{X} \to 2^{\mathscr{X}}$ *is a homeomorphism transforming* $\mathscr{X}$ *onto the set* $\mathbf{S}$ *of all single-element sets* ($\mathscr{X}$ *is supposed to be* $\mathscr{T}_1$).

F is continuous by Theorem 3. The inverse mapping $(x) \to x$ is continuous for the family of all sets (x) such that $x \in G$ (i.e. such that $(x) \subset G$) is open in $\mathbf{S}$ (provided G is open in $\mathscr{X}$).

THEOREM 4. *The union of two continuous functions* $F = F_0 \cup F_1$ *is continuous.*

This follows directly from Theorem 1 and formula (ii).

Remark 1. More precisely, *the union of two continuous functions at* y_0 *is continuous at* y_0. This follows from an analogous statement on semi-continuous functions (established in § 18).

COROLLARY 4a. *The union* $K \cup L$, *considered as a mapping of* $2^{\mathscr{X}} \times 2^{\mathscr{X}}$ *onto* $2^{\mathscr{X}}$, *is continuous.*

THEOREM 5. *Let* E *be closed-open in* $\mathscr{X}$. *Put* $F(K) \equiv K \cap E$. *Then* F *is a continuous mapping of* $2^{\mathscr{X}}$ *into* $2^{\mathscr{X}}$.

Proof. According to (1), we have to show that the set $\underset{K}{E}(K \cap E \subset A)$ is open whenever A is open, and is closed whenever A is closed. Now $(K \cap E \subset A) \equiv (K \subset A \cup (-E))$ and the conclusion follows from Theorem 1 of Section II, since $A \cup (-E)$ is open if A is open and is closed if A is closed.

COROLLARY 5a. *Let* $\mathscr{X} = A_0 \cup A_1$ *where* $A_0 \cap A_1 = 0$ *and* A_0 *and* A_1 *are closed. Then*

$$2^{A_0 \cup A_1} \underset{\text{top}}{=} 2^{A_0} \times 2^{A_1}.$$

Proof. Put $F(K, L) = K \cup L$ where $K \in 2^{A_0}$ and $L \in 2^{A_1}$. F is the required homeomorphism. For, F is continuous by Theorem 4, is one-to-one (since $A_0 \cap A_1 = 0$), is obviously onto and finally the converse mapping is continuous; this means that $X \cap A_j$ is continuous (for $X \in 2^{A_0 \cup A_1}$), what is true by Theorem 5.

Remark 2. The assumption in Theorem 5 for the set E to be open is essential. This is seen on the example where $\mathscr{X} = \mathscr{I}$ and where E is a single-element set. See also § 18, V.

IV. Case of $\mathscr{X}$ regular.

THEOREM 1. *If $\mathscr{X}$ is a regular $\mathscr{T}_1$-space, the sets* $\underset{KL}{E}(K \subset L)$ *and* $\underset{x,K}{E}(x \in K)$ *are closed in* $2^{\mathscr{X}} \times 2^{\mathscr{X}}$, *respectively in* $\mathscr{X} \times 2^{\mathscr{X}}$.

Proof. Let $\mathscr{X}$ be regular. Suppose $K \not\subset L$. As $\mathscr{X}$ is regular, there is an open set G such that

$$K \cap G \neq 0 \quad \text{and} \quad L \subset \mathscr{X} - \bar{G}.$$

Thus,

$$(K \not\subset L) \equiv \bigvee_G (K \cap G \neq 0)(L \subset \mathscr{X} - \bar{G}) \quad (G \text{ open})$$

and consequently (compare § 2, VI (3))

$$\underset{KL}{E}(K \not\subset L) = \bigcup_G [\underset{K}{E}(K \cap G \neq 0) \times \underset{L}{E}(L \subset \mathscr{X} - \bar{G})].$$

The sets $\underset{K}{E}(K \cap G \neq 0)$ and $\underset{L}{E}(L \subset \mathscr{X} - \bar{G})$ being open (according to Theorem 1 of Section II), so is the set in brackets [ ] (relatively to $2^{\mathscr{X}} \times 2^{\mathscr{X}}$). Hence $\underset{KL}{E}(K \not\subset L)$ is open.

This complete the proof of the first part of the theorem. To show the second one, we replace in the preceding argument K by (x).

The converse theorem is also true for $\mathscr{T}_1$-spaces.

THEOREM 2. *Let $\mathscr{X}$ be a $\mathscr{T}_1$-space. If the set* $\underset{x,K}{E}(x \in K)$ *is closed, then the space $\mathscr{X}$ is regular.*

Proof. Let $x_0 \notin K_0$. We have to define U and V open and such that:

$$x_0 \in U, \quad K_0 \subset V, \quad \text{and} \quad U \cap V = 0. \tag{1}$$

Put $\boldsymbol{G} = \underset{x,K}{E}(x \notin K)$. Hence $(x_0, K_0) \in \boldsymbol{G}$. The set $\boldsymbol{G}$ being open by assumption, there exist open sets $U, V, G_1, \ldots, G_n$ such that

$$(x_0, K_0) \in [U \times \boldsymbol{B}(V, G_1, \ldots, G_n)] \subset \boldsymbol{G}.$$

In other terms:

$$x_0 \in U,\ K_0 \in \boldsymbol{B}(V, G_1, \ldots, G_n), \quad \text{i.e.} \quad K_0 \subset V,\ K_0 \cap G_i \neq 0 \text{ for } i = 1, \ldots, n, \tag{2}$$

and

$$(x \in U,\ K \subset V,\ K \cap G_i \neq 0 \text{ for } i = 1, \ldots, n) \Rightarrow (x, K) \in \boldsymbol{G}, \text{ i.e. } x \notin K. \tag{3}$$

It remains to show that $U \cap V = 0$. Suppose on the contrary that $x \in U \cap V$. Let $K = K_0 \cup (x)$. It follows by (2) and (3) that $x \notin K$, which is a contradiction.

THEOREM 3. *If $\mathscr{X}$ is regular, $2^{\mathscr{X}}$ is Hausdorff.*

Proof. Let $K, L \in 2^{\mathscr{X}}$ and $K \neq L$. We have to define two open subsets $\boldsymbol{G}$ and $\boldsymbol{H}$ of $2^{\mathscr{X}}$ such that $K \in \boldsymbol{G}$, $L \in \boldsymbol{H}$, and $\boldsymbol{G} \cap \boldsymbol{H} = 0$.

Clearly, we may assume that $K - L \neq 0$. Let $p \in K - L$. The space $\mathscr{X}$ being regular, there exists an open $U \subset \mathscr{X}$ such that $p \in U$ (hence $K \cap U \neq 0$) and $L \cap \overline{U} = 0$.

One sees easily that the sets

$$\boldsymbol{G} = \underset{F}{E}(F \cap U \neq 0) \quad \text{and} \quad \boldsymbol{H} = \underset{F}{E}(F \cap \overline{U} = 0)$$

are the required sets.

THEOREM 4. *If $\mathscr{X}$ is $\mathscr{T}_1$ and $2^{\mathscr{X}}$ is Hausdorff, then $\mathscr{X}$ is regular*(1).

Proof. Let $x_0 \notin K_0$ where K_0 is closed. We have to define two open sets U and V satisfying conditions (1).

Obviously $K_0 \neq K_0 \cup (x_0)$. As the space $2^{\mathscr{X}}$ is Hausdorff, we have

$$K_0 \in \boldsymbol{B}(G_0, G_1, \ldots, G_m), \quad \big(K_0 \cup (x_0)\big) \in \boldsymbol{B}(H_0, H_1, \ldots, H_n), \tag{4}$$

$$\boldsymbol{B}(G_0, G_1, \ldots, G_m) \cap \boldsymbol{B}(H_0, H_1, \ldots, H_n) = 0. \tag{5}$$

We shall show the existence of an j_0 such that

$$x_0 \in H_{j_0} \quad \text{and} \quad G_0 \cap H_0 \cap H_{j_0} = 0.$$

This will complete the proof since $x_0 \in H_0 \cap H_{j_0}$ and $K_0 \subset G_0$.

Suppose that $x_0 \in H_j \Rightarrow G_0 \cap H_0 \cap H_j \neq 0$ for each $j = 0, 1, \ldots, n$. Put $p_j \in (G_0 \cap H_0 \cap H_j)$ whenever $x_0 \in H_j$, and denote by L the set composed of those p_j. Put $A = K \cup L$. It is easy to verify that

$$A \subset G_0, \quad A \cap G_i \neq 0 \quad (i = 1, \ldots, m), \quad A \subset H_0,$$
$$A \cap H_j \neq 0 \quad (j = 1, \ldots, n).$$

But this means that $A \in \boldsymbol{B}(G_0, \ldots, G_m) \cap \boldsymbol{B}(H_0, \ldots, H_n)$ contrary to (5).

---

(1) For Theorems 3 and 4, compare E. Michael, *loc. cit.*, and O. Frink, *loc. cit.*, p. 577.

COROLLARY 5. *Let $\mathscr{X}$ be a $\mathscr{T}_1$-space. The following conditions are equivalent*:

(i) *$\mathscr{X}$ is regular.*

(ii) *$\underset{KL}{E}(K \subset L)$ is closed in $2^{\mathscr{X}} \times 2^{\mathscr{X}}$,*

(iii) *$\underset{x,K}{E}(x \in K)$ is closed in $\mathscr{X} \times 2^{\mathscr{X}}$,*

(iv) *$2^{\mathscr{X}}$ is Hausdorff.*

V. Case of $\mathscr{X}$ normal.

THEOREM 1. *If $\mathscr{X}$ is normal, the set $\underset{KL}{E}(K \cap L = 0)$ is open in $2^{\mathscr{X}} \times 2^{\mathscr{X}}$; more generally: $\underset{K_1 \dots K_n}{E}(K_1 \cap \dots \cap K_n = 0)$ is open in $(2^{\mathscr{X}})^n$.*

Proof. Suppose that $\mathscr{X}$ is normal. Then

$$(K \cap L = 0) \equiv \bigvee_{GH} (K \subset G)(L \subset H)(G \cap H = 0) \tag{1}$$

where G and H are open. In other words (compare § 2, VI (3)):

$$\underset{KL}{E}(K \cap L = 0) = \bigcup_{G \cap H = 0} \underset{KL}{E}(K \subset G)(L \subset H) = \bigcup_{G \cap H = 0} (2^G \times 2^H).$$

This completes the proof since 2^G and 2^H are open.

In order to prove the second part, we apply instead of (1) the more general formula (§ 14, III):

$$(K_1 \cap \dots \cap K_n = 0) \equiv \bigvee_{G_1 \dots G_n} [(G_1 \cap \dots \cap G_n = 0) \\ (K_i \subset G_i \text{ for } i = 1, \dots, n)] \tag{2}$$

We shall establish now the converse theorem:

THEOREM 2. *If $\mathscr{X}$ is $\mathscr{T}_1$ and $\underset{KL}{E}(K \cap L = 0)$ is open, then $\mathscr{X}$ is normal.*

Proof. Let $K_0 \cap L_0 = 0$. The set $\underset{KL}{E}(K \cap L = 0)$ being supposed to be open, there are in $2^{\mathscr{X}}$ two open sets $\boldsymbol{U}$ and $\boldsymbol{V}$

$$K_0 \in \boldsymbol{U},\ L_0 \in \boldsymbol{V} \quad \text{and} \quad (K \in \boldsymbol{U})(L \in \boldsymbol{V}) \Rightarrow (K \cap L = 0). \tag{3}$$

One can obviously assume that $\boldsymbol{U}$ and $\boldsymbol{V}$ belong to the base of $2^{\mathscr{X}}$ defined in Section I. This means (see I (4)) that there exist two systems of open sets $G_0, \dots, G_m$ and $H_0, \dots, H_n$ such that

$$K_0 \subset G_0,\ K_0 \cap G_i \neq 0\ (i \leqslant m), \quad L_0 \subset H_0,\ L_0 \cap H_j \neq 0\ \ (j \leqslant n) \tag{4}$$

and

$$[(K \subset G_0)(K \cap G_i \neq 0)(L \subset H_0)(L \cap H_j \neq 0)] \Rightarrow (K \cap L = 0). \quad (5)$$

We shall show that $G_0 \cap H_0 = 0$, which will complete the proof.

Suppose, on the contrary, that $p \in G_0 \cap H_0$. Let $a_i \in K_0 \cap G_i$ for $i \leqslant m$ and $b_j \in L_0 \cap H_j$ for $j \leqslant n$. Put

$$K = (p, a_1, \ldots, a_m) \quad \text{and} \quad L = (p, b_1, \ldots, b_n).$$

It follows by (5) that $K \cap L = 0$; but this is a contradiction since $p \in (K \cap L)$.

THEOREM 3. *If $\mathscr{X}$ is normal, $2^{\mathscr{X}}$ is regular.*

Proof. Let $F_0 = \overline{F}_0 \in \boldsymbol{G}$ where $\boldsymbol{G}$ is open (in $2^{\mathscr{X}}$). We have to define two open sets $\boldsymbol{U}$ and $\boldsymbol{V}$ such that

$$F_0 \in \boldsymbol{U}, \quad \boldsymbol{U} \cap \boldsymbol{V} = 0, \quad \text{and} \quad 2^{\mathscr{X}} - \boldsymbol{G} \subset \boldsymbol{V}. \quad (6)$$

We may obviously suppose that $\boldsymbol{G}$ belongs to the base of $2^{\mathscr{X}}$, i.e. that there is a system of open sets $G_0, G_1, \ldots, G_n$ such that $\boldsymbol{G} = \boldsymbol{B}(G_0, G_1, \ldots, G_n)$. That means that

$$F_0 \subset G_0, \quad (7)$$

$$F_0 \cap G_i \neq 0 \quad \text{for} \quad i = 1, \ldots, n. \quad (8)$$

As $\mathscr{X}$ is normal, there are (by (7)) two open sets C_0 and D_0 such that

$$F_0 \subset C_0, \quad C_0 \cap D_0 = 0, \quad \mathscr{X} - G_0 \subset D_0. \quad (9)$$

Put, according to (8), $q_i \in F_0 \cap G_i$ and let C_i and D_i be open sets such that

$$q_i \in C_i, \quad C_i \cap D_i = 0, \quad \mathscr{X} - G_i \subset D_i. \quad (10)$$

We define $\boldsymbol{U}$ and $\boldsymbol{V}$ as follows:

$$(F \in \boldsymbol{U}) \equiv \{F \subset C_0 \text{ and } F \cap C_i \neq 0 \text{ for } i = 1, \ldots, n\},$$

$$(F \in \boldsymbol{V}) \equiv \{\text{either } F \cap D_0 \neq 0 \text{ or } F \subset D_i \text{ for some } i \leqslant n\}.$$

By the first parts of formulas (9) and (10), $F_0 \in \boldsymbol{U}$.

The second parts of (9) and (10) imply $\boldsymbol{U} \cap \boldsymbol{V} = 0$.

Now let $F \in 2^{\mathscr{X}} - \boldsymbol{G}$, i.e. either $F \not\subset G_0$ or $F \cap G_i = 0$ for some $i \leqslant n$. Hence, according to the third parts of (9) and (10) we have either $F \cap D_0 \neq 0$ or $F \subset D_i$ for some $i \leqslant n$. This means that $F \in \boldsymbol{V}$. Thus, (6) is fulfilled.

THEOREM 4. *If $\mathscr{X}$ is $\mathscr{T}_1$ and $2^{\mathscr{X}}$ is regular, then $\mathscr{X}$ is normal.*

Proof. Let $K_0 \cap L_0 = 0$. We have to define an open set G such that

$$K_0 \subset G \text{ and } \bar{G} \cap L_0 = 0, \text{ i.e. } \bar{G} \subset H \text{ where } H = \mathscr{X} - L_0. \quad (11)$$

As $K_0 \subset H$, we have $K_0 \in 2^H$. The set 2^H being open in the (regular) space $2^{\mathscr{X}}$, there are open sets $G_0, \ldots, G_n$ such that

$$K_0 \in \boldsymbol{B}(G_0, \ldots, G_n), \quad \overline{\boldsymbol{B}(G_0, \ldots, G_n)} \subset 2^H, \quad \text{and} \quad G_i \subset G_0. \quad (12)$$

Hence by I (4′) and II (5), we have

$$K_0 \subset G_0, \quad K_0 \cap G_i \neq 0, \quad \text{and} \quad \boldsymbol{B}(\bar{G}_0, \ldots, \bar{G}_n) \subset 2^H.$$

Since $0 \neq G_i \subset G_0$, it follows that $\bar{G}_0 \cap \bar{G}_i = \bar{G}_i \neq 0$, and consequently $\bar{G}_0 \in \boldsymbol{B}(\bar{G}_0, \ldots, \bar{G}_n)$. By (12) $\bar{G}_0 \in 2^H$, i.e. $\bar{G}_0 \subset H$. Thus G_0 can be substituted to G in (11).

COROLLARY 5. *Let $\mathscr{X}$ be a $\mathscr{T}_1$-space. The following conditions are equivalent:*

(i) *$\mathscr{X}$ is normal,*

(ii) *$\underset{KL}{E}(K \cap L = 0)$ is open,*

(iii) *$2^{\mathscr{X}}$ is regular.*

Remark. *For the space $2^{\mathscr{X}}$, regularity is equivalent to complete regularity* (1).

VI. Relations of $2^{\mathscr{X}}$ to lattices and to Brouwerian algebras. It is worthy noticing that many theorems stated in this paragraph (and in § 18) can be extended to distributive lattices or to Brouwerian algebras.

Let us recall that a distributive lattice $\Gamma = (L, \cup, \cap, 0, 1)$ is called a *Brouwerian algebra* (2) if there is in Γ an operation $a - b$ such that

$$(a - b \subset c) \equiv [a \subset (b \cup c)].$$

If $\mathscr{X}$ is a topological space, $2^{\mathscr{X}}$ is a Brouwerian algebra with $\cup$ and $\cap$ having the usual meaning and "$-$" denoting the operation $\overline{A - B}$.

(1) For the Theorems 3, 4 and the above remark, compare the mentioned paper of E. Michael.

(2) See J. C. C. McKinsey and A. Tarski, *On closed elements in closure algebras*, Ann. of Math. 47 (1946), pp. 122–162.

We call[1] Γ a *Wallman lattice*, respectively structurally *regular*, respectively structurally *normal*, if the following conditions are respectively satisfied:

$$(a \not\subset b) \Rightarrow \bigvee_{c} (a \cap c \neq 0)(b \cap c = 0),$$

$$(a \not\subset b) \Rightarrow \bigvee_{cd} (c \cup d = 1)(a \not\subset c)(b \cap d = 0),$$

$$(a \cap b = 0) \Rightarrow \bigvee_{cd} (c \cup d = 1)(a \cap c = 0 = b \cap d).$$

Obviously, if $\mathscr{X}$ is a $\mathscr{T}_1$-space, $2^{\mathscr{X}}$ is a Wallman lattice. $\mathscr{X}$ is regular if and only if $2^{\mathscr{X}}$ is structurally regular (provided $\mathscr{X}$ is $\mathscr{T}_1$). $\mathscr{X}$ is normal if and only if $2^{\mathscr{X}}$ is structurally normal.

Topology in L can be introduced as follows[2]. Consider the ideals

$$I(a) = \mathop{E}_{x}(x \subset a) \quad \text{and} \quad J(a) = \mathop{E}_{x}(x \cap a = 0).$$

We assume that the family of all sets $I(a)$ and of all sets $L - J(a)$ is a closed subbase of L. This topology agrees with the exponential topology of $\mathscr{X}$ when $L = 2^{\mathscr{X}}$.

Let us mention the following theorems (which generalize the corresponding theorems on $2^{\mathscr{X}}$)[3].

Let Γ be a Wallman lattice. Then:

1. *The filter* $\mathop{E}_{x}(a \subset x)$ is closed. Hence L is a $\mathscr{T}_1$-space.

2. $\{\Gamma$ is structurally regular $\} \equiv \{\mathop{E}_{xy}(x \subset y)$ is closed in $L \times L\}$.

3. $\{\Gamma$ is structurally normal$\} \equiv \{\mathop{E}_{xy}(x \cap y = 0)$ is open in $L \times L\}$.

4. If Γ is Brouwerian, then

$$\{\Gamma \text{ is structurally regular}\} \equiv \{\mathop{E}_{xy}(x - y = 0) \text{ is closed}\}.$$

---

(1) See my paper *Mappings of topological spaces into lattices and into Brouwerian algebras*, Bull. Polish Acad. Sc. 12 (1964), pp. 9–16. Cf. H. Wallman, *Lattices and topological spaces*, Annals of Math. 39 (1938), p. 115. Compare also G. Nöbeling, *Grundlagen der analytischen Topologie*, p. 79.

(2) Cf. my mentioned paper and O. Frink, Jr., *Topology in lattices, loc. cit.* p. 576. Compare also B. C. Rennie, *The theory of lattices*, Cambridge 1951.

(3) Cf. my above cited paper and my paper *Characterization of regular lattices with the aid of exponential topology* (Russian), Dokl. Akad. Nauk SSSR, 155 (1964), p. 751.

§ 18. Semi-continuous mappings

I. Definitions. As in § 17, III, let $\mathscr{X}$ and $\mathscr{Y}$ be two topological spaces and F a (set-valued) function assigning to each $y \in \mathscr{Y}$ a set $F(y) \in 2^{\mathscr{X}}$.

F is called *upper* (resp. *lower*) *semi-continuous* if, for each open (resp. closed) $A \subset \mathscr{X}$, the set

$$F^{-1}(2^A) = \mathop{E}_{y}[F(y) \in 2^A] = \mathop{E}_{y}[F(y) \subset A] \tag{1}$$

is open (resp. closed) in $\mathscr{Y}$.

Equivalently, F is u.s.c. (resp. l.s.c.) if, for each closed (resp. open) $A \subset \mathscr{X}$, the set

$$\mathscr{Y} - F^{-1}(2^{\mathscr{X}-A}) = \mathop{E}_{y}[F(y) \cap A \neq 0] \tag{2}$$

is closed (resp. open) in $\mathscr{Y}$.

More precisely, *F is u.s.c. at y_0 if*

$$y_0 \in F^{-1}(2^G) \Rightarrow y_0 \in \mathrm{Int}\big(F^{-1}(2^G)\big) \quad \textit{whenever } G \textit{ is open}. \tag{3}$$

F is l.s.c. at y_0 if

$$y_0 \in \overline{F^{-1}(2^K)} \Rightarrow y_0 \in F^{-1}(2^K) \quad \textit{whenever } K \textit{ is closed}. \tag{4}$$

The next theorem follows easily from these definitions.

THEOREM 1. *F is u.s.c.* (*respectively l.s.c.*) *if and only if F is u.s.c.* (*respectively l.s.c.*) *at each $y \in \mathscr{Y}$.*

By Theorem 1 of § 17, III we have the following theorem.

THEOREM 2. *F is continuous at y_0 if and only if F is both u.s.c. and l.s.c. at y_0.*

Consequently F is continuous if and only if F is both u.s.c. and l.s.c.

THEOREM 3. *Let $D(F)$ denote the set of points of discontinuity of F. Then if F is u.s.c., we have*

$$D(F) = \bigcup_K \{\overline{F^{-1}(2^K)} - F^{-1}(2^K)\} \quad \textit{where } K \textit{ is closed}. \tag{5}$$

If F is l.s.c., then

$$D(F) = \bigcup_G \{F^{-1}(2^G) - \mathrm{Int}[F^{-1}(2^G)]\} \quad \textit{where } G \textit{ is open}. \tag{6}$$

This follows from Corollary 1a of § 17, III.

THEOREM 4. *Let* $f\colon \mathscr{X} \to \mathscr{Y}$ *be continuous and onto. Then the mapping* $f^{-1}\colon \mathscr{Y} \to 2^{\mathscr{X}}$ *is u.s.c. (resp. l.s.c.) if and only if f is a closed (resp. open) mapping*(1).

Like Theorem 2 of § 17, III, this theorem follows immediately from (2) and formula (3) of § 3, II.

THEOREM 5. *If* $F(y)$ *reduces to a single point (denoted by* $f(y)$*), and F is u.s.c. or l.s.c., then f is continuous.*

For

$$\underset{y}{E}[F(y) \subset A] = \underset{y}{E}[f(y) \in A] = f^{-1}(A).$$

II. Examples. Relation to real-valued semi-continuous functions. Remarks.

EXAMPLE 1. Let $f\colon \mathscr{E} \to \mathscr{I}$. By the classical definition (2), f is called upper, resp. lower, semi-continuous at y_0 if $\lim_{n=\infty} y_n = y_0$ implies

$$\limsup_{n=\infty} f(y_n) \leqslant f(y) \quad \text{resp.} \quad f(y) \leqslant \liminf_{n=\infty} f(y_n).$$

In other words, if the conditions $\lim_{n=\infty} y_n = y_0$ and $\lim_{n=\infty} f(y_n) = x_0$ imply $x_0 \leqslant f(y_0)$, resp. $f(y_0) \leqslant x_0$.

Put $F(y) = \underset{x}{E}[0 \leqslant x \leqslant f(y)]$. It is easily seen that F is u.s.c. (l.s.c.) at y_0 if and only if f is such in the sense above defined.

EXAMPLE 2. The function F defined by the condition $F(y) = \mathscr{E} \times (y)$, where $y \in \mathscr{E}$, is not u.s.c. For, denoting by A the hyperbola $y = 1/x$, the set considered in I (2) is not closed.

---

(1) For a stronger statement, see Section III, Theorem 5.

(2) See R. Baire, Ann. di mat. (3) 3 (1899), p. 6. For a more general approach, where $\mathscr{X}$ and $\mathscr{Y}$ are supposed metric, see H. Hahn, *Reelle Funktionen* (1932), p. 148, and for the case where $F(y)$ is supposed to be a closed subset of a compact metric space, see my paper *Les fonctions semi-continues dans l'espace des ensembles fermés*, Fund. Math. 18 (1931), pp. 148–159, and *Topologie*, Vol. II, Chapter IV. The technique consists of using the topological upper and lower limits of sets (see § 29). See also G. Bouligand, Ens. Math. 1932, p. 14. This technique can be generalized by using the Moore-Smith convergence; see the papers quoted at § 20, IX, and also W. L. Strother, *Continuous multi-valued functions*, Bol. Soc. Mat. Sao Paulo 10 (1958), pp. 87–120. Some authors assume that the sets $F(y)$ are compact; see C. Berge, *Espaces topologiques*, p. 110, and Mémorial Sc. Math. 138.

Remark. It is possible to introduce in $2^{\mathscr{X}}$ a topology, called *ϰ-topology* (1), such that the u.s.c. mappings become continuous. This ϰ-topology of $2^{\mathscr{X}}$ is defined by assuming the sets 2^G to form an open base (G open). Obviously, the mapping $F: \mathscr{Y} \to 2^{\mathscr{X}}$ is upper semi-continuous at y_0 relative to the exponential topology if and only if it is continuous at y_0 relative to the ϰ-topology of $2^{\mathscr{X}}$.

Similarly, let us call "*λ-topology*" of $2^{\mathscr{X}}$ its topology obtained by assuming the sets 2^K (K closed) to form a closed subbase. Then the lower semi-continuous mappings are identical with continuous mappings relative to λ-topology.

It should be noted that the ϰ-topology (and λ-topology) is never $\mathscr{T}_1$ (except in the trivial case of $\mathscr{X} = 0$).

III. Fundamental properties.

THEOREM 1. *Suppose that $\mathscr{X}$ is a regular space and F an u.s.c. mapping. Then the set $\underset{xy}{E}[x \in F(y)]$ is closed in $\mathscr{X} \times \mathscr{Y}$.*

Proof. $\mathscr{X}$ being regular, we have the equivalence

$$[x \notin F(y)] \equiv \bigvee_{U \cap V = 0} (x \in U)(F(y) \subset V) \tag{1}$$

where U and V are open. Therefore

$$\underset{xy}{E}[x \notin F(y)] = \bigcup_{U \cap V = 0} [U \times F^{-1}(2^V)]. \tag{2}$$

As F is u.s.c., $F^{-1}(2^V)$ is open in $\mathscr{Y}$, hence $U \times F^{-1}(2^V)$, and consequently $\underset{xy}{E}[x \notin F(y)]$, is open in $\mathscr{X} \times \mathscr{Y}$.

Remark. Without the assumption of regularity of $\mathscr{X}$ the theorem is not true. For, if $\mathscr{Y} = 2^{\mathscr{X}}$ and F is the identity, the set under consideration equals $\underset{xK}{E}(x \in K)$, and the latter set is not closed unless $\mathscr{X}$ is regular (see § 17, IV, Theorem 2).

---

(1) See W. J. Ponomarev, *New space of closed sets and set-valued continuous mappings of bicompacta* (Russian), Matem. Sbornik 48 (1959), pp. 191–212, and a much earlier paper by A. Tychonov, Dokl. Akad. Nauk USSR 3 (1936), p. 49, where the definition of this space appears. See also P. S. Alexandrov, *Some results in topological spaces obtained in the last twenty five years* (Russian), Uspiechi 15 (1960), pp. 25–95.

The ϰ-topology (and the λ-topology considered next) was studied by E. Michael under the name of upper (resp. lower) *semi-finite* topology; see *op. cit.* p. 179.

THEOREM 2. (Generalized Heine condition.) *F is l.s.c. at y_0 if and only if*

$$y_0 \in \bar{B} \Rightarrow F(y_0) \subset \overline{\mathrm{S}F(B)} \quad \textit{whenever} \quad B \subset \mathscr{Y}. \tag{3}$$

Proof. Suppose that F is l.s.c. at y_0. Let $y_0 \in \bar{B}$ and put $K = \overline{\mathrm{S}F(B)}$. We have $B \subset F^{-1}(2^K)$, since

$$y \in B \Rightarrow F(y) \subset \mathrm{S}F(B) \subset K \Rightarrow F(y) \in 2^K \equiv y \in F^{-1}(2^K).$$

Therefore $y_0 \in \overline{F^{-1}(2^K)}$ and by I(4) $y_0 \in F^{-1}(2^K)$, which means that $F(y_0) \in 2^K$, i.e. $F(y_0) \subset K = \overline{\mathrm{S}F(B)}$.

Conversely, let $K = \bar{K}$ and $y_0 \in \overline{F^{-1}(2^K)}$. Put $B = F^{-1}(2^K)$. We have

$$F(B) \subset 2^K, \quad \text{hence} \quad \mathrm{S}F(B) \subset \mathrm{S}2^K = K.$$

Therefore $\overline{\mathrm{S}F(B)} \subset \bar{K} = K$. Now, assuming that (3) is satisfied, we have $F(y_0) \subset \overline{\mathrm{S}F(B)} \subset K$, whence $F(y_0) \in 2^K$ and $y_0 \in F^{-1}(2^K)$. Thus F is l.s.c. at y_0.

COROLLARY 2a. *F is l.s.c. if and only if*

$$\mathrm{S}F(\bar{B}) \subset \overline{\mathrm{S}F(B)} \quad \textit{for each} \quad B \subset \mathscr{Y}. \tag{4}$$

THEOREM 3. (Generalized Cauchy condition.) *F is u.s.c. at y_0 if and only if for each open G containing $F(y_0)$ there is some open H containing y_0 such that $\mathrm{S}F(H) \subset G$, i.e. that $y \in H \Rightarrow F(y) \subset G$*[1].

The proof follows directly from the formulas:

$(F(y_0) \subset G) \equiv y_0 \in F^{-1}(2^G)$ and $(\mathrm{S}F(H) \subset G) \equiv (H \subset F^{-1}(2^G))$.

COROLLARY 3a. *If F is u.s.c. at y_0, then ($\mathscr{X}$ being supposed a $\mathscr{T}_1$-space)*

$$y_0 \in \bar{B} \Rightarrow \mathrm{P}(F(B)) \subset F(y_0), \quad \textit{i.e.} \quad \Big(\bigcap_{y \in B} F(y)\Big) \subset F(y_0). \tag{5}$$

Proof. Put $A = \bigcap_{y \in B} F(y)$ and suppose that $x \in A - F(y_0)$. Put in Theorem 3, $G = \mathscr{X} - (x)$. Then as $y_0 \in \bar{B}$ and H is open, there is $y \in B \cap H$. Hence $A \subset F(y) \subset G$ and therefore $A \subset \mathscr{X} - (x)$, which is a contradiction.

The following theorem is a generalization of Theorem 2 of § 17, II.

---

(1) Compare W. Hurewicz, *Über stetige Bilder von Punktmengen*, Proc. Acad. Amsterdam 29 (1926), pp. 1014–1017.

THEOREM 4. *Let* $\mathscr{X}$ *be a* $\mathscr{T}_1$*-space,* $A \subset \mathscr{X}$ *arbitrary and* F *an u.s.c. function. Then* $\underset{y}{E}\big(A \subset F(y)\big)$ *is closed.*

Proof. Obviously $\big(A \not\subset F(y)\big) \equiv \bigvee_{x \in A} \big(F(y) \subset \mathscr{X}-(x)\big)$. Therefore

$$\underset{y}{E}\big(A \not\subset F(y)\big) = \bigcup_{x \in A} \underset{y}{E}\big(F(y) \subset \mathscr{X}-(x)\big).$$

The set $\mathscr{X}-(x)$ being open and F being u.s.c., $\underset{y}{E}\big(F(y) \subset \mathscr{X}-(x)\big)$ is open, and so is $\underset{y}{E}\big(A \not\subset F(y)\big)$.

THEOREM 5. *Let* $f: \mathscr{X} \to \mathscr{Y}$ *be continuous and* $\mathscr{Y}$ *a* $\mathscr{T}_1$*-space. Then* $\overset{-1}{f}: \mathscr{Y} \to 2^{\mathscr{X}}$ *is l.s.c. (resp. u.s.c.) at* y_0 *if and only if* f *is open (resp. closed) at* y_0.

Proof. Put $F(y) = \overset{-1}{f}(y)$. By § 3, III (16), $SF(B) = f^{-1}(B)$ whenever $B \subset \mathscr{Y}$.

Consider first the case of lower semi-continuity.

Replacing in formula (3) of Theorem 2, $F(y)$ by $f^{-1}(y)$ and $SF(B)$ by $f^{-1}(B)$, we obtain formula (5) of § 13, XIV, which represents a necessary and sufficient condition for f to be open at y_0.

In the case of upper semi-continuity we deduce from Theorem 3, by a similar procedure, the condition stated in Theorem 3 of § 13, XIV, which is necessary and sufficient for f to be closed at y_0.

THEOREM 5a. (Generalization of Theorem 5.) *Let* $f: \mathscr{X} \to \mathscr{Y}$ *be continuous and let* f^{-1} *be restricted to closed subsets of* $\mathscr{Y}$*, i.e.* $f^{-1}: 2^{\mathscr{Y}} \to 2^{\mathscr{X}}$*. Then* f^{-1} *is l.s.c. (resp. u.s.c.) at the closed set* B_0 *if and only if* f *is open (resp. closed) at* B_0(1).

Proof. Put $F = f^{-1}$, $K = \overline{K} \subset \mathscr{X}$, and $G = \mathscr{X}-K$. Note that formula III (13‴) of § 3 is valid when 2^A denotes $\underset{K}{E}\, K \subset A$, i.e.

$$F^{-1}(2^A) = 2^{\mathscr{Y}-f(\mathscr{X}-A)}. \tag{6}$$

According to I (4), (6), § 17, II (3) and § 13, XIV (12), we have:

$$\begin{aligned}(f^{-1} \text{ is l.s.c. at } B) &\equiv \{B \in \overline{F^{-1}(2^K)} \Rightarrow B \in F^{-1}(2^K)\} \\ &\equiv \{B \in \overline{2^{\mathscr{Y}-f(G)}} \Rightarrow B \in 2^{\mathscr{Y}-f(G)}\} \\ &\equiv \{B \in 2^{\overline{\mathscr{Y}-f(G)}} \Rightarrow B \in 2^{\mathscr{Y}-f(G)}\} \\ &\equiv \big\{[B \cap \operatorname{Int}[f(G)] = 0] \Rightarrow [B \cap f(G) = 0]\big\} \\ &\equiv (f \text{ is open at } B).\end{aligned}$$

(1) This theorem answers a question raised by Morton Brown.

Similarly, according to I (3), (6), § 17, II (4), and § 13, XIV (13),

$$\begin{aligned}(f^{-1} \text{ is u.s.c. at } B) &\equiv \{B \in F^{-1}(2^G) \Rightarrow B \in \operatorname{Int}[F^{-1}(2^G)]\} \\ &\equiv \{B \in 2^{\mathscr{Y}-f(K)} \Rightarrow B \in \operatorname{Int}[2^{\mathscr{Y}-f(K)}]\} \\ &\equiv \{B \in 2^{\mathscr{Y}-f(K)} \Rightarrow B \in 2^{\operatorname{Int}[\mathscr{Y}-f(K)]}\} \\ &\equiv \{[B \cap f(K) = 0] \Rightarrow [B \cap \overline{f(K)} = 0]\} \\ &\equiv (f \text{ is closed at } B).\end{aligned}$$

THEOREM 6. *Let $F(y)$ be, for each $y \in \mathscr{Y}$, a one-element set: $F(y) = \{f(y)\}$, where $f: \mathscr{Y} \to \mathscr{X}$ ($\mathscr{X}$ being supposed a $\mathscr{T}_1$-space). Then the semi-continuity of F at y_0 (either upper or lower) implies its continuity at y_0.*

For, in this case, $\mathrm{S}F(B) = \{f(B)\}$, and the conditions stated in Theorems 2 and 3 become the usual Heine and Cauchy conditions of continuity.

The following theorem can easily be shown:

THEOREM 7. *Let $f: \mathscr{X} \to \mathscr{Y}$ and $F: \mathscr{Y} \to 2^{\mathscr{X}}$. Let $f(z_0) = y_0$. If f is continuous at z_0 and F is u.s.c. (resp. l.s.c.) at y_0, then the composed function $H = F \circ f$ is u.s.c. (resp. l.s.c.) at z_0.*

This theorem can also be deduced directly from the remark made in No. II, namely that u.s.c. (l.s.c.) mapping means continuous relative to the $\varkappa$-topology (λ-topology).

IV. Union of semi-continuous mappings.

THEOREM 1. *The union of two u.s.c. functions at y_0 is u.s.c. at y_0.*

Proof. Let $F_j: \mathscr{Y} \to 2^{\mathscr{X}}$ be u.s.c. at y_0 for $j = 0, 1$. Put $F = F_0 \cup \cup F_1$. Let G be open in $\mathscr{X}$ and $y_0 \in F^{-1}(2^G)$. We have to show (see I (3)) that $y_0 \in \operatorname{Int}[F^{-1}(2^G)]$.

By formula (ii) of § 17, III,

$$F^{-1}(2^G) = F_0^{-1}(2^G) \cap F_1^{-1}(2^G); \tag{1}$$

hence $y_0 \in F_j^{-1}(2^G)$. As F_j is u.s.c. at y_0, it follows that $y_0 \in \operatorname{Int}[F_j^{-1}(2^G)]$. Therefore (by § 6, II (1), and (1))

$$\begin{aligned}y_0 \in \operatorname{Int}[F_0^{-1}(2^G)] \cap \operatorname{Int}[F_1^{-1}(2^G)] &= \operatorname{Int}[F_0^{-1}(2^G) \cap F_1^{-1}(2^G)] \\ &= \operatorname{Int}[F^{-1}(2^G)].\end{aligned}$$

THEOREM 2. *The union of two l.s.c. functions at y_0 is l.s.c. at y_0.*

More generally, if each F_t for $t \in T$ (T arbitrary) is l.s.c. at y_0, so is $F = \bigcup_t F_t$.

Proof. Let $K = \overline{K} \subset \mathscr{X}$ and $y_0 \in \overline{F^{-1}(2^K)}$. We have to show (see I (4)) that $y_0 \in F^{-1}(2^K)$.

By formula (iii) of § 17, III,

$$F^{-1}(2^K) = \bigcap_t F_t^{-1}(2^K), \tag{2}$$

and therefore

$$\overline{F^{-1}(2^K)} = \overline{\bigcap_t F_t^{-1}(2^K)} \subset \bigcap_t \overline{F_t^{-1}(2^K)}.$$

Hence, for each $t \in T$, we have $y_0 \in \overline{F_t^{-1}(2^K)}$. As F_t is l.s.c. at y_0, it follows by I (4) that $y_0 \in F_t^{-1}(2^K)$. Thus $y_0 \in \bigcap_t F_t^{-1}(2^K) = F^{-1}(2^K)$.

Remark 1. The second part of Theorem 2 is not true of u.s.c. functions. Such is the case of $F_n(y) = (y^n)$, $0 \leqslant y \leqslant 1$.

Remark 2. Theorems 1 and 2 give the following generalization of Theorem 4 of § 17, III: *The union of two continuous functions at y_0 is continuous at y_0.*

V. Intersection of semi-continuous mappings.

Remark. In connection with formula (3) of § 3, V, let us note that for $F = F_0 \cap F_1$ we have the formula

$$[F_0^{-1}(2^G) \cup F_1^{-1}(2^G)] \subset F^{-1}(2^G) \tag{1}$$

where the sign $\subset$ cannot be replaced by $=$.

However, the following lemma is true.

LEMMA. *Let $\mathscr{X}$ be normal, $F_j\colon \mathscr{Y} \to 2^{\mathscr{X}}$ for $j = 0, 1$, $F = F_0 \cap F_1$, and G open in $\mathscr{X}$. Then*

$$F^{-1}(2^G) = \bigcup_{U_0, U_1} [F_0^{-1}(2^{U_0}) \cap F_1^{-1}(2^{U_1})] \tag{2}$$

where U_0 and U_1 are open and $U_0 \cap U_1 = G$.

Proof. 1. Let $y \in F^{-1}(2^G)$, i.e. $[F_0(y) \cap F_1(y)] \subset G$. Hence the sets $[F_0(y) - G]$ and $[F_1(y) - G]$ are closed and disjoint. $\mathscr{X}$ being normal, there are V_0 and V_1 open and such that

$$F_0(y) - G \subset V_0, \quad F_1(y) - G \subset V_1, \quad \text{and} \quad V_0 \cap V_1 = 0.$$

Put $U_j = V_j \cup G$. Hence U_j is open, $U_0 \cap U_1 = G$, and $F_j(y) \subset U_j$, i.e. $y \in F_j^{-1}(2^{U_j})$.

Thus y belongs to the right-hand side of (2).

2. Conversely, suppose that y belongs to the right-hand side of (2). Hence $F_j(y) \subset U_j$ and consequently $F(y) \subset U_0 \cap U_1$. As $U_0 \cap U_1 = G$, it follows that $y \in F^{-1}(2^G)$.

THEOREM 1. *If $\mathscr{X}$ is normal, the intersection $F = F_0 \cap F_1$ of two u.s.c. functions at y_0 is u.s.c. at y_0.*

Proof. Let G be open in $\mathscr{X}$ and $y_0 \in F^{-1}(2^G)$. We have to show that $y_0 \in \operatorname{Int}[F^{-1}(2^G)]$.

By virtue of (2) there are two open sets U_0 and U_1 such that $y_0 \in F_0^{-1}(2^{U_0}) \cap F_1^{-1}(2^{U_1})$ and $U_0 \cap U_1 = G$. As F_j is u.s.c. at y_0, it follows that

$$y_0 \in \{\operatorname{Int}[F_0^{-1}(2^{U_0})] \cap \operatorname{Int}[F_1^{-1}(2^{U_1})]\}.$$

The latter set being an open subset of $F^{-1}(2^G)$ (according to (2)), y_0 is an interior point of $F^{-1}(2^G)$.

COROLLARY 1a. *If $\mathscr{X}$ is normal, the intersection $K \cap L$ is an u.s.c. mapping of $2^{\mathscr{X}} \times 2^{\mathscr{X}}$ into $2^{\mathscr{X}}$.*

COROLLARY 1b. *Let $\mathscr{X}$ be a $\mathscr{T}_1$-space. Then the following conditions are equivalent* (see also § 17, V, Corollary 5):

(i) *$\mathscr{X}$ is normal,*

(ii) *$K \cap L$ is an u.s.c. mapping of $2^{\mathscr{X}} \times 2^{\mathscr{X}}$ into $2^{\mathscr{X}}$.*

Remark 1. Corollary 1a is also an immediate consequence of the Theorem 1 of § 17, V, according to which the set $\underset{KL}{E}(K \cap L \cap A = 0)$ is open whatever be the closed set A.

Remark 2. As mentioned in Remark 2 to § 17, III, the intersection $K \cap L$ *does not need to be continuous.*

However, one has the following two statements.

THEOREM 2. *Let $F: \mathscr{Y} \to 2^{\mathscr{X}}$ be u.s.c. at y_0 and let $K_0 \subset \mathscr{X}$ be a fixed closed set. Then the mapping $L = K_0 \cap F$ is u.s.c. at y_0 (whatever is the topological space $\mathscr{X}$).*

Proof. Let G be open in $\mathscr{X}$ and $L(y_0) \subset G$. Then we have $F(y_0) \subset [G \cup (\mathscr{X} - K_0)]$. As F is u.s.c. at y_0, there is an open $H \subset \mathscr{Y}$ containing y_0 such that $F(y) \subset [G \cup (\mathscr{X} - K_0)]$ for each $y \in H$. It follows that $[K_0 \cap F(y)] \subset G$, i.e. $L(y) \subset G$ for $y \in H$. By Theorem 3 of Section III, L is u.s.c. at y_0.

THEOREM 3. *Let $F: \mathscr{Y} \to 2^{\mathscr{X}}$ be l.s.c. at y_0 and let $K_0 \subset \mathscr{X}$ be a fixed closed-open set. Then the mapping $L = K_0 \cap F$ is l.s.c. at y_0.*

Proof. Let $B \subset \mathscr{Y}$ and $y_0 \in \bar{B}$. We have to show (compare Theorem 2 of Section III) that $L(y_0) \subset \overline{SL(B)}$, i.e. that $[K_0 \cap F(y_0)] \subset \overline{K_0 \cap SF(B)}$. Now, F being l.s.c. at y_0, we have $F(y_0) \subset \overline{SF(B)}$. Hence

$$[K_0 \cap F(y_0)] \subset [K_0 \cap \overline{SF(B)}] \subset \overline{K_0 \cap SF(B)},$$

since K_0 is open.

Remark(1). For $\mathscr{X}$ *compact*, Theorem 1 can be extended to intersections $\bigcap_{t \in T} F_t$ where T is an arbitrary index-set.

The assumption of compactness of $\mathscr{X}$ is essential, and the following statement, analogous to Corollary 1b, holds:

In order that a topological $\mathscr{T}_1$-space $\mathscr{X}$ be compact it is necessary and sufficient that, for each T, the intersection $\bigcap_{t \in T} K_t$ be an u.s.c. mapping of the product-space $\underset{t \in T}{P} 2^{\mathscr{X}_t}$, where $\mathscr{X}_t = \mathscr{X}$, into $2^{\mathscr{X}}$.

VI. Difference of semi-continuous mappings.

LEMMA. *Let $\mathscr{X}$ be regular, $F_j \colon \mathscr{Y} \to 2^{\mathscr{X}}$, $F = \overline{F_0 - F_1}$, and K closed in $\mathscr{X}$. Then*

$$F^{-1}(2^K) = \bigcap_G \{\mathscr{Y} - [F_1^{-1}(2^G) - F_0^{-1}(2^{\bar{G}})]\} \tag{1}$$

where G is open and $K \subset G$.

Equivalently stated:

$$\mathscr{Y} - F^{-1}(2^K) = \bigcup_G [F_1^{-1}(2^G) - F_0^{-1}(2^{\bar{G}})]. \tag{2}$$

Proof. 1. Let $y \in \mathscr{Y} - F^{-1}(2^K)$, i.e. $\overline{F_0(y) - F_1(y)} \not\subset K$. As K is closed, it follows that $F_0(y) - F_1(y) \not\subset K$. Hence there is an $x \in F_0(y)$ such that $x \notin [K \cup F_1(y)]$. As $\mathscr{X}$ is regular, there is an open G such that

$$[K \cup F_1(y)] \subset G \quad \text{and} \quad x \notin \bar{G}; \quad \text{hence} \quad F_0(y) - \bar{G} \neq 0.$$

---

(1) See R. Engelking, *Quelques remarques concernant les opérations sur les fonctions semi-continues dans les espaces topologiques*, Bull. Acad. Polon. Sci. 11 (1963), pp. 719–726.

Thus

$$K \subset G \quad \text{and} \quad y \in [F_1^{-1}(2^G) - F_0^{-1}(2^{\bar{G}})]. \tag{3}$$

2. Conversely, if (3) is true, then $F_1(y) \subset G$ and $F_0(y) \not\subset \bar{G}$. Hence $F_0(y) - F_1(y) \not\subset \bar{G}$ and therefore $F(y) \not\subset \bar{G}$. Consequently $F(y) \not\subset K$, i.e. $y \in \mathscr{Y} - F^{-1}(2^K)$.

THEOREM 4. *Let $\mathscr{X}$ be regular. If F_0 is l.s.c. at y_0 and F_1 is u.s.c. at y_0, the mapping $F = \overline{F_0 - F_1}$ is l.s.c. at y_0.*

Proof. Let $K = \bar{K}$ and $y_0 \in \overline{F^{-1}(2^K)}$. We have to show that $y_0 \in F^{-1}(2^K)$. According to (1) we have

$$y_0 \in \overline{\bigcap_{G \supset K} [\mathscr{Y} - F_1^{-1}(2^G)] \cup F_0^{-1}(2^{\bar{G}})} \subset \bigcap_{G \supset K} \{\overline{\mathscr{Y} - F_1^{-1}(2^G)} \cup \overline{F_0^{-1}(2^{\bar{G}})}\}.$$

Hence, for each $G \supset K$, we have

$$y_0 \notin \operatorname{Int} F_1^{-1}(2^G) \quad \text{and} \quad y_0 \in \overline{F_0^{-1}(2^{\bar{G}})},$$

which implies by virtue of the semi-continuity assumptions that

$$y_0 \notin F_1^{-1}(2^G) \quad \text{and} \quad y_0 \in F_0^{-1}(2^{\bar{G}}),$$

i.e.

$$y_0 \in \bigcap_{G \supset K} [\mathscr{Y} - F_1^{-1}(2^G)] \cup F_0^{-1}(2^{\bar{G}}).$$

This completes the proof according to (1).

COROLLARY. *Let $\mathscr{X}$ be a $\mathscr{T}_1$-space. Then the mapping $\overline{K - L}$ of $2^{\mathscr{X}} \times 2^{\mathscr{X}}$ into $2^{\mathscr{X}}$ is l.s.c. if and only if $\mathscr{X}$ is regular.*

The sufficiency of the regularity condition follows easily from the preceding theorem. Its necessity follows from the Corollary 5 of § 17, IV. Namely, if the space $\mathscr{X}$ is not regular, then the set

$$\underset{KL}{E}(\overline{K - L} = 0) = \underset{KL}{E}(K - L = 0)$$

is not closed; consequently the mapping $\overline{K - L}$ is not l.s.c.

Remark. *The mapping $\overline{K - L}$ does not need to be continuous* (even in the case $K = \mathscr{X}$).

For example, if $\mathscr{X}$ denotes the (closed) interval $\mathscr{I}$ and L_n is the interval $[1/n, 1]$, then $\overline{\mathscr{X} - L_n}$ is the interval $[0, 1/n]$ and $\operatorname{Lim} L_n = \mathscr{X}$ (compare § 29, VI), hence $\overline{\mathscr{X} - \operatorname{Lim} L_n}$ is void, while $\operatorname{Lim} \overline{\mathscr{X} - L_n}$ is not (it is composed of the point 0).

§ 19. Decomposition space. Quotient topology

I. Definition. Let $\boldsymbol{D}$ be a family of closed, non-void, and disjoint subsets of $\mathscr{X}$ whose union is the whole space $\mathscr{X}$:

$$\boldsymbol{D} \subset 2^{\mathscr{X}} \quad \text{and} \quad \mathscr{X} = \mathrm{S}\boldsymbol{D}. \tag{1}$$

$\boldsymbol{D}$ is called a *decomposition* of $\mathscr{X}$.

The topology in $\boldsymbol{D}$ is defined by means of the following assumption:

$$\boldsymbol{A} \subset \boldsymbol{D} \textit{ is open (in } \boldsymbol{D}\textit{) if and only if } \mathrm{S}\boldsymbol{A} \textit{ is open (in } \mathscr{X}\textit{)}. \tag{2}$$

Equivalently,

$$\boldsymbol{A} \subset \boldsymbol{D} \textit{ is closed if and only if } \mathrm{S}\boldsymbol{A} \textit{ is closed}. \tag{3}$$

$\mathscr{X}$ being assumed a topological space, so is $\boldsymbol{D}$ (a $\mathscr{T}_1$-space).

Consider the equivalence relation ϱ induced by $\boldsymbol{D}$, i.e.

$$x\varrho y \equiv [x \text{ and } y \text{ belong to the same member of } \boldsymbol{D}].$$

Then $\boldsymbol{D} = \mathscr{X}/\varrho$. Owing to this identity, the above defined topology of $\boldsymbol{D}$ is called *quotient topology* (1).

II. Projection. Relationship to bicontinuous mappings. Given a decomposition $\boldsymbol{D}$ of the space $\mathscr{X}$, let us denote by $P(x)$ the member of $\boldsymbol{D}$ containing x:

$$[D = P(x)] \equiv (x \in D \in \boldsymbol{D}). \tag{1}$$

The mapping $P\colon \mathscr{X} \to \boldsymbol{D}$ is called *projection*.

Obviously, we have for $D \in \boldsymbol{D}$ and $A \subset \mathscr{X}$:

$$(D \cap A \neq 0) \equiv [D \in P(A)] \equiv [D \subset \mathrm{S}P(A)], \tag{2}$$

$$\overset{-1}{P}(D) = D, \tag{3}$$

$$P^{-1}(\boldsymbol{A}) = \mathrm{S}\boldsymbol{A} \quad \text{for} \quad \boldsymbol{A} \subset \boldsymbol{D}. \tag{4}$$

It follows, by I (2) and I (3):

THEOREM 1. *$P^{-1}(\boldsymbol{A})$ is open (resp. closed) if and only if $\boldsymbol{A}$ is open (resp. closed).*

---

(1) Cf. J. L. Kelley, *General Topology*, p. 94, N. Bourbaki, *Théorie des ensembles*, Act. Scient., fasc. 1141, § 5, and *Topologie Générale*, fasc. 1142 (1961), § 3, N° 4.

In other words (cf. § 13, XV), *the mapping P is bicontinuous.* The next theorem is its converse.

THEOREM 2. *Given a bicontinuous mapping $f\colon \mathscr{X} \to \mathscr{Y}$, denote by $\boldsymbol{D}$ the (induced) decomposition of $\mathscr{X}$ into the sets $\overset{-1}{f}(y)$ where $y \in \mathscr{Y}$. Then*

$$\mathscr{Y} \underset{\text{top}}{=} \boldsymbol{D},$$

$\overset{-1}{f}$ being the required homeomorphism.

Proof. The diagram

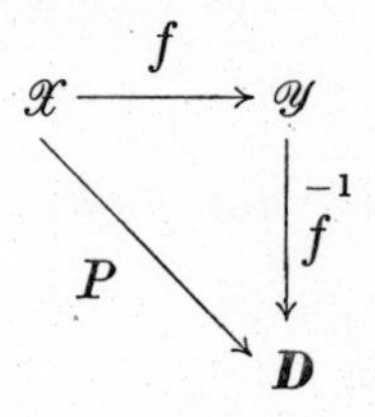

is obviously commutative, i.e. $P(x) = \overset{-1}{f}[f(x)]$. P and f being bicontinuous, so is $\overset{-1}{f}$ (by Theorem 3 of § 13, XV). This implies, by Theorem 2 of § 13, XV, that $\overset{-1}{f}$ is a homeomorphism (since $\overset{-1}{f}$ is obviously one-to-one).

Remark. The preceding thorem can be formulated as follows. *The quotient topology induced by a bicontinuous mapping $f\colon \mathscr{X} \to \mathscr{Y}$ is determined uniquely up to a homeomorphism. Namely*

$$(\mathscr{X}/\varrho) \underset{\text{top}}{=} \mathscr{Y} \quad \textit{where} \quad (x_1 \varrho x_2) \equiv \big(f(x_1) = f(x_2)\big).$$

Thus the study of quotient topology can be reduced to the study of bicontinuous mappings.

In particular, the notion of semi-continuous decomposition (which is used frequently[1]) can be introduced as follows.

Let $f\colon \mathscr{X} \to \mathscr{Y}$ be bicontinuous onto. Denote by $\boldsymbol{D}$ the induced decomposition of $\mathscr{X}$ into sets $f^{-1}(y)$ where $y \in \mathscr{Y}$. According to

---

[1] See: R. L. Moore, *Concerning upper semi-continuous collections of continua...*, Proc. Nat. Acad. Sc. 10 (1924), p. 350; P. Alexandrov, *Über stetige Abbildungen kompakter Räume*, Proc. Akad. Amsterdam 28 (1925), p. 997, and Math. Ann. 96 (1926), p. 555; my paper *Sur les décompositions semi-continues d'espaces métriques compacts*, Fund. Math. 11 (1928), p. 169; G. T. Whyburn, *Analytic Topology*, Chap. VII.

Theorem 2, $\boldsymbol{D}$ may be identified with $\mathscr{Y}$ and f with P (projection of $\mathscr{X}$ in $\boldsymbol{D}$).

DEFINITION. The decomposition $\boldsymbol{D}$ of $\mathscr{X}$ is called *lower* (*upper*) *semi-continuous* if and only if the projection $P\colon \mathscr{X} \to \boldsymbol{D}$ is an open (closed) mapping.

More generally, a member D of $\boldsymbol{D}$ is called *a member of lower* (*upper*) *semi-continuity of* $\boldsymbol{D}$ if and only if the projection P is open (closed) at D.

Of course, *continuity* of $\boldsymbol{D}$ means simultaneous upper and lower semi-continuity.

Since, for $A \subset \mathscr{X}$,

$$P(A) = \mathop{E}_{D \in \boldsymbol{D}} (D \cap A \neq 0) \quad \text{and} \quad P^{-1}P(A) = \mathbf{S}PA,$$

it follows from the definition of open (closed) mappings (see § 13, XIII):

THEOREM 3. $\{\boldsymbol{D}$ *is lower (upper) semi-continuous*$\} \equiv \{$*for each open (closed)* $A \subset \mathscr{X}$, *the union of all* $D \in \boldsymbol{D}$ *which intersect* A *is open (closed)*$\} \equiv \{$*for each closed (open)* $B \subset \mathscr{X}$ *the union of all* $D \in \boldsymbol{D}$ *contained in* B *is closed (open)*$\}$.

THEOREM 4. $\boldsymbol{D}$ *is u.s.c. at* D *if and only if for each open* $G \supset D$ *there is an open* U *such that* $D \subset U \subset G$ *and* U *is the union of some members of* $\boldsymbol{D}$ (1).

THEOREM 5. *If* $\boldsymbol{D}$ *is an u.s.c. decomposition of a normal space* $\mathscr{X}$, *the quotient topology of* $\boldsymbol{D}$ *is normal.*

Consequently, the relation ϱ *induced by* $\boldsymbol{D}$ *is closed.*

This follows from Theorem 2 of § 14, II and Corollary 3a of § 15, IV.

THEOREM 6. *If* $\boldsymbol{D}$ *is l.s.c. and the induced relation* ϱ *is closed, the quotient-topology of* $\boldsymbol{D}$ *is Hausdorff.*

This follows from Theorem 4 of § 15, IV.

THEOREM 7. *The decomposition* $\boldsymbol{D}$ *of a product* $\mathscr{X} \times \mathscr{Y}$ *into sets* $\{x\} \times \mathscr{Y}$, *where* $x \in \mathscr{X}$, *is l.s.c. and* $\boldsymbol{D} \underset{\text{top}}{=} \mathscr{X}$.

Because the mapping $f\colon \mathscr{X} \times \mathscr{Y} \to \mathscr{X}$ given by $f(x, y) = x$ is open (see § 15, II, Theorem 1).

---

(1) This condition is assumed sometimes as definition. See e.g. J. L. Kelley, *loc. cit.* p. 99; G. T. Whyburn, *loc. cit.*

III. Examples and remarks.

EXAMPLE 1. *The decomposition* $\boldsymbol{D}$ *of the plane* $\mathscr{E}^2$ *into vertical lines is not u.s.c.*

For the union of verticals intersecting the hyperbola $y = 1/x$ is not closed.

Note that here the relation ϱ is closed, and $\mathscr{Y} \underset{\text{top}}{=} \mathscr{E}$.

EXAMPLE 2. Let $\mathscr{X}$ be composed of the points $1/n$ $(n = 1, 2, \ldots)$ and of the point 0, and let $\boldsymbol{D}$ have as elements the pair $(0, 1)$ and one-element sets $(1/n)$ where $n > 1$.

$\boldsymbol{D}$ *is u.s.c. but not l.s.c. Here* $\boldsymbol{D} \underset{\text{top}}{=} \mathscr{X}$.

EXAMPLE 3. $\mathscr{X}$ is composed of the points $1/n$ and $1-1/n$ for $n \geqslant 1$. $\boldsymbol{D}$ has as elements the two-elements sets $(1/n, 1-1/n)$ for $n > 2$ and the one-element sets (0), $(1/2)$ and (1).

$\boldsymbol{D}$ *is l.s.c. but not u.s.c.* It is homeomorphic to the space consisting of an infinite sequence of points $p_0, p_1, \ldots$, where each infinite closed set contains both points p_0 and p_1.

This space is not Hausdorff (consequently the relation ϱ is not closed).

Remark. A mapping $F: \mathscr{T} \to 2^{\mathscr{X}}$ may be continuous and such that $\mathscr{X} = \bigcup_{t \in \mathscr{T}} F(t)$ and $F(t) \cap F(t') = 0$, without the decomposition $\boldsymbol{D}$ of $\mathscr{X}$ into the sets $F(t)$ being semi-continuous.

Such is the case where $\mathscr{T}$ is a countable discrete space and $\boldsymbol{D}$ a countable not semi-continuous decomposition of $\mathscr{X}$ (see e.g. Examples 2 and 3).

However, if $\mathscr{T}$ is compact and F is u.s.c., then $\boldsymbol{D}$ is u.s.c. (cf. vol. II, Chapter 4).

IV. Relationship of quotient topology to exponential topology. Generally speaking the quotient topology of $\boldsymbol{D}$ is different from its exponential topology ($\boldsymbol{D}$ being considered as subset of $2^{\mathscr{X}}$). This is seen on Example 2 of Section III.

Clearly, $\boldsymbol{D}$ with the quotient topology is homeomorphic to $\mathscr{X}$, while it is discrete in the exponential topology. P having the same meaning as in the above section, $\overset{1}{P}$ is the identity mapping:

$$\overset{1}{P}\colon \boldsymbol{D} \cap 2^{\mathscr{X}} \to \boldsymbol{D},$$

where $\boldsymbol{D}$ has the quotient and $\boldsymbol{D} \cap 2^{\mathscr{X}}$ the exponential topology.

In the preceding example $\overset{1}{\boldsymbol{P}}$ is not a homeomorphism.

However, the following is true.

THEOREM 1. *The quotient topology of* $\boldsymbol{D}$ *is weaker than its exponential topology, i.e. the mapping* $\overset{1}{P}$ *is continuous.*

Otherwise stated: if $\boldsymbol{E} \subset \boldsymbol{D}$ *and* $\mathrm{S}\boldsymbol{E}$ *is open in* $\mathscr{X}$, *then* $\boldsymbol{E}$ *is open relative to* $\boldsymbol{D}$ *in the exponential topology.*

This follows from the formula $\boldsymbol{E} = \boldsymbol{D} \cap 2^{\mathrm{S}\boldsymbol{E}}$.

THEOREM 2. {*The decomposition* $\boldsymbol{D}$ *is lower* (*upper*) *semi-continuous*} $\equiv$ {*the identity mapping* $\overset{-1}{P}$ *is lower* (*upper*) *semi-continuous*}.

More precisely: {*a member* D *of* $\boldsymbol{D}$ *is a member of lower* (*upper*) *semi-continuity of* $\boldsymbol{D}$} $\equiv$ {$\overset{-1}{P}$ *is lower* (*upper*) *semi-continuous at* D}.

This follows from Theorem 5 of § 18, III, replacing f by P and $\mathscr{Y}$ by $\boldsymbol{D}$ with quotient-topology.

COROLLARY. {*The decomposition* $\boldsymbol{D}$ *is continuous*} $\equiv$ {*the mapping* $\overset{-1}{P}$ *is continuous, hence a homeomorphism*} $\equiv$ {*the quotient topology of* $\boldsymbol{D}$ *is identical with its exponential topology*}.

A member D *of* $\boldsymbol{D}$ *is a member of continuity if and only if* $\overset{-1}{P}$ *is continuous at* D.

CHAPTER TWO

METRIC SPACES

A. RELATIONS TO TOPOLOGICAL SPACES. $\mathscr{L}^*$-SPACES (§§ 20–22)

§ 20. $\mathscr{L}^*$-spaces (provided with the notion of limit)

I. Definition. A set of arbitrary elements is an *$\mathscr{L}^*$-space* if to certain sequences $p_1, p_2, \ldots, p_n, \ldots$ of elements of the space (called *convergent* sequences) there corresponds an element $p = \lim\limits_{n=\infty} p_n$ so that the following conditions are fulfilled([1]):

(i) *if* $\lim\limits_{n=\infty} p_n = p$ *and* $k_1 < k_2 < \ldots$, *then* $\lim\limits_{n=\infty} p_{k_n} = p$,

(ii) *if, for every* n, $p_n = p$, *then* $\lim\limits_{n=\infty} p_n = p$,

(iii) *if the sequence* $p_1, p_2, \ldots$ *does not converge to* p, *it contains a subsequence*([2]) $p_{k_1}, p_{k_2}, \ldots$ *none of whose subsequences converges to* p.

The following two propositions can easily be proved:

THEOREM 1. *Neither the convergence of a sequence nor the limit depend on the initial finite number of terms of this sequence.*

Thus one can add, discard or replace a finite number of terms without making the sequence diverge or converge, and without changing its limit.

THEOREM 2. *Given two sequences with* $\lim\limits_{n=\infty} p_n = p = \lim\limits_{n=\infty} q_n$, *the sequence* $p_1, q_1, p_2, q_2, \ldots$ *converges to* p.

THEOREM 3. *If* $\lim\limits_{n=\infty} p_n = p$ *and the sequence* $\{q_n\}$ *is obtained from the sequence* $\{p_n\}$ *by finite repetition of its elements, then* $\lim\limits_{n=\infty} q_n = p$.

([1]) Abstract spaces having the notion of limit as a primitive term were introduced by M. Fréchet in his *Thesis*, Paris 1906 and Rend. Circ. Mat. di Palermo 22 (1906), *Sur quelques points du calcul functionnel.* For condition (iii) see P. Alexandrov and P. Urysohn, C. R. Paris, 177 (1923), p. 1274, and P. Urysohn *Sur les classes* ($\mathscr{L}$) *de M. Fréchet*, Ens. Math. 25 (1926), pp. 77–83, where the $\mathscr{L}^*$-spaces are denoted by $\mathscr{L}_t$.

([2]) $p_{k_1}, p_{k_2}, \ldots$ is a subsequence of $p_1, p_2, \ldots$, if $k_1 < k_2 < \ldots$.

Proof. By the hypothesis, for every positive m, there exists an r_m such that $p_{r_m} = q_m$ and, for a fixed n, the condition $n = r_m$ is satisfied by, at most, a finite number of indices m. Assume that the sequence $\{q_m\}$ does not converge to p. By (iii) there exists a sequence $k_1 < k_2 < \dots$ such that the sequence $q_{k_1}, q_{k_2}, \dots$ does not contain any sequence convergent to p. Thus no term of the sequence $r_{k_1}, r_{k_2}, \dots$ repeats an infinite number of times and there exists a sequence of indices $l_1 < l_2 < \dots$ such that $r_{k_{l_1}} < r_{k_{l_2}} < \dots$. Let $s_n = r_{k_{l_n}}$. Hence $p_{s_1}, p_{s_2}, \dots$ is a subsequence of $\{p_n\}$. It follows that $p = \lim_{n=\infty} p_{s_n} = \lim_{n=\infty} q_{k_{l_n}}$, since $p_{s_n} = q_{k_{l_n}}$. But $q_{k_{l_1}}, q_{k_{l_2}}, \dots$ is a subsequence of the sequence $q_{k_1}, q_{k_2}, \dots$ and we obtain a contradiction.

Remark. By the asterisk we distinguish the $\mathscr{L}^*$-spaces from the $\mathscr{L}$-spaces of Fréchet which are not subjected to condition (iii). Conditions (i) and (ii) do not characterize the limit sufficiently. For example, the set composed of two elements a and b is an $\mathscr{L}$-space, if we suppose that only the sequences $a, a, a, \dots$ and $b, b, b, \dots$ are convergent (the sequence $a, b, b, b, \dots$ diverges!). See also Sections II and III, and M. Fréchet, *Espaces abstraits*, p. 169.

EXAMPLES. The space $\mathscr{E}$ of real numbers (as well as $\mathscr{E}^n$) is an $\mathscr{L}^*$-space with the usual definition of limit.

The same is true of the space of real-valued functions defined on an arbitrary set, where one assumes that

$$(1) \qquad [\lim_{n=\infty} f_n = f] \equiv \bigwedge_x [\lim_{n=\infty} f_n(x) = f(x)].$$

II. Relation to topological spaces. The closure $\overline{X}$ of a subset X of an $\mathscr{L}^*$-space is defined by the condition: $p \in \overline{X}$ *if and only if* p *is the limit of a sequence of elements of* X.

Consequently, if $\overline{X} = X$, we call X *closed* and $(-X)$ *open*.

THEOREM 1. *Each* $\mathscr{L}^*$*-space satisfies Axioms* 1 *to* 3 *and* 5 *of* § 4, I.

More precisely, condition (i) implies Axiom 1 and (ii) implies Axioms 2 and 5 (compare the argument of § 4, II).

Remark. *An* $\mathscr{L}^*$*-space does not need to be topological.*

Such is the space of real-valued functions considered in § 4,V[1].

---

[1] Necessary and sufficient conditions for a $\mathscr{T}_1$-space to be an $\mathscr{L}^*$-space were given by P. Urysohn in the paper referred to above.

THEOREM 2. *In order that* $p = \lim_{n=\infty} p_n$, *it is necessary and sufficient that, if* X *denotes the set of elements of an arbitrary subsequence* $p_{k_1}, p_{k_2}, \ldots$, *then* $p \in \overline{X}$.

Proof. If $p = \lim_{n=\infty} p_n$, then $p \in \overline{X}$, by (i). Conversely, if the given sequence does not converge to p, it contains, by (iii), a subsequence $p_{k_1}, p_{k_2}, \ldots$ none of whose subsequences converges to p. We can also suppose that the sequence $\{p_{k_n}\}$ does not contain the element p; for, by (ii), it may contain p, at the most, a finite number of times and, in this case, we could replace the sequence p_{k_n} by a sequence which does not contain p. We claim that no sequence $x_1, x_2, \ldots$ (whose elements are different or not) selected from X converges to p. For, in the case when it contains an infinite number of different elements, it contains a subsequence of the sequence $p_{k_1}, p_{k_2}, \ldots$, hence a sequence which does not converge to p; in the case when the sequence $x_1, x_2, \ldots$ contains only a finite number of different elements, there must be one that repeats an infinite number of times. Since this element must be different from p, it follows that the sequence $x_1, x_2, \ldots$ cannot converge to p. Therefore, in either case, $p \notin \overline{X}$.

THEOREM 3. *Let the space under consideration be Hausdorff with a local countable base at each point* p. *Then one can, without affecting its topology, introduce the notion of a limit of a sequence of points and thus endow the space of a* $\mathscr{L}^*$*-structure.*

Namely, we assume that $p = \lim_{n=\infty} p_n$ *if and only if for every neighbourhood* U *of* p *there exists a* k *such that* $p_n \in U$ *for all* $n > k$.

Proof. A sequence cannot have two different limits p and q since the points p and q have disjoint neighbourhoods U and V. Conditions (i) and (ii) are obvious. We shall prove condition (iii). Suppose that the sequence $p_1, p_2, \ldots$ does not converge to p. This means that there is a neighbourhood U of p and a subsequence $p_{k_1}, p_{k_2}, \ldots$ which lies on the outside of U. Therefore no subsequence of this sequence converges to p.

Now let $p \in \overline{A}$ (in the given topology of the space). We have to show that

$$p = \lim_{n=\infty} p_n, \quad \text{where} \quad p_n \in A.$$

Let $U_1, U_2, \ldots$ be a local base of p.

The set $U_1 \cap U_2 \cap \ldots \cap U_n$ being a neighbourhood of p, it contains a point p_n of A; hence $p_n \epsilon (A \cap U_1 \cap \ldots \cap U_n)$. Let U be an arbitrary neighbourhood of p. Then, there exists an n_0 such that $U_{n_0} \subset U$. Therefore, for $n > n_0$,

$$p_n \epsilon U_1 \cap U_2 \cap \ldots \cap U_n \subset U_{n_0} \subset U \quad \text{and} \quad p = \lim_{n=\infty} p_n.$$

Conversely, if $p = \lim_{n=\infty} p_n$, where $p_n \epsilon A$ and if U is a neighbourhood of p, then for sufficiently large values of n, $p_n \epsilon U$, hence $U \cap A \neq 0$. Consequently $p \epsilon \bar{A}$.

It follows that all the theorems concerning $\mathscr{L}^*$-spaces hold true in Hausdorff spaces with local countable base.

III. Notion of continuity.

DEFINITION. Let $f: \mathscr{X} \to \mathscr{Y}$ where $\mathscr{X}$ and $\mathscr{Y}$ are $\mathscr{L}^*$-spaces. The mapping f *is continuous at the point* x *if*

$$[x = \lim_{n=\infty} x_n] \Rightarrow [f(x) = \lim_{n=\infty} f(x_n)]. \tag{1}$$

THEOREM. *The closure being defined as in Section* II, *the above* (*Heine*) *condition is equivalent to* (see § 13,I):

$$[x \epsilon \bar{A}] \Rightarrow [f(x) \epsilon \overline{f(A)}]. \tag{2}$$

Proof. 1. Suppose that (1) is true. Let $x \epsilon \bar{A}$. There exists a sequence $x_1, x_2, \ldots$ such that $x = \lim_{n=\infty} x_n$ and $x_n \epsilon A$. Therefore, by (1), $f(x) = \lim_{n=\infty} f(x_n)$, and since $f(x_n) \epsilon f(A)$, it follows that $f(x) \epsilon \overline{f(A)}$. Thus (2) is true.

2. Suppose that (2) is true. Let $x = \lim_{n=\infty} x_n$. Let A be the set of elements of an arbitrary subsequence $x_{k_1}, x_{k_2}, \ldots$ By Theorem 2 of Section II, $x \epsilon \bar{A}$. Hence, by (2), $f(x) \epsilon \overline{f(A)}$. Since $f(x_{k_1}), f(x_{k_2}), \ldots$ is an arbitrary subsequence of $f(x_1), f(x_2), \ldots$, we have $f(x) = \lim_{n=\infty} f(x_n)$. Thus (1) is true.

The following corollary follows easily.

COROLLARY. f *is a homeomorphism if and only if*

$$[x = \lim_{n=\infty} x_n] \equiv [f(x) = \lim_{n=\infty} f(x_n)]. \tag{3}$$

Remark. The assumption that the space of values is an $\mathscr{L}^*$-space is essential. If the space is an $\mathscr{L}$-space but not an $\mathscr{L}^*$-space, it contains a sequence $y_0, y_1, \ldots$ which does not converge to y_0, but every subsequence of it contains a subsequence which does converge to y_0. Let $f(0) = y_0$ and $f(1/n) = y_n$. Then the function f is continuous in the sense of § 13,I, but does not satisfy Heine's condition.

In general, if the spaces of arguments and of values are $\mathscr{L}$-spaces (but not necessarily $\mathscr{L}^*$-spaces), then the continuity of f (in the sense of § 13, I) is equivalent to the following condition: *If* $x = \lim_{n=\infty} x_n$, *there exists a subsequence* $x_{k_1}, x_{k_2}, \ldots$ *such that* $f(x) = \lim_{n=\infty} f(x_{k_n})$.

IV. Cartesian product of $\mathscr{L}^*$-spaces. Let $\mathscr{X}$ and $\mathscr{Y}$ be $\mathscr{L}^*$-spaces. The set $\mathscr{X} \times \mathscr{Y}$ is given the $\mathscr{L}^*$-space structure by letting the sequence $\mathfrak{z}_n = (x_n, y_n)$ converge to $\mathfrak{z} = (x, y)$, if $\lim_{n=\infty} x_n = x$ and $\lim_{n=\infty} y_n = y$.

Similarly, if $\mathscr{X}_1, \mathscr{X}_2, \ldots$ is an infinite sequence of $\mathscr{L}^*$-spaces, we assume that the sequence $\mathfrak{z}_n = \overset{1}{\mathfrak{z}}_n, \overset{2}{\mathfrak{z}}_n, \ldots$ converges to $\mathfrak{z} = \mathfrak{z}^1, \mathfrak{z}^2, \ldots$ (in the space $\mathscr{X}_1 \times \mathscr{X}_2 \times \ldots$), if $\lim_{n=\infty} \mathfrak{z}_n^i = \mathfrak{z}^i$ for every i, i.e. if the ith term of the variable sequence converges to the ith term of the limit sequence:

$$(\lim_{n=\infty} \mathfrak{z}_n = \mathfrak{z}) \equiv \bigwedge_{i=1}^{\infty} (\lim_{n=\infty} \mathfrak{z}_n^i = \mathfrak{z}^i). \tag{1}$$

Let f be a function which assigns to each point t of a space T a point $\mathfrak{z}$ of the cartesian product $\mathscr{X}_1 \times \mathscr{X}_2 \times \ldots$ (or, less generally, of the product $\mathscr{X} \times \mathscr{Y}$). Let

$$g_i(t) = \big(f(t)\big)^i = \text{the } i\text{th coordinate of the point } \mathfrak{z} = f(t).$$

THEOREM 1. *The necessary and sufficient condition for the function f to be continuous at the point t is that each function g_i where $i = 1, 2, \ldots$, should be continuous at t.*

Proof. If the sequence $t_1, t_2, \ldots$ converges to t, then

$$[f(t) = \lim_{n=\infty} f(t_n)] \equiv \bigwedge_i [\big(f(t)\big)^i = \lim_{n=\infty} \big(f(t_n)\big)^i]. \tag{2}$$

THEOREM 2. *The i-th coordinate $\mathfrak{z}^i$ of a point $\mathfrak{z}$ in the cartesian product is a continuous function of $\mathfrak{z}$.*

In other words, *the projection onto the* ith *axis is a continuous mapping.*

Proof. By (1), the condition $\lim_{n=\infty} \mathfrak{z}_n = \mathfrak{z}$ implies $\lim_{n=\infty} \mathfrak{z}_n^i = \mathfrak{z}^i$.

THEOREM 3. *If the mapping* $f: \mathscr{X} \to \mathscr{Y}$ *is continuous, then its graph* $I = \underset{xy}{E}\big(y = f(x)\big)$ *is a closed set homeomorphic to* $\mathscr{X}$.

Proof. If $\lim_{n=\infty} \big(x_n, f(x_n)\big) = (x, y)$, then $\lim_{n=\infty} x_n = x$ and $\lim_{n=\infty} f(x_n) = y$. The first identity implies that $\lim_{n=\infty} f(x_n) = f(x)$ (by the continuity of f). Hence $y = f(x)$ and $(x, y) \in I$, which proves that $\bar{I} = I$.

The function which carries x to the point $\big(x, f(x)\big)$ is a homeomorphism of $\mathscr{X}$ onto I. For, the condition $\lim_{n=\infty} x_n = x$ implies $\lim_{n=\infty} \big(x_n, f(x_n)\big) = \big(x, f(x)\big)$ and, conversely, the last condition implies $\lim_{n=\infty} x_n = x$.

THEOREM 4. *A finite or countably infinite product of separable spaces is separable.*

Namely, the set $\mathfrak{R}$ considered in the proof of Theorem 6 of § 16,V, is dense in the product space.

V. Countably compact $\mathscr{L}^*$-spaces.

DEFINITION. An $\mathscr{L}^*$-space is *countably compact*(1) (briefly *c. compact*) if every infinite sequence of points $p_1, p_2, \ldots$ contains a convergent subsequence:

$$\lim_{n=\infty} p_{k_n} = p, \quad \text{where} \quad k_1 < k_2 < \ldots . \tag{1}$$

This definition is in agreement with the definition given for topological spaces in § 5, VII (see the forthcoming Remark to Theorem 3).

EXAMPLES. (1) By the classical Bolzano-Weierstrass theorem, the n-dimensional cube $\mathscr{I}^n$ is c. compact. More generally, every closed and bounded subset of the n-dimensional euclidean space $\mathscr{E}^n$ is c. compact (even compact).

(2) The set of ordinal numbers $\alpha < \Omega$ (with the usual notion of limit admitted in the theory of ordinal numbers) is a c. compact space.

---

(1) Compact spaces as defined in § 5, VII will be studied in Chapter IV.

One shows easily (by virtue of I, (iii)):

THEOREM 0. *Let the space be c. compact. If each convergent subsequence of the sequence* $p_1, p_2, \ldots$ *converges to* p, *then* $p = \lim_{n=\infty} p_n$.

THEOREM 1. *Every closed subset* F *of a c. compact space is c. compact.*

Conversely, *a c. compact subset* A *of an arbitrary space* $\mathscr{X}$ *is a closed set.*

Proof. To show the first part of the theorem, let us consider a sequence $p_1, p_2, \ldots$ of points of F. Since the space is c. compact, this sequence has a subsequence which satisfies (1). Since F is closed, it follows that $p \in F$. Hence F is c. compact.

Now let us assume that A is c. compact but not closed. Hence there exists in A a sequence of points converging to a point p which does not belong to A. Then every subsequence of this sequence converges also to p. Thus A is not c. compact.

THEOREM 2 (of Cantor)[1]. *If* $F_1 \supset F_2 \supset \ldots$ *is a decreasing sequence of closed non-empty subsets of a c. compact space, then* $F_1 \cap F_2 \cap \ldots \neq 0$.

Proof. Let us form an infinite sequence $p_1, p_2, \ldots$ by selecting a point p_n from the set F_n. Since the space is c. compact, this sequence contains a convergent subsequence, i.e. formula (1) holds. Since F_n is a closed set and the points p_n, with perhaps the exception of a finite number, belong to F_n, it follows that $p \in F_n$. Hence $p \in (F_1 \cap F_2 \cap \ldots)$.

THEOREM 3 (of Borel)[2]. *Every countable open cover of a countably compact space* $\mathscr{X}$ *contains a finite subcover.*

Proof. Let

$$\mathscr{X} = \bigcup_{n=1}^{\infty} G_n, \quad \text{where } G_n \text{ is open for } n = 1, 2, \ldots \tag{2}$$

We have to show that

$$\mathscr{X} = \bigcup_{n=1}^{k} G_n \tag{3}$$

for a suitable k. Let $F_k = \mathscr{X} - (G_1 \cup \ldots \cup G_k)$. Hence $F_1 \supset F_2 \supset \ldots$.

---

(1) Math. Ann. 17 (1880).

(2) Compare E. Borel, Ann. Ecole Norm. (3) 12 (1895) (Thesis), p. 51.

If we suppose that formula (3) does not hold for any k, we have $F_k \neq 0$ for $k = 1, 2, \ldots$. It follows from Theorem 2 that $\bigcap_{k=1}^{\infty} F_k \neq 0$, i.e. $\bigcup_{k=1}^{\infty}(G_1 \cup \ldots \cup G_k) \neq \mathscr{X}$ and therefore $\bigcup_{n=1}^{\infty} G_n \neq \mathscr{X}$ contrary to (2).

Remark. The theorem of Borel has been deduced from that of Cantor. Similarly, the Cantor theorem can be deduced from the Borel theorem. Thus they are *equivalent* (in spaces satisfying Axiom I of § 4)(1).

Furthermore, in $\mathscr{L}^*$-spaces the *Cantor theorem implies countable compactness*. For, let us suppose that the sequence $p_1, p_2, \ldots$ contains no convergent subsequence. Then the set $F_1 = (p_1, p_2, \ldots)$ is closed and, generally, $F_n = (p_n, p_{n+1}, \ldots)$ is closed. As $F_1 \cap F_2 \cap \ldots = 0$, this is a contradiction to the Cantor theorem.

Consequently, the definition of countable compactness of $\mathscr{L}^*$-spaces is equivalent to the definition of this concept given for topological spaces in § 5, VII.

THEOREM 4. *The cartesian product of a (finite or countably infinite) family of c. compact spaces is c. compact.*

In particular the space $\mathscr{I}^{\aleph_0}$ is c. compact.

Proof. Let $\mathfrak{z}_1, \mathfrak{z}_2, \ldots$ be a sequence of points of the product $\mathscr{X}_1 \times \mathscr{X}_2 \times \ldots$ and let $\mathfrak{z}_n = (\mathfrak{z}_n^1, \mathfrak{z}_n^2, \ldots)$. Since the space $\mathscr{X}_1$ is c. compact, there exists a sequence of integers $1 < k_1 < k_2 < \ldots$ such that the sequence $\mathfrak{z}_{k_1}^1, \mathfrak{z}_{k_2}^1, \ldots$ is convergent. Let $\lim_{n=\infty} \mathfrak{z}_{k_n}^1 = x^1$.

Similarly, since the space $\mathscr{X}_2$ is c. compact, there exists a sequence of integers $1 < l_1 < l_2 < \ldots$ such that the sequence $\mathfrak{z}_{k_{l_1}}^2$, $\mathfrak{z}_{k_{l_2}}^2, \ldots$ Let $\lim_{n=\infty} \mathfrak{z}_{k_{l_n}}^2 = x^2$.

Proceeding step by step we define a sequence of points $x^1 \in \mathscr{X}_1$, $x^2 \in \mathscr{X}_2, \ldots$ (which is finite or infinite according to the number of spaces $\mathscr{X}_1, \mathscr{X}_2, \ldots$). Let $x = (x^1, x^2, \ldots)$. The sequence $\mathfrak{z}_1, \mathfrak{z}_{k_1}$, $\mathfrak{z}_{k_{l_1}}, \ldots$ is convergent to x.

Evidently, $1 < k_1 < k_{l_1} < k_{l_{m_1}} < \ldots$, hence the sequence $\mathfrak{z}_{k_1}^1$, $\mathfrak{z}_{k_{l_1}}^1, \ldots$ converges to x^1 as a subsequence of $\{\mathfrak{z}_{k_n}^1\}$. It follows that

(1) Compare S. Saks, *Sur l'équivalence de deux théorèmes de la Théorie des ensembles*, Fund. Math. 2 (1921), p. 1.

the sequence $\mathfrak{z}_1^1, \mathfrak{z}_{k_1}^1, \mathfrak{z}_{k_{l_1}}^1, \ldots$ converges to x^1. Similarly, the sequence $\mathfrak{z}_{k_{l_1}}^2, \mathfrak{z}_{k_{l_{m_1}}}^2, \ldots$ is a subsequence of $\{\mathfrak{z}_{k_{l_n}}^2\}$ and hence the sequence $\mathfrak{z}_1^2, \mathfrak{z}_{k_1}^2, \mathfrak{z}_{k_{l_1}}^2, \ldots$ converges to x^2. In general, the sequence $\mathfrak{z}_1^n, \mathfrak{z}_{k_1}^n, \mathfrak{z}_{k_{l_1}}^n, \ldots$ converges to x^n. Therefore the sequence $\mathfrak{z}_1, \mathfrak{z}_{k_1}, \mathfrak{z}_{k_{l_1}}, \ldots$ is convergent to x.

THEOREM 5. *The c. compactness of a space is an invariant of continuous mappings.*

Proof. Let f be a continuous mapping of $\mathscr{X}$ onto $\mathscr{Y}$ (they are assumed to be $\mathscr{L}^*$-spaces). Let $y_1, y_2, \ldots$ be a sequence of points of $\mathscr{Y}$ and let $y_n = f(x_n)$. Since $\mathscr{X}$ is c. compact, the sequence $\{x_n\}$ contains a convergent subsequence: $\lim_{n=\infty} x_{k_n} = p$ where $k_1 < k_2 < \ldots$. Since f is continuous, it follows that

$$\lim_{n=\infty} f(x_{k_n}) = f(p), \quad \text{i.e.} \quad \lim_{n=\infty} y_{k_n} = f(p).$$

This shows that $\{y_n\}$ contains a convergent subsequence.

THEOREM 6. *If $\mathscr{X}$ is c. compact and $f: \mathscr{X} \to \mathscr{Y}$ is continuous, then f is a closed mapping.*

In other terms: $(F = \overline{F}) \Rightarrow [f(F) = \overline{f(F)}]$.

Proof. By virtue of Theorem 1, F is c. compact and therefore by Theorem 5, the set $f(F)$ is c. compact and hence closed in $\mathscr{Y}$ (by Theorem 1).

Theorem 6 yields immediately (cf. § 13, XIII) the

COROLLARY 6a. *Any one-to-one continuous mapping of a c. compact space is a homeomorphism.*

Theorem 6 can be sharpened as follows:

THEOREM 7. *Let $\mathscr{Y}$ be c. compact. Then the projection $f: \mathscr{X} \times \mathscr{Y} \to \mathscr{X}$ is closed.*

In other words: *if $\underset{xy}{E}\varphi(x, y)$ is closed, then so is $\underset{x}{E}\underset{y}{\bigvee}\varphi(x, y)$.*

Proof. Let $x_1, x_2, \ldots$ be a convergent sequence of points of $\underset{x}{E}\underset{y}{\bigvee}\varphi(x, y)$ and let $\lim x_n = x_0$. Hence there exists a sequence $y_1, y_2, \ldots$ in $\mathscr{Y}$ such that $\varphi(x_n, y_n)$. $\mathscr{Y}$ being c. compact, this sequence contains a convergent subsequence: $\lim y_{k_n} = y_0$. Since $\underset{xy}{E}\varphi(x, y)$ is closed and $\lim x_{k_n} = x_0$, it follows that $\varphi(x_0, y_0)$ and therefore $x_0 \in \underset{x}{E}\underset{y}{\bigvee}\varphi(x, y)$.

COROLLARY 7a. *Let $\mathscr{Y}$ be compact. If the set* $\underset{xy}{E}\varphi(x, y)$ *is open, so is* $\underset{x}{E}\underset{y}{\wedge}\varphi(x, y)$.

COROLLARY 7b. *Let $\mathscr{Y}$ be a countable union of compact sets. If* $\underset{xy}{E}\varphi(x, y)$ *is* $\boldsymbol{F}_\sigma$, *so is* $\underset{x}{E}\underset{y}{\vee}\varphi(x, y)$. *If* $\underset{xy}{E}\varphi(x, y)$ *is* $\boldsymbol{G}_\delta$, *so is* $\underset{x}{E}\underset{y}{\wedge}\varphi(x, y)$.

THEOREM 8. *Let $\mathscr{Y}$ be c. compact and $f: \mathscr{X} \to \mathscr{Y}$. The set* $\underset{xy}{E}[y = f(x)]$ *is closed if and only if f is continuous.*

Proof. The condition is obviously sufficient (without even the assumption of c. compactness. Compare also Theorem 3 of Section IV.)

To show that it is necessary, put $\lim x_n = x_0$ and choose from the sequence $f(x_1), f(x_2), \ldots$ a convergent subsequence: $\lim f(x_{k_n}) = y_0$. Hence $\langle x_0, y_0\rangle = \lim \langle x_{k_n}, f(x_{k_n})\rangle$. The set $\underset{xy}{E}[y = f(x)]$ being closed, it contains the point $\langle x_0, y_0\rangle$. This means that $y_0 = f(x_0)$. It follows by Theorem 0 that $\lim f(x_n) = f(x_0)$.

Remark. There is a number of further properties of compact spaces which will be shown in Chapter IV and which can be easily proved for c. compact spaces.

VI. Continuous convergence. The set $\mathscr{Y}^{\mathscr{X}}$ as an $\mathscr{L}^*$-space.

DEFINITION 1. Let $f_n: \mathscr{X} \to \mathscr{Y}$, $n = 1, 2, \ldots$. We say [1] that the sequence $f_1, f_2, \ldots$ is *continuously convergent* to f, if

$$\big(\lim_{n=\infty} x_n = x\big) \Rightarrow \big(\lim_{n=\infty} f_n(x_n) = f(x)\big). \tag{0}$$

The continuous convergence implies the convergence in the usual sense. This can be verified by substituting $x_n = x$ in the preceding definition.

Remark. The reverse implication does not hold. For example, if we let $f_n(x) = x^n$ for $0 \leqslant x \leqslant 1$ and $f(x) = 0$ for $x < 1$ and $f(1) = 1$, we obtain $\lim_{n=\infty} f_n(x) = f(x)$. This convergence, however, is not continuous.

The following example shows that a convergence of continuous functions to a *continuous* limit does not need to be continuous.

---

[1] See H. Hahn, *Theorie der reellen Funktionen*, Berlin 1921, p. 238.

The function f_n is defined by the conditions: $f_n(x) = 0$ for $0 \leqslant x \leqslant 1/(n+1)$ and for $1/(n-1) \leqslant x \leqslant 1$, $f_n(1/n) = 1$ and the function f_n is linear in the intervals $(1/(n+1), 1/n)$ and $(1/n, 1/(n-1))$. Then $\lim_{n=\infty} f_n(x) = 0$, but the convergence is not continuous.

DEFINITION 2. If $\mathscr{X}$ and $\mathscr{Y}$ are $\mathscr{L}^*$-spaces, we introduce the *$\mathscr{L}^*$-topology* in $\mathscr{Y}^{\mathscr{X}}$ (see § 13,I) assuming that $f = \lim_{n=\infty} f_n$, if the sequence $\{f_n\}$ is continuously convergent to f.

We have to show that conditions (i) to (iii) of Section I are satisfied.

Let $\lim_{n=\infty} f_n = f$ and $k_1 < k_2 < \ldots$. We have to prove that $\lim_{n=\infty} f_{k_n} = f$; in other words, that the condition

$$\lim_{n=\infty} x_n = x \tag{1}$$

implies

$$\lim_{n=\infty} f_{k_n}(x_n) = f(x). \tag{2}$$

Define the sequence $\{z_m\}$ by $z_m = x_n$ for $k_{n-1} < m \leqslant k_n$ (where $k_0 = 0$). It follows (compare I, Theorem 3) that $\lim_{m=\infty} z_m = \lim_{n=\infty} x_n = x$. Since the sequence $\{f_m\}$ is continuously convergent, it follows that $\lim_{m=\infty} f_m(z_m) = f(x)$, whence $\lim_{n=\infty} f_{k_n}(z_{k_n}) = f(x)$ and, since $z_{k_n} = x_n$, we have (2).

To prove condition (ii), let $f_n = f$ for $n = 1, 2, \ldots$. We have to show that condition (1) implies $\lim f(x_n) = f(x)$. But this is a consequence of the continuity of f.

To prove condition (iii), assume that the sequence $f_1, f_2, \ldots$ is not convergent to f. Hence there exists a sequence $\{x_n\}$ satisfying (1) such that the sequence $\{f_n(x_n)\}$ is not convergent to $f(x)$. Since $\mathscr{Y}$ is an $\mathscr{L}^*$-space, there exists a sequence $k_1 < k_2 < \ldots$ such that the sequence $\{f_{k_n}(x_{k_n})\}$ does not contain any subsequence which converges to $f(x)$. We claim that no subsequence of $\{f_{k_n}\}$ converges to f. For, suppose that there exists a sequence $m_1 < m_2 < \ldots$ such that $\lim f_{k_{m_n}} = f$. Then $\lim f_{k_{m_n}}(x_{k_{m_n}}) = f(x)$, since, by (1), $\lim x_{k_{m_n}} = x$. But $\{f_{k_{m_n}}(x_{k_{m_n}})\}$ is a subsequence of $\{f_{k_n}(x_{k_n})\}$ and we obtain a contradiction with the definition of the sequence $\{k_n\}$.

Remark. The continuity of the functions $f, f_1, f_2, \ldots$ was used only in the proof of condition (ii).

The following theorem can be easily proved.

THEOREM. *Let* $f \in \mathscr{Y}^{\mathscr{X}}$ *and put* $g(f, x) = f(x)$. *Then* $g : \mathscr{Y}^{\mathscr{X}} \times \mathscr{X} \to \mathscr{Y}$ *is continuous.*

VII. Operations on the spaces $\mathscr{Y}^{\mathscr{X}}$ with $\mathscr{L}^*$-topology.

THEOREM 1. $\mathscr{Y}^{\mathscr{X}} \times \mathscr{Z}^{\mathscr{X}} \underset{\text{top}}{=} (\mathscr{Y} \times \mathscr{Z})^{\mathscr{X}}$; *more generally,*

$$\mathscr{Y}_1^{\mathscr{X}} \times \mathscr{Y}_2^{\mathscr{X}} \times \ldots \underset{\text{top}}{=} (\mathscr{Y}_1 \times \mathscr{Y}_2 \times \ldots)^{\mathscr{X}}, \quad \textit{hence} \quad (\mathscr{Y}^{\mathscr{X}})^{\aleph_0} \underset{\text{top}}{=} (\mathscr{Y}^{\aleph_0})^{\mathscr{X}}.$$

Proof. We have to define a homeomorphism h of the Cartesian product $\mathscr{Y}_1^{\mathscr{X}} \times \mathscr{Y}_2^{\mathscr{X}} \times \ldots$ onto the function space $(\mathscr{Y}_1 \times \mathscr{Y}_2 \times \ldots)^{\mathscr{X}}$. Let $h(f) = g$, where

$$f = [f^1, f^2, \ldots], \quad f^i \in \mathscr{Y}_i^{\mathscr{X}}, \quad \text{and} \quad g(x) = [f^1(x), f^2(x), \ldots].$$

It is easily seen that the above defined function g is continuous (compare IV, Theorem 1) and hence belongs to $(\mathscr{Y}_1 \times \mathscr{Y}_2 \times \ldots)^{\mathscr{X}}$. Conversely, to each element g of this space there corresponds an f such that $g = h(f)$; namely $f = [g^1, g^2, \ldots]$. Thus

$$h(\mathscr{Y}_1^{\mathscr{X}} \times \mathscr{Y}_2^{\mathscr{X}} \times \ldots) = (\mathscr{Y}_1 \times \mathscr{Y}_2 \times \ldots)^{\mathscr{X}}.$$

The mapping h is continuous. For, let $\lim f_n = f$. Putting $g_n = h(f_n)$, i.e. $g_n(x) = [f_n^1(x), f_n^2(x), \ldots]$, we have to show that $\lim g_n = g$, i.e. that condition (1) of Section VI implies $\lim g_n(x_n) = g(x)$; in other words that $\lim\limits_{n=\infty} f_n^i(x_n) = f^i(x)$. But the condition $\lim f_n = f$ means that $\lim\limits_{n=\infty} f_n^i = f^i$ for each i, and by virtue of (1) we obtain $\lim\limits_{n=\infty} f_n^i(x_n) = f^i(x)$.

To prove that h is a homeomorphism, it remains to show that the condition $\lim g_n = g$ implies $\lim f_n = f$. Assuming (1), we have $\lim\limits_{n=\infty} g_n(x_n) = g(x)$; hence $\lim\limits_{n=\infty} g_n^i(x_n) = g^i(x)$, i.e. $\lim\limits_{n=\infty} f_n^i(x_n) = f^i(x)$. Therefore $\lim\limits_{n=\infty} f_n^i = f^i$, i.e. $\lim\limits_{n=\infty} f_n = f$.

THEOREM 2. *If* $\mathscr{X} = A \cup B$, *where* A *and* B *are two disjoint closed sets, then*

$$\mathscr{Y}^A \times \mathscr{Y}^B \underset{\text{top}}{=} \mathscr{Y}^{A \cup B}.$$

Proof. With each pair of functions $f \in \mathcal{Y}^A$ and $g \in \mathcal{Y}^B$ let us associate the function $u \in \mathcal{Y}^{A \cup B}$ which is equal to f on A and is equal to g on B. Let $u = h(f, g)$. We deduce that

$$h(\mathcal{Y}^A \times \mathcal{Y}^B) = \mathcal{Y}^{A \cup B}.$$

Let $\lim f_n = f$ and $\lim g_n = g$. Let $x \in A$. Condition (1) yields $\lim f_n(x_n) = f(x)$; hence $\lim u_n(x_n) = u(x)$. The same is true if $x \in B$. It follows that $\lim u_n = u$. This shows that h is continuous.

Conversely, assuming that $\lim u_n = u$ and $x \in A$, we deduce that $\lim f_n(x_n) = f(x)$, whence $\lim f_n = f$. Similarly, we prove that $\lim g_n = g$. It follows that the mapping h is a homeomorphism.

THEOREM 3. $(\mathcal{Y}^{\mathcal{X}})^{\mathcal{T}} \underset{\text{top}}{=} \mathcal{Y}^{\mathcal{X} \times \mathcal{T}}$.

Proof. Let $g(x, t) = f_t(x)$. We shall first show the following theorem.

THEOREM 3′ (1). *The function g is continuous if and only if the function f (which associates with each t the function f_t of the variable x) is continuous, i.e. we have*

$$g \in \mathcal{Y}^{\mathcal{X} \times \mathcal{T}} \equiv f \in (\mathcal{Y}^{\mathcal{X}})^{\mathcal{T}}.$$

Proof. Let $g \in \mathcal{Y}^{\mathcal{X} \times \mathcal{T}}$. For a fixed t, the function g is a continuous function of the variable x, i.e. $f_t \in \mathcal{Y}^{\mathcal{X}}$. We shall show that $f \in (\mathcal{Y}^{\mathcal{X}})^{\mathcal{T}}$. By the continuity of g, the conditions: (1) and $\lim t_n = t$ imply that $\lim g(x_n, t_n) = g(x, t)$, i.e. $\lim\limits_{n=\infty} f_{t_n}(x_n) = f_t(x)$. It follows that the functions f_{t_n} converge continuously to f_t, whence $\lim\limits_{n=\infty} f_{t_n} = f_t$. Therefore the condition $\lim t_n = t$ implies $\lim\limits_{n=\infty} f_{t_n} = f_t$. But this means that the function f is continuous, i.e. $f \in (\mathcal{Y}^{\mathcal{X}})^{\mathcal{T}}$.

Conversely, assuming that the function f is continuous, we see that the condition $\lim t_n = t$ implies that $\lim\limits_{n=\infty} f_{t_n} = f_t$. By the continuous convergence of the sequence f_{t_n} to f_t, (1) implies that $\lim\limits_{n=\infty} f_{t_n}(x_n) = f_t(x)$, i.e. $\lim g(x_n, t_n) = g(x, t)$; but this means that the function g is continuous, i.e. $g \in \mathcal{Y}^{\mathcal{X} \times \mathcal{T}}$.

---

(1) Relations between the continuity of the functions f and g (with a different topology, however, than that of Section VI) were studied by R. H. Fox, *On topologies of function spaces*, Bull. Amer. Math. Soc. 51 (1945), p. 429. See also Vol. II, Chapter IV.

Thus Theorem 3′ has been proved. By putting $g = h(f)$ we can restate it as follows

$$h[(\mathscr{Y}^{\mathscr{X}})^{\mathscr{T}}] = \mathscr{Y}^{\mathscr{X}\times\mathscr{T}}.$$

We have to prove that h is a homeomorphism; we shall first show that it is continuous.

Let $\lim f^n = f$ and $g_n = h(f^n)$. We have to show that $\lim g_n = g$. The condition $\lim f^n = f$ together with $\lim t_n = t$ yields (by the continuous convergence process) $\lim f^n_{t_n}(x_n) = f_t(x)$, which means that $\lim g_n(x_n, t_n) = g(x, t)$. Therefore $\lim g_n = g$.

To show that the inverse of h is continuous, suppose that $\lim g_n = g$, $\lim t_n = t$ and $\lim x_n = x$. It follows that $\lim g_n(x_n, t_n) = g(x, t)$, i.e. $\lim f^n_{t_n}(x_n) = f_t(x)$. Therefore $\lim f^n_{t_n} = f_t$, and $\lim f^n = f$.

VIII. Continuous convergence in the narrow sense. Besides the notion of continuous convergence, it is useful in certain problems (compare § 21, X) to consider the continuous convergence in the narrow sense defined as follows.

DEFINITION. The sequence of functions $\{f_n\}$ is said to be *continuously convergent in the narrow sense*, if the convergence of the sequence $\{f(x_n)\}$ implies the convergence of $\{f_n(x_n)\}$ and the identity $\lim f_n(x_n) = \lim f(x_n)$.

In a manner analogous to that of Section VI, by assuming the convergence in the set $\mathscr{Y}^{\mathscr{X}}$ to be the continuous convergence in the narrow sense, we define in this set an $\mathscr{L}^*$-space structure.

THEOREM 1. *The continuous convergence in the narrow sense implies the continuous convergence (in the set $\mathscr{Y}^{\mathscr{X}}$).*

Proof. Let $\lim x_n = x$. Since the function f is continuous, it follows that $\lim f(x_n) = f(x)$. By the definition of the continuous convergence in the narrow sense, $\lim f_n(x_n) = f(x)$. But this means that the sequence $\{f_n\}$ is continuously convergent.

THEOREM 2. *If the space $\mathscr{X}$ is c. compact, the continuous convergence is equivalent to the continuous convergence in the narrow sense.*

Proof. Suppose that the sequence $\{f_n\}$ is not convergent in the narrow sense; i.e. we have $\lim f(x_n) = y$ but the sequence $\{f_n(x_n)\}$ is not convergent to y. Since $\mathscr{Y}$ is an $\mathscr{L}^*$-space, there exists a sequence $k_1 < k_2 < \ldots$ such that, for any sequence, $m_1 < m_2 < \ldots$, the sequence $\{f_{k_{m_n}}(x_{k_{m_n}})\}$ is not convergent to y.

Since the space $\mathscr{X}$ is c. compact, there exists a sequence $m_1 < m_2 < \ldots$ such that the sequence $\{x_{k_{m_n}}\}$ is convergent. Let $\lim x_{k_{m_n}} = x$. If the sequence $\{f_n\}$ is continuously convergent, then so is the sequence $\{f_{k_{m_n}}\}$. Therefore

$$\lim f_{k_{m_n}}(x_{k_{m_n}}) = f(x).$$

On the other hand, the function f is continuous, whence

$$\lim f(x_{k_{m_n}}) = f(x),$$

and since

$$\lim f(x_{k_{m_n}}) = \lim f(x_n) = y,$$

it follows that $f(x) = y$. Therefore $\lim f_{k_{m_n}}(x_{k_{m_n}}) = y$, contrary to the definition of the sequence $\{k_n\}$.

Remark 1. The hypothesis that $\mathscr{X}$ is countably compact in Theorem 2 is essential. Consider the following example. Let $\mathscr{X}$ be the open interval $0 < x < 1$, $f(x) = 0$, $f_n(x) = x^n$. Obviously, the convergence is continuous, but it is not continuous in the narrow sense. Putting $x_n = 1 - 1/n$, we have $\lim f(x_n) = 0$, while $\lim f_n(x_n) = 1/e$.

Remark 2. If we assume the convergence in $\mathscr{Y}^{\mathscr{X}}$ to be the continuous convergence in the narrow sense, Theorem 3′ of Section VII becomes false. For, let $\mathscr{X}$ be the open interval $0 < x < 1$, let $\mathscr{T} = \mathscr{I}$, and let $g(x, t) = x^{1/t}$ for $t \neq 0$ and $g(x, 0) = 0$. The function g is then continuous, but, if we put $f_t(x) = g(x, t)$, the function f is not continuous since the convergence of the sequence $\{f_{1/n}\}$ to the function f_0 is not continuous in the narrow sense.

THEOREM 3. *If f is a continuous mapping of a c. compact space $\mathscr{T}$, then*

$$\mathscr{Y}^{f(\mathscr{T})} \underset{\text{top}}{\subset} \mathscr{Y}^{\mathscr{T}}.$$

More generally: *If we assume the convergence in the function spaces to be the continuous convergence in the narrow sense, the hypothesis of countable compactness of the space $\mathscr{T}$ can be omitted.*

Proof. Let $\mathscr{X} = f(\mathscr{T})$. With each function $g \in \mathscr{Y}^{\mathscr{X}}$ we associate the composition $u = g \circ f$, i.e. $u(t) = g(f(t))$. Then $u \in \mathscr{Y}^{\mathscr{T}}$.

Let $u = h(g)$. We have to show that h is a homeomorphism under the assumption that the limit $\lim g_n = g$ (respectively, $\lim u_n = u$) means the continuous convergence of the considered functions in the narrow sense.

Suppose that $\lim g_n = g$. We have to show that $\lim u_n = u$, i.e. that $\lim u(t_n) = y$ implies $\lim u_n(t_n) = y$. The first of these conditions says that $g(f(t_n)) = y$. Since $\lim g_n = g$, it follows that $\lim g_n(f(t_n)) = y$, i.e. $\lim u_n(t_n) = y$.

Suppose now that $\lim u_n = u$, i.e. $\lim g_n \circ f = g \circ f$. Let $\lim g(x_n) = y$ where $x_n = f(t_n)$; i.e. $\lim u(t_n) = y$. We have to show that $\lim g_n(x_n) = y$, i.e. that $\lim u_n(t_n) = y$. But this is a direct consequence of the formulas $\lim u_n = u$ and $\lim u(t_n) = y$.

The theorem has been proved in the case of the continuous convergence in the narrow sense. Its validity in the case of the continuous convergence and of the compactness of $\mathscr{T}$ is a consequence of Theorem 2.

IX. Moore-Smith convergence (main definitions) [1]. Given a mapping $f: T \to \mathscr{X}$, it is called a *net* if T is a directed set (see § 3,X). A particular case of a net is an infinite sequence, namely when T is the set of positive integers. For this reason, it may be useful to write f_t instead of $f(t)$.

We say that the net f is *eventually in* $B \subset \mathscr{X}$, if there is a t_0 such that

$$(t_0 < t) \Rightarrow [f_t \in B]. \tag{1}$$

It is said to be *frequently in* B if T is cofinal with a T_0 such that

$$(t \in T_0) \Rightarrow [f_t \in B]. \tag{2}$$

Obviously, f is frequently in B if and only if it is not eventually in $(-B)$.

No assumption has so far been made about $\mathscr{X}$. Now let us suppose that $\mathscr{X}$ is Hausdorff.

DEFINITION. Let $x: T \to \mathscr{X}$. We say that the net x *converges* to a point $p \in X$, and write

$$\lim_{t \in T} x_t = p, \tag{3}$$

if it is eventually in every open G containing p.

---

[1] We refer here to the *General topology*, Chapter 2 of J. Kelley. See also E. H. Moore and H. L. Smith, *A general theory of limits*, Amer. Journ. Math. 44 (1922), pp. 102–121, G. Birkhoff, *Moore-Smith convergence in general topology*, Ann. of Math. 38 (1937), pp. 39–56, E. J. Mc Shane, *Partial orderings and Moore-Smith limits*, Amer. Math. Monthly 59 (1952), pp. 1–11, J. W. Tukey, *Convergence and uniformity in topology*, Ann. of Math. Studies 2 (1940).

The assumption $\mathscr{X}$ to be Hausdorff leads to the conclusion that *a net cannot have two limits.*

The theory of convergence in the general topology can be developed to a large extend on the lines usually adopted for convergence of sequences(1).

§ 21. Metric spaces. General properties

I. Definitions. A set is said to be a *metric space* if to every pair of its elements x, y there is assigned a real number $|x-y|$, called their *distance*, which satisfies the following conditions:

(i) $$|x-y| = 0 \quad \textit{if and only if} \quad x = y;$$

(ii) $$|x-y|+|x-z| \geqslant |y-z| \qquad (\textit{triangle inequality})(2).$$

Substituting, respectively, y and x for z in (ii), we deduce that $|x-y| \geqslant 0$ and $|x-y| = |y-x|$. Hence the distance is a non-negative and symmetric function with respect to two variables(3).

An open *ball* (closed *ball*) with the centre p and radius r is the set of all points x such that $|p-x| < r$ (respectively $|p-x| \leqslant r$).

---

(1) For further theorems on Moore-Smith convergence (formulated also in terms of *filters*), see — besides the mentioned papers: G. Choquet, *Convergences*, Ann. Univ. Grenoble 23 (1948), pp. 55–112, H. J. Kowalsky, Math. Nachr. 12 (1954), pp. 301–340, Jürgen Schmidt, *Eine Studie zum Begriff der Teilfolge*, Jahresber. Deutsche Math. Ver. 63 (1960), pp. 28–50, Z. Frolik, *Concerning topological convergence of sets*, Cechosl. Mat. Journ. 10 (1960), pp. 168–180, J. Flachsmeyer, Math. Nachr. 24 (1962), pp. 1–12; G. Bruns and J. Schmidt, *Zur Aequivalenz von Moore-Smith-Folgen und Filtern*, Math. Nachr. 13 (1955), pp. 169–186; S. Mrówka, *On the convergence of nets of sets*, Fund. Math. 45 (1958), pp. 237–246; G. Helmberg, *On convergence classes of sets*, Proc. Amer. Math. Soc. 13 (1962), pp. 918–921; G. Grimeisen, Math. Ann. 141 (1960), pp. 318–342, and Math. Ann. 144 (1961), pp. 386–417.

(2) This notion is due to M. Fréchet, Rend. Circ. Mat. di Palermo 22 (1906), p. 17. The term "metric space" was introduced by F. Hausdorff, *Grundzüge*, p. 211. For the definition of the text, see M. Fréchet, *Relations entre les notions de limite et de distance*, Trans. Amer. Math. Soc. 19 (1918), p. 54, and A. Lindenbaum, *Contributions à l'étude de l'espace métrique I*, Fund. Math. 8 (1926), p. 211.

(3) For spaces provided with non-symmetric "distance", see W. A. Wilson, *On quasi-metric spaces*, Amer. Journ. of Math. 53 (1931), p. 675.

THEOREM. *Each subset A of a metric space $\mathscr{X}$ is a metric space, the distance in A being the same as in $\mathscr{X}$.*

The proof is immediate.

Remark. If we replace condition (i) by $|x-x| = 0$ (so that the distance of two different points is not necessarily $\neq 0$), we can decompose the space into disjoint subsets by letting two points belong to the same set if and only if their distance is 0 (we say that in this case two points whose distance is 0 are identified). The distance of two sets X and Y of this kind is defined to be $|x-y|$, where $x \in X$ and $y \in Y$. The choice of the points x and y is immaterial, since $|x-x'| = 0$ implies $|x'-y| \leqslant |x-x'|+|x-y| = |x-y|$.

It is easily verified that the sets in question considered as points constitute a metric space.

EXAMPLES. 1) The simplest case is that of the real numbers from which we have taken our notation. The distance of two numbers is defined, as usual, to be the absolute value of their difference. More generally, the n-dimensional euclidean space is a metric space, the distance of two points $x_1, \ldots, x_n$ and $y_1, \ldots, y_n$ being equal to

$$\sqrt{(x_1-y_1)^2+\ldots+(x_n-y_n)^2}.$$

The definition of the ball coincides with that of the n-dimensional ball assumed in geometry.

2) The set of all *bounded functions* $f: \mathscr{I} \to \mathscr{E}$ becomes a metric space, if the distance of two functions is defined to be

$$|f_1-f_2| = \sup|f_1(x)-f_2(x)|,$$

where "sup" denotes the least upper bound (see Section X (1)).

3) The Hilbert space, i.e. the set of all sequences $\mathfrak{x} = (x_1, x_2, \ldots)$ where $|x_1|^2+|x_2|^2+\ldots < \infty$ with the distance

$$|\mathfrak{x}-\mathfrak{y}| = \sqrt{\sum_{n=1}^{\infty}(x_n-y_n)^2},$$

is a metric space (if we assume that $0 \leqslant x_n \leqslant 1/n$, we obtain a set homeomorphic to the Hilbert cube $\mathscr{I}^{\aleph_0}$)([1]).

([1]) See D. Hilbert, Gött. Nachr. 1906, pp. 200 and 439.

4) The set of *square integrable functions* with the distance

$$|f_1-f_2| = \left\{\int_0^1 [f_1(x)-f_2(x)]^2 dx\right\}^{1/2}$$

is a metric space, if we identify two functions when they differ at most in a set of measure zero.

5) A set of an *arbitrary power* can be regarded as a metric space if we define the distance of different points to be equal 1.

II. Topology in metric spaces. The *closure* $\bar{A}$ in a metric space is defined as follows

$$(p \in \bar{A}) \equiv \bigwedge_{\varepsilon} \bigvee_{a} (a \in A)(|a-p| < \varepsilon), \tag{1}$$

in other terms: *every ball with center p contains points of A.*

A topological space is called *metrizable* if it is homeomorphic to a metric space with the closure defined as above.

THEOREM 1. *Each metric space $\mathscr{X}$ is a topological $\mathscr{T}_1$-space with a local countable base at each of its points; it is also an $\mathscr{L}^*$-space.*

For the proof of the fact that $\mathscr{X}$ is $\mathscr{T}_1$, see § 4, II. Next, the family of all balls with center p and rational radius is a local countable base at p. Finally $\mathscr{X}$ is an $\mathscr{L}^*$-space by assuming that

$$(p = \lim_{n=\infty} p_n) \equiv (\lim_{n=\infty} |p-p_n| = 0). \tag{2}$$

Let us note that *in order that p be an interior point of A it is necessary and sufficient that p be the centre of a ball contained in A.*

We shall show later (see Section IV) that each metric space is *normal* (even perfectly normal). Now, let us show the following important theorem.

THEOREM 2. *Each metric separable space contains a countable open base.*

Each metric space with a countable open base is separable.

Proof. A separable space, by definition, contains a countable dense subset S. Let $R_1, R_2, \ldots$ be the sequence of open balls whose centres are points of S and whose radii are rational numbers.

Let G be an open set and $p \in G$. There exists an $\eta > 0$ such that

$$|x-p| < \eta \quad \text{implies} \quad x \in G. \tag{i}$$

S being a dense set, there exists a point $s \in S$ and a rational number ϱ such that

$$|s-p| < \varrho < \eta/2. \tag{ii}$$

If R_n is an open ball with centre s and radius ϱ, then, by (ii), $p \in R_n$. Moreover, if $x \in R_n$, then $|x-s| < \varrho$, hence by (ii), $|x-p| \leqslant |x-s| + |s-p| < \eta$ which proves by (i) that $x \in G$. Therefore $R_n \subset G$.

Thus to each point p of G there is n such that $p \in R_n \subset G$. This means that $R_1, R_2, \ldots$ is a base.

Conversely, if $R_1, R_2, \ldots$ is a base, then by choosing a point p_n in each $R_n (\neq 0)$, we obtain a countable set dense in the space.

Remark 1. *We may assume that $\delta(R_n) < \varepsilon$, where $\varepsilon > 0$ is given.* For, the radii of $R_1, R_2, \ldots$ could be supposed to be less than $\varepsilon/2$.

Remark 2. In the euclidean space, the balls whose radius and coordinates of the centre are rational constitute a base.

III. Diameter. Continuity. Oscillation. The *diameter*, $\delta(X)$, of a set X is the least upper bound of the distances of its points. If $\delta(X)$ is finite, the set X is said to be *bounded*.

The following propositions are easily proved:

$$\{\delta(X) = 0\} \equiv \{X \textit{ is empty or is composed of a single point}\}; \tag{1}$$

$$\textit{if } X \subset Y, \textit{ then } \delta(X) \leqslant \delta(Y); \tag{2}$$

$$\delta(\bar{X}) = \delta(X); \tag{3}$$

$$\textit{if } X \cap Y \neq 0, \textit{ then } \delta(X \cup Y) \leqslant \delta(X) + \delta(Y); \tag{4}$$

$$\textit{if } X \textit{ is countably compact, then } \delta(X) < \infty \textit{ (i.e. } X \textit{ is bounded)}. \tag{5}$$

To prove (5) we observe that in a space which is not bounded there exists an infinite sequence of points $p_1, p_2, \ldots$ such that $|p_n - p_m| > 1$ for each n and $m \neq n$. Obviously the sequence $\{p_n\}$ contains no convergent subsequence.

If f is a mapping of an arbitrary topological space $\mathscr{X}$ onto a metric space $\mathscr{Y}$, the continuity of f at the point p (compare § 13, II, Corollary) can be stated as follows: *to every $\varepsilon > 0$ there corresponds a neighbourhood E of p such that $x \in E$ implies* $|f(x) - f(p)| < \varepsilon$. If we assume that ε is of the form $1/n$, we can reformulate this by saying that *there exists a sequence of neighbourhoods E_n of p such that* $\lim\limits_{n=\infty} \delta[f(E_n)] = 0$. Thus if G_n denotes the union of all open sets $G \subset \mathscr{X}$ such that $\delta[f(G)] < 1/n$, then $C = G_1 \cap G_2 \cap \ldots$ is *the set of points of continuity of* f. *It is a* $\boldsymbol{G}_\delta$*-set.*

We shall generalize this statement by introducing the notion of *oscillation* (which enables us to measure the discontinuity). Given a function f defined for points x of a subset A of $\mathscr{X}$, the oscillation of f at a point p (which may or may not belong to A) is understood to be the greatest lower bound of the numbers $\delta[f(E)]$, where E ranges over open neighbourhoods of p:

$$\omega(p) = \inf \delta[f(E)] \tag{6}$$

("inf" denotes the greatest lower bound).

For example, if $f(x) = \sin 1/x$, then $\omega(0) = 2$.

Obviously, if p does not belong to $\bar{A}$, then $\omega(p) = 0$, since, for $E = 1 - \bar{A}$, the set $f(E)$ is empty and $\delta[f(E)] = 0$.

The set G_n defined as above is readily seen to be the set of points where the oscillation is $< 1/n$. It follows that *the set of points where the oscillation is zero is a* $\boldsymbol{G}_\delta$*-set* (actually this is the set C). Finally $A \cap C$ (i.e. the set of points of continuity of f) is a $\boldsymbol{G}_\delta$-set relative to A.

Observe that if both $\mathscr{X}$ and $\mathscr{Y}$ are metric spaces, the condition for continuity given above can be stated in the classical form (Cauchy definition): For every $\varepsilon > 0$ there exists a $\delta > 0$ such that the condition $|x-p| < \delta$ implies $|f(x)-f(p)| < \varepsilon$ (the neighbourhood E is replaced by the ball with centre p and radius δ).

We have seen previously (§ 15,V, Theorem 2) that if $f: \mathscr{X} \to \mathscr{Y}$ is continuous and $\mathscr{Y}$ Hausdorff, then the set $\mathfrak{I} = \underset{xy}{E}[y = f(x)]$ is closed in $\mathscr{X} \times \mathscr{Y}$. For $\mathscr{Y}$ metric one has the following more precise statement (for the metrization of $\mathscr{X} \times \mathscr{Y}$ see VI (1)).

THEOREM. *Let* $A \subset \mathscr{X}$ *and* $f: A \to \mathscr{Y}$. *If* f *is continuous at the point* x_0 *and* $(x_0, y_0) \in \bar{\mathfrak{I}}$, *then* $y_0 = f(x_0)$.

More generally:

$$\delta\left[\bar{\mathfrak{I}} \cap \left((x) \times \mathscr{Y}\right)\right] \leqslant \omega(x) \tag{7}$$

(*hence if* $\omega(x) = 0$, *there is at most one* y *such that* $(x, y) \in \bar{\mathfrak{I}}$).

Proof. Suppose that the points (x, y) and (x, y^*) belong to the set $\bar{\mathfrak{I}} \cap \left((x) \times \mathscr{Y}\right)$.

Consequently in each neighbourhood E of x and for each $\varepsilon > 0$, there is an x_1 and an x_1^* such that $|f(x_1)-y| < \varepsilon$ and $|f(x_1^*)-y^*| < \varepsilon$. Hence

$$f(x_1) \in f(E) \quad \text{and} \quad f(x_1^*) \in f(E).$$

It follows that

$$\delta[f(E)] \geqslant |f(x_1)-f(x_1^*)| > |y-y^*|-2\varepsilon, \ \omega(x) = \inf \delta[f(E)] \geqslant |y-y^*|,$$

whence formula (7).

IV. The number $\varrho(A, B)$. Generalized ball. Normality of metric spaces.

DEFINITION 1.

$$\varrho(A, B) = \inf |a-b|, \text{ where } a \in A, \ b \in B,$$

the sets A and B being supposed non-empty (¹).

The following formulas hold:

$$\varrho(x, y) = |x-y|; \tag{1}$$

$$\varrho(\bar{A}, \bar{B}) = \varrho(A, B); \tag{2}$$

$$\{\varrho(x, A) = 0\} \equiv \{x \in \bar{A}\}; \tag{3}$$

$$\varrho(A, C) \leqslant \varrho(A, B) + \varrho(B, C) + \delta(B); \tag{4}$$

$$\varrho(A, C) \leqslant \varrho(A, B) + \varrho(A \cup B, C) + \delta(B); \tag{4'}$$

for a fixed A, $\varrho(x, A)$ is a continuous function of x. (5)

Formulas (1)-(3) are obvious; (4) follows from (4′), since

$$\varrho(A \cup B, C) \leqslant \varrho(A, C). \tag{6}$$

Suppose that (4′) is false. Then there exist two pairs of points: x, y and u, z such that $x \in A$, $y \in B$, $u \in A \cup B$, $z \in C$, and

$$|x-y| + |u-z| + \delta(B) < \varrho(A, C). \tag{i}$$

Therefore $|u-z| < \varrho(A, C)$. This implies that $u \notin A$ and hence $u \in B$ and $|y-u| \leqslant \delta(B)$. But

$$\varrho(A, C) \leqslant |x-z| \leqslant |x-y| + |y-u| + |u-z| \leqslant |x-y| + |u-z| + \delta(B),$$

contrary to (i).

(¹) F. Hausdorff calls $\varrho(A, B)$ the lower distance of A to B. See *Mengenlehre*, p. 145.

In order to prove (5), it is obviously sufficient to show that(1)

$$|\varrho(x,A)-\varrho(y,A)| \leqslant |x-y|. \tag{6'}$$

Now, (6') can be established as follows. Let $a \epsilon A$ be such that $|y-a| < \varrho(y,A)+\varepsilon$ (for a given $\varepsilon > 0$). Since $\varrho(x,A) \leqslant |x-a|$ and $|x-a| \leqslant |x-y|+|y-a|$, it follows $\varrho(x,A) < \varrho(y,A)+|x-y| +\varepsilon$. Therefore (as $\varepsilon > 0$ is arbitrary): $\varrho(x,A)-\varrho(y,A) \leqslant |x-y|$. Similarly $\varrho(y,A)-\varrho(x,A) \leqslant |x-y|$. This completes the proof.

It is worthy noticing that uniform continuity can be expressed in terms of the function ϱ. Namely

In order that a function $f\colon \mathcal{X} \to \mathcal{Y}$, where $\mathcal{X}$ and $\mathcal{Y}$ are metric, be uniformly continuous it is necessary and sufficient that(2)

$$[\varrho(A,B)=0] \Rightarrow [\varrho\big(f(A),f(B)\big)=0].$$

The following statement is easily proved:

$$\varrho(A, B \cup C) = \text{either } \varrho(A,B) \text{ or } \varrho(A,C). \tag{7}$$

DEFINITION 2. *A generalized ball with radius r and centre A* (which is supposed to be non-empty) is defined to be the set of all points x such that $\varrho(x,A) < r$. Replacing $<$ with $\leqslant$ we obtain the definition of a generalized *closed* ball.

Since the function $\varrho(x,A)$ is continuous and the set of real numbers $< r$ is open, it follows that the *generalized open ball is an open set* (compare § 13,IV). Moreover, the generalized open ball is easily seen to be the union of open balls with radius r and centres belonging to A.

By (2), replacing A with $\bar{A}$ does not change the ball.

THEOREM 1. *Every metric space is perfectly normal* (see § 14,VI).

More precisely, *if A and B are disjoint closed sets, then the function*

$$f(x) = \frac{\varrho(x,A)}{\varrho(x,A)+\varrho(x,B)} \tag{8}$$

satisfies the conditions:

$$0 \leqslant f(x) \leqslant 1, \quad [f(x)=0] \equiv (x \epsilon A), \quad [f(x)=1] \equiv (x \epsilon B).$$

(1) J. Dieudonné, *Foundations of modern analysis*, New York 1960, p. 32.

(2) See W. A. Efremovitch, *Proximity Geometry* I (Russian), Matem. Sbornik 31 (1952), p. 190.

Here we assume that $\varrho(x, 0) \equiv 1$.

The proof follows immediately from (3) and (5).

By the theorem of Vedenisov (§ 14,VI), it follows that *in a metric space each closed set is a G_δ-set* (and symmetrically, *each open set is an F_σ-set*). This can also be shown directly by putting for $A = \bar{A}$:

$$A = B_1 \cap B_2 \cap \dots \tag{9}$$

where B_n is the open ball with centre A and radius $1/n$.

THEOREM 2. *Every metric space is hereditary normal.*

This follows from Theorem 1 by the Čech theorem mentioned in § 14,VI. It can be also directly deduced from Theorem 1, which implies that every metric space is normal, and from the theorem of Section I by virtue of which every subset of a metric space is metric.

Remarks. As seen, formula (8) gives a direct proof of the *Urysohn lemma* (compare § 14, IV) restricted to metric spaces.

Also the *Tietze extension theorem* can be shown in a more direct way for metric spaces. Namely if $0 \leqslant f(x) \leqslant 1$, a continuous extension of the function f can be obtained by putting

$$f^*(p) = \inf_{x \in F}\left\{f(x) + \frac{|p-x|}{\varrho(p, F)} - 1\right\} \quad \text{for} \quad p \in \mathcal{X} - F\ (^1).$$

It follows that $0 \leqslant f^*(p) \leqslant 1$ (the equality sign can occur).

V. Shrinking mapping [2].

THEOREM. *Every metric space is homeomorphic to a bounded space.*

Proof. We define a "new distance" by the formula

$$\|x-y\| = \frac{|x-y|}{1+|x-y|}.$$

---

[1] See F. Hausdorff, Math. Zft. 5 (1919), p. 296. His proof is reproduced in the first French edition of this book.

[2] Compare "Schränkungstransformation" of H. Hahn, *Theorie der reellen Funktionen I*, Berlin 1921, p. 115; also R. Baire, Acta Math. 30 (1906), p. 6.

We have to show that the distance $\|x-y\|$ satisfies Axioms (i) and (ii) of the definition of metric space. The formula $\|x-y\|=0$ is equivalent to $|x-y|=0$. Moreover,

$$\|x-y\|+\|x-z\| \geqslant \frac{|x-y|}{1+|x-y|+|x-z|}+\frac{|x-z|}{1+|x-y|+|x-z|}$$

$$=\frac{|x-y|+|x-z|}{1+|x-y|+|x-z|}=\frac{1}{1+\dfrac{1}{|x-y|+|x-z|}} \geqslant \frac{1}{1+\dfrac{1}{|y-z|}}=\|y-z\|.$$

The two metric spaces, one with the distance $|x-y|$, and the other with the distance $\|x-y\|$, are homeomorphic, i.e. the equivalence

$$\{\lim_{n=\infty}|x_n-y|=0\} \equiv \{\lim_{n=\infty}\|x_n-y\|=0\}$$

holds. This follows directly from the fact that $u=t/(1+t)$ is a homeomorphism in the domain of non-negative numbers.

Finally, the diameter of the space with respect to the distance $\|x-y\|$ is $\leqslant 1$.

Remark. The preceding theorem could be proved more directly by defining the "new distance" to be equal to $|x-y|$ if $|x-y|\leqslant 1$ and 1 otherwise(1).

VI. Metrization of the cartesian product. Let $\mathscr{X}$ and $\mathscr{Y}$ be two metric spaces. The product $\mathscr{X}\times\mathscr{Y}$ is metrized by the formula

$$|\mathfrak{z}-\mathfrak{z}_1|=\sqrt{|x-x_1|^2+|y-y_1|^2} \quad (\text{where } \mathfrak{z}=(x,y)). \tag{1}$$

It is seen that (compare § 20, IV)

$$\mathfrak{z}=\lim_{n=\infty}\mathfrak{z}_n \textit{ if and only if } \lim_{n=\infty}|\mathfrak{z}-\mathfrak{z}_n|=0.$$

Consequently this definition agrees with the topology of $\mathscr{X}\times\mathscr{Y}$ as introduced in § 15, II.

More generally, let $\mathscr{X}_i$, $i=1,2,\ldots$, be a metric space of diameter $\leqslant 1$. The space $\mathscr{X}_1\times\mathscr{X}_2\times\ldots$ is metrized by the formula

$$|\mathfrak{z}-\mathfrak{y}|=\sum_{i=1}^{\infty}2^{-i}|\mathfrak{z}^i-\mathfrak{y}^i| \quad (\text{where } \mathfrak{z}=(\mathfrak{z}^1,\mathfrak{z}^2,\ldots)). \tag{2}$$

(1) See M. H. Newman, *Elements of the topology of plane sets of points*, Cambridge 1939, p. 47.

The distance so defined satisfies conditions (i) and (ii), since the equivalence $\{|\mathfrak{z}-\mathfrak{y}| = 0\} \equiv \{\mathfrak{z} = \mathfrak{y}\}$ is obvious and the triangle formula follows from the formula

$$|\mathfrak{z}-\mathfrak{y}|+|\mathfrak{z}-\mathfrak{w}| = \sum_{i=1}^{\infty} 2^{-i}[|\mathfrak{z}^i-\mathfrak{y}^i|+|\mathfrak{z}^i-\mathfrak{w}^i|]$$

$$\geqslant \sum_{i=1}^{\infty} 2^{-i}|\mathfrak{y}^i-\mathfrak{w}^i| = |\mathfrak{y}-\mathfrak{w}|.$$

It only remains to prove that $\mathfrak{z} = \lim_{n=\infty} \mathfrak{z}_n$ if and only if $\lim_{n=\infty} |\mathfrak{z}-\mathfrak{z}_n| = 0$, i.e. to prove the equivalence

$$\{\lim_{n=\infty} |\mathfrak{z}-\mathfrak{z}_n| = 0\} \equiv \{\lim_{n=\infty} |\mathfrak{z}^i-\mathfrak{z}_n^i| = 0, \text{ for every } i\}.$$

First, the condition $\lim_{n=\infty} |\mathfrak{z}-\mathfrak{z}_n| = 0$ implies, for every i, $\lim_{n=\infty} |\mathfrak{z}^i-\mathfrak{z}_n^i| = 0$, since $|\mathfrak{z}^i-\mathfrak{z}_n^i| \leqslant 2^i |\mathfrak{z}-\mathfrak{z}_n|$. Conversely, suppose that $\lim_{n=\infty} |\mathfrak{z}^i-\mathfrak{z}_n^i| = 0$ for every i. Given an $\varepsilon > 0$, let m be a positive integer such that $2^{-m} < \varepsilon$. Let j be an integer such that $|\mathfrak{z}^1-\mathfrak{z}_n^1| < \varepsilon, \ldots, |\mathfrak{z}^m-\mathfrak{z}_n^m| < \varepsilon$ for every $n > j$. The formula $|\mathfrak{z}^i-\mathfrak{z}_n^i| < \delta(\mathscr{X}_i) \leqslant 1$ implies that

$$|\mathfrak{z}-\mathfrak{z}_n| \leqslant \sum_{i=1}^{m} 2^{-1}|\mathfrak{z}^i-\mathfrak{z}_n^i| + \sum_{i=m+1}^{m} 2^{-i} < 2\varepsilon.$$

Hence $\lim_{n=\infty} |\mathfrak{z}-\mathfrak{z}_n| = 0$.

Remark 1. There are other definitions of the distance that can be taken in place of (1). For example, nothing will change from the topological point of view, if the distance between the points $\mathfrak{z}$ and $\mathfrak{z}_1$ is defined to be the greater of the two numbers $|x-x_1|$ and $|y-y_1|$, or the sum of these numbers.

Remark 2. If we do not assume that $\delta(\mathscr{X}_i) \leqslant 1$, we define(1)

$$|\mathfrak{z}-\mathfrak{y}| = \sum_{i=1}^{\infty} 2^{-i} \frac{|\mathfrak{z}^i-\mathfrak{y}^i|}{1+|\mathfrak{z}^i-\mathfrak{y}^i|}. \tag{3}$$

THEOREM. *The distance $|x-y|$ is a continuous function of the variable $\mathfrak{z} = (x, y)$.*

---

(1) Compare the formula of Fréchet, *Espaces abstraits*, p. 82.

Proof. Let $\mathfrak{a} = (a, b)$ be a given point and let $|\mathfrak{z}-\mathfrak{a}| < \varepsilon$. It follows that $|x-a| \leqslant |\mathfrak{z}-\mathfrak{a}| < \varepsilon$ and $|y-b| < \varepsilon$. Therefore

$$|x-y| \leqslant |x-a|+|a-b|+|b-y| < |a-b|+2\varepsilon$$

and

$$|a-b| \leqslant |a-x|+|x-y|+|y-b| < |x-y|+2\varepsilon.$$

Hence $\big||x-y|-|a-b|\big| < 2\varepsilon$ which yields the required conclusion (compare Section III).

It is worth noticing the following formulas which can be easily established:

$$\delta(A\times B) = \sqrt{[\delta(A)]^2+[\delta(B)]^2}, \tag{4}$$

$$\delta(\mathop{P}_{n=1}^{\infty} A_n) = \sum_{n=1}^{\infty}\frac{\delta(A_n)}{2^n}. \tag{5}$$

VII. Distance of two sets. The space $(2^{\mathscr{X}})_m$.

DEFINITION. We denote by $(2^{\mathscr{X}})_m$ the space of all *closed* and *bounded* subsets of the metric space $\mathscr{X}$, the distance(¹), $\mathrm{dist}(A, B)$, of two sets $A \neq 0$ and $B \neq 0$ being understood to be the smallest upper bound of the numbers

$$\varrho(x, B) \quad \text{and} \quad \varrho(y, A), \quad \text{where } x \in A \text{ and } y \in B.$$

If $A = 0 \neq B$, then $\mathrm{dist}(A, B) = \delta(\mathscr{X})$ (thus if $\mathscr{X}$ is unbounded, we have to remove the void set from $(2^{\mathscr{X}})_m$; if $\mathscr{X}$ reduces to a single point p, we put $\mathrm{dist}\big(0, (p)\big) = 1$).

THEOREM. *The space $(2^{\mathscr{X}})_m$ is a metric space with respect to the above defined distance.*

Proof. Clearly $\mathrm{dist}(A, B) = 0$ if and only if $A = B$. It only remains to prove the triangle formula.

If $x \in A$ and $y \in B$, then (by IV, (4))

$$\varrho(x, C) \leqslant |x-y|+\varrho(y, C) \leqslant |x-y|+\mathrm{dist}(B, C),$$

whence

$$\begin{aligned}\varrho(x, C) &\leqslant \inf_{y\in B}|x-y|+\mathrm{dist}(B, C)\\ &= \varrho(x, B)+\mathrm{dist}(B, C) \leqslant \mathrm{dist}(A, B)+\mathrm{dist}(B, C).\end{aligned}$$

By symmetry,

$$\varrho(z, A) \leqslant \mathrm{dist}(B, A)+\mathrm{dist}(B, C).$$

(¹) F. Hausdorff, *Grundzüge der Mengenlehre*, Ch. VIII, § 6. Compare also D. Pompeju, Ann. de Toulouse (2) 7(1905).

Therefore

$$\mathrm{dist}(A, C) \leqslant \mathrm{dist}(B, A) + \mathrm{dist}(B, C).$$

Remark 1. *$\varrho(A, B)$ cannot be taken for a distance in this space.* For example, if A is the set consisting of the number 0, B is the interval $1,2$ and C consists of the number 3, then the triangle formula for A, B and C fails (the number $\varrho(A, B)$ is zero, if A and B have a point in common).

Remark 2. As we shall see later (Vol. II), if $\mathscr{X}$ is *compact metric*, then $(2^{\mathscr{X}})_m$ is homeomorphic to $2^{\mathscr{X}}$ with the exponential topology (see § 17,I).

Let us observe that

$$|a-b| = \mathrm{dist}[(a), (b)], \tag{1}$$

$$\{\mathrm{dist}(A, B) \leqslant \varepsilon\} \equiv \{A \subset R_\varepsilon(B) \text{ and } B \subset R_\varepsilon(A)\}, \tag{2}$$

where $R_\varepsilon(X)$ denotes the closed ball with radius ε and centre X,

$$\delta\big((2^{\mathscr{X}})_m\big) = \delta(\mathscr{X}) \quad \text{(if } \mathscr{X} \text{ contains more than one point).} \tag{3}$$

VIII. Totally bounded spaces.

Definition. A metric space is said to be *totally bounded*, if for every $\varepsilon > 0$ it can be decomposed into a finite number of sets of diameter $< \varepsilon$ (1).

Theorem 1. *The space $\mathscr{X}$ is totally bounded if and only if for every $\varepsilon > 0$ there exists a finite set F_ε such that, for each x, $\varrho(x, F_\varepsilon) < \varepsilon$.*

Proof. If the space $\mathscr{X}$ is totally bounded, then

$$\mathscr{X} = A_1^n \cup \ldots \cup A_{k_n}^n, \quad \delta(A_i^n) < 1/n \quad \text{for} \quad i \leqslant k_n.$$

Let p_i^n be a point of A_i^n. The set $F_{1/n}$ composed of the points $p_1^n, \ldots, p_{k_n}^n$ is the required one (for $\varepsilon > 1/n$).

Conversely, if $F_{\varepsilon/2}$ is a set satisfying the condition of the theorem, then the family of balls with radius $\varepsilon/2$ and centre belonging to $F_{\varepsilon/2}$ constitutes the required cover of the space $\mathscr{X}$.

Remark 0. The above condition can also be expressed by means of the notion of distance of sets (Section VII) as follows: *The space is the limit of a sequence of finite sets* (namely, of the sequence $F_1, F_{1/2}, \ldots, F_{1/n}, \ldots$).

---

(1) F. Hausdorff, *Mengenlehre*, p. 108.

This is also a consequence of the following theorem.

THEOREM 2. *If $\mathscr{X}$ is totally bounded, then so is $(2^{\mathscr{X}})_m$.*

Proof. Let F_ε be the set considered above and let $H_{1,\varepsilon}, \ldots, H_{k,\varepsilon}$ be the family of all subsets of F_ε. With each closed set X we associate the set $H_{i,\varepsilon}$ of points p of F_ε such that $\varrho(p, X) < \varepsilon$. Hence $\operatorname{dist}(X, H_{i,\varepsilon}) \leqslant \varepsilon$.

THEOREM 3. *Every totally bounded space is separable.*

To prove this, it suffices to consider the set of points p_i^n, $n = 1, 2, \ldots, i \leqslant k_n$ defined in the proof of Theorem 1.

THEOREM 4. *If G is an open subset of a totally bounded space $\mathscr{X}$, then there exists a sequence of open sets $G_0, G_1, \ldots$ such that*

$$G = G_0 \cup G_1 \cup \ldots, \tag{o}$$

$$G_i \textit{ intersects at most a finite number of sets } G_j, \tag{i}$$

$$\bar{G}_i \subset G, \tag{ii}$$

and

$$\lim_{i=\infty} \delta(G_i) = 0. \tag{iii}$$

Proof. We can assume that $G \neq \mathscr{X}$, for otherwise we put $G_0 = \mathscr{X}$ and $G_i = 0$ for $i > 0$.

Let B_m be the open ball with centre $\mathscr{X} - G$ and radius $1/m$ (compare Section IV). Let $H_{1,m}, \ldots, H_{k_m,m}$ be a family of open subsets such that

$$\mathscr{X} = H_{1,m} \cup \ldots \cup H_{k_m,m} \quad \text{and} \quad \delta(H_{j,m}) < 1/m.$$

Since $G = \bigcup_{m=0}^{\infty} (B_m - \bar{B}_{m+2})$, hence $G = \bigcup_{m=0}^{\infty} \bigcup_{j=1}^{k_m} (B_m - \bar{B}_{m+2}) \cap H_{j,m}$.

By rearranging the terms of this double series into an infinite sequence $G_0, G_1, \ldots$, we obtain a sequence satisfying (o)–(iii).

Condition (i) is a consequence of the formula

$$(B_m - \bar{B}_{m+2}) \cap (B_n - \bar{B}_{n+2}) \subset B_{m+2} - \bar{B}_{m+2} = 0 \quad \text{for} \quad n \geqslant m+2.$$

Condition (ii) follows from the formula

$$\overline{(B_m - \bar{B}_{m+2}) \cap H_{j,m}} \subset \overline{\mathscr{X} - B_{m+2}} \subset G.$$

Finally, condition (iii) follows from the inequality $\delta(H_{j,m}) < 1/m$.

Remark 1. *In an arbitrary metric space, for every $\varepsilon > 0$ there exists a (finite or infinite) closed and discrete set F_ε such that $\varrho(x, F_\varepsilon) < \varepsilon$ for every x.* This can be proved as follows. We order all points of the space into a transfinite sequence $p_0, p_1, \ldots, p_\alpha, \ldots$.

Let $p_{a_0} = p_0$ and, for $\gamma > 0$, let p_{a_γ} be the point with minimal index satisfying $|p_{a_\gamma} - p_{a_\xi}| \geqslant \varepsilon$ for each $\xi < \gamma$ (provided that such a point p_{a_γ} exists). Then F_ε is the set of all p_{a_γ}.

Remark 2. Denote by $\mathscr{X}$ the curve $y = \sin 1/x$, $0 < x \leqslant 1$, with the distance of two points p and q being equal to the diameter of the arc pq. This space is evidently separable and bounded (even complete), but it is not totally bounded. The family of its subsets situated on the x-axis (which are all closed) is uncountable and the distance between any two elements of this family exceeds 1. Therefore this family, and hence also the space $(2^{\mathscr{X}})_m$, is not separable (compare Theorem 2 of Section II).

It is remarkable that the curve $\mathscr{X}$ can be mapped homeomorphically (by projection onto the x-axis) so that the space $(2^{\mathscr{X}})_m$ becomes separable.

It follows that the topological properties of the space $(2^{\mathscr{X}})_m$ are not necessarily topological properties of the space $\mathscr{X}$. This is due to the fact the space $(2^{\mathscr{X}})_m$ is not topologically defined.

THEOREM 5. *If $\{\mathscr{X}_i\}$ is a (finite or countably infinite) sequence of totally bounded spaces such that $\delta(\mathscr{X}_i) \leqslant 1$, then their cartesian product metrized by formula* (2) *of Section* VI, *is totally bounded.*

Proof. Let $\varepsilon > 0$ and let i be an integer such that $2^{-i} < \varepsilon/2$. For every $l \leqslant i$ there exists, by the hypothesis, a finite system of points $r_{l1}, \ldots, r_{lk_l}$ such that each point of the space $\mathscr{X}_l$ lies at a distance $< \varepsilon/2$ from a point of this system. For $n > i$, let r_n be an arbitrary point of $\mathscr{X}_n$.

For a fixed i consider the finite system of sequences of the form

$$\mathfrak{y} = [r_{1j_1}, r_{2j_2}, \ldots, r_{ij_i}, r_{i+1}, r_{i+2}, \ldots]$$

where $j_1 \leqslant k_1, \ldots, j_i \leqslant k_i$.

Given a point $\mathfrak{z} = [x_1, x_2, \ldots]$ of the space $\mathscr{X}_1 \times \mathscr{X}_2 \times \ldots$, there exists, for $l \leqslant i$, a point r_{lj_l} such that $|r_{lj_l} - x_l| < \varepsilon/2$; hence

$$|\mathfrak{y} - \mathfrak{z}| = \sum_{l=1}^{i} 2^{-l} |r_{lj_l} - x_l| + \sum_{n=i+1}^{\infty} 2^{-n} |r_n - x_n| < \frac{\varepsilon}{2} + \frac{\varepsilon}{2} = \varepsilon.$$

Therefore the space $\mathscr{X}_1 \times \mathscr{X}_2 \times \ldots$ is totally bounded.

COROLLARY 5a. *The Hilbert cube $\mathscr{I}^{\aleph_0}$ and consequently, the space $(2^{(\mathscr{I}^{\aleph_0})})_m$ are totally bounded.*

The latter part of the corollary follows from Theorem 2.

IX. Equivalence between countably compact metric spaces and compact metric spaces.

THEOREM 1. *Every countably compact metric space $\mathscr{X}$ is totally bounded, and hence contains a countable open base and furthermore is a Lindelöf space.*

Proof. If the space is not totally bounded, then, by Theorem 1 of Section VIII, there exists an $\varepsilon > 0$ such that for each finite set F there exists a point x with $\varrho(x, F) \geqslant \varepsilon$. This yields a sequence of points $p_1, p_2, \ldots$, such that $|p_n - p_m| \geqslant \varepsilon$ for every n and $m \neq n$. This sequence does not contain any convergent subsequence. Hence $\mathscr{X}$ is not countably compact.

In order to prove that $\mathscr{X}$ is a Lindelöf space, we apply first Theorem 3 of Section VIII, by which every totally bounded space is separable, then Theorem 2 of Section II, by which every metric separable space contains a countable open base, and finally the theorem of Lindelöf (§ 5, XI).

THEOREM 2. *Every countably compact metric space is compact.*

Proof. It is a Lindelöf space by the preceding theorem. Hence every open cover can be reduced to a countable cover and the latter—according to the Borel theorem (§ 20, V)—can be reduced to a finite one.

X. Uniform convergence. Metrization of the space $\mathscr{Y}^{\mathscr{X}}$. Given an arbitrary set $\mathscr{X}$ and a metric space $\mathscr{Y}$ we denote by $\Phi(\mathscr{X}, \mathscr{Y})$ the family of all bounded mappings $f: \mathscr{X} \to \mathscr{Y}$ (i.e. such that the set $f(\mathscr{X})$ is bounded).

THEOREM 1. *$\Phi(\mathscr{X}, \mathscr{Y})$ is a metric space with the distance defined by the formula*(1)

$$|f_1 - f_2| = \sup_{x \in \mathscr{X}} |f_1(x) - f_2(x)|. \tag{1}$$

Proof. First the distance defined above is always finite, for, if x' is a fixed point, then

$$|f_1(x) - f_2(x)| \leqslant |f_1(x) - f_1(x')| + |f_1(x') - f_2(x')| + |f_2(x') - f_2(x)|.$$

(1) Compare M. Fréchet, Rend. di Palermo 22 (1906), p. 36. According to M. Fréchet, this distance between functions was already considered by Weierstrass.

Next the condition $|f_1-f_2| = 0$ is equivalent to $f_1 = f_2$. It remains to prove the triangle inequality. We have

$$|f_1-f_2|+|f_1-f_3| = \sup|f_1(x)-f_2(x)|+\sup|f_1(x)-f_3(x)|$$
$$\geqslant \sup\{|f_1(x)-f_2(x)|+|f_1(x)-f_3(x)|\} \geqslant \sup|f_2(x)-f_3(x)| = |f_2-f_3|.$$

By the definition of convergence in the metric space, the sequence of functions f_n converges to f in the considered function space, if $\lim|f_n-f| = 0$, i.e. if for every $\varepsilon > 0$ there exists an $n(\varepsilon)$ such that, for $n > n(\varepsilon)$, $\sup\limits_{x\in\mathcal{X}}|f_n(x)-f(x)| \leqslant \varepsilon$, hence $|f_n(x)-f(x)| \leqslant \varepsilon$, for every x. This yields the following corollary.

COROLLARY 1a. *The convergence in the space $\Phi(\mathcal{X},\mathcal{Y})$ coincides with the uniform convergence in the usual sense.*

Remark. By Theorem 5 of § 20, V and by III (5), every continuous mapping of a compact metric space into a metric space is bounded, i.e. $\mathcal{Y}^{\mathcal{X}} \subset \Phi(\mathcal{X},\mathcal{Y})$. Therefore formula (1) gives in this case a metric structure to $\mathcal{Y}^{\mathcal{X}}$. Thus, *whenever $\mathcal{X}$ will be assumed to be compact metric*(1) *and $\mathcal{Y}$ to be metric, we shall consider $\mathcal{Y}^{\mathcal{X}}$ as a metric space with distance defined by formula* (1).

If $\mathcal{X}$ is not compact (nor locally compact), one cannot expect $\mathcal{Y}^{\mathcal{X}}$ to be metric. One can however introduce a suitable topology, the so-called *compact-open* topology or the *natural* topology, in which we define the closure of $\Gamma \subset \mathcal{Y}^{\mathcal{X}}$ by putting

$$(f\in\overline{\Gamma}) \equiv [(f|F)\in\overline{\Gamma|F} \text{ for each compact } F \subset \mathcal{X}],$$

or what is the same, by declaring that $\mathcal{Y}^{\mathcal{X}}$ is the inverse limit of $\{\mathcal{Y}^F\}$ with the operation of restriction.

This will be considered in details in Volume II.

Another generalization of $\mathcal{Y}^{\mathcal{X}}$ consists in conferring a topology upon the set of all *partial mappings* $f|F$ with F compact(2).

THEOREM 2. *For each topological space $\mathcal{X}$ the set of continuous bounded functions, i.e. $\mathcal{Y}^{\mathcal{X}} \cap \Phi(\mathcal{X},\mathcal{Y})$, is closed in the space $\Phi(\mathcal{X},\mathcal{Y})$.*

This is a consequence of the following theorem.

---

(1) As will be seen in Chapter IV, the assumption $\mathcal{X}$ to be metric may be omitted.

(2) See my paper *Sur l'espace des fonctions partielles*, Annali di Mat. 40 (1955), pp. 61–67.

THEOREM 2'. *The limit of a uniformly convergent sequence of continuous functions is a continuous function.*

The proof does not differ from the proof of Theorem 3 of § 13,VI (the case of $\mathscr{Y} = \mathscr{E}$).

We also have the following related theorem(1):

THEOREM 2''. *The limit of a continuously convergent sequence of functions* (which need not be continuous) *is a continuous function.*

Proof. By assumption there is f such that $\lim f_n(x) = f(x)$. Suppose that $\lim a_n = a$ and that $\{f(a_n)\}$ is not convergent to $f(a)$. It follows that there exists a neighbourhood G of $f(a)$ and a sequence of integers $k_1 < k_2 < \dots$ such that $f(a_{k_n}) \in \mathscr{Y} - \overline{G}$ for every n. Since $\mathscr{Y} - \overline{G}$ is a neighbourhood of $f(a_{k_n})$, there exists an index m_n such that $f_{m_n}(a_{k_n}) \in \mathscr{Y} - \overline{G}$. We can assume that $m_1 < m_2 < \dots$. Since the sequence $\{f_n\}$ is continuously convergent, the condition $\lim a_{k_n} = a$ (which follows from $\lim a_n = a$) implies the condition

$$\lim f_{m_n}(a_{k_n}) = f(a) \text{ and therefore } f(a) \in \overline{\mathscr{Y} - \overline{G}} \subset \overline{\mathscr{Y} - G} = \mathscr{Y} - G,$$

contrary to the definition of G (for $f(a) \in G$).

THEOREM 3. *The uniform convergence implies the continuous convergence in the narrow sense. Hence, in the domain of continuous functions, the uniform convergence implies the continuous convergence* (compare § 20, VIII, Theorem 1).

In other words, *if $\{f_n\}$ is uniformly convergent, then the condition* $\lim f(x_n) = y$ *implies* $\lim f_n(x_n) = y$.

Proof. Suppose that the sequence $\{f_n(x_n)\}$ is not convergent to y. Hence it contains a subsequence $\{f_{k_n}(x_{k_n})\}$ none of whose subsequences is convergent to y. Since the sequence $\{f_n\}$ is uniformly convergent, then so is the sequence $\{f_{k_n}\}$. It follows that there exists a sequence of integers $m_1 < m_2 < \dots$ such that, for every x,

$$|f_{k_{m_n}}(x) - f(x)| < 1/n.$$

In particular

$$|f_{k_{m_n}}(x_{k_{m_n}}) - f(x_{k_{m_n}})| < 1/n.$$

Comparing this inequality with the identity $\lim f(x_n) = y$ which implies $\lim f(x_{k_{m_n}}) = y$, we obtain that $\lim f_{k_{m_n}}(x_{k_{m_n}}) = y$, contrary to the definition of the sequence $\{k_n\}$.

(1) Compare H. Hahn, *Reelle Funktionen I*, Leipzig 1932, p. 225. The theorem is valid in $\mathscr{L}^*$ topological and regular spaces.

THEOREM 4. *If the space $\mathscr{Y}$ is compact, then the continuous convergence in the narrow sense implies the uniform convergence (hence these two notions of convergence coincide).*

Proof. Suppose that the sequence $\{f_n\}$ is not uniformly convergent to f. Hence there exist an $\varepsilon > 0$, a sequence of integers $k_1 < k_2 < \ldots$, and a sequence of points $x_1, x_2, \ldots$ such that

$$|f_{k_n}(x_n) - f(x_n)| > \varepsilon \quad \text{for every } n. \tag{2}$$

Since $\mathscr{Y}$ is compact, there exists a sequence $m_1 < m_2 < \ldots$ such that the sequence $\{f(x_{m_n})\}$ is convergent; let $\lim f(x_{m_n}) = y$. Since the sequence $\{f_n\}$ is supposed to be uniformly convergent in the narrow sense, then so is the sequence $\{f_{k_{m_n}}\}$. Hence the condition $\lim f(x_{m_n}) = y$ implies that $\lim f_{k_{m_n}}(x_{m_n}) = y$. But, for a sufficiently large n, we have $|f_{k_{m_n}}(x_{m_n}) - f(x_{m_n})| < \varepsilon$, contrary to (2). The contradiction shows that the sequence $\{f_n\}$ is not continuously convergent in the narrow sense.

THEOREM 5. *If the space $\mathscr{X}$ is compact, the continuous convergence implies the uniform convergence (hence these two notions of convergence coincide in the domain of continuous functions).*

Proof. Suppose, as before, that the sequence $\{f_n\}$ is not uniformly convergent to f. Hence we have (2).

Since $\mathscr{X}$ is compact, we can assume that the sequence $\{x_n\}$ is convergent, so $\lim x_n = x$. Since the sequence $\{f_n\}$ is continuously convergent, hence so is the sequence $\{f_{k_n}\}$ (compare § 20, VI, the final remark). Therefore

$$\lim_{n=\infty} f_{k_n}(x_n) = f(x). \tag{3}$$

On the other hand, since $\lim\limits_{m=\infty} f_m(x_n) = f(x_n)$, there exists a sequence of integers $m_1 < m_2 < \ldots$ such that

$$|f_{m_n}(x_n) - f(x_n)| < 1/n, \qquad \text{hence} \qquad |f_{k_n}(x_n) - f_{m_n}(x_n)| > \varepsilon - 1/n,$$

by (2). But this disagrees with (3), since the formula $\lim x_n = x$ implies $\lim f_{m_n}(x_n) = f(x)$.

Remark. If we assume that the space $\mathscr{Y}$ is compact, then the elements of the function space $\mathscr{Y}^{\mathscr{X}}$ are bounded (compare III (5)). Hence *formula* (1) *gives the space $\mathscr{Y}^{\mathscr{X}}$ a metric space structure.* The notion of limit derived from it (according to II, (i)) coincides with that introduced in Section VIII of § 20 (continuous convergence in the narrow sense).

This results easily from Theorem 4.

THEOREM 6. *Let p be a fixed point of a metric space $\mathscr{X}$. Given a bounded set A ($\neq 0$) in $\mathscr{X}$, let f_A be the function defined by*

$$f_A(x) = \varrho(x, A) - |x-p|. \tag{4}$$

Then

$$\operatorname{dist}(A, B) = |f_A - f_B|. \tag{5}$$

Proof. First observe that the function f_A is bounded:

$$|f_A(x)| \leqslant \varrho(p, A) + \delta(A).$$

For, inequality IV (4) yields

$$\varrho(x, A) \leqslant \varrho(p, A) + |x-p| \text{ and } |x-p| \leqslant \varrho(x, A) + \varrho(A, p) + \delta(A).$$

By symmetry we can assume that

$$\operatorname{dist}(A, B) = \sup_{x \in A} \varrho(x, B).$$

Let $\varepsilon > 0$ and let a be a point of A such that $\operatorname{dist}(A, B) \leqslant \varrho(a, B) + \varepsilon$. Since $\varrho(a, A) = 0$, it follows from (4) that

$$\operatorname{dist}(A, B) \leqslant \varrho(a, B) - \varrho(a, A) + \varepsilon \leqslant |f_B - f_A| + \varepsilon. \tag{6}$$

On the other hand, for each $x \in \mathscr{X}$, denote by a_x a point of A such that $\varrho(x, A) \geqslant |x - a_x| - \varepsilon$. By IV (4),

$$\varrho(x, B) \leqslant |x - a_x| + \varrho(a_x, B).$$

Hence

$$\varrho(x, B) - \varrho(x, A) \leqslant \varrho(a_x, B) + \varepsilon \leqslant \operatorname{dist}(A, B) + \varepsilon$$

and similarly

$$\varrho(x, A) - \varrho(x, B) \leqslant \operatorname{dist}(A, B) + \varepsilon.$$

Therefore

$$|f_B - f_A| \leqslant \operatorname{dist}(A, B) + \varepsilon. \tag{7}$$

Formulas (6) and (7) imply (5).

Remark. If the space $\mathscr{X}$ is *bounded*, the function f_A can be defined in a simpler way:

$$f_A(x) = \varrho(x, A).$$

Theorem 6 yields the following theorem.

THEOREM 7. *The space* $(2^X)_m$ *is isometric with a subset of the space* $\mathscr{E}^{\mathscr{X}} \cap \Phi(\mathscr{X}, \mathscr{E})$.

In particular, if the space $\mathscr{X}$ *is compact, the following "topological" (or, more precisely, metrical) inclusions hold*

$$\mathscr{X} \underset{\text{top}}{\subset} (2^{\mathscr{X}})_m \underset{\text{top}}{\subset} \mathscr{E}^{\mathscr{X}},$$

where the space $\mathscr{E}^{\mathscr{X}}$ *is metrized by formula* (1).

The preceding theorem enables us to regard the space $(2^{\mathscr{X}})_m$ as a subset of the space $\mathscr{E}^{\mathscr{X}}$. Conversely, if f is an element of the function space $\mathscr{Y}^{\mathscr{X}}$, then the set of points $(x, f(x))$ of the cartesian product $\mathscr{X} \times \mathscr{Y}$ (which can be identified with f) is a closed subset of $\mathscr{X} \times \mathscr{Y}$ (compare § 15, V, Theorem 2). Hence, if this set is bounded, it is an element of the space $(2^{\mathscr{X} \times \mathscr{Y}})_m$. More precisely, we have the following theorem.

THEOREM 8. *If* $\mathscr{X}$ *is compact, then the following topological inclusion holds*

$$\mathscr{Y}^{\mathscr{X}} \underset{\text{top}}{\subset} (2^{\mathscr{X} \times \mathscr{Y}})_m.$$

Proof. First observe that if $f: \mathscr{X} \to \mathscr{Y}$ and $g: \mathscr{X} \to \mathscr{Y}$, then

$$\operatorname{dist}(f, g) \leqslant |f - g|. \tag{8}$$

To check this, let $p_x = (x, f(x))$ and $q_x = (x, g(x))$. It follows that $p_x \epsilon f$ and $q_x \epsilon g$; hence

$$\varrho(p_x, g) \leqslant |p_x - q_x| = |f(x) - g(x)| \leqslant |f - g|,$$

whence $\sup\limits_{x \epsilon X} \varrho(p_x, g) \leqslant |f - g|$.

By symmetry $\sup\limits_{x \epsilon X} \varrho(q_x, f) \leqslant |f - g|$, whence we obtain (8).

It remains to prove that a sequence of functions $f_n \epsilon \mathscr{Y}^{\mathscr{X}}$ converges uniformly to $f \epsilon \mathscr{Y}^{\mathscr{X}}$ if and only if $\lim\limits_{n=\infty} \operatorname{dist}(f_n, f) = 0$.

Now if the sequence $\{f_n\}$ is uniformly convergent, then

$$\lim_{n=\infty} |f_n - f| = 0, \quad \text{hence} \quad \lim_{n=\infty} \operatorname{dist}(f_n, f) = 0,$$

by (8).

Conversely, let us assume that $\lim\limits_{n=\infty} \operatorname{dist}(f_n, f) = 0$. We shall prove that the sequence $\{f_n\}$ is continuously convergent to f.

By virtue of Theorem 5, this will imply the uniform convergence (the space $\mathscr{X}$ being compact).

Let $\lim x_n = x$. We have to prove that $\lim f_n(x_n) = f(x)$.

Let $q_n = (x_n, f_n(x_n))$. Since $\varrho(q_n, f) \leqslant \operatorname{dist}(f_n, f)$, it follows that $\lim \varrho(q_n, f) = 0$. Hence there exists a sequence of points $p_n \in f$ such that $\lim |q_n - p_n| = 0$. Let $p_n = (x_n^*, f(x_n^*))$. Hence, on one hand, we have

$$\lim |x_n - x_n^*| = 0, \quad \text{i.e.} \quad \lim x_n^* = x \quad \text{and} \quad \lim f(x_n^*) = f(x),$$

and on the other hand,

$$\lim |f_n(x_n) - f(x_n^*)| = 0.$$

This proves that $\lim f_n(x_n) = f(x)$.

The following proposition follows easily from Theorem 6 (and 7):

THEOREM 9 [1]. *Every metric space* $\mathscr{X}$ *is isometric with a subset of* $\mathscr{E}^{\mathscr{X}} \cap \Phi(\mathscr{X}, \mathscr{E})$.

This can be deduced from (4) by taking A to be the set composed of a single point a.

In this case, the proof becomes simpler. By putting

$$f_a(x) = |x - a| - |x - p|,$$

we obtain $|f_a(x)| \leqslant |a - p|$ and $|f_a(x) - f_b(x)| = \big||x - a| - |x - b|\big| \leqslant |a - b|$; hence $|f_a - f_b| \leqslant |a - b|$. On the other hand,

$$f_a(a) - f_b(a) = -|a - p| - |a - b| + |a - p|,$$

and hence $|f_a - f_b| \geqslant |a - b|$. It follows that $|f_a - f_b| = |a - b|$.

XI. Extension of relatively closed or relatively open sets. Let E be a fixed (non empty) subset of the metric space $\mathscr{X}$. To each $X \subset E$ we assign the set

$$F(X) = \mathop{E}_{x}[\varrho(x, X) \leqslant \varrho(x, E - X)]. \tag{1}$$

By assuming that $\varrho(x, 0) = \infty$, we have

$$F(0) = 0, \tag{2}$$

$$F(E) = \mathscr{X}. \tag{3}$$

---

[1] Compare my note in Fund. Math. 25 (1935), p. 543, and K. Kunugui, Proc. Imp. Acad. Japan. 11 (1936), p. 351.

The following relations hold:

$$X \subset F(X), \tag{4}$$

$$\overline{F(X)} = F(X), \tag{5}$$

$$F(X \cup Y) = F(X) \cup F(Y), \tag{6}$$

$$\textit{if } X_1 \cup \ldots \cup X_n = E, \textit{ then } F(X_1) \cup \ldots \cup F(X_n) = \mathscr{X}, \tag{7}$$

$$E \cap F(X) = E \cap \bar{X}, \tag{8}$$

$$\textit{if } X \textit{ is closed in } E, \textit{ then } E \cap F(X) = X. \tag{9}$$

Proof. Inclusion (4) follows from the fact that $\varrho(x, X) = 0$ for $x \in X$. Formula (5) is a consequence of the continuity of the function $\varrho(x, A)$ (compare IV (5)). We shall prove (6).

If $x \in F(X)$, then by (1) and IV (6) we have

$$\varrho(x, X \cup Y) \leqslant \varrho(x, X) \leqslant \varrho(x, E-X) \leqslant \varrho[x, E-(X \cup Y)],$$

therefore $x \in F(X \cup Y)$. It follows that $F(X) \subset F(X \cup Y)$.

Let $x \in F(X \cup Y)$, i.e.

$$\varrho(x, X \cup Y) \leqslant \varrho[x, E-(X \cup Y)], \tag{10}$$

According to IV (7), $\varrho(x, X \cup Y)$ is equal either to $\varrho(x, X)$ or to $\varrho(x, Y)$. Assume that

$$\varrho(x, X \cup Y) = \varrho(x, X). \tag{11}$$

We shall prove that $x \in F(X)$, i.e.

$$\varrho(x, X) \leqslant \varrho(x, E-X)$$

which will complete the proof of (6).

By virtue of (10), (11), and IV (6),

$$\varrho(x, X) \leqslant \varrho[x, E-(X \cup Y)], \quad \varrho(x, X) = \varrho(x, X \cup Y) \leqslant \varrho(x, Y).$$

Therefore by IV (7)

$$\varrho(x, X) \leqslant \varrho[x, E-(X \cup Y) \cup Y] \leqslant \varrho(x, E-X);$$

because $E-(X \cup Y) \cup Y = E-X \cup Y \supset E-X$.

Thus formula (6) is proved; (7) follows from (6) and (3). We shall prove (8).

By (4) and (5), $\overline{X} \subset F(X)$. It remains to prove that $E \cap F(X) \subset \overline{X}$.

Suppose that $x \in [E \cap F(X) - \overline{X}]$. By (1) and IV (3), we have $0 < \varrho(x, X) \leqslant \varrho(x, E-X)$, whence $x \notin E-X$. But $x \in E$, therefore $x \in X$ which gives a contradiction.

Condition (9) follows from (8). For, X being closed in E, we have $X = E \cap \overline{X}$.

Along with the closed set $F(X)$ we consider the *open set* $G(X)$ defined by

$$G(X) = \underset{x}{E}[\varrho(x, X) < \varrho(x, E-X)]. \tag{12}$$

We readily deduce that

$$G(X) = -F(E-X). \tag{13}$$

Relations (2)–(9) together with (13) yield

$$G(E) = \mathscr{X}, \tag{14}$$

$$G(0) = 0, \tag{15}$$

$$G(X \cap Y) = G(X) \cap G(Y), \tag{16}$$

$$\textit{if } X_1 \cap \ldots \cap X_n = 0, \textit{ then } G(X_1) \cap \ldots \cap G(X_n) = 0, \tag{17}$$

$$E \cap G(X) = E - \overline{E-X}, \tag{18}$$

$$\textit{if } X \textit{ is open in } E, \textit{ then } E \cap G(X) = X. \tag{19}$$

The properties of the functions F and G proved above imply directly the following two theorems on extension of relatively closed or relatively open sets.

THEOREM 1. *Given a family $\{X_\iota\}$ of sets closed in E, there exists a family $\{F_\iota\}$ of closed (in $\mathscr{X}$) sets such that $E \cap F_\iota = X_\iota$ and*

$$\textit{the condition } X_{\iota_1} \cup \ldots \cup X_{\iota_n} = E \textit{ implies } F_{\iota_1} \cup \ldots \cup F_{\iota_n} = \mathscr{X}, \tag{20}$$

for every (finite) system of indices $\iota_1, \iota_2, \ldots, \iota_n$.

THEOREM 2. *Given a family $\{X_\iota\}$ of sets open in E, there exists a family $\{G_\iota\}$ of open (in $\mathscr{X}$) sets such that $E \cap G_\iota = X_\iota$ and*

$$\textit{the condition } X_{\iota_1} \cap \ldots \cap X_{\iota_n} = 0 \textit{ implies } G_{\iota_1} \cap \ldots \cap G_{\iota_n} = 0, \tag{21}$$

for every (finite) system of indices $\iota_1, \ldots, \iota_n$.

It suffices to put $F_\iota = F(X_\iota)$ and $G_\iota = G(X_\iota)$.

The following statement is a generalization (in the case of metric spaces) of Theorem 4 of § 14, V.

THEOREM 3. *Condition* (21) *can be replaced by*

$$X_{\iota_1} \cap \ldots \cap X_{\iota_n} = 0 \quad \textit{implies} \quad \bar{G}_{\iota_1} \cap \ldots \cap \bar{G}_{\iota_n} = \bar{X}_{\iota_1} \cap \ldots \cap \bar{X}_{\iota_n}. \tag{22}$$

Proof. By Theorem 2 there exists a family of open sets H_ι such that $E \cap H_\iota = X_\iota$ and that

$$X_{\iota_1} \cap \ldots \cap X_{\iota_n} = 0 \quad \text{implies} \quad H_{\iota_1} \cap \ldots \cap H_{\iota_n} = 0.$$

Substituting $A = X_\iota$ and $B = \mathscr{X} - H_\iota$ in Theorem 3 of § 14, V, we deduce the existence of an open set G_ι such that

$$X_\iota \subset G_\iota \subset H_\iota \quad \text{and} \quad \bar{G}_\iota \cap (\mathscr{X} - H_\iota) \subset \bar{X}_\iota, \quad \text{i.e.} \quad \bar{G}_\iota \subset H_\iota \cup \bar{X}_\iota.$$

Assume that $X_{\iota_1} \cap \ldots \cap X_{\iota_n} = 0$. It follows that

$$\bar{X}_{\iota_1} \cap \ldots \cap \bar{X}_{\iota_n} \subset \bar{G}_{\iota_1} \cap \ldots \cap \bar{G}_{\iota_n} \subset (H_{\iota_1} \cup \bar{X}_{\iota_1}) \cap \ldots \cap (H_{\iota_n} \cup \bar{X}_{\iota_n}).$$

It remains to prove that the right side member of this double inclusion is contained in the left side one. On one hand we have $H_{\iota_1} \cap \ldots \cap H_{\iota_n} = 0$ and, on the other hand, each intersection of n terms of the groups

$$H_{\iota_1}, \ldots, H_{\iota_n} \quad \text{and} \quad \bar{X}_{\iota_1}, \ldots, \bar{X}_{\iota_n},$$

of which at least one member belongs to the second group, is contained in $\bar{X}_{\iota_1} \cap \ldots \cap \bar{X}_{\iota_n}$. This follows from the fact that $E \cap H_\iota = X_\iota$ implies (compare § 5, III) $\bar{E} \cap H_\iota \subset \overline{E \cap H_\iota} = \bar{X}_\iota$, therefore $H_\iota \cap \bar{X}_\varkappa \subset \bar{X}_\iota \cap \bar{X}_\varkappa$.

XII. Refinements of infinite covers. The following statement is—in the case of metric spaces—a generalization of the corollary of § 14, III which concerns normal spaces.

THEOREM. *Each cover of a metric space* $\mathscr{X}$:

$$\mathscr{X} = \bigcup_\iota G_\iota, \quad \textit{where } G_\iota \textit{ is open}, \tag{1}$$

contains a refinement

$$\mathscr{X} = \bigcup_\iota H_\iota \quad \textit{where } H_\iota \textit{ is open and } \bar{H}_\iota \subset G_\iota. \tag{2}$$

Proof. Let us first show that

$$\mathscr{X} = \bigcup_{\iota} F_{\iota} \quad \text{where} \quad F_{\iota} = \overline{F}_{\iota} \subset G_{\iota}. \tag{2'}$$

By Section V, we can assume that the space $\mathscr{X}$ is bounded. Moreover, we can assume that $G_{\iota} \neq \mathscr{X}$ for each ι. For if $G_{\iota_0} = \mathscr{X}$ we can put $F_{\iota_0} = \mathscr{X}$ and $F_{\iota} = 0$ for $\iota \neq \iota_0$.

Let $C_{\iota} = \mathscr{X} - G_{\iota}$. Hence $C_{\iota} \neq 0$ for each ι. Given a point x of the space, let

$$\sigma(x) = \sup \varrho(x, C_{\iota}) \tag{3}$$

and

$$F_{\iota} = \underset{x}{E}[\varrho(x, C_{\iota}) \geqslant \tfrac{1}{2}\sigma(x)]. \tag{4}$$

We shall show that $\overline{F}_{\iota} = F_{\iota}$. The obvious equivalence

$$[\varrho(x, C_{\iota}) \geqslant \tfrac{1}{2}\sigma(x)] \equiv [\varrho(x, C_{\iota}) \geqslant \tfrac{1}{2}\varrho(x, C_{\varkappa}) \text{ for every } \varkappa]$$

yields

$$F_{\iota} = \bigcap_{\varkappa} \underset{x}{E}[\varrho(x, C_{\iota}) \geqslant \tfrac{1}{2}\varrho(x, C_{\varkappa})].$$

It follows that the set F_{ι} is closed, as the intersection of closed sets (by virtue of the continuity of the function ϱ; compare IV, (5)).

By definition of the upper bound, for each x there exists an index ι_0 such that

$$\varrho(x, C_{\iota_0}) \geqslant \tfrac{1}{2} \sup_{\iota} \varrho(x, C_{\iota}).$$

Therefore $x \in F_{\iota_0}$, by (3) and (4). This proves the identity (2').

Formula (1) implies that for each x there exists an ι_0 such that $x \in G_{\iota_0}$; hence (since the set G_{ι_0} is open) $\varrho(x, C_{\iota_0}) > 0$. Therefore $\sigma(x) > 0$. Assuming that $x \in F_{\iota}$, we have by (4)

$$\varrho(x, C_{\iota}) > 0;$$

hence $x \in G_{\iota}$ and $F_{\iota} \subset G_{\iota}$.

Now, denote by H_{ι} according to the normality of $\mathscr{X}$, an open set such that

$$F_{\iota} \subset H_{\iota} \quad \text{and} \quad \overline{H}_{\iota} \cap C_{\iota} = 0.$$

Formula (2) follows.

XIII. G_δ-sets in metric spaces.

THEOREM. *Let $Q \subset \mathcal{X}$ be $\boldsymbol{G}_\delta$. Then there is a continuous $f: Q \to \mathcal{E}^{\aleph_0}$ such that the set $\mathfrak{Z} = \underset{xy}{E}[y = f(x)]$ is closed (in $\mathcal{X} \times \mathcal{E}^{\aleph_0}$).*

Proof. By assumption $Q = G_1 \cap G_2 \cap \ldots$ where G_n is open for $n = 1, 2, \ldots$ Put $F_k = \mathcal{X} - G_k$,

$$f_k(x) = \frac{1}{\varrho(x, F_k)} \quad \text{for} \quad x \in G_n, \tag{1}$$

and for $x \in Q$:

$$f(x) = [f_1(x), f_2(x), \ldots], \quad \text{thus} \quad f: Q \to \mathcal{E}^{\aleph_0}. \tag{2}$$

By IV (3), $\varrho(x, F_k) > 0$ for $x \in G_k$, and by IV (5), ϱ is continuous, and so is f_k by (1). It follows by § 16, II, Theorem 3, that f is continuous.

Now let us suppose that $\mathfrak{Z}$ is not closed. Let

$$\lim_{n=\infty} x_n = x, \quad x_n \in Q, \quad \lim_{n=\infty} f(x_n) = y, \quad (x, y) \notin \mathfrak{Z}. \tag{3}$$

Since f is continuous (on Q), it follows that $x \in \mathcal{X} - Q$. Hence there is a k such that $x \in F_k$ and therefore $\varrho(x, F_k) = 0$. As ϱ is continuous, we have $\varrho(x, F_k) = \lim_{n=\infty} \varrho(x_n, F_k)$ and consequently by (1), $\lim_{n=\infty} f_k(x_n) = \infty$.

This is a contradiction, since we have by (3)

$$\lim_{n=\infty} f(x_n) = y = [y^1, y^2, \ldots], \quad \text{whence} \quad \lim_{n=\infty} f_k(x_n) = y^k.$$

COROLLARY. *Each $\boldsymbol{G}_\delta$ subset of a metric space $\mathcal{X}$ is homeomorphic to a closed subset of $\mathcal{X} \times \mathcal{E}^{\aleph_0}$.*

For Q is homeomorphic to $\mathfrak{Z}$ (by § 15, V, Theorem 1).

Remark 1. If Q is a *difference of two closed sets* $Q = A - B$, one can replace $\mathcal{E}^{\aleph_0}$ by $\mathcal{E}$ in the theorem[1]. For one can put in this case $f(x) = 1/\varrho(x, B)$.

Remark 2. As we shall see later (§ 33, VI), the above corollary, implies that *a $\boldsymbol{G}_\delta$-subset of a complete space is homeomorphic to a complete space.*

---

[1] Compare the paper of W. Sierpiński and myself *Sur les différences de deux ensembles fermés*, Tôhoku Math. Journ. 20 (1921), p. 23.

XIV. Proximity spaces. Uniform spaces (main definitions). In this section we do not assume $\mathscr{X}$ to be metric.

DEFINITION 1(1). *$\mathscr{X}$ is a proximity space in respect to the relation $A\delta B$* defined for subsets of $\mathscr{X}$ and read "*A is close to B*" if the following axioms are fulfilled:

1. the relation δ is symmetric,
2. $[A\delta(B\cup C)]\equiv(A\delta B \text{ or } A\delta C)$,
3. $(p\delta q)\equiv(p=q)$,
4. 0 non-δ $\mathscr{X}$,
5. if A non-δ B, then there are C and D such that

$$A\subset C,\quad B\subset D,\quad C\cap D=0,$$

$$A \text{ non-}\delta\,(-C),\quad B \text{ non-}\delta\,(-D).$$

The natural topology of $\mathscr{X}$ is defined by means of the equivalence

$$(x\in\bar{A})\equiv(x\delta A). \tag{1}$$

One can show that, with this definition, *$\mathscr{X}$ is a completely regular $\mathscr{T}_1$-space.*

Conversely, *each completely regular $\mathscr{T}_1$-space can be considered as a proximity space*, i.e. one can define a relation δ satisfying Axioms 1-5 as well as formula (1).

We can assume, namely, that A non-δ B whenever there exists a continuous $f\colon \mathscr{X}\to\mathscr{I}$ such that $f(A)=0$ and $f(B)=1$ $(A\neq 0\neq B)$.

In a *metric space*, the relation δ can be introduced in the most natural way by means of the equivalence

$$(A\delta B)\equiv[\varrho(A,B)=0]\quad (A\neq 0\neq B). \tag{2}$$

---

(1) See V. Efremovitch, Dokl. Akad. Nauk URSS 76 (1951), p. 341, and *Proximity geometry* (in Russian), Matem. Sbornik 31 (1952), pp. 189–200. See also Yu. Smirnow, *ibid.* pp. 543–574, and *On the dimension of proximity spaces* (in Russian), *ibid.* 38 (1956), pp. 283–302; S. Mrówka, *On the notion of completeness in proximity spaces*, Bull. Acad. Pol. Sc. 4 (1956), p. 477, and *On complete proximity spaces* (in Russian), Dokl. Akad. Nauk URSS 108 (1956), p. 587; J. R. Isbell, *Uniform spaces*, Providence 1964.

The main interest of proximity spaces consists, among others, in the possibility of introducing the concept of *uniformly continuous mappings*. Namely, if $\mathscr{X}$ and $\mathscr{Y}$ are proximity spaces, the mapping $f\colon \mathscr{X} \to \mathscr{Y}$ is called uniformly continuous if

$$A \delta B \Rightarrow f(A) \delta f(B). \tag{3}$$

If $\mathscr{X}$ and $\mathscr{Y}$ are metric, this condition is equivalent to the usual condition adopted in metric spaces (owing to (2); compare IV).

Another concept related to uniform continuity is the following

DEFINITION 2(¹). *$\mathscr{X}$ has an uniform structure relative to a family $\boldsymbol{U}$* (non void) of subsets of $\mathscr{X} \times \mathscr{X}$ if the following axioms are fulfilled for any $V \in \boldsymbol{U}$:

a) the diagonal of $\mathscr{X} \times \mathscr{X}$ is a subset of V,

b) the set $\underset{xy}{E}[(y, x) \in V]$ is a member of $\boldsymbol{U}$,

c) if $V_1 \in \boldsymbol{U}$ and $V_2 \in \boldsymbol{U}$, then $(V_1 \cap V_2) \in \boldsymbol{U}$,

d) if $V \subset Z$, then $Z \in \boldsymbol{U}$,

e) there is a $V_1 \in \boldsymbol{U}$ such that

$$\underset{xy}{E} \bigvee_{z} \big([(x, z) \in V_1] \wedge [(z, y) \in V_1]\big) \subset V.$$

Moreover, let us suppose that:

f) the intersection of members of $\boldsymbol{U}$ is the diagonal of $\mathscr{X} \times \mathscr{X}$.

The induced topology in $\mathscr{X}$ is obtained assuming that each set of the form $\underset{x}{E}[(p, x) \in V]$, where $V \in \boldsymbol{U}$, is a neighbourhood of p. One can show that *$\mathscr{X}$ is a completely regular $\mathscr{T}_1$-space.*

Conversely, *one can give to each completely regular $\mathscr{T}_1$-space a uniform structure*(²).

(¹) A. Weil, *Sur les espaces à structure uniforme et sur la topologie générale*, Paris 1937. Compare also N. Bourbaki, *op. cit.*, Chap. II.

A notion more general than that of proximity spaces and of spaces with uniform structure has been introduced by A. Császár under the name of syntopogene structure. See his paper *Sur une classe de structures topologiques générales*, Revue Math. Pures et Appl. 2 (1957), pp. 399–407, and his book *Fondements de la topologie générale*, Budapest 1960.

(²) N. Bourbaki, *ibidem*.

It is interesting to compare this theorem with the analogous theorem on proximity spaces(¹).

In metric spaces, $\boldsymbol{U}$ can be defined as being composed of sets containing the sets

$$V_\varepsilon = \underset{xy}{E}[|x-y| < \varepsilon] \quad \text{where} \quad \varepsilon > 0. \tag{4}$$

XV. Almost-metric spaces.

DEFINITION 1. A function $\psi\colon \mathscr{X}\times\mathscr{X}\to\mathscr{E}$ is called a *pseudo-distance*(²) if the following axioms are fulfilled (compare I):

α) $\quad \psi(x, x) = 0,$

β) $\quad \psi(x, y)+\psi(x, z) \geqslant \psi(y, z).$

It follows that

$$\psi(x, y) \geqslant 0 \quad \text{and} \quad \psi(x, y) = \psi(y, x).$$

Obviously, if we assume moreover that

$$(x \neq y) \Rightarrow [\psi(x, y) \neq 0],$$

the pseudo-distance becomes a distance.

It is worth noticing that:

(i) if $f\colon \mathscr{X}\to\mathscr{E}$, then $\psi(x, y) = |f(x)-f(y)|$ is a pseudo-distance,

(ii) the sum and the maximum of two pseudo-distances is a pseudo-distance,

(iii) the sum of a convergent series of pseudo-distances is a pseudo-distance,

(iv) if ψ is a pseudo-distance, so is $\overline{\psi}$ defined by the condition

$$\overline{\psi}(x, y) = \begin{cases} \psi(x, y) & \text{if} \quad \psi(x, y) \leqslant 1, \\ 1 & \text{if} \quad \psi(x, y) \geqslant 1. \end{cases}$$

(¹) For different relations between proximity spaces and uniform spaces, see: Yu. Smirnov, *On the completeness of proximity spaces* (Russian), part I, Trudy Mosk. Mat. Ob. 3 (1954), pp. 271–308 and part II, *ibid.* 4 (1955), pp. 421–438, also *On the completeness of uniform spaces and proximity spaces* (Russian), Dokl. Akad. Nauk 91 (1953), p. 1281.

(²) Also called "écart" by Bourbaki (*op. cit.*, Chap. IX, p. 1).

DEFINITION 2 [1]. A space $\mathscr{X}$ is called *almost-metric relative to a family* $\Psi = \{\psi_t\}$ of pseudo-distances ($t \in T$), if

γ) for each pair $x \neq y$ there is a $t \in T$ such that $\psi_t(x, y) \neq 0$.

Denote by Ψ^* the family Ψ enlarged by including in it all the maxima of finite systems of pseudo-distances belonging to Ψ.

THEOREM. *Let us assume that the following "pseudo-balls"*

$$B_{t,p,\varepsilon} = \mathop{E}_{x}[\psi_t(p, x) < \varepsilon] \quad \textit{where} \quad \psi_t \in \Psi^*,\ p \in \mathscr{X},\ \textit{and}\ \varepsilon > 0, \tag{1}$$

form a base of $\mathscr{X}$. *Then* $\mathscr{X}$ *becomes a* $\mathscr{T}_1$ *completely regular space* (this is the *induced* topology in the almost-metric space $\mathscr{X}$).

Proof. Let $F = \overline{F} \subset \mathscr{X}$ and $p \in \mathscr{X} - F$. Then t and ε exist such that $F \cap B_{t,p,\varepsilon} = 0$. Put $\varphi_t = \psi_t/\varepsilon$ and $f(x) = \bar{\varphi}_t(p, \varepsilon)$. Hence $f \in \mathscr{I}^{\mathscr{X}}$, $f(p) = 0$ and $f(x) = 1$ for $x \in F$.

Conversely, we have the following theorem.

THEOREM 2. *In every* $\mathscr{T}_1$ *completely regular space a pseudo-distance can be defined without affecting its topology.*

Namely, $\{f_t\}$ *being a family of functions* $f_t \in \mathscr{I}^{\mathscr{X}}$ *such that, for each* $F = \overline{F}$ *and* $p \in \mathscr{X} - F$, *there is a* t *satisfying the condition*

$$f_t(p) \in \mathscr{I} - \overline{f_t(F)}, \tag{2}$$

then

$$\psi_t(x, y) = |f_t(x) - f_t(y)|. \tag{3}$$

Proof. Obviously $\mathscr{X}$ is almost-metric relative to the family $\{\psi_t\}$. We have to show that the topology induced by this family agrees with the original topology of $\mathscr{X}$.

Let G be open and $p \in G$. Put $F = \mathscr{X} - G$. Let t satisfy condition (2) and put $\varepsilon = \varrho[f_t(p), \overline{f_t(F)}]$. It follows easily that $B_{t,p,\varepsilon} \subset G$ and thus G is open in the topology induced by the family $\{\psi_t\}$.

Conversely, $B_{t,p,\varepsilon}$ is open in the original topology of $\mathscr{X}$ because ψ_t is continuous.

Remark 1. The study of almost-metric spaces can be developed in a way similar to that of metric spaces [2]. One can show, for example, that every closed subset of a completely regular space is the intersection of a family of open sets, the cardinality

---

[1] For the definition and for the theorems which follow see S. Mrówka, *On almost-metric spaces*, Bull. Acad. Pol. Sci. 5 (1957), pp. 123–127.

[2] See S. Mrówka, *ibid.*

of that family being equal to that of the family $\{\psi_t\}$ considered above.

Remark 2. If T is countable and condition γ is satisfied, then the formulas

$$|x-y| = \sum_{n=1}^{\infty} \frac{1}{2^n}\cdot\frac{\psi_n(x, y)}{1+\psi_n(x, y)},$$

as well as

$$|x-y| = \sum_{n=1}^{\infty} \frac{1}{2^n}\bar{\psi}_n(x, y),$$

define a distance in $\mathscr{X}$ (compare (iii) and (iv)).

XVI. Paracompactness of metric spaces.

DEFINITIONS. A family $\boldsymbol{A}$ of sets in a topological space is called *discrete*, if each point of the space has a neighbourhood which has points in common with at most one member of $\boldsymbol{A}$.

A family is *σ-discrete* if it is the union of a countable number of discrete families.

THEOREM 1. *Each open cover $\{G_a\}$ of a metric space has an open σ-discrete refinement.*

Proof. Put

$$G_{a,n} = \mathop{E}_{x}[\varrho(x, -G_a) > 1/2^n]. \tag{0}$$

Obviously (compare IV (3))

$$G_a = G_{a,1} \cup G_{a,2} \cup \ldots \tag{1}$$

and $G_{a,n}$ is open (since ϱ is continuous, compare IV (5)).

Let $x \in G_{a,n}$ and $y \notin G_{a,n+1}$. Hence by IV (6′)

$$|x-y| \geqslant \varrho(x, -G_a) - \varrho(y, -G_a) > 1/2^n - 1/2^{n+1} = 1/2^{n+1},$$

and it follows that

$$\varrho(G_{a,n}, -G_{a,n+1}) \geqslant 1/2^{n+1}. \tag{2}$$

Since the set of indices a is arbitrary, we may assume that a is an ordinal number. Put

$$H_{a_0,n} = G_{a_0,n} - \overline{\bigcup_{a<a_0} G_{a,n+1}}. \tag{3}$$

Therefore, if $a_1 \neq a_2$, then

$$\text{either} \quad H_{a_2,n} \subset -G_{a_1,n+1} \quad \text{or} \quad H_{a_1,n} \subset -G_{a_2,n+1},$$

since $a_1 < a_2$ or $a_2 < a_1$.

Thus, according to (2) and (3), we have

$$\varrho(H_{a_1,n}, H_{a_2,n}) \geqslant 1/2^{n+1} \quad \text{if} \quad a_1 \neq a_2.$$

Hence for each n the family $\boldsymbol{B}_n = \{H_{a,n}\}$ is discrete, and consequently $\{H_{a,n}\}$ is σ-discrete for variable a and n.

It remains to be shown that this family is a cover of the space, i.e. that

$$\mathscr{X} = \bigcup_{a,n} H_{a,n}.$$

Let x_0 be a given point. Denote by a_0 the least index such that $x_0 \epsilon G_a$. By (1) there is an n_0 such that $x_0 \epsilon G_{a_0,n_0}$. Obviously, if $a < a_0$, we have $x_0 \notin G_{a,n_0+2}$. Denote by K the open ball with centre x_0 and radius $1/2^{n_0+2}$. It follows by (2) that

$$K \cap \bigcup_{a<a_0} G_{a,n_0+1} = 0, \quad \text{hence} \quad x_0 \epsilon H_{a_0,n_0}$$

according to (3).

COROLLARY 1a. *Each metric space has a σ-discrete base.*

Proof. Consider for each n the family $\boldsymbol{F}_n$ of all balls of radius less than $1/n$ and denote by $\boldsymbol{A}_n$ a σ-discrete refinement of $\boldsymbol{F}_n$. Then $\boldsymbol{A}_1 \cup \boldsymbol{A}_2 \cup \ldots$ is the required base.

THEOREM 2. *For each σ-locally finite open cover $\boldsymbol{A}$* (compare § 5, VII) *of a metric space, there is a locally finite open refinement.*

Proof. Let

$$\boldsymbol{A} = \boldsymbol{A}_1 \cup \boldsymbol{A}_2 \cup \ldots \quad \text{and} \quad \boldsymbol{A}_n = \{G_{n,t}\}$$

where $t \epsilon T_n$, $G_{n,t}$ is open, and $\boldsymbol{A}_n$ is locally finite.

According to the theorem of Section XII, there are open $H_{n,t}$ such that

$$\overline{H}_{n,t} \subset G_{n,t} \quad \text{and} \quad \bigcup_t H_{n,t} = \bigcup_t G_{n,t}. \tag{4}$$

Put $\boldsymbol{B}_n = \{H_{n,t}\}$ where $t \epsilon T_n$. Since $\bigcup_t G_{n,t} = \mathscr{X}$, it follows that $\boldsymbol{B}_1 \cup \boldsymbol{B}_2 \cup \ldots$ is an (open) cover of the space.

As A_n is locally finite, so is B_n. Therefore, by virtue of the theorem of § 5, VII, the sets

$$K_{n,t} = G_{n,t} - \bigcup_{m<n} \bigcup_{t \in T_m} \overline{H}_{m,t} \tag{5}$$

are open. Thus, to complete the proof, it suffices to show that the family $\{K_{n,t}\}$ with n and t variable, is a locally finite cover of the space.

Let x_0 be a fixed point. Let n_0 be the least index such that $x \in \bigcup H_{n_0,t}$ where $t \in T_{n_0}$. Hence there is a $t_0 \in T_{n_0}$ such that $x_0 \in H_{n_0,t_0}$. As A_m is locally finite, there is a system $V_1, V_2, \ldots, V_{n_0}$ of open sets containing x_0 and such that V_m intersects only a finite number of members of A_m. One sees easily that the set $H_{n_0,t_0} \cap V_1 \cap \ldots \cap V_{n_0}$ intersects only a finite number of sets of the form (5).

Finally, in order to show that $x_0 \in \bigcup_{n,t} K_{n,t}$, denote by n_0 the least index such that $x_0 \in \bigcup G_{n_0,t}$ where $t \in T_{n_0}$. Hence there is a $t_0 \in T_{n_0}$ such that $x_0 \in G_{n_0,t_0}$. By (4) and (5) it follows that $x_0 \in K_{n_0,t_0}$.

COROLLARY 2a (A. H. Stone theorem) (1). *Each metric space is paracompact.*

This follows at once from Theorems 1 and 2, since a discrete family is obviously locally finite.

XVII. Metrization problem. The problem consists in finding necessary and sufficient conditions for a topological space (a regular $\mathcal{T}_1$-space) to be metrizable.

METRIZATION THEOREM OF BING-NAGATA-SMIRNOV (2).

The following conditions on a regular $\mathcal{T}_1$-space $\mathcal{X}$ are equivalent:

---

(1) A. H. Stone, *Paracompactness and product spaces*, Bull. Amer. Math. Soc. 54 (1948), pp. 977–982.

(2) See R. H. Bing, *Metrization of topological spaces*, Canad. Journ. of Math. 3 (1951), pp. 175–186; Jun-iti Nagata, *On a necessary and sufficient condition of metrizability*, Journ. Inst. Polyt. Osaka City Univ. 1 (1950), pp. 93–100; Yu. Smirnov, *A necessary and sufficient condition for metrizability of a topological space* (Russian), Doklady Akad. Nauk URSS 77 (1951), pp. 197–200, and *On metrization of topological spaces* (Russian), Uspiechi Mat. Nauk 6 (1951), pp. 100–111.

See also: P. Alexandrov, *On the metrization of topological spaces*, Bull. Acad. Polon. Sc. 8 (1960), pp. 135–140; A. Archangielskii, *On the metrization of topological spaces*, *ibid.*, pp. 589–595, *New criteria for paracompactness and*

(i) $\mathscr{X}$ *is metrizable,*

(ii) $\mathscr{X}$ *has a σ-discrete base,*

(iii) $\mathscr{X}$ *has a σ-locally finite base.*

Proof. (i) $\Rightarrow$ (ii). This is Corollary 1a of Section XVI.

(iii) $\Rightarrow$ (ii) is obvious.

Thus we have to define in a space $\mathscr{X}$ with a σ-locally finite base $\boldsymbol{B}$ a distance which does not affect the topology of $\mathscr{X}$.

Put

$$\boldsymbol{B} = \boldsymbol{B}_1 \cup \boldsymbol{B}_2 \cup \ldots, \qquad \boldsymbol{B}_n = \{G_{n,t}\}, \qquad t \in T_n, \tag{0}$$

where $\boldsymbol{B}_n$ is locally finite.

It follows, by Theorem 3 of § 14, VI, that $\mathscr{X}$ is perfectly normal. Consequently (by Theorem 1 of § 14, VI) there is for each pair (n, t) a continuous function $f_{n,t}: \mathscr{X} \to \mathscr{I}$ such that

$$(x \in G_{n,t}) \equiv \big(f_{n,t}(x) \neq 0\big). \tag{1}$$

---

metrizability of arbitrary T_1-spaces, Dokl. Akad. Nauk URSS, 141 (1961), pp. 13–15, *Bicompact sets and topology of spaces*, Dokl. Akad. Nauk URSS, 150 (1963), p. 9–12, and *Some metrization theorems*, Uspiechi Mat. Nauk 18 (1963), pp. 139–145; Jun-iti Nagata, *A theorem for metrizability of a topological space*, Proc. Japan. Acad. 33 (1957), pp. 128–130; E. Michael, *Another note on paracompact spaces*, Proc. Amer. Math. Soc. 8 (1957), pp. 822–828.

Historically, the first condition for metrizability was stated in the paper of P. Alexandrov and P. Urysohn, *Une condition nécessaire et suffisante pour qu'une classe (L) soit une classe (D)*, C. R. Paris 177 (1923), p. 1274. Compare also M. Fréchet, *Espaces abstraits*, p. 220, and a condition due to Aronszajn stated in the French edition of this book on p. 138.

See also Chittenden, *On the equivalence of écart and voisinage*, Trans. Amer. Math. Soc. 18 (1917), p. 161.

Let us add that there is an extensive literature concerning the metrization problem for *images of metric spaces* under various mappings such as closed, open or bicontinuous (= metrization of quotient spaces). See e.g. A. V. Martin, *Decompositions and quasi-compact mappings*, Duke Math. Journ. 21 (1954), pp. 463–469; S. Hanai, *On closed mappings*, Proc. Japan. Acad. 30 (1954), pp. 285–288; K. Morita, S. Hanai, *Closed mappings and metric spaces*, *ibid.* 32 (1956), p. 10; K. Morita, *On closed mappings*, *ibid.*, pp. 539–543; A. H. Stone, *Metrizability of decomposition space*, Proc. Amer. Math. Soc. 7 (1956), pp. 690–700.

As $\boldsymbol{B}_n$ is locally finite, each point x is contained in an open $U_n(x)$ which intersects only a finite number of sets $G_{n,t}$. Consequently, on the $U_n(x)$, all the functions $f_{n,t}$, except a finite number, are identically 0.

Thus the following function is well defined:

$$f_n(x, y) = \sum_{t \in T_n} |f_{n,t}(x) - f_{n,t}(y)|, \quad \text{where} \quad x \in \mathscr{X}, y \in \mathscr{X}. \tag{2}$$

Moreover f_n is continuous (compare § 13, V, Theorem 1).

Apply the shrinking mapping (see Section V) to f_n:

$$\bar{f}_n(x, y) = \frac{f_n(x, y)}{1 + f_n(x, y)}. \tag{3}$$

One sees easily that f_n and $\bar{f}_n$ satisfy the triangle inequality (see I (ii)).

Now put

$$|x - y| = \sum_{n=1}^{\infty} \frac{1}{2^n} \bar{f}_n(x, y). \tag{4}$$

We have to prove that this is a distance (i.e. that conditions (i) and (ii) of Section I are fulfilled) and that this distance agrees with the original topology of $\mathscr{X}$.

First, let us observe that, as condition (ii) is satisfied by $\bar{f}_n$, it is also satisfied by $|x-y|$. Next, in order to show that (i) is satisfied by $|x-y|$ and that the distance $|x-y|$ agrees with topology of $\mathscr{X}$, it suffices to show (compare IV (3)) that

$$(x \in \bar{A}) \equiv \big(\varrho(x, A) = 0\big) \quad \text{whenever} \quad x \in \mathscr{X} \text{ and } A \subset \mathscr{X}. \tag{5}$$

The sum of a uniformly convergent series of continuous functions being continuous (compare § 13, VI, 3), and the projection being a continuous mapping, one has the implication from the left member to the right in (5).

On the other hand, if $x \notin \bar{A}$, there is a pair n, t such that

$$x \in G_{n,t} \quad \text{and} \quad A \subset \mathscr{X} - G_{n,t},$$

and it follows from (1)-(4) that, for each $a \in A$,

$$|x - a| \geqslant \frac{1}{2^n} \frac{f_{n,t}(x)}{1 + f_{n,t}(x)} > 0.$$

Hence $\varrho(x, A) > 0$. This completes the proof.

Remark. It follows directly from the above theorem that a regular $\mathcal{T}_1$-space with a countable base is metrizable. In § 22 we shall give a more direct proof of this important statement.

§ 22. Spaces with a countable base

I. General properties. Let $R_1, R_2, \ldots$ denote the base (composed of open non-void sets) of the given $\mathcal{T}_1$-space $\mathcal{X}$.

We have the following theorems.

THEOREM 1. *The property of containing an open countable base is hereditary.*

Namely, if $E \subset \mathcal{X}$, then $E \cap R_1, E \cap R_2, \ldots$ is an open base of E.

THEOREM 2. *If $\mathcal{X}$ is regular and G open, then there is for each $p \in G$ an index n such that $p \in R_n$ and $\bar{R}_n \subset G$.*

For, $\mathcal{X}$ being regular there is an open H such that $p \in H$, $\bar{H} \subset G$. Furthermore there is n such that $p \in R_n \subset H$, whence $\bar{R}_n \subset G$.

THEOREM 3. *Let $f: \mathcal{X} \to \mathcal{Y}$ be continuous and onto. Then, f is a homeomorphism if and only if*

$$(\bar{R}_k \subset R_j) \Rightarrow [f(R_k) \cap \overline{f(\mathcal{X} - R_j)} = 0].$$

Proof. If f is a homeomorphism, then the inclusion $\bar{R}_k \subset R_j$, equivalent to $\bar{R}_k \cap (\mathcal{X} - R_j) = 0$, implies

$$f(\bar{R}_k) \cap f(\mathcal{X} - R_j) = 0; \quad \text{hence} \quad f(R_k) \cap f(\mathcal{X} - R_j) = 0.$$

Therefore $f(R_k) \cap \overline{f(\mathcal{X} - R_j)} = 0$ since the set $f(R_k)$ is open.

Conversely, if f is continuous but not a homeomorphism, two cases can occur: either f is not one-to-one, or there exists an open set G such that $f(G)$ is not open. In the first case, there exist two points $x \neq x'$ with $f(x) = f(x')$. Hence there exists j such that

$$x \in R_j \text{ and } x' \in \mathcal{X} - R_j, \quad \text{therefore} \quad f(x) = f(x') \in f(\mathcal{X} - R_j).$$

In the second case, there exists a point x such that $x \in G$ and $f(x) \in \overline{f(\mathcal{X}) - f(G)}$. Hence there exists j such that

$$x \in R_j \subset G, \quad \text{therefore} \quad f(x) \in \overline{f(\mathcal{X}) - f(R_j)}.$$

In both cases $f(x) \epsilon \overline{f(\mathcal{X}-R_j)}$.

On the other hand, $x \epsilon R_j$, whence there exists, by Theorem 2, a k such that $x \epsilon R_k$ and $\bar{R}_k \subset R_j$. Therefore $f(x) \epsilon f(R_k) \cap \overline{f(\mathcal{X}-R_j)}$ and the proof is completed.

THEOREM 4. *In every space with countable base there exists a sequence of simultaneously open and closed sets such that every open and closed set is the limit (in the sense of § 1, V) of a sequence of certain terms of this sequence* (¹).

This follows immediately from the following theorem which belongs to the theory of sets.

Let $G_1, G_2, \ldots$ be a sequence of subsets of a fixed space $\mathcal{X}$. Let $\boldsymbol{A}$ be the family of all sets X such that X as well as $-X$ is the union of certain terms of this sequence. Then there exists a sequence of sets $A_1, A_2, \ldots$ belonging to $\boldsymbol{A}$ such that each $A \epsilon \boldsymbol{A}$ is of the form

$$A = \operatorname*{Limes}_{k=\infty} A_{i_k}.$$

Proof. Consider all the systems composed of an even number of positive integers $(m_0, \ldots, m_k, n_0, \ldots, n_k)$ for which there exists an $X \epsilon \boldsymbol{A}$ with

$$G_{m_0} \cup \ldots \cup G_{m_k} \subset X \quad \text{and} \quad G_{n_0} \cup \ldots \cup G_{n_k} \subset -X. \tag{1}$$

We order all such systems in an infinite sequence $s_1, s_2, \ldots$.

If $(m_0^i, \ldots, m_{k_i}^i, n_0^i, \ldots, n_{k_i}^i)$ is the ith term of the sequence $\{s_i\}$, let A_i be a set $X \epsilon \boldsymbol{A}$ satisfying (1) (for the index i).

Let $A \epsilon \boldsymbol{A}$. By hypothesis there exist two sequences of integers $m_0, m_1, \ldots$ and $n_0, n_1, \ldots$ such that

$$A = G_{m_0} \cup G_{m_1} \cup \ldots \quad \text{and} \quad -A = G_{n_0} \cup G_{n_1} \cup \ldots.$$

Let $s_{i_k} = (m_0, \ldots, m_k, n_0, \ldots, n_k)$. Since A satisfies (1) for every k, it follows that

$$G_{m_0} \cup \ldots \cup G_{m_k} \subset A_{i_k} \quad \text{and} \quad G_{n_0} \cup \ldots \cup G_{n_k} \subset -A_{i_k},$$

therefore

$$G_{m_0} \cup \ldots \cup G_{m_k} \subset \bigcap_{j=0}^{\infty} A_{i_{k+j}} \quad \text{and} \quad G_{n_0} \cup \ldots \cup G_{n_k} \subset \bigcap_{j=0}^{\infty} (-A_{i_{k+j}}).$$

(¹) See W. Sierpiński, *Sur une propriété des espaces métriques séparables*, Fund. Math. 30 (1938), p. 129. See also de Groot, *A note on 0-dimensional spaces*, Indag. Math. 9 (1947), p. 94.

Hence

$$A \subset \bigcup_{k=0}^{\infty} \bigcap_{j=0}^{\infty} A_{i_{k+j}} \quad \text{and} \quad -A \subset \bigcup_{k=0}^{\infty} \bigcap_{j=0}^{\infty} (-A_{i_{k+j}}),$$

and finally

$$A \subset \bigcup_{k=0}^{\infty} \bigcap_{j=0}^{\infty} A_{i_{k+j}} \subset \bigcap_{k=0}^{\infty} \bigcup_{j=0}^{\infty} A_{i_{k+j}} \subset A, \quad \text{i.e.} \quad A = \operatorname*{Limes}_{k=\infty} A_{i_k}.$$

II. Metrization and introduction of coordinates.

THEOREM 1 (of Urysohn) (¹). *Every regular $\mathcal{T}_1$-space $\mathcal{X}$ with a countable base is homeomorphic to a separable metric space.*

More precisely:

$$\mathcal{X} \underset{\text{top}}{\subset} \mathcal{I}^{\aleph_0}. \tag{0}$$

In other words, *there exists a sequence of functions f^n: $\mathcal{X} \to \mathcal{I}$ such that the function $f = [f^1, f^2, \ldots]$ is a homeomorphism, the distance of two points of $\mathcal{I}^{\aleph_0}$, i.e. of two sequences of real numbers $y = [y^{(1)}, y^{(2)}, \ldots]$, $z = [z^{(1)}, z^{(2)}, \ldots]$ being defined by the formula* (see § 21, VI, (2)):

$$|y-z| = \sum_{n=1}^{\infty} 2^{-n} |y^{(n)} - z^{(n)}|. \tag{i}$$

Proof. Let $R_1, R_2, \ldots$ be the base of the space. Consider all the pairs (R_{k_n}, R_{m_n}) such that $\overline{R}_{k_n} \subset R_{m_n}$. By Corollary 1a of § 14, II, $\mathcal{X}$ is normal; hence by the Urysohn lemma (§ 14, IV), there exists a continuous function f^n such that:

$$0 \leqslant f^n(x) \leqslant 1,$$

$$f^n(x) = 0 \text{ for } x \in R_{k_n} \quad \text{and} \quad f^n(x) = 1 \text{ for } x \notin R_{m_n}. \tag{ii}$$

Since the functions f^n are continuous, then so is the function $f = [f^1, f^2, \ldots]$ (§ 16, II, Theorem 3). To show that f is a homeomorphism, it suffices to show that (see § 13, VIII, (3a)) the condition $p \notin \overline{X}$ implies $f(p) \notin \overline{f(X)}$.

Since $p \in \mathcal{X} - \overline{X}$, there exists an index m_n such that $p \in R_{m_n} \subset \mathcal{X} - \overline{X}$. By virtue of I, Theorem 2, there exists a k_n such that $p \in \overline{R}_{k_n} \subset R_{m_n}$.

(¹) *Zum Metrisationsproblem*, Math. Ann. 94 (1925), p. 310.

Let $x \in X$. It follows that $x \notin R_{m_n}$ and, by (ii), $f^n(x) = 1$. The same formula implies that $f^n(p) = 0$, and, by (i), $|f(x)-f(p)| \geqslant 1/2^n$. This shows that the ball with centre $f(p)$ and radius $1/2^n$ is disjoint with the set $f(X)$. Therefore $f(p) \notin \overline{f(X)}$.

Remark 1. As seen before (§ 21, II, Theorem 2), the converse theorem is also true: *Every metric separable space is a $\mathscr{T}_1$-space with a countable base.*

Among these spaces, *the Hilbert cube $\mathscr{I}^{\aleph_0}$ has the highest topological rank*, i.e. it is itself metric separable and contains topologically all other metric separable spaces. Moreover, *every subset of $\mathscr{I}^{\aleph_0}$ is metric separable.*

Consequently the study of regular $\mathscr{T}_1$-spaces with a countable base is (from the topological point of view) just the study of subsets of $\mathscr{I}^{\aleph_0}$. Equivalently, it is the study of subsets of the Fréchet space $\mathscr{E}^{\aleph_0}$ (since $\mathscr{I}^{\aleph_0}$ and $\mathscr{E}^{\aleph_0}$ are obviously of the same topological rank).

Remark 2. It is interesting to compare the Urysohn theorem (and the above remark) with the following theorem of Banach and Mazur [1].

Every separable metric space is isometric to a subset of the space $\mathscr{E}^{\mathscr{I}}$.

Thus the topological rank of the space $\mathscr{E}^{\mathscr{I}}$ is also highest among separable metric spaces. From the topological point of view, the space $\mathscr{E}^{\mathscr{I}}$, in contrast to $\mathscr{I}^{\aleph_0}$, has the disadvantage of not being compact. From the geometrical point of view, however, it has the advantage of being not only of highest topological rank, but also of highest geometrical rank among separable metric spaces.

Remark 3. For metric separable spaces the inclusion (0) can be proved more directly as follows.

According to § 21, V, we can assume that the space $\mathscr{X}$ has diameter $\leqslant 1$. Let $p_1, p_2, \ldots$ be an (infinite) sequence of points dense in the space. To each point $x \in \mathscr{X}$ we assign the point $h(x)$ of the cube $\mathscr{I}^{\aleph_0}$ with coordinates $|x-p_1|, |x-p_2|, \ldots$:

$$h(x) = [|x-p_1|, |x-p_2|, \ldots], \quad \text{i.e.} \quad h^{(n)}(x) = |x-p_n|. \tag{1}$$

---

(1) S. Banach, *Théorie des opérations linéaires*, p. 187, compare also P. Urysohn, *Sur un espace métrique universel*, Bull. Sc. Math. 151 (1927), pp. 1–38, and W. Sierpiński, *Sur un espace métrique séparable universel*, Fund. Math. 33 (1945).

The function $h^{(n)}$ is continuous (by § 21, IV) and therefore the function h is continuous (by § 16, II, Theorem 3). We shall prove that h is a homeomorphism.

Let

$$\lim_{k=\infty} h(x_k) = h(x). \tag{2}$$

We have to show that $\lim_{k=\infty} x_k = x$.

Suppose that the last identity does not hold. Hence there exists an $\eta > 0$ and a sequence $m_1 < m_2 < \dots$ such that

$$|x_{m_k} - x| > \eta \quad \text{for every } k. \tag{3}$$

Let j be an index such that

$$|x - p_j| < \eta/3. \tag{4}$$

Condition (2) yields $\lim_{k=\infty} h^{(j)}(x_k) = h^{(j)}(x)$, i.e.

$$\lim_{k=\infty} |x_k - p_j| = |x - p_j|.$$

Therefore, if k is sufficiently large, then

$$|x_{m_k} - p_j| < |x - p_j| + \eta/3,$$

hence $|x_{m_k} - x| \leqslant |x_{m_k} - p_j| + |x - p_j| < \eta$, by (4).

Thus we obtain a contradiction with (3).

Remark 4. The assumption made in Theorem 1 for $\mathscr{X}$ to be regular cannot be replaced by the (weaker) assumption to be Hausdorff.

This is seen in the example considered in Remark 2 of § 5, X, of a Hausdorff non-regular space (hence non-metrizable). This space has a countable base.

COROLLARY 1a. *Every separable metric space is homeomorphic to a totally bounded space.*

For the space $\mathscr{I}^{\aleph_0}$ is totally bounded (compare § 21, VIII, Corollary 5a).

Theorem 1 can be sharpened as follows.

THEOREM 2. *The homeomorphisms constitute a dense set in the space* $(\mathscr{I}^{\aleph_0})^{\mathscr{X}}$.

In other words, to each continuous mapping $f: \mathcal{X} \to \mathcal{I}^{\aleph_0}$ and to each $\varepsilon > 0$ there exists a homeomorphism $h: \mathcal{X} \to \mathcal{I}^{\aleph_0}$ such that $|h(x) - f(x)| < \varepsilon$ for every x.

That is to say, *f is the limit of a uniformly convergent sequence of homeomorphisms.*

To prove this, we replace formula (1) by the following one

$$h(x) = [f^{(1)}(x), f^{(2)}(x), \ldots, f^{(n)}(x), |x - p_1|, |x - p_2|, \ldots],$$

where the integer n is chosen so that $1/2^n < \varepsilon$.

COROLLARY 1b. *Every metric separable space is the continuous and one-to-one image of a subset of the Cantor discontinuum $\mathcal{C}$ (as well as of a set of irrational numbers).*

This follows from the Urysohn theorem and from Corollary 6a of § 16, II, according to which $\mathcal{I}^{\aleph_0}$ is a continuous image of $\mathcal{C}$. $\mathcal{C}$ being topologically contained in the set $\mathcal{N}$ of irrational numbers, the proof is completed.

III. Separability of the space $\mathcal{Y}^{\mathcal{X}}$.

THEOREM [1]. *If $\mathcal{X}$ is a compact metric space and $\mathcal{Y}$ is a separable metric space, then the space $\mathcal{Y}^{\mathcal{X}}$ is separable.*

For the inclusion $\mathcal{Y} \underset{\text{top}}{\subset} \mathcal{I}^{\aleph_0}$ implies (compare § 21, X, Theorem 8)

$$\mathcal{Y}^{\mathcal{X}} \underset{\text{top}}{\subset} (\mathcal{I}^{\aleph_0})^{\mathcal{X}} \subset (2^{\mathcal{I}^{\aleph_0} \times \mathcal{X}})_m$$

and the last space is separable since $\mathcal{I}^{\aleph_0} \times \mathcal{X}$ is compact (see § 21, IX, Theorem 2 and VIII, Theorems 2 and 3).

Remark 1. *If a metric space $\mathcal{X}$ is not compact, then the space $\mathcal{I}^{\mathcal{X}}$* (metrized by formula (1) of § 21, X) *is not separable.*

Proof. The space $\mathcal{X}$ contains, by the hypothesis, an infinite countable closed discrete subset $F = (p_1, p_2, \ldots)$. To each point $\mathfrak{z} = [\mathfrak{z}^1, \mathfrak{z}^2, \ldots]$ of the Cantor set $\mathcal{C}$ we assign the function $f_{\mathfrak{z}}$ defined (on F) by the formula $f_{\mathfrak{z}}(p_i) = \frac{1}{2}\mathfrak{z}^i$. Since F is discrete, then the function $f_{\mathfrak{z}}$ is continuous: $f_{\mathfrak{z}} \in \mathcal{I}^F$. According to the Tietze theorem (§ 14, IV), let $f_{\mathfrak{z}}^*$ be an extension of $f_{\mathfrak{z}}$ to the entire space $\mathcal{X}$. It follows that $|f_{\mathfrak{z}}^* - f_{\mathfrak{y}}^*| \geqslant |f_{\mathfrak{z}} - f_{\mathfrak{y}}| = 1$ for $\mathfrak{z} \neq \mathfrak{y}$. The family of functions $f_{\mathfrak{z}}$, where $\mathfrak{z} \in \mathcal{C}$, is uncountable, hence the space $\mathcal{I}^{\mathcal{X}}$ is not separable.

---

(1) See K. Borsuk, *Sur les rétractes*, Fund. Math. 17 (1931), p. 165.

Remark 2. The separability of the space $\mathcal{I}^{\mathcal{I}}$ is a direct consequence of the Weierstrass theorem by which every continuous function is the limit of a uniformly convergent sequence of polynomials with rational coefficients. In other words, this theorem means that the set of polynomials with rational coefficients is dense in the space $\mathcal{I}^{\mathcal{I}}$.

IV. Reduction of closed sets to individual points. Let F be a closed subset of a metric separable space $\mathcal{X}$. By identifying the set F to a point, we obtain again a metric separable space. This can be expressed more precisely as follows.

THEOREM 1 [1]. *There exists a continuous function f which maps the set F onto a single point not belonging to $f(\mathcal{X}-F)$ and which is a homeomorphism on $\mathcal{X}-F$.*

Proof. By the Urysohn theorem, we can assume that $\mathcal{X} \subset \mathcal{E}^{\aleph_0}$, i.e. that the points x of $\mathcal{X}$ are sequences of real numbers $x = [x^1, x^2, x^3, \ldots]$. For brevity, we write $\delta(x) = \varrho(x, F)$. Let $f(x)$ be the point of $\mathcal{E}^{\aleph_0}$ defined by

$$f(x) = [\delta(x), x^1 \cdot \delta(x), x^2 \cdot \delta(x), \ldots].$$

If $x \epsilon F$, then $\delta(x) = 0$ and $f(x) = [0, 0, 0, \ldots]$. Conversely, if $f(x) = [0, 0, 0, \ldots]$, then $\delta(x) = 0$ and $x \epsilon F$.

Since each coordinate of $f(x)$ is a continuous function, it follows that the function f is continuous.

It remains to show that f is a homeomorphism on $\mathcal{X}-F$. If $\lim_{n=\infty} f(x_n) = f(x)$, then $\lim_{n=\infty} \delta(x_n) = \delta(x)$ and $\lim_{n=\infty} x_n^k \cdot \delta(x_n) = x^k \cdot \delta(x)$ for each k. If $x \epsilon \mathcal{X}-F$, then $\delta(x) \neq 0$, hence $\delta(x_n) \neq 0$ for sufficiently large n. It follows that

$$\lim_{n=\infty} x_n^k = \lim_{n=\infty} \frac{x_n^k \cdot \delta(x_n)}{\delta(x_n)} = \frac{\lim_{n=\infty} x_n^k \cdot \delta(x_n)}{\lim_{n=\infty} \delta(x_n)} = \frac{x^k \cdot \delta(x)}{\delta(x)} = x^k.$$

Therefore $\lim_{n=\infty} x_n = x$.

---

[1] See my note *Remarques sur les transformations continues des espaces métriques*, Fund. Math. 30 (1938), p. 48, and F. Hausdorff, *Erweiterung einer stetigen Abbildung*, *ibid.*, p. 40. Compare also W. Nitka, Indag. Math. 21 (1959), p. 36. Note that $f(\mathcal{X})$ does not need to be the quotient-space.

Theorem 1 can be generalized as follows:

THEOREM 2. *Given n disjoint closed subsets $F_1, \ldots, F_n$ of a separable metric space $\mathcal{X}$, there exists a continuous function f which maps $\mathcal{X}$ onto a metric separable space so that the sets $f(F_1), \ldots, f(F_n)$ reduce to n individual (different) points $p_1, \ldots, p_n$ none of whose belongs to $f[\mathcal{X}-(F_1 \cup \ldots \cup F_n)]$ and the function f is a homeomorphism on $\mathcal{X}-(F_1 \cup \ldots \cup F_n)$.*

By applying Theorem 1 step by step, we reduce successively the sets $F_1, F_2, \ldots$ to individual points.

The following theorem is an extension of the Urysohn and Tietze theorems.

THEOREM 3. *Let $F = \overline{F} \subset \mathcal{X} \subset \mathcal{I}^{\aleph_0}$ and $f \in (\mathcal{I}^{\aleph_0})^F$. There exists a continuous mapping $g: \mathcal{X} \to \mathcal{I}^{\aleph_0} \times \mathcal{I} \times \mathcal{I}^{\aleph_0}$ which coincides with f on F and which is a homeomorphism on $\mathcal{X}-F$.*

For, let $f^* \in (\mathcal{I}^{\aleph_0})^{\mathcal{X}}$ be an extension of f (compare the Tietze theorem, § 14, IV) and let [1]

$$g(x) = [f^*(x), \varrho(x, F), x \cdot \varrho(x, F)] \quad \text{where} \quad x \in \mathcal{X}.$$

V. Products of spaces with a countable base. Sets of the first category. We shall assume in this section that *$\mathcal{Y}$ has a countable base.* Denote briefly the vertical section of $\mathcal{X} \times \mathcal{Y}$ by $\mathcal{Y}^x$:

$$\mathcal{Y}^x = (x) \times \mathcal{Y}.$$

THEOREM 1 [2]. *Let $Z \subset \mathcal{X} \times \mathcal{Y}$. If Z is nowhere dense, there is $P \subset \mathcal{X}$ of the first category such that for each $x \in \mathcal{X}-P$ the set $Z \cap \mathcal{Y}^x$ is nowhere dense in $\mathcal{Y}^x$.*

Proof. Let $R_1, R_2, \ldots$ be the open base of $\mathcal{Y}$.

Put

$$E_n = \underset{x}{E}[(x) \times R_n \subset \overline{Z \cap \mathcal{Y}^x}] \quad \text{and} \quad P = E_1 \cup E_2 \cup \ldots \tag{1}$$

Obviously

$$E_n \times R_n \subset \overline{Z}; \quad \text{hence} \quad \overline{E}_n \times \overline{R}_n = \overline{E_n \times R_n} \subset \overline{Z}.$$

---

(1) *Ibidem,* Theorem 1.

(2) See the paper of S. Ulam and myself *Quelques propriétés topologiques du produit combinatoire,* Fund. Math. 19 (1932), p. 248.

The assumption $\mathcal{Y}$ to have a countable base is essential. See *ibid.*, p. 248, footnote 1.

It follows that E_n is nowhere dense; otherwise there would exist an open non-void set $G \subset \bar{E}$ and consequently $0 \neq G \times R_n \subset \bar{E}_n \times \bar{R}_n \subset \bar{Z}$, which is impossible since Z is nowhere dense.

Therefore P is of the first category.

Now let $x \in \mathscr{X} - P$ and suppose that $Z \cap \mathscr{Y}^x$ is not nowhere dense in $\mathscr{Y}^x$. Then there is an n such that $(x) \times R_n \subset \overline{Z \cap \mathscr{Y}^x}$ (since the sequence $(x) \times R_n$, $n = 1, 2, \ldots$, is a base of $\mathscr{Y}^x$). This yields (by (1)) $x \in E_n \subset P$, which is a contradiction.

Corollary 1a. *Let $Z \subset \mathscr{X} \times \mathscr{Y}$. If Z is of the first category, there is $P \subset \mathscr{X}$ of the first category such that $Z \cap \mathscr{Y}^x$ is of the first category in $\mathscr{Y}^x$ for each $x \in \mathscr{X} - P$.*

Proof. Put $Z = Z_1 \cup Z_2 \cup \ldots$ where Z_n is nowhere dense. Apply the preceding theorem to each of the sets Z_n and denote by P_n the corresponding set of first category. Then $P = P_1 \cup P_2 \cup \ldots$ is the required set.

Corollary 1b. *The product $Z = A \times B$ is of the first category in $\mathscr{X} \times \mathscr{Y}$ if and only if either A or B is of the first category.*

Proof. Let Z be of the first category and suppose that neither A nor B is of the first category. Let P denote the set considered in Corollary 1a. Hence $A \not\subset P$. Let $x \in A - P$. It follows that $Z \cap \mathscr{Y}^x$ is of the first category in $\mathscr{Y}^x$. Since $Z = A \times B$, we have $Z \cap \mathscr{Y}^x = (x) \times B$ and consequently B is of the first category (in $\mathscr{Y}$, cf. § 15, VI).

Thus the condition is necessary. Its sufficiency follows from Corollary 2a of § 15, VII.

Corollary 1b can be strengthened as follows.

Corollary 1c [1]. $D(A \times B) = D(A) \times D(B)$.

Proof. By definition, $(a, b) \notin D(A \times B)$ means that $A \times B$ is of the first category at (a, b). In other words there is an open G containing (a, b) and such that $G \cap (A \times B)$ is of the first category. We may assume, of course, that G belongs to the base of $\mathscr{X} \times \mathscr{Y}$; thus

$$G = M \times N, \quad a \in M, \quad b \in N, \quad M \text{ and } N \text{ are open.}$$

---

[1] See R. Sikorski, *On the cartesian products of metric spaces*, Fund. Math. 34 (1947), p. 289.

As $(M\times N)\cap(A\times B) = (M\cap A)\times(N\cap B)$, we conclude that

$$[(a, b)\notin D(A\times B)] \equiv$$

$$\equiv \bigvee_{MN} (a\in M)(b\in N)(M, N \text{ open})[(M\cap A)\times(N\cap B) \text{ of the first cat.}].$$

But by Corollary 1b

$$[(M\cap A)\times(N\cap B) \text{ of the first cat.}] \equiv$$

$$\equiv [\text{either } M\cap A \text{ or } N\cap B \text{ of the first cat.}).$$

Therefore

$$[(a, b)\notin D(A\times B)] \equiv$$

$$\equiv [\text{either } a\in D(A) \text{ or } b\in D(B)] \equiv [(a, b)\notin D(A)\times D(B)].$$

THEOREM 2 (1). *Let* $f: \mathscr{X}\to\mathscr{Y}$ *and* $\mathfrak{J} = \underset{xy}{E}[y = f(x)]$. *Let* $\mathscr{Y}$ *be dense in itself. If* $\mathscr{X}\times\mathscr{Y}-\mathfrak{J}$ *is of the first category at the point* (x, y), *then either* $\mathscr{X}$ *is of the first category at* x *or* $\mathscr{Y}$ *is such at* y.

In other words (see Corollary 1c):

$$D(\mathscr{X}\times\mathscr{Y}-\mathfrak{J}) = D(\mathscr{X}\times\mathscr{Y}). \tag{2}$$

Proof. By assumption, there exist G open in $\mathscr{X}$ and H open in $\mathscr{Y}$ and such that $x\in G$, $y\in H$, and $(G\times H)-\mathfrak{J}$ is of the first category. Suppose that $\mathscr{X}$ is not of the first category at x. Then G is not a set of the first category and it follows by Corollary 1a that there is a point $a\in G$ such that the set

$$\mathscr{Y}^a\cap(G\times H)-\mathfrak{J} = \big((a)\times H\big)-\mathfrak{J}$$

is of the first category in $\mathscr{Y}^a$. Now, this set differs from $\big((a)\times H\big)$ by one point only, namely by the point $\big((a, f(a)\big)$. This point being an accumulation point of $\mathscr{Y}^a$ (since $\mathscr{Y}$ is dense in itself), hence nowhere dense in $\mathscr{Y}^a$, the set $\big((a)\times H\big)$ is of the first category in $\mathscr{Y}^a$. Consequently H is of the first category in $\mathscr{Y}$. But this means that $\mathscr{Y}$ is of the first category at y.

---

(1) See my paper *Sur les fonctions représentables analytiquement et les ensembles de première catégorie*, Fund. Math. 5 (1924), p. 84, and the mentioned paper of S. Ulam and myself, p. 250.

***VI. Products of spaces with a countable base. Baire property.** As in Section V, we assume that $\mathscr{Y}$ *has a countable base.*

THEOREM 1 (1). *Let $Z \subset \mathscr{X}\times\mathscr{Y}$. If Z has Baire property, there is $P \subset \mathscr{X}$ of the first category such that $Z \cap \mathscr{Y}^x$ has the Baire property relative to $\mathscr{Y}^x$ for each $x \in \mathscr{X}-P$.*

Proof. As Z has the Baire property, so $Z = U \cup V$ where U is of the first category and V is a G_δ-set (see § 11, IV, 2). By virtue of Corollary 1a of Section V, there is $P \subset \mathscr{X}$ of the first category such that the set $U \cap \mathscr{Y}^x$ is of the first category in $\mathscr{Y}^x$ for each $x \in \mathscr{X}-P$. Since $V \cap \mathscr{Y}^x$ is obviously a G_δ-set in $\mathscr{Y}^x$, it follows from the identity $Z \cap \mathscr{Y}^x = (U \cap \mathscr{Y}^x) \cup (V \cap \mathscr{Y}^x)$ that the set $Z \cap \mathscr{Y}^x$ has the Baire property relative to $\mathscr{Y}^x$.

THEOREM 2. *The product $Z = A\times B$ has the Baire property without being of the first category if and only if each of the sets A and B has the Baire property without being of the first category.*

Proof. Let us recall (compare § 11, IV, 4) that A has the Baire property if and only if the set $D(A) \cap D(\mathscr{X}-A)$ is nowhere dense; on the other hand, A is of the first category if and only if $D(A) = 0$ (compare § 10, V (7)), which is equivalent to $D(A)$ being nowhere dense (compare § 10, V (11)). Now according to Corollary 1c of Section IV, we have

$$\begin{aligned}
&D(A\times B) \cap D[(\mathscr{X}\times\mathscr{Y})-(A\times B)]\\
&= [D(A)\times D(B)] \cap \{D[(\mathscr{X}-A)\times\mathscr{Y}] \cup D[\mathscr{X}\times(\mathscr{Y}-B)]\}\\
&= \{[D(A)\times D(B)] \cap [D(\mathscr{X}-A)\times D(\mathscr{Y})]\} \cup\\
&\qquad\qquad \cup \{[D(A)\times D(B)] \cap [D(\mathscr{X})\times D(\mathscr{Y}-B)]\}\\
&= \{[D(A) \cap D(\mathscr{X}-A)]\times D(B)\} \cup \{D(A)\times [D(B) \cap D(\mathscr{Y}-B)]\}.
\end{aligned}$$

If we assume that the set $A\times B$ has the Baire property, the first member of this formula, and hence the last one as well, is a nowhere dense set. Therefore $[D(A) \cap D(\mathscr{X}-A)]\times D(B)$ is nowhere dense. It follows by Theorem 2 of § 15, VII, that either $D(A) \cap D(\mathscr{X}-A)$ or $D(B)$ is nowhere dense. Similarly, either $D(A)$ or $D(B) \cap D(\mathscr{Y})-B)$ is nowhere dense. If we now assume that $A\times B$ is not of the first category, then $D(A\times B) \neq 0$ and consequently

(1) See the paper of S. Ulam and myself referred to above.

$D(A) \neq 0 \neq D(B)$ by Corollary 1c of Section IV. It follows, by virtue of § 10, V (11), that neither $D(A)$ nor $D(B)$ is nowhere dense. Thus, both sets

$$D(A) \cap D(\mathscr{X}-A) \quad \text{and} \quad D(B) \cap D(\mathscr{Y}-B) \tag{i}$$

are nowhere dense and therefore both A and B have the Baire property and none of them is of the first category.

Thus our condition is necessary. It is also sufficient. For, as it implies that the sets (i) are nowhere dense, then the sets

$$[D(A) \cap D(\mathscr{X}-A)]\times D(B) \quad \text{and} \quad D(A)\times[D(B) \cap D(\mathscr{Y}-B)]$$

(as well as their union) are nowhere dense, which means that $A\times B$ has the Baire property. Furthermore, as none of the sets A and B is of the first category, $A\times B$ is neither of the first category (by Corollary 1b of Section IV).

Remarks. The analogy between the Baire property and measurability, to which we called attention in § 11, concerns also the statements just established. Namely, if in the theorems of this Section and in the Corollaries 1a and 1b of Section V we replace the sets of the first category by sets of measure zero and the Baire property by measurability, we get well-known theorems on measure theory [1].

THEOREM 3. *Let* $f\colon \mathscr{X} \to \mathscr{Y}$ *and* $\mathfrak{J} = \underset{xy}{E}[y = f(x)]$. *Let* $\mathscr{Y}$ *be dense in itself. If* $\mathfrak{J}$ *has the Baire property, it is of the first category.*

Proof. By V (2) we have

$$D(\mathfrak{J}) = D(\mathfrak{J}) \cap D(\mathscr{X}\times\mathscr{Y}) = D(\mathfrak{J}) \cap D(\mathscr{X}\times\mathscr{Y}-\mathfrak{J}).$$

The latter set is nowhere dense, because $\mathfrak{J}$ has the Baire property (compare § 11, IV, 4). Now $D(\mathfrak{J})$ is never nowhere dense unless it is void (by § 10, V(11)). This completes the proof, because the condition $D(\mathfrak{J}) = 0$ means that $\mathfrak{J}$ is of the first category (by § 10, V (7)).

Remark. If $\mathfrak{J}$ has not been supposed to have the Baire property, it does not need to be of the first category [2].

---

(1) Compare S. Saks, *Theory of the integral*, Monografie Matematyczne 7 (1937), p. 77 (Fubini theorem).

(2) See my above mentioned paper of Fund. Math. 5, p. 85.

B. CARDINALITY PROBLEMS (§§ 23, 24)

The space is assumed to be a regular $\mathscr{T}_1$-space with a countable base. Therefore a distance $|x-y|$ between points of the space can be assumed to be defined. In other words, the space can be supposed separable and metric.

§ 23. Power of the space. Condensation points

I. Power of the space.

THEOREM. *The power of the space is* $\leqslant \mathfrak{c}$.

Proof. There exists a sequence of points $r_1, r_2, \ldots$ such that every point of the space is the limit of a subsequence of this sequence. It follows that the power of the space does not exceed the power of all sequences whose terms belong to a countable set, i.e., the power of the continuum.

II. Dense parts.

THEOREM. *Every subset of the space contains a dense countable subset.*

For a subset of a metric separable space (§ 21, I, Theorem 2 and § 22, I, Theorem 1) is separable.

III. Condensation points.

DEFINITION. A point p is said to be a *condensation point* of the set X, if every neighbourhood of p contains an uncountable set of points of X (1); in other words, if X is not locally countable at the point p (§ 7, IV). The set of condensation points of X will be denoted by X^0.

THEOREM. *The set* $X-X^0$ *is countable.*

Proof. Let $R_1, R_2, \ldots$ be the base of the space. Assign to each $p \in X-X^0$ an index $n(p)$ such that $X \cap R_{n(p)}$ is countable. Then $X-X^0 \subset \bigcup_p X \cap R_{n(p)}$ and this union, as a countable union of countable sets, is countable (by a theorem of the set theory based on the axiom of choice).

(1) E. Lindelöf, Acta Math. 29 (1095), p. 183. Compare also W. H. Young, Quart. Math. Journ. 25 (1903), p. 103.

IV. Properties of the operation X^0.

$$(X \cup Y)^0 = X^0 \cup Y^0; \tag{1}$$

$$X^0 - Y^0 \subset (X - Y)^0; \tag{2}$$

$$\Big(\bigcap_\iota X_\iota\Big)^0 \subset \bigcap_\iota X_\iota^0; \tag{3}$$

$$\bigcup_\iota X_\iota^0 \subset \Big(\bigcup_\iota X_\iota\Big)^0; \tag{4}$$

$$X \subset Y \quad \textit{implies} \quad X^0 \subset Y^0; \tag{5}$$

$$(X - X^0)^0 = 0; \tag{6}$$

$$X^0 \subset X^d \subset \overline{X}; \tag{7}$$

$$X^0 = X^{00} = X^{0d} = \overline{X^0} = (X \cap X^0)^0; \tag{8}$$

$$X \cap X^0 \subset (X \cap X^0)^d. \tag{9}$$

Formulas (1)–(5) follow directly (compare § 7, IV) from the fact that the union of two countable sets is countable and that a subset of a countable set is countable. Formula (6) follows from Section III (for a countable set cannot have a condensation point). Formula (7) follows from the fact that every condensation point is an accumulation point. By (3), we have

$$(X \cap X^0)^0 \subset X^0 \cap X^{00} \subset X^{00}.$$

Since the set X^0 is closed (§ 7, IV), it follows from (7) that

$$X^{00} \subset X^{0d} \subset \overline{X^0} = X^0,$$

and we obtain (8) by the identity $X = (X \cap X^0) \cup (X - X^0)$ which gives by (1) and (6), $X^0 = (X \cap X^0)^0$. Finally, formulas (8) and (7) imply (9):

$$X \cap X^0 \subset X^0 = (X \cap X^0)^0 \subset (X \cap X^0)^d.$$

V. Scattered sets.

THEOREM. *Every scattered space is countable.*

Proof. By IV, (9), the set $\mathscr{X}^0$ is dense in itself and, since the space is scattered, it follows that $\mathscr{X}^0 = 0$ and $\mathscr{X} = \mathscr{X} - \mathscr{X}^0$.

By Section III this difference is countable; hence the proof is completed.

By making use of § 9, VI, Theorem 3 we deduce the *Cantor–Bendixson theorem* (¹): *Every space is the union of two sets, one is perfect and the other is countable* (and scattered).

VI. Unions of scattered sets.

THEOREM. *Every space which is the union of a monotone family of scattered sets is countable* (²).

Proof. Suppose that $\mathcal{X} = \bigcup_\iota C_\iota$, where C_ι are scattered (the index ι ranges over an arbitrary set), and that $\mathcal{X}$ is uncountable. By the preceding theorem, the space contains a set dense in itself which in turn contains (by II) a dense countable subset $D = [p_1, p_2, \ldots]$. Therefore D is dense in itself.

For each p_n, let ι_n be an index such that $p_n \in C_{\iota_n}$ and let

$$S = \bigcup_{n=1}^{\infty} C_{\iota_n}.$$

Therefore $D \subset S$. The set S, as the union of countable sets, is countable and it suffices to prove that $S = \mathcal{X}$.

Suppose that $q \in \mathcal{X} - S$. There exists a C_0 such that $q \in C_0$. It follows that C_0 is not contained in any of the sets C_{ι_n}. Since the family of the sets C_ι is monotone, it follows that $C_{\iota_n} \subset C_0$ for every n. Therefore $S \subset C_0$ and $D \subset S \subset C_0$ which is impossible since C_0 is scattered and D is dense in itself (and non empty).

VII. Points of order $\mathfrak{m}$. The proposition of Section III by which every uncountable space contains condensation points (i.e. points of "uncountable order") can be made more precise as follows: *if the power* $\mathfrak{m} > \aleph_0$ *of the space is not cofinal with* ω (i.e. it is not the sum of a countable series of cardinal numbers less than $\mathfrak{m}$), *then the space contains a point of order* $\mathfrak{m}$ (a point every neighbourhood of which is of the power $\mathfrak{m}$). Also the following proposi-

(¹) G. Cantor, Math. Ann. 21 (1883), p. 575; I. Bendixson, Acta Math. 2 (1883), p. 415; E. Lindelöf, *loc. cit.*

(²) Theorem of W. Sierpiński, *Sur une propriété des ensembles clairsemés*, Fund. Math. 3 (1922), pp. 46–49.

tion holds: *The space is the union of a scattered set and of a sequence of sets such that all the points of the same set are of the same order* (1).

VIII. The concept of effectiveness. This notion is of a metamathematical nature: it concerns ways of proving existence theorems. An existence theorem, i.e. a theorem of the form $\bigvee_x \varphi(x)$, where $\varphi(x)$ is a propositional function (see § 1, III) is said to be *effectively proved*, if an individual a is *defined* and a is proved to satisfy the given theorem, i.e. $\varphi(a)$ is proved (2).

In the topology of regular $\mathcal{T}_1$-spaces with a countable base *we shall understand the base of the space to be well defined* (thus for example one can define a base in the euclidean space to be the sequence of balls whose radius and coordinates of the centre are rational). By using the base of the space one can define other objects (sets, functions and so on).

In problems of power we usually have to prove that the set has a given power. The proof is effective, if the correspondence in question can be well *defined*.

Thus, for example in Section III, the proof that the set $X-X^0$ is countable, is not effective. We have proved that sequences containing this set exist, without defining any particular sequence.

In contrast to this, one can prove effectively that the set $X-X^d$ *is countable*. Namely, if $p \in X-X^d$, there exists an index n such that $p = X \cap R_n$ (where R_n belongs to the base of the space). Let $n(p)$ be the first index with this property. To two different points there correspond different indices. Thus the points of $X-X^d$ have been arranged in a (finite or infinite) sequence.

An effective proof of the countability of the set $X-X^0$ will be given in § 24.

---

(1) See W. Sierpiński, *Sur la décomposition des ensembles de points en parties homogènes*, Fund. Math. 1 (1920). Compare G. Cantor, Acta Math. 7 (1885), p. 118.

(2) Compare the notion of effectiveness due to Borel and Lebesgue and that due to F. Bernstein (who distinguishes between "Existenz" and "Herstellung", Leipz. Ber. 60 (1908)). Many papers of W. Sierpiński were devoted to this subject; see, especially, *Les exemples effectifs et l'axiome du choix*, Fund. Math. 2 (1921). Compare also remarks on effectiveness in a paper of B. Knaster and myself, Fund. Math. 2, p. 251.

§ 24. Powers of various families of sets

I. Families of open sets. Families of sets with the Baire property.

THEOREM 1. *The family of all open sets has a power* $\leqslant \mathfrak{c}$.

Proof. Every open set is the union of certain sets belonging to the base and hence the power of the family of all open sets does not exceed that of all the sequences of elements of the base.

Since every closed set is the complement of an open set, it follows that *the family of all closed sets has a power* $\leqslant \mathfrak{c}$.

Remark. The correspondence between open sets and elements of a subset of a linear continuum can be defined directly as follows.

Let $R_1, R_2, \ldots$ be a base of the space. Assume that each term repeats in this sequence an infinite number of times. Given an open set G, let $k_1, k_2, \ldots$ be the sequence of indices such that $R_{k_n} \subset G$. Let

$$t(G) = \sum_{n=1}^{\infty} \frac{1}{2^{k_n}}, \qquad t(0) = 0 \text{ (}^1\text{)}.$$

In other words, the dyadic expansion of $t(G)$ is $0, a_1, a_2, \ldots$ where $a_n = 1$ or 0 according as to R_n is or is not, contained in G. These expansions always admit an infinite number of units (except when $G = 0$). It follows that the function t is one-to-one:

$$G \neq H \quad \textit{implies} \quad t(G) \neq t(H). \tag{2}$$

We easily deduce that

$$G \subset H \textit{ implies } t(G) \leqslant t(H). \tag{3}$$

THEOREM 2. *Every family of disjoint open sets is (effectively) countable.*

Proof. Given a (non empty) member G of the family, there exists a member of the base such that $R_n \subset G$. Let $n(G)$ be the first index with this property. Since the members of the family are disjoint, it follows that to different sets correspond different indices. Thus the given family has been arranged in a (finite or infinite) sequence.

---

(1) This definition was suggested by Felix Hausdorff.

In particular, every family of non overlapping intervals in a straight line is countable.

THEOREM 3. *Every family of disjoint sets $\{X_\iota\}$ with the Baire property, of which none is of the first category, is (effectively) countable.*

For by § 11, IV, Corollary 3, the sets $\mathrm{Int}[D(X_\iota)]$ are disjoint and non empty by § 10, V, (7).

Remark. It is worth noticing that Theorem 2 can be extended (apart from effectiveness) to spaces which are *products of an arbitrary number of spaces each of which is separable* (¹).

THEOREM 4 (²). *If R is the result of the $(\mathscr{A})$-operation performed on the sets $A_{k_1\ldots k_n}$ with the Baire property, then there exists an index $\mu < \Omega$ such that the set $R - K_\mu$ is of the first category* (here the notation of § 3, XIV, Theorem 3, is used).

Proof. Since $A^\beta_{k_1\ldots k_n} \subset A^\alpha_{k_1\ldots k_n}$ for $\alpha < \beta$, it follows that the sets

$$B^\alpha_{k_1\ldots k_n} = A^\alpha_{k_1\ldots k_n} - A^{\alpha+1}_{k_1\ldots k_n}, \quad \text{where} \quad \alpha < \Omega,$$

are mutually disjoint (for fixed $k_1 \ldots k_n$). By Theorem 3, to every system $r = (k_1 \ldots k_n)$ there corresponds a $\gamma_r < \Omega$ such that the set $B^\alpha_{k_1\ldots k_n}$, for $\alpha > \gamma_r$, is of the first category. Since the family of the systems r is countable, there exists a $\mu < \Omega$ such that $\mu > \gamma_r$ for every r. Hence the set $B^\mu_{k_1\ldots k_n}$ is of the first category for every system $k_1 \ldots k_n$. It follows that the union $\bigcup B^\mu_{k_1\ldots k_n}$ taken over all systems $k_1 \ldots k_n$ is also of the first category. The inclusions

$$R - K_\mu \subset E_\mu - K_\mu \subset T_\mu = \bigcup B^\mu_{k_1\ldots k_n}$$

imply that $R - K_\mu$ is of the first category.

(¹) See E. Marczewski, *Séparabilité et multiplication cartésienne des espaces topologiques*, Fund. Math. 34 (1947), p. 139, and *Remarques sur les produits cartésiens d'espaces topologiques*, C. R. Acad. Sc. URSS 31 (1941), p. 525. Compare also M. Bockstein, *Un théorème de séparabilité pour les produits topologiques*, Fund. Math. 35 (1948), pp. 242–246.

(²) See E. Selivanowski, *Sur les propriétés des constituantes des ensembles analytiques*, Fund. Math. 21 (1933), p. 20; W. Sierpiński, *ibid.* p. 29; E. Szpilrajn-Marczewski, *ibid.* p. 324.

If the Baire property is replaced by the measurability (in the sense of Lebesgue) and the sets of the first category by the sets of measure zero, then Theorem 4 remains valid.

THEOREM 5. *If the sets* $A_{k_1...k_n}$ *have the Baire property in the restricted sense, and if* Z *is a set which contains a single point from each (non empty) difference* $K_a - \bigcup_{\xi<a} K_\xi$, *then* Z *is of the first category on every perfect set.*

Proof. Since $K_a \subset R$, it follows that

$$Z \subset \bigcup_{a \leqslant \mu} (K_a - \bigcup_{\xi<a} K_\xi) \cup (R - K_\mu).$$

By virtue of Theorem 4, we deduce that Z is the union of a countable set and a set of the first category. Relativizing Theorem 4 to an arbitrary perfect set P, we infer that $Z \cap P$ is the union of a countable set and a set of the first category on P. Since the set P is perfect, every point of it is nowhere dense in P. Therefore $Z \cap P$ is of the first category on P.

We shall see later that if R is a non borelian analytic set (in the space $\mathscr{E}$ of real numbers), then the set Z which is of the first category on every perfect set can be supposed to be uncountable. This presents a very remarkable singularity (in the space $\mathscr{E}$).

Another set of the first category on every perfect set can be obtained by selecting a single point from each (non-empty) difference $D_{a+1} - D_a$ where $D_a = -E_a$ (for E_a see § 3, XIV) and where $\{A_{k_1...k_n}\}$ is assumed to be a regular system of closed sets (1).

II. Well ordered monotone families (2).

THEOREM 1. *Every well ordered family of decreasing open sets is (effectively) countable.*

Proof. Let $G_1 \supset G_2 \supset \ldots \supset G_\xi \supset G_{\xi+1} \supset \ldots$ be a transfinite sequence of (different) open sets. If a is not the last index, there exists a point $p_a \in (G_a - G_{a+1})$. Hence there exists an index n such that $p_a \in R_n \subset G_a$. Therefore

(i) $R_n \subset G_a$ and $R_n - G_{a+1} \neq 0$.

Let $n(a)$ be the smallest index satisfying (i).

The indices corresponding to two different transfinite numbers are different. For, the condition $a < \beta$ implies $R_{n(\beta)} \subset G_\beta \subset G_{a+1}$ while, by (i), $R_{n(a)}$ is not contained in G_{a+1}. Thus all the indices a

(1) See N. Lusin and W. Sierpiński, Rend. Acc. Lincei 6, v. VII (1928), p. 214.

(2) See R. Baire, *Thèse*, Ann. di Math. (3) 3 (1899), p. 51.

(with the possible exception of the last one) have been arranged in an infinite sequence.

THEOREM 2. *Every well ordered family of decreasing closed sets is (effectively) countable.*

Proof. Let $F_1 \supset F_2 \supset \ldots \supset F_\xi \supset F_{\xi+1} \supset \ldots$ be the given sequence. If α is not the last index, there exists a point $p_\alpha \epsilon (F_\alpha - F_{\alpha+1})$. Let $n(\alpha)$ be the smallest index such that

$$R_{n(\alpha)} \cap F_\alpha \neq 0 = R_{n(\alpha)} \cap F_{\alpha+1}.$$

For $\alpha < \beta$, we have $F_\beta \subset F_{\alpha+1}$. The condition $R_{n(\beta)} \cap F_\beta \neq 0$ implies, therefore, $R_{n(\beta)} \cap F_{\alpha+1} \neq 0$; hence $n(\alpha) \neq n(\beta)$. It follows that the set of indices α is (effectively) countable.

Since open sets are complements of closed sets, we obtain the following proposition which generalizes the two preceding ones.

THEOREM 3. *Every well ordered family of increasing or decreasing sets which are either all closed or all open is (effectively) countable.*

Remarks. (i) Every set of real numbers well ordered by the "less than" relation is countable. For, a transfinite sequence of real numbers $x_0 < x_1 < \ldots < x_\xi < x_{\xi+1} < \ldots$ determines a well ordered family of increasing closed sets $\{F_\xi\}$, where F_ξ is the half-line $x \leqslant x_\xi$.

(ii) If, in place of well ordering, we assume that each element (except, possibly the last) has an immediate successor, the proof of Theorems 1 and 2 will remain valid. Theorem 3 can also be generalized in this manner.

Other generalizations will be proved in the following sections (III, Theorem 2 and VII, Theorem 1).

III. Resolvable sets. By definiton (§ 12, II), a set E is *resolvable*, if it can be written in the form

$$E = F_1 - F_2 \cup F_3 - F_4 \cup \ldots \cup F_\xi - F_{\xi+1} \cup \ldots$$

where the terms of this transfinite sequence are decreasing closed sets.

By Theorem II, 2, this series is *countable* (in other words, all the indices ξ are less than a number $\alpha < \Omega$).

THEOREM 1. *Resolvable sets are both* $\boldsymbol{F}_\sigma$ *and* $\boldsymbol{G}_\delta$*-sets.*

Proof. The difference of two closed sets is an $\boldsymbol{F}_\sigma$-set as the intersection of a closed set witch an open set (which is $\boldsymbol{F}_\sigma$ by § 21, IV). A resolvable set being composed of countably many differences of closed sets is an $\boldsymbol{F}_\sigma$-set. It is also a $\boldsymbol{G}_\delta$-set, for the complement of a resolvable set is itself resolvable (§ 12, VI, Theorem 1), hence an $\boldsymbol{F}_\sigma$-set.

COROLLARY 1a. *Scattered sets are both $\boldsymbol{F}_\sigma$ and $\boldsymbol{G}_\delta$-sets* (1).

This is because every scattered set is resolvable (§ 12, VI, Theorem 4).

Remarks. (i). Resolvable sets are *effectively $\boldsymbol{F}_\sigma$ and $\boldsymbol{G}_\delta$*, i.e. to every such set we can assign a well determined sequence of closed sets whose union is the given set (as well as a sequence of open sets whose intersection is the given set).

Proof. If E is resolvable, then by § 12, V, (iii),

$$E = \bigcup_{\xi<\alpha} (A_\xi - B_\xi)$$

where A_ξ and B_ξ are terms of the resolution (§ 12, III, 3° (i)).

By II, 2, the sequence A_ξ is effectively countable; hence the indices $\xi < \alpha$ can be assumed to be ordered in a simple infinite sequence $\xi_1, \xi_2, \ldots$ (which is not necessarily increasing). But for every closed set we have defined (§ 21, IV) a sequence of open sets whose intersection is the given set. Therefore

$$B_\xi = \bigcap_{k=1}^{\infty} G_\xi^k \quad \text{and} \quad E = \bigcup_{n,k=1}^{\infty} [A_{\xi_n} - G_{\xi_n}^k].$$

This double series can easily be rearranged in a simple series. Hence a simple infinite series of closed sets whose union is E has been *defined*. This means that E is an *effective $\boldsymbol{F}_\sigma$*.

The complement of a resolvable set being resolvable (§ 12, VI, Theorem 1) we infer that every resolvable set is an effective $\boldsymbol{G}_\delta$.

(ii) We shall see in § 37 that *the converse of Theorem* 1 *is true* in complete spaces, but not in arbitrary spaces. For example, in the space of rational numbers, a set which is both dense and boundary is $\boldsymbol{F}_\sigma$ and $\boldsymbol{G}_\delta$ (since the space is countable), but it is not resolvable, since its boundary is not nowhere dense (it is equal to the entire space, compare § 12, V, (ii)).

(iii) Theorem 1 remains true *in every metric space,* separable or not (see § 30, X, Theorem 5).

(1) Theorem of W. H. Young, *The theory of sets of points*, Cambridge 1906, p. 65.

The following proposition is a generalization of Theorem 3 of Section II.

THEOREM 2. *Every well ordered family of increasing (or decreasing) resolvable sets is countable.*

By using the fact that the characteristic function of a resolvable set is pointwise discontinuous on every closed set (§ 13, VI), Theorem 2 can be derived [1] from the following more general theorem.

THEOREM 2'. *Every well ordered family of real-valued functions*

$$f_1(x) \leqslant f_2(x) \leqslant \ldots \leqslant f_\xi(x) \leqslant f_{\xi+1}(x) \leqslant \ldots \qquad (\xi < \Omega)$$

which are pointwise discontinuous on every closed set is countable.

Proof [2]. It suffices to prove the existence of an index a such that, for every x and $\xi, f_{a+\xi}(x) = f_a(x)$.

For a fixed n, let $R_{n,1}, R_{n,2}, \ldots, R_{n,i}, \ldots$ be the sequence of elements of the base of the space such that there exists an ordinal number $a_{n,i}$ satisfying the condition:

$$x \in R_{n,i} \quad \text{implies} \quad f_{a_{n,i}+\xi}(x) - f_{a_{n,i}}(x) < 1/n \quad \text{for every } \xi.$$

Let $R^n = R_{n,1} \cup R_{n,2} \cup \ldots$ and let $a_n > a_{n,i}$, $i = 1, 2, \ldots$. Hence

$$f_{a_n+\xi}(x) - f_{a_n}(x) < 1/n \quad \text{for} \quad x \in R^n.$$

Consequently, if $a > a_n$, $n = 1, 2, \ldots$, then, for $x \in R^1 \cap R^2 \cap \ldots$,

$$f_{a+\xi}(x) - f_a(x) < 1/n \quad \text{for every } n, \text{ hence} \quad f_{a+\xi}(x) = f_a(x).$$

Therefore it suffices to prove $R^n = \mathscr{X}$, for every n.

Suppose that there exists an n such that the closed set $F = \mathscr{X} - R^n$ is non-empty. Let $[a_1, a_2, \ldots]$ be a countable set dense in F. We shall show that there exists a γ such that

$$\xi \geqslant \gamma \quad \text{implies} \quad f_\xi(a_m) = f_\gamma(a_m) \quad \text{for every } m. \tag{i}$$

For each m there is a γ_m such that $f_\xi(a_m) = f_{\gamma_m}(a_m)$ for $\xi \geqslant \gamma_m$ (compare Remark (i) of Section II); it suffices to take γ greater than all the γ_m's where $m = 1, 2, \ldots$. Moreover, we can assume that $\gamma > a_n$.

---

[1] For a more direct proof, see F. Hausdorff, *Mengenlehre*, p. 171. Compare also Z. Zalcwasser, *Un théorème sur les ensembles qui sont à la fois* $\boldsymbol{F}_\sigma$ *et* $\boldsymbol{G}_\delta$, Fund. Math. 3 (1922), p. 44.

[2] This proof was given by Z. Zalcwasser.

Since $a_m \notin R^n$, there exists in every neighbourhood of a_m a point x and an index $\beta_m > \gamma$ such that $f_{\beta_m}(x) - f_\gamma(x) \geqslant 1/n$; hence there exists a b_m such that

$$|b_m - a_m| < 1/n \quad \text{and} \quad f_{\beta_m}(b_m) - f_\gamma(b_m) \geqslant 1/n.$$

On the other hand, $\gamma > a_n$ implies $f_{a_n}(b_m) \leqslant f_\gamma(b_m)$; hence $f_{\beta_m}(b_m) - f_{a_n}(b_m) \geqslant 1/n$, therefore $b_m \epsilon F$.

Let β be greater than all the β_m's, $m = 1, 2, \ldots$. It follows that

$$f_\beta(b_m) - f_\gamma(b_m) \geqslant 1/n, \qquad m = 1, 2, \ldots. \tag{ii}$$

Since the function f_β is pointwise discontinuous on every closed set, there exists a point of continuity of the partial function $f_\beta | F$. Let $G(\neq 0)$ be an open set in F such that $|f_\beta(x) - f_\beta(x')| < 1/2n$ for x and x' belonging to G. Similarly, there exists a point of continuity of the function $f_\gamma | \bar{G}$ in G. Let $H(\neq 0)$ be an open set in F such that for $x \epsilon H$ and $x' \epsilon H$

$$|f_\beta(x) - f_\beta(x')| < 1/2n \quad \text{and} \quad |f_\gamma(x) - f_\gamma(x')| < 1/2n.$$

For a suitable chosen m we can substitute here a_m for x and b_m for x'. By (i), $f_\beta(a_m) = f_\gamma(a_m)$ and

$$|f_\beta(a_m) - f_\beta(b_m)| < 1/2n \quad \text{and} \quad |f_\beta(a_m) - f_\gamma(b_m)| < 1/2n;$$

hence $|f_\beta(b_m) - f_\gamma(b_m)| < 1/n$, contrary to (ii).

COROLLARY 2a. *Every well ordered family of increasing (or decreasing) scattered sets is countable* (1).

This is because every scattered set is resolvable (§ 12, VI, Theorem 4).

IV. Derived sets of order a (2). The derived set of X of order a is defined by the conditions

$$X^{(1)} = X^d, \quad X^{(a+1)} = (X^a)^d, \quad \text{and} \quad X^{(\lambda)} = \bigcap_{a<\lambda} X^{(a)}$$

if λ is a limit number.

Since the derived set is closed and the intersection of closed sets is also closed, it follows (by transfinite induction) that $X^{(a)}$

(1) A direct proof of this proposition is given in my paper of Fund. Math. 3 (1922), p. 42. An effective proof follows also from the theorem § 23, VI (of W. Sierpiński, see his note referred to above).

(2) G. Cantor, Math. Ann. 17 (1880), p. 357.

is closed for every α. Since every closed set contains its derived set, it follows that the family of derived sets is decreasing:

$$X^{(1)} \supset X^{(2)} \supset \ldots \supset X^{(\alpha)} \supset \ldots .$$

Consequently (by II, Theorem 2), *the family of derived sets of a set X is countable.* In other words, there exists an ordinal number $\beta < \Omega$ such that $X^{(\beta)} = X^{(\beta+1)} = \ldots$.

The set $X^{(\beta)}$ being equal to its derived set, is *perfect.*

In particular, if we put $X = \mathscr{X}$ and $\mathscr{X}^{(0)} = \mathscr{X}$, then (§ 12, I)

$$\mathscr{X} = \bigcup_{\alpha<\beta} (\mathscr{X}^{(\alpha)} - \mathscr{X}^{(\alpha+1)}) \cup \mathscr{X}^{(\beta)}. \tag{i}$$

Each set $\mathscr{X}^{(\alpha)} - \mathscr{X}^{(\alpha+1)}$ is composed of isolated points (§ 23, VIII), hence is countable. It follows that the union $\bigcup_{\alpha<\beta} (\mathscr{X}^{(\alpha)} - \mathscr{X}^{(\alpha+1)})$ is again a countable set.

Remark. Formula (i) is a reformulation of the Cantor–Bendixson theorem (§ 23, V): *By removing a countable set from the space we obtain a perfect set.*

The Cantor–Bendixson theorem was proved originally in this formulation. The advantage of this argument, in contrast to that given in § 23, is the absence of the axiom of choice and the possibility of an *effective* enumeration of the elements of every scattered set. For, the sequence of derived sets and the set $X - X^d$ are effectively countable (§ 23, VIII) ([1]).

V. Logical analysis ([2]). Relations between the following propositions were studied in $\mathscr{L}$-spaces:

(1) *every well ordered family of increasing closed sets is countable*;

(2) *every well ordered family of decreasing closed sets is countable*;

(3) *every set contains a dense countable subset*;

(4) *every scattered set is countable*;

(5) *every uncountable set contains a condensation point.*

([1]) Compare W. Sierpiński, *Une démonstration du théorème sur la structure des ensembles des points*, Fund. Math. 1 (1921), pp. 1–6.

([2]) Compare W. Sierpiński, *Sur l'équivalence de trois propriétés des ensembles abstraits*, Fund. Math. 2 (1921), pp. 179–188, and my paper *Une remarque sur les classes $\mathscr{L}$ de M. Fréchet*, Fund. Math. 3 (1922), pp. 41–43. See also (for the case of metric spaces) W. Gross, *Zur Theorie der Mengen in denen ein Distanzbegriff definiert ist*, Wiener Berichte 123 (1914), p. 801.

It is proved that propositions (4) and (5) are equivalent, that (4) implies (2), and that (3) implies (1). The converse implications hold in $\mathscr{L}$-spaces in which the closure is always a closed set but may fail to be true in more general $\mathscr{L}$-spaces.

VI. Families of continuous functions. Let $f: \mathscr{X} \to \mathscr{Y}$ be continuous. The space $\mathscr{X}$ being separable, there exists a sequence of points $x_1, x_2, \ldots$ such that each point p of $\mathscr{X}$ is the limit of a certain subsequence: $p = \lim x_{k_n}$. Since the function f is continuous, its values are determined by those at the points x_n, for $f(p) = \lim f(x_{k_n})$. In other words, if $f(x_n) = g(x_n)$ for every n, then the functions f and g coincide. It follows that the family $\mathscr{Y}^{\mathscr{X}}$ (of all continuous functions which map the space $\mathscr{X}$ onto subsets of the space $\mathscr{Y}$) has a power not exceeding the power of all (countable) sequences whose terms belong to the space $\mathscr{Y}$, i.e. the power $\mathfrak{c}^{\aleph_0}$, which is equal to $\mathfrak{c}$. Therefore the considered family has power $\leqslant \mathfrak{c}$.

Thus the family of all subsets of $\mathscr{Y}$ which are continuous images of a fixed subset of $\mathscr{X}$ is of the power $\leqslant \mathfrak{c}$. It follows that in a space of the power of the continuum there are as many *topological types* (§ 13, VIII) as subsets, namely $2^{\mathfrak{c}}$.

VII. Structure of monotone families of closed sets(1). Consider a *monotone family* $\boldsymbol{F}$ of closed sets in a separable metric space $\mathscr{X}$. This means that, for every pair A, B of sets, either $A \subset B$ or $B \subset A$. The family $\boldsymbol{F}$ will be considered as ordered and we say that A *precedes* B if $A \subset B$ and $A \neq B$.

The family $\boldsymbol{G}$ of the complements of elements of $\boldsymbol{F}$ is obviously a monotone family of open sets. The function t defined in I (1) establishes, by (2) and (3), a similarity relation between the family $\boldsymbol{G}$ and a subset of the interval $\mathscr{I}$. This can be expressed as follows.

THEOREM 1. *The family* $\boldsymbol{F}$ *is similar to a subset of the interval* $\mathscr{I}$.

That is to say, *it is possible to supply each element* A *of* $\boldsymbol{F}$ *with an index* y *so that* $0 \leqslant y \leqslant 1$ *and that the condition* $y_1 < y_2$ *is equivalent to the relation*

$$A_{y_1} \subset A_{y_2}, \quad A_{y_1} \neq A_{y_2}.$$

(1) For sections VII–X, see my paper *Sur les familles monotones d'ensembles fermés et leurs applications à la théorie des espaces connexes*, Fund. Math. 30 (1938), pp. 17–24. They are applied in *Topology II*.

We shall now prove a more precise theorem.

THEOREM 2. *If J is the set of indices to which the sets A_y correspond, then the following condition can be imposed: the end-points of the intervals contiguous to $\bar{J}$ belong to J; if F contains the first or the last element, then these elements are indexed by 0 and 1.*

Consequently, *if F does not possess gaps and contains the first and the last element, then J is closed. If, moreover, F does not possess leaps, then $J = \mathscr{I}$.*

Theorem 2 is a direct consequence of the following theorem of the general theory of ordered sets.

If F is a linearly ordered set containing a dense countable subset D, then there exists a correspondence φ: $F \to \mathscr{I}$ which preserves the order of the elements. $J = \varphi(F)$ may be assumed to satisfy the additional condition stated in Theorem 2.

The correspondence φ can be defined as follows [1]. Let $d_0, d_1, \ldots$ be a sequence composed of all elements of D and the elements of F which possess a neighbour element. Obviously we can assume that F contains the first element d_0 and the last element d_1.

Let

$$\varphi(d_0) = 0, \quad \varphi(d_1) = 1, \quad \text{and} \quad \varphi(d_n) = \tfrac{1}{2}[\varphi(d_k) + \varphi(d_l)],$$

where (for $n > 1$) d_k, d_l denote the pair of neighbour elements in the system $d_0, \ldots, d_{n-1}$ such that $d_k \prec d_n \prec d_l$. Finally, for $a \in F - D$, let $\varphi(a)$ be the greatest lower bound of the numbers $\varphi(d_n)$, where $a \prec d_n$.

We shall prove that φ is the desired correspondence.

We prove by induction on n that if d_j, d_q are two neighbour elements in the system $d_0, \ldots, d_{n-1}$, there exists s such that $|\varphi(d_j) - \varphi(d_q)| = 1/2^s$. It follows that if $\varphi(d_n) = i/2^m$ and i is odd, then

$$\varphi(d_k) = (i-1)/2^m \quad \text{and} \quad \varphi(d_l) = (i+1)/2^m$$

where k and l are as above.

---

(1) See my paper of Fund. Math. 3 (1921), p. 215.

Let uv be an interval contiguous to $\bar{J}$. Let m be the greatest integer such that the sequence $\varphi(d_0), \varphi(d_1), \ldots$ contains two numbers of the form $i/2^m$ and $(i+1)/2^m$ satisfying the condition

$$i/2^m \leqslant u < v \leqslant (i+1)/2^m.$$

One of the two numbers i or $i+1$, say i, is odd. Let n be the index such that $\varphi(d_n) = i/2^m$. Hence there exists an $l < n$ such that $\varphi(d_l) = (i+1)/2^m$ and the elements d_n and d_l in the system $d_0, \ldots, d_n$ are neighbours. We shal prove that

$$i/2^m = u \quad \text{and} \quad (i+1)/2^m = v.$$

Assume that the opposite is true. Then there exists a d_s with $d_n \prec d_s \prec d_l$. Let s be the smallest possible index; it follows that $\varphi(d_s) = (2i+1)/2^{m+1}$. According to the definition of the number m, $u < (2i+1)/2^{m+1} < v$. But this is a contradiction to the hypothesis that uv is a contiguous interval.

THEOREM 3. *The identities*

$$A_y = \overline{\bigcup_{u<y} A_u}, \tag{1}$$

$$A_y = \bigcap_{z>y} A_z \tag{2}$$

hold for every y, except for a countable set of indices (¹).

Proof. Let $R_1, R_2, \ldots$ be the base of the space $\mathcal{X}$. To every $y \in J$ which does not satisfy (1) we assign an integer $n(y)$ such that

$$A_y \cap R_{n(y)} \neq 0 \quad \text{and} \quad R_{n(y)} \cap \bigcup_{u<y} A_u = 0.$$

If $y \neq y'$ then $n(y) \neq n(y')$. Therefore the set of y which satisfy (1) is countable.

The case (2) is analogous.

THEOREM 4. *The relation*

$$\operatorname{Int}(A_y) = \bigcup_{u<y} \operatorname{Int}(A_u) \tag{3}$$

holds for every y, except for a countable set of indices.

(¹) See W. Sierpiński, Bull. de l'Acad. Polon. des Sci. 1921, p. 62.

Proof. Since the family of the sets $B_y = \overline{\mathscr{X} - A_y}$ is monotone, it follows by Theorem 3 that the family of sets B_y such that

$$B_y \neq \bigcap_{u<y} B_u, \quad \text{i.e. that} \quad \mathrm{Int}(A_y) \neq \bigcup_{u<y} \mathrm{Int}(A_u),$$

is countable. It is sufficient to prove that if $y' < y$ and $B_{y'} = B_y$, then $B_y = \bigcap_{u<y} B_u$. But this follows from the inclusion

$$B_y \subset \bigcap_{u<y} B_u \subset B_{y'}.$$

VIII. Strictly monotone families. We say that a family of (closed) sets is *strictly monotone*, if for every pair $A \neq B$ of its elements either $A \subset \mathrm{Int}(B)$ or $B \subset \mathrm{Int}(A)$. If the elements of the considered families are indexed (according to Theorems 1 and 2 of Section VII), then they are strictly monotone if and only if

the inequality $y_1 < y_2$ *implies the inclusion* $A_{y_1} \subset \mathrm{Int}(A_{y_2})$.

THEOREM. *If* $\boldsymbol{F}$ *is a strictly monotone family of closed sets, then the relations*

$$A_y = \bigcap_{z>y} A_z = \bigcap_{z>y} \mathrm{Int}(A_z); \tag{4}$$

$$A_y = \overline{\bigcup_{u<y} A_u} = \overline{\mathrm{Int}(A_y)}; \tag{5}$$

$$\mathrm{Int}(A_y) = \bigcup_{u<y} \mathrm{Int}(A_u) = \bigcup_{u<y} A_u; \tag{6}$$

$$\overline{\mathscr{X} - A_y} = \bigcap_{u<y} \overline{\mathscr{X} - A_u} = \bigcap_{u<y} (\mathscr{X} - A_u) \tag{7}$$

hold for every y, *except for a countable set of indices.*

Proof. Formula (4) follows from (2) by virtue of the inclusion $A_y \subset \mathrm{Int}(A_z) \subset A_z$. Formula (5) is a consequence of (1); it means that A_y is a closed domain. Formulas (6) and (7) are equivalent and follow from (3) by virtue of $\mathrm{Int}(A_u) \subset A_u \subset \mathrm{Int}(A_y)$.

Combining (4) and (7) we obtain formulas for *the boundary of* A_y:

$$\mathrm{Fr}(A_y) = \bigcap_{z>y} \mathrm{Int}(A_z) - \bigcup_{u<y} A_u \tag{8}$$

and (compare (2) and (7)): there exist two sequences $\{u_n\}$ and $\{z_n\}$ such that

$$u_1 < u_2 < \ldots < y < \ldots < z_2 < z_1$$

and

$$\mathrm{Fr}(A_y) = \bigcap_n \left(\mathrm{Int}(A_{z_n}) - A_{u_n}\right) = \bigcap_n (A_{z_n} \cap \overline{\mathscr{X} - A_{u_n}}). \tag{9}$$

IX. Relations of strictly monotone families to continuous functions.

THEOREM 1. *If* $f \in \mathcal{I}^{\mathcal{X}}$, *then the family* $\boldsymbol{F}$ *of the sets*

$$f^{-1}(0y) = \underset{x}{E}[0 \leqslant f(x) \leqslant y], \quad \textit{where} \quad y \in \mathcal{Y} = f(\mathcal{X}),$$

is strictly monotone and, for each y *which does not have an immediate successor,* $f^{-1}(0y)$ *is the intersection of all the elements following it, i.e.*

$$f^{-1}(0y) = \bigcap_{z>y} f^{-1}(0z),$$

where z *ranges over* $\mathcal{Y}$.

Proof. Let $y_1 < y_2$ be two elements of $\mathcal{Y}$. The obvious inclusion $0y_1 \subset 0y_2 - y_2 = \mathrm{Int}(0y_2)$ implies

$$f^{-1}(0y_1) \subset f^{-1}[\mathrm{Int}(0y_2)] \subset \mathrm{Int}[f^{-1}(0y_2)],$$

because the inclusion $f^{-1}[\mathrm{Int}(B)] \subset \mathrm{Int}[f^{-1}(B)]$ is valid for every continuous function f.

Therefore the family $\boldsymbol{F}$ is strictly monotone.

Assume that y is not the last element of $\mathcal{Y}$ and among the $z > y$ there does not exist the smallest element in $\mathcal{Y}$. Consequently $\mathcal{Y} \cap (0y) = \bigcap_{z>y} \mathcal{Y} \cap (0z)$ and $f^{-1}(0y) = f^{-1}[\bigcap_{z>y} \mathcal{Y} \cap (0z)] = \bigcap_{z>y} f^{-1}(0z)$.

By using (1) and (7) we prove the following theorem.

THEOREM 2. *The relations*:

$$f^{-1}(0y) = \overline{f^{-1}(0y-y)}, \quad f^{-1}(y1) = \overline{f^{-1}(y1-y)},$$
$$f^{-1}(y) = \overline{f^{-1}(0y-y)} \cap \overline{f^{-1}(y1-y)}$$

hold except for a countable set of values of y.

The latter identity is a consequence of the first two ones:

$$f^{-1}(y) = f^{-1}[(0y) \cap (y1)] = f^{-1}(0y) \cap f^{-1}(y1).$$

Theorem 1 has a converse in the following theorem.

THEOREM 3. *If* $\boldsymbol{F}$ *is a strictly monotone family of closed sets indexed according to Theorems* 1 *and* 2 *of Section* VII (*where* $A_1 = \mathcal{X}$), *there exists a function* $f \in \mathcal{I}^{\mathcal{X}}$ *such that, for every index* $y < 1$,

$$A_y = f^{-1}(0y) \quad \text{or} \quad \bigcap_{z>y} A_z = f^{-1}(0y), \tag{10}$$

according as A_y *does, or does not, possess an immediate succesor in* $\boldsymbol{F}$.

Proof. The function f can be defined as follows.

Let $(A_{r_1}, A_{s_1}), (A_{r_2}, A_{s_2}), \ldots$ be the sequence of leaps.

Let

$$\mathcal{X}^* = \mathcal{X} - \bigcup_n [\mathrm{Int}(A_{s_n}) - A_{r_n}].$$

For $x \in \mathcal{X}^*$ let

(i) $f(x) =$ the greatest lower bound of y such that $x \in A_y$;

for $x \in \mathrm{Int}(A_{s_n}) - A_{r_n}$, let

(ii) $f(x) = r_n + (s_n - r_n) \dfrac{\varrho(x, A_{r_n})}{\varrho(x, A_{r_n}) + \varrho[x, \mathcal{X} - \mathrm{Int}(A_{s_n})]}$ [1].

First we shall prove that the function f is continuous on $\mathcal{X}^*$. We note the following double implication

$$[f(x) < y] \to [x \in A_y] \to [f(x) \leqslant y], \tag{11}$$

which is valid for every y belonging to the set J of indices.

Let $\lim x_n = x$, $x_n \in \mathcal{X}^*$, $f(x) = q$ and $\lim f(x_n) = q'$. We must prove that $q = q'$.

In order to prove $q' \geqslant q$, two cases must be considered, according as q is, or is not, the right-hand end-point of an interval contiguous to $\bar{J}$. In the first case, $q \in J$ by VII, Theorem 2. Let r be the left-hand end-point of the contiguous interval in question (then $r \in J$). Since x does not belong to A_r, it follows that, for a sufficiently large n, x_n does not belong to A_r; hence $f(x_n) \geqslant q$ and $q' \geqslant q$.

If q is not a right-hand end-point of a contiguous interval, let $y_1 < y_2 < \ldots$, $\lim y_n = q$ and $y_n \in J$. For a fixed n, $x \notin A_{y_n}$; hence, for sufficiently large values of k, $x_k \notin A_{y_n}$. Therefore $f(x_k) \geqslant y_n$ and $q' \geqslant q$.

The proof of the inequality $q' \leqslant q$ is analogous. If q is the left-hand end-point of a contiguous interval qs, then $x_n \in A_q$ for sufficiently large values of n, since the formulas $x_n \in \mathcal{X}^*$ and $x_n \notin A_q$ imply $x_n \notin \mathrm{Int}(A_s)$, while $x \in A_q \subset \mathrm{Int}(A_s)$. It follows that $f(x_n) \leqslant q$ and $q' \leqslant q$.

---

[1] This formula is an interpolation of the function f defined by the condition (i); i.e. $f(x) \leqslant r_n$ for $x \in A_{r_n}$, $r_n < f(x) < s_n$ for $x \in \mathrm{Int}(A_{s_n}) - A_{r_n}$ and $f(x) \geqslant s_n$ for $x \in \mathcal{X} - \mathrm{Int}(A_{s_n})$.

If $E = 0$, we put $\varrho(x, E) = 1$.

In the contrary case, let $y_1 > y_2 > \ldots$, $\lim y_n = q$, and $y_n \in J$. For a fixed n, $x \in A_{y_{n+1}} \subset \mathrm{Int}(A_{y_n})$. Hence, for sufficiently large values of k, $x_k \in A_{y_n}$, $f(x_k) \leqslant y_n$ and $q' \leqslant q$.

The continuity of the function f on $\mathcal{X}^*$ is thus proved. Now we have to prove that if $\{x_n\}$ is a sequence of elements of $A_{s_{m_n}} - A_{r_{m_n}}$ convergent to x, then $f(x) = \lim f(x_n)$.

Obviously we can assume that $r_{m_1} < r_{m_2} < \ldots$. Letting $q' = \lim r_{m_n}$, we have $q' = \lim f(x_n)$, since $r_{m_n} \leqslant f(x_n) \leqslant s_{m_n} \leqslant r_{m_{n+1}}$. Clearly x does not belong to any of $A_{r_{m_n}}$, hence $f(x) \geqslant q'$. On the other hand, if $q' \leqslant y \in J$, then $x \in \overline{\bigcup_n A_{s_{m_n}}} \subset A_y$, for $s_{m_n} \leqslant q'$. Therefore $f(x) \leqslant y$ and $f(x) \leqslant q'$.

To prove formula (10) we observe that, by (11),

$$A_y \subset f^{-1}(0y) \quad \text{and, for } y < z \in J, \quad f^{-1}(0y) \subset A_z. \tag{12}$$

In the case where y is the left-hand end-point of a contiguous interval, the inclusion $f^{-1}(0y) \subset A_y$ is a direct consequence of (i) and (ii). Otherwise, let $y = \lim z_n$, where $z_1 > z_2 > \ldots$. It follows from (12) that

$$\bigcap_{z>y} A_z = \bigcap_{n=1}^{\infty} A_{z_n} \subset \bigcap_{n=1}^{\infty} f^{-1}(0z_n) \subset \bigcap_{n=2}^{\infty} A_{z_{n-1}} = \bigcap_{z>y} A_z$$

and since $(0y) = \bigcap_{n=1}^{\infty} (0z_n)$, it follows that $f^{-1}(0y) = \bigcap_{n=1}^{\infty} f^{-1}(0z_n) = \bigcap_{z>y} A_z$.

CONSEQUENCES OF THEOREM 3.

$$f^{-1}(0y-y) \subset \mathrm{Int}(A_y) \subset A_y \subset f^{-1}(0y); \tag{13}$$

$$\mathrm{Fr}(A_y) \subset f^{-1}(y) \subset \bigcap_{z>y} A_z - \bigcup_{u<y} A_u \subset \bigcap_{z>y} A_z - \bigcup_{u<y} \mathrm{Int}(A_u); \tag{14}$$

if y does not have an immediate predecessor, then

$$f^{-1}(0y-y) = \bigcup_{u<y} \mathrm{Int}(A_u), \quad \text{i.e.} \quad f^{-1}(y1) = \bigcap_{u<y} \overline{\mathcal{X} - A_u}; \tag{15}$$

if y does not have a neighbour element and $0 < y < 1$, then

$$f^{-1}(y) = \bigcap_{u<y} \overline{\mathcal{X} - A_u} \cap \bigcap_{z>y} A_z; \tag{16}$$

the following relations hold

$$A_y = f^{-1}(0y), \quad \operatorname{Int}(A_y) = f^{-1}(0y-y),$$
$$\mathscr{X} - A_y = f^{-1}(y1-y), \quad \operatorname{Fr}(A_y) = f^{-1}(y), \tag{17}$$

except for a countable set of indices y;

if $\bigcup_{y<1} A_y = A_0 \cup \bigcup_{0<y<1} \operatorname{Fr}(A_y)$, *then* $\operatorname{Fr}(A_y) = f^{-1}(y)$

$$\textit{for } 0 \neq y \neq 1. \tag{18}$$

To prove (13), let $u = f(x) < y$. Hence $x \in f^{-1}(0u)$. If y has an immediate predecessor, then $f^{-1}(0u) \subset \operatorname{Int}(A_y)$. If $u < v < y$ and $v \in J$, then $f^{-1}(0u) \subset A_v \subset \operatorname{Int}(A_y)$. Therefore $f^{-1}(0y-y) \subset \operatorname{Int}(A_y)$. The rest of formula (13) is obvious.

Formula (14) follows from (13) and (10):

$$\operatorname{Fr}(A_y) = A_y - \operatorname{Int}(A_y) \subset f^{-1}(0y) - f^{-1}(0y-y)$$
$$= f^{-1}(y) \subset f^{-1}(0y) \subset A_z$$

and $A_u \subset f^{-1}(0u)$, therefore $A_u \cap f^{-1}(y) = 0$.

To prove (15), let $f(x) < u < y$. Then $x \in f^{-1}(0u-u)$ and, by (13), $x \in \operatorname{Int}(A_u)$. Therefore

$$f^{-1}(0y-y) \subset \bigcup_{u<y} \operatorname{Int}(A_u).$$

The converse inclusion follows from (13).

Formula (16) follows from (15) and (10).

By virtue of (4), (6), (10) and (15), except for a countable set of indices,

$$A_y = \bigcap_{z>y} A_z = f^{-1}(0y) \quad \text{and} \quad \operatorname{Int}(A_y) = \bigcup_{u<y} \operatorname{Int}(A_u) = f^{-1}(0y-y)$$

and this gives (17).

Finally, (18) is derived from (14).

Theorem 3 implies the following corollary.

COROLLARY 3a. *If* $\boldsymbol{F}$ *is a strictly monotone family of closed subsets of* $\mathscr{X}$ *such that each element of* $\boldsymbol{F}$ *which does not have an immediate succesor is the intersection of all the elements following it, then there exists a real valued continuous function defined on* $\mathscr{X}$ *and a set* $J \subset \mathscr{I}$ *such that the family* $\boldsymbol{F}$ *coincides with the family of sets* $f^{-1}(0y)$, *where* $y \in J$.

X. Strictly monotone families of closed order types. We now assume that $\boldsymbol{F}$ is a strictly monotone family consisting of closed subsets of $\mathscr{X}$ ($\mathscr{X} \neq 0$) which does not possess gaps and contains as elements the empty set and the space $\mathscr{X}$. These hypotheses mean that

$$J = \bar{J}, \quad 0 \in J, \quad 1 \in J.$$

In the case where J is uncountable, let P be its perfect kernel (hence $J - P$ is countable, compare § 23, V) and let P^* be the set obtained from P by removing the end-points of its contiguous intervals. Let φ be a non-decreasing continuous function on $\mathscr{I}$ which is increasing on P^* and such that $\varphi(P) = \mathscr{I}$ [1]. Let $\gamma(t)$ and $\Gamma(t)$ be the first and the last y, respectively, such that $\varphi(y) = t$. Let $g(x) = \varphi f(x)$, where f is the function of Section IX, Theorem 3.

In the case where $\bar{\bar{J}} \leqslant \aleph_0$, we put $\varphi(x) = 0$ for $x \in \mathscr{I}$, $\gamma(0) = 0$, $\gamma(1) = 1$. Therefore $g(x) = 0$.

This leads to the following theorem.

THEOREM 1. *The set* $J \cap \varphi^{-1}(t) = \underset{y}{E}(y \in J)[\gamma(t) \leqslant y \leqslant \Gamma(t)]$ *is countable. If* $\gamma(t) \neq \Gamma(t)$ *and* $P \neq 0$, *then* $\gamma(t)$ *and* $\Gamma(t)$ *are end-points of the same interval contiguous to* P. *If* $\gamma(t) = \Gamma(t)$, *then* $\gamma(t)$ *is a point of* P^*. *Therefore, neither does* $\gamma(t)$ *have an immediate predecessor in* J, *nor does* $\Gamma(t)$ *have an immediate successor.*

THEOREM 2. $g^{-1}(t) = \bigcap_{\Gamma(t) < z} A_z \cap \bigcap_{u < \gamma(t)} \overline{\mathscr{X} - A_u}$.

Proof. By (10) and (15),

$$\bigcap_{\Gamma(t) < z} A_z = f^{-1}[0, \Gamma(t)] = \underset{x}{E}[0 \leqslant f(x) \leqslant \Gamma(t)],$$

$$\bigcap_{u < \gamma(t)} \overline{\mathscr{X} - A_u} = f^{-1}[\gamma(t), 1] = \underset{x}{E}[\gamma(t) \leqslant f(x) \leqslant 1].$$

On the other hand,

$$\underset{x}{E}[0 \leqslant f(x) \leqslant \Gamma(t)] \cap \underset{x}{E}[\gamma(t) \leqslant f(x) \leqslant 1]$$

$$= \underset{x}{E}[\gamma(t) \leqslant f(x) \leqslant \Gamma(t)] = \underset{x}{E}[f(x) \in \varphi^{-1}(t)] = f^{-1}\varphi^{-1}(t) = g^{-1}(t),$$

since $\varphi^{-1}(t) = \underset{y}{E}[\gamma(t) \leqslant y \leqslant \Gamma(t)]$ by definitions of γ and Γ.

---

[1] A definition of the function φ does not present any difficulty. See for example § 16, II, Corollary 6a (the definition of the function f).

THEOREM 3. *The relations*

(i) $g^{-1}(0t) \in \boldsymbol{F}$,

(ii) $g^{-1}(t) = f^{-1}(y)$, *where* $t = \varphi(y)$,

hold except for a countable set of values of t.

Proof. Every t, except for a countable set, is of the form $t = \varphi(y)$, where $y \in P^*$. Therefore

$$g^{-1}(0t) = \underset{x}{E}[0 \leqslant \varphi f(x) \leqslant t] = \underset{x}{E}[0 \leqslant f(x) \leqslant y] = f^{-1}(0y),$$

since the conditions $y' \leqslant y$ and $\varphi(y') \leqslant \varphi(y) = t$ are equivalent.

Then (i) follows from (17). (ii) follows from the equivalence

$$\{g(x) = t\} \equiv \{\varphi f(x) = t\} \equiv \{f(x) = y\}.$$

THEOREM 4. *If the family* $\boldsymbol{F}$ *is uncountable, then the function* g *has the finest inverse images of single elements* t *of* $\mathscr{I}$ *among all continuous functions* h *such that* $h(\mathscr{X}) = \mathscr{I}$ *and which satisfy the relation* $h^{-1}(0t) \in \boldsymbol{F}$ *except for a countable set of values of* $t \in \mathscr{I}$.

If the family $\boldsymbol{F}$ *is countable (or finite), then no such function* h *exists.*

Proof. We must prove that the condition $g(x) = g(x')$ implies $h(x) = h(x')$.

Suppose that $h(x) < h(x')$. Hence for every t such that $h(x) < t < h(x')$, we have

$$x \in h^{-1}(0t) \quad \text{and} \quad x' \notin h^{-1}(0t). \tag{19}$$

As the set of such t is uncountable, there exists an uncountable set of t such that $h^{-1}(0t) \in \boldsymbol{F}$. For each such t there exists an index $w \in J$ such that $h^{-1}(0t) = A_w$. To two different values of t there correspond different values of w; for, if we suppose that $t < t'$ and $h^{-1}(0t) = h^{-1}(0t')$, then we can find a t'' such that $t < t'' < t'$ and

$$h^{-1}(t'') \subset h^{-1}(0t') - h^{-1}(0t) = 0,$$

contrary to the hypothesis that $h(\mathscr{X}) = \mathscr{I}$.

Therefore the set W of all w that satisfy

$$A_w = h^{-1}(0t), \quad \text{where} \quad h(x) < t < h(x'), \tag{20}$$

is uncountable. Let W^* be the set obtained from W by removing its first point (if it exists). Then, for $w \in W^*$,

$$x \in \operatorname{Int}(A_w) \quad \text{and} \quad x' \notin A_w. \tag{21}$$

For, let $w' < w$ and, according to (20),

$$A_{w'} = h^{-1}(0t'), \quad \text{where} \quad h(x) < t' < h(x').$$

Therefore $x \in h^{-1}(0t') = A_{w'} \subset \mathrm{Int}(A_w)$ and we obtain the first formula of (21). The second one is a direct consequence of (19) and (20).

Let $g(x) = t_0 = g(x')$, i.e. $x \in g^{-1}(t_0)$ and $x' \in g^{-1}(t_0)$. By virtue of Theorem 2, $x' \in A_z$ for every $z > \Gamma(t_0)$, and $x \in \mathscr{X} - A_u$, i.e. $x \notin \mathrm{Int}(A_u)$ for every $u < \gamma(t_0)$. We deduce from (21) that $w \leqslant \Gamma(t_0)$ and $w \geqslant \gamma(t_0)$, i.e.

$$W^* \subset J \cap \mathop{E}_{y}[\gamma(t_0) \leqslant y \leqslant \Gamma(t_0)].$$

But then the set W^* is countable, by virtue of Theorem 1.

C. PROBLEMS OF DIMENSION (§§ 25–28)

The same assumptions on the space are made as in Section B.

§ 25. Definitions. General properties

I. Definition of dimension[1]. To a space $\mathscr{X}$ we assign an integer $n \geqslant -1$ or ∞ called the *dimension of* $\mathscr{X}$ and denoted by $\dim \mathscr{X}$. The symbol $\dim_p \mathscr{X}$ denotes the dimension of $\mathscr{X}$ at the point p. The following three conditions define the dimension inductively:

1) $\dim \mathscr{X} = -1$ *if and only if the space* $\mathscr{X}$ *is empty*;

2) *if* $\mathscr{X} \neq 0$, *then* $\dim \mathscr{X} = \sup \dim_p \mathscr{X}$ *for* $p \in \mathscr{X}$;

3) $\dim_p \mathscr{X} \leqslant n+1$ *if and only if* p *has arbitrarily small neighbourhoods whose boundaries have dimension* $\leqslant n$.

---

[1] The idea of a definition of dimension was originally due to Henri Poincaré, Revue de métaph. et de mor. 20 (1912), and *Dernières pensées*, p. 65, Paris 1926 (posthumous edition). The definition was made precise by L. E. J. Brouwer, *Über die natürlichen Dimensionsbegriff*, Journ. f. Math. 142 (1913), p. 146, and Proc. Akad. Amsterdam 26 (1923), p. 795. The dimension theory based on a definition rather close to that of Poincaré–Brouwer was created and developed independently by K. Menger and P. Urysohn in a number of papers beginning from 1922. See particularly K. Menger, *Dimensionstheorie*, Leipzig-Berlin, 1928, and P. Urysohn, *Mémoire sur les multiplicités Cantoriennes*, Fund. Math. 7–8 (1925–26).

For a more modern exposition of the dimension theory, see W. Hurewicz and H. Wallman, *Dimension Theory*, Princeton 1941 (Princeton Math. Series N 4).

For a dimension theory based on the notion of homology, see P. Alexandrov, *Dimensionstheorie*, Math. Ann. 106 (1932).

Statement 3) can be formulated equivalently in topological terms as follows:

3′) $\dim_p \mathscr{X} \leqslant n+1$ *if and only if each neighbourhood of* p *contains a neighbourhood of* p *whose boundary is of dimension* $\leqslant n$.

Thus the dimension of the space is defined in a purely topological way. It is an *invariant of homeomorphic mappings of the space.*

EXAMPLES. By definition, a (non-empty) space has *dimension* 0, if every point has arbitrarily small neighbourhoods with empty boundaries. Thus, for example, the set of rational numbers is 0-dimensional, since every interval with irrational end-points has in this space an empty boundary. Also the set of irrational numbers and, in general, *every boundary subset of the straight line is* 0-*dimensional.*

The space of real numbers has dimension $\leqslant 1$, since the boundary of an interval, being composed of two points, has dimension 0. Analogously, the plane has dimension $\leqslant 2$ (since a circumference has dimension $\leqslant 1$) and, in general, the euclidean space $\mathscr{E}^n$ has dimension $\leqslant n$. The proof that the dimension of this space is exactly equal to n is less elementary and we shall return to it in § 28.

II. Dimension of subsets. Given a set $E(\subset \mathscr{X})$ and a point p of E, the condition $\dim_p E \leqslant n+1$ means by 3) that there exists an arbitrarily small open neighbourhood of p *relative to* E whose boundary relative to E has dimension $\leqslant n$. In other words, there exists an arbitrarily small open set G containing p such that $\dim(E \cap \overline{E \cap G} - G) \leqslant n$. For, the boundary of $E \cap G$ relative to E is, by definition,

$$E \cap \overline{E \cap G} - (E \cap G) = E \cap \overline{E \cap G} - G.$$

THEOREM 1. *The dimension of a set does not excede the dimension of the entire space*:

$$\textit{if } p \in E, \textit{ then } \dim_p E \leqslant \dim_p \mathscr{X}.$$

Proof. Proceeding by induction, we can assume that the theorem is true for an n-dimensional space and that $\dim_p \mathscr{X} \leqslant n+1$. Let G be an open neighbourhood of p with $\dim[\mathrm{Fr}(G)] \leqslant n$. We must show that the relative boundary of $E \cap G$ in E has dimension $\leqslant n$. This relative boundary is equal to

$$E \cap \overline{E \cap G} - G \subset \overline{G} - G = \mathrm{Fr}(G).$$

We deduce that

$$\dim(\dot{E} \cap \overline{E \cap G} - G) \leqslant \dim [\mathrm{Fr}(G)] \leqslant n.$$

THEOREM 2. $\dim_p E \leqslant n+1$ *if and only if there exists an arbitrarily small neighbourhood* G *of* p *such that*

$$\dim[E \cap \mathrm{Fr}(G)] \leqslant n.$$

Proof. Suppose that $\dim_p E \leqslant n+1$. Hence there exists a set H open in E such that $\dim(E \cap \bar{H} - H) \leqslant n$. The sets H and $E - \bar{H}$ being separated (§ 6, V, Theorem 3), there exists (by § 14, V, Theorem 1 and § 21, IV, Theorem 2) an open set G such that

$$H \subset G \quad \text{and} \quad \bar{G} \cap E - \bar{H} = 0; \quad \text{hence} \quad E \cap \bar{G} - G \subset E \cap \bar{H} - H$$

and, by Theorem 1,

$$\dim(E \cap \bar{G} - G) \leqslant \dim(E \cap \bar{H} - H) \leqslant n.$$

By § 14, V, 1, the diameter of G is arbitrarily close to that of H. Therefore the condition given by Theorem 2 is necessary.

In order to prove that it is sufficient, let G be the set in question and let $H = E \cap G$. Then

$$E \cap \bar{H} - H = E \cap \overline{E \cap G} - G \subset E \cap \bar{G} - G = E \cap \mathrm{Fr}(G)$$

and

$$\dim(E \cap \bar{H} - H) \leqslant \dim[E \cap \mathrm{Fr}(G)] \leqslant n.$$

Therefore $\dim_p E \leqslant n+1$.

In particular, we have the following theorem.

THEOREM 2_0. $\dim_p E = 0$ *if and only if* p *has arbitrarily small neighbourhoods* G *such that* $E \cap \mathrm{Fr}(G) = 0$.

III. The set $E_{(n)}$.

DEFINITION. p *belongs to the set* $E_{(n)}$, *if* $\dim_p(E \cup p) \leqslant n$, i.e. if p has an arbitrarily small neighbourhood G such that $\dim[E \cap \mathrm{Fr}(G)] \leqslant n-1$.

In particular, if E is the entire space $\mathcal{X}$, then $E_{(n)}$ is the set of points where the space has dimension $\leqslant n$.

The inclusion $E \subset E_{(n)}$ is equivalent to $\dim E \leqslant n$.

By II, Theorem 1, the inclusion $A \subset B$ implies $B_{(n)} \subset A_{(n)}$.

THEOREM 1. $\dim(E \cap E_{(n)}) \leqslant n$.

For, if $p \in E \cap E_{(n)}$, then by definition $\dim_p E \leqslant n$ and, by II, Theorem 1, $\dim_p(E \cap E_{(n)}) \leqslant n$.

Clearly the dimension of $E_{(n)}$ can be greater than n. If, for example, E is composed of a single point, then $E_{(0)} = \mathscr{X}$.

THEOREM 2. *$E_{(n)}$ is a G_δ-set* (1).

Proof. By definition, $p \in E_{(n)} \equiv$ for every k there exists an open set G containing p such that

$$\dim[E \cap \operatorname{Fr}(G)] \leqslant n-1 \quad \text{and} \quad \delta(G) < 1/k.$$

Let H_k be the union of such sets G. Then $E_{(n)} = H_1 \cap H_2 \cap \ldots$.

THEOREM 3. *Given a set E and an integer n, there exists a sequence $D_1, D_2, \ldots$ of open sets such that*

(i) $\dim[E \cap \operatorname{Fr}(D_i)] \leqslant n-1$,

(ii) *the sets $E_{(n)} \cap D_i$, $i = 1, 2, \ldots$, form a base for the set $E_{(n)}$*; in other words, for every point $p \in E_{(n)}$, there exists a set D_i containing p of an arbitrarily small diameter;

(iii) *put* $S = \bigcup_{i=1}^{\infty} \operatorname{Fr}(D_i)$; *then*

$$E_{(n)} \subset (E-S)_{(0)} \quad \textit{and} \quad \dim[E_{(n)}-S] \leqslant 0.$$

Proof. Let k be fixed. For every $p \in E_{(n)}$ let $G(p)$ be an open set containing p such that

$$\dim\{E \cap \operatorname{Fr}[G(p)]\} \leqslant n-1 \quad \text{and} \quad \delta[G(p)] < 1/k.$$

By the Lindelöf theorem, there exists a countable sequence $D_{k,1}, D_{k,2}, \ldots$ of the sets $G(p)$ whose union is equal to the union of the sets $G(p)$. By rearranging this double sequence into a simple sequence, we obtain the desired sequence $\{D_i\}$. As each point p of $E_{(n)}$ is contained in an arbitrarily small set $G(p)$, hence there exists a D_i such that $p \in D_i \subset G(p)$, and condition (ii) follows.

Condition (iii) is its consequence by virtue of the formula

$$(E-S) \cap \operatorname{Fr}(D_i) = 0 = (E_{(n)}-S) \cap \operatorname{Fr}(D_i).$$

---

(1) See K. Menger, *Über die Dimension von Punktmengen*, II. Teil, Mon. f. Math. u. Phys. 34 (1924), p. 141; P. Urysohn, Fund. Math. 8, p. 277; L. Tumarkin, *ibid.* p. 360.

As a particular case of Theorem 3 we have the following theorem.

THEOREM 4. *Every n-dimensional space contains a countable base composed of open sets whose boundaries have dimension* $\leqslant n-1$. *By removing the boundaries from the space, we obtain a* 0*-dimensional (or empty) set.*

THEOREM 5. *The condition* $\dim X \leqslant n$ *implies* $E_{(0)} \subset (E \cup X)_{(n+1)}$.

Proof. If $p \in E_{(0)}$, then there exists a neighbourhood G of p such that $E \cap \mathrm{Fr}(G) = 0$. Therefore $(E \cup X) \cap \mathrm{Fr}(G) = X \cap \mathrm{Fr}(G) \subset X$ and $\dim[(E \cup X) \cap \mathrm{Fr}(G)] \leqslant n$, which proves that $p \in (E \cup X)_{(n+1)}$.

§ 26. 0-dimensional spaces [1]

I. Base of the space. By definition, a space is 0-dimensional, if each of its points lies in an arbitrarily small open neighbourhood which is both closed and open (this is equivalent to the condition that the boundary of the neighbourhood is empty). We deduce the following theorem by virtue of § 25, III, Theorem 4.

THEOREM 1. *Every* 0*-dimensional space contains a countable base composed of sets which are both closed and open.*

Moreover, we can assume that the *diameters of these sets do not exceed a pre-assigned positive number.*

COROLLARIES. *In a* 0*-dimensional space:*

1a. *Every open set (in particular, the entire space) is the union of a sequence of disjoint sets which are both closed and open (and have arbitrarily small diameters);*

1b. *every closed set is the intersection of a sequence of sets which are both closed and open.*

To prove Corollary 1a, let us write according to Theorem 1

$$G = F_1 \cup F_2 \cup \ldots = F_1 \cup (F_2 - F_1) \cup (F_3 - F_1 - F_2) \cup \ldots$$

where G is the given open set and $F_1, F_2, \ldots$ are sets both closed and open. The last member of the formula represents the desired decomposition.

Corollary 1b follows immediately from Corollary 1a.

---

[1] Most theorems of the § 26 will be generalized by induction in the following § 27, where also references will be given.

*THEOREM 1' [1]. *Let Q be a both dense and boundary G_δ-set in a 0-dimensional space. Then Q is the result of the operation* ($\mathscr{A}$) *performed on a regular system* (compare § 3, XIV) *of non-empty sets* $\{A_{k_1\ldots k_n}\}$ *which are closed, open and such that*

1) $\delta(A_{k_1\ldots k_n}) < 1/n$,

2) *two sets* $A_{k_1\ldots k_n}$ *and* $A_{l_1\ldots l_n}$ *corresponding to different systems of* n *indices are disjoint.*

Proof. Let $Q = G_1 \cap G_2 \cap \ldots$ where G_n is open and $G_n \supset G_{n+1}$. Since Q is not closed, there exists an index j_1 such that G_{j_1} is not closed. By Corollary 1a we can assume that

$$G_{j_1} = A_1 \cup A_2 \cup \ldots, \qquad \delta(A_i) < 1,$$

where A_i are non-empty, disjoint closed and open sets.

We proceed by induction. Assuming that the sets $A_{k_1\ldots k_n}$ are defined and satisfy

$$A_{k_1\ldots k_n} \subset G_n \tag{i}$$

for an integer $n \geqslant 1$, we define the sets $A_{k_1\ldots k_n k_{n+1}}$ as follows.

As the set $A_{k_1\ldots k_n}$ is open, there exists an index $j_{n+1} \geqslant n+1$ such that the set $A_{k_1\ldots k_n} \cap G_{j_{n+1}}$ is not closed. For otherwise the set $Q \cap A_{k_1\ldots k_n}$ would be closed contrary to the hypothesis that it is a dense and boundary set in $A_{k_1\ldots k_n}$. We can write as before

$$A_{k_1\ldots k_n} \cap G_{j_{n+1}} = \bigcup_{i=1}^{\infty} A_{k_1\ldots k_n i}, \qquad \delta(A_{k_1\ldots k_n i}) < 1/(n+1),$$

where $A_{k_1\ldots k_n i}$ are non-empty, disjoint closed and open sets. Moreover, we can replace n by $(n+1)$ in (i) by virtue of the inequality $j_{n+1} \geqslant n+1$ which implies that $G_{j_{n+1}} \subset G_{n+1}$.

Thus the sets $A_{k_1\ldots k_n}$ are defined for every n. We proceed to prove that Q coincides with the result of the operation ($\mathscr{A}$) performed on these sets.

Let $p \in Q$. Then $p \in G_{j_1}$. Consequently there exists an index k_1 such that $p \in A_{k_1}$. Suppose that $p \in A_{k_1\ldots k_n}$. Since also $p \in G_{j_{n+1}}$, there exists a k_{n+1} such that $p \in A_{k_1\ldots k_n k_{n+1}}$. It follows that

$$p \in A_{k_1} \cap A_{k_1 k_2} \cap A_{k_1 k_2 k_3} \cap \ldots . \tag{1}$$

On the other hand, (1) and (i) imply $p \in G_1 \cap G_2 \cap \ldots = Q$.

---

[1] This theorem will be used in the chapter on complete spaces (§ 36).

II. Reduction and separation theorems. In (separable) 0-dimensional spaces, Theorem of § 21, XII can be made more precise: we can suppose that the sets F_ι of formula (2') are both *closed and open*. Actually the following theorem holds (compare also the Lindelöf theorem):

THEOREM 1. (Reduction theorem.) (1) *If $G_0, G_1, \ldots$ is a (finite or infinite) sequence of open sets in a 0-dimensional space $\mathscr{X}$, then there exists a sequence of disjoint open sets $H_0, H_1, \ldots$ such that*

$$H_i \subset G_i \quad \text{and} \quad H_0 \cup H_1 \cup \ldots = G_0 \cup G_1 \cup \ldots. \tag{0}$$

Consequently, *if $\mathscr{X} = G_0 \cup G_1 \cup \ldots$, then the sets H_i are both closed and open.*

Proof. Let $G_i = F_{i,0} \cup F_{i,1} \ldots$, where $F_{i,j}$ are both closed and open (compare I, Corollary 1a). We arrange the double sequence $\{i, j\}$ in a simple sequence. Let $n = \varphi(i, j)$ be the integer corresponding to the pair i, j. Let $F^*_{i,j} = F_{i,j} - \bigcup F_{k,l}$, where the union is taken for all pairs k, l such that $\varphi(k, l) < \varphi(i, j)$. Clearly

$$\bigcup_{i,j=0}^{\infty} F^*_{i,j} = \bigcup_{i,j=0}^{\infty} F_{i,j} = \bigcup_{i=0}^{\infty} G_i.$$

Let $H_i = \bigcup_{j=0}^{\infty} F^*_{i,j}$. Then $\bigcup_{i=0}^{\infty} H_i = \bigcup_{i=0}^{\infty} G_i$ and

$$H_i \subset F_{i,0} \cup F_{i,1} \cup \ldots = G_i.$$

As the sets $F^*_{i,j}$ are mutually disjoint, so are the sets H_i.

The next theorem is implied by the reduction theorem.

THEOREM 2. (Separation theorem.) *If $F_0, F_1, \ldots$ is a (finite or infinite) sequence of closed sets in a 0-dimensional space such that $F_0 \cap F_1 \cap \ldots = 0$, then there exists a sequence of sets $E_0, E_1, \ldots$, both closed and open, such that*

$$F_i \subset E_i \quad \text{and} \quad E_0 \cap E_1 \cap \ldots = 0.$$

(1) Compare my paper *Sur les théorèmes de séparation dans la théorie des ensembles*, Fund. Math. 26 (1936), p. 184.

In particular, if A and B are two disjoint closed sets, then there exists a set E, both closed and open, such that $A \subset E$ and $E \cap B = 0$ [1].

Proof. Let us apply the reduction theorem and put

$$G_i = \mathscr{X} - F_i \quad \text{and} \quad E_i = \mathscr{X} - H_i.$$

By hypothesis and by (0),

$$\bigcup_{i=0}^{\infty} G_i = \mathscr{X} - \bigcap_{i=0}^{\infty} F_i = \mathscr{X} = \bigcup_{i=0}^{\infty} H_i.$$

Therefore H_i as well as E_i are both open and closed. Moreover, the last identity implies that $E_0 \cap E_1 \cap \ldots = 0$ and the inclusion $H_i \subset G_i$ gives $F_i \subset E_i$.

The second part of Theorem 2 has the following generalization.

THEOREM 2a. *Given a finite system $A_0, \ldots, A_k$ of disjoint closed subsets of a 0-dimensional space, there exists a system $F_0, \ldots, F_k$ of disjoint sets, both closed and open with*

$$\mathscr{X} = F_0 \cup \ldots \cup F_k \quad \textit{and} \quad A_i \subset F_i.$$

Proof (by induction). The theorem being true for $k = 1$, assume that it is true for $k-1$ and $k \geqslant 2$. There exists a system $F_0, \ldots, F_{k-2}, F^*$ of disjoint sets, both closed and open, with

$$\mathscr{X} = F_0 \cup \ldots \cup F_{k-2} \cup F^*, \quad A_0 \subset F_0, \ldots, A_{k-2} \subset F_{k-2},$$

$$(A_{k-1} \cup A_k) \subset F^*.$$

Applying Theorem 1 to the pair of sets A_{k-1}, A_k and to the set F^* considered as a space, we obtain

$$F^* = F_{k-1} \cup F_k, \quad A_{k-1} \subset F_{k-1}, \quad A_k \subset F_k, \quad F_{k-1} \cap F_k = 0,$$

where the sets F_{k-1} and F_k are both closed and open in F^*, and hence in $\mathscr{X}$ as well.

COROLLARY 1. *Every totally bounded open set G of a 0-dimensional space is of the form*

$$G = F_0 \cup F_1 \cup \ldots, \quad F_i \cap F_j = 0 \quad \textit{for} \quad i \neq j, \quad \lim_{n=\infty} \delta(F_n) = 0 \qquad (1)$$

where the sets F_i are both closed and open.

---

[1] Thus the normality condition in a 0-dimensional space can be stated in a more advantageous form.

COROLLARY 2. *If F is a non-empty closed subset of a 0-dimensional space, then the space can be mapped onto F by a continuous function which is the identity on F* (1).

In other words, *F is a retract of the space.*

This gives (see § 13, V, Theorem 4) the next corollary.

COROLLARY 3. *Every continuous function f defined on a closed subset F of a 0-dimensional space can be extended to the entire space so that its values remain in the set $f(F)$.*

Proof of Corollary 1. According to Corollary 1a of Section I, let $F_0^*, F_1^*, \ldots$ be a sequence of bounded sets, both closed and open, which satisfy the two first identities (1). Being totally bounded, F_n^* is of the form

$$F_n^* = H_{n0} \cup \ldots \cup H_{nk_n}, \quad \text{where} \quad \delta(H_{ni}) < 1/n.$$

Let G_{ni} be an open set with $H_{ni} \subset G_{ni} \subset F_n^*$ and $\delta(G_{ni}) < 1/n$. Hence $F_n^* = G_{n0} \cup \ldots \cup G_{nk_n}$. Applying Theorem 1 to the set F_n^* regarded as a space, we obtain

$$F_n^* = F_{n0} \cup \ldots \cup F_{nk_n}, \quad F_{ni} \cap F_{nj} = 0 \quad \text{for} \quad i \neq j,\ F_{ni} \subset G_{ni},$$

where the sets $F_{n0}, \ldots, F_{nk_n}$ are both closed and open in F_n^*, and hence in the space as well.

As $F_{ni} \subset G_{ni}$, therefore $\delta(F_{ni}) < 1/n$. Thus $F_{00}, \ldots, F_{0k_0}, F_{10}, \ldots, F_{1k_1}, \ldots$ is clearly the desired sequence.

Having proved Corollary 1, we shall now prove Corollary 2.

We can assume that the space is totally bounded, since every separable metric space is homeomorphic to a totally bounded space (§ 22, II, Corollary 1a).

Let $G = \mathcal{X} - F$. According to Corollary 1, let $F_1, F_2, \ldots$ be a sequence of non-empty sets, both closed and open, satisfying (1). Let p_n be a point of F such that

$$\varrho(p_n, F_n) < \varrho(F, F_n) + 1/n. \tag{i}$$

We define

$$f(x) = \begin{cases} x, & \text{for} \quad x \in F, \\ p_n, & \text{for} \quad x \in F_n. \end{cases}$$

(1) Compare W. Sierpiński, Fund. Math. 11 (1928), p. 118.

Since the sets F_n are open and disjoint, the function f is continuous on G. To prove that it is continuous on F, let

$$x_0 \in F, \quad x_0 = \lim_{n=\infty} x_n, \quad x_n \in G; \quad \text{hence} \quad x_n \in F_{k_n}. \tag{ii}$$

The set F_n, being closed and disjoint to F, can contain at most a finite number of terms of the sequence $x_1, x_2, \ldots$. Therefore (compare (1)) $\lim_{n=\infty} \delta(F_{k_n}) = 0$. On the other hand, conditions (ii) imply $\lim_{n=\infty} \varrho(F, F_{k_n}) = 0$ and, by (i), $\lim_{n=\infty} \delta(F_{k_n} \cup p_{k_n}) = 0$. Therefore

$$\lim_{n=\infty} |x_n - p_{k_n}| = 0, \quad \text{i.e.} \quad \lim_{n=\infty} |x_n - f(x_n)| = 0,$$

and

$$\lim_{n=\infty} f(x_n) = \lim_{n=\infty} x_n = x_0 = f(x_0).$$

This proves that the function f is continuous at the point x_0.

COROLLARY 4. *Given a (finite or infinite) sequence of closed sets* $F_0, F_1, \ldots$, *there exists a sequence of open sets* $B_0, B_1, \ldots$ *with*

$$F_i - \bigcap_{m=0}^{\infty} F_m \subset B_i \quad \textit{and} \quad \bigcap_{i=0}^{\infty} B_i = 0. \tag{2}$$

Proof. In Theorem 1 we put

$$G_i = \mathscr{X} - F_i \quad \text{and} \quad B_i = \bigcup_{j \neq i} H_j. \tag{3}$$

Moreover, let

$$S = \bigcup_{i=0}^{\infty} H_i, \quad \text{hence} \quad S = \bigcup_{i=0}^{\infty} G_i. \tag{4}$$

We deduce from (3), (4), and (0) that

$$F_i - \bigcap_{m=0}^{\infty} F_m = \bigcup_{m=0}^{\infty} G_m - G_i = S - G_i = (B_i \cup H_i) - G_i$$
$$= (B_i - G_i) \cup (H_i - G_i) = B_i - G_i \subset B_i.$$

On the other hand, since $B_i = S - H_i$, it follows from (4) that

$$\bigcap_{i=0}^{\infty} B_i = \bigcap_{i=0}^{\infty} (S - H_i) = S - \bigcup_{i=0}^{\infty} H_i = 0.$$

By Corollary 1b of § 22, II every, metric separable space is a continuous image of a subset of the space $\mathcal{N}$ of irrational numbers. This statement can be generalized as follows.

COROLLARY 5 (¹). *Given a family* $\boldsymbol{F}$ *(of sets) of the power of the continuum, there is a* $Z \subset \mathcal{N}$ *such that each member of* $\boldsymbol{F}$ *is a continuous image of* Z.

Proof. Put $\boldsymbol{F} = \{E_z\}$ where $z \in \mathcal{N}$. Denote by N_z the set of elements y of $\mathcal{N}$ such that

$$y^1 = z^1,\ y^3 = z^2,\ y^5 = z^3,\ \dots.$$

As $N_z \underset{\text{top}}{=} \mathcal{N}$, there is, by Corollary 1b of § 22, III, a set $A_z \subset N_z$ such that E_z is its continuous image. Put

$$Z = \bigcup_{z \in \mathcal{N}} A_z.$$

The sets N_z being disjoint and closed in $\mathcal{N}$, A_z is closed in Z, hence — by Corollary 2 — is a continuous image of Z. Therefore E_z is a continuous image of Z.

III. Union theorems for 0-dimensional sets.

THEOREM 1. *If the space can be represented as a (finite or infinite) series of closed sets:* $\mathcal{X} = A_1 \cup A_2 \cup \dots$ *which all, except possibly* A_1, *are 0-dimensional, and* A_1 *has dimension 0 at a given point* p, *then the space has dimension 0 at the point* p.

Proof. Let B be an open ball with centre p. We must find a set G, both closed and open, with $p \in G \subset B$.

The set G will be defined as the union of a series of increasing open sets G_n. The sets G_n will be defined inductively together with increasing open sets H_n so that the following two conditions will be satisfied:

$$\bar{G}_n \cap \bar{H}_n = 0, \tag{i}$$

$$A_n \subset G_n \cup H_n. \tag{ii}$$

Since A_1 has dimension 0 at the point p, there exists a set F_1 containing p, both closed and open relative to A_1. Since A_1 is a closed set, it follows that both F_1 and $A_1 - F_1$ are closed. Moreover we can suppose that F_1 lies in the ball B so that the sets F_1 and $(A_1 - F_1) \cup (\mathcal{X} - B)$ are closed and disjoint. Applying the normality

(¹) W. Sierpiński, Fund. Math. 14 (1929), p. 234.

of the space we deduce the existence of two open sets G_1 and H_1 with

$$F_1 \subset G_1, \quad (A_1 - F_1) \cup (\mathscr{X} - B) \subset H_1, \quad \text{and} \quad \bar{G}_1 \cap \bar{H}_1 = 0.$$

Conditions (i) and (ii) are satisfied for $n = 1$. Suppose that they are satisfied for n. We shall prove them for $n+1$.

Since the sets $A_{n+1} \cap \bar{G}_n$ and $A_{n+1} \cap \bar{H}_n$ are closed and disjoint and A_{n+1} has dimension 0, there exists, by Theorem II, 2, a set F_{n+1}, both closed and open in A_{n+1}, with $A_{n+1} \cap \bar{G}_n \subset F_{n+1}$, $F_{n+1} \cap \cap \bar{H}_n = 0$. As the sets F_{n+1} and $A_{n+1} - F_{n+1}$ are closed (since $A_{n+1} - F_{n+1}$ is closed in A_{n+1} which is closed in the space), the sets $(\bar{G}_n \cup F_{n+1})$ and $\bar{H}_n \cup (A_{n+1} - F_{n+1})$ are closed and disjoint. The space being normal, there exist two open sets G_{n+1} and H_{n+1} with

$$\bar{G}_n \cup F_{n+1} \subset G_{n+1}, \quad \bar{H}_n \cup (A_{n+1} - F_{n+1}) \subset H_{n+1},$$
$$\text{and} \quad \bar{G}_{n+1} \cap \bar{H}_{n+1} = 0.$$

Thus conditions (i) and (ii) are satisfied for $(n+1)$.

Moreover, the sets G_n and H_n being increasing, the condition $G_n \cap H_n = 0$ which follows from (i) implies $G_n \cap H_m = 0$, for every n and m (since $G_n \cap H_{n+k} \subset G_{n+k} \cap H_{n+k} = 0$). Consequently, if we put

$$G = \bigcup_{n=1}^{\infty} G_n \quad \text{and} \quad H = \bigcup_{n=1}^{\infty} H_n,$$

then $G \cap H = 0$. On the other hand, by (ii),

$$\mathscr{X} = \bigcup_{n=1}^{\infty} A_n \subset \bigcup_{n=1}^{\infty} (G_n \cup H_n) = G \cup H.$$

Hence $H = \mathscr{X} - G$. Since G and H are open, hence G is both closed and open.

It remains to show that $p \in G \subset B$. By definition of G_1, $p \in F_1 \subset G_1 \subset G$. By definition of H_1, $\mathscr{X} - B \subset H_1$; hence $\mathscr{X} - H_1 \subset B$ and, since the sets G and H_1 are disjoint, it follows that $G \subset \mathscr{X} - H_1 \subset B$.

COROLLARY 1. *The union of a countable sequence of 0-dimensional closed sets* (or, more generally, 0-*dimensional* $\boldsymbol{F}_\sigma$-*sets*) *is* 0-*dimensional.*

COROLLARY 2. *The union of two 0-dimensional sets one of which is both an* $\boldsymbol{F}_\sigma$ *and a* $\boldsymbol{G}_\delta$ *is 0-dimensional.*

COROLLARY 3. *If a single point is added to a 0-dimensional set, the dimension of the set does not change: if* $\dim A = 0$, *then* $A_{(0)} = \mathscr{X}$.

COROLLARY 4. *If G is an open set, then the condition* $\dim(E \cap G) \leqslant 0$ *implies* $E_{(0)} = (E-G)_{(0)}$.

To prove Corollary 1, it is sufficient to consider the union of the sets in question as a space. Corollary 2 follows from the fact that the sets in question are F_σ-sets relative to their union. Corollary 3 is a particular case of corollary 2. To prove Corollary 4, let $p \in (E-G)_{(0)}$, i.e. $\dim_p(p \cup E-G) = 0$. If the set $p \cup E$ is considered as a space, then the set $E \cap G$ is, in this space, a countable union of closed 0-dimensional sets, and, by Theorem 1, the set $(p \cup E-G) \cup (E \cap G) = p \cup E$ has dimension 0 at the point p, i.e. $p \in E_{(0)}$, and Corollary 4 follows.

Remarks. Theorem 1 may be false, if we suppose only that $\dim_p A_n = 0$. For example, let us divide the interval $\mathscr{I}$ into a sequence of intervals convergent to 0. Let A_1 be the set composed of the point 0 and the even intervals; let A_2 be composed of 0 and the odd intervals. Then clearly $\dim_p A_1 = 0 = \dim A_2$, but $\dim_p(A_1 \cup A_2) = 1$.

The following example shows that instead of assuming A_1 to be closed, one cannot assume A_1 to be an F_σ-set; let A_2 be a sequence of points convergent to the point 0 $(=p)$ and $A_1 = \mathscr{I} - A_2$.

IV. Extension of 0-dimensional sets.

THEOREM 1. *Every 0-dimensional set is contained in a 0-dimensional G_δ-set.*

Proof. By § 25, III, Theorem 3, for every set E there exists an F_σ-set S with $E \cap S = 0$ and $\dim[E_{(0)} - S] \leqslant 0$. Assuming that $\dim E = 0$, we obtain by Section III, Corollary 3, $E_{(0)} = \mathscr{X}$. Hence $(\mathscr{X} - S)$ is a 0-dimensional G_δ-set containing E (1).

THEOREM 2 (2). *Every 0-dimensional space $\mathscr{X}$ is topologically contained in the Cantor discontinuum $\mathscr{C}$.*

Proof. By definition, $\mathscr{C}$ is the set of sequences $[\mathfrak{z}^1, \mathfrak{z}^2, \ldots]$, where $\mathfrak{z}^i$ takes the values 0 or 2 (see § 3, IX). We have to define a homeomorphism $\mathfrak{z}\colon \mathscr{X} \to \mathscr{C}$.

(1) For a proof based on a different idea, see § 35, II, Remark 2.
(2) P. Urysohn, Fund. Math. 7 (1925), p. 77.

Let $R_1, R_2, \ldots, R_i, \ldots$ be a base of the space made up of sets both closed and open (Theorem I, 1). Let $\mathfrak{z}$ be its characteristic function, i.e. $\mathfrak{z}^i(x) = 2$ for $x \in R_i$ and $\mathfrak{z}^i(x) = 0$ otherwise. As the characteristic function of a sequence of sets both closed and open, the function $\mathfrak{z}$ is continuous (§ 16, II, Corollary 6d). To show that it is a homeomorphism, it suffices to prove (§ 13, VIII, (3a)) that the condition $p \in \mathscr{X} - \overline{X}$ implies $\mathfrak{z}(p) \in \mathscr{C} - \overline{\mathfrak{z}(X)}$.

By definition of base, there exists an index i with $p \in R_i \subset \mathscr{X} - \overline{X} \subset \mathscr{X} - X$. Hence $\mathfrak{z}^i(p) = 2$, while for $x \in X$, $\mathfrak{z}^i(x) = 0$. Consequently $|\mathfrak{z}(p) - \mathfrak{z}(x)| \geqslant 1/3^i$ (we identify here the sequence $[\mathfrak{z}^1, \mathfrak{z}^2, \ldots, \mathfrak{z}^i, \ldots]$ with the number $\dfrac{\mathfrak{z}^1}{3} + \dfrac{\mathfrak{z}^2}{3^2} + \ldots + \dfrac{\mathfrak{z}^i}{3^i} + \ldots$). It follows that the number $\mathfrak{z}(p)$ cannot lie in the closure of the set of the numbers $\mathfrak{z}(x)$, i.e. $\mathfrak{z}(p) \in \mathscr{C} - \overline{\mathfrak{z}(X)}$.

COROLLARY 2a. *$\mathscr{C}$ has the highest topological rank among all 0-dimensional spaces.*

Because the Cantor set is itself 0-dimensional as a boundary subset of the straight line (see § 25, I, examples).

Remark. *The set $\mathscr{N}$ of irrational numbers* (of the interval $\mathscr{I}$) also has this property. For, the set $\mathscr{N}$, as the space of all infinite sequences of natural numbers, contains sequences made up of the two numbers 1 and 2, and this set is homeomorphic to $\mathscr{C}$. The set $\mathscr{N}$ is itself 0-dimensional as a boundary subset of the interval. Hence it is topologically contained in $\mathscr{C}$.

We can also say that the space $\mathscr{C}$, as well as $\mathscr{N}$, has the highest topological rank among boundary subsets of the space of real numbers.

COROLLARY 2b. *If* $\dim \mathscr{X}_n = 0$ *for* $n = 1, 2, \ldots$, *then*

$$\dim(\mathscr{X}_1 \times \mathscr{X}_2 \times \ldots) = 0.$$

Because $\mathscr{X}_1 \times \mathscr{X}_2 \times \ldots \subset \mathscr{C}^{\aleph_0} \underset{\text{top}}{=} \mathscr{C}$ by Corollary 6a of § 16, II.

V. Countable spaces.

THEOREM 1. *Every space of power less than the power of continuum is 0-dimensional.*

More generally, *if the space is of order less than $\mathfrak{c}$ at the point p* (see § 23, VII), *the space is 0-dimensional at p.*

Proof. There exists a ball of radius $< \varepsilon$ and centre p whose boundary (the "surface") is empty (since the boundaries of different balls are disjoint). Hence this ball is both closed and open.

THEOREM 2. *Every countable space is topologically contained in the space $\mathcal{R}$ of rational numbers.*

Proof. Let A be a countable space. Since A is topologically contained in $\mathcal{C}$, it can be considered as a subset of the space of real numbers.

By a classical theorem of the theory of ordered sets, the sets $A \cup \mathcal{R}$ and $\mathcal{R}$ have the same order types, i.e. there exists an increasing function f mapping $A \cup \mathcal{R}$ onto $\mathcal{R}$. This function is continuous since, given a sequence $x_1 < x_2 < \ldots$ convergent to x (the points x_n and x are supposed to lie in $A \cup \mathcal{R}$), x is the first point in $A \cup \mathcal{R}$ which succeeds all the x_n and, by the similiarity of the sets $A \cup \mathcal{R}$ and $\mathcal{R}$, the point $f(x)$ has the same property with respect to the sequence $f(x_1), f(x_2), \ldots$ If $f(x) \neq \lim_{n=\infty} f(x_n)$, then there must exist a rational number which is both greater than all the $f(x_n)$ and less than $f(x)$ which is clearly impossible. Thus the function f is continuous. For the same reason the inverse function f^{-1} is continuous. Therefore f is a homeomorphism which maps A onto a subset of $\mathcal{R}$.

By an analogous method we can prove that *the countable dense in itself spaces constitute a single topological type* (*i.e., all are homeomorphic*) (1).

§ 27. *n*-dimensional spaces

I. Union theorems.

THEOREM 1. *The union of an n-dimensional set with a 0-dimensional set has dimension* $\leqslant n+1$.

In other words, the formulas $\dim(\mathcal{X}-Q) = n$ and $\dim Q = 0$ imply $\dim \mathcal{X} \leqslant n+1$.

Proof. By § 26, III, Corollary 3, the set $Q \cup p$ has dimension 0, for every p. Hence there exists an arbitrarily small neighbourhood G of p with $Q \cap \mathrm{Fr}(G) = 0$ (§ 25, II, Theorem 2_0), i.e. $\mathrm{Fr}(G) \subset \mathcal{X} - Q$. Therefore $\dim \mathrm{Fr}(G) \leqslant n$, and $\dim \mathcal{X} \leqslant n+1$.

---

(1) W. Sierpiński, Fund. Math. 1 (1920), p. 11, and Wektor 1915.

THEOREM 2 [1]. *If the space can be represented as a (finite or infinite) series of closed n-dimensional sets:* $\mathscr{X} = A_1 \cup A_2 \cup \ldots$, *then the space itself is n-dimensional.*

Proof. The theorem is true for $n = 0$ (§ 26, III, Theorem 1). Suppose that it is true for $n-1$ and let

$$B_1 = A_1 \quad \text{and} \quad B_k = A_k - (A_1 \cup \ldots \cup A_{k-1}).$$

The sets B_k are then disjoint F_σ-sets and

$$\mathscr{X} = B_1 \cup B_2 \cup \ldots, \qquad \dim B_k \leqslant n.$$

Put $B_k = E \subset E_{(n)}$ and $S_k = E \cap S$ in § 25, III, Theorem 3; we deduce that the set S_k is a countable union of sets of dimension $\leqslant n-1$, closed in B_k and such that $\dim(B_k - S_k) \leqslant 0$. As B_k is an F_σ-set, hence sets closed in B_k are F_σ-sets. Since a countable union of $(n-1)$-dimensional F_σ-sets is, by hypothesis, $(n-1)$-dimensional, it follows that $\dim S_k \leqslant n-1$ and, for the same reason,

$$\dim(S_1 \cup S_2 \cup \ldots) \leqslant n-1.$$

Since the sets B_k are disjoint, it follows that

$$B_i \cap \bigcup_{k=1}^{\infty} (B_k - S_k) = \bigcup_{k=1}^{\infty} (B_i \cap B_k - S_k) = B_i - S_i.$$

It is seen that the set $B_i - S_i$ being the intersection of an F_σ-set with the set $\bigcup_{k=1}^{\infty} (B_k - S_k)$ is an F_σ-set in the latter. Therefore $\bigcup_{k=1}^{\infty} (B_k - S_k)$ regarded as a space is the union of a sequence of 0-dimensional F_σ-sets. Consequently $\dim \bigcup_{k=1}^{\infty} (B_k - S_k) \leqslant 0$.

Therefore the formula $\mathscr{X} = \bigcup_{k=1}^{\infty} B_k = \bigcup_{k=1}^{\infty} S_k \cup \bigcup_{k=1}^{\infty} (B_k - S_k)$ gives a decomposition of the space into the union of a set of dimension $\leqslant n-1$ and a set of dimension $\leqslant 0$. By Theorem 1, the space has dimension $\leqslant n$.

---

[1] See W. Hurewicz, *Normalbereiche und Dimensionstheorie*, Math. Ann. 96 (1927), p. 760, and L. Tumarkin, *Über die Dimension nicht abgeschlossener Mengen*, Math. Ann. (1928), p. 641. For compact spaces, see K. Menger, Monatsh. 34 (1924), p. 147, and P. Urysohn, Fund. Math. 8 (1926), p. 316.

COROLLARY 2a. *The union of a countable sequence of n-dimensional $\boldsymbol{F}_\sigma$-sets is n-dimensional.*

COROLLARY 2b. *The union of two n-dimensional sets, one of which is both $\boldsymbol{F}_\sigma$ and $\boldsymbol{G}_\delta$, has dimension n.*

COROLLARY 2c. *The adjunction of a single point to a (non empty) space does not change its dimension.*

COROLLARY 2d. *Given a set E and an integer n, there exists an $\boldsymbol{F}_\sigma$-set S with*

$$\dim E \cap S \leqslant n-1, \quad E_{(n)} \subset (E-S)_{(0)}, \quad \textit{and} \quad \dim[E_{(n)}-S] \leqslant 0.$$

In particular, *every n-dimensional space is composed of an $(n-1)$-dimensional $\boldsymbol{F}_\sigma$-set and a 0-dimensional $\boldsymbol{G}_\delta$-set.*

COROLLARY 2e. *If G is an open set, then the condition* $\dim(E \cap G) \leqslant n$ *implies $E_{(n)} = (E-G)_{(n)}$.*

COROLLARY 2f. *The conditions $\mathscr{X} = A_1 \cup A_2 \cup \ldots, A_k = \bar{A}_k$,* $\dim_p A_1 \leqslant n$ and $\dim A_k \leqslant n$ *for $k>1$ imply* $\dim_p \mathscr{X} \leqslant n$.

Corollaries 2a–2c are easy consequences of Theorem 2 (compare also the proof of the corollaries of Section III, § 26). By regarding the set $E \cap S$ as a space, where S is the set of § 25, III, Theorem 3, we deduce from Corollary 2a that $\dim E \cap S \leqslant n-1$.

In particular, if E denotes the entire n-dimensional space, then $E_{(n)} = \mathscr{X}$ and $\mathscr{X} = (\mathscr{X}-S)_{(0)}$, and hence $\dim(\mathscr{X}-S) = 0$.

Thus Corollary 2d is proved. By substituting, respectively, $E-G$ and $E \cap G$ for E, we deduce the existence of two $\boldsymbol{F}_\sigma$-sets S and W with

$$\dim(S \cap E-G) \leqslant n-1, \quad (E-G)_{(n)} \subset (E-G-S)_{(0)},$$

$$\dim(W \cap E \cap G) \leqslant n-1, \quad \text{and} \quad \dim(E \cap G-W) \leqslant 0,$$

since by hypothesis $E \cap G \subset (E \cap G)_{(n)}$. The set $E-G-S$ is obtained from $E-G-S \cup (E \cap G)-W$ by subtraction of the open set G whose intersection with the latter is 0-dimensional. It follows from § 26, III, Corollary 4, that

$$\big(E-G-S \cup (E \cap G)-W\big)_{(0)} = (E-G-S)_{(0)}.$$

The sets $S \cap E-G$ and $W \cap E \cap G$ being of dimension $\leqslant n-1$ and also $\boldsymbol{F}_\sigma$-sets relative to E, it follows from Corollary 2a that their union is also of dimension $\leqslant n-1$. Consequently (§ 25,

III, Theorem 5), $(E-G-S \cup (E \cap G)-W)_{(0)} \subset (E-G-S \cup (E \cap G)-W \cup (S \cap E)-G \cup (W \cap E \cap G))_{(n)} = E_{(n)}$; hence $(E-G)_{(n)} \subset (E-G-S)_{(0)} \subset E_{(n)}$ and Corollary 2e follows.

To prove Corollary 2, let $E = \mathscr{X}$ and $G = \mathscr{X} - A_1$. By Corollary 2a, $\dim(A_2 \cup A_3 \cup \ldots) \leqslant n$, whence $\dim G \leqslant n$. The hypothesis $p \in (\mathscr{X}-G)_{(n)}$ implies by virtue of Corollary 2e that $\dim_p \mathscr{X} \leqslant n$.

THEOREM 3 [1]. *In order that a (non-empty) space should have dimension* $\leqslant n$ *it is necessary and sufficient that it should be the union of* $(n+1)$ *0-dimensional sets.*

Proof. The theorem being obvious for $n = 0$, suppose that it is true for $n-1$. By Corollary 2d, an n-dimensional space decomposes itself into the union of a 0-dimensional set and an $(n-1)$-dimensional set. By decomposing the latter into n 0-dimensional sets, we obtain a decomposition of the space into $(n+1)$ 0-dimensional sets.

The fact that the union of $(n+1)$ 0-dimensional sets has dimension $\leqslant n$ is an immediate consequence of Theorem 1.

COROLLARY 3a [2]. *If* $\dim A = n$ *and* $\dim B = m$, *then* $\dim(A \cup B) \leqslant n+m+1$.

II. Separation of closed sets.

THEOREM 1 (compare § 26, II, Theorem 2). *If A and B are two disjoint closed sets in a space of dimension n, then there exists an open set G such that*

$$A \subset G, \quad \bar{G} \cap B = 0 \quad \textit{and} \quad \dim[\mathrm{Fr}(G)] \leqslant n-1 \text{ [3]}.$$

More generally, *if A and B are two disjoint closed sets and E is a set of dimension $n \geqslant 0$ (situated in a space $\mathscr{X}$ of arbitrary dimension), then there exist two closed sets M and N with*

$$\mathscr{X} = M \cup N, \quad A \cap N = 0 = B \cap M, \quad \textit{and} \quad \dim(E \cap M \cap N) \leqslant n-1.$$

---

[1] W. Hurewicz, *loc. cit.*, p. 761, and L. Tumarkin, *loc. cit.*, p. 641.

[2] This corollary can also be derived directly by induction from the definition of dimension (more precisely, from § 25, II, (2)). See K. Menger, *Dimensionstheorie*, p. 114.

[3] W. Hurewicz, *loc. cit.*, p. 763, and L. Tumarkin, *loc. cit.*, p. 653. Compare P. Urysohn, *loc. cit.*, p. 316.

Proof. There exists (§ 22, IV, Theorem 2) a continuous mapping f of the space $\mathscr{X}$ onto a separable metric space $\mathscr{X}^*$ containing two points a and b such that $f^{-1}(a) = A$, $f^{-1}(b) = B$ and f is a homeomorphism on $\mathscr{X} - (A \cup B)$. Therefore

$$\dim f[E-(A \cup B)] = \dim[E-(A \cup B)] \leqslant n.$$

Since

$$f(E) \cup a \cup b = f[E-(A \cup B)] \cup a \cup b,$$

we obtain by I, Corollary 2c,

$$\dim[f(E) \cup a \cup b] \leqslant n.$$

Hence there exists (§ 25, II, Theorem 2) an open set G with

$$a \in G, \quad b \in \mathscr{X}^* - \bar{G}, \quad \text{and} \quad \dim[f(E) \cap \bar{G} - G] \leqslant n-1.$$

Let $M = f^{-1}(\bar{G})$ and $N = f^{-1}(\mathscr{X}^* - G) = \mathscr{X} - f^{-1}(G)$.

Since the function f is continuous, the sets M and N are closed. Since $\mathscr{X}^* = \bar{G} \cup (\mathscr{X}^* - G)$, it follows that $\mathscr{X} = M \cup N$. The condition $a \in G$ implies $(a) \cap (\mathscr{X}^* - G) = 0$; hence

$$A \cap N = f^{-1}(a) \cap f^{-1}(\mathscr{X}^* - G) = f^{-1}[(a) \cap (\mathscr{X}^* - G)] = f^{-1}(0) = 0,$$

and the condition $b \in \mathscr{X}^* - \bar{G}$ implies $B \cap M = 0$.

Finally, the sets $E \cap M \cap N$ and $f(E) \cap \bar{G} - G$ are homeomorphic $\big($for $M \cap N \subset \mathscr{X} - (A \cup B)\big)$; therefore $\dim(E \cap M \cap N) \leqslant n-1$.

To obtain the first part of Theorem 1, we put

$$E = \mathscr{X} \quad \text{and} \quad G = \mathscr{X} - N.$$

Remark. Theorem 1 gives, in the case of an n-dimensional space, a more advantageous condition than the normality property. By § 14, II (7) and § 6, V, Theorem 6, the normality property can also be formulated as follows: Every pair of disjoint closed sets can be separated by a closed set. The following corollary corresponds to this statement.

COROLLARY 1a. *Every pair of disjoint closed sets in a space of dimension n can be separated by a closed set of dimension $\leqslant n-1$.*

The reason for this is that the boundary of G in Theorem 1 is a set of dimension $\leqslant n-1$ *which separates A and B.*

COROLLARY 1b. *The condition of Theorem 1 is necessary and sufficient for a space to be of dimension $\leqslant n$.*

Proof. By Theorem 1 the condition is necessary. Conversely, if we identify A with a given point p and B with the complement of an open neighbourhood of this point, then the condition implies that $\dim_p \mathscr{X} \leqslant n$.

Theorem 1 has the following generalization.

COROLLARY 1c [1]. *If $A_0, \ldots, A_n$ and $B_0, \ldots, B_n$ are two systems of closed sets in a space of dimension n such that $A_i \cap B_i = 0$ for $0 \leqslant i \leqslant n$, then there exist two systems of closed sets $M_0, \ldots, M_n$ and $N_0, \ldots, N_n$ such that, for $i \leqslant n$,*

$$\mathscr{X} = M_i \cup N_i, \quad A_i \cap N_i = 0 = B_i \cap M_i,$$

$$\text{and} \quad \dim(M_0 \cap \ldots \cap M_i \cap N_0 \cap \ldots \cap N_i) \leqslant n-i-1. \quad (1)$$

Proof. The sets $M_0, \ldots, M_i$ and $N_0, \ldots, N_i$ (for a fixed n) will be defined by induction.

By Theorem 1 they exist for $i = 0$. Let them be defined for an integer i such that $0 \leqslant i < n$. We substitute in Theorem 1

$$A = A_{i+1}, B = B_{i+1}, \text{ and } E = M_0 \cap \ldots \cap M_i \cap N_0 \cap \ldots \cap N_i.$$

Hence there exist two closed sets M_{i+1} and N_{i+1} with

$$\mathscr{X} = M_{i+1} \cup N_{i+1}, \quad A_{i+1} \cap N_{i+1} = 0 = B_{i+1} \cap M_{i+1},$$
$$\dim(E \cap M_{i+1} \cap N_{i+1}) \leqslant n-i-2.$$

This proves the corollary.

Remark. Corollary 1c extends to n-dimensional spaces the following property of the n-dimensional cube $\mathscr{I}^n$ (composed of points $(x_1, \ldots, x_n)$ with $0 \leqslant x_i \leqslant 1, 1 \leqslant i \leqslant n$). Let A_i and B_i be the faces of the cube perpendicular to the $\mathscr{X}_i$-axis and let M_i and N_i be two halves of the cube determined by the plane $x_i = \frac{1}{2}$. Then condition (1) is obvious.

The following corollary concerns the "extension of closed sets".

COROLLARY 1d [2]. *If A and B are two closed sets such that $\dim[\mathscr{X} - (A \cup B)] \leqslant n$, then there exist two closed sets P and Q satisfying the conditions:*

$$\mathscr{X} = P \cup Q, \quad P \cap (A \cup B) = A, \quad Q \cap (A \cup B) = B,$$
$$\dim(P \cap Q - A \cap B) \leqslant n-1. \quad (2)$$

---

[1] See S. Eilenberg and E. Otto, *Quelques propriétés caractéristiques de la dimension*, Fund. Math. 31 (1938), p. 151.

[2] See W. Hurewicz, Fund. Math. 24 (1935), p. 146.

Proof. We apply Theorem 1 to the space $\mathscr{X}^* = \mathscr{X} - (A \cap B)$ and substitute $A - B$ for A, $B - A$ for B and $\mathscr{X} - (A \cup B)$ for E. Then

$$\mathscr{X} - (A \cap B) = M \cup N, \qquad A \cap N - B = 0 = B \cap M - A, \tag{3}$$

$$\dim[M \cap N - (A \cup B)] \leqslant n-1. \tag{4}$$

Hence $\dim M \cap N \leqslant n-1$, since $M \cap N \cap (A \cup B) = 0$ by virtue of the formulas

$$M \cap N \cap (A - B) = 0 = M \cap N \cap (B - A) \quad \text{and} \quad M \cap N \cap A \cap B = 0$$

which follow from (3).

Let $P = M \cup (A \cap B)$ and $Q = N \cup (A \cap B)$. By (3), $\mathscr{X} = P \cup Q$.

It follows that $P \cap (A \cup B) = \big(M \cup (A \cap B)\big) \cap (A \cup B) = (M \cap A) \cup (M \cap B) \cup (A \cap B)$.

By (3),

$$M \cap B = (M \cap B - A) \cup (M \cap A \cap B) \subset A, \quad \text{hence} \quad P \cap (A \cup B) \subset A.$$

Since

$$A - B = A - (A \cap B) \subset M \cup N \quad \text{and} \quad (A - B) \cap N = 0,$$

therefore

$$A = (A - B) \cup (A \cap B) \subset M \cup (A \cap B) = P.$$

It follows that $P \cap (A \cup B) = A$.

By symmetry, $Q \cap (A \cup B) = B$. We have $(P \cap Q) - (A \cap B) = \big((M \cap N) \cup (A \cap B)\big) - (A \cap B) \subset M \cap N$ and

$$\dim\big((P \cap Q) - (A \cap B)\big) \leqslant \dim(M \cap N) \leqslant n-1.$$

Finally, since M and N are closed in $\mathscr{X} - (A \cap B)$, it follows that P and Q are closed (in $\mathscr{X}$).

Remark. Concerning an extension of Corollary 3 of § 26, II to an n-dimensional space, we quote the following theorem.

Every continuous function f defined on a closed subset F of a separable metric n-dimensional space can be extended to the entire space so that the set of points of discontinuity has dimension $< n$ and the values remain in the set $f(F)$ (1).

---

(1) See G. Poprugenko, *Sur la dimension de l'espace et l'extension des fonctions continues*, Mon. f. Math. u. Phys. 33 (1931), p. 129. By a remark of E. Otto, the hypothesis that the function is real-valued can be dropped.

III. Decomposition of an n-dimensional space. Condition D_n.

THEOREM 1. [1] (Generalization of Theorem § 26, II, 1.) *If*

$$\mathscr{X} = G_0 \cup G_1 \cup \ldots$$

is a representation of an n-dimensional space as a (finite or infinite) series of open sets, then there exists a sequence of open sets H_0, $H_1, \ldots$ with

$$\mathscr{X} = H_0 \cup H_1 \cup \ldots, \qquad H_i \subset G_i \quad \textit{and} \quad H_{i_0} \cap \ldots \cap H_{i_{n+1}} = 0 \tag{1}$$

for every system of $n+2$ different indices $i_0, \ldots, i_{n+1}$.

Proof. According to Theorem I, 3, the space can be decomposed into $(n+1)$ 0-dimensional sets:

$$\mathscr{X} = Q_0 \cup \ldots \cup Q_n.$$

By Theorem 1 of § 26, II (if Q_j is regarded as a space), Q_j can be decomposed into disjoint sets open in Q_j and contained, respectively, in G_i:

$$Q_j = H_{j0} \cup H_{j1} \cup \ldots \qquad \text{and} \qquad H_{ji} \subset G_i.$$

The sets $H_{j0}, H_{j1}, \ldots$ being disjoint and relatively open in Q_j are mutually separated. Hence, by § 21, XI, Theorem 2, there exist disjoint and open sets $V_{j0}, V_{j1}, \ldots$ with $H_{ji} \subset V_{ji}$. Let

$$H_i = (V_{0i} \cup \ldots \cup V_{ni}) \cap G_i.$$

Then

$$\mathscr{X} = \bigcup_{j=0}^{n} Q_j = \bigcup_{j=0}^{n} \bigcup_{i} H_{ji} = \bigcup_{j=0}^{n} \bigcup_{i} H_{ji} \cap G_i \subset \bigcup_{i} \bigcup_{j=0}^{n} V_{ji} \cap G_i = \bigcup_{i} H_i.$$

Finally, if we suppose that $p \in H_{i_0} \cap \ldots \cap H_{i_{n+1}}$, then there must exist an index $j \leqslant n$ and two different indices i and i' with $p \in V_{ji} \cap V_{ji'}$. But this contradicts the hypothesis that the sets $V_{j0}, V_{j1}, \ldots$ are disjoint.

COROLLARY 1a. *If* $\dim G = n$, *then the following additional condition*

$$G_{i_0} \cap \ldots \cap G_{i_{n+1}} = 0, \qquad \textit{provided} \quad i_0 < i_1 < \ldots < i_{n+1}, \tag{n}$$

can be imposed on the sequence $G_0, G_1, \ldots$ of Theorem § 21, VIII, 4.

---

[1] Theorem 1 in the case of a finite union is due to K. Menger, *Dimensionstheorie*, p. 158. Compare P. Urysohn, *loc. cit.*, p. 292.

For, if the sequence $\{G_i\}$ satisfies the conditions of Theorem 4 in § 21, VIII, then the sequence $\{H_i\}$ also satisfies them (if we put $\mathscr{X} = G$). But the sequence $\{H_i\}$ fulfills condition (n) (after replacing G by H).

DEFINITION. The space *satisfies condition* D_n if the assertion of Theorem 1 holds for every *finite* decomposition into open sets.

By Theorem 1, *every space of dimension* $\leqslant n$ *satisfies condition* D_n. We shall see later (Vol. II) that the converse theorem is also true. For the sake of applications, we shall now prove some properties of spaces satisfying condition D_n. By Theorem 1, these are properties of every space of dimension $\leqslant n$.

THEOREM 2. *If*

$$\mathscr{X} = G_0 \cup \ldots \cup G_m$$

is a finite decomposition of a space satisfying condition D_n *into open sets, then there exist open sets* $H_0, \ldots, H_m$ *such that*

$$\mathscr{X} = H_0 \cup \ldots \cup H_m, \quad \overline{H}_i \subset G_i \quad \text{and} \quad \overline{H}_{i_0} \cap \ldots \cap \overline{H}_{i_{n+1}} = 0 \quad (2)$$

hold for every indices $i_0 < i_1 < \ldots < i_{n+1} \leqslant m$.

Proof. By virtue of § 14, III, Corollary, for the sets $H_0, \ldots, H_m$ of formula (1) there exist open sets $H_0^*, \ldots, H_m^*$ with

$$\mathscr{X} = H_0^* \cup \ldots \cup H_m^* \quad \text{and} \quad \overline{H}_i^* \subset H_i.$$

The sets H_i^* substituted for H_i satisfy formula (2).

Since $\overline{H}_i$ is a closed domain, the following theorem can be deduced from Theorem 2

THEOREM 3. *If* $\mathscr{X} = F_0 \cup \ldots \cup F_m$ *is a finite decomposition of a space satisfying condition* D_n *into closed sets, then for every* $\varepsilon > 0$ *there exist closed domains* $H_0, \ldots, H_m$ *satisfying conditions* (1) *with* G_i *being the generalized open ball with centre* F_i *and radius* ε.

THEOREM 4. *If a space satisfying condition* D_n *is dense in itself and if the open sets* G_i *of the decomposition* $\mathscr{X} = G_0 \cup \ldots \cup G_m$ *are non empty, then the sets* H_i *of formula* (2) *can be subjected to the additional condition:* $H_i \neq 0$ *for* $i = 0, 1, \ldots, m$.

Proof. Since the space is dense in itself, the sets G_i are infinite. Hence we can select a point p_i from each G_i so that the points $p_0, \ldots, p_m$ are mutually distinct. Let

$$G_i^* = G_i - (p_0, \ldots, p_{i-}\ , p_{i+1}, \ldots, p_m).$$

Then $\mathscr{X} = G_0^* \cup \ldots \cup G_m^*$.

By virtue of condition D_n, there exist open sets $H_0, \ldots, H_m$ with

$$\mathscr{X} = H_0 \cup \ldots \cup H_m, \quad \bar{H}_i \subset G_i^* \subset G_i, \quad \bar{H}_{i_0} \cap \ldots \cap \bar{H}_{i_{n+1}} = 0. \quad (3)$$

Since $p_i \notin G_j^*$ for $j \neq i$, hence $p_i \notin H_j$. Therefore, by the first formula of (3), $p_i \in H_i$.

By relativizing condition D_n we are led to the following theorem.

THEOREM 5. *If E is a set satisfying condition D_n in a space (of arbitrary dimension) and $G_0, \ldots, G_m$ are open sets such that $E \subset G_0 \cup \ldots \cup G_m$, then there exist open sets $H_0, \ldots, H_m$ satisfying the conditions*

$$E \subset H_0 \cup \ldots \cup H_m, \quad H_i \subset G_i, \quad \text{and} \quad H_{i_0} \cap \ldots \cap H_{i_{n+1}} = 0 \quad (4)$$

for every system of indices $i_0 < i_1 < \ldots < i_{n+1} \leqslant m$.

If, moreover, E is closed and $G_0 \cup \ldots \cup G_m = \mathscr{X}$, then there exist open sets $Q_0, \ldots, Q_m$ with

$$\mathscr{X} = Q_0 \cup \ldots \cup Q_m, \quad \bar{Q}_i \subset G_i \quad \text{and} \quad E \cap \bar{Q}_{i_0} \cap \ldots \cap \bar{Q}_{i_{n+1}} = 0. \quad (5)$$

Finally, if E is perfect and $G_i \neq 0$ for $i = 0, \ldots, m$, then we can assume that $Q_i \neq 0$ for $i = 0, \ldots, m$.

Proof. The sets $E \cap G_i$ are open in E and $E = E \cap G_0 \cup \ldots \cup E \cap G_m$. By property D_n of E there exist sets $A_0, \ldots, A_m$, open in E, with

$$E = A_0 \cup \ldots \cup A_m, \quad A_i \subset E \cap G_i \subset G_i,$$
$$\text{and} \quad A_{i_0} \cap \ldots \cap A_{i_{n+1}} = 0. \quad (6)$$

By Theorem 2 of § 21, XI, there exist open sets $V_0, \ldots, V_m$ with

$$A_i = E \cap V_i \quad \text{and} \quad V_{i_0} \cap \ldots \cap V_{i_{n+1}} = 0.$$

Then the sets $H_i = V_i \cap G_i$, $i = 0, \ldots, m$, satisfy (4).

In the case where E is closed, let $R_i = H_i \cup (G_i - E)$. Then (4) implies

$$\bigcup_i R_i = \bigcup_i H_i \cup \Big(\bigcup_i G_i - E\Big) \supset \bigcup_i H_i \cup \Big(\bigcup_i G_i - \bigcup_i H_i\Big) = \bigcup_i G_i = \mathscr{X}, \quad (7)$$

$$E \cap R_{i_0} \cap \ldots \cap R_{i_{n+1}} = E \cap H_{i_0} \cap \ldots \cap H_{i_{n+1}} = 0 \quad \text{and} \quad R_i \subset G_i. \quad (8)$$

By (7) and Corollary 1 of § 14, III, there exist open sets $Q_0, \ldots, Q_m$ with

$$\mathscr{X} = Q_0 \cup \ldots \cup Q_m \quad \text{and} \quad \bar{Q}_i \subset R_i.$$

Formulas (8) give (5).

Finally, if E is perfect, we complete formula (6) (according to Theorem 4) with the condition $A_i \neq 0$. Since $A_i \subset H_i \subset R_i$, hence $R_i \neq 0$. We can now assume that $Q_i \neq 0$, for, without affecting conditions (5) we can adjoin to Q_i an arbitrary (non empty) open set whose boundary lies in R_i.

For further applications we shall prove the following theorem.

THEOREM 6. *Let* $K_0, \ldots, K_s$ *be open sets with* $\mathscr{X} = K_0 \cup \ldots \cup K_s$, *and let* $A_0, \ldots, A_l$ *be closed sets with intersection* $A_0 \cap \ldots \cap A_l$ *satisfying condition* D_n. *Then there exist open sets* $G_0, \ldots, G_m$ *with* $\mathscr{X} = G_0 \cup \ldots \cup G_m$ *and such that for each* $i \leqslant m$ *there exists an* $i' \leqslant s$ *with* $G_i \subset K_{i'}$ *and that for each system* $i_0 < i_1 < \ldots < i_{n+1} \leqslant m$ *there exists a* $j \leqslant l$ *with*

$$A_j \cap G_{i_0} \cap \ldots \cap G_{i_{n+1}} = 0.$$

Proof. Let $E = A_0 \cap \ldots \cap A_l$. By Theorem 5 there exist two systems of open sets, $Q_0, \ldots, Q_s$ and $H_0, \ldots, H_s$ with

$$\mathscr{X} = Q_0 \cup \ldots \cup Q_s, \quad Q_i \subset K_i, \quad E \cap Q_{i_0} \cap \ldots \cap Q_{i_{n+1}} = 0,$$

$$E \subset H_0 \cup \ldots \cup H_s, \quad H_i \subset Q_i, \quad H_{i_0} \cap \ldots \cap H_{i_{n+1}} = 0.$$

Then

$$\mathscr{X} = Q_0 \cup \ldots \cup Q_s = \bigcup_{ij} (Q_i - A_j) \cup \bigcap_j A_j,$$

and since

$$\bigcap_j A_j = E \subset \bigcup_i H_i,$$

hence

$$\mathscr{X} = \bigcup_{ij} (Q_i - A_j) \cup \bigcup_i H_i.$$

Let $G_0, \ldots, G_m$ denote the $(s+1)(l+2)$ members of this union.

Let $i_0 < i_1 < \ldots < i_{n+1} \leqslant m$. Suppose that $A_j \cap G_{i_0} \cap \ldots \cap G_{i_{n+1}} \neq 0$, for every $j \leqslant l$. Therefore none of the sets $G_{i_0}, \ldots, G_{i_{n+1}}$ belongs to the system of sets $(Q_i - A_j)$, where $i \leqslant s$ and $j \leqslant l$. Hence they all belong to the system $H_0, \ldots, H_s$. But this contradicts the formula $H_{i_0} \cap \ldots \cap H_{i_{n+1}} = 0$.

IV. Extension of n-dimensional sets.

THEOREM 1. (Generalization of § 26, IV, Theorem 1.) (1) *Every n-dimensional set is contained in an n-dimensional G_δ-set.*

Proof. If E is a set of dimension n, then $E = Q_0 \cup \ldots \cup Q_n$, where the sets Q_i have dimension 0 (Theorem I, 3). By § 26, IV, Theorem 1, the set Q_i is contained in a 0-dimensional G_δ-set: $Q_i \subset Q_i^*$. Therefore $E \subset Q_0^* \cup \ldots \cup Q_n^*$ and this union is an n-dimensional G_δ-set (by the same Theorem I, 3).

V. Dimensional kernel.

THEOREM 1 (2). *The set of points of an n-dimensional space where the space has dimension n, i.e. the set $\mathscr{X}-\mathscr{X}_{(n-1)}$ (called the "dimensional kernel"), has dimension $\geqslant n-1$.*

Proof. By Corollary 2d of Section I, there is an F_σ-set S with

$$\dim S \leqslant n-2 \quad \text{and} \quad \dim(\mathscr{X}_{(n-1)}-S) \leqslant 0.$$

By § 25, III, Theorem 2, $\mathscr{X}-\mathscr{X}_{(n-1)}$ is an F_σ-set, and Corollary 2a of Section I together with the hypothesis that $\dim(\mathscr{X}-\mathscr{X}_{(n-1)}) \leqslant n-2$ imply

$$\dim(\mathscr{X}-\mathscr{X}_{(n-1)} \cup S) \leqslant n-2.$$

But then the identity

$$\mathscr{X} = [\mathscr{X}-\mathscr{X}_{(n-1)} \cup S] \cup [\mathscr{X}_{(n-1)}-S]$$

implies $\dim \mathscr{X} \leqslant n-1$, since the adjoinment of a 0-dimensional set cannot raise the dimension of a set more than by one (Theorem 1 of Section I).

Remark. One can also prove that every (non-empty) set open in the kernel is of dimension $\geqslant n-1$ (3). The closure of the kernel is of dimension n at each point of the kernel (4).

VI. Weakly n-dimensional spaces. If the kernel of an n-dimensional space is not of dimension n (by Theorem V, 1 it is then of dimension $(n-1)$), then the space is said to be *weakly n-dimensional*.

---

(1) A theorem of L. Tumarkin, *loc. cit.*, p. 653.

(2) Theorem of K. Menger, *Das Hauptproblem über die dimensionelle Struktur der Räume*, Proc. Akad. Amsterdam 30 (1926), p. 141.

(3) *Ibid.*

(4) W. Hurewicz, *loc. cit.*, p. 762, and L. Tumarkin, *loc. cit.*, p. 652. Comp. K. Menger, *op. cit.*, Monatsh. 34, p. 144, and P. Urysohn, *loc. cit.*, p. 270.

As an example of a weakly 1-dimensional space, let us consider the set of points $(x, f(x))$ of the plane, where x belongs to the Cantor set $\mathscr{C}$:

$$x = \frac{2}{3^{n_1}} + \frac{2}{3^{n_2}} + \ldots, \qquad n_1 < n_2 < \ldots$$

and where

$$f(x) = \frac{(-1)^{n_1}}{2} + \frac{(-1)^{n_2}}{2^2} + \ldots + \frac{(-1)^{n_k}}{2^k} + \ldots, \qquad f(0) = 0.$$

The kernel of this set is made up of points whose abcissas are end-points of an interval contiguous to $\mathscr{C}$ (1). This is a countable set, hence 0-dimensional.

It can be proved (2) that for every n there exist weakly n-dimensional sets.

VII. Dimensionalizing families. The property of a space being of dimension $\leqslant n$ at a point p is a particular case of the following property: A family $\boldsymbol{F}$ of sets will be said to *dimensionalize* the space at a point p, if p admits arbitrary small neighbourhoods whose boundaries belong to $\boldsymbol{F}$ (3).

In particular, if $\boldsymbol{F}$ is the family of sets of dimension $\leqslant n-1$, the points where the space is dimensionalized by $\boldsymbol{F}$ are those at which the space has dimension $\leqslant n$.

---

(1) See my paper *Une application des images de fonctions à la construction de certains ensembles singuliers*, Mathematica 6 (1932), p. 120. The first example of a set which has dimension 0 everywhere except for a countable set of points was given by W. Sierpiński, Fund. Math. 2 (1921), pp. 81–95.

(2) A theorem of S. Mazurkiewicz, *Sur les ensembles de dimension faible*, Fund. Math. 13 (1929), p. 212.

(3) K. Menger takes a still more general point of view: he calls p an "$\boldsymbol{E}$-point", if p has arbitrarily small neighbourhoods which have the property $\boldsymbol{E}$. See Math. Ann. 95 (1925), p. 281. The term "to dimensionalize" a space, which I owe to B. Knaster, replaces here the term "Unstetigkeitspunkt" of W. Hurewicz; the latter was used by Hurewicz in the paper *Normalbereiche und Dimensionstheorie*, Math. Ann. 96 (1927).

An analogous (but not equivalent) property to that of being a point at which the space is dimensionalized was studied by G. T. Whyburn. He considers points of the form $p = G_1 \cap G_2 \cap \ldots$, where G_n is a neighbourhood of p and $\mathrm{Fr}(G_n) \in \boldsymbol{F}$. See Fund. Math. 16 (1930), p. 169.

Let us assume that the family $\boldsymbol{F}$ satisfies the following conditions (which are fulfilled by the family of sets of dimension $\leqslant n-1$):

(i) $\boldsymbol{F}$ is *hereditary*, i.e. $X \subset Y \in \boldsymbol{F}$ implies $X \in \boldsymbol{F}$;

(ii) $\boldsymbol{F}$ is $\boldsymbol{F}_\sigma$*-additive*, i.e. the relations $X_n \in \boldsymbol{F}$, $S = X_1 \cup X_2 \cup \ldots$ and $X_n = \overline{X}_n \cap S$ imply $S \in \boldsymbol{F}$ ([1]).

As it was shown by W. Hurewicz (in his paper referred to above), a considerable part of the dimension theory can be reduced to the study of 0-dimensional sets and hereditary $\boldsymbol{F}_\sigma$-additive families. This procedure (which is more abstract than that of this book) has the advantage of also being applicable to problems which do not lie in the domain of the dimension theory.

We quote without proofs (these being quite analogous to the proofs of the corresponding theorems of § 27) some fundamental theorems on hereditary $\boldsymbol{F}_\sigma$-additive families $\boldsymbol{F}$.

1. In order that a space be dimensionalized (at each point) by $\boldsymbol{F}$ it is necessary and sufficient that it be composed of a set belonging to $\boldsymbol{F}$ and a set of dimension $\leqslant 0$.

2. In order that a space be dimensionalized by $\boldsymbol{F}$ it is necessary and sufficient that every pair of disjoint closed sets can be separated by a closed set belonging to $\boldsymbol{F}$.

3. If $\boldsymbol{F}$ and $\boldsymbol{F}_1$ are two hereditary $\boldsymbol{F}_\sigma$-additive families, then the family of the unions $X \cup Y$, where $X \in \boldsymbol{F}$ and $Y \in \boldsymbol{F}_1$ is hereditary and $\boldsymbol{F}_\sigma$-additive.

4. The family of sets dimensionalized by $\boldsymbol{F}$ is hereditary and $\boldsymbol{F}_\sigma$-additive.

5. If the space is a countable union of closed sets which all, except possibly one, are dimensionalized by $\boldsymbol{F}$, while the exceptional set is dimensionalized by $\boldsymbol{F}$ at the point p, then the entire space is also dimensionalized by $\boldsymbol{F}$ at the point p.

6. The set of points where the space is not dimensionalized by $\boldsymbol{F}$ (the "kernel" of the space) is an $\boldsymbol{F}_\sigma$-set. It does not belong to $\boldsymbol{F}$ unless it is empty.

([1]) A hereditary $\boldsymbol{F}_\sigma$-additive family is named a "Normalbereich" by W. Hurewicz.

The families satisfying conditions (i) and (ii) which also contain all sets homeomorphic to their elements, were studied by K. Kunugui in his investigations on relations between the notions of dimension and of topological rank ("dimension in the sense of Fréchet"). See Kunugui's thesis, *Sur la Théorie du nombre de dimensions*, Paris 1930, p. 41.

VIII. Dimension of the cartesian product.

THEOREM [1]. $\dim(\mathscr{X}\times\mathscr{Y}) \leqslant \dim\mathscr{X}+\dim\mathscr{Y}$.

Furthermore,

$$(\dim_a\mathscr{X} = 0 = \dim_b\mathscr{Y}) \Rightarrow \dim_{(a,b)}(\mathscr{X}\times\mathscr{Y}) = 0.$$

Proof. Put $\dim_a\mathscr{X} = m$ and $\dim_b\mathscr{Y} = n$. Hence there are two open sets R and S such that

$$a\in R, \quad \delta(R) < \varepsilon/2, \quad \dim[\mathrm{Fr}(R)] \leqslant m-1,$$

$$b\in S, \quad \delta(S) < \varepsilon/2, \quad \dim[\mathrm{Fr}(S)] \leqslant n-1.$$

By § 15, III (3) and § 21, VI (4), we have

$$\mathrm{Fr}(R\times S) = [\mathrm{Fr}(R)\times\bar{S}] \cup [\bar{R}\times\mathrm{Fr}(S)], \quad \delta(R\times S) < \varepsilon. \tag{1}$$

If $m = 0 = n$, then $\dim[\mathrm{Fr}(R)] = -1 = \dim[\mathrm{Fr}(S)]$, i.e. $\mathrm{Fr}(R) = 0 = \mathrm{Fr}(S)$; hence $\mathrm{Fr}(R\times S) = 0$. Consequently

$$\dim_{(a,b)}(\mathscr{X}\times\mathscr{Y}) = 0,$$

which completes the proof of the second part of the theorem.

Thus, the first part has been proved for $\dim\mathscr{X} = 0 = \dim\mathscr{Y}$. Let us proceed by induction.

Suppose that the theorem is true for all pairs $\mathscr{X}, \mathscr{Y}$ such that

$$\dim\mathscr{X}+\dim\mathscr{Y} < k.$$

As

$$\dim[\mathrm{Fr}(R)]+\dim\bar{S} \leqslant (m-1)+n \leqslant k-1,$$

we have by hypothesis

$$\dim[\mathrm{Fr}(R)\times\bar{S}] \leqslant k-1$$

and similarly

$$\dim[\bar{R}\times\mathrm{Fr}(S)] \leqslant k-1.$$

It follows by Theorem 2 of Section I, that

$$\dim\{[\mathrm{Fr}(R)\times\bar{S}] \cup [\bar{R}\times\mathrm{Fr}(S)]\} \leqslant k-1,$$

and by (1) that

$$\dim_{(a,b)}(\mathscr{X}\times\mathscr{Y}) \leqslant k, \quad \text{hence} \quad \dim(\mathscr{X}\times\mathscr{Y}) \leqslant k.$$

---

[1] K. Menger, *Dimensionstheorie*, p. 246.

Remark 1. In the preceding theorem the sign $\leqslant$ cannot be replaced by $=$.

For example if $\mathscr{X}$ denotes the set of points of the Hilbert space (see § 21, I, example 3) all of whose coordinates are rational, we have $\dim \mathscr{X} = 1 = \dim(\mathscr{X} \times \mathscr{X})$ (¹).

A more striking example was given by L. Pontrjagin, namely of *two compact metric spaces, each of dimension* 2 *and whose product has dimension* 3 (²).

Remark 2. If $\mathscr{X}$ is compact and $\dim \mathscr{Y} = 1$, then

$$\dim(\mathscr{X} \times \mathscr{Y}) = \dim \mathscr{X} + 1 \text{ (}^3\text{)}.$$

Remark 3. The second part of the theorem cannot be extended to $n > 0$ (⁴). One sees it easily on the product of $\mathscr{E} \times \mathscr{X}$ where $\mathscr{X}$ is composed of the point 0 and of the intervals $\big(1/(n+1), 1/n\big)$ with n odd.

IX. Continuous and one-to-one mappings of n-dimensional spaces. By Corollary 1b of § 22, II, every metric separable space is a continuous and one-to-one image of a 0-dimensional space.

This theorem can be strengthened as follows.

THEOREM (⁵). *Every metric separable space $\mathscr{X}$ of power $\mathfrak{c}$ is a continuous and one-to-one image of a metric separable space of dimension n for each $n \geqslant 0$, finite or infinite.*

Proof. 1) Consider first the case of $\dim \mathscr{X} = 0$. Let $\mathscr{Y}$ be a space of dimension n (e.g. $\mathscr{Y} = \mathscr{I}^n, n \leqslant \aleph_0$). Denote by $\boldsymbol{G}$ the family of all $\boldsymbol{G}_\delta$-sets contained in $\mathscr{X} \times \mathscr{Y}$. Obviously $\overline{\overline{\boldsymbol{G}}} = \mathfrak{c}$. Consequently we can put $\boldsymbol{G} = \{G_x\}$ where x ranges over $\mathscr{X}$ (thus G_x represents an "universal" function for the $\boldsymbol{G}_\delta$-subsets of $\mathscr{X} \times \mathscr{Y}$; see § 30, XIII).

(¹) See P. Erdös, *The dimension of rational points in Hilbert space*, Ann. of Math. 41 (1940), p. 734. Compare Hurewicz-Wallman, *Dimension Theory*, pp. 13 and 34.

(²) *Sur une hypothèse fondamentale de la théorie de la dimension*, C. R. Paris 190 (1930), p. 1105.

(³) See W. Hurewicz, *Ueber den sogenannten Produktsatz der Dimensionstheorie*, Math. Ann. 102 (1929). p. 306.

(⁴) See also K. Menger, *Bemerkungen ueber dimensionelle Feinstruktur und Produktsatz*, Prace Mat.-Fiz. 37 (1930), p. 78.

(⁵) A. Hilgers, *Bemerkung zur Dimensionstheorie*, Fund. Math. 28 (1937), p. 303.

Let $f\colon \mathscr{X} \to \mathscr{Y}$ be such that

$$[x, f(x)] \in G_x \Rightarrow \big((x)\times\mathscr{Y}\big) \subset G_x. \tag{1}$$

Namely, if $\big((x)\times\mathscr{Y}\big) \subset G_x$, we denote by $f(x)$ an arbitrary point of $\mathscr{Y}$; otherwise, $f(x)$ is a point of $\mathscr{Y}$ such that $[x, f(x)] \notin G_x$.

Since $\mathscr{X}$ is a one-to-one continuous image of the set

$$I = \underset{xy}{E}[y = f(x)],$$

it remains to show that $\dim I = n$.

By the Theorem 5 of Section IV, there is a G_δ-set Z such that

$$I \subset Z \subset \mathscr{X}\times\mathscr{Y} \quad \text{and} \quad \dim Z = \dim I. \tag{2}$$

Hence there is a x_0 such that $Z = G_{x_0}$. It follows that

$$[x_0, f(x_0)] \in G_{x_0}, \quad \text{for} \quad [x_0, f(x_0)] \in I \subset Z.$$

Therefore by (1),

$$(x_0\times\mathscr{Y}) \subset G_{x_0}, \quad \text{whence} \quad \dim\mathscr{Y} \leqslant \dim G_{x_0} \leqslant \dim(\mathscr{X}\times\mathscr{Y}),$$

but (see Section VIII) $\dim(\mathscr{X}\times\mathscr{Y}) \leqslant \dim\mathscr{X} + \dim\mathscr{Y} = 0 + n$, and consequently $\dim G_{x_0} = n$, i.e. $\dim Z = n$. Finally (by (2)) $\dim I = n$.

2) Let $\dim\mathscr{X}$ be arbitrary. According to Corollary 1b of § 22, II, there is a T of dimension 0 and $f\colon T \to \mathscr{X}$, continuous and one-to-one. As just shown, there is I of dimension n and $g\colon I \to T$ continuous and one-to-one. Thus $fg\colon I \to \mathscr{X}$ is continuous and one-to-one.

X. Remarks on the dimension theory in arbitrary metric spaces. In the §§ 25 to 27 the hypothesis of separability of the space under consideration was quite essential. It is worth noticing that some of the results established under the assumption of separability can be extended to arbitrary metric spaces (or even to normal spaces) when an appropriate definition of dimension is adopted.

There are several definitions of dimension which are equivalent for metric separable spaces. This is no longer the case when the space is not assumed to be separable. Thus one has the choice among the following definitions.

First the *inductive dimension*, denoted by $\operatorname{ind} X$, which we have adopted in § 25, I (conditions 1–3). Then the *combinatorial dimension*, denoted by $\dim X$, where $\dim X \leqslant n$ is equivalent to condition D_n (see § 27, III). Finally the "*great*" *inductive dimension* [1], denoted by $\operatorname{Ind} X$, which means that $\operatorname{Ind} 0 = -1$ and that $\operatorname{Ind} X \leqslant n$ whenever for each closed F and each open G containing F there is an open H such that

$$F \subset H, \quad \overline{H} \subset G, \quad \operatorname{Ind} \operatorname{Fr}(H) \leqslant n-1.$$

According to an important theorem of M. Katětov [2], *the identity* $\operatorname{Ind} X = \dim X$ *holds for arbitrary metric spaces*. However for more general spaces the definitions cited above may be different. For example, there is a compact X such that $\dim X = 1$ while $\operatorname{ind} X = 2$ [3].

Related to the above identity are the following inequalities [4]: $\dim X \leqslant \operatorname{ind} X$ valid for compact spaces (Alexandrov) and Lindelöf spaces (Morita, Smirnov); $\dim X \leqslant \operatorname{Ind} X$ valid for normal spaces. There exists [5]: a normal space X for which $\operatorname{ind} X < \operatorname{Ind} X$ (Smirnov), a normal space X for which $0 = \operatorname{ind} X < \operatorname{Ind} X$ (Dowker), and a metric complete space X for which $\operatorname{ind} X = 0$ and $\operatorname{Ind} X = \dim X = 1$.

---

[1] E. Čech, *Sur la dimension des espaces parfaitement normaux*, Bull. Acad. Sc. Bohème 33 (1932), pp. 38–55.

[2] See *On the dimension of metric spaces* (Russian), Dokl. Akad. Nauk URSS 79 (1951), p. 189, and *On the dimension of non-separable spaces I* (Russian), Journ. Math. Tchecoslov. 2 (1952), p. 333–368. See also K. Morita, *Normal families and dimension theory for metric spaces*, Math. Ann. 128 (1954), p. 350–362; H. Dowker and W. Hurewicz, *Dimension of metric spaces*, Fund. Math. 43 (1956), pp. 83–87.

[3] See A. Lunz, Dokl. Akad. Nauk URSS 66 (1949), p. 801. Compare O. Lokutziewskii, *On the dimension of bicompacta* (Russian), *ibid.*, 67 (1949), p. 217, P. Vopenka, *On the dimension of compact spaces*, Journ. Math. Tchecoslov. 8 (1958), pp. 319–326.

[4] P. Alexandrov, *The sum theorem in dimension theory of bicompacta* (Russian), Soobšč. Akad. Nauk Gruz. SSR 2 (1941), pp. 1–6, K. Morita, *On the dimension of normal spaces I*, Jap. Journ. of Math. 20 (1950), pp. 5–36, Yu. M. Smirnov, *Some relations in dimension theory* (Russian), Mat. Sbornik 29 (1951), pp. 157–172, N. B. Vedenisov, *Sur la dimension au sens de E. Čech*, Izv. Akad. Nauk 5 (1941), pp. 211–216.

[5] Yu. M. Smirnov, *loc. cit.*, C. H. Dowker, *Local dimension of normal spaces*, Quart. Journ. 6 (1955), pp. 101–120, P. Roy, *Failure of equivalence of dimension concepts for metric spaces*, Bull. Amer. Math. Soc. 68 (1962), pp. 609–613.

Let us add that there are some fundamental theorems of the dimension theory which do not admit generalizations of that kind. Such as, for example, the theorem stating that the dimension of a subset is never greater than the dimension of the space; it is false for the combinatorial dimension of compact spaces [1] (however, it holds—according to a theorem of Čech [2]—in perfectly normal spaces).

§ 28. Simplexes, complexes, polyhedra

I. Definitions. If $p_0, \ldots, p_n$ is a set of points of euclidean space (of arbitrary dimension), then by the (geometrical open) *simplex* $p_0 \ldots p_n$ we mean the set of points p of the form

$$p = \lambda_0 p_0 + \ldots + \lambda_n p_n, \quad \text{where} \quad \lambda_0 + \ldots + \lambda_n = 1$$
$$\text{and } \lambda_i > 0 \ (i = 0, 1, \ldots, n) \tag{1}$$

and where the points p and p_i are regarded as vectors [3].

---

(1) See P. Alexandrov, *On the dimension of normal spaces*, Proc. R. Soc. A. 189 (1947), p. 26, remark. See also C. H. Dowker, Quart. Journ. 6 (1955).

For various other generalizations of the dimension theory, see: P. Alexandrov, *Contemporary dimension theory*, Uspiechi 6 (1951), p. 43–68, and *ibid.* 15 (1960), p. 78–90; C. H. Dowker, *Inductive dimension of completely normal spaces*, Quart. Journ. 4 (1953), p. 267–281; K. Morita, *On the dimension of normal spaces*, Journ. Math. Soc. Japan 2 (1950), pp. 16–33. B. A. Pasynkov, Dokl. Akad. Nauk, URSS 121 (1958), p. 45; U. V. Proskuriakov, *On the dimension theory of topological spaces* (Russian), Proceed. Univ. Mosk. 148 (1951), p. 219–223; Yu. Smirnov, *On the dimension of proximity spaces* (Russian), Matem. Sb. 38 (1956), p. 283–302; N. B. Vedenisov, *Remarques sur la dimension des espaces topologiques*, Uč. Zapiski Mosk. Univ. 30 (1939), pp. 131–140.

(2) E. Čech, *Contribution à la theorie de la dimension*, Čas. pro Pest. Mat. a Fys. 62 (1933), pp. 277–291.

(3) Given two vectors

$$p = (x_1, \ldots, x_k) \quad \text{and} \quad r = (y_1, \ldots, y_k)$$

in the euclidean space $\mathcal{E}^k$, then, by definition,

$$p + r = (x_1 + y_1, \ldots, x_k + y_k).$$

If λ is a real number, then $\lambda p = (\lambda x_1, \ldots, \lambda x_k)$.

The points $p_0, \ldots, p_n$ are called the *vertices* of the simplex.

The coefficients $\lambda_0, \ldots, \lambda_n$ are the *barycentric coordinates* of the point p with respect to $p_0, \ldots, p_n$ (1).

A simplex of the form $p_{i_0} \ldots p_{i_k}$, where $0 \leqslant k \leqslant n$, is called a *face* of the simplex $S = p_0 \ldots p_n$. Clearly $\bar{S}$ is the union of all faces $p_{i_0} \ldots p_{i_k}$ of S and is made up of points satisfying the condition obtained from (1) by replacing the inequality $\lambda_i > 0$ by $\lambda_i \geqslant 0$.

The set $\bar{S}$ can also be defined to be the smallest *convex* set containing the points $p_0, \ldots, p_n$ (a set is convex if, together with every pair of points, it contains the segment joining them).

If all faces of the simplex S are disjoint, then the simplex is called *non-degenerate*. This condition is equivalent to the *linear independency* of the vertices of S. We recall that the point p is linearly dependent on the points $p_0, \ldots, p_n$, if there exist numbers $\lambda_0, \ldots, \lambda_n$ satisfying the two equalities of (1). In other words, the set of points which depend linearly on $p_0, \ldots, p_n$ is called the *linear manifold* spanned by these points; a simplex is non-degenerate, if none of its vertices belongs to the linear manifold spanned by the remaining vertices.

The number n is called the *geometrical dimension* of a non-degenerate simplex with $(n+1)$ vertices. It is also the geometrical dimension of the linear manifold spanned by $(n+1)$ linearly independent points. We shall see (in Section II) that the geometrical dimension of a simplex coincides with its topological dimension (in the sense of § 25).

In particular, a simplex of dimension 0 is a single point. A (non-degenerate) simplex $p_0 p_1$ is a straight line segment without the end-points; $p_0 p_1 p_2$ is the interior of a triangle (provided that it is non-degenerate).

THEOREM 1. *Every point p of a degenerate simplex $S = p_0 \ldots p_n$ lies on a face $p_{i_0} \ldots p_{i_k}$ with $k < n$* (2).

Proof. Since the simplex S is degenerate, there exist real numbers $\mu_0, \ldots, \mu_n$, not all zero, which satisfy the conditions

$$\mu_0 p_0 + \ldots + \mu_n p_n = 0 \quad \text{and} \quad \mu_0 + \ldots + \mu_n = 0.$$

---

(1) The point p is the centre of gravity of the system $p_0, \ldots, p_n$, when the point p_i has the mass λ_i.

(2) Alexandrov–Hopf, *Topologie I*, p. 607.

We can assume that $\mu_n > 0$ and that λ_n/μ_n is the smallest of the numbers λ_i/μ_i, with μ_i positive ($\lambda_0, \ldots, \lambda_n$ satisfy (1)). Let

$$\lambda_i^* = \lambda_i - \frac{\lambda_n}{\mu_n}\mu_i, \quad \text{for} \quad i = 0, 1, \ldots, n.$$

Then

$$p = \lambda_0^* p_0 + \ldots + \lambda_n^* p_n, \quad \lambda_0^* + \ldots + \lambda_n^* = 1, \quad \text{and} \quad \lambda_i^* \geqslant 0.$$

Since $\lambda_n^* = 0$, hence $p \in \overline{p_0 \ldots p_{n-1}}$, q.e.d.

If we assume that k, in Theorem 1, is the smallest possible integer, we deduce that *every point of the simplex S lies on a non-degenerate face of S*. It follows also that

$$\overline{S} = F_0 + \ldots + F_m, \tag{2}$$

where $F_0, \ldots, F_m$ are *non-degenerate* faces of S.

A (finite or infinite) sequence of points $p_0, p_1, \ldots$ of the space $\mathscr{E}^r$ is said to be *in a general position* (in the space), if every system $p_{i_0}, \ldots, p_{i_k}$ with $i_0 < i_1 < \ldots < i_k$, where $k \leqslant r$, is linearly independent.

THEOREM 2. *Given a sequence of points $a_0, a_1, \ldots$ in the space $\mathscr{E}^r$ (or in the cube $\mathscr{I}^r$) and a sequence of positive numbers $\varepsilon_0, \varepsilon_1, \ldots$, there exists a sequence of points $p_0, p_1, \ldots$ in a general position such that*

$$|p_i - a_i| < \varepsilon_i \quad \textit{for} \quad i = 0, 1, \ldots. \tag{4}$$

Proof by induction. Let $p_0 = a_0$. In order to find p_i, consider all the linear manifolds spanned by the systems $p_{j_0}, \ldots, p_{j_k}$ with $j_0 < \ldots < j_k < i$ and $k < r$. Each manifold being nowhere dense, their union is nowhere dense. Let p_i be a point not in this union which satisfies the condition $|p_i - a_i| < \varepsilon_i$.

Remarks. (i) *If $r = \aleph_0$, then we can assume that the simplexes $p_{i_0} \ldots p_{i_k}$ are non-degenerate (hence disjoint), for every k.*

(ii) *If $r \geqslant 2n+1$ and if L is an n-dimensional linear manifold, then the points $p_0, p_1, \ldots$ can be subjected to the additional condition that the simplexes $p_{i_0} \ldots p_{i_l}$ (where $l \leqslant n$) are disjoint to L.*

To prove this, we denote by $p_{-n-1}, \ldots, p_{-1}$ linearly independent points which span L and find p_i for $i \geqslant 0$ as before.

(iii) *The points $p_0, p_1, \ldots$ can be subjected to the following condition:*

Given $l+1$ linear manifolds $V_0, \ldots, V_l$ spanned by disjoint systems formed, respectively, by $k_0+1, \ldots, k_l+1$ terms of the sequence

$p_0, p_1, \ldots,$ *the geometrical dimension of the linear manifold* $V_0 \cap \ldots \cap V_l$ *(assumed to be non empty) satisfies the relation*

$$\dim(V_0 \cap \ldots \cap V_l) = k_0 + \ldots + k_l - lr. \tag{3}$$

In particular, if $r \geqslant 2n+1$, *then the simplexes* $p_{i_0} \ldots p_{i_k}$ *with* $k \leqslant n$ *are mutually disjoint.*

To find p_i, we consider, together with the manifolds given above, all their intersections as well as the manifolds $\neq \mathcal{E}^r$ spanned by any two of these intersections. We choose p_i outside the union of these manifolds, at a distance $< \varepsilon_i$ from a_i (1).

A *complex* is a finite family of simplexes. The *closure* $\overline{C}$ of a complex C is the complex composed of all the faces of simplexes belonging to C. We say that the complex C is *closed*, if $C = \overline{C}$, i.e. if, together with each simplex $p_0 \ldots p_n$, it contains all the simplexes $p_{i_0} \ldots p_{i_k}$, where $i_0 < \ldots < i_k \leqslant n$.

C is said to be *non-degenerate*, if the simplexes of C are disjoint. In particular, the complex made up by all faces of a non-degenerate simplex is non-degenerate.

The maximum dimension of the simplexes of C is called the *dimension of* C.

A (closed) *polyhedron* C is a set such that there exists a closed non-degenerate complex $C = (R_1, \ldots, R_m)$ with $C = R_1 \cup \ldots \cup R_m$.

A polyhedron is clearly compact.

Every polyhedron is homeomorphic to a polyhedron formed by the union of certain faces of a non-degenerate simplex.

For, if C is a non-degenerate closed complex with vertices $q_0, \ldots, q_m$ and if S is an m-dimensional simplex $p_0 \ldots p_m$, then the union of all the simplexes $p_{i_0} \ldots p_{i_k}$ such that $q_{i_0} \ldots q_{i_k} \in C$ is a polyhedron homeomorphic to the polyhedron $C = S(C)$.

II. Topological dimension of a simplex. Let $S = p_0 \ldots p_n$ be a non-degenerate simplex and let $C = (R_1, \ldots, R_m)$ be a closed non-degenerate complex with

$$\overline{S} = R_1 \cup \ldots \cup R_m. \tag{1}$$

(1) In the proof we use the fact that the dimension of the manifold spanned by the manifolds U and V satisfying $U \cap V \neq 0$ is $\dim U + \dim V - \dim U \cap V$. *Ibidem*, p. 596.

Thus, for example, if $n = 2$, i.e. if S is a triangle, and if we join each vertex of the triangle with the centre of the opposite side, we obtain a decomposition of S into six triangles (the barycentric decomposition). More precisely, the complex $\boldsymbol{C}$ in this case is composed of six 2-dimensional simplexes, twelve 1-dimensional simplexes and seven individual points (hence $m = 25$).

We easily deduce the following three theorems.

THEOREM 1. *If F is a face of S, then, for each $j \leqslant m$, either $R_j \subset F$ or $R_j \cap F = 0$.*

THEOREM 2. *$\overline{F}$ is the union of simplexes belonging to the complex composed of all the R_j with $R_j \subset \overline{F}$.*

THEOREM 3. *For every $\varepsilon > 0$ there exists a closed non-degenerate complex $\boldsymbol{C} = (R_1, \ldots, R_m)$ satisfying* (1) *and such that $\delta(R_j) < \varepsilon$ for $j = 1, 2, \ldots, m$.*

THEOREM 4 (of Sperner) (1). *Let the function $\nu(s)$ assign to each vertex s of $\boldsymbol{C}$ a positive integer such that*

$$\text{if } s \in p_{i_0} \ldots p_{i_k}, \text{ then } \nu(s) \text{ is one of the indices } i_0, \ldots, i_k. \tag{i}$$

Let $\nu(R)$ be the set $[\nu(q_0), \ldots, \nu(q_l)]$, where $R = q_0 \ldots q_l$. Then the number ϱ of the R_j for which $\nu(R_j) = (0, 1, \ldots, n)$ is odd.

Proof by induction. The assertion is evident for $n = 0$, since then $S = p_0 = R$ and $\nu(R) = 0$; hence $\varrho = 1$.

Suppose that the assertion is true for $n-1$. We shall prove it for n.

Let $\boldsymbol{F}$ be the family of $(n-1)$-dimensional simplexes $F \in \boldsymbol{C}$ such that

$$\nu(F) = (0, 1, \ldots, n-1)$$

and let σ be the number of the simplexes F such that $F \in \boldsymbol{F}$ and $F \subset p_0 \ldots p_{n-1}$. By hypothesis and by Theorem 2, σ is odd.

Let us enumerate the simplexes R_j so that $R_1, R_2, \ldots, R_t$ are of dimension n (in the geometrical sense) and $R_{t+1}, \ldots, R_m$ are of dimension $< n$.

---

(1) See E. Sperner, Abh. Math. Seminar Hamburg, 6 (1928), p. 265, and the paper of B. Knaster, S. Mazurkiewicz and myself, Fund. Math. 14 (1929), p. 132.

Let a_j, for $j \leqslant t$, be the number of faces F of R_j such that $F \in \boldsymbol{F}$. One sees easily that

(i) if $\nu(R_j) = (0, \ldots, n)$, then $a_j = 1$;

(ii) if $(0, \ldots, n-1) \subset \nu(R_j) \neq (0, \ldots, n)$, then $a_j = 2$;

(iii) if $(0, \ldots, n-1) \not\subset \nu(R_j)$, then $a_j = 0$.

It follows that

$$\varrho \equiv (a_1 + \ldots + a_t) \pmod 2. \tag{2}$$

If to each R_j with $j \leqslant t$ we assign faces belonging to $\boldsymbol{F}$ (provided they exist), then every $F \in \boldsymbol{F}$ will be assigned to one or two R_j depending on whether F is or is not contained in the simplex $p_0 \ldots p_{n-1}$. Therefore, by (2),

$$a_1 + \ldots + a_t \equiv \sigma \pmod 2; \quad \text{hence} \quad \varrho \equiv \sigma \pmod 2.$$

It follows that the number ϱ is odd.

Remark. In applications of Theorem 4 only the fact $\varrho \neq 0$ will be used. The fact that ϱ is odd is used only in the proof of the theorem.

THEOREM 5. *If $A_0, \ldots, A_n$ are $(n+1)$ closed sets such that each face $p_{i_0} \ldots p_{i_k}$ $(0 \leqslant k \leqslant n)$ of the simplex S satisfies*

$$p_{i_0} \ldots p_{i_k} \subset A_{i_0} \cup \ldots \cup A_{i_k}, \tag{3}$$

then $A_0 \cap \ldots \cap A_n \neq 0$.

Proof. Let r be a positive number. According to Theorem 3 we can assume that the complex $\boldsymbol{C} = (R_1, \ldots, R_m)$ satisfies condition (1) and the inequality

$$\delta(R_j) < 1/r \quad \text{for} \quad j = 1, \ldots, m. \tag{4}$$

To each vertex s of the complex $\boldsymbol{C}$ we assign an integer $\nu(s)$ as follows. By (1), there exists a (unique) simplex $p_{i_0} \ldots p_{i_k}$ containing s. By (3), there exists at least one set $A_{i_0}, \ldots, A_{i_k}$ containing s; let $\nu(s)$ be its index. Therefore $s \in A_{\nu(s)}$. By Theorem 4, since condition (i) is fulfilled, there exists a simplex R_i—which we now denote by $s_o^r \ldots s_n^r$—such that $\nu(s_i^r) = i$. Therefore $s_i^r \in A_i$.

Since $\bar{S}$ is compact, we can assume that the sequence $s_0^1, s_0^2, \ldots, s_0^r, \ldots$ is convergent. Let a be its limit. Condition (4) implies that $a = \lim\limits_{r=\infty} s_i^r$ for $i = 1, \ldots, n$, and the formula $s_i^r \in A_i$ implies $a \in \bar{A}_i = A_i$. Hence $a \in A_0 \cap \ldots \cap A_n$.

Let Q_i be the face of S opposite to p_i, i.e.

$$Q_i = p_0 \dots p_{i-1} p_{i+1} \dots p_n. \tag{5}$$

THEOREM 6. *If $A_0, \dots, A_n$ are $(n+1)$ closed sets such that*

$$\bar{S} = A_0 \cup \dots \cup A_n, \tag{6}$$

$$A_i \cap \bar{Q}_i = 0, \tag{7}$$

then $A_0 \cap \dots \cap A_n \neq 0$.

Moreover, *the hypothesis that the sets A_i are closed can be replaced by the hypothesis that they are open in $\bar{S}$.*

Proof. According to Theorem 5, inclusion (3) is to be proved. Suppose that it does not hold. Then there exists a point a of the simplex $p_{i_0} \dots p_{i_k}$ which is not contained in any of the sets A_{i_0}, $\dots, A_{i_k}$. By (6) there exists an index i such that

$$i \neq i_0, \dots, i \neq i_k, \tag{8}$$

$$a \in A_i. \tag{9}$$

(8) implies that $p_{i_0} \dots p_{i_k} \subset \bar{Q}_i$; hence $a \in \bar{Q}_i$. But the last formula is inconsistent with (7) and (9).

To prove the second part of Theorem 6, suppose that the sets $A_0, \dots, A_n$ are open in $\bar{S}$ and, according to § 14, III, Corollary, denote by $A_0^*, \dots, A_n^*$ closed sets with

$$\bar{S} = A_0^* \cup \dots \cup A_n^* \quad \text{and} \quad A_i^* \subset A_i;$$

$$\text{hence} \quad A_i^* \cap \bar{Q}_i = 0 \ (0 \leqslant i \leqslant n).$$

Since $A_0^* \cap \dots \cap A_n^* \neq 0$, it follows that $A_0 \cap \dots \cap A_n \neq 0$.

THEOREM 7 (fundamental theorem) (1). $\dim \bar{S} = n$.

In other words, $\dim \mathscr{E}^n = n$.

The inequality $\dim \bar{S} \leqslant n$ (or $\dim \mathscr{E}^n \leqslant n$) is obvious (compare § 25, I, examples). We have to prove that $\dim \bar{S} > n-1$.

Suppose that $\dim \bar{S} \leqslant n-1$. Let

$$P_i = \bar{S} - \bar{Q}_i, \tag{10}$$

where Q_i is defined by (5).

---

(1) See L. E. J. Brouwer, *Über den natürlichen Dimensionsbegriff*, Journ. f. Math. 142 (1913), p. 146.

Clearly

$$\bar{S} = P_0 \cup \ldots \cup P_n. \tag{11}$$

By virtue of the inequality $\dim \bar{S} \leqslant n-1$ and Theorem 1 of § 27, III (decomposition theorem), there exist open sets $A_0, \ldots, A_n$ with

$$\bar{S} = A_0 \cup \ldots \cup A_n, \quad A_0 \cap \ldots \cap A_n = 0, \quad \text{and} \quad A_i \subset P_i;$$

$$\text{hence} \quad A_i \cap \bar{Q}_i = 0.$$

But this contradicts Theorem 6.

COROLLARY 7a. *If C is a non-degenerate closed complex of dimension n and C is the polyhedron* $\mathrm{S}(C)$, *then* $\dim C = n$.

COROLLARY 7b. *If $p_0 \ldots p_n$ is a simplex (degenerate or not), then* $\dim p_0 \ldots p_n \leqslant n$.

The second corollary follows from Theorem 7 and formula I (2).

The following theorem related to Theorems 5 and 6 will be applied later.

THEOREM 8. *If $G_1, \ldots, G_n$ are n sets open in $\bar{S}$ with $G_i \cap Q_i = 0$ $(1 \leqslant i \leqslant n)$ such that the union $G_1 \cup \ldots \cup G_n$ separates $\bar{S}$ between the vertex p_0 and the opposite face Q_0, then $G_1 \cap \ldots \cap G_n \neq 0$* (1).

Proof. By hypothesis (compare § 6, V) there exist two closed sets U and V with

$$\bar{S} - (G_1 \cup \ldots \cup G_n) = U \cup V, \quad U \cap V = 0,$$

$$p_0 \in U, \quad \bar{Q}_0 \subset V. \tag{12}$$

According to the normality of $\bar{S}$, let A_0 be a set open in $\bar{S}$ such that

$$U \subset A_0 \quad \text{and} \quad V \subset \bar{S} - \bar{A}_0. \tag{13}$$

For $i > 0$ let

$$A_i = G_i \cup (\bar{S} - \bar{A}_0 - \bar{Q}_i). \tag{14}$$

It follows that (for $i > 0$) $A_i \cap \bar{Q}_i = G_i \cap \bar{Q}_i = 0$, since G_i is open and $G_i \cap Q_i = 0$ by hypothesis. Also, by the last inclusions of (12) and (13), $A_0 \cap \bar{Q}_0 = 0$. Therefore (7) holds for every $i = 0, 1, \ldots, n$.

---

(1) See my note in Ann. Soc. Pol. Math. 16 (1937), p. 219.

On the other hand,

$$A_0 \cup \ldots \cup A_n = A_0 \cup (G_1 \cup \ldots \cup G_n) \cup \big(\bar{S} - \bar{A}_0 - (\bar{Q}_1 \cap \ldots \cap \bar{Q}_n)\big)$$

and clearly $\bar{Q}_1 \cap \ldots \cap \bar{Q}_n = p_0$. By (12) and (13)

$$p_0 \in A_0 \quad \text{and} \quad \bar{S} - (G_1 \cup \ldots \cup G_n) \subset A_0 \cup (\bar{S} - \bar{A}_0).$$

Thus (6) is fulfilled.

Since $A_0 \cap (\bar{S} - A_0 - \bar{Q}_i) = 0$, it follows by Theorem 6 that

$$A_0 \cap \ldots \cap A_n \neq 0 \quad \text{and} \quad G_1 \cap \ldots \cap G_n \neq 0.$$

III. Applications to the fixed point problem. Theorem 5 of Section II enables us to prove easily the following fundamental theorem.

THEOREM 1. (Brouwer fixed point theorem.) [1] *Let $S = p_0 \ldots p_n$ be a non-degenerate simplex and let f be a continuous mapping of $\bar{S}$ onto a subset of $\bar{S}$. Then the mapping f has a fixed point, i.e. there exists a point x_0 with $f(x_0) = x_0$.*

Proof. Let

$$x = \lambda_0 p_0 + \ldots + \lambda_n p_n, \quad \text{where} \quad \lambda_i \geqslant 0 \quad \text{and} \quad \lambda_0 + \ldots + \lambda_n = 1,$$

be a point of $\bar{S}$ and let

$$f(x) = \lambda_0^* p_0 + \ldots + \lambda_n^* p_n,$$
$$\text{where} \quad \lambda_i^* \geqslant 0 \quad \text{and} \quad \lambda_0^* + \ldots + \lambda_n^* = 1.$$

We have to prove that there exists an x such that $\lambda_i^* = \lambda_i$ for $i = 0, \ldots, n$.

Now the sets $A_i = \underset{x}{E}(\lambda_i^* \leqslant \lambda_i)$ satisfy the hypotheses of Theorem 5 of Section II:

(i) they are closed, for the barycentric coordinates λ_i of x are continuous functions of x;

(ii) the condition $x \in p_{i_0} \ldots p_{i_k}$ implies $\lambda_{i_0} + \ldots + \lambda_{i_k} = 1$ and since $\lambda_{i_0}^* + \ldots + \lambda_{i_k}^* \leqslant 1$, there exists an index i_j $(0 \leqslant j \leqslant k)$ such that $\lambda_{i_j}^* \leqslant \lambda_{i_j}$, and hence $x \in A_{i_j}$. Thus $p_{i_0} \ldots p_{i_k} \subset A_{i_0} \cup \ldots \cup A_{i_k}$.

---

[1] For the proof, see the paper of B. Knaster, S. Mazurkiewicz and myself, *Ein Beweis des Fixpunktsatzes für n-dimensionale Simplexe*, Fund. Math. 14 (1929), p. 132.

According to Theorem 5 of Section II, let $x_0 \in A_0 \cap \ldots \cap A_n$. Then $\lambda_i^* \leqslant \lambda_i$ for $i = 0, \ldots, n$ and

$$1 = \lambda_0^* + \ldots + \lambda_n^* \leqslant \lambda_0 + \ldots + \lambda_n = 1.$$

Therefore $\lambda_i^* = \lambda_i$ for $i = 0, \ldots, n$.

COROLLARY 1a. *Let the sphere $\mathscr{S}_n$ be the surface of the solid ball*

$$\mathscr{Q}_{n+1} = \underset{p}{E}[(|p| \leqslant 1)(p \in \mathscr{E}^{n+1})].$$

There exists no continuous mapping f: $\mathscr{Q}_{n+1} \to \mathscr{S}_n$ which is the identity on $\mathscr{S}_n$.

In other words, *$\mathscr{S}_n$ is not a retract of $\mathscr{Q}_{n+1}$.*

Proof. Suppose that there did exist a mapping $f \in \mathscr{S}_n^{\mathscr{Q}_{n+1}}$ such that $f(x) = x$ for $x \in \mathscr{S}_n$.

Since $\mathscr{Q}_{n+1}$ is homeomorphic to the closure of an $(n+1)$-dimensional simplex, by Theorem 1 the function $-f$ has a fixed point x_0, i.e.

$$x_0 = -f(x_0); \quad \text{hence} \quad |x_0| = |-f(x_0)| = 1; \quad \text{therefore} \quad x_0 \in \mathscr{S}_n.$$

But this contradicts our assumption.

IV. Applications to the cubes $\mathscr{I}^n$ and $\mathscr{I}^{\aleph_0}$. Let V_i and W_i, $i = 1, \ldots, n$, be, respectively, the opposite faces $x_i = 0$ and $x_i = 1$ of the cube $\mathscr{I}^n$ (or $\mathscr{I}^{\aleph_0}$).

The following theorem corresponds to Theorem 6 of Section II (and it can be derived from it).

THEOREM 1 [1]. *Let $A_0, \ldots, A_n$ be $(n+1)$ closed sets such that*

$$\mathscr{I}^n = A_0 \cup \ldots \cup A_n, \quad A_{i-1} \cap W_i = 0, \quad A_k \cap V_i = 0 \quad (i \leqslant k).$$

Then $A_0 \cap \ldots \cap A_n \neq 0$.

Proof. Let $S = p_0 \ldots p_n$ be an n-dimensional simplex. Let

$$f(x_1, \ldots, x_n) = \lambda_0 p_0 + \ldots + \lambda_n p_n,$$

$$\lambda_0 = 1 - x_1, \quad \lambda_1 = (1 - x_2)x_1, \quad \lambda_2 = (1 - x_3)x_1 x_2, \ \ldots,$$
$$\lambda_{n-1} = (1 - x_n)x_1 \cdot \ldots \cdot x_{n-1}, \quad \lambda_n = x_1 \cdot \ldots \cdot x_n.$$

---

[1] Compare H. Lebesgue, Fund. Math. 2 (1921), p. 256, and W. Hurewicz, Math. Ann. 101 and Proc. Akad. Amsterdam 31 (1928), p. 917.

Clearly $f(\mathscr{I}^n) = \bar{S}$. Moreover, the boundary B of S and the boundary $V_1 \cup W_1 \cup \ldots \cup V_n \cup W_n$ of $\mathscr{I}^n$ mutually correspond to each other. For if $x_i = 0$, then $\lambda_i = 0$, if $x_i = 1$, then $\lambda_{i-1} = 0$, and conversely, if $\lambda_i = 0$, then one of the equalities

$$x_1 = 0, \ \ldots, \ x_i = 0, \ x_{i+1} = 1$$

holds; this means that (Q_i being defined by II (5))

$$f(V_i) \subset \bar{Q}_i, \quad f(W_i) \subset \bar{Q}_{i-1}, \quad f^{-1}(\bar{Q}_i) \subset V_1 \cup \ldots \cup V_i \cup W_{i+1}.$$

Finally the function f is one-to-one on the interior of $\mathscr{I}^n$, for the condition

$$x_1 \cdot \ldots \cdot x_{i+1} = 1 - (\lambda_0 + \ldots + \lambda_i)$$

implies

$$x_{i+1} = \frac{1 - (\lambda_0 + \ldots + \lambda_i)}{1 - (\lambda_0 + \ldots + \lambda_{i-1})}.$$

Let $A_i^* = f(A_i)$. Then $\bar{S} = A_0^* \cup \ldots \cup A_n^*$ and

$$A_i^* \cap \bar{Q}_i = f[A_i \cap f^{-1}(\bar{Q}_i)]$$
$$\subset f[A_i \cap (V_1 \cup \ldots \cup V_i \cup W_{i+1})] = 0.$$

$A_i^* = \overline{A_i^*}$ (compare § 20, V, Theorem 5). By II, Theorem 6

$$A_0^* \cap \ldots \cap A_n^* \neq 0.$$

Since $A_i^* \cap Q_i = 0$, hence $A_0^* \cap \ldots \cap A_n^* \cap B = 0$ and, the function $f^{-1} | S$ being one-to-one, it follows that $A_0 \cap \ldots \cap A_n \neq 0$.

THEOREM 2. *If $F_1, \ldots, F_n$ and $H_1, \ldots, H_n$ are two systems of closed subsets of $\mathscr{I}^n$ such that*

$$\mathscr{I}^n = F_i \cup H_i \quad \textit{and} \quad F_i \cap V_i = 0 = H_i \cap W_i,$$

then

$$(F_1 \cap H_1) \cap \ldots \cap (F_n \cap H_n) \neq 0.$$

Proof. Let

$$A_0 = H_1, \quad A_1 = F_1 \cap H_2, \ \ldots,$$
$$A_{n-1} = F_1 \cap F_2 \cap \ldots \cap F_{n-1} \cap H_n, \quad A_n = F_1 \cap \ldots \cap F_n,$$

and let $G_i = \mathscr{I}^n - H_i$. Then $G_i \subset F_i$ and

$$A_0 \cup \ldots \cup A_n \supset H_1 \cup (G_1 \cap H_2) \cup \ldots \cup (G_1 \cap \ldots \cap G_{n-1} \cap H_n) \cup$$
$$\cup (G_1 \cap \ldots \cap G_n) = \mathscr{I}^n.$$

The hypotheses of Theorem 1 are fulfilled. Therefore

$$(F_1 \cap H_1) \cap \ldots \cap (F_n \cap H_n) = A_0 \cap \ldots \cap A_n \neq 0.$$

Theorem 2 has the following generalization.

THEOREM 3. *If $F_1, F_2, \ldots$ and $H_1, H_2, \ldots$ are two infinite sequences of closed subsets of $\mathscr{I}^{\aleph_0}$ such that*

$$\mathscr{I}^{\aleph_0} = F_n \cup H_n \quad \text{and} \quad F_n \cap V_n = 0 = H_n \cap W_n,$$

then

$$\bigcap_{n=1}^{\infty} F_n \cap H_n \neq 0.$$

Proof. Let J_n be the set of points $x = [x_1, x_2, \ldots]$ of $\mathscr{I}^{\aleph_0}$ such that $x_{n+1} = x_{n+2} = \ldots = 0$. Obviously the sets J_n and $\mathscr{I}^n$ are homeomorphic and, by this homeomorphism, $V_i \cap J_n$ corresponds for $i \leqslant n$ to the set of x with $x_i = 0$, and conversely. Similarly, the set $W_i \cap J_n$ corresponds to the set of x with $x_i = 1$.

Apply Theorem 2 to J_n. This gives $J_n \subset F_i \cup H_i$, $i \leqslant n$. Hence $(F_1 \cap H_1) \cap \ldots \cap (F_n \cap H_n) \neq 0$ and, by the Cantor theorem (§ 20, V, 2), $\bigcap_{n=1}^{\infty} F_n \cap H_n \neq 0$.

THEOREM 4 (of Hurewicz) (1). *The space $\mathscr{I}^{\aleph_0}$ cannot be represented as a (countable) series of 0-dimensional sets (or, more generally, as a series of sets of finite dimension).*

Proof. Let $Q_1, Q_2, \ldots$ be an infinite sequence of 0-dimensional sets contained in $\mathscr{I}^{\aleph_0}$. We have to prove that

$$\mathscr{I}^{\aleph_0} - \bigcup_{n=1}^{\infty} Q_n \neq 0.$$

The sets V_n and W_n are closed and disjoint. Therefore by § 27, II, Theorem 1, there exist closed sets F_n and H_n with

$$\mathscr{I}^{\aleph_0} = F_n \cup H_n, \quad F_n \cap V_n = 0 = H_n \cap W_n$$
$$\text{and} \quad F_n \cap H_n \cap Q_n = 0.$$

(1) *Über unendlich-dimensionale Punktmengen*, Proc. Akad. Amsterdam 31 (1928), p. 916.

Hence $F_n \cap H_n \subset \mathcal{I}^{\aleph_0} - Q_n$. By Theorem 3 it follows that

$$0 \neq \bigcap_{n=1}^{\infty} F_n \cap H_n \subset \bigcap_{n=1}^{\infty} (\mathcal{I}^{\aleph_0} - Q_n) = \mathcal{I}^{\aleph_0} - \bigcap_{n=1}^{\infty} Q_n.$$

Remarks. (i) Let A_n be the set of points

$$x = [x_1, x_2, \ldots, x_n, 0, 0, 0, \ldots]$$

of $\mathcal{I}^{\aleph_0}$ such that $x_i \leqslant 1/n$ for $i \leqslant n$.

The set $A_0 \cup A_1 \cup \ldots$ is compact, its dimension is infinite, the sets A_n, however, are of a finite dimension [1].

(ii) The continuum hypothesis implies the following theorem [2]:

An (uncountable) space $\mathcal{X}$ which is not a countable union of 0-dimensional subsets contains an uncountable subset every uncountable subset of which is of infinite dimension.

Proof. The family of G_δ-sets is of power $\mathfrak{c}$ by § 24, I, Theorem 1 and § 26, V, Theorem 1. Assuming the continuum hypothesis we order all G_δ-sets of a finite dimension into a transfinite sequence $Q_0, Q_1, \ldots, Q_a, \ldots$ of type Ω. Since each one-point set is a term of this sequence, hence $\mathcal{X} = Q_0 \cup \ldots \cup Q_a \cup \ldots$. On the other hand, every finite dimensional set is a finite union of 0-dimensional sets (§ 27, I, Theorem 3). It follows that no countable union of terms of the sequence $\{Q_a\}$ fills up the entire space $\mathcal{X}$. Consequently there exists an uncountable set of ordinals a such that the set $D_a = Q_a - \bigcup_{\xi<a} Q_\xi \neq 0$. By selecting a point $p_a \epsilon D_a$ we obtain the required set $P = \{p_a\}$.

---

(1) For further theorems concerning spaces of infinite dimension, see a report of Yu. Smirnov, *On dimensional properties of infinite-dimensional spaces*, Proceed. of the Symposium on general topology, Prague, 1961, p. 334, and of the same author *On transfinite dimension*, Mat. Sb. 58 (1962), p. 415. See also in particular: B. Levshenko, *On spaces of infinite dimension*, Dokl. Akad. Nauk URSS 139 (1961), p. 286, E. Skljarenko, *Some remarks on spaces of infinite dimension*, *ibid.* 126 (1959), p. 1200, and *Two theorems on infinitely dimensional spaces*, *ibid.* 143 (1962), p. 1053. C. H. Toulmin, *Shufling ordinals and transfinite dimension*, Proc. London Math. Soc. 3 (1954), pp. 177–196, A. Zareula, Dokl. Akad. Nauk URSS 141 (1961).

(2) Theorem of W. Hurewicz, *Une remarque sur l'hypothèse du continu*, Fund. Math. 19 (1932), p. 8. As Hurewicz observed, the continuum hypothesis is equivalent to the hypothesis that the cube $\mathcal{I}^{\aleph_0}$ satisfies the assertion of the theorem.

To prove this, we notice that the set $P \cap Q_\xi$ is countable for every $\xi < \Omega$, since $p_\alpha \notin Q_\xi$ if $\xi < \alpha$. Hence if $X \subset P$, then $X \cap Q_\xi$ is countable. If we suppose that X is of finite dimension, then by Theorem 1 of § 27, IV, there exists a ξ such that $X \subset Q_\xi$, i.e. $X = X \cap Q_\xi$. Therefore X is countable.

V. Nerve of a system of sets. Let $G_0, \dots, G_n$ be a system of subsets of an arbitrary space $\mathscr{X}$ and let $p_0, \dots, p_n$ be a system of points of an euclidean space (of arbitrary dimension). If the complex N formed by simplexes $p_{i_0} \dots p_{i_k}$ such that $G_{i_0} \cap \dots \cap G_{i_k} \neq 0$ is non-degenerate, then it is called a *nerve* (1) of the system $\{G_0, \dots, G_n\}$. Clearly the nerve is a *closed complex*. Its dimension is equal to the greatest integer k such that there exists a system of indices $i_0 < \dots < i_k$ with $G_{i_0} \cap \dots \cap G_{i_k} \neq 0$.

Two systems have the same nerve if and only if they are combinatorially similar (compare § 14, III).

Let $S = p_0 \dots p_n$ be a non-degenerate simplex. Let (compare II, (10)) P_i be the union of faces of S having p_i as a vertex.

THEOREM 1. *The complex* $\overline{\mathbf{S}}$ *formed by all the faces of* S *is the nerve of the system* $\{P_0, \dots, P_n\}$.

For the intersection $P_{i_0} \cap \dots \cap P_{i_k}$ is the union of the simplexes whose face is $p_{i_0} \dots p_{i_k}$, and hence is non empty if $i_0 < \dots < i_k$.

This can be stated more generally, as follows.

THEOREM 2. *If* $\mathbf{C}$ *is a closed subcomplex of* $\overline{\mathbf{S}}$ *and if* $C = \mathrm{S}(\mathbf{C})$, *then* $\mathbf{C}$ *is a nerve of the system* $\{C \cap P_0, \dots, C \cap P_n\}$.

For the conditions $C \cap P_{i_0} \cap \dots \cap P_{i_k} \neq 0$ and $p_{i_0} \dots p_{i_k} \subset C$ are equivalent.

The following theorem concerns geometrical realization of the nerve of a given system of sets.

THEOREM 3. *Let* $\{G_0, \dots, G_m\}$ *be a given system of sets and let* n *be an integer* $\leqslant m$ *such that no point belongs to any* $(n+2)$ *sets of the system; i.e.*

$$G_{i_0} \cap \dots \cap G_{i_{n+1}} = 0 \quad \textit{for every} \quad i_0 < i_1 < \dots < i_{n+1}. \qquad \text{(n)}$$

(1) See P. Alexandrov, C. R. Paris 184 (1927), p. 317. Compare the notion of "polyèdre réciproque" of H. Poincaré, *Complément à l'analysis situs*, § VII, Rendic. di Palermo 13 (1899).

Then there exists a (non-degenerate, at most n-dimensional) complex in $\mathscr{E}^{2n+1}$ which is a nerve of the given system.

More precisely, *the vertices $p_0, \ldots, p_m$ of the nerve can be subjected to the additional condition: given $a_0, \ldots, a_m$ and $\varepsilon > 0$, then*

$$|p_i - a_i| < \varepsilon, \qquad i = 0, 1, \ldots, m.$$

Proof. We let $r = 2n+1$ in I, Theorem 2 and denote by N the complex formed by the simplexes $p_{i_0} \ldots p_{i_k}$ such that $G_{i_0} \cap \ldots \cap G_{i_k} \neq 0$. By virtue of (n) such simplexes have dimension $\leqslant n$ and the complex N is non-degenerate. Consequently it is a nerve of the system $\{G_0, \ldots, G_m\}$.

VI. Mappings of metric spaces into polyhedra (1). Let $\{G_0, \ldots, G_m\}$ be a system of open subsets of a metric space $\mathscr{X}$ and let $\{p_0, \ldots, p_m\}$ be a system of points of an euclidean space. Let

$$G = G_0 \cup \ldots \cup G_m \quad \text{and} \quad F_i = G - G_i. \tag{1}$$

The *mapping $\varkappa$ corresponding to $\{G_0, \ldots, G_m\}$ and $\{p_0, \ldots, p_m\}$* is the mapping of G defined by

$$\varkappa(x) = \lambda_0(x) \cdot p_0 + \ldots + \lambda_m(x) \cdot p_m \tag{2}$$

where

$$\lambda_i(x) = \frac{\varrho(x, F_i)}{\varrho(x, F_0) + \ldots + \varrho(x, F_m)} \tag{3}$$

(assuming that $\varrho(x, 0) = 1$).

THEOREM 1. *$\varkappa(x)$ is a point of the closure of the (degenerate or not) simplex $S = p_0 \ldots p_m$ and has barycentric coordinates $\lambda_0(x), \ldots, \lambda_m(x)$, i.e.*

$$\lambda_0(x) + \ldots + \lambda_m(x) = 1 \quad \textit{and} \quad \lambda_i(x) \geqslant 0.$$

Proof. We must show that the denominator of (3) is positive. Since F_i is closed in G, hence the following equivalence holds:

$$[\varrho(x, F_i) = 0] \equiv (x \in F_i) \quad \text{for} \quad x \in G. \tag{4}$$

---

(1) Compare P. Alexandrov, C. R. Paris 183 (1926), p. 640, Math. Ann. 98 (1928), p. 635, Ann. of Math. 30 (1928), p. 6, W. Hurewicz, Mon. f. Math. u. Phys. 37 (1930), p. 202, as well as my note *Sur un théorème fondamental concernant le nerf d'un système d'ensembles*, Fund. Math. 20 (1933), pp. 191–196.

If we suppose that $\varrho(x, F_0) + \ldots + \varrho(x, F_m) = 0$, then $\varrho(x, F_i) = 0$ and $x \in F_i$ for every i. Therefore x does not belong to any G_i, contrary to (1).

THEOREM 2. *The mapping $\varkappa$ is continuous on G.*

For the function $\lambda_i(x)$ is continuous on G.

THEOREM 3. *Let Q_i be the face $p_0 \ldots p_{i-1} p_{i+1} \ldots p_m$ and let P_i be the union of faces of the form $p_i p_{i_1} \ldots p_{i_k}$ $(0 \leqslant k \leqslant m)$. Then*

$$\varkappa[G_{i_0} \cap \ldots \cap G_{i_k} - \bigcup_{i \neq i_j} G_i] \subset p_{i_0} \ldots p_{i_k}, \tag{5}$$

$$\varkappa(F_i) \subset \bar{Q}_i, \tag{6}$$

$$\varkappa(G_i) \subset P_i. \tag{7}$$

Proof. Assume that

$$x \in G_{i_j} \quad \text{for} \quad 0 \leqslant j \leqslant k \quad \text{and} \quad x \notin G_i \quad \text{for} \quad i \neq i_j.$$

Then

$$x \notin F_{i_j} \quad \text{and} \quad x \in F_i; \quad \text{therefore} \quad \lambda_{i_j}(x) \neq 0 \quad \text{and} \quad \lambda_i(x) = 0,$$

by (4) and (3). It follows that $\varkappa(x) \in p_{i_0} \ldots p_{i_k}$ and (5) is proved.

By (4), if $x \in F_i$, then $\lambda_i(x) = 0$ and $\varkappa(x) \in \bar{Q}_i$.

Finally, if $x \in G_i$, then $\lambda_i(x) \neq 0$ and $\varkappa(x) \in P_i$.

Remark. If the system $\{G_0, \ldots, G_m\}$ satisfies condition (n), then $\dim\left(\overline{\varkappa(G)}\right) \leqslant n$.

This is an immediate consequence of (5) and of II, Corollary 7b.

THEOREM 4. *If the simplex $S = p_0 \ldots p_m$ is non-degenerate, then inclusions* (5)–(7) *can be replaced by the identities*

$$\varkappa^{-1}(p_{i_0} \ldots p_{i_k}) = G_{i_0} \cap \ldots \cap G_{i_k} - \bigcup_{i \neq i_j} G_i, \tag{8}$$

$$\varkappa^{-1}(\bar{Q}_i) = F_i, \tag{9}$$

$$\varkappa^{-1}(P_i) = G_i. \tag{10}$$

In this case the function $\varkappa$ maps G onto a subset of the polyhedron $N = \mathrm{S}(\boldsymbol{N})$, where $\boldsymbol{N}$ is the nerve of the system $\{G_0, \ldots, G_m\}$, i.e. is made up of the simplexes $p_{i_0} \ldots p_{i_k}$ such that $G_{i_0} \cap \ldots \cap G_{i_k} \neq 0$.

Finally, *if condition* (n) *of Theorem* V, 3 *is fulfilled, then*

$$\dim \boldsymbol{N} = \dim N \leqslant n. \tag{11}$$

Proof. The faces of a non-degenerate simplex are mutually disjoint, hence (8) follows from (5). Let $\bar{Q}_i$ be the union of faces that do not have p_i as a vertex. Then the ith barycentric coordinate of any point of $\bar{Q}_i$ is zero and formula (9) follows. Formula (10) is proved similarly.

To prove the second part of the theorem we observe that, by (8), the condition $G_{i_0} \cap \ldots \cap G_{i_k} = 0$ implies $\varkappa^{-1}(p_{i_0} \ldots p_{i_k}) = 0$. This means that the values of $\varkappa$ do not lie in the simplex $p_{i_0} \ldots p_{i_k}$. Hence they must lie in N.

Remarks. (i) It is easy to prove [1] that every subset of a non-degenerate simplex can be mapped continuously onto a polyhedron being the union of some faces of the simplex without any point leaving the closure of the face to which it belongs. If the given subset is $\varkappa(G)$ and f is the mapping under consideration, then the composed function $g = f \circ \varkappa$ maps G onto a polyhedron and satisfies the inclusion $g^{-1}(p_{i_0} \ldots p_{i_k}) \subset G_{i_0} \cap \ldots \cap G_{i_k}$.

(ii) The right side member of (8) is named (after G. Boole in the algebra of logic) "a constituent of G relative to the system $G_0, \ldots, G_m$". G decomposes into constituents; the constituents are disjoint.

Theorem 4 establishes a correspondence between these constituents and the faces of S.

THEOREM 5. *If $\mathscr{X}$ is a totally bounded space of dimension n, then for every $\varepsilon > 0$ there exists a continuous mapping f of $\mathscr{X}$ onto a subset of an n-dimensional polyhedron such that*

$$\delta[f^{-1}(y)] < \varepsilon \quad \textit{for every} \quad y \in f(\mathscr{X}). \tag{12}$$

More precisely, *let $\mathscr{X}$ be a space with property D_n, let $G_0, \ldots, G_m$ be a system of open sets with $\mathscr{X} = G_0 \cup \ldots \cup G_m$ and let S be a non-degenerate simplex $p_0 \ldots p_m$. Let $\boldsymbol{A}_n$ be the complex formed by the faces of S of dimension $\leqslant n$ and let $A_n = \mathrm{S}(\boldsymbol{A}_n)$; then there exists a function $f \in A_n^{\mathscr{X}}$ such that*

$$f^{-1}(P_i) \subset G_i \quad \textit{for} \quad i = 0, 1, \ldots, m. \tag{13}$$

---

[1] Compare my note quoted above, p. 193.

Proof. By property D_n there exists a system of open sets $G_0^*, \ldots, G_m^*$ with

$$\mathscr{X} = G_0^* \cup \ldots \cup G_m^*, \quad G_i^* \subset G_i \text{ and } G_{i_0}^* \cap \ldots \cap G_{n+1}^* = 0 \tag{14}$$

for $i_0 < \ldots < i_{n+1} \leqslant m$.

The required function f is the mapping $\varkappa$ corresponding to the systems $\{G_0^*, \ldots, G_m^*\}$ and $\{p_0, \ldots, p_m\}$. For, if $\boldsymbol{N}$ is the nerve of the system $\{G_0^*, \ldots, G_m^*\}$, then we have (11). Hence $\boldsymbol{N} \subset \boldsymbol{A}_n$ and $\varkappa(\mathscr{X}) \subset A_n$. On the other hand, (10) implies (13) by virtue of (14).

Thus the second part of Theorem 5 is proved. To prove the first part it is sufficient to consider a cover $\mathscr{X} = G_0 \cup \ldots \cup G_m$ with open sets of diameter $< \varepsilon$.

The converse of Theorem 5 also holds:

THEOREM 6. *If to every cover* $\mathscr{X} = G_0 \cup \ldots \cup G_m$ *with open sets there is a function* $f \in A_n^{\mathscr{X}}$ *satisfying condition* (13), *then the space* $\mathscr{X}$ *has property* D_n.

Proof. Let $G_i^* = f^{-1}(P_i)$. We have to show that conditions (14) are fulfilled.

The first condition of (14) follows from II (11), the second one is a consequence of (13), and the last one is a consequence of the fact that the nerve of the system $\{A_n \cap P_0, \ldots, A_n \cap P_m\}$, and hence of the system $\{G_0^*, \ldots, G_m^*\}$, is of dimension $\leqslant n$ (by Theorem V, 2).

We shall prove the following related theorem:

THEOREM 7. *Suppose that for every* $l > n$ *and for every system of open sets* $\{G_0, \ldots, G_m\}$ *with* $\mathscr{X} = G_0 \cup \ldots \cup G_m$, *whose nerve is of dimension* l, *there exists a function* $f \in A_{l-1}^{\mathscr{X}}$ *satisfying* (13). *Then* $\mathscr{X}$ *has the property* D_n.

Proof. We have to define a system of open sets satisfying (14). Assume that the required system $\{G_0^*, \ldots, G_m^*\}$ exists for a $k \geqslant n$ and for every $l \leqslant k$. We have to prove its existence for $k+1$.

Since $k+1 > n$, there exists, by the hypotheses of Theorem 7, a function $f \in A_k^{\mathscr{X}}$ satisfying (13). Let $H_i = f^{-1}(P_i)$. The nerve of the system $\{H_0, \ldots, H_m\}$, has dimension $\leqslant k$. Therefore there exists a system of open sets $\{G_0^*, \ldots, G_m^*\}$ satisfying the identities (14) and the inclusion $G_i^* \subset H_i$. This, together with (13) yields inclusion (14).

THEOREM 8. *If the points* $p_0, \ldots, p_m$ *of* $\mathscr{E}^n$ *are in general position and condition* (n) *is fulfilled, then, for every* y,

$$\varkappa^{-1}(y) = K_0 \cup \ldots \cup K_m, \quad K_i \subset G_i, \quad K_i \cap K_{i'} = 0 \quad \textit{for} \quad i \neq i',$$

where K_i *are closed (may be empty) in the union* $G = G_0 \cup \ldots \cup G_m$.

Proof. Let $S^* = p_0^* \ldots p_m^*$ be an m-dimensional simplex (in $\mathscr{E}^m$) and let (compare (3))

$$\varkappa^*(x) = \lambda_0(x)\cdot p_0^* + \ldots + \lambda_m(x)\cdot p_m^*.$$

Let A_n^* be the union of simplexes $p_{i_0}^* \ldots p_{i_k}^*$ with $k \leqslant n$ and let

$$f(t) = \beta_0(t)\cdot p_0 + \ldots + \beta_m(t)\cdot p_m, \quad t \in A_n^*,$$

where $\beta_0(t), \ldots, \beta_m(t)$ are the barycentric coordinates of t (with respect to the vertices $p_0^*, \ldots, p_m^*$). Then $\varkappa^*(\mathscr{X}) \subset A_n^*$ and $\varkappa(x) = f[\varkappa^*(x)]$. Consequently $\varkappa^{-1}(y) = \varkappa^{*-1}[f^{-1}(y)]$.

The simplexes $p_{i_0} \ldots p_{i_k}$ with $k \leqslant n$ are non-degenerate, hence the function f is one-to-one on each simplex $p_{i_0}^* \ldots p_{i_k}^*$. Therefore the set $f^{-1}(y)$ is finite. Let P_i^* denote the union of the faces of S^* having p_i^* as a vertex and let $H_i = f^{-1}(y) \cap P_i^*$; then

$$\varkappa^{-1}(y) = \varkappa^{*-1}(H_0) \cup \ldots \cup \varkappa^{*-1}(H_m),$$

since $P_0^* \cup \ldots \cup P_m^* = \overline{S^*}$.

The sets

$$K_0 = \varkappa^{*-1}(H_0), \quad K_1 = \varkappa^{*-1}(H_1 - H_0), \quad \ldots, \quad K_m = \varkappa^{*-1}[H_m - (H_0 \cup \ldots \cup H_{m-1})]$$

are the ones which are required. For

$$K_0 \cup \ldots \cup K_m = \\ = \varkappa^{*-1}\{H_0 \cup (H_1 - H_0) \cup \ldots \cup [H_m - (H_0 \cup \ldots \cup H_{m-1})]\} = \varkappa^{-1}(y),$$

$$K_i \subset \varkappa^{*-1}(H_i) \subset \varkappa^{*-1}(P_i^*) = G_i \quad \text{(by (10))},$$

$$K_i \cap K_{i'} \subset \varkappa^{*-1}[H_i \cap (H_{i'} - H_i)] = 0 \quad \text{for} \quad i < i'.$$

Finally, since H_i is finite, K_i is closed in G.

VII. Approximation of continuous functions by means of kappa functions[1].

THEOREM 1. *If* $\mathscr{X} = G_0 \cup \ldots \cup G_m$ *is a cover of an (arbitrary) metric space* $\mathscr{X}$ *with open subsets,* $f \in (\mathscr{I}^r)^{\mathscr{X}}$, $\varepsilon > 0$ *and* $p_0, \ldots, p_m$ *are points of* $\mathscr{I}^r$ *with*

$$\delta[(p_i) \cup f(G_i)] < \varepsilon \quad \textit{for} \quad i = 0,1,\ldots,m, \tag{15}$$

then the mapping $\varkappa$ *corresponding to the systems* $\{G_0, \ldots, G_m\}$ *and* $\{p_0, \ldots, p_m\}$ *satisfies the condition*

$$|\varkappa(x) - f(x)| < \varepsilon \quad \textit{for every} \quad x \in \mathscr{X}. \tag{16}$$

First, we shall prove that:

THEOREM 2. *If* $p_0 p_j$ *is the longest edge of all edges* $p_0 p_i$ *of the simplex* $p_0 \ldots p_n$ *(which may be degenerate), then*

$$|p_0 - x| \leqslant |p_0 - p_j| \quad \textit{for every} \quad x \in \overline{p_0 \ldots p_n}.$$

Proof of Theorem 2. The centre of gravity of the masses $\lambda_0, \ldots, \lambda_n$ distributed at the points $p_0, \ldots, p_n$ cannot lie on the outside of the sphere with centre p_0 and radius $|p_0 - p_j|$ (this sphere contains all the points $p_0, \ldots, p_n$).

Proof of Theorem 1. Let x be a point of $\mathscr{X}$ and let $i_0, \ldots, i_k$ be the system of all indices such that $x \in G_{i_0}, \ldots, x \in G_{i_k}$. Then

$$x \in [G_{i_0} \cap \ldots \cap G_{i_k} - \bigcup_{i \neq i_j} G_i]; \quad \text{hence} \quad \varkappa(x) \in p_{i_0} \ldots p_{i_k} \tag{17}$$

by virtue of (5).

We can assume that among the points $p_{i_0}, \ldots, p_{i_k}$ the point p_{i_0} lies at the greatest distance from $f(x)$. By substituting the simplex $f(x) p_{i_0} \ldots p_{i_k}$ for $p_0 \ldots p_n$ in Theorem 2, we deduce from (17) that

$$|f(x) - \varkappa(x)| \leqslant |f(x) - p_{i_0}|. \tag{18}$$

Since $x \in G_{i_0}$, hence $f(x) \in f(G_{i_0})$ and

$$|p_{i_0} - f(x)| \leqslant \delta[(p_{i_0}) \cup f(G_{i_0})]. \tag{19}$$

---

[1] Theorems of Section VII will be used to prove (see Vol. II) imbedding theorems in dimension theory; in particular: that every metric separable n-dimensional space is homeomorphic to a subset of $\mathscr{E}^{2n+1}$.

The inequalities (18), (19), and (15) yield (16).

THEOREM 3. *Every function* $f \in (\mathscr{I}^r)^{\mathscr{X}}$ *can be uniformly approximated by* $\varkappa$ *functions.*

More precisely, *if* $\{G_0, \ldots, G_m\}$ *is a system of open sets with*

$$\mathscr{X} = G_0 \cup \ldots \cup G_m \quad \text{and} \quad \delta[f(G_i)] < \varepsilon \quad \text{for} \quad i = 0, \ldots, m, \tag{20}$$

then there exists a system of points $\{p_0, \ldots, p_m\}$ *in* $\mathscr{I}^r$ *in a general position such that the function* $\varkappa$ *corresponding to the systems* $\{G_0, \ldots, G_m\}$ *and* $\{p_0, \ldots, p_m\}$ *satisfies* (16).

Finally, *if* $\mathscr{X}$ *has the property* D_n, *then we can assume that*

$$\dim \overline{\varkappa(\mathscr{X})} \leqslant n. \tag{20'}$$

Proof. Let $a_i \in f(G_i)$ (if $G_i = 0$, then a_i is an arbitrary point of $\mathscr{I}^r$). By I, Theorem 2, there is a system of points $p_0, \ldots, p_m$ in $\mathscr{I}^r$ in a general position which satisfies condition (15). The second part of Theorem 3 is a consequence of Theorem 1.

We shall prove the first part. Let

$$\mathscr{I}^r = H_0 \cup \ldots \cup H_m, \qquad \delta(H_i) < \varepsilon \tag{21}$$

be a cover of the cube $\mathscr{I}^r$ with open sets (e.g. open balls of diameter $< \varepsilon$). Let $G_i = f^{-1}(H_i)$. Then conditions (20) are fulfilled and (16) follows.

If $\mathscr{X}$ has the property D_n, then we replace the system $\{G_0, \ldots, G_m\}$ by a system $\{G_0^*, \ldots, G_m^*\}$ such that

$$\mathscr{X} = G_0^* \cup \ldots \cup G_m^*, \quad G_i^* \subset G_i, \quad G_{i_0}^* \cap \ldots \cap G_{i_{n+1}}^* = 0.$$

Then (20') follows from the remark to Theorem VI, 3.

Remark. If $r \geqslant 2n+1$, then the points $p_0, \ldots, p_m$ can be subjected to the following additional condition.

Given a linear manifold L of dimension $\leqslant n$, all the simplexes $p_{i_0} \ldots p_{i_k}$ with $k \leqslant n$, are disjoint to L (see Section I, Remark (ii)).

Theorem 3 can be expressed more precisely as follows.

THEOREM 4. *If the set* $E = A \cup B$ *has the property* D_n *and the sets* A *and* B *are closed and disjoint, then every function* $f \in (\mathscr{I}^r)^{\mathscr{X}}$, *where* $r \geqslant 2n+1$, *can be uniformly approximated by* $\varkappa$ *functions subjected to the condition*

$$\overline{\varkappa(A)} \cap \overline{\varkappa(B)} = 0.$$

Proof. Let, as in (21), $H_0, \ldots, H_s$ be open sets with

$$\mathscr{I}^r = H_0 \cup \ldots \cup H_s \quad \text{and} \quad \delta(H_j) < \varepsilon \quad \text{for} \quad j \leqslant s. \tag{21'}$$

Let $m = 2s+1$, $K_j = f^{-1}(H_j)$ and

$$G_0 = K_0 - A, \ \ldots, \ G_s = K_s - A, \ G_{s+1} = K_0 - B, \ \ldots, \ G_m = K_s - B.$$

Then $G_0, \ldots, G_m$ are open sets and satisfy conditions (20), since $\delta[f(K_j)] = \delta(H_j) < \varepsilon$. Also, for every $i \leqslant m$,

$$\text{either} \quad G_i \cap A = 0 \quad \text{or} \quad G_i \cap B = 0. \tag{22}$$

Since E has the property D_n, then by § 27, III, Theorem 5, there exist open sets $Q_0, \ldots, Q_m$ with

$$\mathscr{X} = Q_0 \cup \ldots \cup Q_m, \tag{23}$$

$$Q_i \subset G_i, \tag{24}$$

$$E \cap Q_{i_0} \cap \ldots \cap Q_{i_{n+1}} = 0 \quad \text{for} \quad i_0 < \ldots < i_{n+1} \leqslant m. \tag{25}$$

By (20), $\delta[f(Q_i)] < \varepsilon$. Referring to the proof of Theorem 3, we consider a system of points $p_0, \ldots, p_m$ in a general position such that $\delta[p_i \cup f(Q_i)] < \varepsilon$. We infer that the function $\varkappa$ corresponding to the systems $\{Q_0, \ldots, Q_m\}$ and $\{p_0, \ldots, p_m\}$ satisfies (16).

Let N_A and N_B be, respectively, the unions of pairwise disjoint (by (25)) simplexes $p_{i_0} \ldots p_{i_k}$ such that

$$A \cap Q_{i_0} \cap \ldots \cap Q_{i_k} \neq 0, \quad \text{respectively} \quad B \cap Q_{i_0} \cap \ldots \cap Q_{i_k} \neq 0.$$

By (23) and (5)

$$\varkappa(A) \subset N_A \quad \text{and} \quad \varkappa(B) \subset N_B;$$

hence

$$\overline{\varkappa(A)} \subset N_A \quad \text{and} \quad \overline{\varkappa(B)} \subset N_B.$$

Moreover, by (23) and (22), each Q_i is disjoint either to A or to B. Hence

$$N_A \cap N_B = 0 \quad \text{and} \quad \overline{\varkappa(A)} \cap \overline{\varkappa(B)} = 0.$$

Two following generalizations of Theorem 4 will be applied later.

THEOREM 5. *Given $l+1$ closed sets $A_0, \ldots, A_l$ of dimension $\leqslant n$, every function $f \in (\mathscr{I}^r)^{\mathscr{X}}$, where $r \geqslant 2n+1$, can be uniformly approximated by $\varkappa$ functions subjected to the condition*

$$\dim[\overline{\varkappa(A_0)} \cap \ldots \cap \overline{\varkappa(A_l)}] \leqslant \dim(A_0 \cap \ldots \cap A_l). \tag{26}$$

Proof. The sets $H_0, \ldots, H_s$ and $K_0, \ldots, K_s$ have the same meaning as in the proof of Theorem 4. Thus

$$\mathscr{X} = K_0 \cup \ldots \cup K_s \quad \text{and} \quad \delta[f(K_j)] < \varepsilon. \tag{27}$$

Let $E = A_0 \cup \ldots \cup A_l$ and $v = \dim(A_0 \cap \ldots \cap A_l)$. By § 27, III, 6, there exist open sets $G_0, \ldots, G_m$ satisfying (20) such that for every system $i_0 < \ldots < i_{v+1} \leqslant m$ there is a $\nu \leqslant l$ with

$$A_\nu \cap G_{i_0} \cap \ldots \cap G_{i_{v+1}} = 0. \tag{28}$$

Moreover, since E has the property D_n, the system $\{G_0, \ldots, G_m\}$ can be subjected, by § 27, III, Theorem 5, to the condition

$$E \cap G_{i_0} \cap \ldots \cap G_{i_{n+1}} = 0 \tag{29}$$

(by replacing, if necessary, G_i with Q_i satisfying formulas (5) of § 27, III).

If (according to I, Theorem 2) $p_0, \ldots, p_m$ is a system of points in a general position satisfying condition (15), then (16) is fulfilled.

Let, for $\nu = 0, \ldots, l$, N_ν be the union of simplexes $p_{i_0} \ldots p_{i_k}$ such that $A_\nu \cap G_{i_0} \cap \ldots \cap G_{i_k} \neq 0$; then, by (29), $k \leqslant n$. Since the points $p_0, \ldots, p_m$ are in a general position in $\mathscr{I}^r$, whence all these simplexes are disjoint and N_ν is a polyhedron of dimension $\leqslant n$. Therefore the intersection $N_0 \cap \ldots \cap N_l$ is the union of the simplexes $p_{i_0} \ldots p_{i_k}$ such that $A_\nu \cap G_{i_0} \cap \ldots \cap G_{i_k} \neq 0$ for every $\nu \leqslant l$. By (28), $k \leqslant v$ and

$$\dim(N_0 \cap \ldots \cap N_l) \leqslant v.$$

Formula (26) follows, since (by (5))

$$\varkappa(A_\nu) \subset N_\nu \quad \text{and} \quad N_\nu = \overline{N_\nu}; \quad \text{hence} \quad \overline{\varkappa(A_\nu)} \subset N_\nu.$$

THEOREM 6. *Given $l+1$ closed sets $A_0, \ldots, A_l$ of dimensions $k_0, \ldots, k_l$ such that $A_\nu \cap A_\mu = 0$ for $\nu \neq \mu$, every function $f \in (\mathscr{I}^r)^{\mathscr{X}}$, where $r \geqslant k_\nu$ for $\nu = 0, \ldots, l$, can be uniformly approximated by $\varkappa$ functions subjected to the condition*

$$\dim[\overline{\varkappa(A_0)} \cap \ldots \cap \overline{\varkappa(A_l)}] \leqslant k_0 + \ldots + k_l - lr \quad (\textit{or} = -1). \tag{30}$$

Proof. Let H_j and K_j have the same meaning as in the proof of Theorem 4. Then conditions (27) are fulfilled.

Let $\{G_0, \ldots, G_m\}$ be the system of the sets $K_j - \bigcup A_\mu$ such that

$j \leqslant s$ and $\mu \neq \nu \leqslant l$ (hence $m = (s+1)(l+1)-1$). No set G_i, where $i = 0, \ldots, m$, has common points with two different terms of $\{A_0, \ldots, A_l\}$. Therefore, if

$$A_\nu \cap G_{i_0} \cap \ldots \cap G_{i_k} \neq 0 \neq A_{\nu'} \cap G_{i'_0} \cap \ldots \cap G_{i'_k} \qquad \text{and} \qquad \nu \neq \nu',$$

then all the indices $i_0, \ldots, i_k$ are different from all the indices $i'_0, \ldots, i'_{k'}$.

If $p_0, \ldots, p_m$ satisfy Remark (iii) of Section I, and $N_0, \ldots, N_l$ are as in the proof of Theorem 5 and Z_ν is the set of vertices of simplexes of N_ν, then $Z_\nu \cap Z_{\nu'} = 0$. Let $\{G_0, \ldots, G_m\}$ be subjected to condition (29), where n is replaced with r (compare the remark to (29)). Therefore the simplexes in question have dimension $\leqslant r$. Hence by Remark (iii) of Section I,

$$\dim(N_0 \cap \ldots \cap N_l) \leqslant k_0 + \ldots + k_l - lr \ (\text{or} = -1).$$

Therefore (30) follows from the inclusion $\overline{\varkappa(A_\nu)} \subset N_\nu$.

THEOREM 7. *Let $\mathscr{X}$ be a totally bounded space of dimension $\leqslant r$. Given $\eta > 0$, every function $f \in (\mathscr{I}^r)^{\mathscr{X}}$ can be uniformly approximated by $\varkappa$ functions such that, for every y, the set $\varkappa^{-1}(y)$ admits a finite decomposition into disjoint closed sets of diameter $< \eta$.*

Proof. Let H_j and K_j have the same meaning as before and let $V_0, \ldots, V_q$ be open sets with

$$\mathscr{X} = V_0 \cup \ldots \cup V_q \quad \text{and} \quad \delta(V_\nu) < \eta \quad \text{for} \quad \nu = 0, \ldots, q.$$

Let $\{G_0, \ldots, G_m\}$ be the system of all intersections $K_j \cap V_\nu$. Conditions (20) are fulfilled and $\delta(G_i) < \eta$. Since $\dim \mathscr{X} \leqslant r$, we can assume that

$$G_{i_0} \cap \ldots \cap G_{i_{r+1}} = 0 \quad \text{for} \quad i_0 < \ldots < i_{r+1} \leqslant m.$$

Let the points $p_0, \ldots, p_m$ have the same meaning as in the proof of Theorem 5; then (16) is fulfilled.

Moreover, by VI, Theorem 8,

$$\varkappa^{-1}(y) = F_0 \cup \ldots \cup F_m, \quad F_i = \overline{F}_i \subset G_i, \quad \text{and} \quad F_i \cap F_{i'} = 0 \quad \text{for} \quad i \neq i'.$$

Finally $\delta(F_i) < \eta$, since $\delta(G_i) < \eta$.

VIII. Infinite complexes and polyhedra. An infinite countable family of simplexes $R_1, R_2, \ldots$ is called an *infinite complex*, if no point of $\overline{R_i}$ $(i = 1, 2, \ldots)$ is a limit point of a sequence of points selected from different terms of the sequence $\{R_i\}$ (1).

The notions of *closure, closed complex* and *non-degenerate complex* apply to infinite complexes as in the case of finite ones.

Similarly, C is an *infinite polyhedron,* if there exists a non-degenerate closed infinite complex $\boldsymbol{C}$ with $C = \mathrm{S}(\boldsymbol{C})$.

It can be proved that *every open subset of $\mathscr{E}^n$ is an infinite polyhedron* (2).

The notion of *nerve* of an infinite sequence of sets $G_0, G_1, \ldots$ subjected to the condition

$$\textit{Each } G_i \textit{ meets only a finite number of } G_j\textit{'s} \tag{1}$$

is defined as in the case of a finite sequence (Section V).

Theorem V, 1 can be generalized as follows.

THEOREM 1. *If $\boldsymbol{C}$ is a closed, non-degenerate, infinite complex with vertices $p_0, p_1, \ldots$ and P_i is the union of simplexes of $\boldsymbol{C}$ having p_i as a vertex, then $\boldsymbol{C}$ is the nerve of the sequence $P_0, P_1, \ldots$.*

In other words, *the conditions $(p_{i_0} \ldots p_{i_k}) \in \boldsymbol{C}$ and $P_{i_0} \cap \ldots \cap P_{i_k} \neq 0$ are equivalent.*

Proof. $P_{i_0} \cap \ldots \cap P_{i_k}$ is the union of simplexes of $\boldsymbol{C}$ having $p_{i_0} \ldots p_{i_k}$ as a face. Hence on one hand $p_{i_0} \ldots p_{i_k} \subset P_{i_0} \cap \ldots \cap P_{i_k}$ and the condition $(p_{i_0} \ldots p_{i_k}) \in \boldsymbol{C}$ implies $P_{i_0} \cap \ldots \cap P_{i_k} \neq 0$. On the other hand, if $x \in P_{i_0} \cap \ldots \cap P_{i_k}$, then x lies in a simplex $R_j \in \boldsymbol{C}$ having $p_{i_0} \ldots p_{i_k}$ as face; hence $(p_{i_0} \ldots p_{i_k}) \in \boldsymbol{C}$.

Theorem V, 3 has the following extension.

Let $\mathscr{E}^l_{-1}$, respectively $\mathscr{E}^{\aleph_0}_{-1}$, denote the set of points of $\mathscr{E}^l$, respectively of $\mathscr{E}^{\aleph_0}$, whose first coordinate ("the abscissa") is zero. If l is finite, then $\mathscr{E}^l_{-1}$ is isometric to $\mathscr{E}^{l-1}$.

THEOREM 2. *Let $G_0, G_1, \ldots$ be a sequence of sets satisfying* (1), *let $a_0, a_1, \ldots$ be a sequence of points of $\mathscr{E}^{\aleph_0}_{-1}$ and let $\varepsilon_0, \varepsilon_1, \ldots$ be a sequence of positive numbers. Then there exists a non-degenerate*

(1) We can write this (see § 29, III): $\overline{R_i} \cap \mathrm{Ls}\, R_j = 0$.

(2) Theorem of Runge (compare Acta Math. 6 (1884), p. 229). For a proof, see Alexandrov-Hopf, *Topologie I*, p. 143.

infinite complex lying in $\mathscr{E}^{\aleph_0}-\mathscr{E}^{\aleph_0}_{-1}$ *with vertices* $p_0, p_1, \ldots$ *which is a nerve of the system* $G_0, G_1, \ldots$ *and satisfies the condition*

$$|p_i - a_i| < \varepsilon_i \quad \textit{for} \quad i = 0, 1, \ldots. \tag{2}$$

Moreover, *if condition* (n) *of Theorem* V, 3 *is fulfilled, then* $\aleph_0$ *can be replaced by* $2n+1$.

Proof. Let a_i^* be a point whose abscissa is positive $< 1/i$ and such that $|a_i^* - a_i| < \varepsilon_i$. By I, Theorem 2 (compare Remark (i)), there is p_i whose abscissa is positive $< 1/i$, whose distance from a_i^* is so small that (2) is fulfilled and that, moreover, all the simplexes whose vertices belong to $p_0, p_1, \ldots$ are disjoint (in the case where condition (n) is satisfied, we only require the simplexes of dimension $\leqslant n$ to be disjoint).

Let $S_0, S_1, \ldots$ be the complex N made up from the simplexes $p_{i_0} \ldots p_{i_k}$ such that $G_{i_0} \cap \ldots \cap G_{i_k} \neq 0$. We have to show that N is the nerve of the system $G_0, G_1, \ldots$, i.e. that N is non-degenerate. It is sufficient to prove that for every simplex S_0 there exists an i_0 such that

$$\overline{S}_0 \cap \overline{\bigcup_{j>i_0} S_j} = 0. \tag{3}$$

Let η be the smallest abscissa of the vertices of S_0 (hence it is the smallest abscissa of the points of $\overline{S}_0$), and let $1/k \leqslant \eta$. By (1) no p_i can be a vertex of an infinite number of the simplexes S_j. Consequently there exists an i_0 such that none of the points $p_0, \ldots, p_k$, with $j > i_0$, is a vertex of S_j. Since the abscissa of p_j is $< 1/j$, all vertices of S_j, hence also all points of S_j, have abscissas $\leqslant 1/(k+1)$. The same is true for the abscissas of points of the set $\overline{\bigcup_{j>i_0} S_j}$. But $1/(k+1) < \eta$, hence (3) follows.

The notion of the mapping $\varkappa$ extends also to the case of an infinite sequence of open sets $G_0, G_1, \ldots$ subjected to (1).

Let

$$G = G_0 \cup G_1 \cup \ldots \quad \text{and} \quad F_i = G - G_i. \tag{4}$$

Given a sequence of points $p_0, p_1, \ldots$ in $\mathscr{E}^n$ or in $\mathscr{E}^{\aleph_0}$, we put

$$\varkappa(x) = \sum_{i=0}^{\infty} \lambda_i(x) \cdot p_i, \tag{5}$$

where

$$\lambda_i(x) = \frac{\varrho(x, F_i)}{\sum_{j=0}^{\infty} \varrho(x, F_j)}. \tag{6}$$

By (1), if $x \in G$, then $\varrho(x, F_j) = 0$ for sufficiently large j. On the other hand, by (4)

$$\sum_{j=0}^{\infty} \varrho(x, F_j) \neq 0.$$

Hence the summation in (5) is finite.

Consequently

THEOREM 3. *The function $\varkappa$ is continuous on G.*

The following theorem can be proved just as Theorems 3 and 4 of Section VI.

THEOREM 4. *If $\boldsymbol{N}$ is the complex made up from the simplexes $p_{i_0} \dots p_{i_k}$ such that $G_{i_0} \cap \dots \cap G_{i_k} \neq 0$, then*

$$\varkappa(G) \subset N = \mathrm{S}(\boldsymbol{N})$$

and the inclusions VI (5) *and* VI (7) *hold.*

Moreover, *if the complex $\boldsymbol{N}$ is non-degenerate, then it is the nerve of the system $G_0, G_1, \dots$, and conditions* VI (8) *and* VI (10) *hold.*

Finally, *if condition* (n) *of Theorem* V, 3 *is fulfilled, then formula* VI (11) *holds.*

IX. Extension of continuous functions(1). The mapping $\varkappa$ applied to infinite systems can be used to prove the Tietze theorem (§ 14, IV) in a more advantageous form (the space $\mathcal{X}$ is supposed to be metric separable).

THEOREM 1. *Let $F = \overline{F} \subset \mathcal{X}$ and $f \in (\mathcal{E}^r)^F$, where $r \leqslant \aleph_0$. Then there exist a set N which is the union of a sequence of simplexes contained in $\mathcal{E}^r$ and an extension $f^* \in [f(F) \cup N]^{\mathcal{X}}$ of f.*

---

(1) Compare S. Lefschetz, Ann. of Math. 35 (1934), p. 118, and my note *Sur le prolongement des fonctions continues et les transformations en polytopes*, Fund. Math. 24 (1935), p. 259.

More precisely, *suppose that the set $G = \mathscr{X} - F$ is the union of open sets $G_0, G_1, \ldots$ satisfying Theorem* 4 *of* § 21, VIII; i.e. *condition* VIII (1) *and the formulas:*

(ii) $\overline{G}_i \subset G$,

(iii) $\lim\limits_{i=\infty} \delta(G_i) = 0$

hold. Then there exists a sequence of points $p_0, p_1, \ldots$ in $\mathscr{E}^r$ such that the mapping $\varkappa$ corresponding to the systems $(G_0, G_1, \ldots)$ and $(p_0, p_1, \ldots)$ is a continuous extension of f to G.

Proof. We can assume that $G \neq 0 \neq F$ and that the terms of the sequence $G_0, G_1, \ldots$ are non empty.

Let q_i, a_i be a pair of points with

$$q_i \in G_i, \qquad a_i \in F, \tag{1}$$

$$|a_i - q_i| < \varrho(G_i, F) + \frac{1}{i}. \tag{2}$$

Let

$$f_0(q_i) = f(a_i) \quad \text{and} \quad f_0(x) = f(x) \quad \text{for} \quad x \in F. \tag{3}$$

The function f_0 is continuous on $Q = F \cup q_0 \cup q_1 \cup \ldots$. To prove this we observe that the set $q_0, q_1, \ldots$ is discrete by VIII (1), hence we have only to show that the condition

$$\lim_{i=\infty} q_{j_i} = a \in F, \quad \text{where} \quad j_0 < j_1 < \ldots, \tag{4}$$

implies

$$\lim_{i=\infty} f_0(q_{j_i}) = f_0(a). \tag{5}$$

The inequalities

$$|a_{j_i} - a| \leqslant |a_{j_i} - q_{j_i}| + |q_{j_i} - a|$$

and

$$|a_{j_i} - q_{j_i}| < \varrho(G_{j_i}, F) + \frac{1}{j_i} \leqslant |q_{j_i} - a| + \frac{1}{j_i}$$

hold for $a \in F$ (by (2) and (1)). Therefore

$$|a_{j_i} - a| < 2|q_{j_i} - a| + \frac{1}{j_i},$$

hence, by (4), $\lim\limits_{i=\infty} a_{j_i} = a$ and

$$\lim_{i=\infty} f(a_{j_i}) = f(a), \tag{6}$$

since f is continuous on F.

Formulas (6) and (3) imply (5).

Let

$$p_i = f_0(q_i)$$

and let N be the union of simplexes $p_{i_0} \dots p_{i_k}$ such that $G_{i_0} \cap \dots \cap G_{i_k} \neq 0$. Let f^* be the function defined by

$$f^*(x) = \begin{cases} f(x) & \text{for} \quad x \in F, \\ \varkappa(x) & \text{for} \quad x \in G, \end{cases} \tag{7}$$

where $\varkappa$ corresponds to the systems $(G_0, G_1, \dots)$ and $(p_0, p_1, \dots)$.

We have to show that f^* is continuous on F (since G is open).

Let $a \in F$ and $\varepsilon > 0$. Since f_0 is continuous on Q, there exists a neighbourhood H of a such that, for $q_i \in H$,

$$|f_0(q_i) - f(a)| < \varepsilon, \quad \text{i.e.} \quad |p_i - f(a)| < \varepsilon. \tag{8}$$

Let K be the set $H \cap F$ augmented by the points x such that

$$x \in G_i \quad \text{implies} \quad G_i \subset H. \tag{9}$$

We shall prove that K is a neighbourhood of a. Suppose that there exists a sequence of points $\{x_i\}$ and indices $\{m_i\}$ with

$$\lim_{i=\infty} x_i = a, \quad x_i \in G_{m_i} \quad \text{and} \quad G_{m_i} - H \neq 0. \tag{10}$$

By (ii), $a \notin \bar{G}_i$. Hence there is an infinite set of different indices m_i. But then (10) is inconsistent with (iii).

Since K is a neighbourhood of a, it suffices to prove that, given a point $x \in G \cap K$,

$$|f^*(x) - f(a)| < \varepsilon. \tag{11}$$

Let $i_0, \dots, i_k$ be the system of all indices such that $x \in G_{i_0}, \dots, x \in G_{i_k}$, i.e.

$$x \in [G_{i_0} \cap \dots \cap G_{i_k} - \bigcup_{i \neq i_j} G_i]. \tag{12}$$

Hence

$$f^*(x) \in (p_{i_0} \dots p_{i_k}) \tag{13}$$

by (7) and VI (5) (compare VIII, Theorem 4).

Since $x \in K \cap G_{i_m}$ (for $m \leqslant k$), hence by (9), $G_{i_m} \subset H$ and (by (1)) $q_{i_m} \in H$. Therefore (compare (8))

$$|p_{i_m} - f(a)| < \varepsilon, \quad m = 0, 1, \ldots, k. \tag{14}$$

Inequalities (14) together with (13) imply (11) (by virtue of Theorem VII, 2, where the simplex $f(a)p_{i_0} \ldots p_{i_k}$ is substituted for $p_0 \ldots p_n$).

Remark. Theorem 1 remains valid after replacing $\mathscr{E}^r$ with an arbitrary *convex* subset.

THEOREM 2. *If, in the hypothesis of Theorem* 1, $f(F) \subset \mathscr{E}^{\aleph_0}_{-1}$ *and* $r = \aleph_0$, *then* N *can be supposed to be an infinite polyhedron contained in* $\mathscr{E}^{\aleph_0} - \mathscr{E}^{\aleph_0}_{-1}$.

Similarly, *if* $r \geqslant 2n+1$, $f(F) \subset \mathscr{E}^r_{-1}$ *and if the nerve of the sequence* $\{G_i\}$ *has dimension* $\leqslant n$ (i.e. *condition* (n) *of Theorem* V, 3 *is fulfilled*), *then* N *can be supposed to be an infinite polyhedron of dimension* $\leqslant n$ *contained in* $\mathscr{E}^r - \mathscr{E}^r_{-1}$. *Namely*, $N = \mathrm{S}(\boldsymbol{N})$, *where* $\boldsymbol{N}$ *is the nerve of the sequence* $\{G_i\}$.

Proof. We have only to modify the definition of f_0 in Theorem 1 (the first identity of (3)) as follows.

According to VIII, Theorem 2, let $\boldsymbol{N}$ be an infinite complex in $\mathscr{E}^r - \mathscr{E}^r_{-1}$ (where $r = \aleph_0$, respectively $r \geqslant 2n+1$) with vertices $p_0, p_1, \ldots$ which is the nerve of the sequence $G_0, G_1, \ldots$ with

$$\lim_{i=\infty} |p_i - f(a_i)| = 0. \tag{15}$$

Let $f_0(q_i) = p_i$. Condition (4) implies (6) and, by (15), (6) implies $\lim_{i=\infty} p_{j_i} = f(a)$ which is equivalent to (5). Therefore the function f_0 is continuous.

The proof of continuity of f^* does not require modification.

COROLLARY 2a [1]. *If* $\mathscr{X}$ *and* $\mathscr{Y}$ *are separable metric spaces,* F *is a closed subset of* $\mathscr{X}$ *with* $\dim(\mathscr{X} - F) = n$ *and* $f \in \mathscr{Y}^F$, *then the space* $\mathscr{Y}$ *can be imbedded in a space* $\mathscr{Z}$ *such that* $\mathscr{Y}$ *is closed in* $\mathscr{Z}$, $\mathscr{Z} - \mathscr{Y}$ *is an infinite polyhedron of dimension* n, *and* f *has an extension* $f^* \in \mathscr{Z}^{\mathscr{X}}$.

To prove this, we think of $\mathscr{Y}$ as a subset of $\mathscr{E}^{\aleph_0}_{-1}$ and apply the preceding theorem.

---

[1] Compare an analogous theorem of F. Hausdorff for non-separable spaces, Fund. Math. 30 (1938), p. 41.

D. COUNTABLE OPERATIONS. BOREL SETS. *B* MEASURABLE FUNCTIONS

The space considered in §§ 29–32 is supposed to be metric.

§ 29. Lower and upper limits

I. Lower limit (1).

DEFINITION. *The point p belongs to the lower limit* $\operatorname{Li}_{n=\infty} A_n$ *of a sequence of sets $A_1, A_2, \ldots$, if every neighbourhood of p intersects all the A_n from a sufficiently great index n onward.*

Of course, the term "neighbourhood" can be replaced by "open neighbourhood" or by "open ball with centre p".

THEOREM. *The formula $p \in \operatorname{Li}_{n=\infty} A_n$ is equivalent to the existence of a sequence of points $\{p_n\}$ such that*

$$p = \lim_{n=\infty} p_n \quad \textit{and} \quad p_n \in A_n;$$

or to the equality

$$\lim_{n=\infty} \varrho(p, A_n) = 0.$$

Notice that the sequence $\{p_n\}$ is defined from a certain index onwards which is not necessarily equal to 1 (in the case where some A_n are empty).

Proof. If $p \in \operatorname{Li} A_n$ and if S_m is the ball with centre p and radius $1/m$, then there exists an index k_m such that $S_m \cap A_n \neq 0$ for $n \geqslant k_m$. Also we can suppose that $k_m > k_{m-1}$. The sequence $\{p_n\}$, where $p_n \in S_m \cap A_n$ for $k_m \leqslant n < k_{m+1}$, is convergent to p, for $|p_n - p| < 1/m$ for $n \geqslant k_m$. The converse implication is obvious.

---

(1) The idea of lower and upper bound is due to P. Painlevé (compare C. R. Paris 148 (1909), p. 1156, and indications on this subject by L. Zoretti, Journ. de Math. (5) (1905), p. 8). They are called "unterer (oberer) abgeschlossener Limes" by F. Hausdorff. They should not be confused with (see § 1, V):

$$\operatorname{Lim\,inf}_{n=\infty} A_n = \bigcup_n (A_n \cap A_{n+1} \cap A_{n+2} \cap \ldots)$$

and

$$\operatorname{Lim\,sup}_{n=\infty} A_n = \bigcap_n (A_n \cup A_{n+1} \cup A_{n+2} \cup \ldots).$$

EXAMPLES. If A_n is a single point p_n, then

$$\operatorname*{Li}_{n=\infty} A_n = \lim_{n=\infty} p_n$$

provided that the latter limit exists, and $\operatorname*{Li}_{n=\infty} A_n = 0$ otherwise. For $\varrho(p, A_n) = |p - p_n|$.

A sequence of rectangles with a common basis and with altitudes tending to zero has the basis for its lower limit.

II. Formulas. The following formulas hold.

1. $\overline{\mathrm{Li}\, A_n} = \mathrm{Li}\, A_n = \mathrm{Li}\, \bar{A}_n$.
2. $A_n \subset B_n$ *implies* $\mathrm{Li}\, A_n \subset \mathrm{Li}\, B_n$.
3. $\mathrm{Li}\, A_n \cup \mathrm{Li}\, B_n \subset \mathrm{Li}(A_n \cup B_n)$.

3a. $\bigcup_t \mathrm{Li}\, A_n(t) \subset \mathrm{Li}\big(\bigcup_t A_n(t)\big)$.

4. $\mathrm{Li}(A_n \cap B_n) \subset (\mathrm{Li}\, A_n) \cap (\mathrm{Li}\, B_n)$.

4a. $\mathrm{Li}\big(\bigcap_t A_n(t)\big) \subset \bigcap_t \mathrm{Li}\, A_n(t)$.

5. *If* $k_1 < k_2 < \ldots$, *then* $\mathrm{Li}\, A_n \subset \mathrm{Li}\, A_{k_n}$.
6. *If* $A_n = A$, *then* $\mathrm{Li}\, A_n = \bar{A}$.
7. $\mathrm{Li}\, A_n$ *is not affected by changing a finite number of the* A_n.
8. $\bigcap_n A_n \subset \mathrm{Lim\,inf}\, A_n \subset \mathrm{Li}\, A_n$.
9. $\mathrm{Li}(A_n \times B_n) = \mathrm{Li}\, A_n \times \mathrm{Li}\, B_n$.

Proof. If $q \in \overline{\mathrm{Li}\, A_n}$ and if G is an open neighbourhood of q, then there exists a point $p \in G \cap \mathrm{Li}\, A_n$. Hence, from a certain value of n onwards, $G \cap A_n \neq 0$ and $q \in \mathrm{Li}\, A_n$. Also, if H is open, the condition $H \cap A_n \neq 0$ is equivalent to $H \cap \bar{A}_n \neq 0$. Therefore $\mathrm{Li}\, A_n = \mathrm{Li}\, \bar{A}_n$ and formula 1 is proved.

Formulas 2, 5–7, and 9 follow directly from the definition. Formulas 3–4a are consequences of 2 (compare § 4, III). Finally,

$$\bigcap_{n=1}^{\infty} A_n \subset \operatorname*{Li}_{n=\infty} A_n; \qquad \text{hence, by 7,} \qquad \bigcap_{n=m}^{\infty} A_n \subset \operatorname*{Li}_{n=\infty} A_n.$$

This implies 8.

III. Upper limit.

DEFINITION. *The point p belongs to the upper limit* $\operatorname{Ls}_{n=\infty} A_n$ *of a sequence of sets $A_1, A_2, \ldots$, if every neighbourhood of p intersects an infinite set of the terms A_n.*

As in Section I, we prove that this condition is equivalent to *the existence of a sequence of points $\{p_{k_n}\}$ such that*

$$k_1 < k_2 < \ldots, \quad p = \lim_{n=\infty} p_{k_n}, \quad \textit{and} \quad p_{k_n} \in A_{k_n};$$

or, which means the same, to the *condition*

$$\liminf \varrho(p, A_n) = 0.$$

EXAMPLES. Let $\{r_n\}$ be the sequence of all rational points and let A_n be the set composed of the point r_n alone. $\operatorname{Ls} A_n$ is the set of all real numbers. In the example of rectangles considered in Section I, $\operatorname{Ls} A_n = \operatorname{Li} A_n$.

IV. Formulas. The following formulas hold:

1. $\overline{\operatorname{Ls} A_n} = \operatorname{Ls} A_n = \operatorname{Ls} \bar{A}_n$.

2. $A_n \subset B_n$ *implies* $\operatorname{Ls} A_n \subset \operatorname{Ls} B_n$.

3. $\operatorname{Ls}(A_n \cup B_n) = \operatorname{Ls} A_n \cup \operatorname{Ls} B_n$.

3a. $\bigcup_t \operatorname{Ls} A_n(t) \subset \operatorname{Ls}\big(\bigcup_t A_n(t)\big)$.

4. $\operatorname{Ls}(A_n \cap B_n) \subset (\operatorname{Ls} A_n) \cap (\operatorname{Ls} B_n)$.

4a. $\operatorname{Ls}\big(\bigcap_t A_n(t)\big) \subset \bigcap_t \operatorname{Ls} A_n(t)$.

5. *If* $k_1 < k_2 < \ldots$, *then* $\operatorname{Ls} A_{k_n} \subset \operatorname{Ls} A_n$.

6. *If* $A_n = A$, *then* $\operatorname{Ls} A_n = \bar{A}$.

7. $\operatorname{Ls} A_n$ *is not affected by changing a finite number of the* A_n.

8 (¹). $\operatorname{Ls} A_n = \bigcap_n \overline{A_n \cup A_{n+1} \cup \ldots} \subset \overline{\bigcup_n A_n} = \bigcup_n \bar{A}_n \cup \operatorname{Ls} A_n$.

9. $\operatorname{Ls}(A_n \times B_n) \subset \operatorname{Ls} A_n \times \operatorname{Ls} B_n$.

(¹) F. Hausdorff, *Grundzüge der Mengenlehre*, p. 237 (4). An analogous formula for $\operatorname{Li} A_n$ does not exist. See R. Engelking, Bull. Acad. Pol. Sc. 4 (1956), p. 659.

Formula 1 is proved as II, 1. To prove formula 3, let $p = \lim p_{k_n}$ and $p_{k_n} \in (A_{k_n} \cup B_{k_n})$. Then, for an infinite set of indices k_n, either $p_{k_n} \in A_{k_n}$, or $p_{k_n} \in B_{k_n}$. In the first case $p \in \mathrm{Ls}\, A_n$, and in the second case $p \in \mathrm{Ls}\, B_n$.

Formulas 2–4a are direct consequences of 3. Formulas 5–7 and 9 follow easily from the definition.

We shall prove formula 8. Let $p \in \mathrm{Ls}\, A_n$. Then $p = \lim p_{k_n}$ and $p_{k_n} \in A_{k_n}$. Since $k_n \geqslant n$, we have $p_{k_n} \in (A_i \cup A_{i+1} \cup \ldots)$ for $n > i$. Consequently $p \in \overline{A_i \cup A_{i+1} \cup \ldots}$ for each i.

Conversely, if p is not in $\mathrm{Ls}\, A_n$, then there exist a neighbourhood G of p and an index m such that $G \cap A_n = 0$ for $n \geqslant m$. Therefore p is not in $\overline{A_m \cup A_{m+1} \cup \ldots}$

The latter part of formula 8 follows directly from § 4, III, 9.

V. Relations between Li **and** Ls.

0. $\mathrm{Li}\, A_n \subset \mathrm{Ls}\, A_n$.

More precisely:

1. $\mathrm{Li}\, A_n = \bigcap \mathrm{Ls}\, A_{k_n} \subset \bigcup \mathrm{Li}\, A_{k_n} = \mathrm{Ls}\, A_n$,

where $\bigcap$ and $\bigcup$ are taken over all increasing sequences $\{k_n\}$.

Proof. On one hand, by II, 5 and IV, 5

$$\mathrm{Li}\, A_n \subset \bigcap \mathrm{Li}\, A_{k_n} \subset \bigcap \mathrm{Ls}\, A_{k_n}, \qquad \bigcup \mathrm{Li}\, A_{k_n} \subset \bigcup \mathrm{Ls}\, A_{k_n} \subset \mathrm{Ls}\, A_n$$

and, on the other hand:

(i) if $p \notin \mathrm{Li}\, A_n$, then there exist a neighbourhood G of p and a sequence $k_1 < k_2 < \ldots$ such that $G \cap A_{k_n} = 0$ for each n. Therefore $p \notin \mathrm{Ls}\, A_{k_n}$.

(ii) If $p \in \mathrm{Ls}\, A_n$, then there exists a sequence $\{p_{k_n}\}$ such that $p_{k_n} \in A_{k_n}$ and $p = \lim p_{k_n}$. Therefore $p \in \mathrm{Li}\, A_{k_n}$.

2. $\mathrm{Li}\, A_n - \mathrm{Ls}\, B_n \subset \mathrm{Li}(A_n - B_n)$.

Proof. Let $p \in (\mathrm{Li}\, A_n - \mathrm{Ls}\, B_n)$. Then $p = \lim p_n$, $p_n \in A_n$ and, since $p \notin \mathrm{Ls}\, B_n$, it follows that $p_n \notin B_n$ from a sufficiently large value of n onwards. Hence $p_n \in (A_n - B_n)$ and $p \in \mathrm{Li}(A_n - B_n)$.

3. $\mathrm{Li}(A_n \cup B_n) \subset \mathrm{Li}\, A_n \cup \mathrm{Li}\, B_n \cup (\mathrm{Ls}\, A_n \cap \mathrm{Ls}\, B_n)$.

In particular (by II, 3):

4. *If* $\mathrm{Ls}\, A_n \cap \mathrm{Ls}\, B_n = 0$, *then* $\mathrm{Li}(A_n \cup B_n) = \mathrm{Li}\, A_n \cup \mathrm{Li}\, B_n$.

Proof of formula 3. Let $p \in [\mathrm{Li}(A_n \cup B_n) - \mathrm{Ls}\, A_n \cap \mathrm{Ls}\, B_n]$. We must show that $p \in (\mathrm{Li}\, A_n \cup \mathrm{Li}\, B_n)$.

We can assume that $p \notin \mathrm{Ls}\, B_n$. Let $p = \lim p_n$ and $p_n \in (A_n \cup B_n)$; then $p_n \notin B_n$, for sufficiently large value of n. Consequently $p_n \in A_n$ and $p \in \mathrm{Li}\, A_n$.

VI. Limit.

DEFINITION. *The sequence of sets $\{A_n\}$ is said to be convergent to A, if* $\mathrm{Li}\, A_n = A = \mathrm{Ls}\, A_n$. We then write $A = \mathrm{Lim}\, A_n$ [1].

In particular, if A_n consists of a single point p_n, then the sequence $\{A_n\}$ is convergent, either if $\lim p_n$ exists (and then the set $\mathrm{Lim}\, A_n$ consists of the point $\lim p_n$ alone), or if the sequence $\{p_n\}$ does not contain any convergent subsequence (and then $\mathrm{Lim}\, A_n = 0$).

The following formulas hold (in 1–5 and 9 the sequences $\{A_n\}$ and $\{B_n\}$ are supposed to be convergent):

1. $\overline{\mathrm{Lim}\, A_n} = \mathrm{Lim}\, A_n = \mathrm{Lim}\, \overline{A}_n$.
2. $A_n \subset B_n$ *implies* $\mathrm{Lim}\, A_n \subset \mathrm{Lim}\, B_n$.

2a. *The conditions $p_n \in A_n$ and $p = \lim p_n$ imply $p \in \mathrm{Lim}\, A_n$.*

3. $\mathrm{Lim}(A_n \cup B_n) = \mathrm{Lim}\, A_n \cup \mathrm{Lim}\, B_n$.
4. *If* $k_1 < k_2 < \ldots$, *then* $\mathrm{Lim}\, A_{k_n} = \mathrm{Lim}\, A_n$.
5. *If* $A_n = A$, *then* $\mathrm{Lim}\, A_n = \overline{A}$.
6. *Neither the limit nor the convergence is affected by changing a finite number of the terms of the sequence.*
7. *If* $A_1 \subset A_2 \subset \ldots$, *then* $\mathrm{Lim}\, A_n = \overline{\bigcup_n A_n}$.
8. *If* $A_1 \supset A_2 \supset \ldots$, *then* $\mathrm{Lim}\, A_n = \bigcap_n \overline{A_n}$.
9. $\mathrm{Lim}(A_n \times B_n) = \mathrm{Lim}\, A_n \times \mathrm{Lim}\, B_n$.

Formulas 1, 2, 5, 6, and 9 follow directly from the corresponding formulas of Sections II and IV. Formula 2a follows from 2. Formulas 3 and 4 follow from the formulas

$$\mathrm{Ls}(A_n \cup B_n) = \mathrm{Ls}\, A_n \cup \mathrm{Ls}\, B_n$$
$$= \mathrm{Li}\, A_n \cup \mathrm{Li}\, B_n \subset \mathrm{Li}(A_n \cup B_n) \subset \mathrm{Ls}(A_n \cup B_n),$$
$$\mathrm{Lim}\, A_n = \mathrm{Li}\, A_n \subset \mathrm{Li}\, A_{k_n} \subset \mathrm{Ls}\, A_{k_n} \subset \mathrm{Ls}\, A_n = \mathrm{Lim}\, A_n.$$

[1] This should not be confused with $\mathop{\mathrm{Limes}}_{n=\infty} A_n$ in the sense of the general set theory. See § 1, V and Section I, footnote.

The hypothesis of proposition 7 implies that

$$A_n = A_n \cap A_{n+1} \cap \ldots; \quad \text{hence} \quad \bigcup_n A_n = \bigcup_n (A_n \cap A_{n+1} \cap \ldots)$$

and by II, 8 and IV, 8

$$\overline{\bigcup_n A_n} \subset \operatorname{Li} A_n \subset \operatorname{Ls} A_n \subset \overline{\bigcup_n A_n}.$$

Similarly, the hypothesis of proposition 8 implies the identity $A_n = A_n \cup A_{n+1} \cup \ldots$ and

$$\bigcap_n \overline{A_n} \subset \operatorname{Li} \overline{A_n} = \operatorname{Li} A_n \subset \operatorname{Ls} A_n = \bigcap_n \overline{A_n}.$$

VII. Relativization. Given a set E, the lower limit of a sequence A_n of subsets of E relative to E is the set of points $p \in E$ such that for every open set G containing p, the condition $G \cap E \cap A_n \neq 0$ holds for each sufficiently great n. It follows immediately that the relative lower limit with respect to E coincides with the set $E \cap \operatorname{Li} A_n$.

Similarly, the upper limit relative to E is $E \cap \operatorname{Ls} A_n$ and the limit relative to E is $E \cap \operatorname{Lim} A_n$.

VIII. Generalized Bolzano–Weierstrass theorem. *Every sequence of subsets of a separable space contains a convergent subsequence (whose limit may be empty)* (1).

Proof. Let $R_1, R_2, \ldots$ be a base of the space and $A_1, A_2, \ldots$ be the given sequence of sets. Define the sets A_i^n as follows:

1) $A_i^1 = A_i$, for every i,

2) if, for an $n > 1$, there exists a sequence $k_1 < k_2 < \ldots$ with $R_n \cap \operatorname{Ls}_{i=\infty} A_{k_i}^{n-1} = 0$, let $A_i^n = A_{k_i}^{n-1}$ (the sequence $\{k_i\}$ is chosen arbitrarily);

3) if there exists no such sequence, let $A_i^n = A_i^{n-1}$.

We shall prove that the sequence $D_n = A_n^n$ is convergent.

Suppose that $\operatorname{Ls} D_n \neq \operatorname{Li} D_n$. By V, 1, there exists a subsequence $\{D_{j_n}\}$ with $\operatorname{Ls} D_{j_n} \neq \operatorname{Ls} D_n$. The two latter sets being closed, there exists an m such that

$$R_m \cap \operatorname{Ls} D_{j_n} = 0, \tag{i}$$

$$R_m \cap \operatorname{Ls} D_n \neq 0. \tag{ii}$$

(1) See F. Hausdorff, *Mengenlehre*, p. 148, C. Zarankiewicz, Fund. Math. 9 (1927), p. 124, P. Urysohn, Verhandl. Akad. Amsterdam 13 (1927), p. 29. Compare also T. Ważewski, Ann. Soc. Pol. Math. 2 (1923), p. 72, and Fund. Math. 4 (1923), p. 229; R. G. Lubben, Trans. Amer. Math. Soc. 29 (1928), p. 668.

The sequence $\{D_{j_n}\}$ is a subsequence of $\{D_n\}$, and $\{D_n\}$ from the $(m-1)$th term on, is a subsequence of $\{A_i^{m-1}\}$, where $i = 1, 2, \ldots$ It follows by (i) that this is case 2) of the definition, if n is replaced by m. Therefore $R_m \cap \operatorname{Ls}_{i=\infty} A_i^m = 0$. Since $\{D_n\}$ is a subsequence of $\{A_i^m\}$ (except for the first m terms), it follows that $R_m \cap \operatorname{Ls} D_n = 0$, contrary to (ii).

COROLLARY. *In every separable space*

$$\operatorname{Li} A_n = \bigcap{}' \operatorname{Lim} A_{k_n} \quad \textit{and} \quad \operatorname{Ls} A_n = \bigcup{}' \operatorname{Lim} A_{k_n},$$

where $\bigcap'$ and $\bigcup'$ are extended over all convergent subsequences $\{A_{k_n}\}$.

Proof. To prove the first formula, it is sufficient, by V, 1, to show that if $p \notin \operatorname{Li} A_n$, then there exists a convergent sequence $\{A_{k_n}\}$ such that $p \notin \operatorname{Lim} A_{k_n}$. By hypothesis there exists an open neighbourhood G of p and a sequence $\{A_{j_n}\}$ such that $A_{j_n} \cap G = 0$. $\{A_{k_n}\}$ denotes a convergent subsequence of $\{A_{j_n}\}$.

Let us prove the second formula. By V, 1, $\bigcup' \operatorname{Lim} A_{k_n} \subset \operatorname{Ls} A_n$. On the other hand, if $p \in \operatorname{Ls} A_n$, then there exists a subsequence $\{A_{k_n}\}$ such that $p \in \operatorname{Li} A_{k_n}$; as before, there exists a convergent subsequence $\{A_{k_{j_n}}\}$ such that $p \in \operatorname{Lim} A_{k_{j_n}} \subset \bigcup' \operatorname{Lim} A_{k_n}$.

IX. The space $(2^{\mathscr{X}})_L$.

DEFINITION. This symbol denotes the space whose elements are closed subsets of $\mathscr{X}$, the limit being understood in the sense of Definition VI.

THEOREM 1. *$(2^{\mathscr{X}})_L$ is an $\mathscr{L}^*$-space.*

Proof. Conditions (i) and (ii) of § 20, I are fulfilled by VI, 4 and 5. To prove condition (iii) assume that for each sequence $k_1 < k_2 < \ldots$ there exists a sequence $m_1 < m_2 < \ldots$ such that

$$\operatorname{Lim}_{n=\infty} A_{k_{m_n}} = A. \tag{1}$$

We must show that $A = \operatorname{Lim} A_n$, i.e. $\operatorname{Ls} A_n \subset A \subset \operatorname{Li} A_n$.

Let $p \in \operatorname{Ls} A_n$. By V, 1 there exists a sequence $k_1 < k_2 < \ldots$ such that $p \in \operatorname{Li} A_{k_n}$. Let $m_1 < m_2 < \ldots$ be a sequence satisfying (1). Then, by II, 5,

$$p \in \operatorname{Li} A_{k_n} \subset \operatorname{Li} A_{k_{m_n}} = A; \quad \text{hence} \quad \operatorname{Ls} A_n \subset A.$$

Now let $p \in A$. Let $k_1 < k_2 < \dots$ be an arbitrary sequence and let $m_1 < m_2 < \dots$ be the corresponding sequence subjected to (1). Then (compare IV, 5)

$$p \in A = \operatorname*{Lim}_{n=\infty} A_{k_{m_n}} = \operatorname{Ls} A_{k_{m_n}} \subset \operatorname{Ls} A_{k_n}$$

and (by V, 1) $p \in \operatorname{Li} A_n$. Therefore $A \subset \operatorname{Li} A_n$.

Remark 1. The space $(2^{\mathscr{X}})_L$ may fail to be topological.

For instance, let the space $\mathscr{X}$ be composed of 1) the number 0; 2) the numbers $1/n+1/k$, where $n = 2, 3, \dots$, $k = 1, 2, \dots$ and $1/n+1/k < 1/(n-1)$; 3) the number 1; 4) the numbers $1+1/n$.

Let $\boldsymbol{A}$ be the family of all pairs $(1/n+1/k, 1+1/n)$. Hence every number of the form $1+1/n$ is an element of $\bar{\boldsymbol{A}}$. Consequently, the number 1 belongs to $\bar{\bar{\boldsymbol{A}}}$, but not to $\bar{\boldsymbol{A}}$ and $\bar{\bar{\boldsymbol{A}}} \neq \bar{\boldsymbol{A}}$.

Remark 2. It is important to observe that the space $(2^{\mathscr{X}})_L$ is entirely different from the spaces $2^{\mathscr{X}}$ (see § 17, I) and $(2^{\mathscr{X}})_m$ (see § 21, VII). The space $(2^{\mathscr{X}})_m$ is always metric, while $(2^{\mathscr{X}})_L$ does not need to be metrizable (as seen in the above example). We shall show that if $\mathscr{X}$ is separable, then $(2^{\mathscr{X}})_L$ is separable, while $(2^{\mathscr{X}})_m$ may fail to be separable (see § 21, VIII, Remark 2).

THEOREM 2. *If* $A_n \in (2^{\mathscr{X}})_m$ *and* $A \in (2^{\mathscr{X}})_m$, *then the condition* $\lim_{n=\infty} \operatorname{dist}(A_n, A) = 0$ *implies* $\operatorname*{Lim}_{n=\infty} A_n = A$. *Therefore* $(2^{\mathscr{X}})_L$ *is a one-to-one continuous image of* $(2^{\mathscr{X}})_m$ ($\mathscr{X}$ *is assumed to be bounded*) (1).

Proof. Let $\lim_{n=\infty} \operatorname{dist}(A_n, A) = 0$. For every $\varepsilon > 0$ there exists an $n(\varepsilon)$ such that $\operatorname{dist}(A_n, A) < \varepsilon$ for $n > n(\varepsilon)$. In other words

(1) if $x \in A$, then $\varrho(A_n, x) < \varepsilon$ for every $n > n(\varepsilon)$,

(2) there exists a sequence ε_n tending to 0 such that $\varrho(y, A) < \varepsilon_n$ for every $y \in A_n$.

We have to show that $\operatorname{Lim} A_n = A$, i.e. that $A \subset \operatorname{Li} A_n$ and $\operatorname{Ls} A_n \subset A$.

First, let $x \in A$ and let G be a neighbourhood of x. If $\varepsilon > 0$ is small enough, then (by (1)) $G \cap A_n \neq 0$, for every $n > n(\varepsilon)$. Hence $x \in \operatorname{Li} A_n$.

(1) Compare F. Hausdorff, *Mengenlehre*, p. 149.

Next, let $x \in \mathrm{Ls} A_n$. Then $x = \lim_{n=\infty} y_{k_n}$, where $y_{k_n} \in A_{k_n}$. By (2), there is a point $a_{k_n} \in A$ such that $|y_{k_n} - a_{k_n}| < \varepsilon_n$. Since $\lim_{n=\infty} \varepsilon_n = 0$, it follows that $x = \lim_{n=\infty} a_{k_n}$, whence $x \in A$.

THEOREM 3. *If $\mathscr{X}$ is a separable metric space, then $(2^{\mathscr{X}})_L$, as well as each subset of this space, is separable.*

Proof. Let $\mathscr{X}_1$ be a totally bounded space homeomorphic to $\mathscr{X}$ (compare § 22, II, 1a). The space $(2^{\mathscr{X}})_L$ being defined topologically, is homeomorphic to $(2^{\mathscr{X}_1})_L$. Since $\mathscr{X}_1$ is totally bounded, $(2^{\mathscr{X}_1})_m$ is separable (§ 21, VIII, Theorems 2 and 3). As $(2^{\mathscr{X}_1})_L$ and $(2^{\mathscr{X}})_L$ are continuous images of $(2^{\mathscr{X}_1})_m$, they are separable. It follows that every subset of $(2^{\mathscr{X}})_L$, as a continuous image of a separable set, is separable (see § 13, IV, Theorem 3).

Remark 3. The notions of accumulation point and condensation point are invariant under one-to-one continuous mappings (of $\mathscr{L}^*$ spaces). It follows by Theorem 2 that *if A is an uncountable subset of $(2^{\mathscr{X}})_L$, then every element of A, except for a countable set, is a condensation element* (1); *and, moreover, A contains a dense in itself set.* Consequently *every scattered set A is countable* (see § 23, III and V).

THEOREM 4. *If $\mathscr{X}$ is separable metric, then $(2^{\mathscr{X}})_L$ is countably compact.*

This is a consequence of Theorem VIII (together with VI, 1).

THEOREM 5. *If A and B are two disjoint closed subsets of a metric space $\mathscr{X}$, then*

$$(2^{A \cup B})_L \underset{\text{top}}{=} (2^A)_L \times (2^B)_L.$$

Proof. To each pair of closed sets X, Y, where $X \subset A$ and $Y \subset B$, let us assign the union $F(X, Y) = X \cup Y$. It is obvious that the function F is a one-to-one mapping of $(2^A)_L \times (2^B)_L$ onto $(2^{A \cup B})_L$.

The continuity of F is a direct consequence of formula VI, 3. It remains to prove that the inverse of F is continuous, i.e. that $\mathrm{Lim}(X_n \cup Y_n) = X \cup Y$ implies $\mathrm{Lim}\, X_n = X$ and $\mathrm{Lim}\, Y_n = Y$.

---

(1) Compare C. Zarankiewicz, Fund. Math. 11 (1928), p. 129.

By IV, 3,

$$\mathrm{Ls}\, X_n \cup \mathrm{Ls}\, Y_n = \mathrm{Ls}(X_n \cup Y_n) = \mathrm{Lim}(X_n \cup Y_n) = X \cup Y.$$

Hence $\mathrm{Ls}\, X_n = X$ and $\mathrm{Ls}\, Y_n = Y$, because $A \cap B = 0$ and

$$X \subset A, \quad X_n \subset A \quad \text{imply} \quad \mathrm{Ls}\, X_n \subset A,$$
$$Y \subset B, \quad Y_n \subset B \quad \text{imply} \quad \mathrm{Ls}\, Y_n \subset B.$$

Since $\mathrm{Ls}\, X_n \cap \mathrm{Ls}\, Y_n = 0$, it follows by V, 4 that

$$\mathrm{Li}\, X_n \cup \mathrm{Li}\, Y_n = \mathrm{Li}(X_n \cup Y_n) = \mathrm{Lim}(X_n \cup Y_n) = X \cup Y.$$

It follows, as above, that $\mathrm{Li}\, X_n = X$ and $\mathrm{Li}\, Y_n = Y$. Therefore $\mathrm{Lim}\, X_n = X$ and $\mathrm{Lim}\, Y_n = Y$.

*§ 30. Borel sets

I. Equivalences. In § 5, VI we have defined the family of Borel sets to be the smallest family $\boldsymbol{F}$ subjected to the following conditions

1. *Every closed set belongs to* $\boldsymbol{F}$.
2. *If* $X \in \boldsymbol{F}$, *then* $(-X) \in \boldsymbol{F}$.
3. *If* $X_n \in \boldsymbol{F}$, *then* $(X_1 \cap X_2 \cap \ldots) \in \boldsymbol{F}$.

Condition 3 can be replaced by the following condition.

3'. *If* $X_n \in \boldsymbol{F}$, *then* $(X_1 \cup X_2 \cup \ldots) \in \boldsymbol{F}$.

We have also proved that every closed set in a metric space is a $\boldsymbol{G}_\delta$-set and, consequently, every open set is an $\boldsymbol{F}_\sigma$-set (§ 21, IV). By using these facts we shall now prove the following theorem(1).

THEOREM. *The family of Borel sets is the smallest family subjected to conditions* 1, 3, *and* 3'.

Proof. Let $\boldsymbol{F}^*$ be the family in question. We must show that $\boldsymbol{F} = \boldsymbol{F}^*$.

The family $\boldsymbol{F}$ satisfies 1, 3, and 3', so that $\boldsymbol{F}^* \subset \boldsymbol{F}$.

To prove that $\boldsymbol{F} \subset \boldsymbol{F}^*$, let $\boldsymbol{F}^0$ be the family of the complements of sets belonging to $\boldsymbol{F}^*$. It satisfies condition 1, for the complement of a closed set, being open, is a countable union of closed sets;

(1) Compare W. Sierpiński, *Sur les définitions axiomatiques des ensembles mesurables* (*B*), Bull. Acad. Cracovie 1918, p. 29. For a generalization in the set theory, see W. Sierpiński, *Les ensembles boreliens abstraits*, Ann. Soc. Polonaise de Math. 6 (1927), p. 51.

hence it belongs to $\boldsymbol{F}^*$. Applying the de Morgan formulas we see that $\boldsymbol{F}^0$ also satisfies conditions 3 and 3'. Therefore $\boldsymbol{F}^* \subset \boldsymbol{F}^0$ which means that every set of $\boldsymbol{F}^*$ is the complement of a set which also belongs to $\boldsymbol{F}^*$. It follows that $\boldsymbol{F}^*$ satisfies condition 2. Hence $\boldsymbol{F} \subset \boldsymbol{F}^*$.

II. Classification of Borel sets. By a simple argument of the set theory we can prove [1] that the family of Borel sets (the smallest family $\boldsymbol{F}$ satisfying conditions 1, 3, and 3') is a transfinite union (of type Ω)

$$\boldsymbol{F} = \boldsymbol{F}_0 \cup \boldsymbol{F}_1 \cup \ldots \cup \boldsymbol{F}_a \cup \ldots$$

of families $\boldsymbol{F}_0, \boldsymbol{F}_1, \ldots, \boldsymbol{F}_a, \ldots$ such that

(i) $\boldsymbol{F}_0$ is the family of closed sets;

(ii) the sets of the family $\boldsymbol{F}_a$ are countable intersections or unions of sets belonging to $\boldsymbol{F}_\xi$ with $\xi < a$ according to whether a is even or odd (the limit ordinals are understood to be even).

By replacing the term "closed" with "open" in condition I, 1, (compare § 5, VI), we obtain the following classification:

$$\boldsymbol{F} = \boldsymbol{G}_0 \cup \boldsymbol{G}_1 \cup \ldots \cup \boldsymbol{G}_a \cup \ldots$$

where

(i) $\boldsymbol{G}_0$ is the family of open sets;

(ii) the sets of the family $\boldsymbol{G}_a$ are countable unions or intersections of sets belonging to $\boldsymbol{G}_\xi$ with $\xi < a$ depending whether a is even or odd.

III. Properties of the classes $\boldsymbol{F}_a$ and $\boldsymbol{G}_a$. The families $\boldsymbol{F}_a$ with even indices as well as the families $\boldsymbol{G}_a$ with odd indices are countably multiplicative, which means that, given a sequence of sets of the family, its intersection belongs to the same family. The sets belonging to such a family will be said to be of *multiplicative class* a. Similarly, the families $\boldsymbol{F}_a$ with odd indices as well as the families $\boldsymbol{G}_a$ with even indices are additive and form the *additive*

---

[1] Compare F. Hausdorff, *Mengenlehre*, p. 85. See also W. H. Young, Proc. London Math. Soc. (2) 12 (1913), p. 260.

class α. Thus the multiplicative class α (additive class α) consists of the intersections (unions) of sets of classes $< \alpha$ [1].

The classes with finite indices are denoted as follows:

$$\boldsymbol{F}_\sigma, \boldsymbol{F}_{\sigma\delta}, \boldsymbol{F}_{\sigma\delta\sigma}, \ldots, \qquad \boldsymbol{G}_\delta, \boldsymbol{G}_{\delta\sigma}, \boldsymbol{G}_{\delta\sigma\delta}, \ldots .$$

The following properties of $\boldsymbol{F}_\sigma$ and $\boldsymbol{G}_\delta$-sets (compare § 5, V) extend by induction to classes of an arbitrary index.

The complement of a set of class $\boldsymbol{F}_\alpha$ is of class $\boldsymbol{G}_\alpha$. *The finite union* and *finite intersection* of sets of the same class belong to that class. Every set of an additive class α is the union of an *increasing* sequence of sets of indices $\xi < \alpha$. Every set of a multiplicative class α is the intersection of a *decreasing* sequence of sets of indices $\xi < \alpha$. A set is of class $\boldsymbol{F}_\alpha$ (of class $\boldsymbol{G}_\alpha$) *relative* to a set E if and only if it is the intersection of E with a set of class $\boldsymbol{F}_\alpha$ (of class $\boldsymbol{G}_\alpha$). Given a continuous function f defined on $\mathscr{X}$, if Y is of class $\boldsymbol{F}_\alpha$ (of class $\boldsymbol{G}_\alpha$), then so is $f^{-1}(Y)$ (compare § 13, IV (3)–(5)).

Observe that a Borel set of class α belongs to every ($\boldsymbol{F}$ or $\boldsymbol{G}$) class of a *greater index*. This follows by induction from the fact that every open set is an $\boldsymbol{F}_\sigma$-set and every closed set is a $\boldsymbol{G}_\delta$-set.

The cartesian product of two sets of class $\boldsymbol{F}_\alpha$ (of class $\boldsymbol{G}_\alpha$) belongs to the same class. For, the product of two open sets is open and the product of two closed sets is closed and (see § 2, II (8) and (9)):

$$(\bigcup_n A_n)\times(\bigcup_m B_m) = \bigcup_{nm} (A_n\times B_m),$$

$$(\bigcap_n A_n)\times(\bigcap_m B_m) = \bigcap_{nm} (A_n\times B_m).$$

In particular, the product of a set by an axis does not change its class. From this and from the formula (§ 3, VIII, (1))

$$\mathop{P}_i A_i = \bigcap_i (\mathscr{X}_1\times \ldots \times\mathscr{X}_{i-1}\times A_i\times \mathscr{X}_{i+1}\times \ldots)$$

we deduce that the *countable cartesian product* of sets of multiplicative class α is of the multiplicative class α (the analogous theorem for additive classes is not true, even for open sets; compare § 16, V).

---

[1] If α is finite, then the sets of the multiplicative (additive) class α coincide with $\boldsymbol{F}$ sets ($\boldsymbol{O}$ sets) of the class α in the sense of Lebesgue. Compare also W. Sierpiński, *Sur les rapports entre les classifications des ensembles de MM. F. Hausdorff et Ch. de la Vallée Poussin*, Fund. Math. 19 (1932), p. 257.

A countable cartesian product of Borel sets is a Borel set.

THEOREM. *If a subset* Z *of the cartesian product* $\mathcal{X}\times\mathcal{Y}$ *is of class* F_a *or of class* G_a*, then the sets*

$$A_y = \underset{x}{E}[(x, y)\in Z] \quad \textit{and} \quad \underset{x}{E}[(x, x)\in Z] \tag{1}$$

are of the same class (the second conclusion concerns the case where $\mathcal{X} = \mathcal{Y}$).

Proof. The sets $Z\cap \underset{xy}{E}(y = y_0)$ and $Z\cap \underset{xy}{E}(x = y)$ are of class F_a (G_a) relative to $\underset{xy}{E}(y = y_0)$ and $\underset{xy}{E}(x = y)$, respectively; the same is true of the sets (1), by § 15, VI and IV (where we put $\varphi(x, y) \equiv (x, y)\in Z$).

In a separable space, the family of open sets (as well as the family of closed sets) has *power* $\leqslant \mathfrak{c}$ (§ 24, I). The same is true of each Borel class. Since the family of Borel sets is the union of $\aleph_1$ classes, it follows that its power is $\leqslant \mathfrak{c}\cdot\aleph_1 = \mathfrak{c}$. Therefore *in every separable space of power of the continuum there exist non-borelian sets.*

The problem of an effective definition of a non-borelian set in the space of real numbers will be treated in § 38, VI.

The problem of proving (without the continuum hypothesis) the existence of a non-borelian set in every uncountable separable space is open.

IV. Ambiguous Borel sets. A set which belongs to both F_a and G_a is said to be *ambiguous of class* a. For instance, a set both closed and open is ambiguous of class 0; it is ambiguous of class 1, if it is both F_σ and G_δ.

Every Borel set of class a is ambiguous of the class $a+1$.

Clearly the complement of an ambiguous set is ambiguous (of the same class). Therefore the ambiguous sets of class a form a field; thus, the union, intersection and difference of two ambiguous sets of class a are ambiguous of class a.

V. Decomposition of Borel sets into disjoint sets.

THEOREM 1. *Every set of additive class* $a > 0$ *is a countable union of disjoint ambiguous sets of class* a.

Proof. Let $A = A_1\cup A_2\cup\ldots\cup A_n\cup\ldots$. Then

$$A = A_1\cup[A_2-A_1]\cup\ldots\cup[A_n-(A_1\cup\ldots\cup A_{n-1})]\cup\ldots. \tag{0}$$

If each A_n is of multiplicative class $< \alpha$, then it is ambiguous of class α. Hence the terms of (0) are disjoint ambiguous sets of class α.

THEOREM 2. *Every set of the additive class $\alpha > 1$ is a countable union of disjoint sets of multiplicative classes $< \alpha$* (1).

Proof. Consider the decomposition (0). Since each A_n is of a multiplicative class $< \alpha$, then so is the union $A_1 \cup \ldots \cup A_{n-1}$. Therefore the set $1-(A_1 \cup \ldots \cup A_{n-1})$ is of additive class $< \alpha$. By Theorem 1 this set is of the form $\bigcup_{i=1}^{\infty} B_i^n$, where B_i^n are disjoint ambiguous sets of classes $< \alpha$ (for $\alpha > 1$). Thus

$$A = \bigcup_{n=1}^{\infty} \bigcup_{i=1}^{\infty} A_n \cap B_i^n$$

is the desired decomposition, because $A_n \cap B_i^n$ is of a multiplicative class $< \alpha$.

THEOREM 3. *The family of Borel sets is the smallest family that contains:*

(i) *all open sets;*

(ii) *the intersection of its elements;*

(iii) *disjoint unions of its elements.*

Let $\boldsymbol{H}$ be a family satisfying (i)–(iii). We shall prove by induction that every Borel set is in $\boldsymbol{H}$. By (i) and (ii), every $\boldsymbol{G}_\delta$-set, and, by Theorem 2, every $\boldsymbol{G}_{\delta\sigma}$-set is in $\boldsymbol{H}$. Hence every $\boldsymbol{F}_\sigma$-set belongs to $\boldsymbol{H}$.

Now let $\alpha > 1$ and assume that every set of class $< \alpha$ belongs to $\boldsymbol{H}$. By (ii), the sets of multiplicative class α are in $\boldsymbol{H}$ and, by Theorem 2 and (iii), every set of additive class α is in $\boldsymbol{H}$. Consequently all Borel sets of class α are in $\boldsymbol{H}$. This completes the proof.

Remarks. If the space is separable of dimension 0, then Theorem 1 is also true for $\alpha = 0$, i.e. an open set is a countable union of disjoint sets which are both closed and open (§ 26, I, Corollary 1a). It follows that Theorem 2 is valid for $\alpha = 1$, i.e.

---

(1) Theorem of N. Lusin. See W. Sierpiński, *Sur une classification des ensembles mesurables (B)*, Fund. Math. 10 (1927), p. 324.

every F_σ-set in a 0-dimensional space is a countable union of disjoint closed sets.

The latter assertion does not hold in spaces of dimension > 0. For instance, the open interval cannot be decomposed into a countable union of disjoint closed sets (see § 46, Vol. II).

VI. Alternated series of Borel sets.

THEOREM 1. *Let*

$$A_1 \supset A_2 \supset \ldots \supset A_\xi \supset A_{\xi+1} \supset \ldots \supset A_\gamma$$

be a countable transfinite series of ambiguous sets of class a *such that*

$$A_\lambda = \bigcap_{\xi<\lambda} A_\xi, \text{ if } \lambda \text{ is a limit ordinal or if } \lambda = \gamma.$$

Under this condition the set

$$S = A_1 - A_2 \cup A_3 - A_4 \cup \ldots \cup A_{\omega+1} - A_{\omega+2} \cup \ldots$$

is ambiguous of class a.

Proof. By § 12, I (4a),

$$-S = -A_1 \cup A_2 - A_3 \cup \ldots \cup A_\omega - A_{\omega+1} \cup \ldots \cup A_\gamma.$$

Each difference $A_\xi - A_{\xi+1}$ is ambiguous of class a, and the sets S and $-S$ are countable unions of such differences. Hence S is ambiguous of class a.

THEOREM 2. *If the sets* $A_1 \supset A_2 \supset \ldots$ *are ambiguous of class* a *and* $\bigcap_{n=1}^{\infty} A_n = 0$, *then the set* $\bigcup_{n=1}^{\infty} (A_{2n-1} - A_{2n})$ *is ambiguous of class* a.

This is a particular case of Theorem 1.

THEOREM 3. *The countable alternating series*

$$B = B_1 - B_2 \cup B_3 - B_4 \cup \ldots \cup B_{\omega+1} - B_{\omega+2} \cup \ldots$$

of decreasing Borel sets of multiplicative class a *is an ambiguous set of class* $a+1$ [1].

Proof. If λ is a limit ordinal, let $B_\lambda = \bigcap_{\xi<\lambda} B_\xi$. A countable intersection of sets B_ξ is of multiplicative class a, hence it is ambiguous of class $a+1$. It follows from Theorem 1 that B is ambiguous of class $a+1$.

---

[1] The inverse theorem is true in complete spaces (see § 37).

VII. Reduction and separation theorems.

THEOREM 1. (Reduction theorem.) *For every (finite or infinite) sequence* $G_1, G_2, \ldots$ *of sets of additive class* $a > 0$ *there exists a sequence* $H_1, H_2, \ldots$ *of disjoint sets of additive class* $a > 0$ *such that*

$$H_i \subset G_i \quad \text{and} \quad H_1 \cup H_2 \cup \ldots = G_1 \cup G_2 \cup \ldots .$$

Consequently, *if* $\mathcal{X} = G_1 \cup G_2 \cup \ldots$, *then the sets* H_i *are ambiguous of class* a.

Proof. According to V, Theorem 1, let

$$G_i = F_{i,1} \cup F_{i,2} \cup \ldots$$

where $F_{i,j}$ is ambiguous of class a. The rest of the proof is the same as the proof of the reduction theorem of § 26, II.

Similarly, we deduce from Theorem 1 the following theorems.

THEOREM 2. (Separation theorem.) (1) *If* $F_1, F_2, \ldots$ *is a sequence of sets of multiplicative class* $a > 0$ *such that* $F_1 \cap F_2 \cap \ldots = 0$, *then there exists a sequence* $E_1, E_2, \ldots$ *of ambiguous sets of class* a *such that*

$$F_i \subset E_i \quad \text{and} \quad E_1 \cap E_2 \cap \ldots = 0 .$$

In particular, *if* A *and* B *are two disjoint sets of multiplicative class* $a > 0$, *then there exists an ambiguous set* E *of class* a *such that*

$$A \subset E \quad \text{and} \quad E \cap B = 0 . \tag{1}$$

In other words, *if* $A \subset C$ *are two sets of class* $a > 0$ *such that* A *is of the multiplicative class and* C *is of the additive class, then there exists an ambiguous set* E *of class* a *such that*

$$A \subset E \subset C . \tag{2}$$

THEOREM 3. *If* $F_1, F_2, \ldots$ *is a sequence of sets of the multiplicative class* $a > 0$, *then there exists a sequence* $B_1, B_2, \ldots$ *of the additive class* a *such that*

$$F_i - \bigcap_{m=1}^{\infty} F_m \subset B_i \quad \text{and} \quad \bigcap_{i=1}^{\infty} B_i = 0 .$$

(1) See W. Sierpiński, *Sur la séparabilité multiple des ensembles mesurables B*, Fund. Math. 23 (1934), p. 295. For a partiuclar case, see W. Sierpiński, *Sur une propriété des ensembles ambigus*, Fund. Math. 6 (1924), p. 1. Many applications to the general theory of functions are given there.

THEOREM 4. *If $A_1, \ldots, A_k$ is a finite system of disjoint sets of multiplicative class $\alpha > 0$, then there exists a system $F_1, \ldots, F_k$ of disjoint ambiguous sets of class α such that*

$$A_i \subset F_i \quad \textit{and} \quad \mathscr{X} = F_1 \cup \ldots \cup F_k.$$

VIII. Relatively ambiguous sets.

THEOREM 1. *If A is of multiplicative class $\alpha > 0$ and B is ambiguous of class α relative to A, then the set B can be written as $B = A \cap C$, where C is ambiguous of class α (relative to the entire space).*

Proof. By hypothesis, B and $A - B$ are of the multiplicative class α with respect to A. Since A is itself of the multiplicative class α, hence B and $A - B$ are of the multiplicative class α (in the space). By the separation theorem (VII, Theorem 2 (1)), there exists an ambiguous set C of class α such that

$$B \subset C \quad \text{and} \quad C \cap A - B = 0.$$

Therefore

$$B = A \cap B \subset A \cap C \quad \text{and} \quad C \cap A \subset B; \quad \text{hence} \quad B = A \cap C.$$

For further applications, we shall prove the next theorem.

THEOREM 2. *If $\{B_{i_1 \ldots i_n}\}$ is a system of ambiguous sets of class α with respect to A such that*

$$B_{i_1 \ldots i_n} \cap B_{j_1 \ldots j_k} = 0, \quad \textit{if} \quad (i_1 \ldots i_k) \neq (j_1 \ldots j_k) \quad \textit{and} \quad k \leqslant n,$$

then there exists a system $\{C_{i_1 \ldots i_n}\}$ of ambiguous sets of class α with respect to the union

$$A_n = \bigcup C_{i_1 \ldots i_n} \tag{0}$$

(the union is taken over all systems of n indices) such that

$$A_n \textit{ is ambiguous of class } \alpha+1, \tag{1}$$

$$A \cap C_{i_1 \ldots i_n} = B_{i_1 \ldots i_n}, \tag{2}$$

$$C_{i_1 \ldots i_n} \cap C_{j_1 \ldots j_k} = 0, \tag{3}$$

$$C_{i_1 \ldots i_n} = 0, \quad \textit{if} \quad B_{i_1 \ldots i_n} = 0. \tag{4}$$

Moreover, *if A is of multiplicative class $\alpha > 0$ and if*

$$A = \bigcup B_{i_1 \ldots i_n} \quad \textit{for each } n,$$

then A_n coincide with the entire space.

Proof. By hypothesis there exists a system $\{D_{i_1 \dots i_n}\}$ of sets of additive class α (even ambiguous of class α, if A is of multiplicative class $\alpha > 0$) such that

$$A \cap D_{i_1 \dots i_n} = B_{i_1 \dots i_n}, \quad \text{and} \quad D_{i_1 \dots i_n} = 0 \text{ if } B_{i_1 \dots i_n} = 0.$$

Let V_n be the union of all intersections of the form $D_{i_1 \dots i_n} \cap \cap D_{j_1 \dots j_k}$ and let

$$C_{i_1 \dots i_n} = D_{i_1 \dots i_n} - V_n.$$

The set $A_n = \bigcup D_{i_1 \dots i_n} - V_n$ is ambiguous of class $\alpha+1$ as the difference of two sets of the additive class α. The relation

$$A_n \cap D_{i_1 \dots i_n} = D_{i_1 \dots i_n} - V_n = C_{i_1 \dots i_n}$$

implies that $C_{i_1 \dots i_n}$ is of the additive class α with respect to A_n; hence by the relation

$$C_{i_1 \dots i_n} \cap C_{j_1 \dots j_n} = D_{i_1 \dots i_n} \cap D_{j_1 \dots j_n} - V_n = 0$$

we infer that $C_{i_1 \dots i_n}$ is ambiguous of class α with respect to A_n,

$$A \cap D_{i_1 \dots i_n} \cap D_{j_1 \dots j_k} = B_{i_1 \dots i_n} \cap B_{j_1 \dots j_k} = 0$$

gives $A \cap V_n = 0$, and hence

$$A \cap C_{i_1 \dots i_n} = A \cap D_{i_1 \dots i_n} - A \cap V_n = B_{i_1 \dots i_n}.$$

Consider now the case where A is of multiplicative class $\alpha > 0$ and $A = \bigcup B_{i_1 \dots i_n}$. We have

$$B_{i_1 \dots i_n} = \bigcup_k B_{i_1 \dots i_n k} \subset \bigcup_k D_{i_1 \dots i_n k}$$

and there exists (compare VII, Theorem 2 (2)) an ambiguous set $E_{i_1 \dots i_n}$ of class α such that

$$B_{i_1 \dots i_n} \subset E_{i_1 \dots i_n} \subset \bigcup_k D_{i_1 \dots i_n k}.$$

In this case, the sets $C_{i_1 \dots i_n}$ (where $n \geqslant 0$) are defined by induction by letting

1) C be the entire space and $B = A$,

2) if $i_1 \dots i_n$ is a given system and l is the smallest index with $B_{i_1 \dots i_n l} \neq 0$, then

$$C_{i_1 \dots i_n l} = (C_{i_1 \dots i_n} \cap D_{i_1 \dots i_n l}) \cup (C_{i_1 \dots i_n} - E_{i_1 \dots i_n}),$$

3) for $k > l$

$$C_{i_1 \dots i_n k} = (C_{i_1 \dots i_n} \cap D_{i_1 \dots i_n k}) - (C_{i_1 \dots i_n l} \cup \dots \cup C_{i_1 \dots i_n(k-1)}).$$

Remark. Theorems 1 and 2 are also valid for $a = 0$ in a *separable space of dimension* 0.

Theorem 1 can be made more precise as follows [1].

THEOREM 3. *If $B^1, B^2, \ldots$ is a sequence of ambiguous sets of class a relative to a set A, then there exists a set Y of the multiplicative class $a+1$ and a sequence of ambiguous sets $C^1, C^2, \ldots$ of class a relative to Y such that $B^n = A \cap C^n$ and the sequences*

$$B^1, A-B^1, B^2, A-B^2, \ldots, \tag{5}$$

$$C^1, Y-C^1, C^2, Y-C^2, \ldots \tag{6}$$

are similar (in the sense of § 14, III).

Moreover, *if A is of multiplicative class $a > 0$, then Y coincides with the entire space.*

Proof. To each system $i_1, \ldots, i_n$ composed of the numbers 0 and 1 we assign a set $B_{i_1 \ldots i_n}$ by letting

$$B_{i_1 \ldots i_n} = B^1_{i_1} \cap \ldots \cap B^n_{i_n} \quad \text{where} \quad B^n_0 = B^n, \quad B^n_1 = A - B^n. \tag{7}$$

Thus the hypotheses of Theorem 2 are fulfilled [2]. Let the sets $C_{i_1 \ldots i_n}$ and A_n be as in Theorem 2. Let

$$Y = A_1 \cap A_2 \cap \ldots, \tag{8}$$

$$C^1 = Y \cap C_0, \; C^2 = Y \cap (C_{00} \cup C_{10}), \; \ldots, \; C^n = Y \cap \bigcup C_{i_1 \ldots i_{n-1} 0}, \ldots \tag{9}$$

Then $A = B_0 \cup B_1$ and, in general, the sets $B_{i_1 \ldots i_n}$ being constituents of A relative to the sets $B^1, \ldots, B^n$ (compare § 28, VI, Remark (ii)), their union is equal to A (for a fixed n). It follows that (compare (0) and (2))

$$A = \bigcup B_{i_1 \ldots i_n} \subset \bigcup C_{i_1 \ldots i_n} = A_n, \tag{10}$$

and by (8)

$$A \subset A_1 \cap A_2 \cap \ldots = Y. \tag{11}$$

Hence by (9), (2), and (7)

$$A \cap C^n = A \cap Y \cap \bigcup C_{i_1 \ldots i_{n-1} 0} = \bigcup A \cap C_{i_1 \ldots i_{n-1} 0}$$
$$= \bigcup B_{i_1 \ldots i_{n-1} 0} = \bigcup (B^1_{i_1} \cap \ldots \cap B^{n-1}_{i_{n-1}}) \cap B^n = B^n.$$

---

[1] See a paper of T. Posament and myself *Sur l'isomorphie algébro-logique et les ensembles relativement boreliens*, Fund. Math. 22 (1934), p. 285.

[2] In Theorem 2 we can clearly assume that the indices $i_1, \ldots, i_n$ take the values 0 and 1 only.

Since the sets A_n are of the multiplicative class $(\alpha+1)$, so is their intersection Y. Since $C_{i_1\ldots i_n}$ is ambiguous of class α with respect to A_n, then so is $Y \cap C_{i_1\ldots i_n}$ with respect to Y (for $Y \subset A_n$) and the same is true for C^n.

In order to prove that the sequences (5) and (6) are similar, let us notice that, for a fixed n, the sets $B_{i_1\ldots i_n}$ are constituents of A. Therefore each finite intersection of terms of the sequence (5) is the union of some of these sets. Namely, if $l_1 < \ldots < l_n$, then

$$B_{i_{l_1}}^{l_1} \cap \ldots \cap B_{i_{l_n}}^{l_n} = \bigcup B_{j_1 j_2 \ldots j_{l_n}}, \text{ where } j_{l_1} = i_{l_1}, \ldots, j_{l_n} = i_{l_n}. \quad (12)$$

The sequence (6) has a similar property. Formulas (9), (10), (8), and (3) yield

$$C^k = Y \cap \bigcup C_{j_1 \ldots j_{k-1} 0} \cap \bigcup C_{i_1 \ldots i_n} = Y \cap \bigcup C_{j_1 \ldots j_n},$$

where the second union is taken over all systems $j_1 \ldots j_n$ such that $j_k = 0$. This follows from the relation

$$Y \cap C_{j_1 \ldots j_n} = Y \cap A_k \cap C_{j_1 \ldots j_n} = Y \cap \bigcup C_{i_1 \ldots i_k} \cap C_{j_1 \ldots j_k \ldots j_n}$$
$$= Y \cap C_{j_1 \ldots j_k} \cap C_{j_1 \ldots j_k \ldots j_n}.$$

Similarly, by (9), (11), and (0)

$$Y - C^k = Y \cap A_k - \bigcup C_{j_1 \ldots j_{k-1} 0} = Y \cap \bigcup C_{j_1 \ldots j_{k-1} 1},$$

and, as before, $Y - C^k = Y \cap \bigcup C_{j_1 \ldots j_n}$, where $j_k = 1$.

Therefore, if we let $C_0^k = C^k$ and $C_1^k = Y - C^k$, then

$$C_i^k = Y \cap \bigcup C_{j_1 \ldots j_n}, \quad \text{where} \quad j_k = i,$$

and for $l_1 < \ldots < l_n$,

$$C_{i_{l_1}}^{l_1} \cap \ldots \cap C_{i_{l_n}}^{l_n} = Y \cap \bigcup C_{j_1 j_2 \ldots j_{l_n}}, \text{ where } j_{l_1} = i_{l_1}, \ldots, j_{l_n} = i_{l_n}. \quad (13)$$

The conditions $B_{i_1 \ldots i_n} = 0$ and $Y \cap C_{i_1 \ldots i_n} = 0$ are equivalent. For the first one implies the second one by (4); and, by (2) and (11),

$$B_{i_1 \ldots i_n} = A \cap C_{i_1 \ldots i_n} \subset Y \cap C_{i_1 \ldots i_n}.$$

Formulas (12) and (13) imply directly that the sequences (5) and (6) are similar.

Now assume that A is of the multiplicative class $\alpha > 0$. By the first part of (10) and the second part of Theorem 2, the set A_n coincides with the entire space, and hence so does Y (compare (11)).

Remark. The correspondence between the terms of the sequences (5) and (6) (obtained by assigning the nth term of (6) to the nth term of (5)) can be extended to the smallest field containing these terms. To a union (intersection, difference) of two sets we assign the union (intersection, difference) of the corresponding sets. Thus an isomorphism [1] between the two fields can be defined in such a way that *every relation expressed in terms of Boolean algebra which holds in one field does also hold in the other.*

IX. The limit set of ambiguous sets.

THEOREM 1. *If A is an ambiguous set of class $\alpha > 1$, then there exists a sequence of ambiguous sets A_n of classes $< \alpha$ such that*

$$A = \bigcup_{n=0}^{\infty} (A_n \cap A_{n+1} \cap \ldots) = \bigcap_{n=0}^{\infty} (A_n \cup A_{n+1} \cup \ldots),$$

i.e. $A = \operatorname{Limes}_{n=\infty} A_n$ in the sense of the general set theory (compare § 1, V).

In separable 0-dimensional spaces the theorem is also valid for $\alpha = 1$.

Proof. By hypothesis,

$$A = \bigcup_{n=0}^{\infty} K_n, \qquad -A = \bigcup_{n=0}^{\infty} L_n,$$

$$K_n \subset K_{n+1} \quad (\text{hence } K_n = \bigcap_{i=0}^{\infty} K_{n+i}),$$

$$L_n \subset L_{n+1} \quad \big(\text{hence } -L_i = \bigcup_{n=0}^{\infty} (-L_{n+i})\big),$$

where K_n and L_n are of the multiplicative classes $< \alpha$.

By the separation theorem (VII, Theorem 2 (2)) there exists a sequence of ambiguous sets A_n of classes $< \alpha$ such that

$$K_n \subset A_n \subset -L_n.$$

---

[1] See p. 281 of the paper referred to above.

The double identity which is to be proved follows from the formula (compare § 2, IV)

$$A = \bigcup_{n=0}^{\infty} K_n = \bigcup_{n=0}^{\infty}\bigcap_{i=0}^{\infty} K_{n+i} \subset \bigcup_{n=0}^{\infty}\bigcap_{i=0}^{\infty} A_{n+i} \subset \bigcap_{i=0}^{\infty}\bigcup_{n=0}^{\infty} A_{n+i}$$

$$\subset \bigcap_{i=0}^{\infty}\bigcup_{n=0}^{\infty} (-L_{n+i}) = \bigcap_{i=0}^{\infty} (-L_i) = -\bigcup_{i=0}^{\infty} L_i = A .$$

Remark. Theorem 1 follows also from the following lemma (of set theory) which leads to Theorem 2 which is more precise in the case where α is a limit ordinal.

LEMMA [1]. *The conditions*

$$A = \bigcup_{n=1}^{\infty}\bigcap_{m=1}^{\infty} B_{n,m}, \tag{1}$$

$$A = \bigcap_{m=1}^{\infty}\bigcup_{n=1}^{\infty} C_{n,m}, \tag{2}$$

$$\bigcup_{n=1}^{\infty} C_{n,m+1} \subset \bigcup_{n=1}^{\infty} C_{n,m} \tag{3}$$

imply

$$A = \operatorname*{Limes}_{n=\infty} A_n,$$

where

$$A_n = \bigcup_{k=1}^{n} (B_{k,1} \cap \ldots \cap B_{k,n}) \cap (C_{1,k} \cup \ldots \cup C_{n,k}).$$

Proof. We have to prove

(i) $A \subset \bigcup_n A_n \cap A_{n+1} \cap \ldots$ and (ii) $A^c \subset \bigcup_m A_m^c \cap A_{m+1}^c \cap \ldots .$

Condition (i) means that given $p \in A$ there exists an index n_0 such that $p \in A_n$ for $n \geqslant n_0$.

By (1) there exists a k such that

$$p \in B_{k,1} \cap B_{k,2} \cap \ldots .$$

By (2), there exists a sequence of indices $i_1, i_2, \ldots$ such that

$$p \in C_{i_1,1} \cap C_{i_2,2} \cap \ldots .$$

---

[1] W. Sierpiński, *Sur les rapports entre les classifications des ensembles de MM. F. Hausdorff et Ch. de la Vallée Poussin*, Fund. Math. 19 (1932), p. 260.

Let n_0 be the greatest of the numbers k and i_k. Then for $n \geqslant n_0$

$$C_{i_k,k} \subset C_{1,k} \cup \ldots \cup C_{n,k}.$$

Consequently

$$p \in (B_{k,1} \cap \ldots \cap B_{k,n}) \cap C_{i_k,k} \subset (B_{k,1} \cap \ldots \cap B_{k,n}) \cap$$
$$\cap (C_{1,k} \cup \ldots \cup C_{n,k}) \subset A_n.$$

Now let $p \in A^c$. We have to prove that there exists an m_0 such that, for $m \geqslant m_0$, $p \in A_m^c$, where

$$A_m^c = \bigcap_{l=1}^{m} (B_{l,1}^c \cup \ldots \cup B_{l,m}^c) \cup (C_{1,l}^c \cap \ldots \cap C_{m,l}^c).$$

By hypothesis,

$$A^c = \bigcap_n \bigcup_m B_{n,m}^c, \quad A^c = \bigcup_m \bigcap_n C_{n,m}^c, \quad \bigcap_n C_{n,m}^c \subset \bigcap_n C_{n,m+1}^c. \tag{4}$$

The second formula of (4) implies that there exists a k such that

$$p \in \bigcap_n C_{n,k}^c, \quad \text{hence} \quad p \in \bigcap_n C_{n,k+1}^c, \quad p \in \bigcap_n C_{n,k+2}^c, \ldots.$$

The first formula of (4) implies the existence of a sequence of indices $i_1, i_2, \ldots$ such that

$$p \in B_{1,i_1}^c \cap B_{2,i_2}^c \cap \ldots.$$

Let m_0 be the maximum of $i_1, \ldots, i_k$, and let $m \geqslant m_0$. Then

$$p \in B_{l,i_l}^c \subset B_{l,1}^c \cup \ldots \cup B_{l,m}^c \quad \text{for} \quad l \leqslant k,$$

$$p \in (C_{1,l}^c \cap \ldots \cap C_{m,l}^c) \quad \text{for} \quad l \geqslant k.$$

Therefore for each l ($\leqslant m$)

$$p \in (B_{l,1}^c \cup \ldots \cup B_{l,m}^c) \cup (C_{1,l}^c \cap \ldots \cap C_{m,l}^c), \quad \text{i.e.} \quad p \in A_m^c.$$

THEOREM 2. *If A is an ambiguous set of class $\lambda+1$, where λ is a limit ordinal, then there exists a sequence of ambiguous sets A_n of classes $< \lambda$ such that* $A = \operatorname{Limes}_{n=\infty} A_n$.

X. Locally Borel sets. Montgomery operation $\mathcal{M}$. By the general definition of localization (§ 7, IV), a set A is said to be *of class α at the point p*, if there exists a neighbourhood E of p such that $A \cap E$ is a Borel set of class α. The term "neighbourhood" can be replaced by "open neighbourhood" except in the case of the multiplicative class 0 (the case of locally closed set, see § 7, V). For, if $A \cap E$ is of class α and G is the interior of E, then $A \cap G$ is of class α (except for the multiplicative class 0). Also if the space is separable, then open neighbourhoods can be replaced by an open set of a base $R_1, R_2, \ldots$ of the space.

THEOREM 1. *The set B of points of A where A is locally of additive class α (or of multiplicative class $\alpha > 0$) is of the same class.*

In particular, *if a subset A of a separable space $\mathcal{X}$ is of the additive class α (or of the multiplicative class $\alpha > 0$) at each of its points, then A is a set of the same class.*

Proof. Assume first that the space is separable ([1]); in this case the proof is simpler.

Let $R_{n_1}, R_{n_2}, \ldots$ be the sequence of all sets (of the base) such that $A \cap R_{n_k}$ is of the additive class α. Hence the set

$$B = \bigcup_k A \cap R_{n_k}$$

is of the additive class α.

Suppose now that $A \cap R_{n_k}$ is of the multiplicative class $\alpha > 0$. We have

$$B = \bigcup_k A \cap R_{n_k} = A \cap \bigcup_k R_{n_k} = \bigcup_k R_{n_k} - [\bigcup_k R_{n_k} - A]$$
$$= \bigcup_k R_{n_k} - \bigcup_k [R_{n_k} - A \cap R_{n_k}]$$

(by the identity $X \cap Y = X - (X - Y)$).

The set $R_{n_k} - A \cap R_{n_k}$ is of the additive class α, and hence so is $\bigcup_k [R_{n_k} - A \cap R_{n_k}]$ and the proof is completed.

([1]) Compare K. Zarankiewicz, *O zbiorach lokalnie mierzalnych* (B), Wiadomości Matematyczne 30 (1928), p. 115.

The proof for an arbitrary metric space will be based on a general method which will also be applied to other problems [1].

Let $G_0, G_1, \dots, G_\xi, \dots$ be a well ordered family of open sets. Let

$$K_\xi = G_\xi - \bigcup_{\eta<\xi} G_\eta. \tag{1}$$

We call *operation* $\mathscr{M}$ the operation which assigns, to each transfinite sequence $X_0, X_1, \dots, X_\xi, \dots,$ the union

$$S = \bigcup_\xi X_\xi \cap K_\xi. \tag{2}$$

THEOREM 2. *Operation* $\mathscr{M}$ *is additive, multiplicative, and, in the case where*

$$\mathscr{X} = K_1 \cup K_2 \cup \dots \cup K_\xi \cup \dots, \tag{i}$$

it is subtractive.

Proof. The additivity of operation $\mathscr{M}$ is immediate. It means that the condition $X_\xi = \bigcup_\iota X_\xi^\iota$ (where ι runs over an arbitrary set of indices) implies

$$\bigcup_\xi X_\xi \cap K_\xi = \bigcup_\xi (\bigcup_\iota X_\xi^\iota) \cap K_\xi = \bigcup_\iota (\bigcup_\xi X_\xi^\iota) \cap K_\xi. \tag{3}$$

We now prove that operation $\mathscr{M}$ is multiplicative.

Let $X_\xi = \bigcap_\iota X_\xi^\iota$. Observe that by (1)

$$K_\xi \cap K_\eta = 0 \quad \text{for} \quad \xi \neq \eta. \tag{4}$$

It follows that

$$\bigcup_\xi X_\xi \cap K_\xi = \bigcup_\xi \bigcap_\iota X_\xi^\iota \cap K_\xi = \bigcap_\iota \bigcup_\xi X_\xi^\iota \cap K_\xi, \tag{5}$$

by the rule: if $A_{\xi\alpha} \cap A_{\eta\beta} = 0$ for every α, β, and $\eta \neq \xi$, then

$$\bigcup_\xi \bigcap_\alpha A_{\xi\alpha} = \bigcap_\alpha \bigcup_\xi A_{\xi\alpha}.$$

---

[1] This method and its applications are due to D. Montgomery. See *Non-separable metric spaces*, Fund. Math. 25 (1935), p. 527. Compare also my paper *Quelques problèmes concernant les espaces métriques non-séparables*, *ibid.* p. 535.

For a proof based on the paracompactness of metric spaces, see E. Michael, Duke Math. Journ. 21 (1954), p. 163, and K. Nagami, Proc. Japan Acad. 32 (1956), p. 320.

Finally let $X_\xi = \mathscr{X} - X_\xi^c$. Then

$$\bigcup_\xi X_\xi \cap K_\xi = \bigcup_\xi (K_\xi - X_\xi^c) = \bigcup_\xi K_\xi - \bigcup_\xi K_\xi \cap X_\xi^c. \tag{6}$$

For, on one hand

$$K_\xi = (K_\xi \cap X_\xi) \cup (K_\xi \cap X_\xi^c),$$

$$\text{whence} \quad \bigcup_\xi K_\xi = \bigcup_\xi (K_\xi \cap X_\xi) \cup \bigcup_\xi (K_\xi \cap X_\xi^c),$$

and, on the other hand, by (4),

$$(K_\xi \cap X_\xi) \cap (K_\eta \cap X_\eta^c) = 0,$$

$$\text{whence} \quad (\bigcup_\xi K_\xi \cap X_\xi) \cap (\bigcup_\xi K_\xi \cap X_\xi^c) = 0.$$

Formulas (i), (5), and (6) imply the subtractivity of operation $\mathscr{M}$:

$$\bigcup_\xi (X_\xi - Y_\xi) \cap K_\xi = (\bigcup_\xi X_\xi \cap K_\xi) \cap (\bigcup_\xi Y_\xi^c \cap K_\xi)$$

$$= \bigcup_\xi X_\xi \cap K_\xi - \bigcup_\xi Y_\xi \cap K_\xi.$$

Remark. If we drop hypothesis (i), instead of subtractivity we get the following formula

$$\bigcup_\xi X_\xi \cap K_\xi = \bigcup_\xi G_\xi - \bigcup_\xi K_\xi \cap X_\xi^c, \tag{7}$$

which follows from (6) by virtue of the identity

$$\bigcup_\xi K_\xi = \bigcup_\xi G_\xi.$$

THEOREM 3. *The classes* $\boldsymbol{F}_a$ *and* $\boldsymbol{G}_a$ *with* $a > 0$ *are invariant under operation* $\mathscr{M}$.

Proof. Let us agree that $\varrho(X, 0) = 1$ and let, for G_ξ open,

$$X_\xi^n = X_\xi \cap K_\xi \cap \underset{x}{E}[\varrho(x, \mathscr{X} - G_\xi) \geqslant 1/n]. \tag{8}$$

By (1) we get

$$X_\xi \cap K_\xi = \bigcup_{n=1}^{\infty} X_\xi^n, \quad \text{whence} \quad S = \bigcup_\xi \bigcup_{n=1}^{\infty} X_\xi^n$$

by (2). Let

$$S_n = \bigcup_\xi X_\xi^n, \tag{9}$$

therefore

$$S = \bigcup_{n=1}^{\infty} S_n. \tag{10}$$

Also

$$\varrho(X_\eta^n, X_\xi^n) \geqslant \frac{1}{n} \quad \text{if} \quad \eta \neq \xi. \tag{11}$$

For, if $x_\eta \epsilon X_\eta^n$, $x_\xi \epsilon X_\xi^n$, and $\eta < \xi$, then by (8) and (1),

$$\varrho(x_\eta, \mathscr{X}-G_\eta) \geqslant \frac{1}{n} \quad \text{and} \quad x_\xi \epsilon K_\xi \subset \mathscr{X}-G_\eta$$

and $|x_\eta - x_\zeta| \geqslant 1/n$. Formula (11) follows.

It implies that the sets X_ξ^n are both closed and open in S_n. Since $\mathrm{Fr}(G_\xi) \cap \underset{x}{E}[\varrho(x, \mathscr{X}-G_\xi) \geqslant 1/n] = 0$, it follows by (8) and (1) that

$$X_\xi^n = X_\xi \cap [\bar{G}_\xi - \bigcup_{\eta<\xi} G_\eta] \cap \underset{x}{E}[\varrho(x, \mathscr{X}-G_\xi) \geqslant 1/n],$$

which shows that *if X_ξ is closed, then X_ξ^n is closed.* If we suppose that $p \epsilon \bar{S}_n - S_n$, then there exist two points $x_\eta \epsilon X_\eta^n$ and $x_\xi \epsilon X_\xi^n$ such that $|x_\eta - p| < 1/2n$, $|x_\xi - p| < 1/2n$ and $\eta \neq \xi$. Then $|x_\eta - x_\xi| < 1/n$, contrary to (11). It follows that S_n is *closed.*

Thus we have proved (see (10)) that *if X_ξ is closed, S is an $\boldsymbol{F}_\sigma$*-set. It follows by (3) that *if X_ξ is an $\boldsymbol{F}_\sigma$-set, then S is an $\boldsymbol{F}_\sigma$-set.* We deduce from (7) that *if X_ξ is a $\boldsymbol{G}_\delta$-set, then S is a $\boldsymbol{G}_\delta$*-set.

Using formulas (3) and (5) we obtain the required conclusion that *if X_ξ is $\boldsymbol{F}_\alpha$* (or $\boldsymbol{G}_\alpha$), *then so is S.*

Theorem 1 is a particular case (by virtue of Theorem 3) of the following theorem.

THEOREM 4. *Let* P *be a property invariant under operation $\mathscr{M}$. Given a set A, let S be the set of points x of A for which there exists an open set G containing x such that $G \cap A$ has the property* P. *Then the set S has the property* P.

Proof. Consider all open sets G such that $G \cap A$ has the property P and order them in a transfinite sequence $G_0, G_1, \ldots, G_\xi, \ldots$. Let $X_\xi = A \cap G_\xi$. Formula (2) follows from the relation

$$S = \bigcup_\xi A \cap G_\xi = \bigcup_\xi A \cap K_\xi = \bigcup_\xi X_\xi \cap K_\xi.$$

The set X_ξ has the property P. Since P is invariant under operation $\mathscr{M}$, hence S has the property P.

Remark (1). The proof of Theorem 1 leads easily to the conclusion that a subset of a separable space is borelian, if it is locally borelian at each of its points.

However, *there exists a metric non-separable space containing a non-borelian set which is locally borelian at each of its points* (of unbounded class).

As an example, consider the space $\mathscr{X}$ formed by the points (x, a), where $0 \leqslant x \leqslant 1$ and $0 \leqslant a < \Omega$. The distance of the points (x, a) and (x', a') is defined to be $|x-x'|$, if $a = a'$ and 1, if $a \neq a'$ (thus the space is the cartesian product of the interval and the discrete space of numbers $a < \Omega$). Let I_a be the "interval" (x, a), where $0 \leqslant x \leqslant 1$, and let B_a be a Borel set $\subset I_a$ which is not of class a (see Section XIV). The set $S = \bigcup_{a<\Omega} B_a$ is locally borelian, since I_a is open in $\mathscr{X}$; but it is not borelian since, if it were of class a, the set $S \cap I_a = B_a$ would be of the same class.

Theorem 1 enables us to prove, in general metric spaces, the following theorem which has been proved in § 24, III for separable spaces.

THEOREM 5. *Every resolvable set is both F_σ and G_δ.*

Proof. Let

$$E = \bigcup_{\xi<a} (F_\xi - H_\xi), \quad \text{where} \quad F_\xi \supset H_\xi \supset F_\zeta \quad \text{for} \quad \xi < \zeta$$

and $F_\xi = \overline{F}_\xi$, $H_\xi = \overline{H}_\xi$. Assume that the theorem is true for each $a' < a$. To prove it for a we must show, by Theorem 1, that E is locally F_σ and G_δ at each point.

Let $p \in E$. There exists an $a' < a$ such that $p \in F_{a'} - H_{a'}$. Let $G = \mathscr{X} - H_{a'}$. Therefore $p \in G$, G is open and, for $\zeta > a'$, $F_\zeta \subset H_{a'}$, whence $F_\zeta \cap G = 0$ and $G \cap \bigcup_{\zeta>a'} (F_\zeta - H_\zeta) = 0$.

Therefore

$$G \cap E = G \cap \bigcup_{\xi<a'} (F_\xi - H_\xi) \cup (G \cap F_{a'} - H_{a'}).$$

$\bigcup_{\xi<a'} (F_\xi - H_\xi)$ is an F_σ and G_δ-set (by hypothesis), hence so is $G \cap E$. It follows that E is locally an F_σ and G_δ-set at the point p.

(1) This remark is due to E. Szpilrajn-Marczewski, Fund. Math. 21(1933), p. 112.

XI. Evaluation of classes with the aid of logical symbols[1]. We say that a *propositional function* $\varphi(x)$ *is of class* F_a (of class G_a), if the set $\underset{x}{E}\varphi(x)$ is of class F_a (of class G_a). By using formulas proved in the Introduction (§ 1, IV and § 2, V–VI) we prove the following propositions.

THEOREM 1. *If* $\varphi(x)$ *and* $\psi(x)$ *are two propositional functions of class* F_a *(class* G_a*), then the functions* $\varphi(x) \vee \psi(x)$ *and* $\varphi(x) \wedge \psi(x)$ *are of the same class.*

Proof. We have

$$\underset{x}{E}[\varphi(x) \vee \psi(x)] = \underset{x}{E}\varphi(x) \cup \underset{x}{E}\psi(x)$$

and

$$\underset{x}{E}[\varphi(x) \wedge \psi(x)] = \underset{x}{E}\varphi(x) \cap \underset{x}{E}\psi(x)$$

and Borel classes are invariant when taking the union and intersection of sets.

This theorem can be expressed more generally.

THEOREM 1'. *Suppose that* $\varphi(x_1, \ldots, x_n) \equiv \psi(x_{k_1}, \ldots, x_{k_j}) \vee \vee \chi(x_{l_1}, \ldots, x_{l_m})$, *where the indices* $k_1, \ldots, k_j$ *and* $l_1, \ldots, l_m$ *are* $\leqslant n$ *(for instance,* $\varphi(x, y, z) \equiv \psi(x, y) \vee \chi(y, z)$*). If the functions* ψ *and* χ *are of class* F_a *(of class* G_a*), then* φ *is of the same class (the same is true for the conjunction* $\psi \wedge \chi$*).*

Proof.

$$\underset{x_1 \ldots x_n}{E} \varphi(x_1, \ldots, x_n) = \underset{x_1 \ldots x_n}{E} \psi(x_{k_1}, \ldots, x_{k_j}) \cup \underset{x_1 \ldots x_n}{E} \chi(x_{l_1}, \ldots, x_{l_m}).$$

The set $\underset{x_{k_1} \ldots x_{k_j}}{E} \psi(x_{k_1}, \ldots, x_{k_j})$ is of class F_a, hence so is the set $\underset{x_1 \ldots x_n}{E} \psi(x_{k_1}, \ldots, x_{k_j})$ which is obtained from the former one by taking the product with axes (see Section III).

It follows that *the result of performing a finite number of logical disjunctions or conjunctions with propositional functions of class* F_a *(of class* G_a*) is a propositional function of class* F_a *(of class* G_a*).*

---

(1) See the paper of A. Tarski and myself *Les opérations logiques et les ensembles projectifs*, and my paper *Evaluation de la classe borélienne ou projective à l'aide des symboles logiques*, Fund. Math. 17 (1931), pp. 240–272.

THEOREM 2. *If a propositional function $\varphi(x)$ is of class F_a, its negation is of class G_a.*

For $\underset{x}{E}\neg\varphi(x)$ is the complement of $\underset{x}{E}\varphi(x)$.

THEOREM 3. *If $\varphi_n(x)$, $n = 1, 2, \ldots$ is a sequence of propositional functions of classes $< a$, then the function $\underset{x}{\bigvee}\varphi_n(x)$ is of the additive class a and the function $\underset{n}{\bigwedge}\varphi_n(x)$ is of the multiplicative class a.*

Proof. For, we have by § 2, V (1) and (2):

$$\underset{x}{E}\underset{n}{\bigvee}\varphi_n(x) = \bigcup_n \underset{x}{E}\varphi_n(x) \quad \text{and} \quad \underset{x}{E}\underset{n}{\bigwedge}\varphi_n(x) = \bigcap_n \underset{x}{E}\varphi_n(x).$$

In particular, if the functions $\varphi_n(x)$ are of class F_a with a even, then the function $\underset{n}{\bigvee}\varphi_n(x)$ is of class $F_{a\sigma}$ and the function $\underset{n}{\bigwedge}\varphi_n(x)$ is of class F_a. Similarly, if the functions $\varphi_{n,m}(x)$ are of class F_a, then the function $\underset{m}{\bigwedge}\underset{n}{\bigvee}\varphi_{n,m}(x)$ is of class $F_{a\sigma\delta}$, and so on.

Thus *the Theorems* 1–3 *allow us to evaluate the Borel class of a set, if the set is defined by a propositional function obtained from a system of propositional functions whose classes are known by means of logical operations* $\vee, \wedge, \neg, \underset{n}{\bigvee}, \underset{n}{\bigwedge}$, *taken a finite number of times.*

The following rules are often used in applications.

THEOREM 4. *If $\varphi(x)$ is a propositional function of class F_a (of class G_a) and if $f: T \to \mathfrak{X}$ is continuous, then the propositional function $\varphi[f(t)]$ is also of class F_a (of class G_a).*

Proof. Let $A = \underset{x}{E}\varphi(x)$. Then (see § 3, I)

$$\underset{t}{E}\varphi[f(t)] = \underset{t}{E}\{f(t)\in\underset{x}{E}\varphi(x)\} = \underset{t}{E}\{f(t)\in A\} = f^{-1}(A).$$

Since A is of class F_a (of class G_a), then so is $f^{-1}(A)$ (see Section III).

THEOREM 5. *If $\psi(x, y)$ is of class F_a (of class G_a), then the functions $\varphi(x) = \psi(x, y_0)$ (y_0 fixed) and $\chi(x) = \psi(x, x)$ are of class F_a (of class G_a).*

To prove this we substitute

$$Z = \underset{x}{E}\psi(x, y) \quad or \quad Z = \underset{x}{E}\psi(x, x)$$

in the Theorem of Section III.

THEOREM 6. *If $\psi(x)$ is of class F_a (of class G_a), then so is the function $\varphi(x, y) \equiv \psi(x)$.*

For $\underset{xy}{E}\varphi(x, y) = \underset{x}{E}\psi(x)\times\mathcal{Y}$.

XII. Applications. 1) *Evaluation of the Borel class of the sets* $S_a = \underset{t}{E}(\bar{t} < a)$, *where* $a < \Omega$ (compare § 3, XVI).

To each $t\epsilon\mathcal{C}$ and each positive integer n let us assign a point $u\epsilon\mathcal{C}$ defined as follows. If $t^n = 0$, let $u = 0$; if $t^n = 2$, let $u^k = 2$ in the case $t^k = 2$ and $r_n < r_k$, and $u^k = 0$ otherwise. Let the symbol $t^{[n]}$ denote u; then

$$(u = t^{[n]}) \equiv \bigwedge_k \{(u^k = 2) \equiv (t^k = 2 = t^n) \wedge (r_n < r_k)\}. \tag{1}$$

In the notation of § 3, IX,

$$R_{t[n]} = R_t \cap \underset{r}{E}(r_n < r), \quad \text{if} \quad t^n = 2.$$

Therefore, if $\bar{t}$ is an ordinal number, *the sequence* $\overline{t^{[1]}}, \overline{t^{[2]}}, \ldots$ *runs over all ordinal numbers* $< \bar{t}$.

The sets $\underset{t}{E}(t^j = 2)$, $j = 1, 2, \ldots$, are both closed and open in $\mathcal{C}$. Hence so is the set of pairs u, t satisfying the condition in brackets $\{ \}$. It follows that the set $\underset{ut}{E}(u = t^{[n]})$, i.e. the graph of the function $t^{[n]}$, is closed. Since $\mathcal{C}$ is compact, it follows that *the function* $t^{[n]}$ *is continuous* (compare § 20, V, Theorem 8).

THEOREM 1 (1). *If* $a < \Omega$, *the sets* $S_a = \underset{t}{E}(\bar{t} < a)$ *and, consequently, the constituents* $L_a = \underset{t}{E}(\bar{t} = a)$ *are borelian.*

In fact, S_a *is of class* G_a.

Proof. First, the following obvious equivalences hold:

$$(\bar{t} < a+1) \equiv \bigwedge_n (\overline{t^{[n]}} < a), \tag{2}$$

$$(\bar{t} < \lambda) \equiv \bigvee_{\xi<\lambda} (\bar{t} < \xi), \tag{3}$$

where λ is a limit ordinal.

We proceed by induction. The cases $a = 0$ and $a = 1$ are obvious. Assume that the propositional function $\bar{t} < \xi$ is of class G_ξ for

(1) See H. Lebesgue, Journ. de Math. (6) 1 (1905), p. 213; N. Lusin and W. Sierpiński, C. R. Paris 175 (1922), p. 357.

every $\xi < \beta$. Therefore, if $\beta = \alpha+1$, the propositional function $\bar{t} < \alpha$ is of class G_α. Since the function $t^{[n]}$ is continuous, it follows (XI, Theorem 4) that the propositional function $\overline{t^{[n]}} < \alpha$ is also of class G_α. Therefore (by (2) and XI, Theorem 3) the propositional function $\bar{t} < \alpha+1$, i.e. $\bar{t} < \beta$, is of class G_β.

On the other hand, if β is a limit ordinal, the same conclusion follows from (3) and XI, Theorem 3.

Remark. The evaluation of the class of S_α can be done more precisely. Thus it follows easily that the sets S_n are closed if n is finite, S_ω is an F_σ-set, $S_{\omega\cdot 2}$ is an $F_{\sigma\delta\sigma}$-set and so on.

It should be noticed (see § 39, VIII) that the sets S_α are of *non-bounded* classes, i.e. for every $\beta < \Omega$ there exists an S_α which is not of class β.

2) To each irrational number $\mathfrak{z}$ of the interval $\mathscr{I}$ let us assign the set $Z_\mathfrak{z}$ composed of the numbers $r_{\mathfrak{z}1}, r_{\mathfrak{z}2}, \ldots$ (we identify $\mathfrak{z}$ with the sequence $\mathfrak{z}^1, \mathfrak{z}^2, \ldots$, compare § 3, IX, Example 2).

Let $\bar{\mathfrak{z}}$ denote the order type of $Z_\mathfrak{z}$. Let $T_\alpha = \underset{\mathfrak{z}}{E}(\bar{\mathfrak{z}} < \alpha)$.

By an analogous argument we prove that T_α *is of class* G_α.

3) Recall that an order type τ is a *limit type*, if it does not have the last element. Let A be the set of $t \in \mathscr{C}$ such that $\bar{t}$ is a limit type.

In terms of logic (compare the notation of § 3, XVI):

$$(t \in A) \equiv \bigwedge_n \bigvee_k [(r_n \in R_t) \Rightarrow (r_k \in R_t)(r_k < r_n)]$$
$$\equiv \bigwedge_n \bigvee_k [(t^n = 2) \Rightarrow (t^k = 2)(r_k < r_n)].$$

The set $\underset{t}{E}(t^n = 2)$ is both closed and open (in $\mathscr{C}$). Hence the propositional function in brackets [ ] is of class G_0, and *the set A is of class* G_δ.

4) We say that an order type is *even*, if it has the form $\lambda + 2n$, where λ is a limit type and n is an integer $\geqslant 0$. The sums $\lambda + 2n + 1$ will be called *odd* [1].

Let P denote the set of t such that $\bar{t}$ is an even type. Then $P = P_0 \cup P_2 \cup \ldots$, where P_{2n} is the set of t such that $\bar{t}$ has the form $\lambda + 2n$.

---

[1] These concepts are usually considered for well ordered types.

Let us write down the definition of P_2 in logical symbols.

$$(t \in P_2) \equiv \bigvee_{ij} \{(r_i, r_j \in R_t)(r_i < r_j) \wedge \bigwedge_n ((i \neq n \neq j)(r_n \in R_t) \Rightarrow [(r_j < r_n) \bigvee_k (r_j < r_k < r_n)(r_k \in R_t)])\},$$

whence

$$(t \in P_2) \equiv \bigvee_{ij} \{(t^i = 2 = t^j)(r_i < r_j) \wedge \bigwedge_n ((i \neq n \neq j)(t^n = 2) \Rightarrow [(r_j < r_n) \bigvee_k [(r_j < r_k < r_n)(t^k = 2)]])\}.$$

We can easily infer that the set P_2 is a $G_{\delta\sigma}$-set. An analogous definition of $P_4, P_6, \ldots$ shows that they are $G_{\delta\sigma}$-sets. Consequently *P is a $G_{\delta\sigma}$-set.*

Similarly the set of t such that $\bar{t}$ is *odd* is a $G_{\delta\sigma}$-set.

5) *Let Φ be the family of all sequences of points of a metric space $\mathscr{X}$ which satisfy the Cauchy convergence condition* (a sequence $\xi^1, \xi^2, \ldots$ satisfies the Cauchy condition, if for every $\varepsilon > 0$ there exists an index m such that $|\xi^{m+i} - \xi^m| \leqslant \varepsilon$, for each i). We shall prove that *the family Φ is an $F_{\sigma\delta}$-set in the space $\mathscr{X}^{\aleph_0}$.*

By definition,

$$\xi \in \Phi \equiv \bigwedge_k \bigvee_m \bigwedge_i |\xi^{m+i} - \xi^m| \leqslant \frac{1}{k}.$$

Now the propositional function

$$\varphi_{k,m,i}(\xi) \equiv \left\{|\xi^{m+i} - \xi^m| \leqslant \frac{1}{k}\right\}$$

is of class F_0 (for fixed k, m, and i). For, the distance $|x-y|$ is a continuous function of two variables and ξ^n is a continuous function of ξ (compare § 16, II); hence the set

$$\underset{\xi}{E}\left\{|\xi^{m+i} - \xi^m| \leqslant \frac{1}{k}\right\}$$

is closed.

Applying Theorem XI, 3 we infer that the function $\bigwedge_i \varphi_{k,m,i}(\xi)$ is also of class F_0, $\bigvee_m \bigwedge_i \varphi_{k,m,i}(\xi)$ is of class F_σ and, finally, $\bigwedge_k \bigvee_m \bigwedge_i \varphi_{k,m,i}(\xi)$ is of class $F_{\sigma\delta}$. This means that the set Φ is of class $F_{\sigma\delta}$.

By Theorem 4 of Section XI it follows that

If $f_n: T \to \mathscr{X}$, $n = 1, 2, \ldots$, is a sequence of continuous functions the set C of points t for which the Cauchy condition is satisfied by $\{f_n(t)\}$ is an $\boldsymbol{F}_{\sigma\delta}$-set (1).

For, if $\xi(t)$ denotes the sequence $[f_1(t), f_2(t), \ldots]$, then

$$\{t \in C\} \equiv \bigwedge_k \bigvee_m \bigwedge_i \varphi_{k,m,i}[\xi(t)].$$

6) *The family ϑ of sequences dense in themselves is a $\boldsymbol{G}_\delta$-set in the space $\mathscr{X}^{\aleph_0}$.*

For, a sequence $\xi = [\xi^1, \xi^2, \ldots]$ is dense in itself, if for every n, there exists a $\xi^m \neq \xi^n$ arbitrarily close to ξ^n:

$$[\xi \in \vartheta] \equiv \bigwedge_{nk} \bigvee_m \left\{0 < |\xi^n - \xi^m| < \frac{1}{k}\right\}.$$

The propositional function in brackets $\{\ \}$ is obviously of class $\boldsymbol{G}_0$ (for fixed n, m and k). Applying the operator $\bigvee_m$ does not change the class. Therefore the function $\xi \in \vartheta$ is of class $\boldsymbol{G}_\delta$.

XIII. Universal functions (2). Given a family $\boldsymbol{F}$ of sets and a set T, a *universal function with respect to* $\boldsymbol{F}$ is a function F: $T \to \boldsymbol{F}$ onto, i.e.

$$\{X \in \boldsymbol{F}\} \equiv \bigvee_t [X = F(t)].$$

We shall assume subsequently that the space $\mathscr{X}$ (whose subsets are elements of $\boldsymbol{F}$) is *separable metric*. Let $T = \mathscr{N}$ (the set of irrational numbers of the interval $\mathscr{I}$) (3). If $\boldsymbol{F}$ is of power $\leqslant \mathfrak{c}$, there

(1) Compare § 2, VI, example 2. If $\mathscr{X}$ is a complete space, C is the set of points of convergence of the sequence $\{f_n\}$.

(2) See W. Sierpiński, Fund. Math. 14 (1929), p. 82.

(3) Instead of letting the parameter t vary over the entire interval $\mathscr{I}$ (which is often assumed), we restrict its values to irrational numbers in order to avoid inconvenience due to the discontinuity of the function "the nth digit of the dyadic expansion of x". If an irrational number $\mathfrak{z}$ is regarded as a sequence of natural numbers, the nth term of this sequence is a continuous function of $\mathfrak{z}$ (compare § 16, II).

Equally well one could use the Cantor set $\mathscr{C}$ (since it is also a $\aleph_0$th power of a set).

exists obviously a universal function with respect to $\boldsymbol{F}$. Hence a Borel class $\boldsymbol{F}_a$ or $\boldsymbol{G}_a$ can be substituted for $\boldsymbol{F}$.

THEOREM. *For each $a < \Omega$ there exists a universal function G_a with respect to $\boldsymbol{G}_a$ such that the set $\underset{x\mathfrak{z}}{E}[x \in G_a(\mathfrak{z})]$ is a $\boldsymbol{G}_a$-set in the product $\mathcal{X} \times \mathcal{N}$* [1].

Proof. We shall use the following notation. As usual, an irrational number $\mathfrak{z}$ is regarded as a sequence $\mathfrak{z}^{(1)}, \mathfrak{z}^{(2)}, \ldots$ of natural numbers (given e.g. by its expansion into a continuous fraction). Since the space $\mathcal{N}$ is homeomorphic to $\mathcal{N}^{\aleph_0}$ (comp. § 3, IX, 2 and § 16, II, Theorem 6), there is a one-to-one correspondence between irrational numbers $\mathfrak{z}$ and all sequences of irrational numbers $\mathfrak{z}_{(1)}, \mathfrak{z}_{(2)}, \ldots$ such that, for a fixed n, $\mathfrak{z}_{(n)}$ is a continuous function of $\mathfrak{z}$. For example, we can put

$$\mathfrak{z}_{(n)} = [\mathfrak{z}^{(2^{n-1})}, \ldots, \mathfrak{z}^{(2^{n-1}+k\cdot 2^n)}, \ldots].$$

To every transfinite limit ordinal $\lambda < \Omega$ let us assign a sequence $\lambda_1 < \lambda_2 < \ldots$ convergent to λ (its existence follows from the axiom of choice).

Let $R_1, R_2, \ldots$ be a base of the space (containing the empty set). We define the function G_a as follows.

1) $G_0(\mathfrak{z}) = \bigcup_n R_{\mathfrak{z}^n}$;

2) $G_{a+1}(\mathfrak{z}) = \bigcap_n G_a(\mathfrak{z}_{(n)})$ or $\bigcup_n G_a(\mathfrak{z}_{(n)})$, depending whether a is even or odd;

3) $G_\lambda(\mathfrak{z}) = \bigcup_n G_{\lambda_n}(\mathfrak{z}_{(n)})$, if λ is a limit ordinal.

We have to prove that:

(i) *the set $G_a(\mathfrak{z})$ is of class $\boldsymbol{G}_a$*;

(ii) *if X is of class $\boldsymbol{G}_a$, there exists a $\mathfrak{z} \in \mathcal{N}$ such that $X = G_a(\mathfrak{z})$*;

(iii) *the set $\underset{x\mathfrak{z}}{E}[x \in G_a(\mathfrak{z})]$ is of class $\boldsymbol{G}_a$.*

Property (i) is a direct consequence of (iii) (compare the Theorem of Section III).

---

[1] The proof which follows is due essentially to Lebesgue, *op. cit.*, p. 209.

Proof of (ii). First assume that X is of class $\boldsymbol{G}_0$, i.e. an open set. By definition of the base, X has the form $X = \bigcup_n R_{k_n}$. Let $\mathfrak{z}$ be an irrational number with $\mathfrak{z}^1 = k_1, \mathfrak{z}^2 = k_2, \ldots$. Then

$$G_0(\mathfrak{z}) = \bigcup_n R_{\mathfrak{z}^n} = \bigcup_n R_{k_n} = X.$$

Thus condition (ii) is satisfied for $\alpha = 0$.

Assuming it for α, we shall prove it for $\alpha+1$. Let X be a set of class $\boldsymbol{G}_{\alpha+1}$. Then

$$X = \bigcap_n X_n \quad \text{or} \quad X = \bigcup_n X_n$$

(according to as α is even or odd), where X_n are of class $\boldsymbol{G}_\alpha$. By hypothesis there exists a sequence $\{\mathfrak{z}_n\}$ of irrational numbers such that $X_n = G_\alpha(\mathfrak{z}_n)$. By definition of the function $\mathfrak{z}_{(n)}$, there exists a value $\mathfrak{z}$ such that $\mathfrak{z}_n = \mathfrak{z}_{(n)}$ for every n. It follows that

$$G_{\alpha+1}(\mathfrak{z}) = \bigcap_n G_\alpha(\mathfrak{z}_{(n)}) = X \quad \text{or} \quad G_{\alpha+1}(\mathfrak{z}) = \bigcup_n G_\alpha(\mathfrak{z}_{(n)}) = X$$

according to as α is even or odd.

Now suppose that $\lambda = \lim \lambda_n$ and that (ii) holds for each λ_n. If X is a set of class $\boldsymbol{G}_\lambda$, then $X = \bigcup_n X_n$, where X_n is of a class $\boldsymbol{G}_{\alpha_n}$ with $\alpha_n < \lambda$. Since the sequence $\{\lambda_n\}$ is convergent to λ, for each n there exists a k_n such that $\alpha_n \leqslant \lambda_{k_n}$. Therefore X_n is of class $\boldsymbol{G}_{\lambda_{k_n}}$. Hence there exists an irrational number $\mathfrak{z}_{k_n}$ with $X_n = G_{\lambda_{k_n}}(\mathfrak{z}_{k_n})$. If i is different of all k_n, let $\mathfrak{z}_i$ be an irrational number with $G_{\lambda_i}(\mathfrak{z}_i) = 0$. Hence $X = \bigcup_n G_{\lambda_n}(\mathfrak{z}_n)$. As before, let $\mathfrak{z}$ be an irrational number with $\mathfrak{z}_n = \mathfrak{z}_{(n)}$. Then

$$X = \bigcup_n G_{\lambda_n}(\mathfrak{z}_{(n)}) = G_\lambda(\mathfrak{z}).$$

Proof of (iii). First observe that $\underset{x,n}{E}(x \in R_n)$ is open in $\mathscr{X} \times \mathscr{D}$ (where $\mathscr{D}$ is the set of natural numbers). In other words, the propositional function $x \in R_n$ (of two variables) is of class $\boldsymbol{G}_0$. As the function $\mathfrak{z}^n$, for a fixed n, is continuous, then (by Section XI, Theorem 4) the propositional function $x \in R_{\mathfrak{z}^n}$ is also of class $\boldsymbol{G}_0$. The same is true for the propositional function $\bigvee_n (x \in R_{\mathfrak{z}^n})$ which is equivalent to $x \in \bigcup_n R_{\mathfrak{z}^n}$ (see § 2, V). Consequently the propositional function $x \in G_0(\mathfrak{z})$ is of class $\boldsymbol{G}_0$ and the set $\underset{x\mathfrak{z}}{E}[x \in G_0(\mathfrak{z})]$ is open.

Similarly, if the propositional function $x \in G_\alpha(\mathfrak{z})$ is of class α, then so is $x \in G_\alpha(\mathfrak{z}_{(n)})$, for a fixed n, since $\mathfrak{z}_{(n)}$ is a continuous function of $\mathfrak{z}$. It follows that the propositional function $\bigwedge_n [x \in G_\alpha(\mathfrak{z}_{(n)})]$ (for an even α) is of class $G_{\alpha+1}$ (the case of α odd is similar). Since

$$\bigwedge_n [x \in G_\alpha(\mathfrak{z}_{(n)})] \equiv \{x \in \bigcap_n G_\alpha(\mathfrak{z}_{(n)})\} \equiv \{x \in G_{\alpha+1}(\mathfrak{z})\},$$

it follows that condition (iii) is satisfied for $\alpha+1$. Finally, if the propositional function $x \in G_{\lambda_n}(\mathfrak{z})$, for every n, is of class G_{λ_n}, then the function $\bigvee_n [x \in G_{\lambda_n}(\mathfrak{z}_{(n)})]$ is of class G_λ. As before we infer that $\underset{x\mathfrak{z}}{E}[x \in G_\lambda(\mathfrak{z})]$ is of class G_λ.

Remark. An analogous theorem on the classes F_α is true: For every α there exists a universal function F_α such that the set $\underset{x\mathfrak{z}}{E}[x \in F_\alpha(\mathfrak{z})]$ is of class F_α. Namely, $F_\alpha(\mathfrak{z}) = \mathscr{X} - G_\alpha(\mathfrak{z})$.

XIV. Existence of sets of class G_α which are not of class F_α. We shall prove the existence of such sets in the space $\mathscr{N}$ (1). Let $\mathscr{X} = \mathscr{N}$ and consider the set

$$Z_\alpha = \underset{\mathfrak{z}}{E}[\mathfrak{z} \in G_\alpha(\mathfrak{z})]$$

which is the projection, on the $\mathscr{N}$ axis, of the intersection of $\underset{\mathfrak{z}\mathfrak{z}'}{E}[\mathfrak{z} \in G_\alpha(\mathfrak{z}')]$ with *the diagonal* of $\mathscr{N} \times \mathscr{N}$, i.e. with the set $\underset{\mathfrak{z}\mathfrak{z}'}{E}[\mathfrak{z} = \mathfrak{z}']$ (see § 15, IV). Since the set $\underset{\mathfrak{z}\mathfrak{z}'}{E}[\mathfrak{z} \in G_\alpha(\mathfrak{z}')]$ is of class G_α, hence so is Z_α (see the Theorem of Section III.

It remains to prove that Z_α is not of class F_α. If it is, then $\mathscr{N} - Z_\alpha$ is a G_α-set. Since the function G_α is universal, there exists a $\mathfrak{z}_0$ such that $\mathscr{N} - Z_\alpha = G_\alpha(\mathfrak{z}_0)$. By the definition of Z_α, we have

$$\{\mathfrak{z}_0 \in G_0(\mathfrak{z}_0)\} \equiv \{\mathfrak{z}_0 \in Z_\alpha\}$$

and, by the definition of $\mathfrak{z}_0$,

$$\{\mathfrak{z}_0 \in G_\alpha(\mathfrak{z}_0)\} \equiv \{\mathfrak{z}_0 \in (\mathscr{N} - Z_\alpha)\}.$$

This is a contradiction.

---

(1) For $\alpha \leqslant 3$ the existence can be proved more directly. See R. Baire, *Sur la représentation des fonctions discontinues*, Acta Math. 30 (1905) and 32 (1909), and N. Lusin, *Ensembles analytiques*, Paris 1930, p. 97 (an example due to L. Keldysh). For arbitrary α, see also R. Engelking, W. Holsztyński, and R. Sikorski, Coll. Math. 15 (1966), pp. 271–274.

Remark. The second part of the argument is, in fact, a proof of the following theorem of the general set theory.

DIAGONAL THEOREM [1]. *If F is a function which assigns, to each element of a set T, a subset of T, then the set $\underset{t}{E}[t \in T - F(t)]$ is not a value of F.*

XV. Problem of effectiveness [2]. The proof of existence of a set of class $\boldsymbol{G}_a$ which is not of class $\boldsymbol{F}_a$ given in Section XIV *is not effective*. That is, we have not *specified* a function which assigns, to each a, a set with the required property. By inspecting the argument of Section XIII, we see that the non-effectiveness is due to the fact that we have not defined a function which assigns, to each limit ordinal λ, a convergent sequence of ordinals $< \lambda$. We have, in fact, affirmed *the existence*, without determining any specified sequence of this kind. Actually, no effective definition of a sequence convergent to λ is known.

Thus, in order to resolve effectively the problem of existence of $\boldsymbol{G}_a$-sets which are not $\boldsymbol{F}_a$-sets, we shall have to modify condition 3) in the definition of G_a. We shall use for that purpose the function $\bar{\mathfrak{z}}$ defined in Section XII, 2).

Let $\tau(\mathfrak{z}) = \bar{\mathfrak{z}}$, if $\bar{\mathfrak{z}} < \Omega$, and $\tau(\mathfrak{z}) = -1$ otherwise. The function $\tau(\mathfrak{z})$ has two important properties:

(1) To each number $\mathfrak{z}$ it assigns a transfinite number (or -1) so as to cover all ordinals $a < \Omega$.

(2) The propositional function $\varphi_a(\mathfrak{z}) \equiv \{0 < \tau(\mathfrak{z}) < a\}$ is of class $\boldsymbol{G}_a$.

Now let the function G_a be defined by conditions XIII, 1), 2), and the following condition which is to replace condition 3):

$$3')\ \{x \in G_\lambda(\mathfrak{z})\} \equiv \bigvee_n \bigvee_{\xi<\lambda} [0 < \tau(\mathfrak{z}_{(2n)}) = \xi] \wedge [x \in G_\xi(\mathfrak{z}_{(2n+1)})].$$

Assuming that conditions (ii)–(iii) of Section XIII hold for $\xi < \lambda$, we have to prove them for $a = \lambda$ (condition (i) is a consequence of (iii)).

---

[1] This theorem is originally due to G. Cantor. See his proof of the inequality $2^{\mathfrak{m}} > \mathfrak{m}$.

[2] See my paper *Sur l'existence effective des fonctions représentables analytiquement de toute classe de Baire*, C. R. Paris 176 (1923), p. 229. See also W. Sierpiński, *Un exemple effectif d'un ensemble mesurable (B) de classe a*, Fund. Math. 6 (1924), p. 39.

Proof of (ii). Each X of class G_λ is of the form $X = \bigcup_n X_n$, where X_n is of class G_{ξ_n} and $0 < \xi_n < \lambda$. For each ξ_n there exists an irrational number $\mathfrak{y}_n$ such that $\tau(\mathfrak{y}_n) = \xi_n$. Since the function G_{ξ_n} is universal, there exists a $\mathfrak{w}_n$ such that $X_n = G_{\xi_n}(\mathfrak{w}_n)$. By the definition of the function $\mathfrak{z}_{(n)}$, there exists a $\mathfrak{z}$ such that

$$\mathfrak{z}_{(2n)} = \mathfrak{y}_n \quad \text{and} \quad \mathfrak{z}_{(2n+1)} = \mathfrak{w}_n, \quad \text{for every } n.$$

It follows that

$$X_n = G_{\xi_n}(\mathfrak{z}_{(2n+1)}) \quad \text{and} \quad 0 < \tau(\mathfrak{z}_{(2n)}) = \xi_n, \text{ which gives } X = G_\lambda(\mathfrak{z}).$$

Proof of (iii). We have to prove that the propositional function $x \in G_\lambda(\mathfrak{z})$ (of two variables) is of class G_λ. The equivalence

$$[0 < \tau(\mathfrak{z}) = \xi] \equiv \neg \varphi_\xi(\mathfrak{z}) \wedge \varphi_{\xi+1}(\mathfrak{z})$$

shows that the propositional function $[0 < \tau(\mathfrak{z}) = \xi]$ is of class $G_{\xi+1}$; the same is true for $[0 < \tau(\mathfrak{z}_{(2n)}) = \xi]$, since $\mathfrak{z}_{(2n)}$ is a continuous function of $\mathfrak{z}$ (see Theorem 4 of Section XI). For the same reason the propositional function $x \in G_\xi(\mathfrak{z}_{(2n+1)})$ is of class G_ξ, if $\xi < \lambda$. Consequently the conjunction

$$[0 < \tau(\mathfrak{z}_{(2n)}) = \xi] \wedge [x \in G_\xi(\mathfrak{z}_{(2n+1)})]$$

is of class $G_{\xi+1}$. The function $x \in G_\lambda(\mathfrak{z})$ is therefore of class G_λ (by (3')).

Thus the problem of existence of an universal function of class G_a, for each a, is effectively resolved. The definition of the set Z_a such as was given in Section XIV gives an effective solution to the problem of existence of a G_a-set which is not F_a in the space of irrational numbers.

*§ 31. *B* measurable functions

I. Classification. A mapping $f: \mathscr{X} \to \mathscr{Y}$ is said *to be B measurable of class a* (or, briefly, of *class a*) if, for every closed subset $F \subset \mathscr{Y}$, the set $f^{-1}(F)$ is borelian of multiplicative class a (1).

The closed sets are of the multiplicative class 0. Hence the continuous functions are functions of class 0.

(1) See H. Lebesgue, Journ. de math. 1 (1905), p. 166.

A one-to-one mapping f is said to be a (*generalized*) *homeomorphism of class* α, β, if it is of class α and its inverse f^{-1} is of class β (1).

Clearly homeomorphisms of class $0, 0$ coincide with the homeomorphisms in the usual sense.

THEOREM 1. *In order that the characteristic function of a set A be of class α, it is necessary and sufficient that the set A be ambiguous of class α.*

Proof. The characteristic function assumes the values 0 and 1 only. Let $\mathcal{Y}$ be the space composed of 0 and 1. Either of them is then a closed set. If the characteristic function f is of class α, the sets $A = f^{-1}(1)$ and $\mathcal{X} - A = f^{-1}(0)$ are of the multiplicative class α, which means that A is ambiguous of class α.

Conversely, if A is ambiguous of class α, we easily verify that the set $f^{-1}(F)$ is of the multiplicative class α for every closed set F (the space $\mathcal{Y}$ contains, in fact, four closed sets).

By § 30, XIV and III it follows that

THEOREM 2. *In every class α there exist real functions which do not belong to a lower class.*

THEOREM 3. *There exist functions which are not B measurable.*

Theorem 3 for separable spaces follows also from the next theorem.

THEOREM 4. *If $\mathcal{X}$ and $\mathcal{Y}$ are separable, then the family of B measurable functions is of power $\leqslant \mathfrak{c}$.*

Proof. If $R_1, R_2, \ldots$ is a base of $\mathcal{Y}$, then every function f mapping $\mathcal{X}$ into $\mathcal{Y}$ is completely characterized by the sequence of sets $f^{-1}(R_1), f^{-1}(R_2), \ldots$. For, each point y of $\mathcal{Y}$ has the form $y = R_{k_1} \cap R_{k_2} \cap \ldots$ and

$$\{y = f(x)\} \equiv \{x \in f^{-1}(y)\} \equiv \{x \in \bigcap_n f^{-1}(R_{k_n})\}.$$

If the function f is B measurable, the sets $f^{-1}(R_n)$ are borelian. The family of Borel sets is of power $\leqslant \mathfrak{c}$; hence the family of B measurable functions is of power $\leqslant \mathfrak{c}^{\aleph_0} = \mathfrak{c}$.

---

(1) See my paper *Sur le prolongement de l'homéomorphie*, C. R. Paris 197 (1933), p. 1090. Compare also W. Sierpiński, Fund. Math. 21 (1933), p. 66.

Let us state without proof the following theorem.

THEOREM 5. *For each pair of ordinals α, β (less than Ω) there exists a mapping f of $\mathcal{N}$ onto itself which is a homeomorphism of class α, β precisely (i.e. neither the function f is of class $< \alpha$, nor the function f^{-1} is of class $< \beta$)* (1).

II. Equivalences. By using the identity $f^{-1}(\mathcal{Y}-Y) = \mathcal{X}-f^{-1}(Y)$ one can define the functions of class α to be the functions for which the set $f^{-1}(G)$ is of the additive class α, for every open set G.

THEOREM 1. *If $R_1, R_2, \ldots$ is a base of $\mathcal{Y}$, then f is of class α if and only if each set $f^{-1}(R_n)$ is of additive class α.*

Proof. As $G = R_{k_1} \cup R_{k_2} \cup \ldots$, therefore

$$f^{-1}(G) = f^{-1}(R_{k_1}) \cup f^{-1}(R_{k_2}) \cup \ldots .$$

It follows also that *if the sets $f^{-1}(R_n)$ are borelian, for $n = 1, 2, \ldots$, then the function f is B measurable.* In fact, its class is α, where $\alpha > \alpha_n$ and where $f^{-1}(R_n)$ is of class α_n.

In particular, if $\mathcal{Y}$ is the space of *real numbers*, the functions of class α can be defined by assuming that the sets $\underset{x}{E}\{a < f(x) < b\}$ are of additive class α, for every a and b (which can be assumed to be rational).

THEOREM 2. *If the space $\mathcal{Y}$ is discrete and the function f is of class α, the set $f^{-1}(Y)$ is ambiguous of class α, for every $Y \subset \mathcal{Y}$.*

For, every set Y is both closed and open in $\mathcal{Y}$.

THEOREM 3 (2). *If the space $\mathcal{Y}$ is separable, then a function f is of class α if and only if for every $\varepsilon > 0$ there exists a sequence $Z_1, Z_2, \ldots$ of sets of additive class α such that*

$$\mathcal{X} = Z_1 \cup Z_2 \cup \ldots \quad \textit{and} \quad \delta[f(Z_n)] < \varepsilon \quad \textit{for} \quad n = 1, 2, \ldots .$$

(1) For a proof see my note *Sur une généralisation de la notion d'homéomorphie*, Fund. Math. 22 (1934), p. 219.

(2) See H. Lebesgue, *op. cit.*, p. 172 (real domain). For the general case, see B. Gagaeff, *Sur les suites convergentes de fonctions mesurables B*, Fund. Math. 18 (1932), p. 183. See also P. Veress, *Ueber kompakte Funktionenmengen und Bairesche Klassen*, Fund. Math. 7 (1925), p. 244, where numerous applications of this theorem are given.

Proof. Since $\mathscr{Y}$ is separable, there exists (see § 21, II, Theorem 2 and Remark 1) a sequence $B_1, B_2, \ldots$ of open balls such that

$$\mathscr{Y} = B_1 \cup B_2 \cup \ldots \quad \text{and} \quad \delta(B_n) < \varepsilon.$$

It suffices to put $Z_n = f^{-1}(B_n)$.

Assume now that the condition of the theorem is satisfied. Then

$$\mathscr{X} = Z_1^k \cup Z_2^k \cup \ldots \quad \text{and} \quad \delta[f(Z_n^k)] < \frac{1}{k},$$

where Z_n^k is of the additive class a. Let G be an open set in $\mathscr{Y}$. We shall show that $f^{-1}(G)$ is the union of the sets Z_n^k such that $f(Z_n^k) \subset G$, which will prove that $f^{-1}(G)$ is of the additive class a.

Firstly, the conditions $x \in Z_n^k$ and $f(Z_n^k) \subset G$ imply $f(x) \in G$ and $x \in f^{-1}(G)$. Secondly, the condition $f(x) \in G$, equivalent to $x \in f^{-1}(G)$, implies that, for sufficiently large k, if $|y - f(x)| < 1/k$, then $y \in G$ (since G is open). Let n be such that $x \in Z_n^k$. Since $\delta[f(Z_n^k)] < 1/k$, it follows that $f(Z_n^k) \subset G$.

III. Composition of functions.

THEOREM 1. *If f is a function of class a and Y is a set of class β, then the set $f^{-1}(Y)$ is of class $a+\beta$ (which is multiplicative or additive according to the class of Y).*

The proof is by transfinite induction (on β). We use the identities

$$f^{-1}\Big(\bigcup_n Y_n\Big) = \bigcup_n f^{-1}(Y_n), \quad f^{-1}\Big(\bigcap_n Y_n\Big) = \bigcap_n f^{-1}(Y_n)$$

and the implication $\beta_n < \beta$ implies $a+\beta_n < a+\beta$.

In particular if f is *continuous*, i.e. $a = 0$, then $f^{-1}(Y)$ is of class β.

THEOREM 2. *If $f\colon \mathscr{X} \to \mathscr{Y}$ is of class a and $g\colon \mathscr{Y} \to \mathscr{Z}$ is of class β, then the function $h = g \circ f$ is of class $a+\beta$.*

Proof. By § 3, IV (1), we have

$$h^{-1}(F) = f^{-1}[g^{-1}(F)].$$

If the set F is closed, then $g^{-1}(F)$ is of the multiplicative class β, and $f^{-1}[g^{-1}(F)]$ is, by Theorem 1, of class $a+\beta$.

In particular if g is *continuous*, then the functions $g \circ f$ and $f \circ g$ are of the same class as f.

IV. Partial functions.

THEOREM 1. *If $\{E_n\}$ is a sequence of sets of the additive class α such that $\mathscr{X} = E_1 \cup E_2 \cup \ldots$ and the partial function $f_n = f|E_n$ is of class α on E_n, then the function f is of class α (on the entire space).*

Proof. Let G be open in $\mathscr{Y}$. Then (§ 3, III, (15)):

$$f^{-1}(G) = f_1^{-1}(G) \cup f_2^{-1}(G) \cup \ldots ;$$

$f_n^{-1}(G)$ is of the additive class α with respect to E_n which itself is of the additive class α; therefore $f^{-1}(G)$ is of the additive class α as the union of sets of this class.

THEOREM 2. *If M and N are two sets of the multiplicative class α such that $\mathscr{X} = M \cup N$ and if the partial functions $f|M$ and $f|N$ are of class α, then the function f is of class α.*

The proof is analogous to the proof above. We replace the open set G by the closed set F and infinite union by union of two terms.

THEOREM 3. *If f is of class α, then so is $f|E$, for every E.*

This is an immediate consequence of § 3, III, 14.

THEOREM 4. *For every finite system $F_1, \ldots, F_k$ of disjoint sets of the multiplicative class $\alpha > 0$ and for every finite space $\mathscr{Y} = (y_1, \ldots, y_k)$ there exists a function $f: \mathscr{X} \to \mathscr{Y}$ of class α such that $f(x) = y_i$ for $x \in F_i$.*

Proof. By § 30, VII, Theorem 4, there exists a system $A_1, \ldots, A_k$ of disjoint ambiguous sets of class α such that

$$\mathscr{X} = A_1 \cup \ldots \cup A_k, \qquad F_i \subset A_i.$$

We define $f(x) = y_i$ for $x \in A_i$, $i = 1, \ldots, k$. By Theorem 1, the function f is of class α.

V. Functions of several variables. Consider the *cartesian product* $\mathscr{X} \times \mathscr{Y}$ of two spaces $\mathscr{X}$ and $\mathscr{Y}$.

Obviously every function $f: \mathscr{X} \to \mathscr{Z}$ of a single variable can be regarded as a function of two variables $g: \mathscr{X} \times \mathscr{Y} \to \mathscr{Z}$, if we put $g(x, y) = f(x)$.

THEOREM 1. *If f is of class α and $g(x, y) = f(x)$, then g is of class α.*

Proof. The abscissa x of the point (x, y) is its continuous function (see § 15, II). By the composite function theorem (Section III, Theorem 2), f is a function of class α.

THEOREM 2. *If the function f of the variables x and y is continuous in x and of class a in y, then it is of class $a+1$ in the variable (x, y)* [1].

In particular, *if the function f is continuous in each variable separately, it is of class* 1.

Consequently, a *function of n variables, which is continuous in each variable, is of class $n-1$.*

The proof of Theorem 2 is simpler, if $\mathscr{X}$ is assumed to be separable. In order to prove it in this particular case, let us prove first the following proposition.

Let $r_1, r_2, \ldots$ be a dense sequence of points in $\mathscr{X}$ and g be a continuous function. In order that the point $g(x)$ should belong to a closed set F it is necessary and sufficient that for every n there exist a k such that $|x-r_k| < 1/n$ and $g(r_k) \epsilon B_n$, where B_n is the generalized open ball with centre F and radius $1/n$ (see § 21, IV):

$$\{g(x) \epsilon F\} \equiv \bigwedge_n \bigvee_k [|x-r_k| < 1/n] \wedge [g(r_k) \epsilon B_n]. \tag{i}$$

Proof. If $r_{k_1}, r_{k_2}, \ldots$ is a sequence convergent to x, then $\lim_{m=\infty} g(r_{k_m}) = g(x)$ and $|r_{k_m}-x| < 1/n$, $|g(r_{k_m})-g(x)| < 1/n$, for sufficiently large values of m. Supposing that $g(x) \epsilon F$, we get $g(r_{k_m}) \epsilon B_n$ and the right-hand member of (i) is fulfilled. Conversely, if for every n there exists a k_n such that $|x-r_{k_n}| < 1/n$ and $g(r_{k_n}) \epsilon B_n$, whence $\varrho[g(r_{k_n}), F] < 1/n$, then $\lim_{n=\infty} r_{k_n} = x$ and $\lim_{n=\infty} g(r_{k_n}) = g(x)$. Since $\lim_{n=\infty} \varrho[g(r_{k_n}), F] = 0$, it follows that $\varrho[g(x), F] = 0$, and hence $g(x) \epsilon F$.

This completes the proof of (i). Replace now $g(x)$ by $f(x, y)$ in formula (i):

$$\{f(x, y) \epsilon F\} \equiv \bigwedge_n \bigvee_k [|x-r_k| < 1/n] \wedge [f(r_k, y) \epsilon B_n].$$

Hence

$$f^{-1}(F) = \bigcap_n \bigcup_k \{[(\underset{x}{E}\, |x-r_k| < 1/n) \times \mathscr{Y}] \cap [\mathscr{X} \times \underset{y}{E}\, f(r_k, y) \epsilon B_n]\}. \tag{ii}$$

---

[1] See H. Lebesgue, *loc. cit.*, p. 201, and my note *Sur la théorie des fonctions dans les espaces métriques*, Fund. Math. 17 (1931), p. 278. Elementary examples show that a function of two variables may be *discontinuous*, but, at the same time, it may be continuous in each variable separately.

Put $f_k(y) = f(r_k, y)$. Hence f_k is a function of class a. Thus the set $\underset{y}{E}[f(r_k, y)\in B_n]$ is of the additive class a. Note that $\underset{x}{E}[|x-r_k| < 1/n]$ is an open ball. By the method of evaluation of the class of a propositional function (§ 30, XI) it follows that the propositional function $\{f(x, y)\in F\}$ of two variables and the set $f^{-1}(F)$ are of the multiplicative class $a+1$.

We now pass to the general case of $\mathscr{X}$ arbitrary metric (1).

For $a = 0$, we apply the following equivalence in place of (i):

$$\{g(x)\in F\} \equiv \bigwedge_n \bigvee_{x'} [|x-x'| < 1/n] \wedge [g(x')\in B_n], \tag{iii}$$

which can be proved in an analogous way.

In formula (ii) we must replace $\bigcup_k$ by $\bigcup_{x'}$ and r_k by x'. Now the set in brackets $\{\ \}$ is open (since $a = 0$), the (uncountable) disjunction $\bigcup_{x'}$ leads again to an open set and, finally, $f^{-1}(F)$ is a G_δ-set.

The proof of Theorem 2 for $a = 0$ is completed.

Now let $a > 0$.

Firstly, consider a closed set F and a sequence of points such that $\lim y_n = y$. Let B_n be the open ball with centre F and radius $1/n$. We shall prove that $y\in F$ *if and only if for every* n *there exists a* k *such that* $y_{n+k}\in B_n$:

$$\{y\in F\} \equiv \bigwedge_n \bigvee_k (y_{n+k}\in B_n). \tag{iv}$$

On one hand, if $y\in F$, the points y_m with sufficiently large values of m satisfy $|y_m-y| < 1/n$ and $y_m\in B_n$. Hence the right member of (i) is true. On the other hand, if $y\notin F$, then by the formula $F = \bigcap_n \bar{B}_n$ there exists an m such that $y\notin\bar{B}_m$. The condition $y = \lim y_n$ implies that, from an index $n > m$ onwards, the points y_{n+k} lie on outside of $\bar{B}_m$ which shows that the right-hand member of (iv) fails.

Thus formula (iv) is proved. Now we shall prove an auxiliary theorem on the operation $\mathscr{M}$ (see § 30, X; using the same notation).

---

(1) See D. Montgomery, *op. cit.*, Fund. Math. 25, and my note in the same issue. See also R. Engelking, Prace Matematyczne (to appear).

AUXILIARY THEOREM. *Let* P *be a property of sets which is invariant under operation* $\mathscr{M}$ *and under the cartesian product with* $\mathscr{X}$. *Let* $\mathscr{X} = p_0, p_1, \ldots, p_\zeta, \ldots$ (*the space* $\mathscr{X}$ *is well ordered*) *and let* $G_0, G_1, \ldots, G_\xi, \ldots$ *be a transfinite sequence of open sets such that* $\mathscr{X} = \bigcup_\xi G_\xi$. *Let* $\xi(x)$ *be the minimal index with* $x \in G_{\xi(x)}$ *and let* $w(x) = p_{\xi(x)}$. *Let* $\varphi(x, y)$ *be a propositional function of two variables running over* $\mathscr{X}$ *such that the set* $\underset{y}{E}\,\varphi(x, y)$ *has the property* P *for every* x.

Under these hypotheses the set $\underset{xy}{E}\,\varphi[w(x), y]$ *also has the property* P.

Proof. By the definition of $w(x)$,

$$\{p_\xi = w(x)\} \equiv \{x \in G_\xi - \bigcup_{\eta<\xi} G_\eta\}. \tag{1}$$

Therefore

$$\varphi[w(x), y] \equiv \bigvee_\xi \varphi(p_\xi, y) \wedge [p_\xi = w(x)] \equiv \bigvee_\xi \varphi(p_\xi, y) \wedge [x \in (G_\xi - \bigcup_{\eta<\xi} G_\eta)].$$

It follows that

$$\begin{aligned}\underset{xy}{E}\,\varphi[w(x), y] &= \underset{xy}{E} \bigvee_\xi \varphi(p_\xi, y) \wedge [x \in (G_\xi - \bigcup_{\eta<\xi} G_\eta)] \\ &= \bigcup_\xi \{\underset{xy}{E}\,\varphi(p_\xi, y) \cap \underset{xy}{E}\, x \in (G_\xi - \bigcup_{\eta<\xi} G_\eta)\} \\ &= \bigcup_\xi \{\underset{xy}{E}\,\varphi(p_\xi, y) \cap \underset{xy}{E}\,(x \in G_\xi) - \bigcup_{\eta<\xi} \underset{xy}{E}\,(x \in G_\eta)\}.\end{aligned}$$

Let $\underset{xy}{E}\,\varphi(p_\xi, y) = X_\xi$ and $\underset{xy}{E}\,(x \in G_\xi) = G_\xi^*$. Then

$$\underset{xy}{E}\,\varphi[w(x), y] = \bigcup_\xi (X_\xi \cap G_\xi^* - \bigcup_{\eta<\xi} G_\eta^*) \tag{2}$$

which shows that the set $\underset{xy}{E}\,\varphi[w(x), y]$ is derived from $\{X_\xi\}$ by means of the operation $\mathscr{M}$.

Since $X_\xi = \mathscr{X} \times \underset{y}{E}\,\varphi(p_\xi, y)$, X_ξ has the property P. Since $G_\xi^* = G_\xi \times \mathscr{X}$, G_ξ^* is open. Since the property P is invariant under the operation $\mathscr{M}$, the required conclusion follows from (2).

Thus the auxiliary theorem is proved. Now let f be a function continuous in x and of class $\alpha > 0$ in y. We must prove that, for every closed set F, the set $\underset{xy}{E}\,[f(x, y) \in F]$ is of multiplicative class $(\alpha+1)$.

For a fixed n, let G_ξ be the open ball with centre p_ξ and radius $1/n$. Let $w_n(x)$ denote the function $w(x)$ defined by (1). Since $x \in G_{\xi(x)}$ and $w_n(x) = p_{\xi(x)} \in G_{\xi(x)}$, it follows that

$$|w_n(x) - x| < 1/n, \quad \text{and hence} \quad \lim_{n=\infty} w_n(x) = x. \tag{3}$$

By the continuity of f in x, condition (3) yields

$$\lim_{n=\infty} f[w_n(x), y] = f\{[\lim_{n=\infty} w_n(x)], y\} = f(x, y).$$

By virtue of (iv)

$$\{f(x, y) \in F\} \equiv \bigwedge_{n=1}^{\infty} \bigvee_{k=1}^{\infty} \{f[w_{n+k}(x), y] \in B_n\}. \tag{4}$$

We apply the auxiliary theorem to the property P of being of the additive class α, and to the propositional function

$$\varphi_n(x, y) \equiv \{f(x, y) \in B_n\}.$$

As the additive class $\alpha > 0$ is invariant under the operation $\mathcal{M}$ (see § 30, X, Theorem 3) and under the product by $\mathcal{X}$ (see § 30, III), it follows that the set

$$\underset{xy}{E}\, \varphi_n[w_m(x), y] = \underset{xy}{E}\{f[w_m(x), y] \in B_n\}$$

is of the additive class α for every m and n. Hence $f^{-1}(F)$ is of multiplicative class $(\alpha+1)$, by (4).

Remark. *A function $f\colon \mathcal{X} \times \mathcal{Y} \to \mathcal{Z}$ of class 1 in each variable separately need not be B measurable* (even measurable in the sense of Lebesgue) [1].

Let A be a non-borelian set situated on a circumference in the euclidean plane $\mathcal{E} \times \mathcal{E}$ (or a set which is not surface measurable in the sense of Lebesgue and which has at most two points in

---

[1] W. Sierpiński, *Sur un problème concernant les ensembles mesurables superficiellement*, Fund. Math. 1 (1920), p. 114, and *Funkcje przedstawialne analitycznie*, Lwów 1925, p. 68.

common with every vertical and every horizontal line). The characteristic function of A is not B measurable (see Section I), while it is of class 1 with respect to each variable; in fact, it is zero everywhere except at most two points.

VI. Complex and product functions. A pair of functions $f: \mathcal{T} \to \mathcal{X}$, $g: \mathcal{T} \to \mathcal{Y}$ define a "complex" function $h: \mathcal{T} \to \mathcal{X} \times \mathcal{Y}$ (compare § 3, III).

THEOREM 1. *Let the spaces $\mathcal{X}$ and $\mathcal{Y}$ be separable* (1). *The function h is of class a if and only if the functions f and g (the "coordinates") are of class a.*

Proof. The condition is necessary. If G is an open subset of $\mathcal{X}$, then $G \times \mathcal{Y}$ is open. Since the function h is of class a, the set $\underset{t}{E}[h(t) \in G \times \mathcal{Y}]$ is of the additive class a. By virtue of the equivalence $\{f(t) \in G\} \equiv \{h(t) \in G \times \mathcal{Y}\}$, it coincides with $f^{-1}(G)$. Therefore the function f is of class a.

The condition is sufficient. Let $R_1, R_2, \ldots$ be a base of $\mathcal{X}$ and $S_1, S_2, \ldots$ a base of $\mathcal{Y}$. The double sequence $R_m \times S_n$ is a base of $\mathcal{X} \times \mathcal{Y}$ (see § 15, I, Theorem 3). It suffices (see Section II, Theorem 1) to show that the set $h^{-1}(R_m \times S_n)$ is of the additive class a. By § 3, III (23), we have

$$h^{-1}(R_m \times S_n) = f^{-1}(R_m) \cap g^{-1}(S_n).$$

By hypothesis, the functions f and g are of class a, so the sets $f^{-1}(R_m)$ and $g^{-1}(S_n)$ are of the additive class a. Therefore their intersection $h^{-1}(R_m \times S_n)$ is of the additive class a.

The preceding theorem can be extended to countable products:

THEOREM 1'. *Let $\mathcal{X}_1, \mathcal{X}_2, \ldots$ be a sequence of separable spaces and $\mathfrak{z}: \mathcal{T} \to \mathcal{X}_1 \times \mathcal{X}_2 \times \ldots$, i.e. the function $\mathfrak{z}$ represents a sequence of functions $\mathfrak{z}_1, \mathfrak{z}_2, \ldots$. The function $\mathfrak{z}$ is of class a if and only if each function $\mathfrak{z}_i$ is of this class.*

Proof. As before we prove that the condition is necessary by using the equivalence

$$\{\mathfrak{z}_1(t) \in G\} \equiv \{\mathfrak{z}(t) \in (G \times \mathcal{X}_2 \times \mathcal{X}_3 \times \ldots)\}.$$

(1) It would be interesting to know if the assumption of separability can be dropped.

To prove that it is sufficient, denote by R^i_m, $m = 1, 2, \ldots$ a base of $\mathscr{X}_i$. The sets of the form

$$R^1_{k_1} \times R^2_{k_2} \times \ldots \times R^n_{k_n} \times \mathscr{X}_{n+1} \times \mathscr{X}_{n+2} \times \ldots$$

constitute a base of $\mathscr{X}_1 \times \mathscr{X}_2 \times \ldots$ (§ 16, I). Then

$$\mathfrak{z}^{-1}(R^1_{k_1} \times \ldots \times R^n_{k_n} \times \mathscr{X}_{n+1} \times \mathscr{X}_{n+2} \times \ldots)$$
$$= \mathfrak{z}_1^{-1}(R^1_{k_1}) \cap \ldots \cap \mathfrak{z}_n^{-1}(R^n_{k_n}) \cap \mathscr{T} \cap \mathscr{T} \cap \ldots$$

by formula (9) of § 3, VIII.

The first n factors of the product are of the additive class a, hence the total set is of this class. This completes the proof.

Let $f_i\colon \mathscr{X}_i \to \mathscr{Y}_i$ and let $\mathfrak{z} = [\mathfrak{z}^1, \mathfrak{z}^2, \ldots]$ be a variable point of $\mathscr{X}_1 \times \mathscr{X}_2 \times \ldots$; then putting $\mathfrak{y}(\mathfrak{z}) = [f_1(\mathfrak{z}^1), f_2(\mathfrak{z}^2), \ldots]$ we get the product function

$$\mathfrak{y}\colon \mathscr{X}_1 \times \mathscr{X}_2 \times \ldots \to \mathscr{Y}_1 \times \mathscr{Y}_2 \times \ldots$$

(compare § 3, VIII).

THEOREM 2. *If each function $f_i\colon \mathscr{X}_i \to \mathscr{Y}_i$ is of class a, then so is $\mathfrak{y}$. If the function $g\colon \mathscr{Y}_1 \times \mathscr{Y}_2 \times \ldots \to \mathscr{Z}$ is of class β, then the function $g \circ \mathfrak{y}\colon \mathscr{X}_1 \times \mathscr{X}_2 \times \ldots \to \mathscr{Z}$ is of class $a+\beta$* (provided that the spaces $\mathscr{Y}_i$ are separable).

If a separable space $\mathscr{Y}_i$ is an image of a space $\mathscr{X}_i$ under a mapping of class a, then the space $\mathscr{Y}_1 \times \mathscr{Y}_2 \times \ldots$ is an image of $\mathscr{X}_1 \times \mathscr{X}_2 \times \ldots$ under a mapping of class a.

Proof. Since the functions $\mathfrak{z}^i$ are continuous and the coordinates $f_i(\mathfrak{z}^i)$ of $\mathfrak{y}(\mathfrak{z})$ are functions of class a of $\mathfrak{z}$, it follows that the function $\mathfrak{y}$ is of the same class. Moreover, the function $g \circ \mathfrak{y}$ is of class $a+\beta$.

THEOREM 3. *If f_i, $i = 1, 2, \ldots$, is a mapping of class a of a space $\mathscr{X}_i$ onto a subset of a separable space $\mathscr{Y}_i$, then the set*

$$\mathfrak{Z} = \underset{\mathfrak{z}}{E}[f_1(\mathfrak{z}^1) = f_2(\mathfrak{z}^2) = \ldots], \quad \textit{where} \quad \mathfrak{z} \in (\mathscr{X}_1 \times \mathscr{X}_2 \times \ldots),$$

is of the multiplicative class a in $\mathscr{X}_1 \times \mathscr{X}_2 \times \ldots$.

Moreover, *the function f^* defined by the condition*

$$f^*(\mathfrak{z}) = f_1(\mathfrak{z}^1) \quad \textit{for} \quad \mathfrak{z} \in \mathfrak{Z}$$

is a mapping of class a of $\mathfrak{Z}$ onto $f^*(\mathfrak{Z}) = f_1(\mathscr{X}_1) \cap f_2(\mathscr{X}_2) \cap \ldots$.

Finally, *if the functions f_i are homeomorphisms of class α, β then so is the function f^* (provided that $\mathscr{X}_i$ are separable).*

The proof is quite analogous to that of Theorems 4 and 5 of § 16, IV.

THEOREM 4. *In order that the characteristic function of a sequence of sets $A_1, A_2, \ldots$* (compare § 3, VII) *should be of class α .it is necessary and sufficient that each of the sets be ambiguous of class α.*

This is a direct consequence of Theorems 1′ and I, 1.

EXAMPLES. (i) If $\mathscr{Y}$ is separable and the functions $f_1\colon \mathscr{X}_1 \to \mathscr{Y}$, $f_2\colon \mathscr{X}_2 \to \mathscr{Y}$ are of class α, and $h(x_1, x_2) = |f_1(x_1) - f_2(x_2)|$, h is of class α.

We apply Theorem 2 to $g(y_1, y_2) = |y_1 - y_2|$, the distance being a continuous function.

(ii) If f_1 and f_2 are two real valued functions of class α, then $f_1(x_1) \pm f_2(x_2), f_1(x_1) \cdot f_2(x_2), f_1(x_1) : f_2(x_2)$ define functions of the same class.

VII. Graph of $f\colon \mathscr{X} \to \mathscr{Y}$. Let $I = \underset{xy}{E}[y = f(x)]$.

THEOREM 1. *If f is of class α, then I is of multiplicative class α.*

If $\mathscr{Y}$ is separable, this theorem is a particular case of Theorem VI, 3, for $\mathscr{X}_1 = \mathscr{Y}$, $\mathscr{X}_2 = \mathscr{X}$, $f_1 = \text{identity}$, $f_2 = f\,(i = 1, 2)$.

To give a more direct proof (based on a different idea), let us consider the function $h(x, y) = |y - f(x)|$. Then

$$I = \underset{xy}{E}[h(x, y) = 0] = h^{-1}(0)$$

(where 0 is the number zero). Since the function h is of class α (see VI, example (i)), the set $h^{-1}(0)$ is of the multiplicative class α [1].

---

[1] F. Hausdorff, *Mengenlehre*, p. 269. For the case of a real function, see W. Sierpiński, *Sur les images des fonctions représentables analytiquement*, Fund. Math. 2 (1921), p. 78. In the general case, I gave given another proof (in my paper of Fund. Math. 17, p. 277 referred to above) based on the "separation of variables formula"

$$(y \neq y') \equiv \bigvee_n (y \in \mathscr{Y} - R_n) \wedge (y' \in R_n),$$

where $R_1, R_2, \ldots$ is a base of the space.

The proof in the case where $\mathcal{Y}$ is an arbitrary metric space will be based on the auxiliary theorem of Section V (¹). Replace there $\mathcal{X}$ by $\mathcal{Y}$ and G_ξ by the open ball with centre p_ξ and radius $1/n$ (for a fixed n). As in the proof of Theorem V, 2, let $w_n(y)$ be the function $w(y)$ defined by V (1). The inequality V (3) implies the equivalence

$$[a = b] \equiv \bigwedge_n |w_n(a) - b| \leqslant \frac{1}{n},$$

whence

$$[y = f(x)] \equiv \bigwedge_n |w_n(y) - f(x)| \leqslant \frac{1}{n}$$

and therefore

$$I = \underset{xy}{E}[y = f(x)] = \bigcap_{n=1}^{\infty} \underset{xy}{E}[|w_n(y) - f(x)| \leqslant 1/n]. \tag{1}$$

In the auxiliary theorem let

$$\varphi(x, y) \equiv [|y - f(x)| \leqslant 1/n]. \tag{2}$$

For every y, the set $\underset{x}{E}\varphi(x, y)$ is of the multiplicative class a (since f is of class a). The multiplicative class $a > 0$ is invariant under operation $\mathcal{M}$ (by § 30, X, Theorem 3) and under multiplication by $\mathcal{Y}$. It follows that the set $\underset{xy}{E}\varphi[x, w_n(y)]$ is of multiplicative class a. Therefore (by (1) and (2)) I is of the same class.

It is well known that if $a = 0$, i.e. if f is continuous, I is closed (by § 15, V, 2). If $a = 1$, I is a G_δ-set. The converse theorems are, however, not true.

THEOREM 2. *If f is of class a and A is of class β in $\mathcal{X} \times \mathcal{Y}$, the projection P of $I \cap A$ on the $\mathcal{X}$ axis is of class $a + \beta$; this class is multiplicative or additive according to the class of A (provided that $\mathcal{X}$ and $\mathcal{Y}$ are separable).*

Proof. Let $h(x) = [x, f(x)]$. Then h is of class a (by VI, Theorem 1), whence $P = h^{-1}(A)$ is of class $a + \beta$ (by III, Theorem 1).

(¹) See D. Montgomery, *op. cit.*, p. 532. See also R. Engelking, *loc. cit.*

VIII. Limit of functions[1].

THEOREM 1. *The limit of a convergent sequence of functions of class a is of class $a+1$.*

Proof. Let $f(x) = \lim_{n=\infty} f_n(x)$. By V (iv)

$$\{f(x)\in F\} \equiv \bigwedge_n \bigvee_k \{f_{n+k}(x)\in B_n\},$$

whence

$$f^{-1}(F) = \mathop{E}_x [f(x)\in F] = \bigcap_n \bigcup_k \mathop{E}_x [f_{n+k}(x)\in B_n] = \bigcap_n \bigcup_k f_{n+k}^{-1}(B_n).$$

Since the functions f_n are of class a, the set $f_{n+k}^{-1}(B_n)$ is of the additive class a. Consequently the set $f^{-1}(F)$ is of the multiplicative class $a+1$.

Thus, in particular, the limit of a sequence of continuous functions is of class 1. The limit of a sequence of functions of finite classes is of class $\omega+1$.

THEOREM 2. *The limit of a uniformly convergent sequence of functions of class a is of class a.*

Proof. By the uniform convergence there exists an (increasing) sequence m_n such that $|f(x)-f_{m_n+k}(x)|<1/n$ for every x and $k\geqslant 0$.

We shall show that

$$\{f(x)\in F\} \equiv \bigwedge_n \bigwedge_k \{f_{m_n+k}(x)\in \overline{B}_n\}.$$

For brevity, let us denote $y=f(x)$ and $y_n=f_n(x)$. If $y\in F$, then $y_{m_n+k}\in\overline{B}_n$, for every n and k, since $|y-y_{m_n+k}|<1/n$. Conversely, if the right-hand member of the equivalence is satisfied, then $y_{m_n}\in\overline{B}_n$ for every n, whence $\varrho(y_{m_n}, F)\leqslant 1/n$. Since $\varrho(y, F)$ is a continuous function of y (§ 21, IV, (5)), the condition $y=\lim_{n=\infty} y_{m_n}$ implies $\varrho(y,F)=0$, hence $y\in F$.

Thus

$$f^{-1}(F) = \bigcap_n \bigcap_k f_{m_n+k}^{-1}(\overline{B}_n).$$

As the set $f_{m_n+k}^{-1}(\overline{B}_n)$ is of the multiplicative class a, so $f^{-1}(F)$ is of the same class. Therefore the function f is of class a.

---

[1] See F. Hausdorff, *Mengenlehre*, p. 267.

In particular, the limit of a uniformly convergent sequence of continuous functions is continuous (compare § 21, X, Theorem 2'). The limit of a uniformly convergent sequence of functions of finite (increasing) classes is of class ω.

Remark. The uniform convergence is by no means a necessary condition for the function $f(x) = \lim_{n=\infty} f_n(x)$ to be of class a. A necessary and sufficient condition (provided that the space $\mathcal{Y}$ is separable) is: *for every $\varepsilon > 0$ there exists an arbitrary large index n such that the set $\underset{x}{E}\{|f(x)-f_n(x)| < \varepsilon\}$ is of additive class a* (1).

The condition is necessary, since the function

$$\varphi_n(x) = |f(x)-f_n(x)|$$

is of class a (by VI, example (i)).

We shall prove that it is sufficient. The condition implies the existence of an increasing sequence $k_1 < k_2 < \ldots$ of integers such that the sets $E_n = \underset{x}{E}\{|f(x)-f_{k_n}(x)| < \varepsilon\}$ are of the additive class a. Since the sequence $\{f_n(x)\}$ is convergent, we have $\mathcal{X} = E_1 \cup E_2 \cup \ldots$ and since each f_{k_n} is of class a, there exists (by II, Theorem 3) a double sequence of sets Z_i^n of the additive class a such that

$$\mathcal{X} = \bigcup_{i=1}^{\infty} Z_i^n = \bigcup_{n,i=1}^{\infty} E_n \cap Z_i^n \quad \text{and} \quad \delta[f_{k_n}(Z_i^n)] < \varepsilon.$$

Since the sets $E_n \cap Z_i^n$ are of the additive class a, it is sufficient (by virtue of Theorem 3 of Section II) to show that

$$\delta[f(E_n \cap Z_i^n)] \leqslant 3\varepsilon.$$

Let x_1 and x_2 be two points of $E_n \cap Z_i^n$. Then

$$|f(x_1)-f_{k_n}(x_1)| < \varepsilon, \quad |f(x_2)-f_{k_n}(x_2)| < \varepsilon, \quad |f_{k_n}(x_1)-f_{k_n}(x_2)| < \varepsilon,$$

whence

$$|f(x_1)-f(x_2)| < 3\varepsilon.$$

(1) This condition is due to E. Szpilrajn-Marczewski (compare B. Gagaeff, *loc. cit.*, p. 187). For other necessary and sufficient conditions, see *ibid.* and H. Hahn, *Reelle Funktionen*, p. 309.

THEOREM 3. *If $\mathscr{Y}$ is separable, every function f of class $\alpha > 0$ is the limit of a uniformly convergent sequence of functions f_n of class α such that the sets $f_n(\mathscr{X})$ are discrete* (1).

Moreover, *if $\mathscr{Y}$ is totally bounded, the sets $f_n(\mathscr{X})$ can be assumed to be finite.*

Proof. Since $\mathscr{Y}$ is separable (respectively totally bounded), for every $\varepsilon > 0$ there exists a discrete (respectively finite) set I such that each point of the space lies at a distance $< \varepsilon$ from a point of I (see § 21, VIII, Theorem 1 and Remark 1). We order I into a (finite or infinite) sequence of distinct elements $y_1, y_2, \ldots$ and let

$$A_k = \underset{x}{E}\{|f(x) - y_k| \leqslant \varepsilon\}, \qquad B_k = \underset{x}{E}\{|f(x) - y_k| \geqslant 2\varepsilon\}.$$

The sets A_k and B_k are disjoint and of the multiplicative class α. By the separation theorem (§ 30, VII, (1)), there exists an ambiguous set F_k of class α such that $A_k \subset F_k$ and $F_k \cap B_k = 0$. It follows

$$\mathscr{X} = A_1 \cup A_2 \cup \ldots, \qquad \text{whence} \qquad \mathscr{X} = F_1 \cup F_2 \cup \ldots .$$

The function g defined by the following conditions

$$g(x) = \begin{cases} y_1, & \text{if} \quad x \in F_1, \\ y_k, & \text{if} \quad x \in F_k - (F_1 \cup \ldots \cup F_{k-1}) \end{cases}$$

is of class α. This follows from the fact that the set $g^{-1}(y_k)$, equal to $F_k - (F_1 \cup \ldots \cup F_{k-1})$, is of the additive class α (it is even ambiguous of class α) and, for every G, the set $g^{-1}(G)$ is of the additive class α, for the values of g form a countable set.

Moreover, $|f(x) - g(x)| < 2\varepsilon$, for every x, since the condition $g(x) = y_k$ implies $x \in F_k \subset \mathscr{X} - B_k$.

Now we define the function f_n to be equal to g assuming that the number ε is equal to $1/n$.

The two auxiliary theorems which follow will be used in the proof of Theorem 6 (the first is a particular case of it) (2).

(1) For the case of real functions, see Ch. de la Vallée-Poussin, *Intégrale de Lebesgue...*, p. 118; S. Kempisty, Fund. Math. 2 (1921), p. 135; W. Sierpiński, Fund. Math. 6 (1924), p. 4, and, for the general case, S. Banach, *Über analytisch darstellbare Operationen in abstrakten Räumen*, Fund. Math. 17 (1931), p. 287.

(2) For the results of section VIII which follow, compare S. Banach, *op. cit.*, p. 283.

THEOREM 4. *Every function f of class $\alpha > 1$ whose values form a finite set of points $E = (y_1, \ldots, y_k)$ is the limit of a sequence of functions f_n of classes $< \alpha$ whose values are in E.*

Moreover, *if $\alpha = \lambda + 1$, where λ is a limit ordinal, the functions f_n are of classes $< \lambda$.*

Proof. The set $A_i = f^{-1}(y_i)$ is ambiguous of class α. By § 30, IX, Theorems 1 and 2,

$$A_i = \operatorname*{Limes}_{n=\infty} A_{in}, \tag{1}$$

where A_{in} is ambiguous of class $< \alpha$ (respectively, of class $< \lambda$). Let

$$F_{1n} = A_{1n}, \; F_{2n} = A_{2n} - A_{1n}, \; \ldots, \; F_{kn} = A_{nk} - (A_{1n} \cup \ldots \cup A_{k-1,n}).$$

For a fixed n these sets are disjoint and of multiplicative classes $< \alpha$ ($< \lambda$ respectively). Hence there exists (by IV, Theorem 4) a function f_n of class $< \alpha$ ($< \lambda$ respectively) defined on $\mathscr{X}$ whose values are in E and such that $f_n(x) = y_i$ for $x \in F_{in}$.

We shall prove that $f(x) = \lim_{n=\infty} f_n(x)$. Let x_0 be a fixed point and $x_0 \in A_i$, say (since $\mathscr{X} = A_1 \cup \ldots \cup A_k$), i.e. $f(x_0) = y_i$. Since $x_0 \notin A_l$ for $l \neq i$, (1) implies that, if n is sufficiently large, then

$$x_0 \in A_{in} \quad \text{and} \quad x_0 \notin A_{ln}.$$

Therefore

$$x_0 \in F_{in}, \quad \text{whence} \quad f_n(x_0) = y_i = f(x_0).$$

THEOREM 5. *If $\{f_n\}$ and $\{g_n\}$ are two sequences of functions of classes $< \alpha$, each assuming a finite number of values and such that*

$$\lim_{n=\infty} f_n(x) = f(x), \quad \lim_{n=\infty} g_n(x) = g(x), \tag{2}$$

then there exists a sequence $\{h_n\}$ of functions of classes $< \alpha$, each assuming a finite number of values and such that

$$\lim_{n=\infty} h_n(x) = g(x), \quad |h_n(x) - f_n(x)| \leqslant c, \tag{3}$$

where c is a number satisfying $|f(x) - g(x)| < c$, for every x ($\mathscr{Y}$ is separable).

Proof. Let $d_n(x) = |g_n(x) - f_n(x)|$. Then d_n is of class $< \alpha$ (Section VI, (i)) and assumes a finite number of values. Hence the set

A_n of x such that $d_n(x) \leqslant c$ is ambiguous of class $< \alpha$ (see II, Theorem 2). Let

$$h_n(x) = \begin{cases} g_n(x), & \text{if} \quad x \in A_n, \\ f_n(x), & \text{if} \quad x \in \mathcal{X} - A_n. \end{cases}$$

The function h_n is of class $< \alpha$ (by IV, Theorem 2).

By the definition of h_n, the second part of (3) is satisfied.

To prove that the first part of (3) holds, let x_0 be a fixed point. By (2), if n is sufficiently large, then

$$|g_n(x_0) - f_n(x_0)| \leqslant c, \quad \text{so} \quad x_0 \in A_n \text{ and } h_n(x_0) = g_n(x_0).$$

Therefore $\lim\limits_{n=\infty} h_n(x_0) = \lim\limits_{n=\infty} g_n(x_0) = g(x_0)$.

THEOREM 6. *If $\mathcal{Y}$ is separable, every function f of class $\alpha > 1$ is the limit of a sequence of functions of classes $< \alpha$.*

Moreover, *if $\alpha = \lambda + 1$, where λ is a limit ordinal, these functions are of classes $< \lambda$.*

Proof. We can assume that $\mathcal{Y}$ is totally bounded (see § 22, II, Corollary 1a).

By Theorem 3,

$$f(x) = \lim_{m=\infty} f_m(x)$$

where the functions $f_1, f_2, \ldots$ are of class α, assume a finite number of values and the convergence is uniform. We can assume that

$$|f_{m+1}(x) - f_m(x)| < \frac{1}{2^m} \tag{4}$$

replacing, if necessary, the sequence $\{f_m\}$ by a partial sequence.

By Theorem 4,

$$f_m(x) = \lim_{n=\infty} f_{mn}(x),$$

where f_{mn} is of class $< \alpha$ (respectively $< \lambda$) and assumes a finite number of values. By induction on m we shall define a double sequence of functions h_{mn} of classes $< \alpha$ (respectively $< \lambda$) assuming a finite number of values such that

$$\lim_{n=\infty} h_{mn}(x) = f_m(x) \quad \text{and} \quad |h_{m+1,n}(x) - h_{mn}(x)| \leqslant 1/2^m. \tag{5}$$

Let $h_{1n}(x) = f_{1n}(x)$. Assume that the sequence $h_{m1}, h_{m2}, \ldots$ is defined. The existence of a sequence $h_{m+1,1}, h_{m+1,2}, \ldots$ with the required properties is provided by Theorem 5, if f is replaced by f_m, f_n by h_{mn}, g by f_{m+1}, g_n by $f_{m+1,n}$, and c by $1/2^m$.

We are going to prove that

$$\lim_{n=\infty} h_{nn}(x) = f(x), \tag{6}$$

which will complete the proof of Theorem 6.

Let x_0 be a given point and $\varepsilon > 0$. Let m be an integer such that $1/2^{m-1} < \varepsilon$ and

$$|f_m(x_0) - f(x_0)| < \varepsilon. \tag{7}$$

Finally, let $n_0 > m$ be an integer such that for $n > n_0$

$$|h_{mn}(x_0) - f_m(x_0)| < \varepsilon. \tag{8}$$

It follows that for $n > n_0$

$$|h_{nn}(x_0) - f(x_0)| \leqslant \{|h_{nn}(x_0) - h_{n-1,n}(x_0)| + \ldots + |h_{m+1,n}(x_0) - h_{mn}(x_0)|\}$$
$$+ |h_{mn}(x_0) - f_m(x_0)| + |f_m(x_0) - f(x_0)|$$
$$< \left(\frac{1}{2^{n-1}} + \ldots + \frac{1}{2^m}\right) + \varepsilon + \varepsilon < 3\varepsilon$$

by (5), (8), and (7). Formula (6) follows.

Remark. Theorem 6 is not, in general, true for $\alpha = 1$. For instance, if $\mathscr{Y}$ is the space composed of two elements 0 and 1, then the characteristic function of a single point of the space $\mathscr{X}$ of real numbers is of class 1; however, it is not a limit of any convergent sequence of continuous functions (with values in $\mathscr{Y}$).

But this theorem remains valid for $\alpha = 1$ in the two important cases:

(i) if $\mathscr{X}$ is a (separable) space of dimension 0;

(ii) if $\mathscr{Y} = \mathscr{I}$ (or $\mathscr{E}$) or, more generally, if $\mathscr{Y} = \mathscr{I}^{\mathfrak{m}}$ $(\mathfrak{m} \leqslant \aleph_0)$.

For, if $\dim \mathscr{X} = 0$, Theorem 3 is valid for $\alpha = 0$ and Theorem 4 is valid for $\alpha = 1$. In case (ii) we have the following theorem:

THEOREM 7. *If* $\mathscr{Y} = \mathscr{I}$ *(or, more generally,* $\mathscr{Y} = \mathscr{I}^{\mathfrak{m}}$, $\mathfrak{m} \leqslant \aleph_0$*), then every function of class* 1 *is the limit of a sequence of continuous functions.*

Proof. By virtue of § 20, IV, Theorem 1, we can confine ourselves to the case where $\mathscr{Y}$ is the interval. First consider the case where the function f assumes only a *finite set* of values $y_1, \ldots, y_k$. The set $f^{-1}(y_i)$ is an $\boldsymbol{F}_\sigma$-set and we can write

$$A_i = f^{-1}(y_i) = F_{i1} \cup F_{i2} \cup \ldots, \qquad F_{in} \subset F_{i,n+1}, \qquad F_{in} = \overline{F}_{in}. \quad (9)$$

By the Tietze extension theorem (§ 14, IV), for each n there exists a continuous function f_n defined on $\mathscr{X}$ such that $f_n(x) = y_i$ for $x \in F_{in}$, $1 \leqslant i \leqslant k$, and that the values of f_n remain between the lower and upper bounds of f.

Let x_0 be given. Since $\mathscr{X} = A_1 \cup \ldots \cup A_k$, we may put $x_0 \in A_i$. By (9), if n is sufficiently large, then $x_0 \in F_{in}$, whence $f_n(x_0) = y_i$, i.e. $f_n(x_0) = f(x_0)$. Therefore $f(x) = \lim\limits_{n=\infty} f_n(x)$.

Now let us consider the general case. According to Theorem 3 let $\{f_m\}$ be a uniformly convergent to f sequence of functions of class 1 each assuming a finite set of values. We can assume that (4) holds and $f_0(x) = 0$.

The difference $f_m(x) - f_{m-1}(x)$ is a function of class 1 (see VI, example (ii)). By the above argument, there exists a double sequence of continuous functions g_{mn} such that

$$\lim_{n=\infty} g_{mn}(x) = f_m(x) - f_{m-1}(x) \quad \text{and} \quad |g_{mn}(x)| \leqslant 1/2^{m-1}.$$

Let $h_{mn}(x) = g_{1n}(x) + \ldots + g_{mn}(x)$.

Formulas (5) hold and (6) follows as in the case of Theorem 6.

IX. Analytic representation. The family of *analytic representable functions* is defined [1] to be the smallest family of functions which contains

1) all continuous functions;

2) the limits of convergent sequences of functions belonging to it.

This family can be divided into classes Φ_α, where $\alpha < \Omega$, as follows:

(1) The class Φ_0 consists of continuous functions.

(2) The class Φ_α $(\alpha > 0)$ consists of the limits of convergent sequences of functions of classes Φ_ξ with $\xi < \alpha$.

---

[1] See R. Baire, *Thèse*, Ann. di Math. (3), 3 (1899), p. 68.

LEBESGUE-HAUSDORFF THEOREM [1]. *If $\mathscr{Y}$ is the interval $\mathscr{I}$ (or, more generally, $\mathscr{Y} = \mathscr{I}^{\mathfrak{m}}$, where $\mathfrak{m} \leqslant \aleph_0$), then the class Φ_a coincides with the family of B measurable functions of class a, respectively $a+1$, according as a is finite or infinite* [2].

Proof by induction. The assertion is obvious for $a = 0$. Assume that it is true for every $\xi < a$.

By Theorems VIII, 7 and 1, Φ_1 coincides with the family of functions of class 1. By Theorems VIII, 6 and 1, the family of functions of (finite) class $n+1$ coincides with the family of limits of convergent sequences of functions of class n and hence of the class Φ_n (by the induction assumption). Therefore this family coincides with Φ_{n+1} (by (2)).

Thus the theorem is proved for $a < \omega$. Assume now that $a = \lambda$, where λ is a limit ordinal.

Each function $f \epsilon \Phi_\lambda$ has the form

$$f(x) = \lim_{m=\infty} f_m(x), \quad \text{where} \quad f_m \epsilon \Phi_{\xi_m} \text{ and } \xi_m < \lambda.$$

By the hypothesis, f_m is then *B* measurable of class ξ_m+1, hence of class λ. By Theorem VIII, 1, f is of class $\lambda+1$.

Conversely, if f is of class $\lambda+1$, then by Theorem VIII, 6

$$f(x) = \lim_{m=\infty} f_m(x), \quad \text{where} \quad f_m \text{ is of class } \xi_m < \lambda.$$

Therefore $f_m \epsilon \Phi_{\xi_m}$ and $f \epsilon \Phi_\lambda$.

Thus we have proved that Φ_λ is the family of *B* measurable functions of class $\lambda+1$. By finite induction as above we prove that $\Phi_{\lambda+n}$ is the family of *B* measurable functions of class $\lambda+n+1$.

Remark. In arbitrary (metric separable) spaces $\mathscr{Y}$ we take, instead of continuous functions, the functions of class 1 as the starting point in the definition of analytic representable functions. Thus we obtain a transfinite sequence

$$\Phi_1^*, \Phi_2^*, \ldots, \Phi_a^*, \ldots \; (a < \Omega),$$

where Φ_1^* is the family of *B* measurable functions of class 1 and Φ_a^* with $a > 1$ is defined by condition (2) as above.

---

(1) H. Lebesgue, *op. cit.*, p. 168, and F. Hausdorff, *Mengenlehre*, Chapter 9.

(2) Conversely, the class a of *B* measurable functions coincides with Φ_{a-1} if a is infinite and is not a limit ordinal. See F. Hausdorff, *op. cit.*

By an analogous proof to that of the preceding theorem we obtain

BANACH THEOREM (1). *If $\mathscr{Y}$ is a separable space, the class Φ_a^*, where $1 \leqslant a < \Omega$, coincides with the family of B measurable functions of class a, respectively $a+1$, according to as a is finite or infinite.*

X. Baire theorems on functions of class 1 (2).

Let $f: \mathscr{X} \to \mathscr{Y}$. Let us recall that the set D of points of discontinuity of f satisfies the formula

$$D = \bigcup_G \{f^{-1}(G) - \operatorname{Int}[f^{-1}(G)]\} = \bigcup_F \{\overline{f^{-1}(F)} - f^{-1}(F)\}, \quad (1)$$

where G runs over the family of open sets and F over that of closed sets in $\mathscr{Y}$ (see § 13, III, (3)).

If $\mathscr{Y}$ is *separable*, formula (1) can be replaced by a formula which involves only countable union (see § 13, III, (4)):

$$D = \bigcup_n \{f^{-1}(R_n) - \operatorname{Int}[f^{-1}(R_n)]\} = \bigcup_n \{\overline{f^{-1}(S_n)} - f^{-1}(S_n)\}, \quad (2)$$

where $S_n = \mathscr{Y} - R_n$ and $R_1, R_2, \ldots$ is an open base of $\mathscr{Y}$.

THEOREM 1. *The set D of points of discontinuity of a B measurable function of class 1 is of the first category (provided that the space $\mathscr{Y}$ is separable).*

Proof. The set S_n in formula (2) is closed, hence $f^{-1}(S_n)$ is a G_δ-set, by hypothesis. Therefore $\overline{f^{-1}(S_n)} - f^{-1}(S_n)$ is an F_σ-set. It is also a boundary set, for it has the form $\overline{X} - X$ (§ 8, IV, Theorem 1), hence a set of the first category (§ 10, II). It follows that D is of the first category.

COROLLARY 1a. *If f is a function of class 1 and A is an arbitrary subset of the space $\mathscr{X}$, then the set of points of discontinuity of the partial function $f|A$ is of the first category in A (provided that $\mathscr{Y}$ is separable).*

---

(1) S. Banach, *op. cit.*, p. 285.

(2) See the Thesis of R. Baire. The functions of class 1 play an important rôle in applications. For example, the semi-continuous, monotone (more generally, of bounded variation), derived functions, all belong to this class. See also the Chapter IV on compact spaces of Vol. II.

THEOREM 2. *A function pointwise discontinuous on every closed set is B measurable of class* 1.

Proof. Let F be a closed subset of $\mathscr{Y}$. We have to show that $f^{-1}(F)$ is a G_δ-set.

Let

$$\mathscr{Y}-F = F_1 \cup F_2 \cup \ldots, \quad \text{where} \quad \overline{F}_n = F_n.$$

We are going to apply theorem 1^0 of § 12, III to the sets $f^{-1}(F)$ and $f^{-1}(F_n)$ (n fixed). It asserts that if E and H are two sets such that the equation $X = \overline{X \cap E} \cap \overline{X \cap H}$ is satisfied only for $X = 0$, then there exists a resolvable set D such that $E \subset D$ and $H \cap D = 0$; by Theorem 5 of § 30, X, D is a G_δ-set.

Therefore suppose that $X = \overline{X \cap f^{-1}(F)} \cap \overline{X \cap f^{-1}(F_n)}$ and $X \neq 0$. Since X is closed, there exists a point p of continuity of the partial function $g = f|X$. Hence (see § 3, III, (14))

$$p \in X \subset \overline{X \cap f^{-1}(F)} = \overline{g^{-1}(F)} \quad \text{and} \quad g(p) \in \overline{gg^{-1}(F)}$$

by the continuity of g.

Since $g(p) = f(p)$ and $\overline{gg^{-1}(F)} \subset \overline{F} = F$, it follows that $f(p) \in F$. Similarly the inclusion $X \subset \overline{X \cap f^{-1}(F_n)}$ implies $f(p) \in F_n$.

But this is impossible, for $F_n \cap F = 0$.

Therefore $X = 0$. Hence for each n there exists a G_δ-set D_n such that

$$f^{-1}(F) \subset D_n \subset \mathscr{X} - f^{-1}(F_n).$$

It follows that

$$f^{-1}(F) \subset \bigcap_n D_n \subset \bigcap_n [\mathscr{X} - f^{-1}(F_n)] = \mathscr{X} - \bigcup_n f^{-1}(F_n)$$

$$= \mathscr{X} - f^{-1}(\bigcup_n F_n) = \mathscr{X} - f^{-1}(\mathscr{Y} - F) = \mathscr{X} - [\mathscr{X} - f^{-1}(F)] = f^{-1}(F),$$

and hence $f^{-1}(F) = \bigcap_n D_n$. Since each D_n is a G_δ-set, it follows that $f^{-1}(F)$ is a G_δ-set.

Remarks. 1) The pointwise discontinuity of f on every closed set is equivalent to *the existence of a point of continuity of the partial function $f|A$ on every non-empty closed set A.*

This is, in fact, the last condition which was used in the proof of Theorem 2.

2) The term *closed* in Theorem 2 can be replaced by *perfect*. For, every isolated point is a point of continuity.

It readily follows (see § 9, V, Theorem 5) that if the set of points of discontinuity of a function is *scattered* (in particular, if it is finite), then the function is of class 1.

3) For complete spaces (see § 34, VII), the assertion of Theorem 1 is *equivalent* to the hypothesis of Theorem 2, so that each of them *characterizes* the functions of class 1. In non-complete spaces, however, the first of them is not sufficient (if the continuum hypothesis $\aleph_1 = \mathfrak{c}$ is assumed), and the second one is not necessary for a function to be of class 1.

The reason for this is that on one hand, there exists a separable space E of power $\aleph_1$ (see § 40, IV) such that every function defined on E satisfies the assertion of the corollary. As the family of real functions defined on E is of power $\mathfrak{c}^{\aleph_1}$ and the family of B measurable functions is of power $\mathfrak{c}$ (see Section I), hence the inequality $\mathfrak{c} < \mathfrak{c}^{\aleph_1}$ (which follows from the continuum hypothesis) implies the existence of functions which are not B measurable and which satisfy the assertion of the corollary.

On the other hand, if E is a set which is an $\boldsymbol{F}_\sigma$ and $\boldsymbol{G}_\delta$ but is not resolvable into an alternate series of decreasing closed sets (as, for example, a set which is both dense and boundary in the space of rational numbers, see § 24, III, Remark (ii)), then the characteristic function of E is of class 1 (Section I), but does not satisfy the hypothesis of Theorem 2 (see § 13, VI, Corollary).

4) Every function f of class 1 is *effectively* of class 1 (in the sense of § 23, VIII). In fact, given a closed set $F \subset \mathcal{Y}$, the proof of Theorem 2 enables us to define a sequence of open sets $G_n \subset \mathcal{X}$ so that $f^{-1}(F) = G_1 \cap G_2 \cap \ldots$ (for every resolvable set D is an effective $\boldsymbol{G}_\delta$, see § 24, III, Remark (i)). If f is a real function, this also enables us to *define* a sequence of continuous functions convergent to f (so that f is an effectively analytically representable function of class 1) [1].

---

[1] We can use, for example, the procedure described by F. Hausdorff, *Mengenlehre*, Chapter 9; compare Ch. de la Vallée-Poussin, *Intégrales de Lebesgue*..., p. 107 ("Problème de Baire"); and my paper of Fund. Math. **3** (1922), p. 100, where a solution of this problem is presented without using transfinite numbers.

5) *The words "B measurable function" in Theorem 1 can be replaced by "analytically representable functions". Then the assumption that the space $\mathscr{Y}$ is separable can be dropped* (1).

Theorem 1 so modified is more general than the one above. For, if $\mathscr{Y}$ is separable, every B measurable function of class 1 becomes analytically representable after imbedding $\mathscr{Y}$ in the Hilbert space.

To prove the modification of Theorem 1, let

$$f(x) = \lim_{n=\infty} f_n(x), \tag{3}$$

where f_n is continuous. Let D denote the set of points of discontinuity of f and let $\omega(x)$ be the oscillation of f at x. Let $E(\varepsilon)$ be the set of x such that $\omega(x) \geqslant \varepsilon$. Since

$$D = E(1) \cup E(1/2) \cup E(1/3) \cup \ldots,$$

it is sufficient to prove that $E(\varepsilon)$ is of the first category for every $\varepsilon > 0$.

For a fixed ε, let $E = E(\varepsilon)$. Let A_k be the set of x such that

$$|f_m(x) - f_k(x)| \leqslant \varepsilon/4, \quad \text{for every} \quad m > k. \tag{4}$$

As the functions $f_1, f_2, \ldots$ are continuous, hence A_k is closed. Therefore $\mathrm{Fr}(A_k)$ is nowhere dense.

On the other hand, since the sequence $f_1, f_2, \ldots$ is convergent, it follows that

$$\mathscr{X} = A_1 \cup A_2 \cup \ldots, \qquad \text{whence} \qquad E = (E \cap A_1) \cup (E \cap A_2) \cup \ldots.$$

It remains to prove that

$$E \cap A_k \subset \mathrm{Fr}(A_k), \quad \text{i.e.} \quad E \cap \mathrm{Int}(A_k) = 0.$$

Let $x \in \mathrm{Int}(A_k)$. Since f_k is continuous, there exists an open set G such that

$$x \in G \subset A_k \quad \text{and} \quad \delta[f_k(G)] < \varepsilon/4. \tag{5}$$

(1) See my paper *Sur les fonctions représentables analytiquement et les ensembles de première catégorie*, Fund. Math. 5 (1923), p. 75.

Let $x', x'' \in G$. By (3)–(5)

$$|f(x')-f_k(x')| \leqslant \varepsilon/4, \quad |f(x'')-f_k(x'')| \leqslant \varepsilon/4,$$
$$|f_k(x')-f_k(x'')| \leqslant \varepsilon/4,$$

whence $|f(x')-f(x'')| \leqslant 3\varepsilon/4$ and $\delta[f(G)] < \varepsilon$. It follows that $\omega(x) < \varepsilon$ and $x \notin E$.

6) *If $\mathscr{X}$ and $\mathscr{Y}$ are $\mathscr{T}_1$-spaces, f is a function pointwise discontinuous on every closed set* (1) *and F is a set which is both closed and open in $\mathscr{Y}$, then the set $f^{-1}(F)$ is resolvable.*

Proof. According to § 12, V, (i), it is sufficient to show that, for every closed set X, the condition $X = \overline{X \cap f^{-1}(F)} \cap \overline{X - f^{-1}(F)}$ implies $X = 0$. The proof is completely analogous to that of Theorem 2 in what concerns the identity $X = 0$ (F_n should be replaced by $\mathscr{Y} - F$).

7) *The separability of a space is invariant under mappings f which are pointwise discontinuous on every G_δ-set.*

Proof. Suppose that the space $\mathscr{Y} = f(\mathscr{X})$ is not separable. Hence it contains (by § 21, VIII, Remark 1) a closed discrete set of power $\aleph_1$:

$$\mathscr{Y}^* = (y_1, y_2, \ldots, y_\xi, \ldots), \quad \text{where} \quad \xi < \Omega.$$

Let $\mathscr{X}^* = f^{-1}(\mathscr{Y}^*)$ and $F_\xi = (y_1, y_2, \ldots, y_\xi)$.

By Theorem 2, f is of class 1, so $\mathscr{X}^*$ is a G_δ-set. By the hypothesis, the partial function $g = f|\mathscr{X}^*$ is pointwise discontinuous on every closed set (even on every G_δ-set) in $\mathscr{X}^*$. Since F_ξ is both closed and open in $\mathscr{Y}^*$, it follows by Remark 6 that the set $A_\xi = g^{-1}(F_\xi)$ is resolvable into sets closed in $\mathscr{X}^*$. By letting ξ vary, we obtain a well ordered uncountable family of increasing resolvable sets A_ξ. According to Theorem 2 of § 24, III, $\mathscr{X}^*$ is not separable. Hence $\mathscr{X}$ is not separable.

8) Under the continuum hypothesis we obtain a more general theorem:

---

(1) If $\mathscr{X}$ is complete, these mappings coincide with the mappings of class 1 (see § 34, VII).

The separability of a space is invariant under B measurable mappings.

Proof. If the space $\mathscr{Y} = f(\mathscr{X})$ is not separable, we can consider the discrete set $\mathscr{Y}^*$ as in the preceding proof and, to each ξ, assign an x_ξ such that $f(x_\xi) = y_\xi$. The set $A = (x_1, x_2, \ldots, x_\xi, \ldots)$ is not countable, so, by the continuum hypothesis, its power is $\mathfrak{c}$. If $\mathscr{X}$ is separable, it follows that A contains a set E which is non-borelian in A (see § 30, III). Let

$$E = (x_{\xi_1}, x_{\xi_2}, \ldots, x_{\xi_\gamma}, \ldots) \quad \text{and} \quad F = (y_{\xi_1}, y_{\xi_2}, \ldots, y_{\xi_\gamma}, \ldots).$$

Since the function $g = f|A$ is B measurable and F is closed in $\mathscr{Y}^*$ (since it is discrete), it follows that $E = g^{-1}(F)$ is a Borel set. Thus we obtain a contradiction.

9) *Every function defined on a countable space is of class* 1.

For, every subset of a countable space is an F_σ-set.

*§ 32. Functions which have the Baire property

I. Definition. Let f: $\mathscr{X} \to \mathscr{Y}$. *The function f has the Baire property, if, for every closed subset F of $\mathscr{Y}$, the set $f^{-1}(F)$ has the Baire property.*

Clearly the complement of a set which has the Baire property does also have this property. Therefore, in this definition, the term *closed* can be replaced by *open.*

On the other hand, the Baire property of sets is invariant under countable union and intersection (§ 11, III). It follows that *if the function f has the Baire property and X is a Borel set, then $f^{-1}(X)$ has the Baire property.*

Every Borel set has the Baire property. It follows that *every B measurable function has the Baire property.* This is one of the most important properties common to all B measurable functions, and hence to all analytically representable functions (see also Section II).

Remark. *If X has the Baire property, the set $f^{-1}(X)$ may not have the Baire property*, even if f is continuous and X is nowhere dense. For instance, let $\mathscr{X} = \mathscr{C}$, $\mathscr{Y} = \mathscr{I}$, $f(x) = x$ and let X be a subset of $\mathscr{X}$ which does not have the Baire property with respect to $\mathscr{X}$.

II. Equivalences (1).

THEOREM. *A necessary and sufficient condition for a function f to have the Baire property is the existence of a set P of the first category (in $\mathscr{X}$) such that the partial function $f|(\mathscr{X}-P)$ is continuous* (it is said then that f is *continuous apart from sets of the first category*). The space $\mathscr{Y}$ is assumed to be *separable*.

Proof. The condition is necessary. We have to define a set P of the first category such that the function $g=f|(\mathscr{X}-P)$ is continuous, i.e. given an arbitrary open set H in $\mathscr{Y}$, the set $g^{-1}(H)$ is to be open relative to $\mathscr{X}-P$.

Let $S_1, S_2, \dots$ be a base in $\mathscr{Y}$. Therefore $H = S_{k_1} \cup S_{k_2} \cup \dots$. By hypothesis, $f^{-1}(S_n)$ has the Baire property. It follows that

$$f^{-1}(S_n) = G_n - P_n \cup R_n,$$

where G_n is open and P_n and R_n are of the first category (see § 11, I). Let

$$P = (P_1 \cup R_1) \cup (P_2 \cup R_2) \cup \dots .$$

The formula $g^{-1}(H) = f^{-1}(H) - P$ (§ 3, III, (14)) gives

$$g^{-1}(H) = \bigcup_n [f^{-1}(S_{k_n})] - P = \bigcup_n [(G_{k_n} - P_{k_n} \cup R_{k_n}) - P].$$

Since $P_{k_n} \cup R_{k_n} \subset P$, it follows that

$$(G_{k_n} - P_{k_n} \cup R_{k_n}) - P = G_{k_n} - P.$$

Therefore $g^{-1}(H) = (\bigcup_n G_{k_n}) - P$ and since $\bigcup_n G_{k_n}$ is open, $g^{-1}(H)$ is open in $\mathscr{X}-P$.

The condition is sufficient. Let P be a set of the first category such that the function $g=f|(\mathscr{X}-P)$ is continuous. The continuity of g means (§ 13, IV, (3)) that if H is an arbitrary open set, then the set $g^{-1}(H) = f^{-1}(H) - P$ is open in $\mathscr{X}-P$, i.e. it is of the form $G-P$, where G is open (in $\mathscr{X}$). Then

$$f^{-1}(H) = f^{-1}(H) - P \cup \big(f^{-1}(H) \cap P\big) = G - P \cup \big(f^{-1}(H) \cap P\big). \quad (1)$$

Since P and also $f^{-1}(H) \cap P$ is of the first category, decomposition (1) shows that the set $f^{-1}(H)$ has the Baire property.

Remark. By this theorem and the conclusions of Section I we see that *every B measurable function and, therefore, every analy-*

---

(1) See my paper *La propriété de Baire dans les espaces métriques*, Fund. Math. 16 (1930), p. 391. Compare O. Nikodym, *Sur la condition de Baire*, Bull. Acad. Pol. 1929, p. 595.

tically representable function is continuous apart from sets of the first category (*$\mathscr{Y}$ being separable*) [1].

Note also that *the Baire property of a given set and of its characteristic function are equivalent.*

III. Operations on functions which have the Baire property.

THEOREM 1 (on composition of functions). *If the function $f: \mathscr{X} \to \mathscr{Y}$ has the Baire property and the function $g: \mathscr{Y} \to \mathscr{Z}$ is B measurable, then the function $h = g \circ f$ has the Baire property.*

Proof. As $h^{-1}(F) = f^{-1}[g^{-1}(F)]$ and $g^{-1}(F)$ is B measurable, it follows that the set $f^{-1}[g^{-1}(F)]$ has the Baire property.

Remark. In the *reverse* case, when f is B measurable and g has the Baire property, the function $g \circ f$ may fail to have the Baire property. For instance consider the example of Section I and denote by g the characteristic function of X (regarded as a subset of the space $\mathscr{Y}$).

THEOREM 2. *The limit of a convergent sequence of functions with the Baire property still has this property.*

This is a consequence of the formula (compare § 31, VIII, Theorem 1):

$$f^{-1}(F) = \bigcap_n \bigcup_k [f_{n+k}^{-1}(B_n)],$$

where B_n is the ball with centre F and radius $1/n$. The sets $f_{n+k}^{-1}(B_n)$ have the Baire property, hence every set obtained from them by taking countable union and intersection has it.

Remark. If $\mathscr{Y}$ is separable, then the fact that the limit of a sequence of functions which are continuous apart from sets of the first category is a function of the same kind can be proved as follows. Let P_n be a set of the first category such that the partial function $f_n|(\mathscr{X}-P_n)$ is continuous and let $P = P_1 \cup P_2 \cup \ldots$. Then $f_n|(\mathscr{X}-P)$ is continuous for every n. If f is the limit of f_n, then $f|(\mathscr{X}-P)$ is of class 1 (on $\mathscr{X}-P$). Since the points of discontinuity of a function of class 1 form a set of the first category (§ 31, X, Corollary 1a), the set R of points of discontinuity of the function $f|(\mathscr{X}-P)$ is of the first category in $\mathscr{X}-P$, hence also in $\mathscr{X}$. Thus the function $f|(\mathscr{X}-P-R)$ is continuous and the set $P \cup R$ is of the first category.

---

(1) This statement is due to R. Baire, C. R. Paris 129 (1899), p. 1010. Compare my paper *Sur les fonctions représentables analytiquement et les ensembles de première catégorie*, Fund. Math. 5 (1924), p. 82.

THEOREM 3 (on functions of several variables). *If the function f has the Baire property and if we put $g(x, y) = f(x)$, then the function g has the Baire property (on the product $\mathscr{X} \times \mathscr{Y}$).*

For $g^{-1}(G) = f^{-1}(G) \times \mathscr{Y}$ and the Baire property is invariant under product with an axis (§ 15, VII, Corollary 2a).

Remark. By using this invariance, we infer from formula (ii) of § 31, V, in the case where $\mathscr{X}$ is separable, that

If the function $f: \mathscr{X} \times \mathscr{Y} \to \mathscr{Z}$ is continuous in x and has the Baire property with respect to y, then it has the Baire property (on $\mathscr{X} \times \mathscr{Y}$).

The same conclusion can be obtained without assuming the separability. We shall use, as in § 31, V, the invariance of the Baire property under the operation $\mathscr{M}$ (see § 30, X) (¹).

We shall first prove that the property of being of the first category is invariant under $\mathscr{M}$.

With the notation of § 30, X, assume that the sets X_ξ are of the first category and consider the sets X_ξ^n defined by formula § 30, X (8). Since X_ξ^n is open in the union $S_n = \bigcup_\xi X_\xi^n$ (ibidem (11)), S_n is of the first category (by § 10, III). Therefore $S = S_1 \cup S_2 \cup \ldots$ is of the first category, which means that the property of being of the first category is invariant under $\mathscr{M}$.

Since the property of being a G_δ-set is invariant under $\mathscr{M}$ (§ 30, X, Theorem 3) and the operation $\mathscr{M}$ is additive (§ 30, X, Theorem 2), the invariance of the Baire property follows (for, the Baire property means that the given set is the union of a G_δ-set and a set of the first category).

The case of *complex functions* can be treated just as in § 31, VI. We obtain the following conclusions.

THEOREM 4. *If $f: \mathscr{T} \to \mathscr{X}_1 \times \mathscr{X}_2 \times \ldots$ where $\mathscr{X}_i$ are separable spaces, then the function f has the Baire property if and only if each coordinate of f has it.*

Combining Theorems 1, 3, and 4, we infer the following theorem, analogous to Theorem 2 of § 31, VI.

THEOREM 5. *If each of the functions $f_i: \mathscr{X}_i \to \mathscr{Y}_i$ has the Baire property and the function $g: \mathscr{Y}_2 \times \mathscr{Y}_2 \times \ldots \to \mathscr{Z}$ is B measurable,*

(¹) For a proof based on a different idea, see the paper of R. Engelking referred to above.

then the composite function $g \circ \mathfrak{y}$ *has the Baire property.* The spaces $\mathcal{Y}_i$ are assumed to be *separable.*

By following the argument of § 31, VII, we show

THEOREM 6. *If* f *has the Baire property, then the set* $I = \underset{xy}{E}[y = f(x)]$ *has it also.*

Remarks. 1. The converse theorem is not true. The graph of the equation $y = f(x)$ may have the Baire property without the function f having it. We shall return to this question in § 40, IV.

2. By § 22, VI, Theorem 3, *if* $\mathcal{Y}$ *is separable and dense in itself, then the set* I *is of the first category.*

IV. Functions which have the Baire property in the restricted sense. We say that the function f *has the Baire property in the restricted sense,* if, for every closed set F, the set $f^{-1}(F)$ *has the Baire property in the restricted sense.*

This means that, *for every set* E, *the partial function* $f|E$ *has the Baire property relative to* E; in other words (by virtue of the theorem of Section II) *that the partial function* $f|E$ *is continuous apart from sets of the first category relative to* E. The space $\mathcal{Y}$ is assumed to be separable.

Let us prove that this condition is necessary. Let $g = f|E$. Then (§ 3, III, (14)) $g^{-1}(F) = E \cap f^{-1}(F)$. By hypothesis the set $f^{-1}(F)$ has the Baire property in the restricted sense. So $g^{-1}(F)$ has the Baire property relative to E. This means that the function g has the Baire property (relative to E).

The condition is sufficient. Because, if, for every E, $g^{-1}(F)$ has the Baire property relative to E, then the set $f^{-1}(F)$ has the Baire property in the restricted sense. Therefore the function f has the Baire property in the restricted sense.

It is clear that E can be assumed to vary over *perfect* sets or over *closed* sets (compare § 11, VI).

Since every Borel set has the Baire property in the restricted sense (§ 11, VI), it follows that *every B measurable function has this property.* However, there exist functions with the Baire property in the restricted sense which are not B measurable (see § 39, II).

The Theorems 1, 2, and 4 of Section III can be proved for the Baire property in the restricted sense in a quite similar way.

Note also that the Baire property in the restricted sense of a set is equivalent to that of its *characteristic function.*

Remark. The continuum hypothesis implies that the statements 3, 5, and 6 applied to the Baire property in the restricted sense do not hold (see § 40, VIII).

V. Relations to the Lebesgue measure [1].

THEOREM. *If f is a real valued function with the Baire property defined on a separable space $\mathscr{X}$, then there exists a set Z such that $\mathscr{X}-Z$ is of the first category and $f(Z)$ has measure zero*: $mf(Z) = 0$.

Proof. Firstly assume that the function f is continuous.

Let $R = r_1, r_2, \ldots$ be a countable set dense in $\mathscr{X}$. Since f is continuous, for each k and n, there exists a ball $B_{k,n}$ such that

$$r_k \in B_{k,n} \text{ and } \delta[f(B_{k,n})] < \frac{1}{2^k n}, \text{ so that } m_e f(B_{k,n}) < \frac{1}{2^k n},$$

where m_e is the exterior measure.

Let

$$Z = \bigcap_{n=1}^{\infty} \bigcup_{k=1}^{\infty} B_{k,n}.$$

Then, for each n,

$$Z \subset \bigcup_{k=1}^{\infty} B_{k,n}, \quad \text{so} \quad f(Z) \subset \bigcup_{k=1}^{\infty} f(B_{k,n})$$

and consequently

$$m_e f(Z) \leqslant \bigcup_{k=1}^{\infty} m_e f(B_{k,n}) \leqslant \frac{1}{n}, \quad \text{whence} \quad mf(Z) = 0.$$

On the other hand,

$$\mathscr{X}-Z = \bigcup_{n=1}^{\infty} \{\mathscr{X} - \bigcup_{k=1}^{\infty} B_{k,n}\} \quad \text{and} \quad R \subset \bigcup_{k=1}^{\infty} B_{k,n},$$

which proves that, for a fixed n, $\bigcup_{k=1}^{\infty} B_{k,n}$ is open and dense. It follows that $\mathscr{X} - \bigcup_{k=1}^{\infty} B_{k,n}$ is nowhere dense and, consequently, $\mathscr{X}-Z$ is of the first category.

The case when f is a function with the Baire property can be reduced to the preceding one. For, by the Theorem of Section II, $\mathscr{X} = P \cup (\mathscr{X}-P)$, where f is continuous on $\mathscr{X}-P$ and P is of the first category. By the above argument, there exists a set Z such that $mf(Z) = 0$ and $\mathscr{X}-P-Z$ is of the first category. Therefore the set $\mathscr{X}-Z \subset (\mathscr{X}-P-Z) \cup P$ is of the first category.

---

[1] Compare W. Sierpiński, Fund. Math. 11 (1928), p. 302. An application of this theorem will be given in Chapter III, § 40.

CHAPTER THREE

COMPLETE SPACES

§ 33. Definitions. General properties

I. Definitions. We say that a sequence of points $p_1, p_2, \ldots, p_n, \ldots$ of a metric space is a *Cauchy sequence*, if for every $\varepsilon > 0$ there exists an n such that $|p_n - p_k| < \varepsilon$ for every $k > n$; in other words, if $\lim_{n=\infty} \delta(E_n) = 0$, where E_n denotes the set $(p_n, p_{n+1}, \ldots)$.

A metric space is said to be *complete* (¹), if every Cauchy sequence is convergent; i.e. if there exists a point p of the space such that $p = \lim_{n=\infty} p_n$.

Remark. The notion of complete space is not topological: The open interval $0 < x < 1$ is not complete, but it is homeomorphic to the space of real numbers which is complete (by the classical Cauchy theorem on the equivalence of the Cauchy condition to the convergence of a sequence of real numbers).

We distinguish between the notion of complete space in the *metric sense* (defined above) and that of complete space in the *topological sense*: namely, of a space homeomorphic to a complete space in the metric sense.

Complete separable spaces are also called *Polish spaces* (by Bourbaki).

II. Convergence and Cauchy sequences.

THEOREM 1. *Every convergent sequence is a Cauchy sequence.* If $p = \lim_{n=\infty} p_n$, then for every $\varepsilon > 0$ there exists an n such that

$$|p_k - p| < \frac{\varepsilon}{2} \quad \text{for} \quad k \geqslant n.$$

Hence $|p_n - p_k| < \varepsilon$.

(¹) This notion is due to M. Fréchet (Thesis). Compare F. Hausdorff, *Grundzüge...*, p. 315. It should be remarked that the spaces which satisfy axiom I of R. L. Moore (*Foundations...*, p. 464) are complete; see J. H. Roberts, Bull. Amer. Math. Soc. 38 (1932), p. 835. Compare also W. Sierpiński, Fund. Math. 6 (1924), p. 106.

THEOREM 2. *If a Cauchy sequence contains a convergent subsequence, then the entire sequence is convergent* (to the limit of the subsequence under consideration).

Proof. Let $p = \lim_{j=\infty} p_{i_j}$. Let ε and n be the same as in the definition of a Cauchy sequence and let m be an index $> n$ such that $|p_{i_m} - p| < \varepsilon$. Since $i_m \geqslant m > n$, it follows that $|p_{i_m} - p_k| < 2\varepsilon$ for $k > n$. Therefore $|p_k - p| < 3\varepsilon$, and so $\lim_{k=\infty} p_k = p$.

THEOREM 3. *Every compact metric space is complete.*

This follows from Theorem 2 of § 21, IX.

III. Cartesian product.

THEOREM 1. *The cartesian product $\mathscr{X} \times \mathscr{Y}$ of two complete spaces is complete, if it is metrized by the formula*

$$|\mathfrak{z} - \mathfrak{z}_1| = \sqrt{|x - x_1|^2 + |y - y_1|^2},$$

where $\mathfrak{z} = (x, y)$ and $\mathfrak{z}_1 = (x_1, y_1)$ (compare § 21, VI, (1)).

Proof. If $\mathfrak{z}_1, \mathfrak{z}_2, \ldots$ is a Cauchy sequence, then $x_1, x_2, \ldots$ and $y_1, y_2, \ldots$ are Cauchy sequences, for

$$|x_n - y_k| \leqslant |\mathfrak{z}_n - \mathfrak{z}_k| \quad \text{and} \quad |y_n - y_k| \leqslant \mathfrak{z}_n - \mathfrak{z}_k|.$$

Since $\mathscr{X}$ and $\mathscr{Y}$ are complete, there are x and y such that

$$x = \lim_{n=\infty} x_n \quad \text{and} \quad y = \lim_{n=\infty} y_n, \quad \text{whence} \quad (x, y) = \lim_{n=\infty} \mathfrak{z}_n.$$

The preceding theorem is valid for every finite number of spaces. It also extends to countable products.

THEOREM 2. *The product $\mathscr{X}_1 \times \mathscr{X}_2 \times \ldots$ of a sequence of complete spaces is a complete space, if it is metrized by the formula*

$$|\mathfrak{z} - \mathfrak{y}| = \sum_{i=1}^{\infty} 2^{-i} \frac{|\mathfrak{z}^i - \mathfrak{y}^i|}{1 + |\mathfrak{z}^i - \mathfrak{y}^i|},$$

where $\mathfrak{z}$ denotes the sequence $\mathfrak{z}^1, \mathfrak{z}^2, \ldots$ and $\mathfrak{y}$ denotes the sequence $\mathfrak{y}^1, \mathfrak{y}^2, \ldots$ (see § 21, VI, Remark 2).

Proof. Let $\mathfrak{z}_1, \mathfrak{z}_2, \ldots$ be a Cauchy sequence. Let i be an arbitrary fixed index. For every $\varepsilon > 0$ (with $1 - 2^i\varepsilon > 0$) there exists an n such that $|\mathfrak{z}_n - \mathfrak{z}_k| < \varepsilon$, for every $k > n$. Then

$$\frac{|\mathfrak{z}_n^i - \mathfrak{z}_k^i|}{1 + |\mathfrak{z}_n^i - \mathfrak{z}_k^i|} < 2^i\varepsilon, \quad \text{so that} \quad |\mathfrak{z}_n^i - \mathfrak{z}_k^i| < \frac{2^i\varepsilon}{1 - 2^i\varepsilon} = \frac{1}{1/(2^i\varepsilon) - 1}.$$

The last number tends to zero with ε. It follows that $\mathfrak{z}_1^i, \mathfrak{z}_2^i, \ldots, \mathfrak{z}_n^i, \ldots$ is a Cauchy sequence. Since $\mathcal{X}_i$ is a complete space, there exists a $\mathfrak{z}^i = \lim_{n=\infty} \mathfrak{z}_n^i$, and hence $\mathfrak{z} = \lim_{n=\infty} \mathfrak{z}_n$.

In particular, *the euclidean space* $\mathcal{E}^n$, *the space* $\mathcal{E}^{\aleph_0}$ *of sequences of real numbers, the n-dimensional cube* $\mathcal{I}^n$, *the Hilbert cube* $\mathcal{I}^{\aleph_0}$, *the space* $\mathcal{N}$ *of irrational numbers between* 0 *and* 1 (as the $\aleph_0$-power of the space $\mathcal{D}$ of natural numbers) *are topologically complete* (i.e. homeomorphic to complete spaces) (¹).

IV. The space $(2^{\mathcal{X}})_m$.

THEOREM. *If* $\mathcal{X}$ *is complete, then the space* $(2^{\mathcal{X}})_m$ *is complete* (²).

Proof. Let $A_1, A_2, \ldots$ be a Cauchy sequence of elements of $(2^{\mathcal{X}})_m$. That is, for every $\varepsilon > 0$ there exists an $n(\varepsilon)$ such that

$$n > n(\varepsilon) \quad \text{implies} \quad \operatorname{dist}(A_n, A_{n(\varepsilon)}) < \varepsilon. \tag{1}$$

Let $L = \operatorname{Ls}_{n=\infty} A_n$. We shall prove that

$$\lim_{n=\infty} \operatorname{dist}(L, A_n) = 0. \tag{2}$$

It suffices to prove that $\operatorname{dist}(L, A_{n(\varepsilon)}) \leqslant 2\varepsilon$, for then (1) will imply that $\operatorname{dist}(L, A_n) \leqslant 3\varepsilon$ for $n > n(\varepsilon)$. We have to show that

(i) if $p \in L$, then $\varrho(p, A_{n(\varepsilon)}) \leqslant 2\varepsilon$;

(ii) if $q \in A_{n(\varepsilon)}$, then $\varrho(q, L) \leqslant 2\varepsilon$.

Let R be the (generalized) closed ball with centre $A_{n(\varepsilon)}$ and radius ε. By (1), $A_n \subset R$ for $n > n(\varepsilon)$ (compare § 21, VII, (2)), and as $L \subset \overline{A_n \cup A_{n+1} \cup \ldots}$ (compare § 29, IV, 8), therefore $L \subset R$, which proves proposition (i).

To prove (ii), let $n(\varepsilon/2^k) = n_k$ (we can assume that $n_k > n_{k-1}$) and consider the sequence $q_{n_0}, q_{n_1}, \ldots, q_{n_k}, \ldots$ defined by induction as follows. Choose a point q_{n_k} in A_{n_k}, so that $q_{n_0} = q$ and $|q_{n_k} - q_{n_{k-1}}| < \varepsilon/2^{k-1}$; this is possible by (1).

$q_{n_0}, q_{n_1}, \ldots$ is a Cauchy sequence, because $|q_{n_m} - q_{n_k}| < \varepsilon/2^{k-1}$ for $m > k$. Since the space is complete, this sequence converges to a point l of the space. By the definition of L, l belongs to L. Moreover, since $|q_{n_k} - q_{n_0}| < 2\varepsilon$ for every k, we have $|q - l| \leqslant 2\varepsilon$, and (ii) follows.

(¹) The space $\mathcal{N}$, when metrized so that it becomes complete, is also called the "0-*dimensional Baire space*".

(²) Theorem of H. Hahn, *Reelle Funktionen*, p. 124, Leipzig 1932.

V. Function space. We have seen in § 21, X, Theorem 1, that the family $\Phi(\mathscr{X}, \mathscr{Y})$ of bounded functions $f\colon \mathscr{X} \to \mathscr{Y}$ can be given a metric, if the distance is defined by the formula

$$|f_1 - f_2| = \sup_{x \in \mathscr{X}} |f_1(x) - f_2(x)|. \tag{0}$$

THEOREM 1. *If $\mathscr{Y}$ is complete, then so is $\Phi(\mathscr{X}, \mathscr{Y})$.*

Proof. Suppose that $|f_n - f_k| < \varepsilon$ for $k > n$. Then, for a fixed x, $|f_n(x) - f_k(x)| < \varepsilon$ and $f_1(x), f_2(x), \ldots$ is a Cauchy sequence. Let $f(x) = \lim f_n(x)$. Since the sequence $\{f_n\}$ is uniformly convergent, the functions f_n converge to f in the sense of the distance defined by (0) and as easily seen, $f \in \Phi(\mathscr{X}, \mathscr{Y})$.

THEOREM 2. *If $\mathscr{Y}$ is complete, the space of continuous bounded functions is complete.*

This is because the limit of a uniformly convergent sequence of continuous functions is continuous (§ 21, X, Theorem 2).

The same is true for the space of functions of class α (compare § 31, VIII, Theorem 2), the space of B measurable functions, and the space of functions with the Baire property (compare § 32, III, Theorem 2).

VI. Complete metrization of G_δ-sets. It follows directly from the definition of complete spaces that

A closed subset of a complete space is itself a complete space (with the same distance).

By the corollary of § 21, XIII, every G_δ-subset of a metric space $\mathscr{X}$ is homeomorphic to a closed subset of the product $\mathscr{X} \times \mathscr{E}^{\aleph_0}$. Now, if $\mathscr{X}$ is complete, then $\mathscr{X} \times \mathscr{E}^{\aleph_0}$, as a product of two complete spaces, is complete, and every closed subset of it is complete. This leads to the following theorem.

ALEXANDROV THEOREM [1]. *Every G_δ-subset of a complete space is homeomorphic to a complete space (i.e. it is complete in the topological sense).*

---

[1] P. Alexandrov, *Sur les ensembles de la première classe et les espaces abstraits*, C. R. Paris 178 (1924), p. 185. A simple proof is due to F. Hausdorff, *Die Mengen G_δ in vollständigen Räumen*, Fund. Math. 6 (1924), p. 146. For the converse theorem, see § 35, III.

Remark 1. The procedure used in the proof of the corollary of § 21, XIII, allows us to define directly a "new distance" in a G_δ-set so that this set becomes a complete space. In fact, given a set $Q = G_1 \cap G_2 \cap \ldots$, where G_i is open, we define

$$f_i(x) = \frac{1}{\varrho(x, \mathscr{X} - G_i)}.$$

The new distance between the points x and y of Q is then

$$\|x-y\| = |x-y| + \sum_{i=1}^{\infty} 2^{-i} \frac{|f_i(x)-f_i(y)|}{1+|f_i(x)-f_i(y)|}. \tag{1}$$

Remark 2. This is a particular case of the following theorem on *complete metrization* [1]:

Let $\mathscr{X}$ be metric and $f_m: \mathscr{X} \to \mathscr{E}$ continuous for $m = 1, 2, \ldots$. Let us make the following assumption:

If $x_1, x_2, \ldots$ is a Cauchy sequence such that the sequence $f_i(x_1)$, $f_i(x_2), \ldots$ is bounded (for each $i = 1, 2, \ldots$), then the sequence $x_1, x_2, \ldots$ is convergent. (2)

Under this assumption, $\mathscr{X}$ becomes a complete space relative to the new distance $\|x-y\|$ defined by (1), *and its topology remains unaffected, i.e.*

$$(\lim_{i=\infty} \|x_i - x\| = 0) \equiv (\lim_{i=\infty} |x_i - x| = 0). \tag{3}$$

A still more general theorem, having important applications, is obtained from the above theorem supposing that one has defined in $\mathscr{X}$ a notion of convergence $x_i \to x$ such that

$$(x_i \to x) \Rightarrow (\lim_{i=\infty} |x_i - x| = 0),$$

and that the continuity of f_m, as well as the convergence considered in (2), are to be understood in terms of the convergence $x_i \to x$. Under these assumptions, the theorem remains true when $x_i \to x$ is substituted for $\lim_{i=\infty} |x_i - x| = 0$ in (3).

---

(1) See my papers: *Un théorème sur les espaces complets et ses applications à l'étude de la connexité locale*, Bull. Acad. Polon. des Sci. 3 (1955), pp. 75–80, and *Sur une méthode de métrisation complète de certains espaces d'ensembles fermés*, Fund. Math. 43 (1956), pp. 114–138.

VII. Imbedding of a metric space in a complete space. It has been shown (§ 21, X, Theorem 9) that every metric space $\mathscr{X}$ is isometric to a subset of the space of real-valued bounded continuous functions defined on $\mathscr{X}$. This function space is complete, since the space of real numbers is complete (see V, Theorem 2). This gives the following theorem.

HAUSDORFF THEOREM. *Every metric space is isometric to a subset of a complete space.*

This theorem can also be proved as follows [1].

Consider the Cauchy sequences $\mathfrak{z} = [\mathfrak{z}^1, \mathfrak{z}^2, \ldots, \mathfrak{z}^i, \ldots]$ of points of the space $\mathscr{X}$ and regard them as points of a space $\tilde{\mathscr{X}}$; define the distance of two points $\mathfrak{z}$ and $\mathfrak{y}$ of $\tilde{\mathscr{X}}$ by the formula

$$|\mathfrak{z}-\mathfrak{y}| = \lim_{i=\infty} |\mathfrak{z}^i - \mathfrak{y}^i|, \tag{0}$$

assuming that two sequences with distance zero are identified (compare § 21, I, Remark). We are going to show that, with this definition, *$\tilde{\mathscr{X}}$ is complete and contains a subset isometric to $\mathscr{X}$.*

First we verify that $\tilde{\mathscr{X}}$ is a metric space. The inequality

$$|\mathfrak{z}^i - \mathfrak{y}^i| \leqslant |\mathfrak{z}^i - \mathfrak{z}^j| + |\mathfrak{z}^j - \mathfrak{y}^j| + |\mathfrak{y}^j - \mathfrak{y}^i|$$

gives

$$|\mathfrak{z}^i - \mathfrak{y}^i| - |\mathfrak{z}^j - \mathfrak{y}^j| \leqslant |\mathfrak{z}^i - \mathfrak{z}^j| + |\mathfrak{y}^j - \mathfrak{y}^i|.$$

Now, if $\mathfrak{z}$ and $\mathfrak{y}$ are Cauchy sequences then $|\mathfrak{z}^1 - \mathfrak{y}^1|$, $|\mathfrak{z}^2 - \mathfrak{y}^2|$, ... is a Cauchy sequence (of numbers), and the limit $\lim_{i=\infty} |\mathfrak{z}^i - \mathfrak{y}^i|$ exists. Thus the distance is defined for any pair of elements of $\tilde{\mathscr{X}}$.

The triangle law holds, since the inequality $|\mathfrak{z}^i - \mathfrak{y}^i| + |\mathfrak{z}^i - \mathfrak{w}^i| \geqslant |\mathfrak{y}^i - \mathfrak{w}^i|$ yields in the limit $|\mathfrak{z}-\mathfrak{y}| + |\mathfrak{z}-\mathfrak{w}| \geqslant |\mathfrak{y}-\mathfrak{w}|$.

If, for each i, $\mathfrak{z}^i = x$ and $\mathfrak{y}^i = y$, then clearly $|x-y| = |\mathfrak{z}-\mathfrak{y}|$. Thus the space $\mathscr{X}$ is isometric to the subset of $\tilde{\mathscr{X}}$ which consists of sequences composed of identical elements.

It remains to prove that $\tilde{\mathscr{X}}$ is complete.

---

[1] F. Hausdorff, *Grundzüge...*, p. 315. The proof which we are going to reproduce here after Hausdorff is a generalization of the well-known Cantor–Méray procedure of defining irrational numbers.

Let $\mathfrak{z}_1, \mathfrak{z}_2, \ldots, \mathfrak{z}_n, \ldots$ be a sequence of elements of $\tilde{\mathscr{X}}$. Then there exists a sequence of positive integers $m_1, m_2, \ldots, m_n, \ldots$ such that

$$|\mathfrak{z}_n^{m_n} - \mathfrak{z}_n^i| < \frac{1}{n} \quad \text{for} \quad i > m_n. \tag{1}$$

Suppose that $\mathfrak{z}_1, \mathfrak{z}_2, \ldots$ is a Cauchy sequence. We shall prove that $\mathfrak{z} = [\mathfrak{z}_1^{m_1}, \mathfrak{z}_2^{m_2}, \ldots]$ is also a Cauchy sequence, i.e. $\mathfrak{z} \in \tilde{\mathscr{X}}$, and that $\mathfrak{z} = \lim_{n=\infty} \mathfrak{z}_n$.

Let $\varepsilon > 0$. There exists an index $q = q(\varepsilon) > 1/\varepsilon$ such that $|\mathfrak{z}_q - \mathfrak{z}_{q+k}| < \varepsilon$ for every k. By (0), there exists a sequence of integers j_k such that

$$|\mathfrak{z}_q^i - \mathfrak{z}_{q+k}^i| < \varepsilon \quad \text{for} \quad i > j_k. \tag{2}$$

The inequality

$$|\mathfrak{z}_q^{m_q} - \mathfrak{z}_{q+k}^{m_{q+k}}| \leqslant |\mathfrak{z}_q^{m_q} - \mathfrak{z}_q^i| + |\mathfrak{z}_q^i - \mathfrak{z}_{q+k}^i| + |\mathfrak{z}_{q+k}^i - \mathfrak{z}_{q+k}^{m_{q+k}}|$$

implies, by (1) and (2), that, if i is greater than m_q, j_k, and m_{q+k}, then

$$|\mathfrak{z}_q^{m_q} - \mathfrak{z}_{q+k}^{m_{q+k}}| < \frac{1}{q} + \varepsilon + \frac{1}{q+k} < 3\varepsilon, \tag{3}$$

and this proves that $\mathfrak{z}$ is a Cauchy sequence.

The inequality

$$|\mathfrak{z}_n^i - \mathfrak{z}_i^{m_i}| \leqslant |\mathfrak{z}_n^i - \mathfrak{z}_n^{m_n}| + |\mathfrak{z}_n^{m_n} - \mathfrak{z}_i^{m_i}|$$

implies, by (1) and (3), that if $i > m_n$, $n > q$, and $i > q$, then

$$|\mathfrak{z}_n^i - \mathfrak{z}_i^{m_i}| < 1/n + 6\varepsilon < 7\varepsilon. \tag{4}$$

Thus for every $\varepsilon > 0$ there exists an index q such that, for $n > q$, (4) holds, provided that i is sufficiently large. By letting i increase to infinity, we deduce from (0) that

$$|\mathfrak{z}_n - \mathfrak{z}| \leqslant 7\varepsilon, \quad \text{so that} \quad \lim_{n=\infty} \mathfrak{z}_n = \mathfrak{z}.$$

Remark. If the points x of the space $\mathscr{X}$ are identified with the sequences $[x, x, x, \ldots]$, then $\mathscr{X}$, regarded as a subset of $\tilde{\mathscr{X}}$, is *dense* in $\tilde{\mathscr{X}}$. In fact, if $\mathfrak{z} = [\mathfrak{z}^1, \mathfrak{z}^2, \ldots]$ is an element of $\tilde{\mathscr{X}}$, then $\mathfrak{z} = \lim_{n=\infty} \mathfrak{z}_n$, where $\mathfrak{z}_n = [\mathfrak{z}^n, \mathfrak{z}^n, \mathfrak{z}^n, \ldots]$. For, by (0), $|\mathfrak{z}_n - \mathfrak{z}| = \lim_{i=\infty} |\mathfrak{z}^n - \mathfrak{z}^{n+i}|$ and $\mathfrak{z}$ is a Cauchy sequence, so $\lim_{n=\infty} |\mathfrak{z}_n - \mathfrak{z}| = 0$.

In particular, *if $\mathscr{X}$ is separable, then $\tilde{\mathscr{X}}$ is separable.*

§ 34. Sequences of sets. Baire theorem

We assume that the space considered in § 34 is complete (but not necessarily separable).

I. The coefficient $\alpha(A)$. We denote by $\alpha(A)$ the greatest lower bound of the numbers ε such that A can be decomposed into a finite union of sets of diameter $< \varepsilon$. The condition $\alpha(A) = 0$ means therefore that A is totally bounded (§ 21, VIII).

THEOREM [1]. *Every sequence $A_1 \supset A_2 \supset \dots$ of non-empty closed subsets of the space $\mathscr{X}$ such that $\lim_{n=\infty} \alpha(A_n) = 0$ is convergent in the space $(2^{\mathscr{X}})_m$ to the non-empty set $A_1 \cap A_2 \cap \dots$.*

Proof. First suppose that $\{A_n\}$ is not a Cauchy sequence. Then there exists an $\varepsilon > 0$ and an increasing sequence of integers $\{k_n\}$ such that $\operatorname{dist}(A_n, A_{k_n}) > \varepsilon$. It follows that there exists a sequence of points $\{p_n\}$ such that $p_n \in A_n$ and $\varrho(p_n, A_{k_n}) > \varepsilon$, for the inclusion $A_n \supset A_{k_n}$ implies $\varrho(A_n, x) = 0$ for every $x \in A_{k_n}$.

Let m be an integer with $\alpha(A_m) < \varepsilon$ and let

$$A_m = Z_1 \cup \dots \cup Z_l, \quad \text{where} \quad \delta(Z_i) < \varepsilon.$$

The points $p_m, p_{m+1}, p_{m+2}, \dots$ belong to A_m, so one of the sets $Z_1, \dots, Z_l$ (say Z_1) contains an infinite sequence $p_{i_1}, p_{i_2}, \dots$ Let j be an index such that $i_j > k_{i_1}$. Then $A_{k_{i_1}} \supset A_{i_j}$ and $p_{i_j} \in A_{k_{i_1}}$. The formulas $p_{i_1} \in Z_1$, $p_{i_j} \in Z_1$ and $\delta(Z_1) < \varepsilon$ imply $|p_{i_1} - p_{i_j}| < \varepsilon$, whence $\varrho(p_{i_1}, A_{k_{i_1}}) < \varepsilon$, contrary to the definition of the sequence $\{p_n\}$.

The contradiction proves that $\{A_n\}$ is a Cauchy sequence. Since the space $(2^{\mathscr{X}})_m$ is complete, the sequence $\{A_n\}$ is (by § 33, IV (2)) convergent to $\operatorname{Ls} A_n$, which coincides with the intersection $A_1 \cap A_2 \cap \dots$, for the sequence is decreasing (§ 29, VI, 8).

COROLLARY. *Let $\{F_\iota\}$ be a family of closed sets such that*

(i) *every finite intersection of the sets F_ι is non-empty;*

(ii) *there are sets F_ι with arbitrarily small $\alpha(F_\iota)$.*

Then the intersection of all the sets F_ι is non-empty.

---

[1] See my paper *Sur les espaces complets*, Fund. Math. 15 (1930), p. 303. By a theorem of Chapter IV, the set $A_1 \cap A_2 \cap \dots$ is *compact*.

Proof. Let $\varkappa_n$ be an index such that $\alpha(F_{\varkappa_n}) < 1/n$ and let $P = F_{\varkappa_1} \cap F_{\varkappa_2} \cap \dots$. Then $\alpha(P) = 0$, i.e. P is totally bounded, hence separable (§ 21, VIII, Theorem 3). By the Lindelöf theorem (§ 5, XI) applied to the set P as to a space (see § 21, II, Theorem 2), there exists a sequence of indices $\iota_1, \iota_2, \dots$ such that

$$\bigcup_{n=1}^{\infty} P - F_{\iota_n} = \bigcup_{\iota} P - F_{\iota}, \quad \text{hence} \quad \bigcap_{n=1}^{\infty} P \cap F_{\iota_n} = \bigcap_{\iota} P \cap F_{\iota} = \bigcap_{\iota} F_{\iota}.$$

Let $A_n = F_{\varkappa_1} \cap \dots \cap F_{\varkappa_n} \cap F_{\iota_1} \cap \dots \cap F_{\iota_n}$.
We infer from the preceding theorem that

$$\bigcap_{n=1}^{\infty} P \cap F_{\iota_n} = \bigcap_{n=1}^{\infty} A_n \neq 0, \quad \text{hence} \quad \bigcap_{\iota} F_{\iota} \neq 0.$$

II. Cantor theorem[1].

If $A_1 \supset A_2 \supset \dots$ *is a sequence of non-empty closed sets such that* $\lim_{n=\infty} \delta(A_n) = 0$, *then the set* $A_1 \cap A_2 \cap \dots$ *consists of a single point.*

This is an immediate consequence of the theorem of Section I and of the inequality $\alpha(A_n) \leqslant \delta(A_n)$.

Remark 1. The Cantor theorem can be proved more directly. Namely, by selecting a point p_n from each A_n, we obtain a Cauchy sequence. The limit of this sequence is a point of each A_n (since A_n is closed), so it belongs to $A_1 \cap A_2 \cap \dots$

Remark 2. The Cantor theorem *characterizes* the complete spaces (among metric spaces). For, consider a Cauchy sequence $p_1, p_2, \dots$. If $E_n = (p_n, p_{n+1}, \dots)$, then $\lim_{n=\infty} \delta(E_n) = 0$. By the Cantor theorem, there exists a point

$$p \in \bar{E}_1 \cap \bar{E}_2 \cap \dots, \quad \text{so that} \quad |p - p_n| \leqslant \delta(\bar{E}_n) = \delta(E_n)$$

and $p = \lim_{n=\infty} p_n$.

---

[1] Compare G. Cantor, Math. Ann. 17 (1880) and F. Hausdorff, *Grundzüge*..., p. 318, where this theorem is in the form stated here.

III. Application to continuous functions. *If f is a continuous function defined on the space $\mathscr{X}$ and $F_1 \supset F_2 \supset \ldots$ is a sequence of non-empty closed sets such that $\lim_{n=\infty} \alpha(F_n) = 0$ (in particular, if $\lim_{n=\infty} \delta(F_n) = 0$), then*

$$f\left(\bigcap_{n=1}^{\infty} F_n\right) = \bigcap_{n=1}^{\infty} f(F_n).$$

Proof. The inclusion

$$f(F_1 \cap F_2 \cap \ldots) \subset f(F_1) \cap f(F_2) \cap \ldots$$

is always true (§ 3, III, (2a)). Conversely, let

$$y \in f(F_1) \cap f(F_2) \cap \ldots \quad \text{and} \quad A_n = F_n \cap f^{-1}(y).$$

By the theorem of Section I it follows that

$$A_1 \cap A_2 \cap \ldots \neq 0, \text{ whence } f^{-1}(y) \cap F_1 \cap F_2 \cap \ldots \neq 0,$$

and $y \in f(F_1 \cap F_2 \cap \ldots)$.

IV. Baire theorem(1).

THEOREM. *Every set of the first category is a boundary set.*

Proof. Let $E = N_1 \cup N_2 \cup \ldots$ be a countable union of nowhere dense sets and let B_0 be an arbitrary (closed) ball. We have to prove that $B_0 - E \neq 0$.

Consider the sequence $B_0 \supset B_1 \supset \ldots \supset B_n \supset \ldots$ of closed balls defined by induction as follows. Since N_n is nowhere dense, there exists a closed ball, B_n, such that

$$B_n \subset B_{n-1} - N_n \quad \text{and} \quad \delta(B_n) < 1/n.$$

By the Cantor theorem, there exists a point $p \in \bigcap_{n=0}^{\infty} B_n$. Since

$$\bigcap_{n=0}^{\infty} B_n \subset \bigcap_{n=1}^{\infty} (-N_n) = -\bigcup_{n=1}^{\infty} N_n,$$

it follows that $p \in B_0 - E$.

COROLLARY 1. *The complement of a first category set is not of the first category (unless the space is empty).*

(1) *Thèse*, Ann. di Math. (3) 3 (1899), p. 65.

COROLLARY 2. *The space is not of the first category at any point.*

Otherwise there would exist a non-empty open set of the first category.

COROLLARY 3. *Every point of a dense in itself complete space is a condensation point.*

Otherwise there would exist a countable open set G. Since each set composed of a single point is nowhere dense, then G would be of the first category.

COROLLARY 4. *Every countable complete space is scattered* (1).

Proof. A space that is not scattered contains a non-empty perfect subset (§ 9, VI, Theorem 3). This set, regarded as a space, is complete and dense in itself, and, by Corollary 3, must be uncountable.

COROLLARY 5. *If $\mathscr{X} = Z_1 \cup Z_2 \cup \ldots$ is a representation of a complete space $\mathscr{X}$ as a countable union of closed sets, and E is such that, for each n, $E \cap Z_n$ is scattered, then E is scattered.*

Proof. Let D be a subset of E, dense in itself. The scattered set $D \cap Z_n$ is a boundary set in D (see § 9, VI, 1). Hence

$$\overline{D-Z_n} = \overline{D}, \quad \text{therefore} \quad \overline{D} \cap Z_n \subset \overline{D} = \overline{D-Z_n} \subset \overline{\overline{D}-Z_n},$$

which proves that $\overline{D} \cap Z_n$ is a boundary set in $\overline{D}$ (compare § 8, VI). Since it is also closed, it is nowhere dense in $\overline{D}$.

The identity $\overline{D} = (\overline{D} \cap Z_1) \cup (\overline{D} \cap Z_2) \cup \ldots$ implies that $\overline{D}$ is of the first category on itself, so it must be empty (by Corollary 2, when $\overline{D}$ is regarded as a space). Therefore $D = 0$.

COROLLARY 6. *If E is a (non-empty) dense in itself subset of a complete space $\mathscr{X}$ and f is a continuous function defined on the space such that the partial function $f|E$ is one-to-one, then the set $f(\mathscr{X})$ is uncountable.*

Proof. Suppose, conversely, that $f(\mathscr{X}) = (y_1, y_2, \ldots)$. Let $Z_n = f^{-1}(y_n)$. Then $\mathscr{X} = Z_1 \cup Z_2 \cup \ldots$ and $Z_n = \overline{Z}_n$. The set $E \cap Z_n$ contains one point at the most, since the conditions $x \in E \cap Z_n$ and $x' \in E \cap Z_n$ imply $f(x) = y_n = f(x')$ and the function $f|E$ is one-to-one. By Corollary 5, the set E must be scattered which contradicts the hypothesis.

---

(1) Compare also the theorem of § 36, V.

COROLLARY 7. *Let $E_{\alpha,\beta}$, where $\alpha<\Omega$ and $\beta<\Omega$, be a system of sets with the Baire property in a complete, separable, and uncountable space $\mathscr{X}$. If $E_{\alpha,\beta}\cap E_{\alpha,\beta'}=0$ for $\beta\neq\beta'$, then there exists a transfinite sequence $\{\gamma_\alpha\}$, $\alpha<\Omega$, such that*

$$\overline{\overline{\mathscr{X}-\bigcup_{\alpha<\Omega}E_{\alpha,\gamma_\alpha}}}>\aleph_0. \tag{1}$$

Proof. We can assume that the space is perfect (by removing, if necessary, its scattered part).

By a transfinite induction we shall define two sequences $\{\gamma_\alpha\}$ and $\{p_\alpha\}$. By § 24, I, Theorem 3, if α is fixed and β is sufficiently large, the set $E_{\alpha,\beta}$ is of the first category. For a fixed α, the sets $E_{\alpha,\beta}$ are disjoint, so there exists an ordinal β (which we now denote by γ_α) such that $p_\xi\in\mathscr{X}-E_{\alpha,\gamma_\alpha}$ for every $\xi<\alpha$. The sets $E_{1,\gamma_1}, E_{2,\gamma_2}, \ldots, E_{\alpha,\gamma_\alpha}$ are of the first category and the p_ξ are accumulation points of the space. Hence by the Baire theorem

$$\mathscr{X}-\Big(\bigcup_{\xi\leqslant\alpha}E_{\xi,\gamma_\xi}\cup\bigcup_{\xi<\alpha}p_\xi\Big)\neq 0.$$

Let p_α be a point of this set.

Let P be the set of the p_α with $\alpha<\Omega$. Since $p_\alpha\neq p_\xi$, for $\xi<\alpha$, it follows that $\overline{\overline{P}}=\aleph_1$. As $P\cap E_{\alpha,\gamma_\alpha}=0$, formula (1) follows.

Remarks. 1. The proof of the Baire theorem can be made *effective* in the following sense (compare § 23, VIII). If we are given

(i) a base $R_1, R_2, \ldots$ of the space;

(ii) a sequence $N_1, N_2, \ldots$ of nowhere dense sets;

(iii) a ball B_0,

then we can *define* a point p which belongs to $B_0-(N_1\cup N_2\cup\ldots)$.

For $n>0$, we define $B_n=\overline{R}_{k_n}$, where k_n is the smallest index such that

$$\overline{R}_{k_n}\subset R_{k_{n-1}}-N_n \quad\text{and}\quad \delta(R_{k_n})<1/n.$$

Then the intersection $B_0\cap B_1\cap B_2\cap\ldots$ consists of just one point which we denote by p.

2. *The Baire theorem is equivalent to the hypothesis that every sequence of arbitrary sets X_n satisfies the condition*

$$\overline{\mathscr{X}-\bigcup_{n=1}^{\infty}X_n\cap\overline{\mathscr{X}-\overline{X}_n}}=\mathscr{X}.$$

This is because a set is nowhere dense if and only if it has the form $X \cap \overline{\mathscr{X}-\overline{X}}$ (compare § 8, II).

3) The continuum hypothesis implies that Corollary 7 fails to be true if we drop the assumption that $E_{\alpha,\beta}$ has the Baire property. Because it implies the existence of a system of sets $E_{n,\beta}$, $n = 1, 2, \ldots$, $\beta < \Omega$, in the interval $\mathscr{I}$ such that $E_{n\beta} \cap E_{n\beta'} = 0$ for $\beta \neq \beta'$ and, for every infinite sequence $\gamma_1, \gamma_2, \ldots$,

$$\overline{\overline{\mathscr{I} - \bigcup_{n=1}^{\infty} E_{n,\gamma_n}}} \leqslant \aleph_0 \text{ (}^1\text{)}.$$

V. Applications to G_δ-sets.

THEOREM 1. *If $Q_1, Q_2, \ldots$ are dense G_δ-sets, then so is the set $Q_1 \cap Q_2 \cap \ldots$.*

Proof. Each of the sets $\mathscr{X} - Q_n$ is a boundary F_σ-set, and hence of the first category. The union $(\mathscr{X}-Q_1) \cup (\mathscr{X}-Q_2) \cup \ldots$ is also of the first category. It follows that its complement $Q_1 \cap Q_2, \cap \ldots$ is dense.

THEOREM 2. *Every G_δ-set of the first category is nowhere dense.*

Proof. If Q is not nowhere dense, there exists an open set $G \neq 0$ such that the set $Q \cap G$ is dense in G. Consequently, if Q is a G_δ-set, $G-Q$ is a boundary F_σ-set, and so a set of the first category. Suppose that Q is of the first category. Then the open set G, as the union of two first category sets, is of the first category, contrary to Corollary IV, 2.

Since a G_δ-subset of a complete space is homeomorphic to a complete space (§ 33, VI), all the statements of Section IV can be *relativized* to G_δ-sets. Thus we have:

THEOREM 3. *If Q is a G_δ-set, then every subset of it which is of the first category in Q is a boundary set in Q.*

Every countable G_δ-set is scattered.

VI. Applications to F_σ and G_δ sets.

THEOREM [2]. *Every set which is both F_σ and G_δ is resolvable into a (transfinite) alternate series of decreasing closed sets.*

Proof. Let E be a set both F_σ and G_δ and let X be an arbitrary non-empty closed set. We have to show (compare § 12, V, 1°)

(1) S. Braun and W. Sierpiński, Fund. Math. 19 (1932), p. 1.

(2) F. Hausdorff, *Grundzüge der Mengenlehre*, p. 462.

that either $X \cap E$ or $X-E$ is not a boundary set in X. Otherwise, the sets $X \cap E$ and $X-E$ would be of the first category in X (compare § 10, II), so their union $X = (X \cap E) \cup (X-E)$ would be of the first category on itself, contrary to Corollary 1 of Section IV (X being regarded as a space).

The converse theorem was proved in § 30, X, Theorem 5: Therefore, *in complete spaces, the sets both* $\boldsymbol{F}_\sigma$ *and* $\boldsymbol{G}_\delta$ *coincide with the resolvable sets.*

Remarks. 1. We infer (compare § 12, V and VII) that a subset E of a complete space is both $\boldsymbol{F}_\sigma$ and $\boldsymbol{G}_\delta$ if and only if, for every non-empty closed set F:

(i) the boundary of $F \cap E$ relative to F is nowhere dense in F (in other words, it is not equal to F);

(ii) $F \cap E$ is the union of an open set and a set nowhere dense in F (in other words, it is the difference of a closed set and a set nowhere dense in F);

(iii) $F \cap E$ is locally closed at one of its points (or it is empty).

2. It follows from (iii) that *every* $\boldsymbol{F}_\sigma$ *and* $\boldsymbol{G}_\delta$ *set which is homogeneous in the space is the difference of two closed sets* (for it is locally closed at any of its points, comp. § 7, V, Corollary).

3. In complete separable spaces, *every well ordered family of increasing* (or *decreasing*) $\boldsymbol{F}_\sigma$ *and* $\boldsymbol{G}_\delta$ *sets is countable* (compare § 24, III, 2).

This statement does not extend to $\boldsymbol{G}_\delta$-sets; nor the hypothesis that the space is complete can be dropped (see § 40, III).

Effectiveness problems. 1. *Every* $\boldsymbol{F}_\sigma$ and $\boldsymbol{G}_\delta$ set is an *effective* $\boldsymbol{F}_\sigma$ and $\boldsymbol{G}_\delta$ set (compare § 24, III, Remark (i)).

2. *The axiom of choice* can be applied effectively in the domain of sets that are both $\boldsymbol{F}_\sigma$ and $\boldsymbol{G}_\delta$. That is, we can determine a function f defined for the non-empty sets X that are both $\boldsymbol{F}_\sigma$ and $\boldsymbol{G}_\delta$ so that $f(X) \in X$.

Proof. By the preceding statement, it suffices to prove this for effective $\boldsymbol{G}_\delta$-sets. In other words, to every sequence of open sets $G_1, G_2, \ldots$ such that $G_1 \cap G_2 \cap \ldots \neq 0$, we must assign a point p that belongs to each G_n, $n = 1, 2, \ldots$.

Let $R_1, R_2, \ldots$ be a base of the space, and let k_1 be the smallest index such that

$$\bar{R}_{k_1} \subset G_1, \quad R_{k_1} \cap \bigcap_{n=1}^{\infty} G_n \neq 0 \quad \text{and} \quad \delta(R_{k_1}) < 1.$$

In general, let k_m be the smallest index such that

$$\bar{R}_{k_m} \subset G_m \cap R_{k_{m-1}}, \quad R_{k_m} \cap \bigcap_{n=1}^{\infty} G_n \neq 0, \quad \text{and} \quad \delta(R_{k_m}) < 1/m.$$

By the Cantor theorem, the intersection $\bar{R}_{k_1} \cap \bar{R}_{k_2} \cap \ldots$ consists of just one point. This is the point p we choose.

3. Every countable family of disjoint $\boldsymbol{F}_\sigma$-sets which cover the space is *effectively countable.*

Proof. The complement of each of the sets is an $\boldsymbol{F}_\sigma$-set, hence each of them is both $\boldsymbol{F}_\sigma$ and $\boldsymbol{G}_\delta$. By the preceding statement, we can select a point from each of them to obtain a scattered set of points E (see Section IV, Corollary 5). By § 24, IV, E is effectively countable and the assertion follows.

VII. Applications to functions of class 1. By Theorem 1 of § 31, X, if f is a B measurable function of class 1 which maps a metric space $\mathscr{X}$ onto a subset of a separable space $\mathscr{Y}$, then the set of points of discontinuity of f is of the first category in $\mathscr{X}$. If $\mathscr{X}$ is complete, this must be a boundary set. In other words, the function f is pointwise discontinuous. Combining this statement with Theorem 2 of § 31, X, we obtain

THEOREM [1]. *If $\mathscr{X}$ is complete and $\mathscr{Y}$ separable, then the function f is B measurable of the class 1 if and only if it is pointwise discontinuous on every closed set.*

Remarks. It is easily seen that the condition of pointwise discontinuity can be replaced by that of the existence of a continuity point of $f|A$ for every closed set $A \neq 0$, and that the term *closed* can be replaced by *perfect* (see § 31, X, Remarks 1 and 2).

Further, this term can be replaced by $\boldsymbol{G}_\delta$-set. For every $\boldsymbol{G}_\delta$-subset A of $\mathscr{X}$ is topologically complete (§ 33, VI). If the function f is of class 1, $f|A$ is of class 1, and so it is pointwise discontinuous.

---

[1] Compare R. Baire, *Thèse*, Ann. di Mat. (3) 3 (1899).

COROLLARY 1. *Under the same hypotheses on $\mathscr{X}$ and $\mathscr{Y}$, if the set of points of discontinuity of the function f is countable, then f is of class* 1.

Proof. Every non-empty perfect set is uncountable (Section IV, Corollary 3), and therefore must contain a point of continuity of f.

COROLLARY 2. *Every well ordered family of (real valued) functions of class* 1 *defined on a complete space such that*

$$f_1(x) \leqslant f_2(x) \leqslant \ldots \leqslant f_\xi(x) \leqslant f_{\xi+1}(x) \leqslant \ldots$$

is countable.

This is a direct consequence of the Theorem 2′ of § 24, III.

VIII. Applications to existence theorems. The Baire theorem is also important because it often enables the *existence* of elements satisfying a given property **P** to be proved. Namely, if the set of elements of a complete space which do not satisfy the property **P** decomposes into a countable union of nowhere dense sets, it is then of the first category and cannot fill the space up, i.e. there exists an element which does possess the property **P** (even there exists a dense set of such elements).

Thus, for example, the Menger-Nöbeling theorem says that every separable metric n-dimensional space $\mathscr{X}$ is homeomorphic to a subset of the cube $\mathscr{I}^{2n+1}$. A simple proof of it is based on the fact that, in the function space $(\mathscr{I}^{2n+1})^{\mathscr{X}}$, the set of functions that are not homeomorphisms is of the first category (¹).

In particular this method has proved fruitful in problems of existence of functions (of real variable) enjoying certain singularities (²).

Consider, for example, a proof of the *existence of a continuous function without a derivative* (³). In fact, we shall prove the existence of a function with an additional singularity; namely, the

(¹) See W. Hurewicz, Sgb. Preuss. Akad. 24 (1933), p. 757, and my paper of Fund. Math. 28 (1937), p. 334. For other applications to the dimension theory, see Vol. II and K. Borsuk, Fund. Math. 28 p. 91.

(²) See numerous papers of Auerbach, Banach, Kaczmarz, Mazurkiewicz in Studia Math. 3 (1931); of H. Steinhaus, ibid. 1 (1929), pp. 51–83; of S. Saks, Fund. Math. 10 (1927), pp. 186–196 and 19 (1932), pp. 211–219; of W. Orlicz, Bull. Acad. Polon. 1932, pp. 221-228.

(³) S. Banach, *Über die Baire'sche Kategorie gewisser Funktionenmengen*, Studia Math. 3 (1931), p. 174.

existence of a continuous function that does not at any point have two right derived finite numbers. Let **P** denote the property in question.

Consider the space Φ of continuous functions $f: \mathscr{E} \to \mathscr{E}$ which are periodic of period 1 (compare § 33, V). The condition that the function f does not possess the property **P**, i.e. that it has two finite right derived numbers at one point at least, is equivalent to the existence of a positive integer n and a point $x \in \mathscr{I}$ such that

$$\left|\frac{f(x+h)-f(x)}{h}\right| \leqslant n \quad \text{for every} \quad h > 0. \tag{1}$$

The set of functions f that fail to have the property **P** is therefore equal to the union $N_1 \cup N_2 \cup \ldots$, where N_n is the set of functions f for which there exists an x satisfying (1).

The space Φ is complete (compare § 33, V, Theorem 2), so it is sufficient to prove that the set N_n is nowhere dense (in Φ).

Now N_n is closed; since, if $\{f_k\}$ uniformly converges to f and $\{x_k\}$ is a sequence of points satisfying (1), then $f \in N_n$ and its corresponding x is any accumulation point of $\{x_k\}$.

The fact that N_n is closed becomes also evident, if we write its definition in logical symbols:

$$N_n = \underset{f}{E} \underset{x}{\bigvee} \underset{h>0}{\bigwedge} \left\{\left|\frac{f(x+h)-f(x)}{h}\right| \leqslant n\right\}. \tag{2}$$

Thus N_n is a projection (parallel to the $\mathscr{I}$ axis) of the closed set

$$\underset{fx}{E} \underset{h>0}{\bigwedge} \left\{\left|\frac{f(x+h)-f(x)}{h}\right| \leqslant n\right\} = \bigcap_{h>0} \underset{fx}{E} \left\{\left|\frac{f(x+h)-f(x)}{h}\right| \leqslant n\right\}.$$

Since the $\mathscr{I}$ axis is compact, the projection of a closed set is closed (see § 20, V, Theorem 7).

Since N_n is closed, in order to prove that N_n is nowhere dense it suffices to prove that it is a boundary set; as the set of piecewise linear functions is dense in Φ, it is sufficient to prove that, given a piecewise linear function f, there exists a function in $\Phi - N_n$ which is arbitrarily close to f. But this is an elementary fact.

Thus the existence of continuous functions without a derivative has been proved. Furthermore, it has been proved that the functions which do not enjoy this singularity form a first category set (thus they are "exceptional" in the entire set of continuous functions).

§ 35. Extension of functions

I. Extension of continuous functions.

THEOREM 1. *If f is a continuous function defined on a subset A of a (metric) space $\mathscr{X}$ with values in a complete space $\mathscr{Y}$, then there exists a continuous extension of f to a G_δ-set.*

Namely, *there is an extension of f to the set A^* of points p of $\bar{A}$ with $\omega(p) = 0$, i.e. such that the oscillation of f at p is zero* (see § 21, III).

Proof. Let us assign to every point p of A^* a sequence $\{p_n\}$ such that $p_n \in A$ and $\lim\limits_{n=\infty} p_n = p$ and denote by E_n the set $(p_n, p_{n+1}, \ldots)$. Then $\omega(p) = 0$ implies $\lim\limits_{n=\infty} \delta[f(E_n)] = 0$. So $f(p_1), f(p_2), \ldots$ is a Cauchy sequence. As $\mathscr{Y}$ is complete, let

$$f^*(p) = \lim_{n=\infty} f(p_n), \quad \text{i.e.} \quad f^*(p) = \bigcap_G \overline{f(G)},$$

where G is a variable neighbourhood of p (compare § 34, I, the corollary).

Thus f^* is defined on the set A^* (which is a G_δ, by § 21, III). If $x \in A$, then $f(x) = f^*(x)$. It remains to prove that f^* is continuous for $p \in A^*$, i.e. that $\omega^*(p) = 0$, where ω^* is the oscillation of f^*.

Now if G is an open subset of $\mathscr{X}$, then

$$f^*(G) \subset \overline{f(G)}, \quad \text{whence} \quad \delta[f^*(G)] \leqslant \delta[\overline{f(G)}] = \delta[f(G)].$$

Therefore (compare § 21, III)

$$\omega^*(p) = \inf \delta[f^*(G)] \leqslant \inf \delta[f(G)] = \omega(p) = 0,$$

where G runs over open sets containing p.

Remarks. (i) The hypothesis that $\mathscr{Y}$ is complete is essential, as is seen on the example $\mathscr{X} = \mathscr{I}$, $A = \mathscr{Y} =$ the set of rational numbers of $\mathscr{X}$ and $f(x) = x$, for $x \in A$.

(ii) From the preceding theorem we deduce that

If $\mathscr{X}$ is a metric separable space of power $\mathfrak{c}$, then there exists a real valued function g on $\mathscr{X}$, which is discontinuous on every set of power $\mathfrak{c}$ (1).

The proof is based on the following lemma of the general set theory.

(1) A theorem of Sierpiński and Zygmund, Fund. Math. 4 (1923), p. 316.

LEMMA 1. *Let Φ be a family of mappings of subsets of a given set $\mathcal{X}$ onto subsets of a given set $\mathcal{Y}$. If $\mathcal{X}$, $\mathcal{Y}$, and Φ are of power $\mathfrak{m}$, there exists a mapping g of $\mathcal{X}$ onto a subset of $\mathcal{Y}$ which does not coincide with any mapping of the family Φ on any set of power $\mathfrak{m}$.*

Proof. Let us order the elements of $\mathcal{X}$ and of Φ into two transfinite sequences $\{x_a\}$ and $\{f_a\}$ of the smallest possible order types and define g by transfinite induction assuming that $g(x_a) \neq f_\xi(x_a)$, for every $\xi < a$. Thus the proof is completed.

Let Φ be the family of real-valued continuous functions defined on G_δ-subsets of $\mathcal{X}$. By § 30, III and § 24, VI, Φ has power $\mathfrak{c}$. Then g is the required function. Would it be continuous on a subset E of power $\mathfrak{c}$, there would exist a continuous function f defined on a G_δ containing E such that $f(x) = g(x)$ for $x \in E$. But this is impossible, since $f \in \Phi$.

(iii) We quote without proof the following theorem [1]:

Let $\mathcal{X}$ and $\mathcal{Y}$ be two separable complete spaces let $A \subset \mathcal{X}$ and let $f: A \to f(A)$ be open. Then f has a continuous open extension $f^*: A^* \to f^*(A^*)$ where A^* is a G_δ-set.

Applications to the problem of power. The following statements 2–4 will be used to prove Theorem 5 [2].

LEMMA 2. *Given a function which assigns, to each $\xi < \omega_\delta$, a set F_ξ of power $\aleph_\delta$, there is a G_ξ, where $\xi < \omega_\delta$, such that*

$$G_\xi \cap G'_\xi = 0, \quad \text{for} \quad \xi \neq \xi', \quad G_\xi \subset F_\xi, \quad \overline{\overline{G}}_\xi = \aleph_\delta. \tag{1}$$

Proof. Let us order all ordinals $\xi < \omega_\delta$ into a transfinite sequence $\{a_\eta\}$ of type ω_δ in which each term repeats $\aleph_\delta$ times. To each $\eta < \omega_\delta$ let us assign an element p_η of F_{a_η} so that

$$p_\eta \neq p_{\eta'} \quad \text{for} \quad \eta' < \eta. \tag{2}$$

---

(1) See S. Mazurkiewicz, *Sur les transformations intérieures*, Fund. Math. 19 (1932), p. 198. For *closed* mappings, see I. Wainstein, Utch. Zapiski Mosc. Univ. (1952), pp. 3–53. See also F. Hausdorff, Fund. Math. 23 (1934).

(2) A theorem announced by A. Lindenbaum, Ann. Soc. Pol. Math. 10 (1932), p. 114 and 15 (1937), p. 185, and proved by W. Sierpiński, *Deux théorèmes sur les familles des transformations*, Fund. Math. 34 (1947), p. 30.

In that direction, see Z. Waraszkiewicz, *Une famille indénombrable de continus plans dont aucun n'est l'image continue d'un autre*, Fund. Math. 18 (1932), p. 118.

Let G_ξ be the set of p_η such that $a_\eta = \xi$.

Suppose that $p_\eta \epsilon G_\xi \cap G_{\xi'}$. Then $a_\eta = \xi$ and $a_\eta = \xi'$, whence $\xi = \xi'$, which proves that the first condition of (1) is fulfilled.

To prove the second one, let $p_\eta \epsilon G_\xi$. Then $a_\eta = \xi$. On the other hand, by the definition of the sequence $\{p_\eta\}$, $p_\eta \epsilon F_{a_\eta}$. Consequently $p_\eta \epsilon F_\xi$, which proves that $G_\xi \subset F_\xi$.

The set of η such that $a_\eta = \xi$ is, for a fixed ξ, of power $\aleph_\delta$, and so is the set of all the p_η satisfying (2), that is, the set G_ξ.

LEMMA 3. *If $\mathscr{X}$ is an infinite set of power $\mathfrak{m}$, there exists a family $\boldsymbol{F}$ of power $2^\mathfrak{m}$ of subsets of $\mathscr{X}$ such that the conditions $X \epsilon \boldsymbol{F}$, $Y \epsilon \boldsymbol{F}$, and $X \neq Y$ imply $\overline{\overline{X - Y}} = \mathfrak{m}$.*

Moreover: *Given a function which assigns a subset A_x of $\mathscr{X}$ of power $\mathfrak{m}$ to every $x \epsilon \mathscr{X}$, the family $\boldsymbol{F}$ can be subjected to a more precise condition: the formulas $X \epsilon \boldsymbol{F}$, $Y \epsilon \boldsymbol{F}$, and $X \neq Y$ imply $\overline{\overline{A_x \cap X - Y}} = \mathfrak{m}$, for every $x \epsilon \mathscr{X}$.*

Proof. The identity $\mathfrak{m} = \mathfrak{m}^2$ implies directly that $\mathscr{X}$ has the form

$$\mathscr{X} = \bigcup_{x \epsilon \mathscr{X}} M_x, \quad \overline{\overline{M_x}} = \mathfrak{m}, \quad \text{and} \quad M_x \cap M_{x'} = 0, \quad \text{for} \quad x \neq x'.$$

Since $\mathfrak{m} = 2\mathfrak{m}$,

$$M_x = P_x \cup Q_x, \quad \overline{\overline{P_x}} = \mathfrak{m} = \overline{\overline{Q_x}}, \quad \text{and} \quad P_x \cap Q_x = 0.$$

For $X \subset \mathscr{X}$, let

$$F(X) = \bigcup_{x \epsilon X} P_x \cup \bigcup_{x \epsilon X^c} Q_x, \quad \text{where} \quad X^c = \mathscr{X} - X,$$

and denote by $\boldsymbol{F}$ the family of sets $F(X)$, where X runs over the family of all subsets of $\mathscr{X}$.

It is easily seen that, if $x_0 \epsilon X - Y$, then

$$P_{x_0} \subset F(X) - F(Y), \quad \text{and hence} \quad \overline{\overline{F(X) - F(Y)}} = \mathfrak{m}$$

and

$$Q_{x_0} \subset F(Y) - F(X), \quad \text{so that} \quad \overline{\overline{F(Y) - F(X)}} = \mathfrak{m}.$$

The first part of Lemma 3 has been proved.

To prove the second part, let, according to Lemma 2,

$$A_x^* \cap A_{x'}^* = 0 \quad \text{for} \quad x \neq x', \quad A_x^* \subset A_x, \quad \text{and} \quad \overline{\overline{A_x^*}} = \mathfrak{m}.$$

We have to define a family $\boldsymbol{F}^*$ of power $2^\mathfrak{m}$ of subsets of $\mathscr{X}$ such that the conditions $X \epsilon \boldsymbol{F}^*$, $Y \epsilon \boldsymbol{F}^*$, $X \neq Y$ and $x \epsilon \mathscr{X}$ imply $\overline{\overline{A_x^* \cap X - Y}} = \mathfrak{m}$.

The sets $\mathscr{X}$ and A_x^* are of power $\mathfrak{m}$, so for every $x \in \mathscr{X}$ there exists a one-to-one mapping f_x of $\mathscr{X}$ onto A_x^*. To every $X \in \boldsymbol{F}$ let us assign the set $S(X) = \bigcup_{x \in \mathscr{X}} f_x(X)$ and denote by $\boldsymbol{F}^*$ the family of all sets $S(X)$, where X belongs to $\boldsymbol{F}$.

Since $f_x(X) \subset A_x^*$ and $A_x^* \cap A_{x'}^* = 0$ for $x \neq x'$, it follows that

$$A_{x_0}^* \cap S(X) = A_{x_0}^* \cap \bigcup_x f_x(X) = A_{x_0}^* \cap f_{x_0}(X) = f_{x_0}(X),$$

so that

$$A_{x_0}^* \cap S(X) - S(Y) = f_{x_0}(X) - f_{x_0}(Y). \tag{3}$$

The function f_{x_0} is one-to-one, therefore

$$f_{x_0}(X) - f_{x_0}(Y) = f_{x_0}(X - Y),$$

and hence

$$\overline{\overline{f_{x_0}(X) - f_{x_0}(Y)}} = \overline{\overline{f_{x_0}(X - Y)}} = \overline{\overline{X - Y}}. \tag{4}$$

If the sets $S(X)$ and $S(Y)$ are different, then $X \neq Y$, so $\overline{\overline{X - Y}} = \mathfrak{m}$ and, by (3) and (4),

$$\overline{\overline{A_{x_0}^* \cap S(X) - S(Y)}} = \mathfrak{m}.$$

The same identity shows that the condition $X \neq Y$ implies $S(X) \neq S(Y)$. It follows that $\overline{\overline{\boldsymbol{F}^*}} = \overline{\overline{\boldsymbol{F}}} = 2^{\mathfrak{m}}$.

LEMMA 4 [1]. *If $\mathscr{X}$ is an infinite set of power $\mathfrak{m}$ and Φ is a family of power $\leqslant \mathfrak{m}$ of mappings of subsets of $\mathscr{X}$ onto subsets of power $\mathfrak{m}$ of $\mathscr{X}$, then there exists a family $\boldsymbol{F}$ of power $2^{\mathfrak{m}}$ of subsets of $\mathscr{X}$ such that the conditions $X \in \boldsymbol{F}$, $Y \in \boldsymbol{F}$, and $X \neq Y$ imply $\overline{\overline{f(X) - Y}} = \mathfrak{m}$, for every function $f \in \Phi$.*

Proof. Let $\mathfrak{m} = \aleph_\delta$ and let us order the elements of Φ into a transfinite sequence $f_0, f_1, \ldots, f_\alpha, \ldots$ $(\alpha < \omega_\delta)$ so that each element of Φ repeats $\mathfrak{m}$ times. For a fixed f, let us assign, to each value y of f, an x such that $f(x) = y$ and, for $f = f_\alpha$, let A_α be the set of x which are chosen. Then $\overline{\overline{A_\alpha}} = \mathfrak{m}$ and the partial function $g_\alpha = f_\alpha | A_\alpha$ is one-to-one.

[1] See my paper *Sur l'extension de deux théorèmes topologiques à la Théorie des ensembles*, Fund. Math. 34 (1947), p. 34. Compare also the papers of A. Lindenbaum and W. Sierpiński referred to above.

There exists a transfinite sequence $p_0, p_1, .., p_\alpha, \ldots$ $(\alpha < \omega_\delta)$ such that $p_\alpha \in A_\alpha$ and

(i) $p_\alpha \neq p_\xi$ for $\xi < \alpha$,

(ii) $p_\alpha \neq g_\eta(p_\xi)$ for $\xi < \alpha$ and $\eta < \alpha$,

(iii) $g_\eta(p_\alpha) \neq p_\xi$ for $\xi < \alpha$ and $\eta < \alpha$.

This is because the set of p_ξ with $\xi < \alpha$, the set of $g_\eta(p_\xi)$ with $\xi < \alpha$ and $\eta < \alpha$ as well as the set $g_\eta^{-1}(p_\xi)$ are each, for a fixed α, of power $< \aleph_\delta$, while $\overline{\overline{A_\alpha}} = \aleph_\delta$.

The terms of the sequence $\{p_\alpha\}$, $\alpha < \omega_\delta$, are mutually distinct (by (i)), so their set P has power $\mathfrak{m}$. Moreover, if we let $A_\alpha^* = P \cap A_\alpha$, then $\overline{\overline{A_\alpha^*}} = \mathfrak{m}$, for each A_α repeats $\mathfrak{m}$ times in the sequence $\{A_\xi\}$, $\xi < \omega_\delta$. Therefore we can replace $\mathcal{X}$ by P and A_x by A_α^* in Lemma 3. We deduce the existence of a family $\boldsymbol{F}$ of power $2^{\mathfrak{m}}$ of subsets of P such that the conditions $X \in \boldsymbol{F}$, $Y \in \boldsymbol{F}$, and $X \neq Y$ imply $\overline{\overline{A_\alpha^* \cap X - Y}} = \mathfrak{m}$, for every $\alpha < \omega_\delta$. We shall prove that this implies $\overline{\overline{f_\alpha(X) - Y}} = \mathfrak{m}$, which will complete the proof.

The set $A_\alpha^* \cap X - Y$ is of power $\aleph_\delta$; so let $\{p_{\beta_\eta}\}$, where $\alpha < \eta < \omega_\delta$ and $\eta \leqslant \beta_\eta$, be a sequence of elements of this set. Since $p_{\beta_\eta} \in A_\alpha$, the function g_α is defined for p_{β_η} and $g_\alpha(p_{\beta_\eta}) \in f_\alpha(X)$. Since this function is one-to-one, the set of all $g_\alpha(p_{\beta_\eta})$, when η varies, has power $\mathfrak{m}$. It only remains to prove that $g_\alpha(p_{\beta_\eta}) \notin Y$.

Suppose the contrary is true. Since $Y \subset P$, there exists a $\zeta < \omega_\delta$ such that $g_\alpha(p_{\beta_\eta}) = p_\zeta$. Therefore $p_\zeta \in Y$ and, since $p_{\beta_\eta} \notin Y$, it follows that $\zeta \neq \beta_\eta$.

Two cases are to be considered:

1) $\beta_\eta < \zeta$; hence $\alpha < \zeta$. Now the conditions $p_\zeta = g_\alpha(p_{\beta_\eta})$, $\beta_\eta < \zeta$, and $\alpha < \zeta$ are inconsistent, by (ii).

2) $\zeta < \beta_\eta$. Then the following conditions are inconsistent, by (iii):

$$g_\alpha(p_{\beta_\eta}) = p_\zeta, \quad \zeta < \beta_\eta, \quad \text{and} \quad \alpha < \beta_\eta.$$

Thus, in either case, we are led to a contradiction.

THEOREM 5. *In every complete separable space $\mathcal{X}$ of power $\mathfrak{c}$ there exists a family $\boldsymbol{F}$ of power $2^{\mathfrak{c}}$ of sets none of which is a continuous image of the other.*

Moreover, *the conditions $X \in \boldsymbol{F}$, $Y \in \boldsymbol{F}$, and $X \neq Y$ imply $\overline{\overline{f(X) - Y}} = \mathfrak{c}$, for every continuous function f defined on $\mathcal{X}$ and assuming a set of values of power $\mathfrak{c}$.*

Proof. Let Φ be the family of continuous mappings of $\boldsymbol{G}_\delta$-sets onto subsets of power $\mathfrak{c}$ of $\mathscr{X}$. Then $\overline{\overline{\Phi}} = \mathfrak{c}$ by § 30, III and § 24, VI. The family $\boldsymbol{F}$ of Lemma 4 is the required one.

To prove this, observe that the condition $\overline{\overline{f(X)-Y}} = \mathfrak{c}$ implies $\overline{\overline{X}} = \mathfrak{c}$. Therefore, no element of $\boldsymbol{F}$ can be mapped onto any other element by a function which assumes less than $\mathfrak{c}$ values. On the other hand, if f is a continuous function defined on X and assuming $\mathfrak{c}$ values, then (by Theorem 1) there exists a continuous extension f^* of f to a $\boldsymbol{G}_\delta$-set containing X. Therefore $f^* \epsilon \Phi$ and, consequently, $\overline{\overline{f^*(X)-Y}} = \mathfrak{c}$ for $X \epsilon \boldsymbol{F}$, $Y \epsilon \boldsymbol{F}$, and $X \neq Y$. Since $f^*(X) = f(X)$, we obtain the required conclusion.

THEOREM 6. *There exists an uncountable subset P of the interval $\mathscr{I}$ such that $\mathscr{I}$ is not a continuous image of P.*

We shall derive this theorem from the following statement which will be proved with aid of the continuum hypothesis.

THEOREM 7 (1). *Let $\boldsymbol{F}$ be a family of power $\leqslant \aleph_1$ of uncountable subsets of a complete separable (uncountable) space $\mathscr{X}$. Then there exists an uncountable subset P of $\mathscr{X}$ such that no element of $\boldsymbol{F}$ is a continuous image of P.*

More precisely, *if $Y \epsilon \boldsymbol{F}$, then $Y-f(P) \neq 0$, for every continuous function f (defined on a subset of $\mathscr{X}$ containing P or not).*

Proof. Consider all pairs f, Y, where $Y \epsilon \boldsymbol{F}$ and f is a continuous function defined on a $\boldsymbol{G}_\delta$ whose set of values contains Y. Let us order such pairs into a transfinite sequence of type Ω: $\{f_\alpha, Y_\alpha\}$, $\alpha < \Omega$ (the existence of such a sequence follows from the fact that the family of $\boldsymbol{G}_\delta$-sets and the family of continuous functions defined on $\boldsymbol{G}_\delta$-sets are of power $\mathfrak{c}$, so, by the continuum hypothesis, of power $\aleph_1$).

For a fixed α, let us order the set Y_α into a transfinite sequence $\{y_{\alpha,\beta}\}$, where $\beta < \Omega$, and let $E_{\alpha,\beta} = f_\alpha^{-1}(y_{\alpha,\beta})$. The function f_α is continuous, hence the set $E_{\alpha,\beta}$ is closed in the set of arguments of f_α; hence it has the Baire property. Moreover, the condition $\beta \neq \beta'$ implies $E_{\alpha,\beta} \cap E_{\alpha,\beta'} = 0$, and Corollary 7 of § 34, IV can be applied. We deduce the existence of a sequence γ_α, where $\alpha < \Omega$, and of an uncountable set P such that $P \cap E_{\alpha,\gamma_\alpha} = 0$, for every α.

(1) See W. Sierpiński, *Un théorème concernant les transformations continues des ensembles linéaires*, Fund. Math. 19 (1932), p. 205.

Let $Y \epsilon \boldsymbol{F}$ and let f be a continuous function. Suppose that $Y \subset f(P)$. According to Theorem 1, let f^* be a continuous extension to a $\boldsymbol{G}_\delta$ of the partial function $f|P$. Therefore $Y \subset f(P) \subset f^*(P)$. It follows that the pair f^*, Y belongs to the sequence $\{f_a, Y_a\}$. Let $f^* = f_a$ and $Y = Y_a$.

Since $P \cap E_{a,\gamma_a} = 0$, i.e. $P \cap f_a^{-1}(y_{a,\gamma_a}) = 0$, hence $y_{a,\gamma_a} \notin f_a(P)$ and $y_{a,\gamma_a} \epsilon Y_a - f_a(P) \subset Y - f(P)$.

Remarks. 1. Theorem 6 follows from 7 if we let $\mathscr{X} = \mathscr{I}$ and $\boldsymbol{F} = (\mathscr{I})$. The proof of Theorem 6 does not require the continuum hypothesis, for if we assume that $\mathfrak{c} > \aleph_1$, Theorem 6 is obvious.

Without the continuum hypothesis, however, we are unable to prove the existence of a set P of power $\mathfrak{c}$ such that the interval is not a continuous image of P.

2. In Theorem 7, the inequality $Y - f(P) \neq 0$ can be replaced by the formula $\overline{Y - f(P)} = \aleph_1$.

Proof. Let us decompose each $Y \epsilon \boldsymbol{F}$ into $\aleph_1$ disjoint subsets of power $\aleph_1$ and denote by $\boldsymbol{F}^*$ the family obtained from $\boldsymbol{F}$ after replacing each Y by the subsets. By applying Theorem 7 to the family $\boldsymbol{F}^*$ we obtain the assertion.

3. If we replace the family of continuous functions by a family of power $\leqslant \aleph_1$ (or even $\leqslant \aleph_0$) of arbitrary functions, Theorem 6 becomes false. In fact, the following theorem holds (it is equivalent to the continuum hypothesis): *there exists a countable family Φ of mappings of $\mathscr{I}$ such that, for every uncountable subset P of $\mathscr{I}$, there exists in Φ a mapping which maps P onto $\mathscr{I}$* (1).

4. A notion analogous to the topological type is that of *continuity type*. Two sets are said to have the same continuity type, if each of them is a continuous image of the other (2).

By Theorem 5, *in a complete separable space of power $\mathfrak{c}$ there exist $2^{\mathfrak{c}}$ different continuity types* (3).

---

(1) See S. Braun and W. Sierpiński, Fund. Math. 19 (1932), p. 2.

(2) See W. Sierpiński, *Sur les images continues des ensembles de points*, Fund. Math. 14 (1929), p. 234.

(3) For other problems concerning the continuity types, see: Z. Waraszkiewicz, *Sur une famille des types de continuité qui remplit un intervalle*, Fund. Math. 18 (1932), p. 309, W. Sierpiński, *op. cit.*, and N. Aronszajn, *Sur les invariants des transformations continues d'ensembles*, Fund. Math. 19 (1932), p. 92.

II. Extension of homeomorphisms.

LAVRENTIEV THEOREM [1]. *Every homeomorphism between two subsets A and B of complete spaces $\mathscr{X}$ and $\mathscr{Y}$, respectively, can be extended to two G_δ-sets situated in these spaces.*

Proof. Let f be a homeomorphism defined on A such that $f(A) = B$. Let $g = f^{-1}$. Let f^* be a continuous extension of f to a G_δ-set A^* (see Section I, Theorem 1). Similarly, let g^* be an extension of g to B^*. Let

$$I = \underset{xy}{E}[y = f^*(x)] \quad \text{and} \quad J = \underset{xy}{E}[x = g^*(y)].$$

Let A_1 and B_1 be the projections of $I \cap J$ on the $\mathscr{X}$ axis and on the $\mathscr{Y}$ axis, respectively. Clearly f^* is a homeomorphism of A_1 onto B_1.

We shall prove that A_1 and B_1 are G_δ-sets. Let $h(x) = [x, f^*(x)]$. Then $A_1 = h^{-1}(J)$. Since h is continuous and J is closed in $\mathscr{X} \times B^*$ (§ 15, V, Theorem 2), hence a G_δ in $\mathscr{X} \times \mathscr{Y}$, it follows that $h^{-1}(J)$ is a G_δ in A^* (§ 13, IV, (5)), hence also in $\mathscr{X}$. Similarly, B_1 is a G_δ in $\mathscr{Y}$.

Compare the following corollary to Theorem 1 of Section I.

COROLLARY. *Let f be a homeomorphism which maps a subset A of a metric space $\mathscr{X}$ (not necessarily complete) into a complete space $\mathscr{Y}$. Then f can be extended to a homeomorphism defined on a G_δ-set.*

Proof. Let $\mathscr{X}^*$ be a complete space which contains $\mathscr{X}$ topologically (see § 33, VII). The homeomorphism f can be extended to a set A_1 which is a G_δ relative to $\mathscr{X}^*$. Thus $A \subset A_1 \subset \mathscr{X}^*$. Then $A_2 = A_1 \cap \mathscr{X}$ is the required set. It is a G_δ (in $\mathscr{X}$), it contains A, and the extended function is a homeomorphism on A_2.

Remark 1. The set $f(A_2)$ need not be a G_δ. Consider for example: $\mathscr{X} = A =$ the set of rational numbers of the interval $\mathscr{I}$, $\mathscr{Y} = \mathscr{I}$ and $f(x) = x$.

Remark 2. The Lavrentiev theorem implies the following theorem which was proved in a different way in § 27, IV (Theorem 1): *every n-dimensional set* (lying in a metric separable space) *is contained in an n-dimensional G_δ-set; it is therefore contained topologically in a complete n-dimensional space* [2].

---

(1) C. R. Paris 178 (1924), p. 187 and *Contribution à la théorie des ensembles homéomorphes*, Fund. Math. 6 (1924), p. 149.

(2) As we shall see in Volume II, the term *complete* can be replaced by *compact*.

The proof (1) reduces to the case $n = 0$ (compare § 27, IV, Theorem 1). Now if A is a 0-dimensional set, there exists a homeomorphism h of A onto a subset of the Cantor set $\mathscr{C}$ (§ 26, IV, Theorem 2). By the preceding corollary, h can be extended to a G_δ-set B. But B is homeomorphic to a subset of $\mathscr{C}$, hence $\dim B = 0$.

III. Topological characterization of complete spaces.

THEOREM. *Every set (situated in an arbitrary metric space $\mathscr{X}$) homeomorphic to a complete space $\mathscr{Y}$ is a G_δ* (2).

Proof. Assume in the corollary of Section II that f maps A onto $\mathscr{Y}$. Then the extension f^* of f coincides with f (because f^* is one-to-one). The set of arguments of f (i.e. the set A) is therefore identical to the set of arguments of f^* and the latter is a G_δ-set (3).

Remarks. This theorem and the theorem of § 33, VI, imply that a *subset of a complete space is homeomorphic to a complete space* (i.e. is *topologically complete*) *if and only if it is a G_δ-set*. It follows that *every set homeomorphic to a G_δ-subset of a complete space is a G_δ* (4).

The complete separable spaces — from the topological point of view — *coincide with the G_δ-subsets of the Hilbert cube $\mathscr{I}^{\aleph_0}$*.

For, by the Urysohn theorem, every metric separable space is homeomorphic to a subset of $\mathscr{I}^{\aleph_0}$ (§ 22, II); moreover, if the space is complete, the subset in question is, as we have proved, a G_δ.

IV. Intrinsic invariance of various families of sets. According to § 13, IX, a property **P** of sets is said to be an *intrinsic invariant* with respect to complete spaces, if every subset of a complete

(1) See L. Tumarkin, Math. Ann. 98 (1928), p. 653.

(2) This statement yields a particular case of the invariance of the Borel class (see Section IV).

(3) For a more direct proof, see W. Sierpiński, *Sur les ensembles complets d'un espace (D)*, Fund. Math. 11 (1928), p. 203.

(4) Theorem of S. Mazurkiewicz, *Über Borelsche Mengen*, Bull. Acad. Cracovie 1916, pp. 490–494. See also W. Sierpiński, *Sur l'invariance topologique des ensembles G_δ*, Fund. Math. 8 (1926), p. 135. The theorem can also be deduced directly from the Lavrentiev theorem without referring to the theorem of § 33, VI; see M. Lavrentiev, *loc. cit.*

space which is homeomorphic to a subset with the property **P** of some complete space has also the property **P**.

THEOREM [1]. *Let* **P** *be an intrinsic invariant such that if* X *has the property* **P**, *then every* G_δ *in* X *has the property* **P**. *Under this hypothesis the following properties are intrinsic invariants:*

(i) *the property* $\mathbf{P}_\sigma$ *of being the union of a countable family of sets with the property* **P**;

(ii) *the property* $\mathbf{P}_\delta$ *of being the intersection of a countable family of sets with the property* **P**;

(iii) *the property* $\mathbf{P}_\varrho$ *of being the difference of two sets with the property* **P**;

(iv) *the property* $\mathbf{P}_c$ *of being the complement of a set with the property* **P**, *under the additional hypothesis that a set with the property* **P**, *after the adjunction of an* F_σ*-set, still has the property* **P**.

Proof. The invariance of the property $\mathbf{P}_\sigma$ is obvious (it does not depend on the hypothesis that relative G_δ-sets have the property **P**).

To prove (ii), put $A = A_1 \cap A_2 \cap \dots$, where A_n has the property **P** and A is homeomorphic to B. By the Lavrentiev theorem, the homeomorphism f of A onto B can be extended to a G_δ-set A^*. It follows that $A = A \cap A^*$, and hence

$$A = \bigcap_{n=1}^{\infty} (A_n \cap A^*) \quad \text{and} \quad B = f(A) = \bigcap_{n=1}^{\infty} f(A_n \cap A^*),$$

since the function f is one-to-one.

Since A^* is a G_δ, the set $A_n \cap A^*$ has the property **P** and so has it $f(A_n \cap A^*)$. It follows that B has the property $\mathbf{P}_\delta$.

The proofs of (iii) and (iv) are analogous. Let $A = A_1 - A_2$, and let B, f, and A^* be as before. Then

$$A = A \cap A^* = (A_1 \cap A^*) - (A_2 \cap A^*), \quad \text{whence}$$
$$B = f(A) = f(A_1 \cap A^*) - f(A_2 \cap A^*),$$

which completes the proof of (iii).

Now let $A = A_1^c$ (the complement of A_1). Then

$$A = A^* \cap A_1^c = A^* - (A_1 \cap A^*)$$

and

$$B = f(A) = f(A^*) - f(A_1 \cap A^*) = \{[f(A^*)]^c \cup f(A_1 \cap A^*)\}^c.$$

---

[1] M. Lavrentiev, Fund. Math. 6, *op. cit.*

The set $f(A^*)$ is a G_δ, hence $[f(A^*)]^c$ is an F_σ. The set in the brackets $\{\}$ therefore has the property **P** and B has the property $\mathbf{P}_c$.

(i), (ii), (iv) and the intrinsic invariance of G_δ-sets (proved in Section III) imply the following statement:

COROLLARY 1 (1). *The property of being a Borel set of class G_α with $\alpha > 0$ (G_δ, $G_{\delta\sigma}$ and so on) and that of class F_α with $\alpha > 1$ ($F_{\sigma\delta}$, $F_{\sigma\delta\sigma}$ an so on) are intrinsic invariants.*

Remark 1. We can assume that A lies in a complete space, while B (which is homeomorphic to A) lies in an arbitrary metric space $\mathscr{X}$. For, if $\mathscr{X}^*$ is the "completion" of $\mathscr{X}$ (compare § 33, VII, then, by Corollary 1, B is of class G_α (of class F_α) in the space $\mathscr{X}^*$, hence also in $\mathscr{X}$, which contains B.

COROLLARY 2. *The Baire property in the restricted sense is an intrinsic invariant.*

For a set A has this property if and only if each Z closed in A is the union of a Borel set and a set of the first category in Z (§ 11, VI).

Remark 2. The invariance of $\mathbf{P}_\varrho$ can be used to prove the intrinsic invariance of *developments into alternate* (finite or transfinite) *series* of sets of class G_α (or of class F_α) (2). For example, the property of being the difference of two G_δ-sets is an intrinsic invariant (3).

(1) This corollary was proved in some particular cases without employing the Lavrentiev theorem (under certain additional hypotheses however). Besides the papers of S. Mazurkiewicz and W. Sierpiński referred to above, see W. Sierpiński, *Sur une propriété des ensembles $F_{\sigma\delta}$* and *Sur une définition topologique des ensembles $F_{\sigma\delta}$*, Fund. Math. 6 (1924), pp. 21–29; of the same author: C. R. Paris, 171 (1920), p. 24 (the case of $G_{\delta\sigma}$, $G_{\delta\sigma\delta}$, ...) and S. Mazurkiewicz, *Sur l'invariance de la notion d'ensemble $F_{\sigma\delta}$*, Fund. Math. 2 (1921), p. 104.

For generalizations of Corollary 1 obtained by replacing homeomorphisms by mappings related to open and to closed mappings, see F. Hausdorff, *Ueber innere Abbildungen*, Fund. Math. 23 (1934), pp. 279–291 and A. D. Taimanov, *On open images of Borel sets* (Russian), Matem. Sb. 37 (1955), pp. 293–300, *On closed mappings* (Russian), Matem. Sb. 36 (1955), pp. 349–352, and 52 (1960), pp. 579–588.

See also L. Keldysh, *On open mappings of A-sets*, Dokl. Ak. N. URSS, 49 (1945), p. 646, and I. Wainstein, *On closed mappings*, Utch. Zapiski Mosc. Univ. 1952, pp. 3–53.

(2) Compare "small" Borel classes, § 37.

(3) W. Sierpiński, *Sur quelques invariants d'Analysis Situs*, Fund. Math. 3 (1922), p. 119.

V. Applications to topological ranks.

THEOREM 1. *In every complete separable space of power* $\mathfrak{c}$ *there is a family of* $2^{\mathfrak{c}}$ *sets whose topological ranks are incomparable* (1).

This theorem is a consequence of Theorem I, 5. It can be derived more directly from the following theorem of set theory.

Let Φ be a family of mappings of subsets of a given set $\mathcal{X}$ onto subsets of $\mathcal{X}$. We say that two sets X and Y (contained in $\mathcal{X}$) are *incomparable* (with respect to Φ), if no mapping of Φ maps one of them onto a subset of the other.

THEOREM 2. *If the mappings of* Φ *are one-to-one and* $\overline{\overline{\mathcal{X}}} = \overline{\overline{\Phi}} = \mathfrak{m}$, *then there exists a family* $\boldsymbol{F}$ *of* $2^{\mathfrak{m}}$ *subsets of* $\mathcal{X}$ *which are incomparable with each other.*

Proof. Consider the set $\mathcal{X}$ and the family Φ augmented with the inverses of functions belonging to Φ. Order these two sets into transfinite sequences $\{x_a\}$ and $\{f_a\}$ of the smallest possible order type. Let $P = \{p_a\}$ be a transfinite sequence defined by induction so that p_a is different from all the p_ξ and all the $f_\xi(p_\eta)$ with $\xi < a$ and $\eta < a$. Let $\boldsymbol{F}$ be a family of subsets of P such that $\overline{\overline{\boldsymbol{F}}} = 2^{\mathfrak{m}}$ and $\overline{\overline{X - Y}} = \mathfrak{m}$, for every pair of distinct sets of $\boldsymbol{F}$ (compare I, Theorem 3, first part).

Suppose that a function $f_\gamma \in \Phi$ maps X onto Y_1, where $X \in \boldsymbol{F}$ and $Y_1 \subset Y \in \boldsymbol{F}$. Let $f_\gamma^{-1} = f_\xi$. We shall prove that the formulas $p_a \in X - Y$ and $\xi < a$ imply $a < \gamma$, from which follows $\overline{\overline{X - Y}} < m$, contrary to the hypothesis.

Let $f_\gamma(p_a) = p_\eta$, hence $p_a = f_\xi(p_\eta)$. By definition of p_a, $\eta \geqslant a$. On the other hand, $p_\eta \in Y$ and $p_a \notin Y$, and hence $\eta \neq a$. The formulas $a < \eta$ and $p_\eta = f_\gamma(p_a)$ imply $\eta \leqslant \gamma$, hence $a < \gamma$.

To prove Theorem 1, let Φ be the family of homeomorphisms defined on G_δ-subsets of $\mathcal{X}$. If we assume that $X \in \boldsymbol{F}$, $X \neq Y \in \boldsymbol{F}$ and that X is topologically contained in Y, then, by the Lavrentiev theorem, there exists a homeomorphism f defined on a G_δ containing X such that $f(X) \subset Y$. But this is impossible, since $f \in \Phi$ and X and Y are incomparable with respect to Φ.

(1) See my paper *Sur la puissance des „nombres de dimension" au sens de M. Fréchet,* Fund. Math. 8 (1926), p. 201. See also W. Sierpiński, Fund. Math. 19 (1932), p. 70, and S. Banach, *ibid.* p. 10.

Remarks. In an analogous direction, one shows *the existence of a transfinite sequence of power* $> \mathfrak{c}$ *of sets whose topological ranks increase as well as of an transfinite sequence of power* $\mathfrak{c}$ *of sets whose topological ranks decrease* (1).

VI. Extension of B measurable functions.

THEOREM (2). *Let* $\mathcal{X}$ *be an arbitrary metric space, let* $\mathcal{Y}$ *be a complete separable space and let* A *be a subset of* $\mathcal{X}$. *Every function* f *of class* α *defined on* A *can be extended (without altering its class) to a set* A^* *of the multiplicative class* $(\alpha+1)$.

If, in particular, A *is of multiplicative class* $\alpha>0$, *then* f *can be extended to the entire space* (3).

Proof. The theorem is proved for $\alpha=0$ in Section I; so let $\alpha>0$. By § 31, VIII, Theorem 3, there exists a sequence of functions $\{f_n\}$ of class α defined on A which uniformly approximate f so that the sets $f_n(A)$ are discrete. Therefore we can put

$$|f_n(x)-f_k(x)|<\frac{1}{2^k} \quad \text{for every} \quad n\geqslant k \tag{1}$$

and $f_n(A)=[y_n^1, y_n^2, \ldots, y_n^i, \ldots]$, a finite or infinite sequence of distinct elements.

Let $B_{i_1\ldots i_n}=[f_1^{-1}(y_1^{i_1})]\cap\ldots\cap[f_n^{-1}(y_n^{i_n})]$.

Each of the sets $f_k^{-1}(y_k^i)$ is ambiguous of class α with respect to A (since the point y_k^i is isolated in $f_k(A)$ and hence constitutes a set both closed and open). Therefore the hypotheses of Theorem 2 of § 30, VIII are fulfilled.

It follows that for every n there exists a set A_n of the multiplicative class $(\alpha+1)$ and a system $\{C_{i_1\ldots i_n}\}$ of ambiguous sets of class α with respect to A_n such that

(i) $C_{i_1\ldots i_n}\cap C_{j_1\ldots j_k}=0$ for $(i_1\ldots i_k)\neq(j_1\ldots j_k)$ and $k\leqslant n$;

(ii) $A_n=\bigcup C_{i_1\ldots i_n}$;

(iii) $A\cap C_{i_1\ldots i_n}=B_{i_1\ldots i_n}$;

(iv) if $B_{i_1\ldots i_n}=0$, then $C_{i_1\ldots i_n}=0$.

---

(1) See W. Sierpiński, *loc. cit.*, and a paper of W. Sierpiński and myself *Sur un problème de M. Fréchet concernant les dimensions des ensembles linéaires*, Fund. Math. 8 (1926), p. 193.

(2) See my paper *Sur les théorèmes topologiques de la théorie des fonctions de variables réelles*, C. R. Paris 197 (1933), p. 19.

(3) For the second part of the theorem restricted to the real valued functions, see F. Hausdorff, *Mengenlehre*, p. 244.

Moreover (by the second part of the theorem referred to above), if A is of the multiplicative class α, then $A_n = \mathscr{X}$, since the formula

$$A = f_k^{-1} f_k(A) = f_k^{-1}(y_k^1) \cup f_k^{-1}(y_k^2) \cup \ldots$$

is valid for each k and implies $A = \bigcup B_{i_1 \ldots i_n}$ for every n.

Thus $A^* = A_1 \cap A_2 \cap \ldots$ is of the multiplicative class $\alpha+1$; moreover, if A is of the multiplicative class α (hence $A_n = \mathscr{X}$), then $A^* = \mathscr{X}$.

Extend each function f_n to A_n assuming that $f_n(x) = y_n^{i_n}$, if $x \in C_{i_1 \ldots i_n}$. The set of values of the function f_n is countable. In order to prove that this function is of class α (on A_n), it suffices to show that the set $f_n^{-1}(y_n^i)$ is of the additive class α with respect to A_n, for fixed n and i. This follows directly from the formula $f_n^{-1}(y_n^i) = \bigcup C_{i_1 \ldots i_{n-1} i}$, where the union is taken over all systems of $(n-1)$ indices.

Now we shall prove that the (extended) functions f_n are uniformly convergent on A^*; in fact, that condition (1) is valid for every $x_0 \in A^*$.

Let $x_0 \in C_{i_1 \ldots i_n}$ and $x_0 \in C_{j_1 \ldots j_k}$ for $k \leqslant n$. By (i), we have $i_1 = j_1, \ldots, i_k = j_k$. Let $x \in B_{i_1 \ldots i_n}$; then

$$x \in f_n^{-1}(y_n^{i_n}), \quad \text{whence} \quad f_n(x) = y_n^{i_n} = f_n(x_0).$$

The relation $i_k = j_k$ implies that

$$x \in f_k^{-1}(y_k^{j_k}), \quad \text{so that} \quad f_k(x) = y_k^{j_k} = f_k(x_0).$$

As a point of A, x satisfies (1). It follows that x_0 satisfies (1), since $f_n(x) = f_n(x_0)$ and $f_k(x) = f_k(x_0)$.

The space $\mathscr{Y}$ is complete, so let $f(x) = \lim f_n(x)$. Thus the function f is defined for every $x \in A^*$ and is of class α as a uniformly convergent sequence of functions of class α (§ 31, VIII, Theorem 2).

COROLLARY [1]. *Under the same hypotheses on $\mathscr{X}$ and $\mathscr{Y}$, every function of class α defined on A can be extended to the entire space to give a function of class $(\alpha+1)$.*

---

[1] For the case $\mathscr{Y} = \mathscr{E}$, see W. Sierpiński, *Sur l'extension des fonctions de Baire définies sur les ensembles linéaires quelconques*, Fund. Math. 16 (1930), p. 81 and G. Alexits, *Über die Erweiterung einer Baireschen Funktion*, Fund. Math. 15 (1930), p. 51.

Proof. By the first part of the preceding theorem, the function f can be extended to a set E of the multiplicative class $(\alpha+1)$ without altering the class of f (if $\alpha = 0$, see Section I). By the second part of the same theorem, the extension of f can be further extended to the entire space to give a function of class $(\alpha+1)$.

VII. Extension of a homeomorphism of class α, β. The following statements enable us to generalize those of Sections II–IV under the assumption that the considered spaces are separable.

THEOREM [1]. *Every homeomorphism of class* α, β *between arbitrary subsets* A *and* B *of complete separable spaces* $\mathscr{X}$ *and* $\mathscr{Y}$, *respectively, can be extended to two subsets of classes* $\alpha+\beta+1$ *and* $\beta+\alpha+1$ *of these spaces.*

Proof. We return to the proof of the Lavrentiev theorem. Let f be a homeomorphism of class α, β of A onto B, let f^* be an extension of class α of f defined on a set A^* of multiplicative class $\alpha+1, A \subset A^*$, (see Section VI) and let $I = \mathop{E}_{xy}[y = f^*(x)]$. Let $g = f^{-1}$ and let B^*, g^* and J have an analogous meaning with respect to g. Finally let A_1 and B_1 be the projections of the set $I \cap J$ on the $\mathscr{X}$ and $\mathscr{Y}$ axes, respectively. Since J is of the multiplicative class $\beta+1$ (see § 31, VII, Theorem 1), A_1 is of the multiplicative class $\alpha+\beta+1$ with respect to A^* (§ 31, VII, Theorem 2), and hence with respect to $\mathscr{X}$. This completes the proof.

Now consider the particular case where A is of multiplicative class α. Let $A^* = A$ and $f^* = f$. It follows that $A_1 = A$ and $B_1 = B$. Now I is of the multiplicative class α, and B_1 is of the multiplicative class $\beta+\alpha$ with respect to B^*, and hence with respect to $\mathscr{Y}$. Thus we obtain the following statements:

COROLLARIES. 1. *The property of being a set of the multiplicative class* $\alpha > 0$ *is invariant under homeomorphisms of class* $\alpha, 0$.

2. *If* A *is of the multiplicative class* $\alpha > 0$ *and* f *is a homeomorphism of class* α, β, *then* $f(A)$ *is of multiplicative class* $\beta+\alpha$.

3. *Every homeomorphism of class* $1, \beta$ *defined on a complete separable space maps the space onto a set of the multiplicative class* $\beta+1$.

---

[1] See my paper *Sur le prolongement de l'homéomorphie*, C. R. Paris 197 (1933), p. 1090.

*§ 36. Relations of complete separable spaces to the space $\mathcal{N}$ of irrational numbers

I. Operation ($\mathscr{A}$). Assume that to every system $n_1, \dots n_k$ of positive integers there corresponds a closed (empty or not) subset $F_{n_1 \dots n_k}$ of a complete space $\mathscr{X}$.

As usual, denote by $\mathfrak{z}$ a variable sequence $[\mathfrak{z}^1, \mathfrak{z}^2, \dots]$ of positive integers. We can identify such a sequence with the irrational number $\frac{1|}{|\mathfrak{z}^1} + \frac{1|}{|\mathfrak{z}^2} + \dots$, so we can assume that $\mathfrak{z}$ runs over the space $\mathcal{N}$ (compare § 3, XIV and IX).

We shall subsequently assume that

$$F_{\mathfrak{z}^1 \dots \mathfrak{z}^k \mathfrak{z}^{k+1}} \subset F_{\mathfrak{z}^1 \dots \mathfrak{z}^k}, \tag{1}$$

$$\lim_{k=\infty} \delta(F_{\mathfrak{z}^1 \dots \mathfrak{z}^k}) = 0. \tag{2}$$

Let Z be the set of $\mathfrak{z}$ such that $F_{\mathfrak{z}^1 \dots \mathfrak{z}^k} \neq 0$ for every k. Then the set $F_{\mathfrak{z}^1} \cap F_{\mathfrak{z}^1 \mathfrak{z}^2} \cap F_{\mathfrak{z}^1 \mathfrak{z}^2 \mathfrak{z}^3} \cap \dots$ reduces to a single point (§ 34, II) which we denote by $f(\mathfrak{z})$. Consequently

$$f(Z) = \bigcup_{\mathfrak{z}} \bigcap_{k=1}^{\infty} F_{\mathfrak{z}^1 \dots \mathfrak{z}^k}, \tag{3}$$

so that $f(Z)$ is the result of the operation ($\mathscr{A}$) performed on the system $\{F_{n_1 \dots n_k}\}$.

Let $\mathcal{N}_{n_1 \dots n_k}$ denote the set of $\mathfrak{z}$ such that $\mathfrak{z}^1 = n_1, \dots, \mathfrak{z}^k = n_k$. The sets $\mathcal{N}_{n_1 \dots n_k}$ are both *closed and open* in $\mathcal{N}$ and constitute a *base in* $\mathcal{N}$ (compare § 5, XI, Example 4).

The following inclusion holds:

$$f(Z \cap \mathcal{N}_{n_1 \dots n_k}) \subset F_{n_1 \dots n_k}. \tag{4}$$

For, if $\mathfrak{z} \in Z$, then $f(\mathfrak{z}) \in F_{\mathfrak{z}^1 \dots \mathfrak{z}^k}$ for every k. If $\mathfrak{z} \in \mathcal{N}_{n_1 \dots n_k}$, then $\mathfrak{z}^1 = n_1, \dots, \mathfrak{z}^k = n_k$. Therefore $f(\mathfrak{z}) \in F_{n_1 \dots n_k}$ and (4) follows.

(a) *The set Z is closed in $\mathcal{N}$.*

Proof. Let $\mathfrak{z} \in \mathcal{N} - Z$. There exists a k such that $F_{\mathfrak{z}^1 \dots \mathfrak{z}^k} = 0$. By (4), $f(Z \cap \mathcal{N}_{\mathfrak{z}^1 \dots \mathfrak{z}^k}) = 0$, hence $Z \cap \mathcal{N}_{\mathfrak{z}^1 \dots \mathfrak{z}^k} = 0$. As the set $\mathcal{N}_{\mathfrak{z}^1 \dots \mathfrak{z}^k}$ is open and contains $\mathfrak{z}$, it follows that every point of $\mathcal{N} - Z$ belongs to an open set disjoint to Z. Therefore Z is closed (in $\mathcal{N}$).

(b) *The function f is continuous* (*on* Z).

Proof. Let $\mathfrak{z} \in Z$ and $\varepsilon > 0$. According to (2), let k be an index such that $\delta(F_{\mathfrak{z}^1 \dots \mathfrak{z}^k}) < \varepsilon$. By (4), $\delta[f(Z \cap \mathcal{N}_{\mathfrak{z}^1 \dots \mathfrak{z}^k})] < \varepsilon$. Since

$\mathscr{N}_{\mathfrak{z}^1 \ldots \mathfrak{z}^k}$ is a neighbourhood of $\mathfrak{z}$, it follows that the function f is continuous at $\mathfrak{z}$.

The following proposition is obvious:

(c) *If* $F_{\mathfrak{z}^1 \ldots \mathfrak{z}^k} \neq 0$, *for every* $\mathfrak{z}$ *and* k, *then* $Z = \mathscr{N}$.

(d) *If* $\{F_{\mathfrak{z}^1 \ldots \mathfrak{z}^k}\}$ *is a dyadic system, i.e. the set* $F_{\mathfrak{z}^1 \ldots \mathfrak{z}^k}$ *is empty except for the systems* $\mathfrak{z}^1 \ldots \mathfrak{z}^k$ *composed of the digits* 1 *and* 2, *then the set* Z *is homeomorphic to the Cantor set* $\mathscr{C}$.

For, in this case, Z is the set of sequences composed of the digits 1 and 2.

(e) *If two sets* $F_{\mathfrak{z}^1 \ldots \mathfrak{z}^k}$ *and* $F_{\mathfrak{y}^1 \ldots \mathfrak{y}^k}$ *indexed by two different systems of indices are always disjoint, then the function* f *is one-to-one.*

Moreover,

$$f(Z) = \bigcap_{k=1}^{\infty} \bigcup_{\mathfrak{z}} F_{\mathfrak{z}^1 \ldots \mathfrak{z}^k} \tag{5}$$

so that the set $f(Z)$ is an $F_{\sigma\delta}$ and

$$f(Z \cap \mathscr{N}_{n_1 \ldots n_k}) = f(Z) \cap F_{n_1 \ldots n_k}. \tag{6}$$

Proof. If $\mathfrak{z} \neq \mathfrak{y}$, there exists an index k such that $\mathfrak{z}^k \neq \mathfrak{y}^k$. Since $f(\mathfrak{y}) \in F_{\mathfrak{y}^1 \ldots \mathfrak{y}^k}$, it follows that $f(\mathfrak{y}) \notin F_{\mathfrak{z}^1 \ldots \mathfrak{z}^k}$ and $f(\mathfrak{z}) \neq f(\mathfrak{y})$.

Formula (5) follows directly from Theorem 2 of § 3, XIV and from (3).

To prove (6) we have to show, by (4), that the right-hand member is contained in the left-hand one.

Let $p \in f(Z) \cap F_{n_1 \ldots n_k}$. Then

$$p = f(\mathfrak{z}) \in F_{\mathfrak{z}^1 \ldots \mathfrak{z}^k}, \quad \mathfrak{z} \in Z \quad \text{and} \quad p \in F_{n_1 \ldots n_k}.$$

By hypothesis, the formula $p \in F_{\mathfrak{z}^1 \ldots \mathfrak{z}^k} \cap F_{n_1 \ldots n_k}$ implies that $\mathfrak{z}^1 = n_1, \ldots, \mathfrak{z}^k = n_k$. Therefore $\mathfrak{z} \in \mathscr{N}_{n_1 \ldots n_k}$ and $p \in f(Z \cap \mathscr{N}_{n_1 \ldots n_k})$.

(f) *If in statement* (e), *we assume that the sets* $F_{\mathfrak{z}^1 \ldots \mathfrak{z}^k}$ *are open* (*all being closed*), *then the function* f *is a homeomorphism of* Z *onto* $f(Z)$.

Proof. Suppose that $\lim_{m=\infty} f(\mathfrak{z}_m) = f(\mathfrak{z})$. We have to prove that $\lim_{m=\infty} \mathfrak{z}_m = \mathfrak{z}$, i.e. $\lim_{m=\infty} \mathfrak{z}_m^k = \mathfrak{z}^k$, for every k; in other words, that, for a fixed k, $\mathfrak{z}_m^k = \mathfrak{z}^k$, if m is sufficiently large.

Now the set $F_{\mathfrak{z}^1 \ldots \mathfrak{z}^k}$ is open, and hence the formula $\lim_{m=\infty} f(\mathfrak{z}_m) = f(\mathfrak{z}) \in F_{\mathfrak{z}^1 \ldots \mathfrak{z}^k}$ implies that $f(\mathfrak{z}_m) \in F_{\mathfrak{z}^1 \ldots \mathfrak{z}^k}$, if m is sufficiently large.

On the other hand, $f(\mathfrak{z}_m)\in F_{\mathfrak{z}_m^1\ldots\mathfrak{z}_m^k}$. Therefore the hypothesis of (e) implies that $\mathfrak{z}_m^1=\mathfrak{z}^1,\ldots,\mathfrak{z}_m^k=\mathfrak{z}^k$.

(g) *If* $\mathscr{X}=\bigcup_{i=1}^{\infty}F_i$ *and* $F_{\mathfrak{z}^1\ldots\mathfrak{z}^k}=\bigcup_{i=1}^{\infty}F_{\mathfrak{z}^1\ldots\mathfrak{z}^k i}$, *for every* $\mathfrak{z}$ *and* k, *then* $f(Z)=\mathscr{X}$.

Proof. Given a point p, there exists an index n_1 such that $p\in F_{n_1}$. We proceed by induction. Assume that $p\in F_{n_1\ldots n_k}$. By hypothesis, there exists an index n_{k+1} such that $p\in F_{n_1\ldots n_k n_{k+1}}$. Let $\mathfrak{z}$ be the sequence $n_1, n_2, n_3, \ldots$. Then $p=f(\mathfrak{z})$.

For further applications we shall prove the following three propositions.

(h) *If* $\mathscr{X}=\overline{\bigcup_{i=1}^{\infty}F_i}$ *and* $F_{\mathfrak{z}^1\ldots\mathfrak{z}^k}=\overline{\bigcup_{i=1}^{\infty}F_{\mathfrak{z}^1\ldots\mathfrak{z}^k i}}$, *for every* $\mathfrak{z}$ *and* k, *then*

$$\overline{f(Z)}=\mathscr{X} \quad \text{and} \quad \overline{f(Z)\cap F_{\mathfrak{z}^1\ldots\mathfrak{z}^k}}=F_{\mathfrak{z}^1\ldots\mathfrak{z}^k}. \tag{7}$$

Proof. Let $p\in F_{n_1\ldots n_k}$ and let $\varepsilon>0$. We have to prove that there exists a $\mathfrak{z}\in Z$ such that $f(\mathfrak{z})\in F_{n_1\ldots n_k}$ and $|f(\mathfrak{z})-p|\leqslant\varepsilon$.

By hypothesis, there exists an index n_{k+1} and a point p_1 such that

$$p_1\in F_{n_1\ldots n_k n_{k+1}} \quad \text{and} \quad |p_1-p|<\varepsilon/2.$$

In general, there exists a sequence of integers $n_{k+1}, n_{k+2}, \ldots$ and a sequence of points $p_1, p_2, \ldots$ such that

$$p_j\in F_{n_1\ldots n_k\ldots n_{k+j}} \quad \text{and} \quad |p_j-p_{j-1}|<\varepsilon/2^j.$$

Let $\mathfrak{z}$ be the sequence $n_1, n_2, \ldots$. Thus $\mathfrak{z}^m=n_m$, for every m. It follows that

$$f(\mathfrak{z})\in F_{\mathfrak{z}^1\ldots\mathfrak{z}^m}=F_{n_1\ldots n_m} \quad \text{and} \quad \lim_{j=\infty}|f(\mathfrak{z})-p_j|=0$$

(by (2)). Therefore $f(\mathfrak{z})=\lim\limits_{j=\infty}p_j$ and since $|p_j-p|<\varepsilon$, it follows that $|f(\mathfrak{z})-p|\leqslant\varepsilon$.

Thus the second identity of formula (7) is proved. The first one is its particular case, if we let $F=\mathscr{X}$ for $k=0$.

(i) *If the hypotheses of statements* (c), (e), *and* (h) *are fulfilled and the set* $F_{n_1\ldots n_k}$ *is nowhere dense in* $F_{n_1\ldots n_{k-1}}$ (*for every system of indices* $n_1\ldots n_{k-1}$), *then the function* f *maps every open set in* $\mathscr{N}$ *onto a set of the first category on itself.*

Proof. Since the sets $\mathcal{N}_{n_1 \dots n_k}$ form a base in $\mathcal{N}$, it is sufficient to prove that $f(\mathcal{N}_{n_1 \dots n_k})$ is of the first category on itself.

By (c), $Z = \mathcal{N}$, hence, by (4), $f(\mathcal{N}_{n_1 \dots n_k}) \subset F_{n_1 \dots n_k}$, whence

$$f(\mathcal{N}_{n_1 \dots n_k}) = \bigcup_{i=1}^{\infty} f(\mathcal{N}_{n_1 \dots n_k i}) \subset \bigcup_{i=1}^{\infty} F_{n_1 \dots n_k i}.$$

Since each of the sets $F_{n_1 \dots n_k i}$ is nowhere dense in $F_{n_1 \dots n_k}$ (by hypothesis), their union is of the first category in $F_{n_1 \dots n_k}$. So $f(\mathcal{N}_{n_1 \dots n_k})$ is of the first category in $F_{n_1 \dots n_k}$. On the other hand, by (7), (6), and (c),

$$F_{n_1 \dots n_k} = \overline{f(Z) \cap F_{n_1 \dots n_k}} = \overline{f(Z \cap \mathcal{N}_{n_1 \dots n_k})} = \overline{f(\mathcal{N}_{n_1 \dots n_k})}.$$

The set $f(\mathcal{N}_{n_1 \dots n_k})$ is of the first category on itself, since it is of the first category on its closure (compare § 10, IV, Theorem 2).

We complete statement (i) by the following one:

(j) *In the space $\mathcal{C}$ there exists a system $\{F_{n_1 \dots n_k}\}$ of perfect sets satisfying the hypotheses of Theorem* (i).

Proof. Observe first that given a point p of the product $\mathcal{C} \times \mathcal{C}$, there exists a perfect set $P_p \subset \mathcal{C} \times \mathcal{C}$ containing p and nowhere dense in the product $\mathcal{C} \times \mathcal{C}$ (and homeomorphic to $\mathcal{C}$).

Now since $\mathcal{C} \times \mathcal{C}$ is homeomorphic to $\mathcal{C}$ (§ 16, II, Theorem 6), every point of $\mathcal{C}$ lies in a perfect set, nowhere dense in $\mathcal{C}$. Therefore, if $p_1, p_2, \dots$ is a dense sequence of points in $\mathcal{C}$, there exists a sequence $F_1, F_2, \dots$ of non-empty perfect sets, nowhere dense in $\mathcal{C}$ such that

$$F_n \cap F_m = 0 \quad \text{for} \quad n \neq m \quad \text{and} \quad p_n \in (F_1 \cup \dots \cup F_n).$$

It follows that $\mathcal{C} = \overline{F_1 \cup F_2 \cup \dots}$

If we now replace $\mathcal{C}$ by F_n, we obtain sets $F_{n_1 n_2}$, then $F_{n_1 n_2 n_3}$ and so on. We can assume, moreover, that

$$\delta(F_{n_1 \dots n_k}) < 1/k.$$

The system $\{F_{n_1 \dots n_k}\}$ satisfies the hypotheses of Theorem (i).

II. Mappings of the space $\mathcal{N}$ onto complete spaces.

THEOREM 1 [1]. *Every complete separable space is (effectively [2]) a continuous image of the space $\mathcal{N}$.*

---

[1] This theorem will be extended in § 37 to Borel sets.

[2] i.e. the following proof permits to define a continuous function f which maps $\mathcal{N}$ onto $\mathcal{X}$ (compare § 23, VIII).

Proof. Let $R_1, R_2, \ldots$ be a base in $\mathcal{X}$ such that $R_i \neq 0$ and $\delta(R_i) < 1$ (compare § 21, II). Let $F_i = \bar{R}_i$. Then

$$\mathcal{X} = \bigcup_{i=1}^{\infty} F_i \quad \text{and} \quad \delta(F_i) \leqslant 1. \tag{1}$$

Proceed by induction. Assume that a non-empty closed set $F_{n_1 \ldots n_k}$ has been defined and is now regarded as a space. There exists an infinite sequence of non-empty closed sets $F_{n_1 \ldots n_k i}$, $i = 1, 2, \ldots$, such that

$$F_{n_1 \ldots n_k} = \bigcup_{i=1}^{\infty} F_{n_1 \ldots n_k i} \quad \text{and} \quad \delta(F_{n_1 \ldots n_k i}) \leqslant \frac{1}{k+1}. \tag{2}$$

By (b), (c), and (g), $\mathcal{X}$ is a continuous image of $\mathcal{N}$.

THEOREM 2. *Every complete separable space of dimension* 0 *is homeomorphic to a closed subset of* $\mathcal{N}$.

Proof. Since $\dim \mathcal{X} = 0$, every open set is the union of an infinite sequence of disjoint sets (empty or not) which are both closed and open and are of arbitrarily small diameters (§ 26, I, Corollary 1a). Thus, as in the preceding proof, we obtain formulas (1) and (2), where $F_{n_1 \ldots n_k}$ are closed, open and disjoint for a fixed k.

By (a), (f) and (g), $\mathcal{X}$ is homeomorphic to a closed subset Z of $\mathcal{N}$.

COROLLARY 2a. *Every* G_δ*-set in the space* $\mathcal{N}$ *is homeomorphic to a closed set.*

For, by virtue of § 33, VII, every G_δ-set contained in $\mathcal{N}$ is homeomorphic to a complete space of dimension 0.

THEOREM 3 (of S. Mazurkiewicz) [1]. *Every* G_δ*-set which is both dense and boundary in a complete, separable, and* 0*-dimensional space is homeomorphic to the space* $\mathcal{N}$.

Proof. By Theorem 1′ of § 26, I, this G_δ-set is the result of the operation ($\mathcal{A}$) performed on a system $\{F_{n_1 \ldots n_k}\}$ of non-empty, closed and open sets, disjoint for a fixed k and satisfying conditions (1) and (2) of Section I. By (3), (c) and (f), this set is homeomorphic to $\mathcal{N}$.

---

[1] *Teorja zbiorów* G_δ, Wektor 1918. Compare also L. E. J. Brouwer, *On linear inner limiting sets*, Proc. Akad. Amsterdam 20 (1917), p. 1191, and P. Alexandrov and P. Urysohn, *Ueber nulldimensionale Mengen*, Math. Ann. 98 (1927), p. 89.

COROLLARY 3a. *Every G_δ-set Q, both dense and boundary in the space $\mathscr{E}$ of real numbers, is homeomorphic to $\mathscr{N}$.*

Proof. Let D be a countable dense subset of $\mathscr{E}-Q$ such that the set $\mathscr{E}-Q-D$ is dense (such a D exists, since $\mathscr{E}-Q$ is dense and dense in itself). It follows that Q is dense and boundary in $\mathscr{E}-D$. Moreover, $\mathscr{E}-D$ is a G_δ, and hence it is a topologically complete space of dimension 0 (§ 33, VI). By Theorem 3, Q is homeomorphic to $\mathscr{N}$.

III. One-to-one mappings.

THEOREM. *The interval $\mathscr{I}$, as well as the n-dimensional cube $\mathscr{I}^n$ (n is finite or $\aleph_0$), can be obtained from $\mathscr{N}$ by a homeomorphism of class* 0, 1.

Proof. Let N be the Cantor set $\mathscr{C}$ with the left-hand end points of the contiguous intervals removed. Let t be the function which carries the number

$$x = \frac{c_1}{3^1}+\frac{c_2}{3^2}+\ldots+\frac{c_m}{3^m}+\ldots \qquad (c_m = 0 \text{ or } 2)$$

to

$$t(x) = \frac{c_1}{2^2}+\frac{c_2}{2^3}+\ldots+\frac{c_m}{2^{m+1}}+\ldots.$$

Then t is a continuous one-to-one mapping of N onto $\mathscr{I}$ (compare § 16, II, Corollary 6a).

The inverse function t^{-1} admits only a countable set of discontinuity points (these are the points represented by finite dyadic fractions), hence it is of class 1 (§ 34, VII, Corollary 1).

Since N is dense and boundary in $\mathscr{C}$, the sets N and $\mathscr{N}$ are homeomorphic (Theorem 3 of Section II). Let s be a homeomorphism such that $s(\mathscr{N}) = N$. The function $f = t \circ s$ satisfies the theorem for $n = 1$.

By assigning the sequence $f(x_1), f(x_2), \ldots$ to a sequence $x_1, x_2, \ldots$ of irrational numbers, we define a continuous mapping g of $\mathscr{N}^n$ onto $\mathscr{I}^n$.

The inverse mapping g^{-1} carries every point $y_1, y_2, \ldots$ of $\mathscr{I}^n$ to the point $f^{-1}(y_1), f^{-1}(y_2), \ldots$ of $\mathscr{N}^n$. Therefore the mapping g^{-1} is of class 1, since f^{-1} is of class 1 (§ 31, VI, Theorem 3). As $\mathscr{N}^n$ is homeomorphic to $\mathscr{N}$ (§ 16, II, Theorem 6), let h be a homeomorphism such that $h(\mathscr{N}) = \mathscr{N}^n$. Then $g \circ h$ is the required function.

Remark. The preceding theorem can be generalized so as to replace $\mathcal{I}^n$ by an arbitrary $F_{\sigma\delta}$ which consists of condensation points only (and lies in a complete separable space) [1].

COROLLARY [2]. *Every separable metric space $\mathcal{X}$ can be obtained from a subset E of $\mathcal{N}$ by a homeomorphism of class* $0, 1$.

If, moreover, $\mathcal{X}$ is complete, then E is closed (and hence topologically complete, separable, and 0-dimensional).

Proof. $\mathcal{X}$ can be regarded as a subset Q of $\mathcal{I}^{\aleph_0}$ (§ 22, II).

If $\mathcal{X}$ is complete, then Q is a G_δ (compare § 35, III). Now if f is the function provided by the preceding theorem (for $n = \aleph_0$), the set $f^{-1}(Q)$ is a G_δ in $\mathcal{N}$. It is therefore homeomorphic to a closed subset of $\mathcal{N}$ (Corollary 2a, Section II).

IV. Decomposition theorems.

THEOREM 1. *Every complete, separable, 0-dimensional uncountable space is composed of an infinite countable set and a set homeomorphic to $\mathcal{N}$.*

Proof. By the Cantor-Bendixson theorem (§ 23, V), the space is composed of a countable set and a non-empty perfect set P. Let D be a countable dense set in P. The set P is a complete, separable, and 0-dimensional space. Moreover, every point of P is its condensation point (§ 34, IV, Corollary 3), so the set $P-D$ is dense in P. Being a G_δ-set, dense and boundary in P, the set $P-D$ is homeomorphic to $\mathcal{N}$, by Theorem 3 of Section II.

Remark. It follows that every *uncountable G_δ subset of $\mathcal{E}$ becomes homeomorphic to $\mathcal{N}$ after deleting a suitably chosen countable set* [3]. For, after deleting the rational points we obtain a 0-dimensional G_δ-set which then becomes homeomorphic to $\mathcal{N}$ after removing a countable set from it.

THEOREM 2. *Every complete separable uncountable space is composed of a countable set and a set obtained from $\mathcal{N}$ by a homeomorphism of class* $0, 1$.

Proof. Let $\mathcal{Y}$ be the considered space and let $\mathcal{X}$ be a complete separable 0-dimensional space such that $\mathcal{Y}$ can be obtained from $\mathcal{X}$

---

(1) F. Hausdorff, *Die schlichten stetigen Bilder des Nullraums*, Fund. Math. 29 (1937), p. 153.

(2) This corollary will be extended in § 37 to Borel sets.

(3) Theorem of S. Mazurkiewicz, *loc. cit.*

by a homeomorphism f of class 0, 1 (see Section III, Corollary). By the preceding theorem, $\mathscr{X} = D \cup N$, where D is countable and N is homeomorphic to $\mathscr{N}$. Then $\mathscr{Y} = f(D) \cup f(N)$ is the required decomposition.

V. Relations to the Cantor discontinuum $\mathscr{C}$.

THEOREM. *Every continuous mapping f of a complete separable space $\mathscr{X}$ is a homeomorphism on a set A homeomorphic to $\mathscr{C}$, provided that f assumes an uncountable set of values.*

Moreover (1), *every non-empty, complete, and dense in itself space* (which may fail to be separable) *contains $\mathscr{C}$ topologically.*

Proof. For every value y of f, let x_y be a point such that $y = f(x_y)$. As $\mathscr{X}$ is separable and the set of the x_y is uncountable, let D be a (non-empty) dense in itself subset of this set (compare § 23, V).

Let $p_0 \neq p_2$ be two points of D and let F_0 and F_2 be two (closed) balls with centres p_0 and p_2 such that

$$\delta(F_0) < 1, \quad \delta(F_2) < 1, \tag{i}$$

$$f(F_0) \cap f(F_2) = 0. \tag{ii}$$

The existence of the balls F_0 and F_2 follows from the continuity of f. If they did not exist, we could define (by taking the balls smaller and smaller) two sequences $\{r_n\}$ and $\{s_n\}$ such that

$$\lim_{n=\infty} r_n = p_0, \quad \lim_{n=\infty} s_n = p_2, \quad \text{and} \quad f(r_n) = f(s_n).$$

Then we would have

$$\lim_{n=\infty} f(r_n) = \lim_{n=\infty} f(s_n), \quad \text{and hence} \quad f(p_0) = f(p_2) \text{ and } p_0 = p_2,$$

since f is one-to-one on D.

Since D is dense in itself, there exist two points p_{00} and p_{02} of D in the interior of F_0 and two balls F_{00} and F_{02} such that

$$\delta(F_{00}) < 1/2, \quad \delta(F_{02}) < 1/2, \quad f(F_{00}) \cap f(F_{02}) = 0$$

and $F_{00} \cup F_{02} \subset F_0$.

By letting F_{20} and F_{22} have an analogous meaning and by continuing this procedure step by step, we obtain a dyadic system $\{F_{c_1 \ldots c_k}\}$ of non-empty closed sets satisfying (1) and (2) of Section I. Moreover,

(1) Compare W. H. Young, Leipziger Ber. 55 (1903), p. 287.

(i) $f(F_{c_1 \ldots c_k}) \cap f(F_{d_1 \ldots d_k}) = 0$, if $(c_1 \ldots c_k) \neq (d_1 \ldots d_k)$, which implies that $F_{c_1 \ldots c_k} \cap F_{d_1 \ldots d_k} = 0$.

By Section I, (d) and (e), it follows that the set $A = \bigcap_{k=1}^{\infty} \bigcup F_{c_1 \ldots c_k}$, where the union is taken over all systems of k digits $c_1 \ldots c_k$, is a continuous one-to-one image of $\mathscr{C}$: $A = g(\mathscr{C})$ (1). By § 20, V, Corollary 6a, it follows that A is homeomorphic to $\mathscr{C}$.

To prove that the partial function $f|A$ is a homeomorphism it is sufficient to show that it is one-to-one. Now, given two points $x_1 \neq x_2$ of A, there exist two different systems of indices $c_1 \ldots c_k$ and $d_1 \ldots d_k$ such that $x_1 \in F_{c_1 \ldots c_k}$ and $x_2 \in F_{d_1 \ldots d_k}$. It follows from (i) that $f(x_1) \neq f(x_2)$.

To prove the second part of the theorem it suffices to let $D = \mathscr{X}$ and $f(x) = x$ in the preceding argument.

COROLLARY 1. *Every uncountable continuous image of a complete separable space contains $\mathscr{C}$ topologically. It is therefore of the power of continuum.*

COROLLARY 2. *Every uncountable complete separable space contains a G_δ-set homeomorphic to $\mathcal{N}$.*

This is a consequence of the preceding corollary, of Theorem 1 of Section II and of the fact that the set $\mathscr{C}$ with the ends of the contiguous intervals removed is homeomorphic to $\mathcal{N}$.

COROLLARY 3. *The set of values of a continuous function f defined on a complete separable space $\mathscr{X}$ is uncountable if and only if there exists a dense in itself set $E \subset \mathscr{X}$ such that the partial function $f|E$ is one-to-one.*

Proof. The condition is necessary by the theorem of this section. It is sufficient by § 34, IV, Corollary 6.

Remark (2). The theorem can be extended to arbitrary metric spaces as follows. We say that a function f is *locally constant at*

---

(1) The function g can be defined directly by letting

$$g(c) = F_{c_1} \cap F_{c_1 c_2} \cap F_{c_1 c_2 c_3} \cap \ldots, \quad \text{if} \quad c = \frac{c_1}{3} + \frac{c_2}{9} + \frac{c_3}{27} + \ldots.$$

Compare F. Hausdorff, *Mengenlehre*, p. 134.

(2) W. Hurewicz, *Ein Satz über stetige Abbildungen*, Fund. Math. 23 (1934), p. 54.

the point x_0, if there exists a neighbourhood E of x_0 such that the partial function $f|E$ is constant. Then the following theorem holds.

Let f be a continuous mapping of a metric space $\mathscr{X}$. If there exists an uncountable subset A of $\mathscr{X}$ such that the partial function $f|A$ is not constant at any point, then $\mathscr{X}$ contains a perfect set P such that $f|P$ is a homeomorphism.

It follows that:

Every continuous mapping of a metric separable space which assumes an uncountable set of values is a homeomorphism on a perfect set.

*§ 37. Borel sets in complete separable spaces(1)

I. Relations of Borel sets to the space $\mathscr{N}$.

THEOREM 1. *Every Borel set is a continuous image of the space* $\mathscr{N}$.

Proof. Every closed subset of a complete space is complete, and hence, by § 36, II, Theorem 1, a continuous image of $\mathscr{N}$. Therefore it is sufficient to prove (compare the Theorem of § 30, I) that if $f_1, f_2, \ldots$ is a sequence of continuous mappings of $\mathscr{N}$ onto subsets $f_1(\mathscr{N}), f_2(\mathscr{N}), \ldots$ of a space $\mathscr{X}$, then the sets

$$S = f_1(\mathscr{N}) \cup f_2(\mathscr{N}) \cup \ldots \quad \text{and} \quad P = f_1(\mathscr{N}) \cap f_2(\mathscr{N}) \cap \ldots$$

are continuous images of $\mathscr{N}$.

Let N_n be the set of irrational numbers of the interval $n-1$, n and let $f(x) = f_n(x-n+1)$ for $x \in N_n$, $n = 1, 2, \ldots$. Then f is a continuous function on $\mathscr{N}^* = N_1 \cup N_2 \cup \ldots$ and $f(\mathscr{N}^*) = S$. Clearly $\mathscr{N}^*$ is homeomorphic to $\mathscr{N}$.

To prove that P is a continuous image of $\mathscr{N}$, consider the space $\mathscr{N}^{\aleph_0}$ (of sequences $\mathfrak{z} = [\mathfrak{z}^1, \mathfrak{z}^2, \ldots]$, where $\mathfrak{z}^n \in \mathscr{N}$), and the subset $\mathfrak{Z}$ of $\mathscr{N}^{\aleph_0}$ which consists of the $\mathfrak{z}$ such that $f_1(\mathfrak{z}^1) = f_2(\mathfrak{z}^2) = \ldots$. If $\mathfrak{z} \in \mathfrak{Z}$, let $f^*(\mathfrak{z}) = f_1(\mathfrak{z}^1)$. By § 16, IV, 5, $\mathfrak{Z}$ is a closed subset of $\mathscr{N}^{\aleph_0}$, and therefore a continuous image of $\mathscr{N}$. By § 16, IV, Theorem 3, f^* is a continuous mapping of $\mathfrak{Z}$ onto P.

(1) Compare N. Lusin, *Ensembles analytiques*, Chapter II, and my paper *Sur une généralisation de la notion d'homéomorphie*, Fund. Math. 22 (1934), p. 206. See also § 39, Sections IV and V. For complete spaces without the assumption of separability, see A. H. Stone, *Non-separable sets*, Rozprawy Matematyczne 28 (1962).

THEOREM 2. *Every Borel set is a one-to-one continuous image of a complete separable 0-dimensional space* (1).

Proof. Every open set is a topologically complete space (§ 33, VI) and hence satisfies the assertion of the theorem (by § 36, III, Corollary). Therefore it is sufficient to prove (compare § 30, V, Theorem 3) that if $\mathscr{X}_1, \mathscr{X}_2, \ldots$ is a sequence of complete separable 0-dimensional spaces and $f_1, f_2, \ldots$ is a sequence of one-to-one continuous mappings of the spaces $\mathscr{X}_1, \mathscr{X}_2, \ldots$ onto subsets $f_1(\mathscr{X}_1)$, $f_2(\mathscr{X}_2), \ldots$ of a space $\mathscr{X}$, then:

(i) the set $P = f_1(\mathscr{X}_1) \cap f_2(\mathscr{X}_2) \cap \ldots$ satisfies the theorem;

(ii) the set $S = f_1(\mathscr{X}_1) \cup f_2(\mathscr{X}_2) \cup \ldots$ also satisfies the theorem under the additional hypothesis that the sets $f_1(\mathscr{X}_1), f_2(\mathscr{X}_2), \ldots$ are mutually disjoint.

To prove (i), consider the subset $\mathfrak{Z}$ of the countable product $\mathscr{X}_1 \times \mathscr{X}_2 \times \ldots$ consisting of the $\mathfrak{z} = [\mathfrak{z}^1, \mathfrak{z}^2, \ldots]$ such that $f_1(\mathfrak{z}^1) = f_2(\mathfrak{z}^2) = \ldots$ and let $f^*(\mathfrak{z}) = f_1(\mathfrak{z}^1)$, for $\mathfrak{z} \in \mathfrak{Z}$. The product $\mathscr{X}_1 \times \mathscr{X}_2 \times \ldots$ is complete (§ 33, III), separable (§ 16, V, Theorem 6) and 0-dimensional (§ 26, IV, Corollary 2b) and so is $\mathfrak{Z}$ as its closed subset (by § 16, IV, Theorem 5). Moreover, f^* is a one-to-one continuous mapping of $\mathfrak{Z}$ onto P (§ 16, IV, 3–4).

In order to prove (ii), assume that $\mathscr{X}_n$ is, according to § 36, II, Theorem 2, a closed subset of the space N_n of irrational numbers of the interval $(n-1), n$. The functions f_n, $n = 1, 2, \ldots$, define a one-to-one continuous mapping of the union $\mathscr{X}_1 \cup \mathscr{X}_2 \cup \ldots$ onto S. This union is closed in the space of positive irrational numbers and thus constitutes a topologically complete separable 0-dimensional space.

Theorems 1 and 2 imply (separately) the following important theorem:

THEOREM 3 (of Alexandrov-Hausdorff) (2). *Every uncountable Borel set contains the Cantor set $\mathscr{C}$ topologically. Its power is therefore* $\mathfrak{c}$.

This theorem can be derived from Theorem 1 and Corollary 1 of § 36, V, or from Theorem 2 and Theorem 1 of § 36, IV.

---

(1) If a Borel set consists of condensation points only, it is a one-to-one continuous image of $\mathscr{N}$. See W. Sierpiński, *Sur les images biunivoques de l'ensemble de tous les nombres irrationnels,* Mathematica 2 (1924), p. 18.

(2) P. Alexandrov, C. R. Paris 162 (1916), p. 323. F. Hausdorff, Math. Ann. 77 (1916), p. 430.

II. Characterization of the Borel class with aid of generalized homeomorphisms.

Theorem 2 of Section I can be made more precise as follows.

THEOREM 1. *For every set A of multiplicative (respectively ambiguous) class $(a+1)>0$ lying in the space $\mathscr{X}$ there exists a complete separable space $\mathscr{T}$ and a homeomorphism f of class 0, a which maps $\mathscr{T}$ onto $\mathscr{X}$ so that the set $f^{-1}(A)$ is a $\boldsymbol{G}_\delta$ (respectively an $\boldsymbol{F}_\sigma$ and $\boldsymbol{G}_\delta$).*

Moreover, *if $a+1>1$, $\mathscr{T}$ can be assumed to be of dimension* 0.

Proof. We shall first prove the theorem for ambiguous classes.

Suppose that the theorem is true for each ambiguous set of class $<\beta$ $(\beta>1)$. We shall prove it for any ambiguous set A of class β.

By § 30, IX, Theorem 1, A has the form

$$A=\bigcup_{n=1}^{\infty}(A_n\cap A_{n+1}\cap\ldots)=\bigcap_{n=1}^{\infty}(A_n\cup A_{n+1}\cup\ldots),$$

where A_n is ambiguous of class a_n and $0<a_n<\beta$.

If $\beta>2$, there exists, by hypothesis, a closed subset C_n of $\mathscr{N}$ and a homeomorphism f_n of class $0, a_n$ defined on C_n such that $f_n(C_n)=A_n$. If $\beta=2$, the same conclusion follows from the corollary of § 36, III (where the space $\mathscr{X}$ can be replaced by A_n, since A_n is a $\boldsymbol{G}_\delta$).

Since $\mathscr{X}-A_n$ is ambiguous of class a_n, there exists a closed subset D_n of the set of irrational numbers of the interval $(1,2)$ which is related to $\mathscr{X}-A_n$ just as C_n is to A_n. Thus if we let $\mathscr{X}_n=C_n\cup D_n$, then the function f_n defined on the entire set $\mathscr{X}_n$ is continuous, one-to-one and fulfills the formulas $f_n(C_n)=A_n$ and $f_n(D_n)=\mathscr{X}-A_n$. The function f_n^{-1} is of class a_n, since the partial functions $f_n^{-1}|A_n$ and $f_n^{-1}|(\mathscr{X}-A_n)$ are, by hypothesis, of class a_n and A_n is ambiguous of class a_n (compare § 31, IV, Theorem 2).

It should be noticed that $\mathscr{X}_n$ is topologically complete, separable and 0-dimensional and that the sets C_n and D_n are closed and open in this space. Let $E=\mathscr{X}_1\times\mathscr{X}_2\times\ldots$ and let $\mathscr{T}$ be the set of the $\mathfrak{z}\in E$ such that $f_1(\mathfrak{z}^1)=f_2(\mathfrak{z}^2)=\ldots$. We define f by the formula: $f(\mathfrak{z})=f_1(\mathfrak{z}^1)$ for $\mathfrak{z}\in\mathscr{T}$ (compare § 16, IV).

Since $\mathcal{T}$ is closed in E and E is topologically complete, separable, and 0-dimensional (§ 33, III, § 16, V, Theorem 6, § 25, IV, Corollary 2b), so is $\mathcal{T}$.

By § 16, IV, $f(\mathcal{T}) = \mathcal{X}$ and by § 31, VI, Theorem 3, f is a homeomorphism of class $0, \beta$, if β is a limit ordinal, and of class $0, \alpha$, if $\beta = \alpha+1$.

We shall prove that $f^{-1}(A)$ is an $\boldsymbol{F}_\sigma$ and $\boldsymbol{G}_\delta$. We have

$$f^{-1}(A) = \bigcup_{n=1}^{\infty} [f^{-1}(A_n) \cap f^{-1}(A_{n+1}) \cap \dots]$$

$$= \bigcap_{n=1}^{\infty} [f^{-1}(A_n) \cup f^{-1}(A_{n+1}) \cup \dots].$$

Since the function f is continuous, it is sufficient to prove that the set $f^{-1}(A_n)$ is closed and open in $\mathcal{T}$. Now, if $\mathfrak{z} \in \mathcal{T}$, then

$$\{f(\mathfrak{z}) \in A_n\} \equiv \{f_n(\mathfrak{z}^n) \in A_n\} \equiv \{\mathfrak{z}^n \in C_n\},$$

and since C_n is closed and open in $\mathcal{X}_n$, it follows that the set

$$f^{-1}(A_n) = \mathcal{T} \cap \underset{\mathfrak{z}}{E}\{\mathfrak{z}^n \in C_n\}$$

is closed and open in $\mathcal{T}$, for

$$\underset{\mathfrak{z}}{E}\{\mathfrak{z}^n \in C_n\} = \mathcal{X}_1 \times \dots \times \mathcal{X}_{n-1} \times C_n \times \mathcal{X}_{n+1} \times \dots .$$

Having proved the theorem for ambiguous sets, we now proceed to prove it in the case of multiplicative classes.

If A is of the multiplicative class $\alpha+1$ (>0), then we have $A = A_1 \cap A_2 \cap \dots$, where A is ambiguous of class $\alpha+1$. As it has just been proved, there exist a complete separable space $\mathcal{X}_n$, a $\boldsymbol{G}_\delta$-set (even a $\boldsymbol{G}_\delta$ and $\boldsymbol{F}_\sigma$-set) C_n and a homeomorphism f_n of class $0, \alpha$ such that

$$f_n(\mathcal{X}_n) = \mathcal{X} \quad \text{and} \quad f_n^{-1}(A_n) = C_n.$$

Let $\mathcal{T}$ and f be defined as above. Then

$$f(\mathcal{T}) = f_1(\mathcal{X}_1) \cap f_2(\mathcal{X}_2) \cap \dots = \mathcal{X},$$

$$f^{-1}(A) = f^{-1}(A_1) \cap f^{-1}(A_2) \dots,$$

$$f^{-1}(A_n) = \mathcal{T} \cap (\mathcal{X}_1 \times \dots \times \mathcal{X}_{n-1} \times C_n \times \mathcal{X}_{n+1} \times \dots),$$

which proves that $f^{-1}(A_n)$, and hence $f^{-1}(A)$, is a $\boldsymbol{G}_\delta$ in $\mathcal{T}$.

Remarks. (i) The class of f^{-1} cannot be *reduced*. This is because if f^{-1} is of class ξ, then the set $A = ff^{-1}(A)$ is ambiguous of class $\xi+1$ since $f^{-1}(A)$ is ambiguous of class 1 (§ 31, III, 1). Consequently, if A is of class $\alpha+1$ but not of class α, then $\xi \geqslant \alpha$.

(ii) Theorem 1 can be generalized so as to replace $\alpha+1$ by $\alpha+\beta$ and $\boldsymbol{G}_\delta$ (respectively $\boldsymbol{F}_\sigma$ and $\boldsymbol{G}_\delta$) by the multiplicative (respectively ambiguous) class β [1].

COROLLARY 1a. *Every set A of the multiplicative class $\alpha+1>0$ can be obtained from a suitably defined complete separable space by means of a homeomorphism of class $0, \alpha$.*

If $\alpha+1>1$, this space can be assumed to have dimension 0.

The reason for this is that the set $f^{-1}(A)$ of Theorem 1 is a $\boldsymbol{G}_\delta$, so it is topologically complete.

Corollary 1a and Corollary 3 of § 35, VII imply:

COROLLARY 1b. *A set A is of the multiplicative class $\alpha+1>0$ if and only if it can be obtained from a complete separable space by means of a homeomorphism of class $0, \alpha$.*

COROLLARY 1c. *Every uncountable Borel set A of multiplicative class $\alpha+1>1$ contains a countably infinite set D such that the set $A-D$ can be obtained from $\mathcal{N}$ by means of a homeomorphism of class $0, \alpha$.*

Proof. According to Corollary 1a, let $\mathcal{T}$ be a complete separable 0-dimensional space and f a homeomorphism of class $0, \alpha$ such that $f(\mathcal{T}) = A$. Let E be a countable subset of $\mathcal{T}$ such that $\mathcal{T}-E$ is homeomorphic to $\mathcal{N}$ (§ 36, IV, Theorem 1) and $D = f(E)$.

THEOREM 2. *For every pair of uncountable sets A and B of the multiplicative classes $\alpha+1>2$ and $\beta+1>2$, respectively, there exists a homeomorphism f of A onto B of class α, β* [2].

Proof. According to Corollary 1c let D and E be two countable infinite sets and f_1 and f_2 be two homeomorphisms of classes $0, \alpha$ and $0, \beta$, respectively, such that $f_1(\mathcal{N}) = A-D$ and $f_2(\mathcal{N}) = B-E$. We let $f(x) = f_2 f_1^{-1}(x)$, for $x \in A-D$ and let $f|D$ be a one-to-one mapping onto E.

---

(1) See my paper referred to above, Fund. Math. 22 (1934), p. 217.

(2) This relation between A and B, called by G. W. Mackey *Borel isomorphism,* has interesting applications in group theory. See of this author *Les ensembles boréliens et les extensions des groupes,* Journ. de Math. 36 (1957), p. 171.

The function f so defined is of class a on $A-D$ (§ 31, III, Theorem 2) and of class 1 (§ 31, X, Remark 9), and so of class a as well, on D. The sets D and $A-D$ are $G_{\delta\sigma}$-sets with respect to A. Hence they are of the additive class a and it follows that f is of class a on the entire set A (by § 31, IV, Theorem 1).

By symmetry, f^{-1} is of class β on B.

Remarks. (i) It follows that, in the domain of Borel subsets of complete separable spaces, *an equivalence in the sense of set theory (equal powers) can always be obtained by means of a generalized homeomorphism.*

(ii) In particular one proves [1] that: *For each pair of complete separable spaces* [2] *of the same power there exists a homeomorphism of class* 1, 1.

III. Resolution of ambiguous sets into alternate series.

THEOREM. *A necessary and sufficient condition for a set A to be ambiguous of class $a+1$ is that it should be resolvable into an alternate countably transfinite series of decreasing sets of the multiplicative class a:*

$$A = B_1 - B_2 \cup B_3 - B_4 \cup \ldots \cup B_\omega - B_{\omega+1} \cup \ldots . \tag{1}$$

Proof. The condition is sufficient by § 30, VI, Theorem 3. To prove that it is necessary, consider, as in Theorem 1 of Section II, a complete separable space $\mathcal{T}$ and a homeomorphism f of class $0, a$ which maps $\mathcal{T}$ onto the space $\mathcal{X}$ (containing A) so that $f^{-1}(A)$ is an F_σ and G_δ. By the theorem of § 34, VI, the set $f^{-1}(A)$ has the form

$$f^{-1}(A) = F_1 - F_2 \cup F_3 - F_4 \cup \ldots \cup F_\omega - F_{\omega+1} \cup \ldots,$$

where the sets F_ξ are closed and decreasing.

Since the function f is one-to-one, it follows that

$$A = ff^{-1}(A) = f(F_1) - f(F_2) \cup \ldots \cup f(F_\omega) - f(F_{\omega+1}) \cup \ldots$$

and the decreasing sets $B_\xi = f(F_\xi)$ are of the multiplicative class a because the function f^{-1} is of class a.

---

[1] See my paper in Fund. Math. 22 (1934), p. 212.

[2] Hence also for each pair of G_δ-sets. As for the $F_{\sigma\delta}$-sets, see the remark of § 36, III.

IV. Small Borel classes. The preceding theorem leads to a natural classification of ambiguous sets of a given class α (where α is not a limit ordinal). Namely, if series (1) is of type β, we say that A belongs to the *small class* $\boldsymbol{F}_\alpha^\beta$. Similarly, if the complement of A is resolvable into a series of type β, we say that A belongs to the *small class* $\boldsymbol{G}_\alpha^\beta$.

Thus it is seen that the classification of ambiguous sets of class α (for a fixed α) into "small classes" is quite analogous to the classification of Borel sets into the classes $\boldsymbol{F}_\alpha$ (or $\boldsymbol{G}_\alpha$).

Thus, for example, the sets of small classes $\boldsymbol{F}_1^1, \boldsymbol{F}_1^2, \boldsymbol{F}_1^3, \ldots$ and $\boldsymbol{G}_1^1, \boldsymbol{G}_1^2, \boldsymbol{G}_1^3, \ldots$ are of the form

$$F,\ F_1 - F_2,\ (F_1 - F_2) \cup F_3,\ \ldots,$$

$$\mathscr{X} - F,\ (\mathscr{X} - F_1) \cup F_2,\ (\mathscr{X} - F_1) \cup (F_2 - F_3),\ \ldots, \text{ respectively.}$$

We have shown that for every α there exists a Borel set (in the space of irrational numbers) that is of class α, but not of class $< \alpha$ (§ 30, XIV). By an analogous method one can prove the existence of sets that belong to a given small class, but not to a class of a lower index [1].

Remark. The definition of the classes $\boldsymbol{G}_\alpha^\beta$ can be expressed in a more natural way with the aid of the division of sets, where $X : Y = X \cup Y^c$. $\boldsymbol{G}_\alpha^\beta$ is then the class of sets that can be resolved into an "alternate intersection" of type β of increasing sets of the additive class α:

$$A = (G_1 : G_2) \cap (G_3 : G_4) \cap \ldots \cap (G_\omega : G_{\omega+1}) \cap \ldots,$$

where

$$G_1 \subset G_2 \subset \ldots \subset G_\omega \subset \ldots .$$

This is because the formula

$$A^c = (B_1 - B_2) \cup \ldots \cup (B_\omega - B_{\omega+1}) \cup \ldots$$

implies

$$A = (B_1 - B_2)^c \cap \ldots \cap (B_\omega - B_{\omega+1})^c \cap \ldots$$
$$= (B_1^c : B_2^c) \cap \ldots \cap (B_\omega^c : B_{\omega+1}^c) \cap \ldots .$$

---

[1] M. Lavrentiev, *Sur les sous-classes de la classification de M. Baire*, C. R. Paris 1925. Compare also N. Lusin, *op. cit.*, p. 123 and W. Sierpiński, *Sur l'existence de diverses classes d'ensembles*, Fund. Math. 14 (1929), p. 82.

*§ 38. Projective sets

The space is assumed to be complete and separable.

I. Definitions. *The projective sets of class* 0 are understood to be the Borel sets. *The projective sets of class* $2n+1$ are continuous images of the projective sets of class $2n$ (lying in the same space); *the projective sets of class* $2n$ are the complements of the projective sets of class $2n-1$ (1).

In particular, the projective sets of class 1, i.e. the continuous images of Borel sets, are called *analytic sets*, or ***A*** *sets* (2); their complements, i.e. the projective sets of class 2 are called *analytic complements*, or ***CA*** *sets* (3).

II. Relations between projective classes. Given a family $\boldsymbol{X}$ of sets, let $\boldsymbol{CX}$ denote the family of the complements of sets of $\boldsymbol{X}$ and let $\boldsymbol{PX}$ denote the family of continuous images of sets belonging to $\boldsymbol{X}$. The following formulas are obvious:

1. $\boldsymbol{CCX} = \boldsymbol{X}$.

2. $\boldsymbol{X} \subset \boldsymbol{PX} = \boldsymbol{PPX}$.

3. $\boldsymbol{X} \subset \boldsymbol{Y}$ *implies* $\boldsymbol{CX} \subset \boldsymbol{CY}$ *and* $\boldsymbol{PX} \subset \boldsymbol{PY}$.

Moreover, if $\boldsymbol{L}_n$ denotes the nth projective class, then

4. $\boldsymbol{L}_{2n+1} = \boldsymbol{PL}_{2n}$, $\boldsymbol{PL}_{2n+1} = \boldsymbol{L}_{2n+1}$.

5. $\boldsymbol{L}_{2n} = \boldsymbol{CL}_{2n-1}$, $\boldsymbol{CL}_{2n} = \boldsymbol{L}_{2n-1}$.

6. $\boldsymbol{CL}_0 = \boldsymbol{L}_0$ *and* $\boldsymbol{PL}_0 = \boldsymbol{L}_1 = \boldsymbol{A}$.

---

(1) By means of countable operations $\bigcup_{n=1}^{\infty}$ and $\bigcap_{n=1}^{\infty}$ this classification can easily be extended to transfinite ordinals ($< \Omega$) thus copying the classification of Borel sets.

(2) Or "Souslin sets" (F. Hausdorff, *Mengenlehre*, p. 177).

(3) The idea of the analytic set was introduced by Souslin and Lusin. See *Sur une définition des ensembles mesurables B sans nombres transfinis*, C. R. Paris 164 (1917), p. 88. The theory of analytic sets has been developed mainly by Lusin and Sierpiński. The notion of the projective set is due to Lusin; see *Sur les ensembles projectifs de M. Henri Lebesgue*, C. R. Paris 180 (1925), p. 1570. Compare L. Kantorovitch and E. Livenson, *Memoir on the analytical operations and projective sets*, Fund. Math. 18 (1932), p. 214 and 20 (1933), p. 54.

We shall prove that

7. $\boldsymbol{L}_{2n} \subset \boldsymbol{L}_{2n+k}$ *and* $\boldsymbol{L}_{2n+1} \subset \boldsymbol{L}_{2n+2+k}$, *if* $n \geqslant 0$ *and* $k = 1, 2, \ldots$

Clearly it is sufficient to prove that

$$\text{(i) } \boldsymbol{L}_{2n} \subset \boldsymbol{L}_{2n+1}, \qquad \text{(ii) } \boldsymbol{L}_n \subset \boldsymbol{L}_{n+2}, \qquad \text{(iii) } \boldsymbol{L}_{2n+1} \subset \boldsymbol{L}_{2n+4}.$$

Inclusion (i) is a direct consequence of 2 and 4.

Inclusion (ii) holds for $n = 0$ by virtue of 2, 3, and 6. For $n > 0$ we have either $\boldsymbol{L}_n = \boldsymbol{PCL}_{n-2}$, or $\boldsymbol{L}_n = \boldsymbol{CPL}_{n-2}$ (if we let $\boldsymbol{L}_{-1} = \boldsymbol{L}_0$). Assume the induction hypothesis $\boldsymbol{L}_{n-2} \subset \boldsymbol{L}_n$. Then, by 3, $\boldsymbol{PCL}_{n-2} \subset \boldsymbol{PCL}_n$ and $\boldsymbol{CPL}_{n-2} \subset \boldsymbol{CPL}_n$ and (ii) follows.

Finally inclusion (i) yields $\boldsymbol{CL}_{2n+1} = \boldsymbol{L}_{2n+2} \subset \boldsymbol{L}_{2n+3}$, and therefore by 3 and 1: $\boldsymbol{L}_{2n+1} = \boldsymbol{CCL}_{2n+1} \subset \boldsymbol{CL}_{2n+3} = \boldsymbol{L}_{2n+4}$.

III. Properties of projective sets. 1. *If P_k is a projective set of class n situated in a complete separable space $\mathscr{X}_k$, $k = 1, 2, \ldots$, then the (finite or infinite) product $P_1 \times P_2 \times \ldots$ is a projective set of class n in the space $\mathscr{X}_1 \times \mathscr{X}_2 \times \ldots$.*

2. *If P and Q are two projective sets of class n situated in complete separable spaces $\mathscr{X}$ and $\mathscr{Y}$, respectively, and f is a continuous function defined on P, then the set $f^{-1}(Q)$ is of class n.*

Moreover, *if n is odd, then $f(P)$ is of class n; if n is even, then $f(P)$ is of class $n+1$.*

3 [1]. *The property of being a projective set of class n is countably additive and multiplicative; in other words, if $P_1, P_2, \ldots$ are projective sets of class n, then so are the sets $P_1 \cup P_2 \cup \ldots$ and $P_1 \cap P_2 \cap \ldots$ as well.*

We shall prove these properties by finite induction.

Proof. For $n = 0$ (the case of Borel sets), properties 1, 3, and the first part of 2 are satisfied (§ 30, III). Also the second part of 2 is satisfied. Since, assuming that $\mathscr{Y}$ is uncountable (otherwise $f(P)$ would be countable and hence of class 1), P is a continuous image of a Borel subset N of $\mathscr{Y}$; $P = g(N)$ (compare § 36, V, Corollary 2 and § 37, I, 1). It follows that the set $f(P) = fg(N)$ is of class 1.

Assume now that $n > 0$ and that the three properties (which we now denote by $1_{n-1}, 2_{n-1}, 3_{n-1}$) are satisfied for $n-1$.

---

[1] Compare W. Sierpiński, *Sur les familles inductives et projectives d'ensembles*, Fund. Math. 13 (1929), p. 228–239.

(a) *Case of n odd.* Let $P_k = p_k(R_k)$, where R_k is a subset of $\mathscr{X}_k$ of class $(n-1)$ and p_k is a continuous function defined on R_k.

1. The function which carries the point $(x_1, x_2, \ldots)$ to $[p_1(x_1), p_2(x_2), \ldots]$ is a continuous mapping of $R_1 \times R_2 \times \ldots$ onto $P_1 \times P_2 \times \ldots$ (see § 16, V, 8).

Since the first set is of (even) class $(n-1)$ by 1_{n-1}, the second one is of class n.

2. As before, let $P = p(R)$, where $R \subset \mathscr{X}$. Let T be a subset of $\mathscr{Y}$ of class $n-1$ and q a continuous function defined on T such that $Q = q(T)$.

We have the following obvious equivalences (where $x \in \mathscr{X}$, $x^* \in R$, and $y \in T$):

$$\{x \in f^{-1}(Q)\} \equiv \{f(x) \in Q\} \equiv \bigvee_{x^*} \{[x = p(x^*)]\,[fp(x^*) \in Q]\}$$
$$\equiv \bigvee_{x^* y} \{[x = p(x^*)]\,[fp(x^*) = q(y)]\},$$

and hence

$$f^{-1}(Q) = \mathop{E}_{x} \bigvee_{x^* y} \{[x = p(x^*)]\,[fp(x^*) = q(y)]\}.$$

Thus $f^{-1}(Q)$ is a continuous image by a projection (§ 2, V, Theorem) of the set M of points (x, x^*, y) which satisfy the condition in the brackets $\{\ \}$. Since the functions p, $f \circ p$, and q are continuous and since p is defined on R and q is defined on T, the set M is closed in the product $\mathscr{X} \times R \times T$. By 1_{n-1}, $\mathscr{X} \times R \times T$ is of class $(n-1)$, so M is the intersection of a closed set (hence a set of class $(n-1)$) and a set of class $n-1$. It follows from 3_{n-1} that M is of class $(n-1)$ in $\mathscr{X} \times \mathscr{X} \times \mathscr{Y}$ and $f^{-1}(Q)$, as a continuous image of M, is of class n by the last part of 2_{n-1}.

Moreover, the formula $f(P) = fp(R)$ implies, by 2_{n-1}, that $f(P)$ is of class n in $\mathscr{Y}$, as a continuous image of R.

3. Now let $P_k = p_k(R_k)$ and $R_k \subset \mathscr{X}$. We have

$$\{x \in \bigcup_{k=1}^{\infty} P_k\} \equiv x \in \bigcup_{k=1}^{\infty} p_k(R_k)\} \equiv \bigvee_{k} \{x \in p_k(R_k)\}$$
$$\equiv \bigvee_{k} \bigvee_{x^*} \{x = p_k(x^*)\} \equiv \bigvee_{x^*} \bigvee_{k} \{x = p_k(x^*)\}.$$

Hence

$$\bigcup_{k=1}^{\infty} P_k = \mathop{E}_{x} \bigvee_{x^*} \bigvee_{k} \{x = p_k(x^*)\}.$$

This proves that $\bigcup_{k=1}^{\infty} P_k$ is a projection of

$$M = \underset{xx^*}{E} \bigvee_k \{x = p_k(x^*)\} = \bigcup_{k=1}^{\infty} \underset{xx^*}{E} \{x = p_k(x^*)\}.$$

Since the function p_k, defined on R_k, is continuous, the set $M_k = \underset{xx^*}{E}\{x = p_k(x^*)\}$ is closed in the product $\mathscr{X} \times R_k$, which is of class $(n-1)$ by 1_{n-1}. It follows that M_k, as the intersection of a closed set and a set of class $(n-1)$, is itself of class $(n-1)$, by 3_{n-1}. By the same proposition, the set $M = \bigcup_{k=1}^{\infty} M_k$ is of class $n-1$ and, by 2_{n-1}, $\bigcup_{k=1}^{\infty} P_k$ is of class n as a continuous image (a projection) of M.

On the other hand, the condition $x \in \bigcap_{k=1}^{\infty} P_k$ means that there exists a sequence of points $x_1, x_2, \ldots$ such that $x = p_k(x_k)$. Denoting as usual, by $\mathfrak{z} = [\mathfrak{z}^1, \mathfrak{z}^2, \ldots]$ a variable point of $\mathscr{X}^{\aleph_0}$, we have

$$\{x \in \bigcap_{k=1}^{\infty} P_k\} \equiv \bigvee_{\mathfrak{z}} \bigwedge_k \{x = p_k(\mathfrak{z}^k)\}.$$

Consequently, $\bigcap_{k=1}^{\infty} P_k$ is a projection of the set

$$\underset{x\mathfrak{z}}{E} \bigwedge_k \{x = p_k(\mathfrak{z}^k)\} = \bigcap_{k=1}^{\infty} \underset{x\mathfrak{z}}{E} \{x = p_k(\mathfrak{z}^k)\}.$$

It is sufficient therefore (by virtue of 3_{n-1}) to show that the set $\underset{x\mathfrak{z}}{E}\{x = p_k(\mathfrak{z}^k)\}$ is of class $(n-1)$ (in the space $\mathscr{X} \times \mathscr{X}^{\aleph_0}$). And this follows, as before, from the fact that the function p_k (considered as a function of $\mathfrak{z}$) is continuous (since $\mathfrak{z}^k$ is a continuous function of $\mathfrak{z}$; compare § 16, II) and is defined on the set

$$\underset{\mathfrak{z}}{E}\{\mathfrak{z}^k \in R_k\} \doteq \mathscr{X} \times \ldots \times \mathscr{X} \times R_k \times \mathscr{X} \times \mathscr{X} \times \ldots$$

which is of class $n-1$ by 1_{n-1}.

(b) *Case of an even* $n > 0$. Since P is of class n, its complement, P^c, is of odd class $n-1$.

Proposition 3 follows directly from 3_{n-1} and the de Morgan formulas:

$$(\bigcup_k P_k)^c = \bigcap_k P_k^c \quad \text{and} \quad (\bigcap_k P_k)^c = \bigcup_k P_k^c.$$

By virtue of the identity $P\times\mathscr{Y} = \mathscr{X}\times\mathscr{Y}-(P^c\times\mathscr{Y})$ (compare § 2, II, (3)) and of 1_{n-1}, the set $P\times\mathscr{Y}$ is of class n and so is the set $\mathscr{X}_1\times \ldots \times\mathscr{X}_{k-1}\times P_k\times\mathscr{X}_{k+1}\times \ldots$. Proposition 1_n follows from 3_n and the identity (compare § 3, VIII, (1));

$$P_1\times P_2\times \ldots = (P_1\times\mathscr{X}_2\times\mathscr{X}_3\times \ldots)\cap(\mathscr{X}_1\times P_2\times\mathscr{X}_3\times \ldots)\cap \ldots .$$

Finally we shall prove 2. By the extension theorem (§ 35, I, 1) there exists a G_δ-set P^* containing P and a continuous extension f_1 of f to P^*; that is $f = f_1|P$. Then (compare § 3, III, (14)):

$$f^{-1}(Q) = P\cap f_1^{-1}(Q) \quad \text{and} \quad f_1^{-1}(Q) = f_1^{-1}(\mathscr{Y}-Q^c) = P^*-f_1^{-1}(Q^c).$$

By 2_{n-1}, $f_1^{-1}(Q^c)$ is of class $n-1$; therefore by 3_n, the set $f_1^{-1}(Q) = P^*\cap[\mathscr{X}-f_1^{-1}(Q^c)]$ is of class n. By taking the intersection of this set with P and applying 3_n, we infer that $f^{-1}(Q)$ is a projective set of class n.

Moreover, if N denotes (as in the case $n = 0$) a Borel subset of $\mathscr{Y}$ which can be mapped continuously onto $\mathscr{X}$: $\mathscr{X} = g(N)$, then the set $g^{-1}(P)$ is of class n in $\mathscr{Y}$ and hence the set $f(P) = fg[g^{-1}(P)]$ is of class $n+1$.

Thus the proof of propositions 1–3 is completed.

4. *If f is a B measurable function defined on a set P of class n, then the set $I = \underset{xy}{E}\{y = f(x)\}$ is of class n.*

Proof. I is borelian with respect to the product $P\times\mathscr{Y}$ (§ 31, VII, Theorem 1). As the intersection of a Borel set with the set $P\times\mathscr{Y}$, which is of class n (by 1), I is therefore of class n (by 3).

5. *Proposition 2 remains valid, if we assume that f is an arbitrary B measurable function.*

Proof. By § 35, VI, there exists a B measurable function f_1 defined on the entire space $\mathscr{X}$ such that $f = f_1|P$. It follows that (§ 3, III, (14)) $f^{-1}(Q) = P\cap f_1^{-1}(Q)$. Thus it is sufficient to prove that $f_1^{-1}(Q)$ is of class n. Now this is the case, if n is odd, for $f_1^{-1}(Q)$ is the projection of $J\cap(\mathscr{X}\times Q)$, where $J = \underset{xy}{E}\{y = f_1(x)\}$, on the $\mathscr{X}$ axis. If n is even, the same conclusion follows from the identity $f_1^{-1}(\mathscr{Y}-Q) = \mathscr{X}-f_1^{-1}(Q)$.

Moreover, $f(P)$ is the projection of the set $\underset{xy}{E}\{y = f(x)\}$ on the $\mathscr{Y}$ axis.

Remark. Thus we see that an equivalent definition of the projective class is the following one: A subset P of $\mathscr{X}$ is of class $2n+1$, if there exists (in $\mathscr{X}$ or in some other complete separable space) a set R of class $2n$ and a B measurable function f defined on R such that $P = f(R)$.

The following proposition concerns the relativization of projective sets.

6. *Let E be a G_δ-subset of the space $\mathscr{X}$. A set $A \subset E$ is projective of class n relative to E if and only if it is of class n (relative to $\mathscr{X}$).*

Proof by induction. If $n = 0$, the theorem is obvious: A is a Borel set in E (which is a Borel set) if and only if A is a Borel set.

Assume that $n > 0$ and that the theorem is true for $(n-1)$.

Let n be odd. If A is of class n in E, then A is a continuous image of a subset of E of class $n-1$. By 2, A is therefore of class n in $\mathscr{X}$. Similarly, if A is of class n, then A is of class n in E.

Let n be even. If A is of class n in E, then $A = E - B$, where B is of class $(n-1)$ in E, and hence in $\mathscr{X}$. Since $\mathscr{X} - B$ is of class n, it follows that the set $A = E \cap (\mathscr{X} - B)$ is of class n.

Conversely, if A is of class n, then $A = \mathscr{X} - B$, where B is of class $(n-1)$. Therefore $E \cap B$ is of class $(n-1)$ and, consequently, $E \cap B$ is of class $(n-1)$ in E. Since $A \subset E$, it follows that $A = E - B = E - (E \cap B)$, which proves that A is of class n in E.

IV. Projections.

THEOREM. *The projective sets of the odd class n situated in a space $\mathscr{X}$ coincide with the projections of sets of class $(n-1)$ situated in the product $\mathscr{X} \times \mathscr{X}$* [1].

Proof. On one hand, projections of sets of class $(n-1)$ are sets of class n (by III, 2). On the other hand, every set P of class n is a continuous image of a set R of class $n-1$, i.e. $f(R) = P$, where $R \subset \mathscr{X}$. Thus P is a projection of the set $\underset{xx^*}{E}\{x^* = f(x)\}$ which, by III, 4, is of class $n-1$ (in $\mathscr{X} \times \mathscr{X}$).

Remark 1. The theorem remains true *after replacing $\mathscr{X} \times \mathscr{X}$ by $\mathscr{X} \times \mathscr{N}$.*

---

[1] This proposition motivates the term "projective set".

Proof. By Theorem 1 of § 36, II, $\mathscr{X}$ is a continuous image of $\mathscr{N}$, i.e. $\mathscr{X}=f(\mathscr{N})$. If P is a set of (even or odd) class n, then so is the set $P_1=f^{-1}(P)$ (by III, 2). Consequently, P is a continuous image of a subset of $\mathscr{N}$ of class n; and this subset, if n is odd, is a continuous image of a subset of $\mathscr{N}$ of class $(n-1)$. It follows that P is the projection on the $\mathscr{X}$ axis of a subset of $\mathscr{X}\times\mathscr{N}$ of class $(n-1)$.

Remark 2. In particular, *the analytic sets coincide with the projections* (*on* the $\mathscr{X}$ axis) *of closed subsets of* $\mathscr{X}\times\mathscr{N}$ (1).

Proof. By Theorem 1 of § 37, I, every Borel set is a continuous image of $\mathscr{N}$; so, also, is every analytic set. Consequently, if A is an analytic set, there exists a continuous function f defined on $\mathscr{N}$ such that A is the projection on the $\mathscr{X}$ axis of the closed subset $\underset{x\mathfrak{z}}{E}\{x=f(\mathfrak{z})\}$ of the space $\mathscr{X}\times\mathscr{N}$.

Conversely, a projection of a closed set is an analytic set (by Section III, 2).

V. Universal functions. According to the definition of § 30, XIII, a function L_n which assigns, to each irrational number $\mathfrak{z}$, a projective set $L_n(\mathfrak{z})$ of class n (in the given space $\mathscr{X}$) is said to be *universal* with respect to the class $\boldsymbol{L}_n$, if $\boldsymbol{L}_n$ coincides with the set of values of the function L_n.

THEOREM. *For every* $n>0$ *there exists a universal function* L_n *such that the set* $\underset{x\mathfrak{z}}{E}\{x\in L_n(\mathfrak{z})\}$ *is of class* n (2).

Proof. According to the remark of Section XIII of § 30, let F be a universal function with respect to the family of closed subsets of the space $\mathscr{X}\times\mathscr{N}$ such that

$$\underset{x\mathfrak{z}\mathfrak{z}^*}{E}\{(x\mathfrak{z}^*)\in F(\mathfrak{z})\} \text{ is closed in the space } \mathscr{X}\times\mathscr{N}\times\mathscr{N}.$$

Now every analytic subset of $\mathscr{X}$ is a projection of a closed subset of $\mathscr{X}\times\mathscr{N}$. Thus if we let $L_1(\mathfrak{z})$ be the projection of $F(\mathfrak{z})$ on $\mathscr{X}$, we obtain a universal function with respect to $\boldsymbol{L}_1$ $(=\boldsymbol{A})$. That is

$$\{x\in L_1(\mathfrak{z})\}\equiv\bigvee_{\mathfrak{z}^*}\{(x\mathfrak{z}^*)\in F(\mathfrak{z})\}.$$

(1) Compare the paper of Szpilrajn-Marczewski and myself, *Sur les cribles fermés et leurs applications*, Fund. Math. 18 (1931), p. 160.

(2) Compare N. Lusin, *Ensembles analytiques*, p. 290.

The set

$$\underset{x\mathfrak{z}}{E}\{x \in L_1(\mathfrak{z})\} \equiv \underset{x\mathfrak{z}}{E} \underset{\mathfrak{z}^*}{\bigvee} \{(x\mathfrak{z}^*) \in F(\mathfrak{z})\}$$

is analytic as a projection of the set $\underset{x\mathfrak{z}\mathfrak{z}^*}{E}\{(x\mathfrak{z}^*) \in F(\mathfrak{z})\}$.

Proceed by induction. If n is even, let $L_n(\mathfrak{z}) = \mathcal{X} - L_{n-1}(\mathfrak{z})$. The function L_n is universal with respect to $\boldsymbol{L}_n$ and the set

$$\underset{x\mathfrak{z}}{E}\{x \in L_n(\mathfrak{z})\} = \underset{x\mathfrak{z}}{E}\{x \notin L_{n-1}(\mathfrak{z})\} = \mathcal{X}\times\mathcal{N} - \underset{x\mathfrak{z}}{E}\{x \in L_{n-1}(\mathfrak{z})\}$$

is of class n.

If $n > 1$ is odd, let L^*_{n-1} be a universal function with respect to the family of subsets of $\mathcal{X}\times\mathcal{N}$ of class $n-1$ such that the set $\underset{x\mathfrak{z}\mathfrak{z}^*}{E}\{(x\mathfrak{z}^*) \in L^*_{n-1}(\mathfrak{z})\}$ is of class $n-1$ (in $\mathcal{X}\times\mathcal{N}\times\mathcal{N}$). As in the case $n = 1$ we prove that the function L_n defined by the equivalence

$$\{x \in L_n(\mathfrak{z})\} \equiv \underset{\mathfrak{z}^*}{\bigvee} \{(x\mathfrak{z}^*) \in L^*_{n-1}(\mathfrak{z})\}$$

is the required one.

VI. Existence theorem [1]**.**

THEOREM. *For each $n > 0$ the space $\mathcal{N}$ of irrational numbers of the interval $\mathcal{I}$ contains a projective set of class n which is not of class $< n$.*

$\mathcal{N}$ also contains non-projective sets.

Proof. Consider the universal function L_n of Section V (where $\mathcal{X} = \mathcal{N}$). If $n \geqslant 1$ is odd, let

$$E_n = \underset{\mathfrak{z}}{E}\{\mathfrak{z} \in L_n(\mathfrak{z})\} \quad \text{and} \quad E_{n+1} = \mathcal{N} - E_n = \underset{\mathfrak{z}}{E}\{\mathfrak{z} \notin L_n(\mathfrak{z})\}.$$

As E_n is a projection of the set $\underset{x\mathfrak{z}}{E}\{x \in L_n(\mathfrak{z})\} \cap \underset{x\mathfrak{z}}{E}(x = \mathfrak{z})$, which is of class n, so E_n is of class n. Hence E_{n+1} is of class $(n+1)$.

On the other hand, by the diagonal theorem (§ 30, XIV), E_{n+1} is not of class n, and hence is not of a class $< n$ (since n is odd). It follows that E_n is not of class $< n$, for, otherwise, E_{n+1} would be of class $< n+1$.

Now think of each E_n as placed in the interval $(n-1, n)$. The set $E_\infty = E_1 \cup E_2 \cup \ldots$ is not projective.

Remarks. (i). The procedure used in Sections V and VI permits the definition of the sets E_n and E_∞ *effectively* (compare § 23, VIII).

---

[1] *Ibidem.*

The existence of non-projective sets can be established non-effectively in a simpler way. Since the family of Borel set has power of continuum and the family of all continuous images of a set is also of this power (§ 24, VI), it follows that, for each n, *the class* $\boldsymbol{L}_n$ *has power* $\mathfrak{c}$, and so does the union $\boldsymbol{L}_1 \cup \boldsymbol{L}_2 \cup \ldots$. But the family of all subsets of a complete separable (uncountable) space is of power $2^{\mathfrak{c}} > \mathfrak{c}$; so the existence of non-projective sets follows.

(ii) One can prove [1] that, in the space $\mathcal{I}^{\mathcal{I}}$ (of continuous functions) there exists a *linear* set of class $2n$ that is not of a lower class (a set is linear if it contains, together with f and g, every function of the form $\lambda f + \mu g$).

(iii) It arises the problem of the *Lebesgue measurability* of the set E_3. Note that this set is determined explicitly [2].

VII. Invariance.

THEOREM. *If P is a projective set of class n, then every set homeomorphic to P* (situated in the same space or in some other complete separable space) *is projective of class n.*

Proof. The theorem is obvious, if n is odd, since every continuous image of P is then of class n. If $n = 0$ (the case of Borel sets) the theorem holds by virtue of Corollary 1 of § 35, IV. If $n > 0$ is even, the property $\boldsymbol{L}_{n-1}$ fulfils the hypothesis of the theorem of § 35, IV; that is, it is an invariant of homeomorphisms and, moreover: (i) every $\boldsymbol{G}_\delta$-set with respect to a set of class $(n-1)$ is of class $(n-1)$; (ii) the union of a set of class $(n-1)$ with an $\boldsymbol{F}_\sigma$-set is of class $(n-1)$. It follows that the property of being the complement of a set of class $(n-1)$, i.e. of being of class n, is an intrinsic invariant [3].

---

[1] See a paper of S. Banach and myself *Sur la structure des ensembles linéaires*, Studia Math. 4 (1933), p. 95. For further results see V. Klee, *On the borelian and projective types of linear subspaces*, Math. Scand. 6 (1958), p. 189.

[2] See my paper on this subject *Sur le problème de la mesurabilité des ensembles définissables*, Congr. Int. Math. Zürich 1932, vol. II, p. 117. Compare also § 39, I, (ii).

[3] For the invariance of the $\boldsymbol{CA}$ class, see P. Alexandrov, *Sur les ensembles complémentaires aux ensembles* (A), Fund. Math. 5 (1924), pp. 160–165.

For some extension of the theorem to mappings more general than homeomorphisms, see the papers referred to in footnote to Corollary 1 of § 35, IV.

VIII. Projective propositional functions(1). A propositional *function* $\varphi(x)$ *is said to be of class* $\boldsymbol{L}_n$, if the set $\underset{x}{E}\,\varphi(x)$ is of class $\boldsymbol{L}_n$.

We shall establish the following rules of "projective calculus":

THEOREM 1. *If* $\varphi(x)$ *is projective of class* $\boldsymbol{L}_n$, *its negation* $\neg\varphi(x)$ *is of class* $\boldsymbol{CL}_n$.

For

$$\underset{x}{E}[\neg\,\varphi(x)] = [\underset{x}{E}\,\varphi(x)]^c.$$

THEOREM 2. *If* $\varphi(x)$ *is of class* $\boldsymbol{L}_n$ *in the space* $\mathscr{X}$, $\varphi(x)$ *is of class* $\boldsymbol{L}_n$ *in the space* $\mathscr{X}\times\mathscr{Y}$.

The reason for this is (Section III, 1) that $\underset{xy}{E}\,\varphi(x) = [\underset{x}{E}\,\varphi(x)]\times\mathscr{Y}$.

THEOREM 3. *If* $\varphi_1(x), \varphi_2(x), \ldots$ *is a (finite or infinite) sequence of functions of class* $\boldsymbol{L}_n$ *(n is fixed)* (2), *then the functions*

$$\bigvee_k \varphi_k(x) \text{ and } \bigwedge_k \varphi_k(x) \text{ are also of class } \boldsymbol{L}_n.$$

This is a consequence of Section III, 3.

THEOREM 4. *If* $\psi(x, y)$ *is of class* $\boldsymbol{L}_n$ *(in* $\mathscr{X}\times\mathscr{Y}$*), then the propositional function* $\bigvee_y \psi(x, y)$ *is of class* $\boldsymbol{PL}_n$ *and* $\bigwedge_y \psi(x, y)$ *is of class* $\boldsymbol{CPCL}_n$ *(in* $\mathscr{X}$*).*

Proof. The set $\underset{x}{E}\bigvee_y \psi(x, y)$ is a projection (compare § 2, V, Theorem) of $\underset{xy}{E}\,\psi(x, y)$, so it is of class $\boldsymbol{PL}_n$. Moreover

$$\bigwedge_y \psi(x, y) \equiv \neg\,[\bigvee_y \neg\,\psi(x, y)].$$

THEOREM 5. *If* $\psi(x, y)$ *is of class* $\boldsymbol{L}_n$, *then the functions* $\varphi(x) = \psi(x, y_0)$, *for a fixed* y_0, *and* $\chi(x) = \psi(x, x)$ *are of class* $\boldsymbol{L}_n$.

Proof. The functions $\psi(x, y) \wedge (y = y_0)$ and $\psi(x, y) \wedge (x = y)$ are of class $\boldsymbol{L}_n$, so the set $\underset{xy}{E}\big(\psi(x, y) \wedge (y = y_0)\big)$ is of class $\boldsymbol{L}_n$ with respect to $\underset{xy}{E}(y = y_0)$ and the set $\underset{xy}{E}\big(\psi(x, y) \wedge (x = y)\big)$ is of class $\boldsymbol{L}_n$ with respect to the diagonal $\underset{xy}{E}(x = y)$ (compare III, 6).

(1) Compare § 1, § 2, and § 30, XI. See papers on this subject of A. Tarski and myself in Fund. Math. 17 (1931), pp. 241–272.

(2) This hypothesis can be dropped by extending the classes $\boldsymbol{L}_n$ to infinite indices (compare Section I, footnote).

Consequently (§ 15, IV and VI), the sets $\underset{x}{E}\,\psi(x, y_0)$ and $\underset{x}{E}\,\psi(x, x)$ are of class L_n (in $\mathscr{X}$). This completes the proof.

Remark 1. Theorems 1–4 show that *the logical operations*: $\neg$, $\vee$, $\wedge$, $\bigvee_x$, $\bigwedge_x$, *performed on projective propositional functions always lead to projective propositional functions.* Together with the rules of § 30, XI, they permit, at the same time, the *evaluation* of the Borel or projective class of a propositional function and therefore the class of a set that it defines.

This explains the fact that, by using the logical notation, one is led in a natural way to the idea of projective set.

Remark 2. Given a propositional function $\psi(x, y)$ of two variables, in order to evaluate the class of $\bigvee_y \psi(x, y)$ we can use either Theorem 4, by which the set $\underset{x}{E} \bigvee_y \psi(x, y)$ is a *projection* of $\underset{xy}{E}\,\psi(x, y)$, or formula § 2, V (1), by which $\underset{x}{E} \bigvee_y \psi(x, y)$ is the *union* $\bigcup_y \underset{x}{E}\,\psi(x, y)$. The second method (where y is regarded as an *index*) is more advantageous in the case when y runs over a countable set, as well as in the case when the set $\underset{x}{E}\,\psi(x, y)$ is open for each y (compare § 15, II, Theorem 3 and § 20, V, Corollary 7b).

Analogous remarks concern the operator $\wedge$. Thus, for instance, in § 34, VIII, (2), h is regarded as an index and not as a point of a space.

Application. Rectilinear accessibility. A point x is said to be rectilinearly accessible with respect to a set $M \subset \mathscr{X}$ (we write $x \in M_a$), if there exists a rectilinear segment xy in the space $\mathscr{X}$ which has at most the point x in common with M (and which does not reduce to this point). The condition that the point z belongs to the segment xy can be expressed by the formula $|x-z|+|z-y| = |x-y|$ (where $x, y, z \in \mathscr{X}$). It follows that

$$\{x \in M_a\} \equiv (x \in M) \wedge \bigvee_y \{(y \neq x) \wedge$$
$$\wedge \bigwedge_z [(|x-z|+|z-y| = |x-y|) \wedge (z \neq x) \Rightarrow \neg (z \in M)]\}$$
$$\equiv (x \in M) \wedge \bigvee_y \{(y \neq x) \wedge$$
$$\wedge \bigwedge_z [(|x-z|+|z-y| \neq |x-y|) \vee (z = x) \vee \neg (z \in M)]\}.$$

Thus the propositional function $\varphi(x) \equiv \{x \in M_a\}$ is the result of logical operations performed on the functions

$$\beta(x) \equiv \{x \in M\}, \quad \gamma(x, y) \equiv \{x = y\},$$
$$\delta(x, y, z) \equiv \{|x-z| + |z-y| = |x-y|\}.$$

The second and third functions are obviously of class $\boldsymbol{F}_0$. The class of the first function depends on the hypotheses on M. If M is an $\boldsymbol{F}_\sigma$, then by applying the rules of § 30, XI, we readily see that the function in the brackets [ ] is of class $\boldsymbol{G}_\delta$. If the space is compact, the result of the operation $\bigwedge_z$ performed on this function is still of class $\boldsymbol{G}_\delta$ (by § 20, V, Corollary 7b). The conjunction with $(y \neq x)$, which is of class $\boldsymbol{G}_0$, is again of class $\boldsymbol{G}_\delta$. Then, applying the operation $\bigvee_y$, we get a function of class $\boldsymbol{A}$ (Theorem 4). Its conjunction with the function $(x \in M)$, which is of class $\boldsymbol{F}_\sigma$, is the function $\varphi(x)$. Therefore $\varphi(x)$ is of class $\boldsymbol{A}$.

In other words, *if M is an $\boldsymbol{F}_\sigma$-subset of a compact space, the set of its rectilinearly accessible points is analytic* (1).

More generally, if M is a set of class $\boldsymbol{L}_{2n+1}$ in a complete space, then M_a is of class $\boldsymbol{L}_{2n+3}$.

Remark 3. The class of M_a cannot be reduced. In fact, there exists (in $\mathscr{E}^2$) a closed set M such that M_a is not borelian (2).

IX. Invariance of projective classes with respect to the sieve operation and the operation ($\mathscr{A}$). According to § 3, XV, a *sieve* is any operation W which assigns a set $W_r \subset \mathscr{X}$ to each $r \in \mathscr{R}_0$. The set A of all points x such that there exists an infinite sequence $r_1 < r_2 < \ldots$ with $x \in W_{r_1} \cap W_{r_2} \cap \ldots$ is said to be *sieved* by W.

In logical symbols (where $\mathfrak{z}$ varies in the space $\mathscr{I}^{\aleph_0}$):

$$x \in A \equiv \bigvee_{\mathfrak{z}} \bigwedge_k (\mathfrak{z}^k < \mathfrak{z}^{k+1}) \wedge (x \in W_{\mathfrak{z}^k})$$
$$\equiv \bigvee_{\mathfrak{z}} \bigwedge_k [(\mathfrak{z}^k < \mathfrak{z}^{k+1}) \wedge \bigvee_r (r = \mathfrak{z}^k) \wedge (x \in W_r)]. \tag{1}$$

(1) Theorem of Nikodym and Urysohn. For the proof see p. 255 of the paper of mine referred to earlier.

(2) See: O. Nikodym, Fund. Math. 7 (1925), p. 250 and 11 (1928); Ann. Soc. Pol. Math. 7 (1929), p. 79; P. Urysohn, Proc. Acad. Amsterdam 28 (1925), p. 984, and N. Lusin, Fund. Math. 12 (1928), p. 158. This is no more true of arcwise accessibility. See S. Mazurkiewicz, Fund. Math. 26 (1936), p. 153.

We shall assume subsequently that the $\mathcal{X}$ is a complete separable space.

The principal result concerning sieves is that the *projective class* $\boldsymbol{L}_n$ *is an invariant of the sieve operation, if* $0 \neq n \neq 2$ (Theorems 1 and 5).

THEOREM 1. *If all the sets of a sieve* W *are of odd class* n, *then the set* A *sieved by this sieve is also of class* n.

This is an easy consequence of formula (1). Since the existential quantifier $\bigvee_r$ is countable, the propositional function in the brackets [ ], of the variables x and $\mathfrak{z}$, is of class $\boldsymbol{L}_n$. Therefore A is of class $\boldsymbol{PL}_n = \boldsymbol{L}_n$.

We shall apply this theorem to a particularly important case, when C is the sieve defined in § 3, XVI. The set sieved by C is then the complement of the set $L = \underset{t}{E}(\bar{t} < \Omega)$ (for the notations see § 3, XVI). Since the sets C_r are analytic (they are even both closed and open), we obtain the following corollary.

COROLLARY 1a (1). *The set* $L = \underset{t}{E}(\bar{t} < \Omega)$ *is of class* $\boldsymbol{CA}$.

More directly:

$$[x \in (\mathscr{C} - L)] \equiv \bigvee_{\mathfrak{z}} \bigwedge_{k} [(\mathfrak{z}^k < \mathfrak{z}^{k+1}) \wedge \bigvee_{l} (r_l = \mathfrak{z}^k) \wedge (t^l = 2)]. \tag{2}$$

Consider now two sieves W and Z where $W_r \subset \mathcal{X}$ and $Z_r \subset \mathcal{Y}$, and let

$$M_x = \underset{r}{E}(x \in W_r) \quad \text{and} \quad N_y = \underset{r}{E}(y \in Z_r). \tag{3}$$

We shall prove the following statement.

LEMMA 2 (2). *If the sets* W_r *are of class* $2n$ *and the sets* Z_r *are of class* $(2n-1)$, *then the set* P *composed of all pairs* xy *such that* M_x *is similar to a subset of* N_y *is of class* $2n-1$ (*in the space* $\mathcal{X} \times \mathcal{Y}$).

Proof. Notice first that if M and N are two subsets of $\mathscr{R}_0$, then the condition that M is similar to a subset of N is equivalent to the existence of two sequences $\mathfrak{z} = [\mathfrak{z}^1, \mathfrak{z}^2, \ldots]$ and $\mathfrak{y} = [\mathfrak{y}^1, \mathfrak{y}^2, \ldots]$ of real numbers (of the interval $\mathscr{I}$) such that

---

(1) See N. Lusin and W. Sierpiński, Journ. de Math. 1923, p. 53.

(2) Compare "deuxième principe" of N. Lusin, *Ensembles analytiques* and my note *Sur la géométrisation des types d'ordre dénombrable*, Fund. Math. 28 (1937), p. 167.

(i) the sequence $\mathfrak{z}^1, \mathfrak{z}^2, \ldots$ contains all elements of M;

(ii) the sequence $\mathfrak{y}^1, \mathfrak{y}^2, \ldots$ is entirely contained in N;

(iii) the condition $\mathfrak{z}^i > \mathfrak{z}^j$ implies $\mathfrak{y}^i > \mathfrak{y}^j$.

In logical symbols,

$$\bigvee_{\mathfrak{z}\mathfrak{y}} \{\bigwedge_r [(r \in M) \Rightarrow \bigvee_n (r = \mathfrak{z}^n)] \wedge \bigwedge_k (\mathfrak{y}^k \in N) \wedge \bigwedge_{ij} [(\mathfrak{z}^i > \mathfrak{z}^j) \Rightarrow (\mathfrak{y}^i > \mathfrak{y}^j)]\}.$$

Replace M by M_x and N by N_y. Therefore (by the definition of M_x and N_y) the following equivalences hold:

$$(r \in M_x) \equiv (x \in W_r),$$

$$(\mathfrak{y}^k \in N_y) \equiv (y \in Z_{\mathfrak{y}^k}) \equiv \bigvee_r [(r = \mathfrak{y}^k) \wedge (\mathfrak{y} \in Z_r)].$$

Replacing $(\alpha \Rightarrow \beta)$ by $(\neg\alpha \vee \beta)$, we finally obtain

$$[(xy) \in P] \equiv \bigvee_{\mathfrak{z}\mathfrak{y}} \{\bigwedge_r [(x \in W_r^c) \vee \bigvee_n (r = \mathfrak{z}^n)] \wedge$$

$$\wedge \bigwedge_k \bigvee_r [(r = \mathfrak{y}^k) \wedge (y \in Z_r)] \wedge \bigwedge_{ij} [(\mathfrak{z}^i \leqslant \mathfrak{z}^j) \vee (\mathfrak{y}^i > \mathfrak{y}^j)]\}.$$

The sets W_r^c and Z_r are of class $(2n-1)$ and the operations $\bigwedge_r$, $\bigvee_n$, $\bigwedge_k$, $\bigvee_r$, and $\bigwedge_{ij}$ are countable. It follows that the propositional function in the brackets $\{\ \}$ (of the variables x, y, $\mathfrak{z}$, and $\mathfrak{y}$) is of class $(2n-1)$. Thus P is of class $\boldsymbol{PL}_{2n-1} = \boldsymbol{L}_{2n-1}$. The proof of the lemma is therefore completed.

Now apply the lemma to the particular case when the sieves W and Z coincide with the sieve C of § 3, XVI.

Let τ and σ be two order types. We write

$$\tau \prec\!\!\!\sim \sigma \tag{4}$$

if a set T of type τ is similar to a subset of a set S of type σ (in the particular case, when τ and σ are well order types, the relation $\tau \prec\!\!\!\sim \sigma$ clearly coincides with $\tau \leqslant \sigma$).

The lemma implies the following corollary.

COROLLARY 2a. *The set* $\underset{tu}{E}(\bar{t} \prec\!\!\!\sim \bar{u})$ *is analytic.*

For, if $W_r = C_r$, $\bar{t}$ is by definition (§ 3, XVI) the order type of the set M_t defined by (3).

THEOREM 3 ([1]). *If all the sets of a sieve* W *are of even class* $n \geqslant 4$, *then the sieved set* A *is also of class* n.

([1]) For Theorems 3 and 4, see my paper *Les ensembles projectifs et l'opération* (*A*), C. R. Paris 203 (1936), p. 211.

Proof. Apply Lemma 2 to the case when $Z_r = C_r$ (compare § 3, XVI) and, for brevity, denote by $\mu(x)$ the order type of the set M_x. Hence the set $P = \underset{xt}{E}[\mu(x) \prec \bar{t}]$ is of class $(n-1)$.

On the other hand, since $\bar{t}$ varies over all countable order types a (§ 3, XVI), it follows by the definition of the sieve that

$$[x \in (\mathscr{X} - A)] \equiv [\mu(x) < \Omega] \equiv \bigvee_a [\mu(x) \leqslant a < \Omega]$$

$$\equiv \bigvee_t [\mu(x) \leqslant \bar{t} < \Omega] \equiv \bigvee_t [(x, t) \in P] \wedge (\bar{t} < \Omega).$$

By Corollary 1a, the set $\underset{t}{E}(\bar{t} < \Omega)$ is of class 2, and hence of class $(n-1)$ (see II, 7). It follows that $\underset{xt}{E}[(x, t) \in P] \wedge (\bar{t} < \Omega)$ is also of class $(n-1)$. The set $\mathscr{X} - A$, as a set of class $\boldsymbol{PL}_{n-1} = \boldsymbol{L}_{n-1}$, is of class $(n-1)$. Therefore A is of class n (by II, 5).

THEOREM 4. *The projective classes $\boldsymbol{L}_n$, for $0 \neq n \neq 2$ are invariants of operation $(\mathscr{A})$.*

Proof[1]. Let $A = \bigcup_{\mathfrak{z}} \bigcap_m A_{\mathfrak{z}^1 \ldots \mathfrak{z}^m}$. We can assume that the system $\{A_{\mathfrak{z}^1 \ldots \mathfrak{z}^m}\}$ is regular (compare § 3, XIV), as otherwise it can be made regular by replacing it with

$$\overset{*}{A}_{\mathfrak{z}^1 \ldots \mathfrak{z}^m} = A_{\mathfrak{z}^1} \cap A_{\mathfrak{z}^1 \mathfrak{z}^2} \cap \ldots \cap A_{\mathfrak{z}^1 \mathfrak{z}^2 \ldots \mathfrak{z}^m},$$

without affecting the projective class of the sets involved.

By the theorem of § 3, XV, the set A is then sieved by the sieve consisting of sets belonging to the system $\{A_{\mathfrak{z}^1 \ldots \mathfrak{z}^m}\}$ which are all of class n. By Theorems 1 and 3, A is of class $\boldsymbol{L}_n$.

Remark. If n is odd, Theorem 4 can be proved more directly. In fact, we have

$$(x \in A) \equiv \bigvee_{\mathfrak{z}} \bigwedge_m (x \in A_{\mathfrak{z}1 \ldots \mathfrak{z}m}).$$

In order to prove that the propositional function is of class n it suffices to show that, for a fixed m, the function $x \in A_{\mathfrak{z}1 \ldots \mathfrak{z}m}$ is of class n (in the space $\mathscr{X} \times \mathscr{N}$). But this follows from

$$(x \in A_{\mathfrak{z}1 \ldots \mathfrak{z}m}) \equiv \bigvee_{k_1} \ldots \bigvee_{k_m} (\mathfrak{z}^1 = k_1) \wedge \ldots \wedge (\mathfrak{z}^m = k_m) \wedge (x \in A_{k_1 \ldots k_m}).$$

[1] For a different proof, see J. W. Addison and S. C. Kleene, *A note on function quantification,* Proc. Amer. Math. Soc. 8 (1957), p. 1002.

COROLLARY 4a. *The operation ($\mathscr{A}$) and the subtraction performed on the family of sets which are of classes both* **PCA** *and* **CPCA** *do not lead out of this family.*

Consequently, *the smallest family of sets which contains all the closed sets and which is closed with respect to the operation ($\mathscr{A}$) and the subtraction, is contained in the classes* **PCA** *and* **CPCA** (¹).

Lemma 2 concerned the relation $\tau \precsim \sigma$; the following lemma concerns the identity $\tau = \sigma$.

LEMMA 5. *If the sets W_r and Z_r are of classes both $2n$ and $(2n-1)$, then the set of all pairs xy such that M_x is similar to N_y is of class $(2n-1)$.*

The proof is quite analogous to that of Lemma 2. Namely the property of M of being similar to N is expressed as follows

$$\bigvee_{\mathfrak{z}\mathfrak{y}} \{\bigwedge_r [(r \in M) \Rightarrow \bigvee_n (r = \mathfrak{z}^n)] \wedge \bigwedge_r [(r \in N) \Rightarrow \bigvee_n (r = \mathfrak{y}^n)] \wedge$$
$$\wedge \bigwedge_n (\mathfrak{z}^n \in M) \wedge (\mathfrak{y}^n \in N) \wedge \bigwedge_{ij} [(\mathfrak{z}^i > \mathfrak{z}^j) \Rightarrow (\mathfrak{y}^i > \mathfrak{y}^j)]\}.$$

It follows that

$$[(xy) \in T] \equiv \bigvee_{\mathfrak{z}\mathfrak{y}} \{\bigwedge_r [(x \in W_r^c) \vee \bigvee_n (r = \mathfrak{z}^n)] \wedge$$
$$\wedge \bigwedge_r [(x \in Z_r^c) \vee \bigvee_n (r = \mathfrak{y}^n)] \wedge$$
$$\wedge \bigwedge_n \bigvee_{rs} [(r = \mathfrak{z}^n) \wedge (x \in W_r) \wedge (s = \mathfrak{y}^n) \wedge (\mathfrak{y} \in Z_s)] \wedge$$
$$\wedge \bigwedge_{ij} [(\mathfrak{z}^i \leqslant \mathfrak{z}^j) \vee (\mathfrak{y}^i > \mathfrak{y}^j)]\}.$$

An easy logical calculus shows that the set T is of class $(2n-1)$.

COROLLARY 5a. *The set $\underset{tu}{E}(\bar{t} = \bar{u})$ is analytic.*

Proof. Substitute the sieve C of § 3, XVI, for $\dot{W}$ and Z.

COROLLARY 5b. *The set $L_\tau = \underset{t}{E}(\bar{t} = \tau)$ is analytic, for every countable order type τ.*

(¹) The study of this family of sets, which constitutes a natural generalization of the sets **A** and **CA** was proposed by N. Lusin. See Fund. Math. 5 (1924), p. 165, footnote 3. Compare O. Nikodym, Fund. Math. 14 (1929), p. 145.

Corollary 4a was announced by Kantorovitch and Livensohn in the C. R. Paris 190 (1930), p. 1115 and in Fund. Math. 18 (1932), p. 217. Compare also indications of N. Lusin in his *Compte rendu sur la théorie descriptive des fonctions, présenté à l'Académie des Sciences d'URSS*, 1935, p. 61.

Proof. For a fixed τ, let u_0 be an element of $\mathscr{C}$ such that $\bar{u}_0 = \tau$. Since the propositional function $\bar{t} = \bar{u}$ (of two variables) is of class $\boldsymbol{A}$, so is the propositional function $\bar{t} = \bar{u}_0$ of the single variable t (compare VIII, Theorem 5). The assertion follows.

Remark. Quite recently Professor Dana Scott has obtained a stronger result, namely: *L_τ is a Borel set for every τ* (1).

X. Transfinite induction. We shall prove that, besides the operations considered above, the transfinite induction does not lead out of the projective sets either — under very general hypotheses (2).

Observe first that every operation F which assigns, to each family $\boldsymbol{X}$ of subsets of a space $\mathscr{X}$, a subset $F(\boldsymbol{X})$ of $\mathscr{X}$ determines a transfinite sequence of sets $A_0, A_1, \ldots, A_\alpha, \ldots$ defined as follows:

$$A_0 = F(0) \quad \text{and} \quad A_\alpha = F(\boldsymbol{X}_\alpha) \quad \text{for} \quad \alpha > 0;$$

$\boldsymbol{X}_\alpha$ denotes here the family of sets $\{A_\xi\}$ with $\xi < \alpha$.

The problem arises: under what conditions is the set $\underset{tx}{E}(x \in A_{\bar{t}})$, where $\bar{t} < \Omega$, projective in $\mathscr{C} \times \mathscr{X}$?

The capital German letters will be used to denote subsets of the product $\mathscr{C} \times \mathscr{X}$. Let

$$\mathfrak{Z}^t = \underset{x}{E}[(t, x) \in \mathfrak{Z}] \quad \text{where} \quad t \in \mathscr{C}, \tag{1}$$

and as usual

$$L = \underset{t}{E}(\bar{t} < \Omega). \tag{2}$$

THEOREM 1. *Let R_t be an operation which assigns, to each $t \in L$ and each $\mathfrak{B} \subset \underset{u}{E}(\bar{u} < \bar{t}) \times \mathscr{X}$, a set $R_t(\mathfrak{B}) \subset \mathscr{X}$.*

Then there exists a unique set $\mathfrak{Z} \subset L \times \mathscr{X}$ such that

$$\mathfrak{Z}^t = R_t[\mathfrak{Z} \cap \underset{ux}{E}(\bar{u} < \bar{t})] \quad \textit{for every} \quad \bar{t} < \Omega. \tag{3}$$

---

(1) See *Invariant Borel sets,* Fund. Math. 56 (1964), pp. 117–128. See also C. Ryll-Nardzewski, *On Borel measurability of orbits, ibid.*, pp. 129–130. For a partial result see S. Hartman *Zur Geometrisierung der abzählbaren Ordnungstypen,* Fund. Math. 29 (1937), p. 210.

(2) See my paper *Les ensembles projectifs et l'induction transfinie,* Fund. Math. 27 (1936), p. 269.

Proof. Consider the transfinite sequence $\mathfrak{A}_0, \mathfrak{A}_1, \ldots, \mathfrak{A}_\alpha, \ldots$ $(\alpha < \Omega)$ of subsets of $L \times \mathscr{X}$ defined by induction as follows:

$$\mathfrak{A}_0 = \underset{tx}{E}(t = 0) \wedge [x \in R_0(0)], \tag{4}$$

$$\mathfrak{A}_\alpha = \bigcup_{\beta<\alpha} \mathfrak{A}_\beta \cup \mathfrak{R}_\alpha, \text{ where } \mathfrak{R}_\alpha = \underset{tx}{E}(\bar{t} = \alpha) \wedge [x \in R_t(\bigcup_{\beta<\alpha} \mathfrak{A}_\beta)]. \tag{5}$$

It is easily proved by induction that

$$\mathfrak{A}_\alpha \subset \underset{u}{E}(\bar{u} \leqslant \alpha) \times \mathscr{X}. \tag{6}$$

Moreover $\mathfrak{A}_0 \subset \mathfrak{A}_1 \subset \ldots \subset \mathfrak{A}_\alpha \subset \ldots$.
Let $\mathfrak{Z} = \mathfrak{A}_0 \cup \mathfrak{A}_1 \cup \ldots \cup \mathfrak{A}_\alpha \cup \ldots$ By (6) we have

$$\mathfrak{A}_\alpha^t = \mathfrak{A}_{\bar{t}}^t \quad \text{for} \quad \bar{t} \leqslant \alpha \tag{7}$$

$$\mathfrak{A}_\alpha^t = 0 \quad \text{for} \quad \bar{t} \text{ non} \leqslant \alpha. \tag{8}$$

It follows that

$$\mathfrak{Z}^t = \bigcup_{\alpha<\Omega} \mathfrak{A}_\alpha^t = \mathfrak{A}_{\bar{t}}^t = R_t(\bigcup_{\beta<\bar{t}} \mathfrak{A}_\beta), \tag{9}$$

and since $\bigcup_{\beta<\bar{t}} \mathfrak{A}_\beta = \mathfrak{Z} \cap \underset{ux}{E}(\bar{u} < \bar{t})$, formula (3) follows.

The uniqueness of $\mathfrak{Z}$ can easily be proved by induction.

With the same notation, we have the following statement (which gives an explicit definition of $\mathfrak{Z}$, defined implicitly by condition (3)).

THEOREM 2. *If $\mathfrak{G}$ is a function defined on $\mathscr{C}$ which assumes all the sets $\mathfrak{A}_\alpha$, $\alpha < \Omega$, as values, then the following equivalences hold (where $t, v, z \in \mathscr{C}$, $x \in \mathscr{X}$):*

$$\{(v, x) \in \mathfrak{Z}\} \equiv (\bar{v} < \Omega) \wedge$$

$$\wedge \bigvee_z \bigwedge_t \{(\bar{t} \leqslant \bar{v}) \Rightarrow [\mathfrak{G}^t(z) = R_t(\mathfrak{G}(z) \cap \underset{ux}{E}(\bar{u} < \bar{t}))]\} \wedge \{x \in \mathfrak{G}^v(z)\}$$

$$\equiv (\bar{v} < \Omega) \wedge \bigwedge_z \{[\bigwedge_t (\bar{t} \leqslant \bar{v}) \Rightarrow [\mathfrak{G}^t(z) = R_t(\mathfrak{G}(z) \cap \underset{ux}{E}(\bar{u} < \bar{t}))]] \Rightarrow$$

$$\Rightarrow [x \in \mathfrak{G}^v(z)]\}.$$

Proof. Observe first that the set $\mathfrak{A}_\alpha^t$ can be defined by the conditions

$$\mathfrak{A}_\alpha^t = \begin{cases} R_t[\mathfrak{A}_\alpha \cap \underset{ux}{E}(\bar{u} < \bar{t})] & \text{for} \quad \bar{t} \leqslant \alpha, \\ 0 \quad \text{otherwise}. \end{cases}$$

For, by virtue of (5),

$$\mathfrak{A}_\alpha \cap \underset{ux}{E}(\bar{u} < \bar{t}) = \bigcup_{\beta < \bar{t}} \mathfrak{A}_\beta$$

and the assertion follows by (7) and (9).

Consequently, $\mathfrak{Z}$ as the union of the $\mathfrak{A}_\alpha$, is the set of points v, x such that $\bar{v} < \Omega$ and such that there exists a $z \in \mathscr{C}$ satisfying the conditions:

(i) $\mathfrak{G}^t(z) = R_t[\mathfrak{G}(z) \cap \underset{ux}{E}(\bar{u} < \bar{t})]$ for $\bar{t} \leqslant \bar{v}$;

(ii) $x \in \mathfrak{G}^v(z)$

($\mathfrak{G}(z) = \mathfrak{A}_\alpha$ of course).

This, when translated into logical symbols, gives the first equivalence.

Since the set $\mathfrak{A}_\alpha$ is defined uniquely by conditions (7) and (8), the condition for (v, x) to belong to $\mathfrak{Z}$ (where $\bar{v} < \Omega$) can be restated as follows: if z satisfies (i), condition (ii) is fulfilled. This leads to the second equivalence.

THEOREM 3. *Let $\mathfrak{G}$ be a function defined on $\mathscr{C}$, universal for the projective sets of class $\boldsymbol{L}_n$ $(n \geqslant 1)$ situated in $\mathscr{C} \times \mathscr{X}$. Suppose that the sets*

$$\underset{ztx}{E}[(t, x) \in \mathfrak{G}(z)] \quad \textit{and} \quad \underset{ztx}{E}\{x \in R_t[\mathfrak{G}(z) \cap \underset{ux}{E}(\bar{u} < \bar{t})]\}$$

are projective of class $\boldsymbol{L}_n$. Then the set $\mathfrak{Z}$ of Theorem 1 is projective (of classes $\boldsymbol{L}_{n+4}$ and $\boldsymbol{CL}_{n+4}$).

Proof. Firstly, the function $\mathfrak{G}$ satisfies the hypothesis of Theorem 2; i.e. the sets $\mathfrak{A}_\alpha$ defined by (4) and (5) are of class $\boldsymbol{L}_n$. By an induction argument, it is sufficient to prove that the set $\mathfrak{R}_\alpha$ is of class $\boldsymbol{L}_n$, assuming that $\mathfrak{A}_\beta$ is of class $\boldsymbol{L}_n$ for $\beta < \alpha$. Now since $\bigcup_{\beta<\alpha} \mathfrak{A}_\beta$ is of class $\boldsymbol{L}_n$, there exists a z_0 such that $\mathfrak{G}(z_0) = \bigcup_{\beta<\alpha} \mathfrak{A}_\beta$, and since $\bigcup_{\beta<\alpha} \mathfrak{A}_\beta \subset \underset{ux}{E}(\bar{u} < \alpha)$, it follows that

$$(\bar{t} = \alpha) \wedge [x \in R_t(\bigcup_{\beta<\alpha} \mathfrak{A}_\beta)] \equiv (\bar{t} = \alpha) \wedge \{x \in R_t[\mathfrak{G}(z_0) \cap \underset{ux}{E}(\bar{u} < \bar{t})]\}.$$

By hypothesis, the propositional function in the brackets { } is of class $\boldsymbol{L}_n$ (compare VIII, 5) and $\underset{t}{E}(\bar{t} = \alpha)$ is a Borel set (§ 30, XII). It follows that $\mathfrak{R}_\alpha$ is of class $\boldsymbol{L}_n$.

By virtue of the identity

$$\{A = B\} \equiv \bigwedge_x \Big[\big((x \in A) \wedge (x \in B)\big) \vee \big((x \notin A) \wedge (x \notin B)\big)\Big] \qquad (10)$$

we infer that the set

$$\underset{ztvx}{E}\big[\mathfrak{G}^t(z) = R_t\big(\mathfrak{G}(z) \cap \underset{ux}{E}(\bar{u} < \bar{t})\big)\big]$$

is projective of class $(n+2)$ or $(n+3)$ according as n is even or odd. Now replace $\bar{t} \leqslant \bar{v}$ by $\bar{t} \prec \bar{v}$ (compare IX (4)). Since the set $\underset{tv}{E}(\bar{t} \prec \bar{v})$ is projective (even analytic, compare IX, Corollary 2a) and $\underset{v}{E}(\bar{v} < \Omega)$ is a $\boldsymbol{CA}$ set (IX, Corollary 1a), it follows from Theorem 2 that $\mathfrak{Z}$ is projective (of classes $\boldsymbol{L}_{n+4}$ and $\boldsymbol{CL}_{n+4}$).

Applications. 1) *The Lebesgue set, universal with respect to analytically representable functions, is projective* (1).

This set can be defined as follows. For $t = \dfrac{t^{(1)}}{3} + \dfrac{t^{(2)}}{9} + \ldots$ let

$$t_{(n)} = \frac{t^{(1\cdot 2^n)}}{3} + \frac{t^{(3\cdot 2^n)}}{9} + \frac{t^{(5\cdot 2^n)}}{27} + \ldots . \qquad (11)$$

Let A be a plain Borel set $\subset \mathscr{C}\times\mathscr{E}$ such that all closed subsets and all open subsets of $\mathscr{E}$ occur as vertical sections of A. Consider the subset $\mathfrak{Z}$ of $\mathscr{C}\times\mathscr{C}\times\mathscr{E}$ defined by the conditions:

$$\mathfrak{Z}^0 = A, \qquad \mathfrak{Z}^{t,y} = \operatorname*{Lim\,sup}_{n=\infty} \mathfrak{Z}^{t^{[n]}, y_{(n)}} \quad \text{for} \quad 0 < \bar{t} < \Omega \text{ and } y \in \mathscr{C} \ (2),$$

$$\mathfrak{Z}^t = 0 \quad \text{otherwise.}$$

In this case, the set $R_t(\mathfrak{B})$ can be defined as follows: if $\mathfrak{B}$ is a subset of $\underset{u}{E}(\bar{u} < \bar{t})\times\mathscr{C}\times\mathscr{E}$, then $R_t(\mathfrak{B})$ is the plane subset of $\mathscr{C}\times\mathscr{E}$ defined by the conditions:

$$R_0(0) = A, \qquad R_t(\mathfrak{B}) = \underset{yz}{E}\big[x \in \operatorname*{Lim\,sup}_{n=\infty} \mathfrak{B}^{t[n], y_{(n)}}\big] \quad \text{for} \quad t \neq 0.$$

(1) For a definition of this set, see the paper of Lebesgue in Journal de Math., 1905, Ch. VIII. The problem whether this set is projective was raised by N. Lusin. Its solution has been given in my paper *Sur un problème concernant l'induction transfinie*, C. R. Paris 202 (1936), p. 1239.

(2) For the definition of $t^{[n]}$ see § 30, XII. According to (1), $\mathfrak{Z}^{t,y}$ denotes the set $\underset{x}{E}[(t, y, x) \in \mathfrak{Z}]$. By definition (§ 1, V): $\operatorname*{Lim\,sup}_{n=\infty} A_n = \bigcap_{n=0}^{\infty} \bigcup_{k=0}^{\infty} A_{n+k}$.

Let $\mathfrak{G}$ be a universal function with respect to all analytic subsets of $\mathscr{C}\times\mathscr{C}\times\mathscr{E}$ such that the set $\underset{ztyx}{E}[(t, y, x)\,\epsilon\,\mathfrak{G}(z)]$ is analytic (compare Section V). It remains to be proved that the set $\underset{ztyx}{E}\big[(y, x)\,\epsilon\,R_t\big(\mathfrak{G}(z)\cap\underset{uyx}{E}(\bar{u}<\bar{t})\big)\big]$ is analytic (in $\mathscr{C}^3\times\mathscr{E}$). Now we have (if $t \neq 0$):

$$(y, x)\,\epsilon\,R_t\big(\mathfrak{G}(z)\cap\underset{uyx}{E}(\bar{u}<\bar{t})\big)$$

$$\equiv x\,\epsilon\,\underset{n=\infty}{\mathrm{Lim\,sup}}\,\mathfrak{G}^{t^{[n]},y_{(n)}}(z) \equiv \bigwedge_n \bigvee_k x\,\epsilon\,\mathfrak{G}^{t^{[n+k]},y_{(n+k)}}(z)$$

$$\equiv \bigwedge_n \bigvee_k [(t^{[n+k]}, y_{(n+k)}, x)\,\epsilon\,\mathfrak{G}(z)].$$

Since the functions $t^{[n]}$ and $y_{(n)}$ are continuous, for a fixed n (compare § 30, XII), the required assertion follows.

2) Now let us modify the definition of $R_t(\mathfrak{B})$ as follows: let $\mathscr{X}=\mathscr{C}$; then $R_0(0)=A=$ an open set such that A^y is a universal function with respect to open sets (compare § 30, XIII, footnote 3);

$$R_t(\mathfrak{B}) = \underset{yx}{E}\{x\,\epsilon\bigcup_{n=1}^{\infty}\mathfrak{B}^{t^{[n]},y_{(n)}}\},$$

or

$$\underset{yx}{E}\{x\,\epsilon\bigcap_{n=1}^{\infty}\mathfrak{B}^{t^{[n]},y_{(n)}}\},$$

according to as $\bar{t}$ is even or odd (in the sense of § 30, XII, 4).

The set $\mathfrak{Z}$, whose definition can be deduced from that of $R_t(\mathfrak{B})$, is a subset of the space $\mathscr{C}^3$ such that, for each $\bar{t}<\Omega$, $\mathfrak{Z}^t$ is a Borel set of class $\boldsymbol{G}_{\bar{t}}$ and that that the function $\mathfrak{Z}^{t,y}$ (as a function of y) is universal with respect to the family of subsets of $\mathscr{C}$ of class $\boldsymbol{G}_{\bar{t}}$. As in the preceding example, the fact that $\mathfrak{Z}$ is projective follows from the definition of $R_t(\mathfrak{B})$. The only additional premise here is the property of the set of all t such that $\bar{t}$ is an even order type of beind a Borel set (as proved in § 30, XII, 4).

3) Let t^* be the point of $\mathscr{C}$ such that the set $\underset{r_n}{E}(t^{*n}=2)$ is obtained from $\underset{r_n}{E}(t^n=2)$ (¹) by omitting the last element (provided it exists; if not, we let $t^*=t$). It is easily seen that t^* is a B measurable function of t (compare § 39, V, Theorem 2).

(¹) This set was denoted by R_t in § 3, XVI.

$\mathfrak{Z}$ is defined by the conditions:

$$\mathfrak{Z}^0 = A, \quad \mathfrak{Z}^{t,y} = \bigcup_{n=1}^{\infty} \mathfrak{Z}^{t[n],y(n)} \quad \text{or} \quad \mathcal{E} - \mathfrak{Z}^{t*,y}$$

according to as $\bar{t}$ is even or odd.

The corresponding set $R_t(\mathfrak{B})$ is defined as in example 2 by replacing $\bigcap_{n=1}^{\infty} \mathfrak{B}^{t[n],y(n)}$ with $\mathcal{E} - \mathfrak{B}^{t*,y}$. As before, $\mathfrak{Z}$ is projective.

In examples 1–3, the set $\mathfrak{Z}^t$ depends solely on $\bar{t}$ (not on t). Thus if we let $A_a = \mathfrak{Z}^t$, where t is any number such that $\bar{t} = a$, we obtain a transfinite sequence of Borel sets (precisely of class a in examples 2 and 3). In the example which follows, $\mathfrak{Z}^t$ depends on t:

4) Let:

$$\mathfrak{Z}^0 = A, \mathfrak{Z}^{t,y} = \bigcap_{n=1}^{\infty} \mathfrak{Z}^{t*,y(n)}, \quad \text{if } \bar{t} \text{ is odd},$$

$$x \in \mathfrak{Z}^{t,y} \equiv \bigvee_{n=1}^{\infty} (\bar{y}_{(2n+1)} < \bar{t}) \wedge (x \in \mathfrak{Z}^{y(2n+1),y(2n)}),$$

$$\text{if } \bar{t} \text{ is even and } 0 < \bar{t} < \Omega.$$

Here $R_t(\mathfrak{B})$ can be defined as follows:

$$\{(y, x) \in R_t(\mathfrak{B})\} \equiv \Big\{\big[(t = 0) \wedge \big((y, x) \in A\big)\big] \vee \big[(\bar{t} \text{ is odd}) \wedge \bigwedge_n (x \in \mathfrak{B}^{t*,y(n)})\big]$$

$$\vee \big[(\bar{t} \text{ is even and } > 0) \wedge \bigvee_n (\bar{y}_{(2n+1)} + 1 \leqslant \bar{t}) \wedge (x \in \mathfrak{B}^{y(2n+1),y(2n)})\big]\Big\}.$$

As before we infer that the set $\mathfrak{Z}$ is projective (¹).

XI. Hausdorff operations. Given a sequence of sets $A_1, A_2, \ldots$ and a set B of infinite sequences of positive integers (regarded as a subset of the space $\mathcal{N}$), we say that H *is obtained from the sequence* $\{A_n\}$ *by means of the Hausdorff operation with basis* B (²), *if*

$$(x \in H) \equiv \bigvee_{\mathfrak{z}} \{(\mathfrak{z} \in B) \bigwedge_n (x \in A_{\mathfrak{z}n})\}. \tag{1}$$

(¹) See my paper *Sur l'existence effective des fonctions représentables analytiquement de toute classe de Baire*, C. R. Paris 176 (1923), p. 229.

(²) F. Hausdorff, *Mengenlehre*, p. 87. A. Kolmogoroff, *Opérations sur des ensembles*, Rec. Math. Moscow 35 (1928), p. 418. See also L. Kantorovitsch and E. Livensohn, *Memoir on the Analytical Operations and Projective Sets* (I), Fund. Math. 18 (1932), p. 214, and A. A. Liapounov, *R-sets* (Russian), Trudy Stieklov Inst., Moscow 40 (1953), *On operations which admit transfinite indices* (Russian), Trudy Mosc. Math. Soc. 6 (1957), pp. 196–230 (where numerous bibliographical references can be found).

Thus, for example,

$$\operatorname*{Lim\,sup}_{n=\infty} A_n = \bigcap_{n=0}^{\infty} \bigcup_{k=0}^{\infty} A_{n+k}$$

is a Hausdorff operation. Its basis consists of sequences of positive integers $\mathfrak{z} = [\mathfrak{z}^1, \mathfrak{z}^2, \ldots]$ containing an infinite set of terms different from each other.

$\bigcup_n$ and $\bigcap_n$ are Hausdorff operations. More generally, the operation $(\mathscr{A})$ can be reduced to a Hausdorff operation (¹).

THEOREM 1. *If the sets A_n as well as the basis B are projective of odd class k, then so is the set H.*

Proof. We have

$$(x \in A_{\mathfrak{z}^n}) \equiv \bigwedge_m [(m = \mathfrak{z}^n) \Rightarrow (x \in A_m)].$$

By virtue of (1)

$$(x \in H) \equiv \bigvee_{\mathfrak{z}} \{(\mathfrak{z} \in B) \wedge \bigwedge_{mn} [(m = \mathfrak{z}^n) \Rightarrow (x \in A_m)]\}.$$

We infer from this that H is of class $\boldsymbol{PL}_k = \boldsymbol{L}_k$.

We return now to the evaluation of the projective class of the set $\mathfrak{Z}$ defined by a transfinite induction (see Section X). When this definition is obtained by means of a Hausdorff operation, it is often more convenient to use a method different from that given in Section X. The following method due to J. v. Neumann (²) permits, in particular, to prove that the set $\mathfrak{Z}$ considered in examples 1, 2, and 4 of Section X is the difference of two analytic sets.

In order to simplify the notation we shall confine ourselves to the example 1.

We shall use the following notation. If r denotes a variable finite system of positive integers $k_1, \ldots, k_n$ $(n \geqslant 0)$, consider the functions γ which assign, to each r, an integer $\gamma(r) \geqslant 0$. r can be regarded as a finite binary fraction (compare § 3, XV)

$$r = \frac{1}{2^{k_1}} + \frac{1}{2^{k_1+k_2}} + \ldots + \frac{1}{2^{k_1+\ldots+k_n}}, \quad 0 \leqslant r < 1.$$

(¹) Compare F. Hausdorff, *op. cit.*, p. 93.

(²) See a paper of J. v. Neumann and mine *On some analytic sets defined by transfinite induction*, Annals of Mathematics 38 (1937), pp. 521–525.

For some applications, see T. Tugué, *Sur les fonctions qui sont définies par l'induction transfinie*, Journ. Math. Soc. Japan 7 (1955), pp. 94–122.

Let Γ be the set of the functions γ. We endow Γ with the topological space structure by assuming that the sequence $\gamma_1, \gamma_2, \ldots$ is convergent to γ, if $\lim_{n=\infty} \gamma_n(r) = \gamma(r)$ for every r. *The space* Γ *is complete separable*; in fact it is homeomorphic to the space of irrational numbers. This can be proved by ordering the numbers r into an infinite sequence and assigning, to each element γ of Γ, the element $\mathfrak{z} = \{\gamma(r_1), \gamma(r_2), \ldots\}$ of $\mathcal{N}$.

Let $\mathfrak{y}_r(\gamma) = \mathfrak{y}_{k_1,\ldots,k_n}(\gamma)$ denote the infinite sequence

$$\gamma(k_1, \ldots, k_n, 1),\ \gamma(k_1, \ldots, k_n, 2), \ldots .$$

In particular, $\mathfrak{y}_0(\gamma) = \{\gamma(1), \gamma(2), \ldots\}$.

Finally, let $\overline{\gamma, r} = \{\gamma(k_1), \ldots, \gamma(k_1, \ldots, k_n)\}$. Hence $\overline{\gamma, 0} = 0$ (the empty set).

We easily check that, for a fixed r, $\gamma(r)$ and $\mathfrak{y}_r(\gamma)$, considered as functions of the variable γ, are continuous.

Let the symbols $t^{[n]}$ and $t_{(n)}$ (where $t \in \mathscr{C}$) have the same meaning as in Section X (example 1). Define the symbols $t^{[r]}$ and $t_{(r)}$ by finite induction as follows:

$$t^{[0]} = t = t_{(0)},$$

$$t^{[k_1,\ldots,k_n,k_{n+1}]} = (t^{[k_1,\ldots,k_n]})^{[k_{n+1}]},$$

$$t_{(k_1,\ldots,k_n,k_{n+1})} = (t_{(k_1,\ldots k_n)})_{(k_{n+1})}.$$

It is easily proved that $t^{[\overline{\gamma,r}]}$ and $t_{[\overline{\gamma,r}]}$ are, for a fixed r, continuous functions of t and γ and that

$$(t^{[k_1]})^{[k_2,\ldots,k_n]} = t^{[k_1,k_2,\ldots,k_n]}, \qquad (t_{(k_1)})_{(k_2,\ldots,k_n)} = t_{(k_1,k_2,\ldots,k_n)}.$$

THEOREM 2. *Let* $A \subset \mathscr{C}\times\mathscr{E}$ *and* $B \subset \mathcal{N}$ *be two analytic sets. The set* $\mathfrak{Z} \subset \mathscr{C}\times\mathscr{C}\times\mathscr{E}$ *defined by the conditions*:

$$\mathfrak{Z}^0 = A, \quad \mathfrak{Z}^{t,y} = \bigcup_{\mathfrak{z}} \bigcap_{n} \mathfrak{Z}^{t^{[\mathfrak{z}^n]}, y_{(\mathfrak{z}^n)}}, \quad \textit{if} \quad 0 < \bar{t} < \Omega,\ y \in \mathscr{C},\ \mathfrak{z} \in B,$$

$$\mathfrak{Z}^t = 0 \quad \textit{otherwise},$$

is the difference of two analytic sets.

More precisely, *the following equivalence holds*:

$$[(tyx) \in \mathfrak{Z}] \equiv (\bar{t} < \Omega) \wedge \bigvee_{\gamma} \{\gamma(0) \neq 0) \wedge$$

$$\wedge \bigwedge_{r} [(\gamma(r) \neq 0) \wedge (t^{[\overline{\gamma,r}]} = 0) \Rightarrow (y_{(\overline{\gamma,r})}, x) \in A] \wedge$$

$$\wedge [(\gamma(r) \neq 0) \wedge (t^{[\overline{\gamma,r}]} \neq 0) \Rightarrow \mathfrak{y}_r(\gamma) \in B]\}.$$

As soon as this equivalence is proved, we shall deduce that $\mathfrak{Z}$ is the intersection of a **CA** set with an analytic set. For, on one hand, the set $\underset{t}{E}(\bar{t} < \Omega)$ is **CA** (by IX, Corollary 1a) and, on the other hand, the propositional function in the brackets $\{\ \}$ is analytic, since the quantifier $\bigwedge_r$ is countable and $\gamma(r)$, $t^{[\overline{\gamma, r}]}$, $y_{(\overline{\gamma,r})}$ and $\mathfrak{y}_r(\gamma)$ are continuous functions of γ (for a fixed r).

We shall now prove the equivalence under consideration.

1) Let $(tyx) \in \mathfrak{Z}$. We shall show by transfinite induction that there exists a function $\gamma \in \Gamma$ satisfying the condition in the brackets $\{\ \}$ and, moreover, assuming a preassigned value $p \neq 0$ for $r = 0$.

First let $\bar{t} = 0$, i.e. $t = 0$. Hence $(yx) \in A$. Let γ be the function defined by: $\gamma(0) = p$ and $\gamma(r) = 0$ for $r \neq 0$. Clearly the condition in the brackets is fulfilled.

Assume now that $\alpha > 0$ and that our hypothesis is fulfilled for each t such that $\bar{t} < \alpha$.

Let $\bar{t} = \alpha$. The formula $(tyx) \in \mathfrak{Z}$ implies the existence of an infinite sequence belonging to B such that, for every n,

$$x \in \mathfrak{Z}^{t^{[\mathfrak{z}^n]}, y_{(\mathfrak{z}^n)}}, \qquad \text{i.e.} \qquad (t^{[\mathfrak{z}^n]}, y_{(\mathfrak{z}^n)}, x) \in \mathfrak{Z}.$$

Since $\overline{t^{[\mathfrak{z}^n]}} < \bar{t}$, there exists for each n, by hypothesis, a function $\gamma_n \in \Gamma$ such that

(i) $\gamma_n(0) = \mathfrak{z}^n$,

(ii) if $\gamma_n(r) \neq 0$ and $t^{[\mathfrak{z}^n][\overline{\gamma_n, r}]} = 0$, then $(y_{(\mathfrak{z}^n)(\overline{\gamma_n, r})}, x) \in A$,

(iii) if $\gamma_n(r) \neq 0$ and $t^{[\mathfrak{z}^n][\overline{\gamma_n, r}]} \neq 0$, then $\mathfrak{y}_r(\gamma_n) \in B$.

Define γ as follows: $\gamma(0) = p$ and, for $r = (k_1, \ldots, k_m)$ (where $m \neq 0$), $\gamma(k_1, \ldots, k_m) = \gamma_{k_1}(k_2, \ldots, k_m)$. It follows that

$$\begin{aligned}\overline{\gamma, r} = \overline{\gamma, (k_1, \ldots, k_m)} &= \{\gamma(k_1), \gamma(k_1, k_2), \ldots, \gamma(k_1, k_2, \ldots, k_m)\} \\ &= \{\mathfrak{z}^{k_1}, \gamma_{k_1}(k_2), \ldots, \gamma_{k_1}(k_2, \ldots, k_m)\} = \{\mathfrak{z}^{k_1}, \overline{\gamma_{k_1}, k_2, \ldots, k_m}\}\end{aligned}$$

and

$$\begin{aligned}\mathfrak{y}_r(\gamma) &= \{\gamma(k_1, \ldots, k_m, 1), \gamma(k_1, \ldots, k_m, 2), \ldots\} \\ &= \{\gamma_{k_1}(k_2, \ldots, k_m, 1), \gamma_{k_1}(k_2, \ldots, k_m, 2), \ldots\} = \mathfrak{y}_{k_2, \ldots, k_m}(\gamma_{k_1}).\end{aligned}$$

It is easily proved that the condition in the brackets is fulfilled.

2) Assume, on the other hand, that γ satisfies the condition in the brackets. We have to prove that $(tyx)\in\mathfrak{Z}$.

First let $t=0$. Since $\gamma(0)\neq 0$ and $\overline{t^{[\gamma,0]}}=t^{[0]}=t$, we have $(y_{(\overline{\gamma,0})}, x)\in A$. But $y_{(\overline{\gamma,0})}=y_{(0)}=y$, hence $(y,x)\in A=\mathfrak{Z}^0$, i.e. $(tyx)\in\mathfrak{Z}$.

Assume now that $a>0$ and that our hypothesis is satisfied for each t such that $\bar{t}<a$.

Let $\bar{t}=a$, and let $\mathfrak{z}=\mathfrak{y}_0(\gamma)=\{\gamma(1),\gamma(2),\ldots\}$. Since $\gamma(0)\neq 0$ and $\overline{t^{[\gamma,0]}}=t\neq 0$, it follows that $\mathfrak{y}_0(\gamma)\in B$, i.e. $\mathfrak{z}\in B$. It remains to be proved that, for every n,

$$x\in\mathfrak{Z}^{t^{[\mathfrak{z}^n]},y_{(\mathfrak{z}^n)}},\qquad \text{i.e.}\qquad (t^{[\mathfrak{z}^n]},y_{(\mathfrak{z}^n)},x)\in\mathfrak{Z}.$$

We let $\gamma_{k_1}(k_2,\ldots,k_m)=\gamma(k_1,k_2,\ldots,k_m)$, $m\neq 0$. It is easy to prove that conditions (i)–(iii) are fulfilled for every n. Since $\overline{t^{[\mathfrak{z}^n]}}<a$, we infer (by the hypothesis) that $(t^{[\mathfrak{z}^n]},y_{(\mathfrak{z}^n)},x)\in\mathfrak{Z}$.

§ 39. Analytic sets

The space $\mathscr{X}$ is assumed to be complete and separable.

I. General theorems. We begin by recalling some properties of analytic sets established in § 38. By definition, the analytic sets are continuous images of Borel sets (§ 38, I). Since every Borel set is a continuous image of the space $\mathscr{N}$, the analytic sets can be defined as *continuous images of* $\mathscr{N}$ (§ 38, IV) (¹).

The property of being an analytic set is *invariant* with respect to the countable operations: union, intersection, and cartesian product. It is also invariant with respect to the operation $(\mathscr{A})$ and with respect to B measurable mappings (of this set onto a subset of the same space or of some other complete separable space).

(¹) For a more general approach to the notion of analytic sets (which do not need to be metric), see G. Choquet, *Theory of capacities*, Ann. Inst. Fourier 5 (1953–1954), pp. 131–294, and *Ensembles K-analytiques et K-Susliniens. Cas général et cas métrique, ibid.* 9 (1959), pp. 75–81; Z. Frolik, *On the descriptive theory of sets* and *On bianalytic spaces*, Czechosl. Math. Journ. 13 (1963), pp. 335–359 and pp. 561–572. See also V. Shneider, *Descriptive theory of sets in topological spaces*, Dokl. Akad. Nauk SSSR, 50 (1945), p. 77; M. Sion, *On analytic sets in topological spaces*, Trans. Amer. Math. Soc. 96 (1960), pp. 341–354; D. W. Brassler and M. Sion, *The current theory of analytic sets*, Canad. J. Math. 16 (1964), pp. 207–230, and C. A. Rogers, *Analytic sets in Hausdorff spaces*, Mathematica 11 (1964), pp. 1–8.

By Corollary 1 of § 36, V, we have:

THEOREM 0. *Every uncountable analytic set contains $\mathscr{C}$ topologically* (1).

Thus, for the analytic sets the continuum hypothesis is fulfilled: *their power is either finite, or* $\aleph_0$, *or* $\mathfrak{c}$.

We shall see (compare Section II), that every analytic set, and hence every ***CA*** set as well, has the Baire property in the restricted sense and is Lebesgue measurable (if it is contained in $\mathscr{E}$).

Remarks. (i) The above properties show that the analytic sets have certain "regularity" which accounts for their applications. As one of the applications of the analytic set theory we quote the important theorem on inversion of continuous functions (Section V, Theorem 3).

Numerous problems of an elementary character will be seen to lead to analytic or ***CA*** sets. There are also important and very simple examples of sets that are projective but not borelian. Thus, in the space $2^{\mathscr{I}}$ (of closed subsets of the interval), the set of closed uncountable sets is analytic, but non-borelian (2); the set of differentiable functions is a non-borelian ***CA*** set in the space $\mathscr{E}^{\mathscr{I}}$ (3); the functions f of two variables x and y for which there exists a y such that the partial derivative $f'_x(x, y)$ exists for every $x \in \mathscr{I}$ form a ***PCA*** set in the space $\mathscr{E}^{\mathscr{I} \times \mathscr{I}}$ (which is not of a lower class) (4).

(ii) It is essential to notice that the projective sets of higher classes do not, in general, have the same "regularity" as the analytic sets. In fact, the hypothesis of *existence of non-measurable sets of class* ***PCA*** *and* ***CPCA*** (*in the space of real numbers*) is non-contradictory with the system of axioms of the set theory (5).

(1) See M. Souslin, C. R. Paris 164 (1917).

(2) W. Hurewicz, *Zur Theorie der analytischen Mengen*, Fund. Math. 15 (1930), pp. 4–17.

(3) S. Mazurkiewicz, *Über die Menge der differenzierbaren Funktionen*, Fund. Math. 27 (1936), p. 244.

(4) S. Mazurkiewicz, *Eine projektive Menge der Klasse PCA im Funktionalraum*, Fund. Math. 28 (1937), p. 7.

(5) See K. Gödel, *The consistency of the axiom of choice and of the generalized continuum hypothesis*, Proc. Nat. Acad. Sci. 24 (1938), p. 556, and Annals of Math. Studies 3, Princeton 1951.

See also J. W. Addison, *Some consequences of the axiom of constructibility*, Fund. Math. 46 (1959), pp. 338–357; P. S. Novikov, *On the consistency of*

The same can be said about the *existence of uncountable* $\boldsymbol{CA}$ *sets that do not contain perfect subsets.*

(iii) As was proved by W. Hurewicz [1], the result of the operation $(\mathscr{A})$ performed on closed subsets of a metric separable (not necessarily complete) space contains a perfect set (provided that it is uncountable).

Since every analytic set is the result of the operation $(\mathscr{A})$ performed on closed sets (Section II), the Hurewicz theorem is a generalization of the theorem which states that every uncountable analytic set contains a perfect set.

For another generalization, see Section II, Corollary 3.

THEOREM 1. *Given an uncountable analytic set* $A \subset \mathscr{X}$, *every analytic set* A^* *can be obtained from* A *by a mapping of class* 1 [2].

Proof. Clearly it is sufficient to prove the theorem for $A^* = \mathscr{N}$.

Since A is analytic and uncountable, it contains a set homeomorphic to $\mathscr{C}$, and also a set N homeomorphic to $\mathscr{N}$. Let $f(x) = x$ for $x \in N$. Since N is a $\boldsymbol{G}_\delta$-set in the space $\mathscr{X}$ which contains A (compare § 35, III), N is topologically complete and there exists a function f^* of class 1 defined in the entire space $\mathscr{X}$ with values belonging to N and such that f^* coincides with f on N (§ 35, VI, Corollary). It follows that $f^*(A) = N$.

THEOREM 2. *Given two uncountable analytic sets of dimension* 0, *one of them is a continuous image of the other* [3].

THEOREM 3. *Every analytic set is a one-to-one continuous image of a* $\boldsymbol{CA}$ *set* [4].

---

some statements of the descriptive set theory (Russian), Trudy Mat. Inst. Stiekl. 38 (1951), pp. 279–316; a paper of mine *Ensembles projectifs et ensembles singuliers*, Fund. Math. 35 (1948), pp. 131–140, and a paper of W. Sierpiński and myself *Sur l'existence des ensembles projectifs non mesurables*, Académie Bulgare 61 (1941), pp. 207–212.

[1] *Relativ perfekte Teile von Punktmengen und Mengen* (A), Fund. Math. 12 (1928), pp. 78–109.

[2] W. Sierpiński, *Sur les images de Baire des ensembles linéaires*, Fund. Math. 15(1930), p. 195.

[3] For a proof, see W. Sierpiński, *Sur les images continues des ensembles analytiques linéaires ponctiformes*, Fund. Math. 14 (1929), p. 345.

[4] S. Mazurkiewicz, *Sur une propriété des ensembles* $C(A)$, Fund. Math. 10 (1927), p. 172.

More precisely, *if f is a continuous function defined on a closed subset F of $\mathcal{N}$, then F contains a* ***CA****-set C such that $f(C) = f(F)$ and that the partial function $f \mid C$ is one-to-one*(¹).

Proof. Let us consider the lexicographical ordering of the irrational numbers $\mathfrak{z} = [\mathfrak{z}^1, \mathfrak{z}^2, \ldots]$

$$[\mathfrak{z} \prec \mathfrak{y}] \equiv (\mathfrak{z}^1 \leqslant \mathfrak{y}^1) \wedge \bigwedge_k [(\mathfrak{z}^k \leqslant \mathfrak{y}^k) \vee \bigvee_{i<k} (\mathfrak{z}^i < \mathfrak{y}^i)]. \tag{1}$$

It is easily seen that the propositional function $[\mathfrak{z} \prec \mathfrak{y}]$ is of class $\boldsymbol{F}_0$. Moreover, in every non-empty closed set X there exists the first element. In fact, this is the point $\mathfrak{p}(X) = X_1 \cap X_2 \cap \ldots$, where X_1 is the set of the $\mathfrak{z} \in X$ such that $\mathfrak{z}^1$ assumes the minimum value; and X_n, with $n > 1$, is the set of the $\mathfrak{z} \in X_{n-1}$ such that $\mathfrak{z}^n$ assumes the minimum value. By the Cantor theorem (§ 34, II), the intersection of the sets X_n reduces to a single point.

Now let C be the set of the points $\mathfrak{p}[f^{-1}(x)]$, where x runs over $f(F)$; that is

$$[\mathfrak{z} \in C] \equiv [\mathfrak{z} \in F] \wedge \bigwedge_{\mathfrak{y}} \{[f(\mathfrak{z}) = f(\mathfrak{y})] \Rightarrow [\mathfrak{z} \prec \mathfrak{y}]\}. \tag{2}$$

C is a ***CA*** set, since the propositional function in the brackets $\{\}$ is borelian. Moreover, the partial function $f \mid C$ is one-to-one, since the conditions $\mathfrak{z} \in C$, $\mathfrak{y} \in C$ and $f(\mathfrak{z}) = f(\mathfrak{y})$ imply $\mathfrak{z} \prec \mathfrak{y} \prec \mathfrak{z}$, and hence $\mathfrak{z} = \mathfrak{y}$. Finally we have $f(C) = f(F)$, since $\mathfrak{z}$ can be replaced by $\mathfrak{p}[f^{-1}(x)]$ in the left side member of (2).

II. Analytic sets as results of the operation $(\mathcal{A})$. It is easily seen that the set $\mathcal{N}_{n_1 \ldots n_k}$ of irrational numbers $\mathfrak{z} = \frac{1|}{|\mathfrak{z}^1} + \frac{1|}{|\mathfrak{z}^2} + \ldots$ such that $\mathfrak{z}^1 = n_1, \ldots, \mathfrak{z}^k = n_k$ (compare § 36, I) satisfies the following propositions:

$$\mathcal{N} = \bigcup_{n=1}^{\infty} \mathcal{N}_n, \quad \mathcal{N}_{n_1 \ldots n_k} = \bigcup_{n=1}^{\infty} \mathcal{N}_{n_1 \ldots n_k n}, \tag{1}$$

$$\mathcal{N}_{n_1 \ldots n_k} \cap \mathcal{N}_{m_1 \ldots m_k} = 0, \quad \text{if} \quad (n_1 \ldots n_k) \neq (m_1 \ldots m_k), \tag{2}$$

$$(\mathfrak{z}) = \bigcap_{k=1}^{\infty} \mathcal{N}_{\mathfrak{z}^1 \ldots \mathfrak{z}^k}, \tag{3}$$

$$\mathcal{N} = \bigcup_{\mathfrak{z}} \bigcap_{k=1}^{\infty} \mathcal{N}_{\mathfrak{z}^1 \ldots \mathfrak{z}^k} = \bigcap_{k=1}^{\infty} \bigcup_{\mathfrak{z}} \mathcal{N}_{\mathfrak{z}^1 \ldots \mathfrak{z}^k}. \tag{4}$$

(¹) For a generalization of this statement, see Section V, 5.

Formula (4) shows that the set $\mathcal{N}$ is the result of the operation $(\mathscr{A})$ performed on the system of sets $\mathcal{N}_{n_1 \dots n_k}$.

$$\lim_{k=\infty} \delta(\mathcal{N}_{\mathfrak{z}^1 \dots \mathfrak{z}^k}) = 0. \tag{5}$$

If f is a continuous function defined on $\mathcal{N}$, then (6)

$$f(\mathfrak{z}) = \bigcap_{k=1}^{\infty} f(\mathcal{N}_{\mathfrak{z}^1 \dots \mathfrak{z}^k}) = \bigcap_{k=1}^{\infty} \overline{f(\mathcal{N}_{\mathfrak{z}^1 \dots \mathfrak{z}^k})}.$$

We shall prove (6). By (3) we have

$$f(\mathfrak{z}) = f\Big(\bigcap_{k=1}^{\infty} \mathcal{N}_{\mathfrak{z}^1 \dots \mathfrak{z}^k}\Big) \subset \bigcap_{k=1}^{\infty} f(\mathcal{N}_{\mathfrak{z}^1 \dots \mathfrak{z}^k}) \subset \bigcap_{k=1}^{\infty} \overline{f(\mathcal{N}_{\mathfrak{z}^1 \dots \mathfrak{z}^k})},$$

and by (5), the continuity of f implies that $\lim_{k=\infty} \delta[f(\mathcal{N}_{\mathfrak{z}^1 \dots \mathfrak{z}^k})] = 0$, whence $\lim_{k=\infty} \delta[\overline{f(\mathcal{N}_{\mathfrak{z}^1 \dots \mathfrak{z}^k})}] = 0$ and $\delta\Big(\bigcap_{k=1}^{\infty} \overline{f(\mathcal{N}_{\mathfrak{z}^1 \dots \mathfrak{z}^k})}\Big) = 0$. It follows that the set $\bigcap_{k=1}^{\infty} \overline{f(\mathcal{N}_{\mathfrak{z}^1 \dots \mathfrak{z}^k})}$ reduces to a single point.

THEOREM. *A set is analytic if and only if it is the result of the operation $(\mathscr{A})$ performed on a regular system of closed sets.*

Proof. If A is analytic, then there exists a continuous function f such that $A = f(\mathcal{N})$. Thus A is the result of the operation $(\mathscr{A})$ performed on the sets $\overline{f(\mathcal{N}_{n_1 \dots n_k})}$, since we have, by (6),

$$A = f(\mathcal{N}) = \bigcup_{\mathfrak{z}} f(\mathfrak{z}) = \bigcup_{\mathfrak{z}} \bigcap_{k=1}^{\infty} \overline{f(\mathcal{N}_{\mathfrak{z}^1 \dots \mathfrak{z}^k})}.$$

Conversely, operation $(\mathscr{A})$ performed on closed sets (even on analytic sets) always leads to analytic sets (§ 38, IX, Theorem 4).

COROLLARY 1. *The analytic sets have the Baire property in the restricted sense.*

For, this property belongs to closed sets and is invariant under operation $(\mathscr{A})$, by the corollary of § 11, VII.

For the same reason, *the analytic sets are Lebesgue measurable* (if they are sets of real numbers, complex numbers and so on).

The complement of a set has the Baire property, or is measurable, whenever the set itself has the Baire property, or is measurable, respectively. It follows that the ***CA*** sets have these properties.

The same is true for the sets which are obtained from closed sets by means of subtraction and operation $(\mathscr{A})$.

Remark 1. By Corollary 1, the set E_1 of § 38, VI, is non-borelian and has the Baire property in the restricted sense(1).

COROLLARY 2. *Every analytic set A is sieved by a sieve composed of closed sets.*

In other words, *there exists a family of closed sets W_r, where $r \in \mathscr{R}_0$, such that the condition $x \in A$ is equivalent to the existence of an infinite sequence of numbers $r_1 < r_2 < \ldots$ satisfying the condition $x \in W_{r_1} \cap W_{r_2} \cap \ldots$.*

This is a consequence of the theorem (by § 3, XV).

Remark 2. If $A \subset \mathscr{I}$ is defined by means of a sieve W, then an element $f(W) \in \mathscr{I} - A$ can be defined *effectively* (provided $A \neq \mathscr{I}$)(2).

COROLLARY 3. *Every projective* ***PCA*** *set E is the union of $\aleph_1$ Borel sets.*

It follows that *if $\overline{\overline{E}} > \aleph_1$, then E contains a perfect set* $(\neq 0)$.

Proof. Assume first that E is analytic. Hence it is the result of operation $(\mathscr{A})$ performed on a system of closed sets. Using the notation of § 3, XIV, we have

$$E = \bigcap_{a<\Omega} E_a = \bigcup_{a<\Omega} K_a. \tag{7}$$

It is easily proved by transfinite induction that E_a and K_a are Borel sets. Thus the first part of the corollary is proved in the case when E is an $\boldsymbol{A}$ or $\boldsymbol{CA}$ set(3).

Assume now that E is a $\boldsymbol{PCA}$ set and let $E = f(C)$, where C is a $\boldsymbol{CA}$ set and f is a continuous function. We have

$$C = \bigcup_{a<\Omega} B_a, \quad \text{hence} \quad E = f(C) = \bigcup_{a<\Omega} f(B_a).$$

(1) See N. Lusin and W. Sierpiński, *Sur un ensemble non mesurable B*, Journ. de Math. 1923, p. 53.

(2) See N. Lusin and P. Novikov, *Choix effectif d'un point dans un complémentaire analytique arbitraire, donné par un crible*, Fund. Math. 25 (1935), p. 559, and Y. Sampei, Comm. Math. Univ. S. Pauli 9 (1961), p. 91.

(3) Compare Theorem VIII, 3.

Since the B_a are borelian, $f(B_a)$ is analytic and therefore it is the union of $\aleph_1$ Borel sets. It follows that E is also the union of $\aleph_1$ Borel sets.

The second part of the corollary follows from the first one. For, if $\overline{\overline{E}} > \aleph_1$, then at least one of the members of the decomposition of E is uncountable and contains a non-empty perfect subset (§ 37, I, Theorem 3).

Remark 3. We see thus that the uncountable ***PCA*** sets can be either of power $\aleph_1$ or of power $\mathfrak{c}$. We do not know, however, if there exist ***PCA*** sets, or more generally, projective sets, of power $\aleph_1$.

No analogous theorem is known for ***CPCA*** sets.

COROLLARY 4. *Every uncountable complete separable space $\mathscr{X}$ can be decomposed into $\aleph_1$ non-empty disjoint Borel sets.*

Proof. Let E be an analytic non-borelian subset of $\mathscr{X}$ (it exists by § 38, VI). Applying formula (7) and letting

$$A_a = K_a - \bigcup_{\xi<a} K_\xi, \quad \text{we get} \quad E = \bigcup_{a<\Omega} A_a.$$

Similarly, if we let

$$L_a = \mathscr{X} - E_a \quad \text{and} \quad B_a = L_a - \bigcup_{\xi<a} L_\xi,$$

we have

$$\mathscr{X} - E = \bigcup_{a<\Omega} L_a = \bigcup_{a<\Omega} B_a.$$

Consequently

$$\mathscr{X} = \bigcup_{a<\Omega} A_a \cup \bigcup_{a<\Omega} B_a. \tag{8}$$

A_a and B_a are Borel sets. The members of decomposition (8) are mutually disjoint. There are uncountably many indices a such that $A_a \neq 0$, for, otherwise, E would be a Borel set (as a countable union of Borel sets).

Remark 4. A decomposition of any uncountable complete separable space $\mathscr{X}$ into $\aleph_1$ non-empty disjoint Borel sets can also be obtained by using the following Hausdorff theorem [1]. *In the space $\mathscr{X}$ there is a transfinite sequence of different $\boldsymbol{G}_\delta$-sets $G_0 \subset G_1 \subset \ldots \subset G_a \subset \ldots$ $(a < \Omega)$ such that $\mathscr{X} = \bigcup_{a<\Omega} G_a$.*

If we let $F_a = G_a - \bigcup_{\xi<a} G_\xi$, we have $\mathscr{X} = \bigcup_{a<\Omega} F_a$, where F_a are non-empty and disjoint $\boldsymbol{F}_{\sigma\delta}$ sets.

---

(1) *Summen von $\aleph_1$ Mengen*, Fund. Math. 26 (1936), p. 248.

The advantage of this decomposition, in contrast to (8), is that its members are of class $F_{\sigma\delta}$, while the members of (8) are of non-bounded classes. On the other hand, decomposition (8) is *effective* (both the set E and the members can be defined), while the existence of the sequence $\{G_a\}$ is not established effectively.

We mention that the problem of decomposing a space into $\aleph_1$ non-empty and disjoint G_δ-sets is open(1).

III. First separation theorem.

THEOREM. *If A and B are two disjoint analytic sets, there exist a Borel set E such that*

$$A \subset E \quad \textit{and} \quad E \cap B = 0\,(^2).$$

LEMMA(3). *Let us say that a pair of sets A, B is "B separable", if it fulfills the assertion of the theorem. If the sets $P = P_1 \cup P_2 \cup \ldots$ and $Q = Q_1 \cup Q_2 \cup \ldots$ are not B separable, then there exists a pair P_n, Q_m which is not B separable.*

Proof. Let Z_{nm} be a Borel set such that

$$P_n \subset Z_{nm} \subset Q_m^c \; (= \mathscr{X} - Q_m).$$

It follows that

$$P_n \subset \bigcap_{m=1}^{\infty} Z_{nm} \subset \bigcap_{m=1}^{\infty} Q_m^c = \Big(\bigcup_{m=1}^{\infty} Q_m\Big)^c = Q^c,$$

hence

$$P = \bigcup_{n=1}^{\infty} P_n \subset \bigcup_{n=1}^{\infty} \bigcap_{m=1}^{\infty} Z_{nm} \subset Q^c,$$

which shows that the Borel set $\bigcup_{n=1}^{\infty} \bigcap_{m=1}^{\infty} Z_{nm}$ separates the sets P and Q.

---

(1) Compare W. Sierpiński, *Sur deux conséquences d'un théorème de Hausdorff*, Fund. Math. 33 (1945), p. 272.

(2) Theorem of N. Lusin, *Sur les ensembles analytiques*, Fund. Math. 10 (1927), p. 52. It should be noticed that the analogous theorem for CA sets is false. See a simple example of W. Sierpiński, *Sur deux complémentaires analytiques non separables B*, Fund. Math. 17 (1931), p. 296, and another of N. Lusin, *Ensembles analytiques*, p. 220, 260, 263. See also P. Novikov, Fund. Math. 17, p. 25.

Concerning the separation theorems for higher projective classes and their relations to the axiom of constructibility of Gödel, see J. W. Addison, *Separation principles in the hierarchies of classical and effective set theory*, Fund. Math. 46 (1958), pp. 123–135. See also P. S. Novikov, *loc. cit.*

(3) W. Sierpiński, *Zarys teorii mnogości*, II, p. 180.

Proof of the theorem. Suppose the contrary is true, i.e. the disjoint analytic sets A and B are not B separable. Let f and g be two continuous functions mapping $\mathcal{N}$ onto A and onto B: $A = f(\mathcal{N})$, $B = g(\mathcal{N})$. It follows that (compare II (1)):

$$A = f(\mathcal{N}_1) \cup f(\mathcal{N}_2) \cup \ldots \quad \text{and} \quad B = g(\mathcal{N}_1) \cup g(\mathcal{N}_2) \cup \ldots .$$

By the lemma, some two sets $f(\mathcal{N}_{n_1})$ and $g(\mathcal{N}_{m_1})$ are not B separable. Since, by II (1),

$$f(\mathcal{N}_{n_1 \ldots n_k}) = \bigcup_{n=1}^{\infty} f(\mathcal{N}_{n_1 \ldots n_k n}) \quad \text{and} \quad g(\mathcal{N}_{m_1 \ldots m_k}) = \bigcup_{m=1}^{\infty} g(\mathcal{N}_{m_1 \ldots m_k m}),$$

we deduce, by induction, the existence of two infinite sequences of indices $n_1, n_2, \ldots$ and $m_1, m_2, \ldots$ such that, for each k, the sets $f(\mathcal{N}_{n_1 \ldots n_k})$ and $g(\mathcal{N}_{m_1 \ldots m_k})$ are not B separable.

Let

$$\mathfrak{z} = \frac{1|}{|n_1} + \frac{1|}{|n_2} + \ldots \quad \text{and} \quad \mathfrak{y} = \frac{1|}{|m_1} + \frac{1|}{|m_2} + \ldots .$$

Since $f(\mathfrak{z}) \in A$, $g(\mathfrak{y}) \in B$ and $A \cap B = 0$, we have $|f(\mathfrak{z}) - g(\mathfrak{y})| > 0$. Since the functions f and g are continuous, we have for a sufficiently large k (compare II (5)):

$$\delta[f(\mathcal{N}_{n_1 \ldots n_k})] + \delta[g(\mathcal{N}_{m_1 \ldots m_k})] < |f(\mathfrak{z}) - g(\mathfrak{y})| .$$

Since

$$f(\mathfrak{z}) \in f(\mathcal{N}_{n_1 \ldots n_k}) \quad \text{and} \quad g(\mathfrak{y}) \in g(\mathcal{N}_{m_1 \ldots m_k})$$

(compare II (3)), it follows that

$$\overline{f(\mathcal{N}_{n_1 \ldots n_k})} \cap \overline{g(\mathcal{N}_{m_1 \ldots m_k})} = 0$$

and hence the sets $f(\mathcal{N}_{n_1 \ldots n_k})$ and $g(\mathcal{N}_{m_1 \ldots m_k})$ are B separable. But this contradicts the hypothesis.

COROLLARY 1 ([1]). *If the sets A and $\mathcal{X} - A$ are analytic, then A is a Borel set.*

For, if we let $B = \mathcal{X} - A$ in the preceding theorem, we get $E = A$.

Corollary 1 implies that the *family of sets that are both* **A** *and* **CA** *coincides with the family of Borel sets.*

([1]) M. Souslin, *loc. cit.*

COROLLARY 2 (simultaneous separation). *If $A_1, A_2, \ldots$ is a sequence of disjoint analytic sets, then there exists a sequence of disjoint Borel sets $B_1, B_2, \ldots$ such that $A_n \subset B_n$.*

Proof. By the separation theorem, there exists a system of Borel sets E_{nm} such that

$$A_n \subset E_{nm} \quad \text{and} \quad E_{nm} \cap A_m = 0 \quad (\text{for } n \neq m).$$

Let

$$B_1 = \bigcap_{m=2}^{\infty} E_{1m} \quad \text{and} \quad B_n = \bigcap_{m \neq n} E_{nm} - (B_1 \cup \ldots \cup B_{n-1}).$$

The inclusion $B_m \subset E_{mn}$ implies $A_n \cap B_m \subset A_n \cap E_{mn} = 0$ for $m < n$. Therefore $A_n \subset B_n$.

For another generalization of the separation theorem, see Section X.

IV. Applications to Borel sets.

THEOREM[1]. *Every one-to-one continuous image of a Borel set is a Borel set.*

Proof. Every Borel set is composed of a countable set and of a one-to-one continuous image of the set $\mathcal{N}$ (§ 37, II, Corollary 1c). It suffices therefore to prove that if f is a one-to-one continuous mapping of $\mathcal{N}$, then $f(\mathcal{N})$ is a Borel set.

Since the function f is one-to-one, it follows from II (2) that the sets of the system $\{f(\mathcal{N}_{n_1 \ldots n_k})\}$ are disjoint for a fixed k. By Corollary 2 of Section III, for each k there exists a system of disjoint Borel sets $B_{n_1 \ldots n_k}$ such that $f(\mathcal{N}_{n_1 \ldots n_k}) \subset B_{n_1 \ldots n_k}$.

Let us set $B^*_{n_1} = B_{n_1} \cap \overline{f(\mathcal{N}_{n_1})}$ and

$$B^*_{n_1 \ldots n_k} = B_{n_1 \ldots n_k} \cap \overline{f(\mathcal{N}_{n_1 \ldots n_k})} \cap B^*_{n_1 \ldots n_{k-1}}.$$

We shall prove the inclusion

$$f(\mathcal{N}_{n_1 \ldots n_k}) \subset B^*_{n_1 \ldots n_k} \subset \overline{f(\mathcal{N}_{n_1 \ldots n_k})}. \tag{1}$$

This formula is obvious for $k = 1$. Assume it for $k-1$. By II (1), we have

$$f(\mathcal{N}_{n_1 \ldots n_k}) \subset f(\mathcal{N}_{n_1 \ldots n_{k-1}}) \subset B^*_{n_1 \ldots n_{k-1}},$$

and formula (1) follows.

---

[1] M. Souslin, *loc. cit.* For generalizations of this theorem, see Sections V and VIII.

Formula (1) implies by virtue of II (6) that

$$\bigcap_{k=1}^{\infty} f(\mathcal{N}_{\mathfrak{z}^1\ldots\mathfrak{z}^k}) = \bigcap_{k=1}^{\infty} B^*_{\mathfrak{z}^1\ldots\mathfrak{z}^k} = \bigcap_{k=1}^{\infty} \overline{f(\mathcal{N}_{\mathfrak{z}^1\ldots\mathfrak{z}^k})},$$

and, since $f(\mathcal{N}) = \bigcup_{\mathfrak{z}} \bigcap_{k=1}^{\infty} f(\mathcal{N}_{\mathfrak{z}^1\ldots\mathfrak{z}^k})$, by II (6), it follows that

$$f(\mathcal{N}) = \bigcup_{\mathfrak{z}} \bigcap_{k=1}^{\infty} B^*_{\mathfrak{z}^1\ldots\mathfrak{z}^k}.$$

Now, the system $\{B^*_{n_1\ldots n_k}\}$ is regular, i.e. $B^*_{n_1\ldots n_k} \subset B^*_{n_1\ldots n_{k-1}}$ and, for a fixed k, it is composed of disjoint sets.

It follows that (compare § 3, XIV, Theorem 2)

$$\bigcup_{\mathfrak{z}} \bigcap_{k=1}^{\infty} B^*_{\mathfrak{z}^1\ldots\mathfrak{z}^k} = \bigcap_{k=1}^{\infty} \bigcup_{\mathfrak{z}} B^*_{\mathfrak{z}^1\ldots\mathfrak{z}^k}.$$

For a fixed k, the union $\bigcup_{\mathfrak{z}} B^*_{\mathfrak{z}^1\ldots\mathfrak{z}^k}$ is taken over a countable system of Borel sets; hence $\bigcup_{\mathfrak{z}} B^*_{\mathfrak{z}^1\ldots\mathfrak{z}^k}$ is a Borel set. It follows that $f(\mathcal{N})$ is also a Borel set.

Remark 1. The preceding theorem and Corollary 1a of § 37, II imply that the *Borel subsets of a complete separable space coincide with one-to-one continuous images of* 0-*dimensional complete separable spaces.*

Remark 2. If B is a Borel set and g is a one-to-one continuous mapping, then the space $\mathcal{Y}$ containing $g(B)$ can be assumed to be an *arbitrary metric* space (*complete separable or not*). For, as a continuous image of a separable set, $g(B)$ is separable. The set $\overline{g(B)}$ (which is also separable) can be completed to a space $\widetilde{\overline{g(B)}}$ (see § 33, VII). By the preceding theorem, $g(B)$ is borelian in $\widetilde{\overline{g(B)}}$, so it is borelian in $\overline{g(B)}$. Hence $g(B)$ is the intersection of $\overline{g(B)}$ with a Borel set in $\mathcal{Y}$. It follows that $g(B)$ is a Borel set in $\mathcal{Y}$.

On the other hand, the assumptions that the space $\mathcal{X}$ (containing B) is *complete* and *separable* cannot be dropped. For instance, let B be a non-borelian subset of the interval $\mathcal{I} = \mathcal{Y}$ and let

(i) $\mathcal{X} = B$ (case when $\mathcal{X}$ is not complete),

(ii) $\mathcal{X}$ is composed of the points of the interval $\mathcal{I}$ with the distance of any two different points defined to be equal to one (case when $\mathcal{X}$ is not separable).

In both cases, the set B, borelian in $\mathcal{X}$, is mapped by the identity mapping onto a non-borelian set (in $\mathcal{Y}$).

V. Applications to B measurable functions.

THEOREM 1. *If f is a one-to-one B measurable function defined on a Borel set E, then $f(E)$ is a Borel set.*

Proof. Since the function f is one-to-one, the projection of the set $I = \underset{xy}{E}\{y = f(x)\}$ to the $\mathcal{Y}$ axis is a one-to-one continuous mapping of I onto $f(E)$. By § 38, III, Theorem 4, I is a Borel set, and so is $f(E)$, by the Theorem of Section IV.

THEOREM 2. *If f is a function defined on an analytic set A such that the set $I = \underset{xy}{E}\{y = f(x)\}$ is analytic, then f is B measurable.*

Consequently (see § 38, III, Theorem 4), *if A is borelian and I analytic, then I is borelian.*

Proof. We must show that, for every closed subset F of $\mathcal{Y}$, the set $f^{-1}(F)$ is borelian in A. Now the set $f^{-1}(F)$ is analytic, since it is a projection of the analytic set $I \cap (\mathcal{X} \times F)$. For the same reason the set $f^{-1}(\mathcal{Y} - F) = A - f^{-1}(F)$ is analytic. By the separation theorem of Section III there exists a Borel set E such that $f^{-1}(F) \subset E$ and $E \cap A - f^{-1}(F) = 0$, whence $f^{-1}(F) = E \cap A$. This proves that $f^{-1}(F)$ is a Borel set in A.

THEOREM 3. *If f is a one-to-one B measurable function defined on an analytic set A, then the inverse function f^{-1} is B measurable on the set $f(A)$; i.e. the function f is a generalized homeomorphism* (in the sense of § 31, I)(1).

Proof. In the preceding theorem we replace x by y, y by x and A by $f(A)$. Since

$$\underset{xy}{E}\{x = f^{-1}(y)\} = \underset{xy}{E}\{y = f(x)\} = I$$

and since the sets I and $f(A)$ are analytic (compare § 38, III, 4 and 5), the theorem follows.

THEOREM 4. *If f is a one-to-one B measurable function defined on a Borel set E and P is a projective subset of E of class $\mathbf{L}_n$ $(n \geqslant 0)$, then the set $f(P)$ is also of class $\mathbf{L}_n$.*

(1) This theorem is originally due to Lebesgue (for $A = \mathcal{I}$), Journ. de Math. 1905, *op. cit.* The analysis of the Lebesgue's proof, which was not correct, suggested the idea of analytic sets to Souslin. A correct proof is due to Lusin and Souslin.

Proof. By Theorems 3 and 1, the function f^{-1} is B measurable on $f(E)$, so by § 38, III, 5, the set $f(P) = (f^{-1})^{-1}(P)$ is of class $\boldsymbol{L}_n$.

THEOREM 5. *If f is a continuous function on a Borel set E, then there exists a* ***CA*** *subset C of E such that the partial function $f \mid C$ is one-to-one and $f(C) = f(E)$.*

Proof. Being a Borel set, E is a one-to-one continuous image of a closed set $F \subset \mathcal{N}$ (see § 37, II, Corollary 1a): $E = g(F)$. Let $h(\mathfrak{z}) = fg(\mathfrak{z})$, where $\mathfrak{z} \in F$. By I, 3, F contains a ***CA*** subset H such that the partial function $h \mid H$ is one-to-one and $h(H) = h(F)$. Since the function g is one-to-one and continuous, the set $C = g(H)$ is a ***CA*** set by Theorem 4.

Moreover,

$$f(C) = fg(H) = h(H) = h(F) = fg(F) = f(E).$$

Finally, if $x \neq x'$, $x \in C$ and $x' \in C$, then

$$x = g(\mathfrak{z}), \quad x' = g(\mathfrak{z}'), \quad \mathfrak{z} \neq \mathfrak{z}', \quad \mathfrak{z} \in H, \quad \text{and} \quad \mathfrak{z}' \in H.$$

The inequality $\mathfrak{z} \neq \mathfrak{z}'$ implies

$$f(x) = h(\mathfrak{z}) \neq h(\mathfrak{z}') = f(x'),$$

so that the function $f \mid C$ is one-to-one.

Remarks. (i) For every α there exists a one-to-one continuous function f such that the inverse function f^{-1} is not of class α (see § 31, I, 5)[1].

(ii) In Theorem 3, the assumption of *completeness* of the space $\mathcal{X}$ is essential. For instance, let Z be a non-borelian subset of the interval $\mathcal{I}$, and let Z^* be its complement placed in the interval 1,2, let $\mathcal{X} = Z \cup Z^*$, $\mathcal{Y} = \mathcal{I}$, $f(x) = x$ for $x \in Z$, and $f(x) = x - 1$, for $x \in Z^*$. The function f^{-1} is not B measurable (it has not even the Baire property, if the set Z does not have the Baire property).

Example (ii) of Section IV (Remark) shows that the assumption of separability of $\mathcal{X}$ is essential.

---

(1) Compare W. Sierpiński, *Sur l'inversion des fonctions représentables analytiquement*, Fund. Math. 3 (1922), p. 26.

(iii) The assumption of separability of $\mathcal{Y}$ in Theorem 1 is not necessary. It is of course sufficient to show that $f(E)$ is separable. Suppose it is not. Then it contains a discrete uncountable set B closed in $f(E)$. Let $A \subset E$ be such that $f(A) = B$ and $f|A$ is one-to-one. As each subset of B is closed in $f(E)$, so each subset of A is borelian in E, hence in $\mathcal{X}$. Since A is borelian uncountable, hence $\overline{\overline{A}} = \mathfrak{c}$. But then A contains non-borelian subsets(¹).

(iv) *Application to metric groups.* Let $\mathcal{X}$ and $\mathcal{Y}$ be two complete spaces which are topological groups (compare § 13, XII). Let f be an additive function (i.e. $f(x+x') = f(x)+f(x')$) which maps $\mathcal{X}$ onto $\mathcal{Y}$. It can be proved that the function f is continuous, whenever it has the Baire property (in particular, if it is B measurable)(²). It follows that *if the groups $\mathcal{X}$ and $\mathcal{Y}$ are complete and separable, then every one-to-one continuous additive function f such that $f(\mathcal{X}) = \mathcal{Y}$ is a homeomorphism*(³). For, if $f\colon \mathcal{X} \to \mathcal{Y}$ is additive, so is f^{-1} and since f^{-1} is B measurable, it is continuous.

(v) Theorem 5 has been generalized to arbitrary CA sets(⁴).

VI. Second separation theorem.

THEOREM(⁵). *If A and B are two analytic sets, then there exist two CA sets, D and H, such that*

$$A-B \subset D, \quad B-A \subset H, \quad \textit{and} \quad D \cap H = 0. \tag{1}$$

(¹) See A. H. Stone, *Non-separable Borel sets*, Rozprawy Matematyczne 28 (1962), p. 32, Theorem 16.

(²) S. Banach, *Théorie des opérations linéaires*, Mon. Mat. (1932), p. 23.

(³) S. Banach, *Über metrische Gruppen*, Studia Math. 3 (1931), p. 111. For numerous applications of this theorem to differential equations, see the Banach monograph referred to above, Chapter III. Compare, of the same author, *Sur les fonctionnelles linéaires II*, Studia Math. 1 (1929), p. 238 and J. Schauder, *Über die Umkehrung linearer, stetiger Funktionaloperationen*, Studia Math. 2 (1930), p. 1.

(⁴) See M. Kondô, *Sur l'uniformisation des complémentaires analytiques et les ensembles projectifs de la seconde classe*, Japan. J. Math. 15 (1938), pp. 197–230. For simpler proofs, see Y. Sampei, *On the uniformization of the complement of an analytic set*, Comment. Math. Univ. St. Paul. 10 (1960), pp. 57–62, and Y. Suzuki, *On the uniformization principle*, Proceed. Symp. on the Foundations of Mathematics, Katada (Japan) 1962, pp. 137–144.

(⁵) See N. Lusin, *Ensembles analytiques*, p. 210 ("deuxième principe").

Proof. Let the sets W_r be as in Corollary 2 of II. Let us order the set $M_x = \underset{r}{E}(x \in W_r)$ by the relation $r > s$. We have the following equivalence:

$$\{x \in A\} \equiv \{\text{the set } M_x = \underset{r}{E}(x \in W_r) \text{ is not well ordered}\}.$$

Similarly, there exists a family of closed sets Z_r, with $r \in \mathscr{R}_0$, such that

$$\{x \in B\} \equiv \{\text{the set } N_x = \underset{r}{E}(x \in Z_r) \text{ is not well ordered}\}.$$

Set $M_x \precsim N_x$, if the set M_x is similar to a part of N_x. Let

$$D = B^c \cap [\underset{x}{E}(M_x \precsim N_x)]^c \quad \text{and} \quad H = A^c \cap [\underset{x}{E}(N_x \precsim M_x)]^c,$$

where X^c, as usual, is the complement of X.

Formula (1) holds. For, if $x \in A - B$, then M_x is not well ordered while N_x is; hence we cannot have $M_x \precsim N_x$ and therefore

$$x \in [\underset{x}{E}(M_x \precsim N_x)]^c.$$

Since $x \in B^c$, it follows that $x \in D$.

So $A - B \subset D$ and, by the symmetry, $B - A \subset H$.

On the other hand, if $x \in B^c \cap A^c$ both M_x nad N_x are well ordered. Hence they are comparable, i.e. either $M_x \precsim N_x$ or $N_x \precsim M_x$. Consequently, we cannot have $x \in [\underset{x}{E}(M_x \precsim N_x)]^c \cap [\underset{x}{E}(N_x \precsim M_x)]^c$. It follows that $D \cap H = 0$.

To prove that D and H are $\boldsymbol{CA}$ sets, it suffices to show that the set $\underset{x}{E}(M_x \precsim N_x)$ is analytic. But this is a particular case of § 38, IX, 2 (for $n = 1$).

COROLLARY 1 (simultaneous separation). *If $A_1, A_2, \ldots$ is a sequence of analytic sets, then there exists a sequence $C_1, C_2, \ldots$ of disjoint $\boldsymbol{CA}$ sets such that*

$$A_n - (A_1 \cup \ldots \cup A_{n-1} \cup A_{n+1} \cup \ldots) \subset C_n.$$

Proof. According to the preceding theorem, let

$$A_n - \bigcup_{k \neq n} A_k \subset D_n, \quad \bigcup_{k \neq n} A_k - A_n \subset H_n, \quad D_n \cap H_n = 0.$$

Let $C_n = D_n \cap \bigcap_{k \neq n} H_k$. Thus C_n are disjoint $\boldsymbol{CA}$ sets. Moreover,

$$A_n - \bigcup_{k \neq n} A_k \subset A_n - A_k \subset \bigcup_{k \neq n} A_n - A_k \subset H_k,$$

for every $k \neq n$. Therefore $A_n - \bigcup_{k \neq n} A_k \subset \bigcap_{k \neq n} H_k$ and the required inclusion follows.

COROLLARY 2. *If $A_{n_1 \dots n_k}$ is a system of analytic sets such that*

$$A_{n_1 \dots n_k} = A_{n_1 \dots n_k 1} \cup A_{n_1 \dots n_k 2} \cup \dots,$$

then there exists a regular system $\{C_{n_1 \dots n_k}\}$ of $\boldsymbol{CA}$ sets, disjoint for a fixed k, such that

$$A_{n_1 \dots n_k} - \bigcup' A_{m_1 \dots m_k} \subset C_{n_1 \dots n_k}, \tag{i}$$

where the union is taken over all systems $(m_1 \dots m_k) \neq (n_1 \dots n_k)$.

Proof. According to Corollary 1, let, for a fixed k, $\{D_{n_1 \dots n_k}\}$ be a system of disjoint $\boldsymbol{CA}$ sets such that

$$A_{n_1 \dots n_k} - \bigcup' A_{m_1 \dots m_k} \subset D_{n_1 \dots n_k}.$$

Let

$$C_{n_1} = D_{n_1} \quad \text{and} \quad C_{n_1 \dots n_k} = D_{n_1 \dots n_k} \cap C_{n_1 \dots n_{k-1}} \quad \text{for} \quad k > 1.$$

We must show that

$$A_{n_1 \dots n_k} - \bigcup' A_{m_1 \dots m_k} \subset C_{n_1 \dots n_{k-1}}.$$

But this follows from inclusion (i) assumed for $k-1$ and from the following formulas assumed by hypothesis:

$$A_{n_1 \dots n_k} \subset A_{n_1 \dots n_{k-1}},$$

$$A_{m_1 \dots m_{k-1}} = A_{m_1 \dots m_{k-1} 1} \cup A_{m_1 \dots m_{k-1} 2} \cup \dots \subset \bigcup' A_{m_1 \dots m_k}.$$

For other generalizations of the second separation theorem, see Section X.

The following section gives applications of this theorem.

VII. Order of value of a B measurable function. A point y is said to be a value of *order* 1 of the function f, written as $y \in Z_f$, if there exists a unique x such that $y = f(x)$. The following theorem is a generalization of that of Section IV.

THEOREM 1 [1]. *The set of values of order* 1 *of a B measurable function defined on a Borel set is a* **CA** *set.*

Proof. First we shall prove the theorem in the case when f is continuous. Every (uncountable) Borel set, after removing from it a suitably chosen countable set, is a one-to-one continuous image of the set $\mathcal{N}$ (§ 37, II, Corollary 1c). Thus it suffices to show that if f is a continuous function defined on $\mathcal{N}$, then the set Z of its values of order 1 is a **CA** set.

By definition we have

$$Z = \bigcup_{\mathfrak{z}} [f(\mathfrak{z}) - f(\mathcal{N} - \mathfrak{z})].$$

By II and § 34, III,

$$\mathfrak{z} = \bigcap_{k=1}^{\infty} \mathcal{N}_{\mathfrak{z}^1 \ldots \mathfrak{z}^k} \quad \text{and} \quad f\Big[\bigcap_{k=1}^{\infty} \mathcal{N}_{\mathfrak{z}^1 \ldots \mathfrak{z}^k}\Big] = \bigcap_{k=1}^{\infty} f(\mathcal{N}_{\mathfrak{z}^1 \ldots \mathfrak{z}^k}).$$

It follows that

$$f(\mathfrak{z}) - f(\mathcal{N} - \mathfrak{z}) = \bigcap_{k=1}^{\infty} f(\mathcal{N}_{\mathfrak{z}^1 \ldots \mathfrak{z}^k}) - f\Big[\bigcup_{k=1}^{\infty} (\mathcal{N} - \mathcal{N}_{\mathfrak{z}^1 \ldots \mathfrak{z}^k})\Big]$$
$$= \bigcap_{k=1}^{\infty} [f(\mathcal{N}_{\mathfrak{z}^1 \ldots \mathfrak{z}^k}) - f(\mathcal{N} - \mathcal{N}_{\mathfrak{z}^1 \ldots \mathfrak{z}^k})].$$

Let

$$A_{n_1 \ldots n_k} = f(\mathcal{N}_{n_1 \ldots n_k}) \quad \text{and} \quad B_{n_1 \ldots n_k} = f(\mathcal{N} - \mathcal{N}_{n_1 \ldots n_k}).$$

Therefore $Z = \bigcup_{\mathfrak{z}} \bigcap_{k=1}^{\infty} (A_{\mathfrak{z}^1 \ldots \mathfrak{z}^k} - B_{\mathfrak{z}^1 \ldots \mathfrak{z}^k})$.

By II (1) and (2), $B_{n_1 \ldots n_k} = \bigcup' A_{m_1 \ldots m_k}$, where the union $\bigcup'$ is taken over the systems $(m_1 \ldots m_k) \neq (n_1 \ldots n_k)$. On the other hand, by II (1),

$$A_{n_1 \ldots n_k} = A_{n_1 \ldots n_k 1} \cup A_{n_1 \ldots n_k 2} \cup \ldots$$

Thus Corollary 2 of Section VI can be applied. Let

$$C^*_{n_1 \ldots n_k} = C_{n_1 \ldots n_k} \cap \bar{A}_{n_1 \ldots n_k} - B_{n_1 \ldots n_k},$$

where the $C_{n_1 \ldots n_k}$ have the same meaning as in Corollary 2. Consequently,

$$A_{n_1 \ldots n_k} - B_{n_1 \ldots n_k} \subset C^*_{n_1 \ldots n_k} \subset \bar{A}_{n_1 \ldots n_k} - B_{n_1 \ldots n_k}, \tag{1}$$

[1] N. Lusin, *op. cit.*, p. 257.

and the sets $C^*_{n_1\dots n_k}$ are $\boldsymbol{CA}$, disjoint for a fixed k. Moreover, the system $\{C^*_{n_1\dots n_k}\}$ is regular, since $\{C_{n_1\dots n_k}\}$ is regular and the inclusion $\mathscr{N}_{n_1\dots n_k}\subset\mathscr{N}_{n_1\dots n_{k-1}}$ (compare II (1)) yields

$$A_{n_1\dots n_k}=f(\mathscr{N}_{n_1\dots n_k})\subset f(\mathscr{N}_{n_1\dots n_{k-1}})=A_{n_1\dots n_{k-1}}\subset\bar{A}_{n_1\dots n_{k-1}},$$
$$B_{n_1\dots n_k}=f(\mathscr{N}-\mathscr{N}_{n_1\dots n_k})\supset f(\mathscr{N}-\mathscr{N}_{n_1\dots n_{k-1}})=B_{n_1\dots n_{k-1}},$$

and we obtain the formula

$$C^*_{n_1\dots n_k}=C_{n_1\dots n_k}\cap\bar{A}_{n_1\dots n_k}-B_{n_1\dots n_k}\subset$$
$$\subset C_{n_1\dots n_{k-1}}\cap\bar{A}_{n_1\dots n_{k-1}}-B_{n_1\dots n_{k-1}}=C^*_{n_1\dots n_{k-1}}.$$

Double inclusion (1) yields

$$Z\subset\bigcup_{\mathfrak{z}}\bigcap_{k=1}^{\infty}C^*_{\mathfrak{z}^1\dots\mathfrak{z}^k}\subset\bigcup_{\mathfrak{z}}\bigcap_{k=1}^{\infty}(\bar{A}_{\mathfrak{z}^1\dots\mathfrak{z}^k}-B_{\mathfrak{z}^1\dots\mathfrak{z}^k}). \tag{2}$$

It follows from II (6) that

$$\bigcap_{k=1}^{\infty}(\bar{A}_{\mathfrak{z}^1\dots\mathfrak{z}^k}-B_{\mathfrak{z}^1\dots\mathfrak{z}^k})=\overline{\bigcap_{k=1}^{\infty}f(\mathscr{N}_{\mathfrak{z}^1\dots\mathfrak{z}^k})}\cap\bigcap_{k=1}^{\infty}B^c_{\mathfrak{z}^1\dots\mathfrak{z}^k}$$
$$=\bigcap_{k=1}^{\infty}f(\mathscr{N}_{\mathfrak{z}^1\dots\mathfrak{z}^k})\cap\bigcap_{k=1}^{\infty}B^c_{\mathfrak{z}^1\dots\mathfrak{z}^k}$$
$$=\bigcap_{k=1}^{\infty}(A_{\mathfrak{z}^1\dots\mathfrak{z}^k}-B_{\mathfrak{z}^1\dots\mathfrak{z}^k}),$$

and hence

$$\bigcup_{\mathfrak{z}}\bigcap_{k=1}^{\infty}(\bar{A}_{\mathfrak{z}^1\dots\mathfrak{z}^k}-B_{\mathfrak{z}^1\dots\mathfrak{z}^k})=\bigcup_{\mathfrak{z}}\bigcap_{k=1}^{\infty}(A_{\mathfrak{z}^1\dots\mathfrak{z}^k}-B_{\mathfrak{z}^1\dots\mathfrak{z}^k})=Z.$$

It follows that the three members of double inclusion (2) are equal. Since the system $\{C^*_{n_1\dots n_k}\}$ is regular and disjoint, for a fixed k, formula (7') of § 3, XIV, implies that

$$Z=\bigcup_{\mathfrak{z}}\bigcap_{k=1}^{\infty}C^*_{\mathfrak{z}^1\dots\mathfrak{z}^k}=\bigcap_{k=1}^{\infty}\bigcup_{\mathfrak{z}}C^*_{\mathfrak{z}^1\dots\mathfrak{z}^k}=\bigcap_{k=1}^{\infty}\bigcup_{n_1\dots n_k}C^*_{n_1\dots n_k}$$

is a $\boldsymbol{CA}$ set.

Consider now the case when g is a B measurable function defined on a Borel set E. Since the sets E and $g(E)$ lie in complete separable spaces $\mathscr{X}$ and $\mathscr{Y}$, the set $I=\underset{xy}{E}\{y=g(x)\}$ is borelian in $\mathscr{X}\times\mathscr{Y}$. Clearly the set of values of order 1 of g coincides with the set of values of order 1 of the projection of I (which is a continuous mapping of I) on the $\mathscr{Y}$ axis; and the latter was shown to be a $\boldsymbol{CA}$ set.

Remarks. (i) In the case considered here, the formula

$$\{y \in Z_f\} \equiv \bigcup_x [y = f(x)] \cap \bigcap_{xx'} \{[y = f(x) = f(x')] \Rightarrow (x = x')\}$$

would give a less precise result. However, if f is defined on an $\boldsymbol{F}_\sigma$ subset of a compact space, Z_f is by this formula a difference of two $\boldsymbol{F}_\sigma$ sets (this result cannot be improved).

(ii) CONVERSE THEOREM. *If C is a* **CA** *subset of $\mathcal{X}$, then there exists a continuous function f defined on a closed subset of $\mathcal{N}$ such that $C = Z_f$.*

Proof. Let F be a closed subset of $\mathcal{N}$ contained in the interval $0, 1/2$ and f a one-to-one continuous mapping of F onto $\mathcal{X}$ (compare § 36, III, Corollary). It suffices to define $f(\mathfrak{z})$ for $1/2 < \mathfrak{z} < 1$ so as to map the part of $\mathcal{N}$ contained in the interval $1/2, 1$ onto $\mathcal{X} - C$.

A value y of the function f is said to be of *uncountable order*, if the set $f^{-1}(y)$ is uncountable. Symbolically

$$(y \in A_f) \equiv [\overline{\overline{f^{-1}(y)}} > \aleph_0].$$

THEOREM 2 (1). *The set of values of uncountable order of a B measurable function f defined on an analytic set is an analytic set.*

Let $I = \underset{xy}{E}[y = f(x)]$. Since the set I is analytic (§ 38, III, 4), Theorem 2 reduces to the following theorem.

THEOREM 3. *If $\mathcal{X}$ and $\mathcal{Y}$ are complete separable spaces and I is an analytic subset of $\mathcal{X} \times \mathcal{Y}$, then the set A of the points y for which the set $\underset{x}{E}[(x, y) \in I]$ is uncountable is analytic.*

Proof. The analytic set I admits a parametric representation on the set $\mathcal{N}$ of irrational numbers, i.e. there exist two continuous functions g and h defined on $\mathcal{N}$ such that, to each point (x, y) of I, there corresponds a $\mathfrak{z}$ satisfying the conditions $x = g(\mathfrak{z})$ and $y = h(\mathfrak{z})$. Thus the condition that the set $\underset{x}{E}[(x, y) \in I]$ is uncountable is equivalent (§ 36, V, Corollary 3) to the existence of a dense in itself sequence of points of $\mathcal{N}$ such that

(i) $h(\mathfrak{z})$ is identically equal to y;

(ii) the function g is one-to-one.

(1) Compare S. Mazurkiewicz and W. Sierpiński, *Sur un problème concernant les fonctions continues*, Fund. Math. 6 (1924), p. 161, my paper in Fund. Math. 17, p. 161, and S. Saks, Fund. Math. 19 (1932), p. 218.

We can write this as follows:

$$\{y \in A\} \equiv \bigvee_{\xi} \bigwedge_{n} \{[y = h(\xi^n)] \wedge \bigwedge_{m \neq n} [g(\xi^n) \neq g(\xi^m)]\},$$

where ξ runs over the set of sequences, dense in themselves and belonging to $\mathcal{N}$. This set of sequences is topologically complete, since it is a G_δ in the complete space $\mathcal{N}^{\aleph_0}$ (§ 30, XII, (6)). It follows that the set

$$\underset{y\xi}{E} \bigwedge_{n} \{[y = h(\xi^n)] \wedge \bigwedge_{m \neq n} [g(\xi^n) \neq g(\xi^m)]\}$$

is a G_δ and its projection, A, is analytic.

Remarks. 1. Conversely, *for every analytic set A there exists a continuous function f defined on $\mathcal{N}$, such that $A = A_f$.*

Proof. Define

$$g\left(\frac{1|}{|\mathfrak{z}^1} + \frac{1|}{|\mathfrak{z}^2} + \frac{1|}{|\mathfrak{z}^3} + \ldots\right) = \frac{1|}{|\mathfrak{z}^1} + \frac{1|}{|\mathfrak{z}^3} + \frac{1|}{|\mathfrak{z}^5} + \ldots.$$

Then $g(\mathcal{N}) = \mathcal{N}$ and the set $g^{-1}(\mathfrak{z})$ is uncountable, for any $\mathfrak{z}$. Let h be a continuous mapping of $\mathcal{N}$ onto A. Then it is sufficient to let $f(\mathfrak{z}) = hg(\mathfrak{z})$.

2. Even if the function f is continuous and $\mathcal{X} = \mathcal{Y} = \mathcal{I}$, the set A_f need not be borelian.

3. Saks (*loc. cit.*) derived an interesting application of Theorem 3. Let: $\mathcal{X} = \mathcal{I}$, $\mathcal{Y} = \mathcal{E}^{\mathcal{I}}$, $I =$ the set of the "points" (x,y) of the product $\mathcal{X} \times \mathcal{Y}$ such that the right derivative of the function y at the point x is equal to $+\infty$. The set I is analytic (even an $F_{\sigma\delta}$), since

$$\{(x,y) \in I\} \equiv \bigwedge_{n} \bigvee_{m} \bigwedge_{0<h\leqslant 1/m} \left\{\frac{y(x+h)-y(x)}{h} \geqslant n\right\}.$$

Theorem 3 therefore implies that the *set of continuous functions that have an infinite right derivative at uncountably many points is analytic* (in the space $\mathcal{Y}$).

Moreover, this set is not of the first category at any point (*ibid.* p. 215), so — by the Baire property (compare § 11, IV, Corollary 2) — *its complement is of the first category* (and non-empty, by a theorem of Besicovitch, *ibid.* p. 212). Compare this result with § 34, VIII.

Denote, respectively, by I_f, D_f, and C_f the sets of the points $y \in \mathcal{Y}$ such that:

(i) $f^{-1}(y)$ contains an isolated point,

(ii) $f^{-1}(y)$ is countable (finite or infinite) and non-empty,

(iii) $f^{-1}(y)$ contains a point which is not its condensation point.

The following corollaries generalize the theorem of Section IV.

COROLLARIES 1–3 (¹). *If f is a B measurable function defined on a Borel set, then I_f, D_f, and C_f are* **CA** *sets.*

Proof. 1. Let $R_1, R_2, \ldots$ be a base of the space $\mathcal{X}$. Let f_n be the partial function $f \mid R_n$. Then

$$I_f = Z_{f_1} \cup Z_{f_2} \cup \ldots .$$

2. As in Theorem 1, the proof reduces to the case where f is a continuous function defined on the space $\mathcal{N}$. The non-empty countable sets (in the complete space $\mathcal{N}$) are characterized, among closed sets F, by the following two conditions: (i) F contains an isolated point; (ii) F is not uncountable. Hence

$$D_f = I_f - A_f.$$

3. The function f_n is as in 1; then

$$C_f = D_{f_1} \cup D_{f_2} \cup \ldots .$$

COROLLARY 4 (²). *If the function f is B measurable and the set E is a Borel set, while $f(E)$ is not, then $f(E) - C_f$ is not a Borel set.*

Otherwise the set $f(E) = C_f \cup [f(E) - C_f]$ would be a **CA** set, by Corollary 3, and the analytic set $f(E)$ (see § 38, III, 5) would be a Borel set (Section III, Corollary 1).

The following corollary follows directly from Corollary 2.

COROLLARY 5. *If f is a continuous function defined on a complete separable space $\mathcal{X}$ such that $f(\mathcal{X}) = D_f$ (i.e. all values of f are of countable order), then $f(\mathcal{X})$ is a Borel set.*

Moreover, *$\mathcal{X}$ is the union of a series of Borel sets: $\mathcal{X} = B_1 \cup \cup B_2 \cup \ldots$ such that the partial functions $f \mid B_n$ are homeomorphisms* (³).

(¹) S. Braun, *Quelques théorèmes sur les cribles boreliens*, Fund. Math. 20 (1933), pp. 168–172.

(²) *Ibid.* Compare also N. Lusin, *loc. cit.*, p. 171.

(³) For a proof, see N. Lusin, *loc. cit.*, p. 237, and H. Hahn, *Reelle Funktionen*, p. 381 (42.5.3).

VIII. Constituents of CA sets (1). Let A be an analytic set and — according to Corollary 2 of Section II — let $W = \{W_r\}$ be a sieve composed of closed sets (or, more generally, of Borel sets) which sieves A. As in § 3, XV, let

$$M_x = \underset{r}{E}(x \in W_r), \tag{1}$$

$$\mu(x) = \text{the order type of } M_x, \tag{2}$$

$$A_\tau = \underset{x}{E}[\mu(x) = \tau]. \tag{3}$$

Since A is sieved by the sieve W, hence

$$\mathscr{X} - A = \bigcup_{a<\Omega} A_a. \tag{4}$$

The sets A_a with $a < \Omega$ are called the *constituents of the set* $\mathscr{X} - A$ *determined by the sieve* W.

Let us order the binary fractions $r \in R_0$ into an infinite sequence $r_1, r_2, \ldots, r_n, \ldots$ considered to be fixed from now on. Let us recall the definitions of the sieve C and constituents L_a introduced in § 3, XVI. The set C_{r_n} has been defined by the equivalence

$$(t \in C_{r_n}) \equiv (t^{(n)} = 2), \tag{5}$$

where $t = [t^{(1)}, t^{(2)}, \ldots, t^{(n)}, \ldots]$ is a point of the Cantor set $\mathscr{C}$. Let

$$R_t = \underset{r}{E}(t \in C_r), \tag{6}$$

$$\bar{t} = \text{the order type of } R_t, \tag{7}$$

$$L_\tau = \underset{t}{E}(\bar{t} = \tau) \quad \text{and} \quad L = \underset{t}{E}(\bar{t} < \Omega) = \bigcup_{a<\Omega} L_a. \tag{8}$$

Equivalence (5) means that the *characteristic function of the sequence* $C_{r_1}, C_{r_2}, \ldots$ *is the identity.*

THEOREM 1. *Let f be the characteristic function of the sequence $W_{r_1}, W_{r_2}, \ldots$, i.e. $f(x)$ is a point of the Cantor set $\mathscr{C}$ such that*

$$[f^{(n)}(x) = 2] \equiv (x \in W_{r_n}). \tag{9}$$

(1) Compare N. Lusin, *Ensembles analytiques*, p. 188.

Then the following formulas hold:

$$M_x = R_{f(x)}, \tag{10}$$

whence

$$\mu(x) = \overline{f(x)}, \tag{11}$$

$$A_\tau = f^{-1}(L_\tau) \tag{12}$$

and

$$\mathscr{X} - A = f^{-1}(L). \tag{13}$$

Proof. By (1), (9), (5), and (6),

$$(r_n \in M_x) \equiv (x \in W_{r_n}) \equiv [f^{(n)}(x) = 2] \equiv [f(x) \in C_{r_n}] \equiv [r_n \in R_{f(x)}],$$

and formula (10) follows; by (2) and (7), we obtain formula (11).

Formula (12) follows from (8), (11), and (3):

$$[x \in f^{-1}(L_\tau)] \equiv [f(x) \in L_\tau] \equiv [\overline{f(x)} = \tau] \equiv [\mu(x) = \tau] \equiv (x \in A_\tau).$$

Finally, (13) follows from (4), (12), and (8):

$$\mathscr{X} - A = \bigcup_{\alpha<\Omega} A_\alpha = \bigcup_{\alpha<\Omega} f^{-1}(L_\alpha) = f^{-1}(\bigcup_{\alpha<\Omega} L_\alpha) = f^{-1}(L).$$

We shall derive from Theorem 1 the following two theorems.

THEOREM 2. (Lusin–Sierpiński theorem.)(1) *The set* $L = \underset{t}{E}(\bar{t} < \Omega)$ *is not analytic* (by Corollary 1a of § 38, IX, it is a $\boldsymbol{CA}$ set).

Proof. The space $\mathscr{N}$ of irrational numbers contains an analytic set A whose complement is not analytic (§ 38, VI). Let $W = \{W_r\}$ be a sieve composed of closed sets that sieves A. Since the set W_{r_n} is closed, its characteristic function (and the function $f^{(n)}$ defined by formula (9)) is of class 1 (§ 31, I, Theorem 1). Hence the characteristic function f of the sequence $W_{r_1}, W_{r_2}, \ldots$ is also of class 1 (§ 31, VI, Theorem 1′). If we suppose that L is an analytic set, then $f^{-1}(L)$ must be analytic (§ 38, III, 5). But, by (13), $f^{-1}(L) = \mathscr{N} - A$, and $\mathscr{N} - A$ is not analytic by the hypothesis.

THEOREM 3. *The constituents* A_α *are Borel sets* (2).

---

(1) Journ. de Math. 1923, p. 53. The proof given here can be found in my paper *Sur les suites analytiques d'ensembles*. Fund. Math. 29 (1937), p. 57.

(2) By the remark to Corollary 5b of § 38, IX, this is true of A_τ for any order type τ.

Proof. The sets L_a are Borel sets for $a < \Omega$ (by § 30, XII, Theorem 1) and the function f is of class 1; hence A_a is a Borel set by virtue of (12) (compare § 31, III, Theorem 1).

LEMMA 4. *The sets* $\underset{tx}{E}[\bar{t} \prec\!\!\!\!\sim \mu(x)]$, $\underset{tx}{E}[\mu(x) \prec\!\!\!\!\sim \bar{t}]$ *and* $\underset{tx}{E}[\bar{t} = \mu(x)]$ *are analytic.*

This is a direct consequence of § 38, IX, Lemmas 2 and 5.

THEOREM 5. (Covering theorem.)(1) *Let E and A be two analytic disjoint sets, i.e.*

$$E \subset \mathscr{X} - A = \bigcup_{a<\Omega} A_a.$$

There exists an index $a_0 < \Omega$ *such that*

$$E \subset \bigcup_{a<a_0} A_a. \tag{14}$$

Proof. Suppose, on the contrary, that for each $a < \Omega$ there is a $\xi \geqslant a$ such that $E \cap A_\xi \neq 0$; so there exists an $x \in E$ such that $\mu(x) = \xi$, and hence $\mu(x) \geqslant a$. It follows that the inequality $\bar{t} < \Omega$ is equivalent to the existence of an $x \in E$ such that $\mu(x) \geqslant \bar{t}$, therefore such that $\bar{t} \prec\!\!\!\!\sim \mu(x)$ (since $\mu(x)$ is an ordinal number, if $x \in E$). That is,

$$(\bar{t} < \Omega) \equiv \bigvee_x (x \in E) \wedge [\bar{t} \prec\!\!\!\!\sim \mu(x)].$$

The propositional functions $x \in E$ and $\bar{t} \prec\!\!\!\!\sim \mu(x)$ are of class **A** (the first one by hypothesis and the second one by Lemma 4); hence so are the propositional function $\bar{t} < \Omega$. But this contradicts Theorem 2 which asserts that the set $L = \underset{t}{E}(\bar{t} < \Omega)$ is not analytic.

Remark. Theorem 5 leads to a proof of Corollary 1 of Section III that does not involve the first separation theorem.

If the sets A and $\mathscr{X} - A$ are analytic, we have by Theorem 5:

$$\mathscr{X} - A \subset \bigcup_{a<a_0} A_a, \quad \text{hence} \quad \mathscr{X} - A = \bigcup_{a<a_0} A_a.$$

Each A_a is a Borel set (Theorem 3), hence so is the set

$$\bigcup_{a<a_0} A_a = \mathscr{X} - A.$$

(1) Theorem of N. Lusin. Compare *Ensembles analytiques*, p. 183.

Theorems 5 and 3 imply the following corollary.

COROLLARY 5a. *The set* $\mathcal{X}-A = \bigcup_{\alpha<\Omega} A_\alpha$ *is analytic (or, what amounts to the same, borelian) if and only if the values of the function* μ *form an at most countable set of ordinal numbers; i.e.* $A_\alpha = 0$, *if* α *is sufficiently large.*

THEOREM 6. *There exists an* $\alpha_0 < \Omega$ *such that the set* $\bigcup_{\xi\geqslant\alpha_0} A_\xi$ *is of the first category.*

Proof. Since $\mathcal{X}-A$ is a **CA** set, it has the Baire property (II, Corollary 1). Hence it contains a Borel set E (a $\boldsymbol{G}_\delta$) such that $\mathcal{X}-A-E$ is of the first category (§ 11, IV, 2). According to (14) we have

$$\bigcup_{\xi\geqslant\alpha_0} A_\xi = \mathcal{X}-A-\bigcup_{\alpha<\alpha_0} A_\alpha \subset \mathcal{X}-A-E$$

and the assertion follows.

THEOREM 7. *If a set* Z *can be ordered into a sequence* $x_1, x_2, \ldots, x_\alpha, \ldots$ *of type* Ω *such that* $\mu(x_1) < \mu(x_2) < \ldots < \mu(x_\alpha) < \ldots < \Omega$, *then* Z *is of the first category on every perfect set*(1).

Proof. The preceding argument can be relativized to a perfect set P given in advance. Thus, for a fixed P, there exists an α_0 such that the set of the x_ξ with $\xi \geqslant \alpha_0$ (which belong to P) is of the first category on P. The set of the x_α with $\alpha < \alpha_0$ is countable, so it is of the first category on P (since P is perfect). It follows that $Z \cap P$ is of the first category on P.

THEOREM 8. *The sets* L_α $(\alpha < \Omega)$ *are of non-bounded Borel classes, i.e. for each* $\beta < \Omega$ *there exists an* L_α *which is not of class* β.

To prove this, it suffices, by (12), to establish the existence of a **CA** set whose constituents are of non-bounded classes. Now, we shall prove

THEOREM 9(2). *If* F *is an universal function* (defined on $\mathcal{N}$) *with respect to the family of analytic subsets of* $\mathcal{N}$ *such that the set* $A = \underset{\mathfrak{z}\mathfrak{y}}{E}[\mathfrak{y}\in F(\mathfrak{z})]$ *is analytic* (compare § 38, V) *and if* $W = \{W_r\}$ *is a sieve (formed by closed sets) that sieves* A, *then the constituents* A_α *which correspond to this sieve are of non-bounded classes.*

(1) For another construction of sets of the first category on every perfect set, see § 24, I, Theorem 5.

(2) See N. Lusin and W. Sierpiński, C. R. Paris 189 (1929), p. 794.

Proof. Suppose, on the contrary, that the constituents are all of a Borel class β. Let B be a Borel subset of $\mathcal{N}$ which is not of class $\beta+1$. There exists a $\mathfrak{z}_0$ such that $F(\mathfrak{z}_0) = \mathcal{N}-B$. Let

$$V = \mathfrak{z}_0 \times \mathcal{N} \quad \text{and} \quad D = \mathfrak{z}_0 \times B.$$

Then

$$D = \mathfrak{z}_0 \times [\mathcal{N}-F(\mathfrak{z}_0)] = V-(V \cap A) = V-A.$$

Since D is a Borel set, there is, by Theorem 5, an α_0 such that

$$D \subset \bigcup_{\alpha<\alpha_0} A_\alpha, \quad \text{whence} \quad V-A \subset \bigcup_{\alpha<\alpha_0} V \cap A_\alpha$$

and

$$V-A = \bigcup_{\alpha<\alpha_0} V \cap A_\alpha,$$

for (compare (4)) $V \cap A_\alpha \subset V-A$. By hypothesis, the sets A_α are of class β. It follows that the set $V-A$ (i.e. the set D, hence also the set $\mathcal{N}-F(\mathfrak{z}_0)$) is of class $\beta+1$. But then B would be of class $\beta+1$, contrary to hypothesis.

IX. Projective class of a propositional function that involves variable order types(1). Let τ be a variable which runs over the set $\mathcal{T}$ of countable order types. If t is an element of the Cantor set $\mathcal{C}$, let $\bar{t}$ denote its order type according to § 3, XVI.

DEFINITION 1. *A propositional function $\varphi(\tau_1, \ldots, \tau_n)$ of n variables is said to be of projective class $\boldsymbol{L}_n$, if the propositional function $\varphi(\bar{t}_1, \ldots, \bar{t}_n)$ is of this class, i.e. if the set* $\underset{t_1 \ldots t_n}{E}\, \varphi(\bar{t}_1, \ldots, \bar{t}_n)$ *is of class* $\boldsymbol{L}_n$.

EXAMPLES. The propositional functions: "τ is a limit type", "τ is an even type", "τ is an odd type", "$\tau = \alpha$", where α is a fixed ordinal $< \Omega$, are borelian, since the corresponding sets are borelian (by § 30, XII, 3), 4), and 1)). Similarly, the relations (propositional functions of two variables) "$\tau \precsim \sigma$" and "$\tau = \sigma$" are analytic (by § 38, IX, Corollaries 2a and 5a). The propositional function "$\tau < \Omega$" is of class $\boldsymbol{CA}$ (it is not analytic, compare VIII, 2).

---

(1) See my papers *Sur la géométrisation des types d'ordre dénombrables* and *Les suites transfinies d'ensembles et les ensembles projectifs*, Fund. Math. 28 (1937). For some applications, see Y. Sampei, *On set-theoretical operations*, Comment. Univ. S. Pauli (Tokyo) 9 (1961).

We shall prove in that direction the following two theorems.

THEOREM 1. *The addition $\tau = \sigma + \varrho$, regarded as a propositional function of three variables, is an analytic operation.*

Proof. Let T, S, and R be three subsets of $\mathscr{R}_0 = (r_1, r_2, \ldots)$ (the set of binary fractions) of the order types τ, σ, and ϱ, respectively. Let R^* be the set obtained from R by addition of 2 to each element of R. Since S and R^* are disjoint, $S \cup R^*$ is of order type $\sigma + \varrho$. The condition for a set T to be similar to $S \cup R^*$ can be expressed as follows: There exist two sequences $\mathfrak{z} = [\mathfrak{z}^1, \mathfrak{z}^2, \ldots]$ and $\mathfrak{y} = [\mathfrak{y}^1, \mathfrak{y}^2, \ldots]$ of real numbers such that

(i) each element of $\mathfrak{z}$ belongs to T and each element of $\mathfrak{y}$ belongs to $S \cup R^*$;

(ii) each element of T occurs in the sequence $\mathfrak{z}$ and each element of $S \cup R^*$ occurs in $\mathfrak{y}$;

(iii) the conditions $\mathfrak{z}^i < \mathfrak{z}^j$ and $\mathfrak{y}^i < \mathfrak{y}^j$ are equivalent.

That is:

$$(\tau = \sigma + \varrho) \equiv \bigvee_{\mathfrak{z}\mathfrak{y}} \bigwedge_n \{(\mathfrak{z}^n \in T) \wedge [(\mathfrak{y}^n \in S) \vee ((\mathfrak{y}^n - 2) \in R)]\} \wedge$$
$$\wedge \bigwedge_k \{[(r_k \in T) \Rightarrow \bigvee_n (r_k = \mathfrak{z}^n)] \wedge [(r_k \in S) \Rightarrow \bigvee_n (r_k = \mathfrak{y}^n)] \wedge$$
$$\wedge [(r_k \in R) \Rightarrow \bigvee_n (r_k = \mathfrak{y}^n - 2)]\} \wedge \bigwedge_{ij} \{(\mathfrak{z}^i < \mathfrak{z}^j) \equiv (\mathfrak{y}^i < \mathfrak{y}^j)\}.$$

As $\bar{t}$ is the order type of R_t (compare § 3, XVI), then the equality $\bar{t} = \bar{s} + \bar{r}$ is equivalent to the expression obtained from the preceding equivalence by replacing T with R_t, S with R_s and R with R_r. From the equivalence

$$(\mathfrak{z}^n \in R_t) \equiv \bigvee_k (r_k = \mathfrak{z}^n) \wedge (r_k \in R_t) \equiv \bigvee_k (r_k = \mathfrak{z}^n) \wedge (t^{(k)} = 2)$$

and from two analogous equivalences concerning $(\mathfrak{y}^n \in R_s)$ and $[(\mathfrak{y}^n - 2) \in R_r]$, we easily infer that the propositional function (of the variables $\mathfrak{z}$, $\mathfrak{y}$, t, s, and r) which follows the quantifier $\bigvee_{\mathfrak{z}\mathfrak{y}}$ is borelian. Hence the propositional function $\bar{t} = \bar{s} + \bar{r}$ is analytic.

THEOREM 2. *The multiplication $\tau = \sigma \cdot \varrho$ is an analytic operation.*

Proof. In notation of the preceding proof we have

$$(\tau = \sigma\cdot\varrho) \equiv \bigvee_{\mathfrak{z}\mathfrak{y}\mathfrak{w}} \bigwedge_{n} [(\mathfrak{z}^n \in T) \wedge (\mathfrak{y}^n \in S) \wedge (\mathfrak{w}^n \in R)] \wedge$$

$$\wedge \bigwedge_{k} [(r_k \in T) \Rightarrow \bigvee_{n} (r_k = \mathfrak{z}^n)] \wedge$$

$$\wedge \bigwedge_{km} [(r_k \in S) \wedge (r_m \in R) \Rightarrow \bigvee_{n} (r_k = \mathfrak{y}^n) \wedge (r_m = \mathfrak{w}^n)] \wedge$$

$$\wedge \bigwedge_{ij} [(\mathfrak{z}^i < \mathfrak{z}^j) \equiv (\mathfrak{w}^i < \mathfrak{w}^j) \vee (\mathfrak{w}^i = \mathfrak{w}^j) \wedge (\mathfrak{y}^i < \mathfrak{y}^j)].$$

An argument completely analogous to the preceding one proves that the propositional function $\bar{t} = \bar{s}\cdot\bar{r}$ is analytic.

Definition 1 can be extended to propositional functions which, besides variable order types, involve variable points of some complete separable spaces.

DEFINITION 2. *A propositional function* $\varphi(\tau_1, \ldots, \tau_n, x_1, \ldots, x_n)$ *is said to be of class* $\mathbf{L}_n$, *if the propositional function* $\varphi(\bar{t}_1, \ldots, \bar{t}_n, x_1, \ldots, x_n)$ *is of class* $\mathbf{L}_n$.

Thus, for example, in notation of Section VIII, the propositional functions $\tau \prec \mu(x)$ and $\tau = \mu(x)$ are analytic (VIII, Lemma 4), and so is the propositional function $x \in A_\tau$, i.e. the set $\underset{tx}{E}(x \in A_{\bar{t}})$.

In order to evaluate the projective class of a propositional function involving order type variables, we use the following rules, which can easily be derived from Theorems 1–5 of § 38, VIII.

(i) *If* $\varphi(\tau, x)$ *is of class* $\mathbf{L}_n$, *its negation* $\neg\varphi(\tau, x)$ *is of class* $\mathbf{CL}_n$.

(ii) *If* $\varphi(\tau, x)$ *is of class* $\mathbf{L}_n$ *in* $\mathcal{T}\times\mathcal{X}$, *then* $\varphi(\tau, x)$ *is of class* $\mathbf{L}_n$ *in* $\mathcal{T}^2\times\mathcal{X}$ *and in* $\mathcal{T}\times\mathcal{X}\times\mathcal{Y}$ *as well.*

(iii) *If* $\varphi_1(\tau, x)$, $\varphi_2(\tau, x), \ldots$ *is a (finite or infinite) sequence of functions of class* $\mathbf{L}_n$ *(n is fixed), then the functions* $\bigvee_k \varphi_k(\tau, x)$ *and* $\bigwedge_k \varphi_k(\tau, x)$ *are of class* $\mathbf{L}_n$.

(iv) *If* $\varphi(\tau, \sigma, x, y)$ *is of class* $\mathbf{L}_n$, *then the functions* $\bigvee_\sigma \varphi(\tau, \sigma, x, y)$ *and* $\bigvee_y \varphi(\tau, \sigma, x, y)$ *are of class* $\mathbf{PL}_n$ *and the functions* $\bigwedge_\sigma \varphi(\tau, \sigma, x, y)$ *and* $\bigwedge_y \varphi(\tau, \sigma, x, y)$ *are of class* $\mathbf{CPCL}_n$.

(v) *If* $\varphi(\tau, \sigma, x, y)$ *is of class* $\mathbf{L}_n$, *then so are the following functions:* $\varphi(\tau, \sigma_0, x, y)$, $\varphi(\tau, \sigma, x, y_0)$, $\varphi(\tau, \tau, x, y)$, *and* $\varphi(\tau, \sigma, x, x)$.

We complete this list by another two rules.

(vi) *If $\varphi(\sigma, x)$ is of class $\boldsymbol{L}_n$, then the functions*

$$\psi(\tau, x) \equiv \bigvee_{\sigma<\tau} \varphi(\sigma, x) \quad \textit{and} \quad \chi(\tau, x) \equiv \bigwedge_{\sigma<\tau} \varphi(\sigma, x)$$

are also of class $\boldsymbol{L}_n$.

Proof. Clearly it is sufficient to prove this theorem for a function $\varphi(\sigma)$ which does not depend on x. In § 30, XII, (1) we defined a sequence $\{t^{[k]}\}$ of continuous functions such that the sequence $\overline{t^{[1]}}, \overline{t^{[2]}}, \ldots$ covers all ordinal numbers $< \bar{t}$ (under the condition $t \neq 0$). Consequently, the existence of a $\sigma < \bar{t}$ (where $t \neq 0$) such that $\varphi(\sigma)$, is equivalent to the existence of a k such that $\varphi(\overline{t^{[k]}})$. In other words, we have (for $\tau = \bar{t}$)

$$\psi(\tau) \equiv \psi(\bar{t}) \equiv (t \neq 0) \wedge \bigvee_{k=1}^{\infty} \varphi(\overline{t^{[k]}}).$$

Since the function $t^{[k]}$ is continuous, we infer from (iii) that ψ is of class $\boldsymbol{L}_n$. By applying de Morgan law we deduce that the function χ is also of class $\boldsymbol{L}_n$.

Let μ be a function defined in the space $\mathscr{X}$ or in the set $\mathscr{T}$ and whose values belong to $\mathscr{T}$. We say that the function μ is of class $\boldsymbol{L}_n$, if the propositional function $\sigma = \mu(\xi)$ is of class $\boldsymbol{L}_n$. With this definition, we have the following rule.

(vii) *If Φ is a set of order types of class $\boldsymbol{L}_n$ $(n > 0)$ and if the function μ is analytic, then the set*

$$\mu^{-1}(\Phi) = \underset{\xi}{E}[\mu(\xi) \in \Phi]$$

is of class $\boldsymbol{L}_n$.

Consequently, *if $\varphi(\tau, x)$ is of class $\boldsymbol{L}_n$ $(n > 0)$, then the propositional function $\varphi[\mu(\xi), x]$ is also of class $\boldsymbol{L}_n$.*

Proof. We have

$$[\mu(\xi) \in \Phi] \equiv \bigvee_{\sigma} [\sigma = \mu(\xi)] \wedge (\sigma \in \Phi).$$

Thus, if Φ is of odd class $\boldsymbol{L}_n$, then the set $\mu^{-1}(\Phi)$ is of class $\boldsymbol{PL}_n = \boldsymbol{L}_n$. If Φ is of even (> 0) class $\boldsymbol{L}_n$, its complement, Φ^c, is of odd class $\boldsymbol{L}_{n-1}$, and hence $\mu^{-1}(\Phi^c)$ is of class $\boldsymbol{L}_{n-1}$. Therefore the set $\mu^{-1}(\Phi) = [\mu^{-1}(\Phi^c)]^c$ is of class $\boldsymbol{CL}_{n-1} = \boldsymbol{L}_n$.

In the proof of the second part of (vii) we can confine ourselves to the case when φ is a function of a single variable τ. Observe that, by virtue of the identity $\varphi(\sigma) \equiv \big[\sigma \in \underset{\tau}{E}\, \varphi(\tau)\big]$, we have

$$\varphi[\mu(\xi)] \equiv \big[u(\xi) \in \underset{\tau}{E}\, \varphi(\tau)\big].$$

Let $\Phi = \underset{\tau}{E}\, \varphi(\tau)$. By hypothesis, the set $\underset{\tau}{E}\, \varphi(\tau)$ is of class $\boldsymbol{L}_n$, so, as we have just proved, the propositional function $\mu(\xi) \in \Phi$ is of class $\boldsymbol{L}_n$. Hence $\varphi[\mu(\xi)]$ is of class $\boldsymbol{L}_n$.

(viii) *The composition of two analytic functions is an analytic function.*

Proof. If the function μ_1 is analytic, the propositional function $\sigma = \mu_1(\tau)$ is analytic. Hence, by (vii) (second part), if the function μ_2 is analytic, then the propositional function $\sigma = \mu_1[\mu_2(\xi)]$ is analytic. This means that the composed function $\mu_1 \circ \mu_2$ is analytic.

EXAMPLE. The function $(\tau \cdot 2 + 1)2^n$ (n is fixed) is analytic since it is the composition of the functions $\tau \cdot 2$, $\tau + 1$, and $\tau \cdot 2^n$ (we shall use this function in Section X).

Remark. It is easily seen that the variables τ and x (in rules (i) to (viii)) can be replaced by *complex* variables (which vary in the cartesian product). Also the values of μ may be complex.

In the study of propositional functions defined for *ordinal numbers*, it is useful to introduce the following concept of a *relatively analytic* propositional function: a propositional function $\varphi(\alpha)$ defined for $\alpha < \Omega$ is relatively analytic, if there exists an analytic propositional function $\varphi^*(\tau)$ defined for $\tau \in \mathscr{T}$ such that

$$\varphi^*(\tau) \wedge (\tau < \Omega) \equiv \varphi(\tau).$$

More generally, a function $\varphi(\alpha_1, \ldots, \alpha_n)$ is relatively analytic, if it has the form

$$\varphi(\tau_1, \ldots, \tau_n) \equiv \varphi^*(\tau_1, \ldots, \tau_n) \wedge (\tau_1 < \Omega) \wedge \ldots \wedge (\tau_n < \Omega)$$

where φ^* is analytic.

Propositional functions $\varphi(\alpha_1, \ldots, \alpha_n)$, where $\alpha_1 < \Omega, \ldots, \alpha_n < \Omega$, which are both of class $\boldsymbol{CA}$ and relatively analytic will be called *elementary*. The reason for this name is the fact that the propositional functions considered in the theory of ordinal numbers are usually *elementary*.

For instance, the relations $\alpha = \beta$, $\alpha \leqslant \beta$, $\alpha < \beta$, $\alpha = \beta + \gamma$, $\alpha = \beta\gamma$ are elementary(1).

In order to prove that the identity $\alpha = \beta$ is elementary, we observe that the following equivalence holds:

$$(\alpha = \beta) \equiv (\alpha \text{ non} \prec\!\!\!\sim \beta) \wedge (\beta \text{ non} \prec\!\!\!\sim \alpha).$$

Since the relation $\tau \prec\!\!\!\sim \alpha$ is analytic and the relation $\tau < \Omega$ is of class $\boldsymbol{CA}$, it follows that the relation

$$(\tau \text{ non} \prec\!\!\!\sim \sigma) \wedge (\sigma \text{ non} \prec\!\!\!\sim \tau) \wedge (\tau, \sigma < \Omega),$$

hence also the relation $\alpha = \beta$, is of class $\boldsymbol{CA}$. Moreover, the relation $\alpha = \beta$ is relatively analytic, since the relation $\tau = \sigma$ (in the entire domain of order types) is analytic.

Transfinite induction does not lead out of elementary functions. For we have the following theorem(2).

THEOREM. *Let* α *be a given ordinal* $< \Omega$ *and* $\varkappa$ *a function which assigns an ordinal* $< \Omega$ *to each* $\xi < \Omega$. *Assume that the function* $\varkappa$ *is elementary* (*i.e. the relation* $\lambda = \varkappa(\xi)$ *is elementary*). *Then the function* μ *defined for* $\xi < \Omega$ *by the conditions*

1) $\mu(0) = \alpha$,

2) $\mu(\xi + 1) = \varkappa[\mu(\xi)]$,

3) $\mu(\zeta) = \lim_{\eta < \zeta} \mu(\eta)$, *if* ζ *is a limit number* ($\neq 0$),

is elementary.

Thus, for example, the relation $\alpha = \beta^\gamma$ is elementary.

X. Reduction theorems(3).

THEOREM 1. *If* $U^1, U^2, \ldots$ *is an infinite sequence of* $\boldsymbol{CA}$ *sets, then there exists a sequence* $V^1, V^2, \ldots$ *of disjoint* $\boldsymbol{CA}$ *sets such that*

$$V^n \subset U^n \quad \textit{and} \quad \bigcup_{n=1}^{\infty} V^n = \bigcup_{n=1}^{\infty} U^n. \tag{1}$$

Consequently, *if* $\mathscr{X} = U^1 \cup U^2 \cup \ldots$, *then* V^n *are Borel sets.*

---

(1) For a proof and for more details, see my paper referred to above, Fund. Math. 28, pp. 179–185.

(2) *Ibidem*, p. 181.

(3) See my paper *Sur les théorèmes de séparation dans la Théorie des ensembles*, Fund. Math. 26 (1936), pp. 183–191.

Proof. Let $U^n = \bigcup_{\alpha<\Omega} A_\alpha^n$ be a resolution of U^n into constituents (corresponding to a sieve formed by closed sets). According to VIII (3) and Lemma 4, there exists a function μ_n which assigns to each $x \in \mathcal{X}$ a countable order type so that:

(i) $A_\alpha^n = \underset{x}{E}[\mu_n(x) = \alpha]$, for $\alpha < \Omega$;

(ii) the function μ_n is analytic in the sense of Section IX, i.e. the set $\underset{xt}{E}[\mu_n(x) = \bar{t}]$ is analytic.

Let

$$\gamma_n(x) = [\mu_n(x)\cdot 2+1]\cdot 2^n \tag{2}$$

and

$$V^n = U^n \cap \bigcap_{k\neq n} \underset{x}{E}[\gamma_k(x) \text{ non } \preccurlyeq \gamma_n(x)], \tag{3}$$

where the relation τ non $\preccurlyeq \sigma$ means that a set of order type τ is not similar to any subset of a set of order type σ (compare § 38, IX, (4)).

The function γ_n is analytic, since it is the composition of two analytic funtions (see IX, (viii), example). Since the relation $\tau \preccurlyeq \sigma$ is analytic (IX, examples, Definition 1), so is $\gamma_k(x) \preccurlyeq \gamma_n(x)$, for fixed k and n (by IX, (vii)). It follows that V^n is a ***CA*** set.

Observe that $V^n \cap V^k = 0$, if $n \neq k$. For, if $x \in V^n \cap V^k$, then

$$\gamma_n(x) < \Omega, \quad \gamma_k(x) < \Omega, \quad \gamma_k(x) \text{ non } \preccurlyeq \gamma_n(x) \text{ non } \preccurlyeq \gamma_k(x).$$

Now, the relation $\alpha \preccurlyeq \beta$ means, for ordinal numbers, that $\alpha \leqslant \beta$. Therefore $\gamma_k(x) = \gamma_n(x)$. On the other hand, in the domain of ordinal numbers, the condition

$$(\alpha\cdot 2+1)2^n = (\beta\cdot 2+1)2^k$$

implies $n = k$ (and $\alpha = \beta$)(1).

It follows that $x \in V^n \cap V^k$ implies $n = k$. In other words, the sets $V^1, V^2, \ldots$ are disjoint.

It remains to prove that

$$U^n \subset \bigcup_{m=1}^{\infty} V^m. \tag{3a}$$

(1) See, e.g., W. Sierpiński, *Cardinal and ordinal numbers*, Warszawa 1958, p. 330.

Let $x \in A^n_\alpha$; then $\mu_n(x) = \alpha < \Omega$ and $\gamma_n(x) < \Omega$. Let m_0 be the index m of the smallest ordinal $\gamma_m(x) < \Omega$. Hence, for $k \neq m_0$:

$$\gamma_k(x) \text{ non} \prec \gamma_{m_0}(x), \quad \text{whence} \quad x \in \bigcap_{k \neq m_0} \underset{x}{E}[\gamma_k(x) \text{ non} \prec \gamma_{m_0}(x)].$$

Finally $x \in U^{m_0}$, i.e. $\mu_{m_0}(x) < \Omega$. For, if τ were not an ordinal number, $(\tau \cdot 2 + 1) \cdot 2^n$ would not be one either. Therefore the inequality $\mu_{m_0}(x) < \Omega$ is a consequence of the inequality $\gamma_{m_0}(x) < \Omega$.

Thus we have proved that $x \in V^{m_0}$ and inclusion (3a) follows.

To prove the second part of the theorem, observe that since the sets $V^1, V^2, \ldots$ are disjoint, the condition $\mathscr{X} = V^1 \cup V^2 \cup \ldots$ implies

$$\mathscr{X} - V^n = \bigcup_{k \neq n} V^k.$$

Thus $\mathscr{X} - V^n$ is a ***CA*** set, as a countable union of ***CA*** sets. Hence V^n is a Borel set, since it is an ***A*** and ***CA*** set.

Remarks. 1. The function $\varrho(\alpha, n) = (\alpha \cdot 2 + 1) 2^n$ transforms the set of all pairs (α, n) into a transfinite sequence of type Ω in a one-to-one manner. The definition of V^n can be restated as follows: a point x belongs to V^n, whenever it belongs to a constituent A^n_α such that the "rank" $\varrho(\alpha, n)$ of (α, n) is lower than the rank of any other pair (β, k) with $x \in A^k_\beta$.

2. In the proof of Theorem 1 an essential role is played by the *analycity* of the relation $x \in A^n_\tau$, i.e. of the set $\underset{xt}{E}(x \in A^n_t)$.

The reduction theorem implies the following two theorems. They generalize the separation theorems of Sections III and VI. Their proof is identical to that of Theorem 2 and Corollary 4 of § 26, II.

THEOREM 2. (First generalized separation theorem.) (¹) *If $A_1, A_2, \ldots$ is a sequence of analytic sets such that $\bigcap_{n=1}^{\infty} A_n = 0$, then there exists a sequence $B_1, B_2, \ldots$ of Borel sets such that*

$$A_n \subset B_n \quad \textit{and} \quad \bigcap_{n=1}^{\infty} B_n = 0. \tag{4}$$

(¹) Theorem of P. Novikov, C. R. Acad. Sc. URSS 2 (1934), p. 165.

THEOREM 3. (Second generalized separation theorem.) *If A_1, $A_2, \ldots$ is a sequence of analytic sets, then there exists a sequence $B_1, B_2, \ldots$ of* **CA** *sets such that*

$$A_n - \bigcap_{m=1}^{\infty} A_m \subset B_n \quad \textit{and} \quad \bigcap_{n=1}^{\infty} B_n = 0. \tag{5}$$

The reduction theorem is also valid in the domain of **PCA** sets:

THEOREM 4. *If $U^1, U^2, \ldots$ is an infinite sequence of* **PCA** *sets, then there exists a sequence $V^1, V^2, \ldots$ of disjoint* **PCA** *sets satisfying conditions* (1)(¹).

Just as in the case of **CA** sets, the reduction theorem for **PCA** sets yields two generalized separation theorems(²):

THEOREM 5. *If $A_1, A_2, \ldots$ is a sequence of* **CPCA** *sets such that $\bigcap_{n=1}^{\infty} A_n = 0$, then there exists a sequence $B_1, B_2, \ldots$ of sets which are both* **PCA** *and* **CPCA** *and which satisfy conditions* (4).

THEOREM 6. *If $A_1, A_2, \ldots$ is a sequence of* **CPCA** *sets, then there exists a sequence $B_1, B_2, \ldots$ of* **PCA** *sets satisfying conditions* (5).

THEOREM 7(³). *If $A_1, A_2, \ldots$ is a sequence of analytic sets (resp.* **CPCA** *sets) such that* $\operatorname{Lim\,sup} A_n = 0$, *then there exists a sequence $D_1, D_2, \ldots$ of Borel sets (resp. of sets which are both* **PCA** *and* **CPCA**) *such that*

$$A_n \subset D_n \quad \textit{and} \quad \operatorname{Lim\,sup} D_n = 0. \tag{8}$$

Proof. Let $A_n^* = A_n \cup A_{n+1} \cup \ldots$. By hypothesis, we have

$$\bigcap_{n=1}^{\infty} A_n^* = \bigcap_{n=1}^{\infty} \bigcup_{k=0}^{\infty} A_{n+k} = \operatorname{Lim\,sup} A_n = 0.$$

Theorem 2 (resp. 5) can be applied. We obtain

$$A_n^* \subset B_n \quad \text{and} \quad \bigcap_{n=1}^{\infty} B_n = 0.$$

(¹) For a proof, see my paper referred to above, Fund. Math. 26 and also Fund. Math. 28 (1937), p. 136.

(²) Theorems of P. Novikov, *Sur la séparabilité des ensembles projectifs de seconde classe*, Fund. Math. 25 (1935), Theorem VI, p. 465 and Theorem VII, p. 466.

(³) Proved for analytic sets by A.A. Liapunov, C. R. Acad. Sc. URSS, 2 (1934), p. 276.

Let

$$D_1 = B_1 \quad \text{and, in general,} \quad D_n = D_{n-1} \cap B_n. \tag{9}$$

We have

$$A_n \subset A_n^* \subset B_n \quad \text{and} \quad A_n^* \subset A_{n-1}^*.$$

Inclusion (8) is proved by induction. First,

$$A_1 \subset A_1^* \subset B_1 = D_1.$$

Next, assuming that $A_{n-1}^* \subset D_{n-1}$, we have $A_n^* \subset D_{n-1}$ and, since $A_n^* \subset B_n$, we obtain

$$A_n^* \subset D_{n-1} \cap B_n = D_n, \quad \text{and hence} \quad A_n \subset A_n^* \subset D_n.$$

Moreover, since $D_n \subset D_{n-1}$, we have

$$\operatorname{Lim\,sup} D_n = \bigcap_{n=1}^{\infty} \bigcup_{k=0}^{\infty} D_{n+k} = \bigcap_{n=1}^{\infty} D_n \subset \bigcap_{n=1}^{\infty} B_n = 0.$$

Finally, if B_n are Borel sets (resp. ***PCA*** and ***CPCA*** sets), then so are D_n, by virtue of (9).

XI. *A* and *CA* functions.** We say that a real valued function f is an ***A *function* (resp. ***CA*** *function*), if the set $\underset{x}{E}[f(x) > c]$ is an ***A*** set (resp. a ***CA*** set) for every c(1).

It is easily seen that ***A*** functions and ***CA*** functions are Lebesgue *measurable* and *have the Baire property in the restricted sense.* The functions which are both ***A*** and ***CA*** coincide with the B measurable functions. The *characteristic function* of an ***A*** set (***CA*** set) is an ***A*** function (***CA*** function).

THEOREM 1. *The limit of a convergent sequence of* ***A*** (*or* ***CA***) *functions is an* ***A*** (*or* ***CA***) *function.*

Proof. The condition $f(x) = \lim_{n=\infty} f_n(x)$ implies the equivalence

$$\{f(x) > c\} \equiv \bigvee_n \bigwedge_k [f_{n+k}(x) > c + 1/n],$$

and therefore

$$\underset{x}{E}\{f(x) > c\} = \bigcup_n \bigcap_k \underset{x}{E}[f_{n+k}(x) > c + 1/n].$$

(1) Compare L. Kantorovitch, *Sur les fonctions du type* (U), C. R. Paris 192, p. 1267. It is clear that functions of arbitrary projective class may be defined similarly.

THEOREM 2. *If f is an **A** or **CA** function, then the set* $\underset{xy}{E}[y = f(x)]$ *is the difference of two analytic sets.*

Proof. If $\{r_n\}$ denotes the sequence of rational numbers, we have the equivalence

$$[y \neq f(x)] \equiv \bigvee_n \{[y < r_n < f(x)] \vee [f(x) < r_n < y]\}.$$

THEOREM 3. *If f is a* (bounded) *B measurable function of two variables x and t, and $g(x) = \sup_t f(x, t)$, then g is an **A** function, and if $h(x) = \inf_t f(x, t)$, then h is a **CA** function*[1].

Proof. The equivalence

$$[\sup_t f(t) \leqslant c] \equiv \bigwedge_t [f(t) \leqslant c]$$

yields

$$[g(x) \leqslant c] \equiv \bigwedge_t [f(x, t) \leqslant c], \quad \text{i.e.} \quad [g(x) > c] \equiv \bigvee_t [f(x, t) > c],$$

which proves that the set $\underset{x}{E}[g(x) > c]$ is analytic, as a projection of the Borel set $\underset{xt}{E}[f(x, t) > c]$. Similarly, the equivalence

$$[h(x) < c] \equiv \bigvee_t [f(x, t) < c]$$

gives

$$[h(x) \leqslant c] \equiv \bigwedge_n [h(x) < c + 1/n] \equiv \bigwedge_n \bigvee_t [f(x, t) < c + 1/n]$$

and we infer that the set $\underset{x}{E}[h(x) \leqslant c]$ is analytic.

Under the same assumptions, $\limsup_{t=a} f(x, t)$ *is an **A** function and* $\liminf_{t=a} f(x, t)$ *is a **CA** function.*

For $\limsup g(t) = \lim_{n=\infty} m_n$, where $m_n = \sup g(t)$ for $0 < |a - t| < 1/n$. It follows by Theorems 1 and 3 that $\limsup_{t=a} f(x, t)$ is an **A** function. The case of liminf is similar.

---

[1] F. Hausdorff, *Mengenlehre*, p. 274.

EXAMPLE. *The Dini upper (or lower) partial derivatives of a B measurable function g of two variables are* $\boldsymbol{A}$ *(or* $\boldsymbol{CA}$*) functions* (they can assume also infinite values)(¹).

Proof. If we put

$$f(x,y,t)=\frac{g(x+t,y)-g(x,y)}{t},$$

we have

$$\limsup_{t=+0}\frac{g(x+t,y)-g(x,y)}{t}=\limsup_{t=0}f(x,y,t),$$

and f is a B measurable function (provided that $t>0$).

§ 40. Totally imperfect spaces and other singular spaces

The spaces considered in § 40 are assumed to be metric and separable.

I. Totally imperfect spaces. A space which contains no set homeomorphic to the Cantor discontinuum $\mathscr{C}$ is called *totally imperfect.*

For complete separable spaces, the assumption that a set E is totally imperfect means that E *contains no non-empty perfect set*; hence that E contains no analytic uncountable set, for every analytic uncountable set contains $\mathscr{C}$ topologically, see § 39, I.

THEOREM 1 (of F. Bernstein)(²). *In every uncountable complete separable space there exists a set Z which, together with its complement, is totally imperfect and has power of continuum.*

This theorem follows directly from the following lemma of the general set theory.

LEMMA 2. *Let R be a set of power $\mathfrak{c}$ and let $\boldsymbol{M}$ be a family of power $\leqslant\mathfrak{c}$ of subsets of R each of which has power $\mathfrak{c}$. Then R contains a subset Z such that both Z and $R-Z$ have power $\mathfrak{c}$ and both contain at least one element from each set belonging to the family $\boldsymbol{M}$.*

(¹) M. Neubauer, *Ueber die partiellen Derivierten unstetiger Funktionen*, Mon. f. Math. u. Phys. 38 (1931), p. 139.

(²) Leipz. Ber. 60 (1908), p. 329. Compare P. Mahlo, ibid. 63 (1911), p. 346 and A. Schönflies, *Entwicklung der Mengenlehre* I, Leipzig 1913, p. 361. For a proof, see a paper of W. Sierpiński and myself, Fund. Math. 8 (1926), p. 193.

The problem of existence of totally imperfect separable spaces was originally raised by L. Scheeffer, Acta Math. 5 (1884), p. 287.

Proof. The proof will be based on the well-order theorem: Let $\Omega_{\mathfrak{c}}$ be the smallest ordinal number of power $\mathfrak{c}$. Assume that the elements of $\boldsymbol{M}$ are arranged into a transfinite sequence of type $\Omega_{\mathfrak{c}}$:

$$M_1, M_2, \ldots, M_\omega, M_{\omega+1}, \ldots \tag{1}$$

whose terms are not necessarily different.

Similarly, let

$$x_1, x_2, \ldots, x_\omega, x_{\omega+1}, \ldots \tag{2}$$

be a sequence of type $\Omega_{\mathfrak{c}}$ with distinct terms, composed of all elements of R.

We define by transfinite induction two sequences $\{p_a\}$ and $\{q_a\}$ ($a < \Omega_{\mathfrak{c}}$) as follows.

(i) p_1 is the first term of sequence (2) contained in M_1, and q_1 is the first term of sequence (2) such that $p_1 \neq q_1$ and $q_1 \in M_1$.

(ii) Let $a > 1$ and let S_a be the set of all the p_ξ and q_ξ with $\xi < a$. Define p_a to be the first term of (2) contained in $M_a - S_a$ (such a term exists, since S_a is of power $< \mathfrak{c}$); similarly, let q_a be the first term of (2) contained in $M_a - S_a - p_a$.

Define Z to be the set of all the p_a with $a < \Omega_{\mathfrak{c}}$.

Remarks. 1. In the case where R is an interval, *the set Z is not Lebesgue measurable.* For Z, as well as the complement of Z, have the interior measure zero, since they are totally imperfect.

2. The proof of the existence of uncountable totally imperfect sets given here is not *effective*. That is, the existence was proved, without specifying any individual set Z (sequences (1) and (2) were not defined). No effective proof is known (even in the case of the space of real numbers). This is one of the fundamental problems connected with the concept of effectiveness (1).

3. *If $\mathscr{X}$ is complete and uncountable, there exists an uncountable (totally imperfect) set $P \subset \mathscr{X}$ each of whose continuous images (situated in $\mathscr{X}$) is a totally imperfect set.*

For, if we suppose that $\mathfrak{c} > \aleph_1$, then every set of power $\aleph_1$ has this property. If we assume the continuum hypothesis, it is sufficient to apply Theorem 7 of § 35, I, to the family $\boldsymbol{F}$ of non-empty perfect subsets of $\mathscr{I}^{\aleph_0}$.

(1) Compare remarks of Bernstein on this matter, *loc. cit.*

THEOREM 3. *If $\mathscr{X}$ is a complete, separable and dense in itself space and Z is a totally imperfect subset of $\mathscr{X}$, then $\mathscr{X}-Z$ is not of the first category at any point of $\mathscr{X}$.*

Proof. Suppose that G is a non-empty open set such that $G-Z$ is of the first category. Let B be an $\boldsymbol{F}_\sigma$ of the first category such that $G-Z\subset B$. The set $G-B$ is dense in G (§ 34, IV), and hence dense in itself (§ 9, V, Theorem 3). As a dense in itself $\boldsymbol{G}_\delta$-set, $G-B$ is uncountable (§ 34, V, Theorem 3) and must contain a non-empty perfect set (§ 36, V). But then the set Z, which contains $G-B$, is not totally imperfect.

THEOREM 4. *If the space $\mathscr{X}$ is complete, separable and dense in itself, then every totally imperfect subset of $\mathscr{X}$ with the Baire property is of the first category.*

Consequently, *if the sets Z and $\mathscr{X}-Z$ are totally imperfect, they do not have the Baire property.*

This is a consequence of Theorem 3 and Corollary 2 of § 11, IV.

II. Spaces that are always of the first category. We say that a space is always of the first category, if every dense in itself subset of it is of the first category on itself.

THEOREM 1. *In a complete separable space the following four properties are equivalent:*

(i) *to be a totally imperfect $\boldsymbol{B}_r$ set (i.e. a set with the Baire property in the restricted sense);*

(ii) *to be a set of the first category on every perfect set;*

(iii) *to be a set always of the first category;*

(iv) *to be a set each of whose subsets is a $\boldsymbol{B}_r$.*

Proof. (i) $\Rightarrow$ (ii). This is a consequence of I, Theorem 4.

(ii) $\Rightarrow$ (iii). If E has property (ii) and if X is a dense in itself subset of E, then $E\cap\overline{X}$ is of the first category on $\overline{X}$. It follows that (§ 10, IV, Theorem 2) X is of the first category on itself.

(iii) $\Rightarrow$ (iv). Clearly property (iii) implies property $\boldsymbol{B}_r$ and every subset of a set with property (iii) also has property (iii).

(iv) $\Rightarrow$ (i). By Theorems 1 and 4 of Section I, every non-empty perfect set contains a set that does not have the Baire property.

The existence of uncountable sets that are always of the first category was stated in § 24, I, Theorem 5 and in § 39, VIII, Theorem 7. We shall prove it in a more direct way in Section III.

THEOREM 2. *The property of being always of the first category is countably additive.*

Proof. Let $\mathscr{X} = E_1 \cup E_2 \cup \ldots$, where the sets E_n are always of the first category. Let $\mathscr{X}^*$ be a complete space containing $\mathscr{X}$. If P is a perfect subset of $\mathscr{X}^*$, the sets $P \cap E_n$ are of the first category on P, according to (ii); and so too is their union $\mathscr{X} \cap P$.

III. Rarefied spaces (or λ spaces)(1). We say that a space is *rarefied, if every countable subset of it is a* $\boldsymbol{G_\delta}$.

THEOREM 1. *Every rarefied space is always of the first category.*

Proof. Let W be a dense in itself set and let D be a countable set, dense in W, i.e. $D \subset W \subset \overline{D}$. Since $\overline{W}-D$ is a boundary $\boldsymbol{F_\sigma}$-set in $\overline{W}$, it is of the first category in $\overline{W}$. Hence $\overline{W} = (\overline{W}-D) \cup D$ is also of the first category in $\overline{W}$. It follows that W is of the first category on itself (§ 10, IV, Theorem 2).

Remark. The converse theorem does not hold. Compare Section V, p. 523 and Section VIII, Theorem 6.

THEOREM 2. *Every uncountable complete separable space contains an uncountable rarefied set*; therefore *it contains a family of power* $2^{\aleph_1}$ *of uncountable rarefied sets* (since every subset of a rarefied set is rarefied).

It is sufficient to prove the assertion for the space $\mathscr{N}$, since every uncountable complete separable space contains $\mathscr{N}$ topologically.

We shall prove first that *the space* $\mathscr{N}$ *contains an uncountable transfinite sequence of increasing (and distinct)* $\boldsymbol{G_\delta}$*-sets*:

$$Q_0 \subset Q_1 \subset \ldots \subset Q_\omega \subset Q_{\omega+1} \subset \ldots ^{(2)}. \tag{1}$$

Proof. Let $Q_0 = 0$. Let $a > 0$ and suppose that the sets Q_ξ with $\xi < a$ are $\boldsymbol{G_\delta}$ of (Lebesgue) measure zero. Hence $\bigcup_{\xi<a} Q_\xi$ is still of measure zero. Since (by a theorem of measure theory) every set of

(1) See my paper *Sur une famille d'ensembles singuliers*, Fund. Math. 21 (1933), p. 127.

(2) This proposition is due to Z. Zalcwasser. For a proof without using measure theory see W. Sierpiński, C. R. Soc. Sc. de Varsovie, 1934. Compare also F. Hausdorff, *Summen von* $\aleph_1$ *Mengen*, Fund. Math. 26 (1936), p. 248.

measure zero is contained in a G_δ-set of measure zero, there exists a G_δ-set Q_a such that

$$Q_a \supset \bigcup_{\xi<a} Q_\xi \quad \text{and} \quad Q_a \neq \bigcup_{\xi<a} Q_\xi.$$

Clearly we can assume that the sets Q_a are contained in $\mathcal{N}$.

Thus we have shown that a sequence (1) exists. Let $p_{a+1} \in Q_{a+1} - Q_a$. We shall show that the set E of the points $p_1, p_2, \ldots, p_{\omega+1}, p_{\omega+2}, \ldots$ is rarefied. Let $D = (p_{\xi_1}, p_{\xi_2}, \ldots, p_{\xi_n}, \ldots)$ be a countable subset of E and let a be an ordinal greater than all ξ_n. Then $D \subset Q_a \cap E$, and hence $D = Q_a \cap E \cap D = E \cap [Q_a - (Q_a \cap E - D)]$. Since the set $Q_a \cap E$ is countable (for, if $\beta > a$, then p_β does not belong to Q_a), it follows that $Q_a - (Q_a \cap E - D)$ is a G_δ-set. This proves that D is a G_δ relative to E(1).

Application. Let

$$\mathfrak{z} = [\mathfrak{z}^1, \mathfrak{z}^2, \ldots] \quad \text{and} \quad \mathfrak{y} = [\mathfrak{y}^1, \mathfrak{y}^2, \ldots]$$

be two irrational numbers (i.e. two sequences of positive integers). Define: $\mathfrak{z} \prec \mathfrak{y}$, if $\mathfrak{z}^i < \mathfrak{y}^i$ from an index i on:

$$\{\mathfrak{z} \prec \mathfrak{y}\} \equiv \bigvee_n \bigwedge_k \{\mathfrak{z}^{n+k} < \mathfrak{y}^{n+k}\}.$$

THEOREM 3. *Every set $\mathfrak{G}$ of sequences which is well ordered by the relation $\mathfrak{z} \prec \mathfrak{y}$ into type Ω(2) is rarefied (in $\mathcal{N}$).*

Proof. Let $\mathfrak{D}$ be a countable subset of $\mathfrak{G}$:

$$\mathfrak{G} = [\mathfrak{z}_0, \mathfrak{z}_1, \ldots, \mathfrak{z}_\omega, \mathfrak{z}_{\omega+1}, \ldots], \qquad \mathfrak{D} = [\mathfrak{z}_{\xi_1}, \mathfrak{z}_{\xi_2}, \ldots, \mathfrak{z}_{\xi_n}, \ldots].$$

Let a be an ordinal greater than all the ξ_n and let $\mathfrak{H}$ be the set $(\mathfrak{z}_0, \mathfrak{z}_1, \ldots, \mathfrak{z}_a)$. Since $\mathfrak{H}$ contains $\mathfrak{D}$ and since $\mathfrak{H} - \mathfrak{D}$ is countable, it is sufficient to show that $\mathfrak{H}$ is a G_δ in $\mathfrak{G}$, i.e. that $\mathfrak{G} - \mathfrak{H}$ is an F_σ in $\mathfrak{G}$.

(1) This argument is due to Sierpiński, *Sur un ensemble linéaire non dénombrable qui est de première catégorie sur tout ensemble parfait*, C. R. Soc. Sc. de Varsovie, 1933, p. 102.

(2) The existence of a set of this kind follows easily from the Zermelo theorem.

The following equivalences are obvious:

$\{\mathfrak{z} \in (\mathfrak{G} - \mathfrak{H})\} \equiv \{(\mathfrak{z} \in \mathfrak{G}) \wedge (\mathfrak{z}_a \prec \mathfrak{z})\}$ and $\{\mathfrak{z}_a \prec \mathfrak{z}\} \equiv \bigvee_n \bigwedge_k \{\mathfrak{z}_a^{n+k} < \mathfrak{z}^{n+k}\}$,

whence

$$\mathfrak{G} - \mathfrak{H} = \mathfrak{G} \cap \bigcup_n \bigcap_k \underset{\mathfrak{z}}{E} \{\mathfrak{z}_a^{n+k} < \mathfrak{z}^{n+k}\}.$$

Since the set $\underset{\mathfrak{z}}{E}\{m < \mathfrak{z}^i\} = \underset{\mathfrak{z}}{E}\{m+1 \leqslant \mathfrak{z}^i\}$ is closed, so also is the set $\underset{\mathfrak{z}}{E}\{\mathfrak{z}_a^{n+k} < \mathfrak{z}^{n+k}\}$. It follows that $\mathfrak{G} - \mathfrak{H}$ is an $\boldsymbol{F}_\sigma$ in $\mathfrak{G}$ (1).

IV. Mappings.

THEOREM 1. *The properties of being a totally imperfect set and to be a rarefied space are invariant under any one-to-one mapping f whose inverse f^{-1} is a continuous mapping.*

Proof. If the space $\mathcal{Y} = f(\mathcal{X})$ contains a set C homeomorphic to $\mathscr{C}$, then $\mathcal{X}$ contains the set $f^{-1}(C)$ which contains $\mathscr{C}$ topologically (§ 36, V).

On the other hand, if $\mathcal{X}$ is rarefied and P is a countable subset of $\mathcal{Y}$, then $f^{-1}(P)$ is a $\boldsymbol{G}_\delta$-set in $\mathcal{X}$, since it is countable, and the set $P = ff^{-1}(P)$ is a $\boldsymbol{G}_\delta$-set, since the function f^{-1} is continuous. Therefore $\mathcal{Y}$ is rarefied.

THEOREM 2. *If f is an arbitrary function defined on a space $\mathcal{X}$ which is either totally imperfect, or rarefied, then so also is the set $I = \underset{xy}{E}\{y = f(x)\}$.*

This is because the projection parallel to the $\mathcal{Y}$ axis is a one-to-one continuous mapping of I onto $\mathcal{X}$.

THEOREM 3. *Every subset Z of power $\aleph_1$ of a complete separable space $\mathcal{Y}$ is a one-to-one continuous image of a rarefied space contained in $\mathcal{N} \times \mathcal{Y}$; hence, of a set which has the Baire property in the restricted sense.*

Proof. According to III, Theorem 2, $\mathcal{N}$ contains a rarefied set E of power $\aleph_1$. Let f be a one-to-one mapping of E onto Z. The set $I = \underset{xy}{E}\{y = f(x)\}$ is rarefied (by Theorem 2) and Z is a one-to-one continuous image of I.

---

(1) This is, in fact, the argument that was used by Lusin to prove the existence of uncountable sets with property (ii). See *Sur l'existence d'un ensemble non dénombrable qui est de première catégorie sur tout ensemble parfait*, Fund. Math. 2 (1921), pp. 155–157.

THEOREM 4. *An arbitrary function f defined on a space $\mathcal{X}$ that is always of the first category has the Baire property in the restricted sense.* Moreover, *if $A \subset \mathcal{X}$, then the set D of discontinuity points of the partial function $f \mid A$ is of the first category on A* (¹).

Proof. There exists a countable subset E of A such that $A-E$ is dense in itself (§ 23, V). Moreover, since D contains no isolated points of A (§ 13, III), it follows that $D \subset (A-E) \cup E \cap A^d$. Since the set $E \cap A^d$ consists of a sequence of accumulation points, it is of the first category in A. Also, by hypothesis, $A-E$ is of the first category in A; it follows that so is D.

Assuming the *continuum hypothesis* we have the following three propositions:

THEOREM 5. *There exist rarefied spaces of every, finite or infinite dimension* (²).

Proof. By § 27, IX, for every rarefied space $\mathcal{X}$ of power $\mathfrak{c}$ there exists a space $\mathcal{X}^*$ of a dimension given in advance such that $\mathcal{X}$ is a one-to-one image of $\mathcal{X}^*$. By Theorem 1, $\mathcal{X}^*$ is rarefied.

Remarks. (i) *The set* $I = \underset{xy}{E}\{y = f(x)\}$ *may have the Baire property in the restricted sense without the function f having it, even in the unrestricted sense* (³).

This follows from Theorem 2 by virtue of the continuum hypothesis. In fact, let A be a subset of the interval $\mathcal{I}$ without the Baire property, let E be an uncountable rarefied space, let F be a closed subset of E such that the sets F and $E-F$ are uncountable and let f be a one-to-one function such that $f(\mathcal{I}) = E$ and $f(A) = F$.

(ii) *Neither the Baire property in the restricted sense nor that of being a rarefied space are invariant under one-to-one continuous mappings.*

This follows from Theorem 3 by virtue of the continuum hypothesis. For, if A is a set without the Baire property (in the unrestricted sense, compare § 11, IVa), then there exists a rarefied set E such that A is a one-to-one continuous image of E.

(¹) Compare Remark 3 of § 31, X.

(²) See S. Mazurkiewicz and E. Szpilrajn-Marczewski, *Sur la dimension de certains ensembles singuliers*, Fund. Math. 28 (1937), p. 306.

(³) Compare § 32, III, p. 403, Remark 1. See W. Sierpiński, *La propriété de Baire des fonctions et de leurs images*, Fund. Math. 11 (1928), p. 306.

***V. Property λ'.**

DEFINITION (1).We say that a subset E of a space $\mathcal{X}$ *has property* λ' if, for every countable set $X \subset \mathcal{X}$, the set $E \cup X$ is rarefied. In other words: if every countable set $X^* \subset E \cup X$ is a G_δ relative to $E \cup X$.

THEOREM 1. *Theorem* III, 2, *remains valid after replacing the term "rarefied" by* λ'.

Proof. The sets Q_α in formulas III (1) can be subjected to the following additional condition (compare § 39, II, Corollary 4, Remark):

$$\mathcal{N} = \bigcup_{\alpha<\Omega} Q_\alpha.$$

Then the corresponding set E has property λ'(2).

To prove this, we observe that, since $X^* \subset E \cup X$ is countable, there exists an $\alpha < \Omega$ such that $X^* \subset Q_\alpha$. Hence $X^* \subset Q_\alpha \cap (E \cup X)$ and

$$X^* = Q_\alpha \cap (E \cup X) \cap X^* = (E \cup X) \cap \{Q_\alpha - [Q_\alpha \cap (E \cup X) - X^*]\}.$$

Since $Q_\alpha \cap E$ and $Q_\alpha \cap (E \cup X)$ are countable, the set in the brackets $\{\ \}$ is a G_δ. It follows that X^* is a G_δ relative to $E \cup X$.

THEOREM 2. *A subset E of a space $\mathcal{X}$ has property λ' if and only if for every countable set $D \subset \mathcal{X}$ there exists a G_δ-set Q such that* $Q \cap E \subset D \subset Q$.

Proof. If E has property λ', $E \cup D$ is rarefied and $D = Q \cap (E \cup D)$, where Q is a suitably chosen G_δ-set. Hence $Q \cap E \subset D \subset Q$.

Conversely, let D and X be two countable sets and let Q be a G_δ-set such that

$$D \subset E \cup X \quad \text{and} \quad Q \cap E \subset D \subset Q. \tag{1}$$

Let us put

$$Q^* = (Q - X) \cup D = Q - (X - D).$$

Thus Q^* is a G_δ-set and we have

$$Q^* \cap (E \cup X) = (Q - X) \cap (E \cup X) \cup \big(D \cap (E \cup X)\big) = D,$$

(1) W. Sierpiński, *Sur une propriété additive d'ensembles*, C. R. Soc. Sc. Varsovie 30 (1937), p. 257.

(2) W. Sierpiński, *Sur deux conséquences d'un théorème de Hausdorff*, Fund. Math. 33 (1945), p. 269.

since

$$(Q-X) \cap (E \cup X) \subset Q \cap E \subset D \quad \text{and} \quad D \cap (E \cup X) = D,$$

by (1). The formula $D = Q^* \cap (E \cup X)$ proves that $E \cup X$ is rarefied. It follows that E has property λ'.

THEOREM 3. *Property λ' is countably additive.*

Proof. Let $E_1, E_2, \ldots$ be a (finite or infinite) sequence of subsets of $\mathscr{X}$ with property λ'. Let D be a countable subset of $\mathscr{X}$. We have to define a G_δ-set Q such that $Q \cap (E_1 \cup E_2 \cup \ldots) \subset D \subset Q$.

By hypothesis, there exists a sequence $Q_1, Q_2, \ldots$ of G_δ-sets such that $Q_n \cap E_n \subset D \subset Q_n$. Define $Q = Q_1 \cap Q_2 \cap \ldots$. Then

$$Q \cap (E_1 \cup E_2 \cup \ldots) \subset (Q_1 \cap E_1) \cup (Q_2 \cap E_2) \cup \ldots \subset D \subset Q.$$

Remarks. The property of being a rarefied set is not additive(1). In fact, *a rarefied set may cease to be rarefied after adjunction of a countable set.*

This will now be proved. Define the set $\mathfrak{C}$ (compare Section III, Remark) as follows. Let $\mathfrak{y}_0, \mathfrak{y}_1, \ldots, \mathfrak{y}_\alpha, \ldots$ $(\alpha < \gamma)$ be a transfinite sequence composed of all irrational numbers, where γ is the smallest ordinal number corresponding to the power $\mathfrak{c}$; the irrational numbers are regarded as sequences of positive integers. Define $\mathfrak{z}_0 = \mathfrak{y}_0$. Let $\alpha > 0$ and assume that the $\mathfrak{z}_\xi$ with $\xi < \alpha$ are defined. Define $\mathfrak{z}_\alpha$ to be the first term of the sequence $\{\mathfrak{y}_\beta\}$, $\beta < \gamma$, satisfying the relation $\mathfrak{z}_\xi \prec \mathfrak{z}_\alpha$ for every $\xi < \alpha$, provided that such a term exists; otherwise the set $\mathfrak{C}$ is of type α.

One can prove(2) that $\mathfrak{C}$ so defined is rarefied, while $\mathfrak{C} \cup \mathscr{R}$ (where $\mathscr{R}$ is the set of rational numbers) is not.

The set $\mathfrak{C}$ is also an example of *a rarefied set that is not a λ' set.* The existence of such sets follows also from the topological invariance of the property of being a rarefied set, for *there exist sets homeomorphic to λ' sets that are not λ' sets*(3).

---

(1) Compare F. Rothberger, *Sur un ensemble toujours de première catégorie qui est dépourvu de la propriété λ*, Fund. Math. 32 (1939), p. 306.

(2) W. Sierpiński, *Sur un ensemble à propriété λ*, Fund. Math. 32 (1939), p. 306.

(3) W. Sierpiński, *Sur la non-invariance topologique de la propriété λ′*, Fund. Math. 33 (1945), p. 264.

Finally we remark that, by virtue of II, Theorem 2, *the set* $\mathfrak{C} \cup \mathcal{R}$ *is always of the first category, but it is not rarefied*(1).

THEOREM 4. *If every uncountable subset of a set* $E \subset \mathcal{X}$ *has infinite dimension*(2), *then the set* E *has property* λ' (3).

Proof. Given an arbitrary countable set D, there exists a 0-dimensional G_δ-set Q containing D. Let

$$Q^* = Q - \big((Q \cap E) - D\big).$$

Since $\dim(Q \cap E) = 0$, hence $\overline{\overline{Q \cap E}} \leqslant \aleph_0$ by hypothesis. Q^* is therefore a G_δ-set.

Since $D \subset Q$, we have

$$Q^* = (Q - E) \cup (Q \cap D) = (Q - E) \cup D \supset D$$

and $Q^* \cap E = Q^* \cap D \subset D$.

It follows by Theorem 2 that E has property λ'.

VI. σ spaces.

DEFINITION. $\mathcal{X}$ *is a* σ *space, if every* F_σ *subset of* $\mathcal{X}$ *is a* G_δ*-set*(4).

The existence of uncountable σ spaces (contained in $\mathcal{N}$) will be deduced from the continuum hypothesis. The proof of this is quite analogous to that of existence of rarefied spaces (Section III, Theorem 2). We consider first all G_δ-sets of measure zero as being ordered into a transfinite sequence of type Ω:

$$K_0, K_1, \ldots, K_\omega, K_{\omega+1}, \ldots. \tag{1}$$

There are uncountably many non-empty differences of the form $K_\alpha - \bigcup_{\xi<\alpha} K_\xi$. For, otherwise, we would have, from a certain index α on,

$$\bigcup_{\xi<\alpha} K_\xi = \bigcup_{\xi<\alpha+1} K_\xi = \bigcup_{\xi<\alpha+2} K_\xi = \ldots$$

(1) The existence of sets with this property was first proved by Lusin with aid of the continuum hypothesis; see *Sur les ensembles toujours de première catégorie*, Fund. Math. 21 (1933), p. 119; then by Rothberger without the continuum hypothesis in the paper referred to above.

(2) About the existence of such sets E see § 28, IV, Remark 2.

(3) See Mazurkiewicz and Szpilrajn-Marczewski, *loc. cit.*, p. 307.

(4) See W. Sierpiński, *Sur l'hypothèse du continu*, Fund. Math. 5 (1924), p. 184, and E. Szpilrajn-Marczewski, *Sur un problème de M. Banach*, Fund. Math. 15 (1930), p. 212.

and, since each one-point G_δ-set occurs in sequence (1), we would have $\bigcup_{\xi<\alpha} K_\xi = \mathcal{N}$ which is impossible, for the union $\bigcup_{\xi<\alpha} K_\xi$ has measure zero. We shall show that:

Every set E that contains a single point from each non-empty set $K_\alpha - \bigcup_{\xi<\alpha} K_\xi$ *is a* σ *set.*

We shall prove first that *every Lebesgue measurable set H contained in E is countable.*

Since the set $E \cap K_\alpha$ is countable, for every α, there exists no uncountable G_δ-set of measure zero contained in E. Hence E has the interior measure zero and therefore H has measure zero. Consequently there exists a K_α with $H \subset K_\alpha$, i.e. $H \subset E \cap K_\alpha$. Since $E \cap K_\alpha$ is countable, so also is H.

Now let B be an arbitrary F_σ-set contained in the interval $\mathcal{I} = 01$. We must show that $B \cap E$ is a G_δ-set relative to E. Let $\mathcal{I} - B = U \cup V$, where U is an F_σ-set and V has measure zero. Then $E - B = (U \cap E) \cup (V \cap E)$. As a set of measure zero, $V \cap E$ is countable and, consequently, $E - B$ is an F_σ-set in E and $B \cap E$ is a G_δ-set. This completes the proof.

THEOREM 1. *Every Borel set in a* σ *space is both* F_σ *and* G_δ.

Proof. We have to prove that the countable union and intersection of ambiguous sets of class 1 in a σ space are ambiguous sets of class 1. Now if $X_1, X_2, \dots$ are both F_σ and G_δ, their union $X_1 \cup X_2 \cup \dots$ is an F_σ and, since the space is a σ space, it is a G_δ; their intersection is a G_δ, i.e. the complement of an F_σ, hence of a G_δ; it is therefore an F_σ.

THEOREM 2. *Every finitely dimensional* σ *space is* 0*-dimensional.*

Proof. Suppose $\mathcal{X}$ is a σ space of finite dimension $n > 0$, and let S be an F_σ-set such that (compare § 27, I, Corollary 2d):

$$\dim S \leqslant n-1 \quad \text{and} \quad \dim(\mathcal{X} - S) = 0.$$

The set $\mathcal{X} - S$ is a G_δ and, by Theorem 1, it is an F_σ. Thus $\mathcal{X}$ decomposes into the union of two F_σ-sets, S and $\mathcal{X} - S$, both of dimension $\leqslant n-1$. It follows from § 27, I, Theorem 2, that $\dim \mathcal{X} \leqslant n-1$, which is a contradiction.

Theorem 2 together with Theorem IV, 5 implies

THEOREM 3. *There exist rarefied spaces that are not* σ *spaces.*

Clearly, *every* σ *space is rarefied.*

The next theorem easily follows from Theorem 1.

THEOREM 4. *Every B measurable function defined on a σ space is of class* 1.

Remark. The set E used in the proof of existence of uncountable σ spaces is clearly not Lebesgue measurable. Thus *there exist non-measurable sets that have the Baire property in the restricted sense* (because every rarefied space has this property, hence every σ space has it).

More generally, if every Lebesgue measurable subset of a set E is countable, then E has the Baire property in the restricted sense (1).

Let us mention that the converse problem, that of the existence of a measurable set (of measure zero) without the Baire property (even in the unrestricted sense), is easily solved without the continuum hypothesis.

VII. ν spaces, concentrated spaces, property C.

DEFINITION 1. $\mathscr{X}$ is a ν space, if *every nowhere dense set is countable* (2).

The existence of uncountable ν spaces (contained in $\mathscr{N}$) follows from the *continuum hypothesis*. We shall first prove the existence of an uncountable set $E \subset \mathscr{N}$ which has the following property (called *property* L):

PROPERTY L. *For every nowhere dense subset N of the interval $\mathscr{I}$*, $\overline{\overline{N \cap E}} \leqslant \aleph_0$.

Proof. We think of closed nowhere dense subsets of $\mathscr{I}$ as being ordered into a transfinite sequence of type Ω:

$$F_0, F_1, \ldots, F_\omega, F_{\omega+1}, \ldots.$$

Since $\mathscr{I}$ is not of the first category, it follows (as in the proof in Section VI) that there are uncountably many non-empty differences $F_\alpha - \bigcup_{\xi<\alpha} F_\xi$.

Let E_0 be the set containing a single point from each of these differences and let $E = E_0 \cap \mathscr{N}$. Then $\overline{\overline{E}} > \aleph_0$.

---

(1) See S. Saks, *Sur un ensemble non mesurable jouissant de la propriété de Baire*, Fund. Math. 11 (1928), p. 277. Compare N. Lusin, *Sur une question concernant la propriété de Baire*, Fund. Math. 9 (1927), p. 117.

(2) N. Lusin, C. R. Paris 158 (1914), p. 1259. See also P. Mahlo, *loc. cit.*

Moreover, if N is a nowhere dense subset of $\mathscr{I}$, there exists an index a with $F_a = \overline{N}$. Since $\overline{\overline{E_0 \cap F_a}} \leqslant \aleph_0$ and $N \cap E \subset E_0 \cap F_a$, it follows that $\overline{\overline{N \cap E}} \leqslant \aleph_0$. This completes the proof.

Now it follows that E is an uncountable ν space, since clearly property ν follows from property L.

Remark. The converse implication does not hold (this can be seen by considering a subset of $\mathscr{C}$ with property ν). One can prove, however[1], that *every (metric separable) ν space is homeomorphic to a set with property* L.

THEOREM 1. *Every ν space is totally imperfect.*

For the Cantor set $\mathscr{C}$ contains a perfect subset nowhere dense in $\mathscr{C}$.

THEOREM 2. *Every subset with the Baire property of a ν space is the union of a G_δ-set and a countable set.*

For it is the union of a G_δ-set and a set of the first category (§ 11, IV, 2) and the latter is countable by hypothesis.

We have therefore

THEOREM 3. *Every function with the Baire property defined on a ν space is of the second class*[2].

DEFINITION 2. A space $\mathscr{X}$ is said to be *concentrated about a set A*, if, for every neighbourhood G of A, the set $\mathscr{X} - G$ is countable[3].

THEOREM 4. *$\mathscr{X}$ is a ν space if and only if it is concentrated about every set dense in it; or, if and only if it is concentrated about every countable dense set*[4].

Proof. Let $\mathscr{X}$ be a ν space. Let $\overline{A} = \mathscr{X}$ and let E be a neighbourhood of A. If G denotes the interior of E, we have $A \subset G$, whence $\overline{G} = \mathscr{X}$, which proves that $\mathscr{X} - G$ is nowhere dense, hence countable. Since $\mathscr{X} - E \subset \mathscr{X} - G$, the set $\mathscr{X} - E$ is countable.

---

(1) See a paper of Sierpiński and myself *Sur les ensembles qui ne contiennent aucun sous-ensemble indénombrable non-dense*, Fund. Math. 26 (1936), p. 137.

(2) G. Poprougénko, *Sur un problème de M. Mazurkiewicz*, Fund. Math. 15 (1930), p. 285.

(3) After S. Besicovitch (Acta Math. 62 (1934), p. 289), a subset E of a space $\mathscr{X}$ is concentrated with respect to a set $A \subset \mathscr{X}$ if, for every neighbourhood G of A, the set $E - G$ is countable. Clearly, if E is concentrated with respect to A, then so is $E \cup A$.

(4) E. Szpilrajn-Marczewski, Fund. Math. 31 (1938), p. 218.

Assume now that N is nowhere dense and that A is a countable subset of $\mathscr{X}-\overline{N}$, dense in $\mathscr{X}-\overline{N}$. Then $\overline{A} = \overline{\mathscr{X}-\overline{N}} = \mathscr{X}$. Since $\mathscr{X}-\overline{N}$ is a neighbourhood of A, its complement, the set $\overline{N}$, is countable by hypothesis. It follows that N is countable.

THEOREM 5. *Every space $\mathscr{X}$ concentrated about a countable set A, hence also every ν space, has the following property, called property* C'':

If, for every $x \in \mathscr{X}$, there exists a sequence of open sets $G_{x,n}$, where $n = 1, 2, \ldots$, such that $x \in G_{x,n}$, then there exists an infinite sequence x_1, $x_2, \ldots$ such that

$$\mathscr{X} = G_{x_1,1} \cup G_{x_2,2} \cup \ldots \text{ [1]}. \tag{1}$$

Proof. Let $A = [x_1, x_3, x_5, \ldots]$. Since $G_{x_1,1} \cup G_{x_3,3} \cup \ldots$ is a neighbourhood of A and since $\mathscr{X}$ is concentrated about A, the set $\mathscr{X}-(G_{x_1,1} \cup G_{x_3,3} \cup \ldots)$ is countable. Let us arrange it in a sequence $x_2, x_4, x_6, \ldots$ (we can assume that neither this set nor the set A is empty). Then $x_1, x_2, x_3, x_4, \ldots$ is the required sequence.

THEOREM 6. *Property* C'' *is invariant with respect to continuous mappings.*

Proof. Let $\mathscr{X}$ be a space with property C'', let f be a continuous mapping of $\mathscr{X}$ onto $\mathscr{Y}$ and let $H_{y,n}$ be a system of open sets in $\mathscr{Y}$ such that $y \in H_{y,n}$. We must define a sequence of points $y_1, y_2, \ldots$ such that

$$\mathscr{Y} = H_{y_1,1} \cup H_{y_2,2} \cup \ldots.$$

Since the function f is continuous, the sets $f^{-1}(H_{y,n})$ are open. Let $G_{x,n} = f^{-1}(H_{f(x),n})$. Since $f(x) \in H_{f(x),n}$, we have $x \in G_{x,n}$. Hence there exists a sequence $x_1, x_2, \ldots$ satisfying (1). It follows that

$$\mathscr{Y} = H_{f(x_1),1} \cup H_{f(x_2),2} \cup \ldots.$$

THEOREM 7. *Property* C'' *implies the following property* C[2]:

---

[1] F. Rothberger, *Eine Verschärfung der Eigenschaft C*, Fund. Math. 30 (1938), p. 50.

[2] The problem of the existence of uncountable sets with property C was originally raised by Borel. These sets coincide with "the sets whose asymptotic measure is lower than any series given in advance" of Borel. See *Sur la classification des ensembles de mesure nulle*, Bull. Soc. Math. de France 47 (1919), p. 123. It is not known whether the existence of (uncountable) sets with property C can be proved without the continuum hypothesis. Numerous bibliographical references concerning property C can be found in a paper of W. Sierpiński, *Sur le rapport de la propriété* (C) *à la théorie générale des ensembles*, Fund. Math. 29 (1937), p. 91.

For every sequence $\{\lambda_n\}$ of positive numbers there exists a sequence $\{A_n\}$ of sets such that

$$\mathscr{X} = A_1 \cup A_2 \cup \ldots \quad \textit{and} \quad \delta(A_n) < \lambda_n. \tag{2}$$

It follows that *every subset of $\mathscr{E}$ with property* C″ *has Lebesgue measure zero; the same is true for every continuous image* ($\subset \mathscr{E}$) *of a ν set* (by Theorems 5 and 6)(¹).

Proof. Let $G_{x,n}$ be an open ball with centre x and diameter $< \lambda_n$. By hypothesis there exists a sequence $x_1, x_2, \ldots$ satisfying (1). By letting $A_n = G_{x_n,n}$, we get $\delta(A_n) < \lambda_n$, and the assertion follows.

Remark. The last part of Theorem 7 remains valid after replacing continuous mapping by *mapping with the Baire property.*

For, if f is a function with the Baire property, then there exists a set A such that $\mathscr{X} - A$ is of the first category, therefore countable, and the partial function $f \mid A$ is continuous (§ 32, II). Since A has property ν, $f(A)$ has measure zero. Hence $f(\mathscr{X})$ has measure zero.

This statement can also be proved easily by using Theorem V of § 32.

THEOREM 8. *Every space $\mathscr{X}$ with property* C, *hence every continuous image of a ν space, has dimension* 0(²).

Proof. Let x_0 be a fixed point of $\mathscr{X}$ and let $f(x) = |x - x_0|$. We shall first prove that the set $f(\mathscr{X})$ has property C. Let $\{\lambda_n\}$ be a given sequence and let $\{A_n\}$ be a sequence satisfying (2). Then

$$f(\mathscr{X}) = f(A_1) \cup f(A_2) \cup \ldots \quad \text{and} \quad \delta[f(A_n)] \leqslant \delta(A_n),$$

because

$$|f(x) - f(x')| = \big||x - x_0| - |x' - x_0|\big| \leqslant |x - x'|.$$

Since $f(\mathscr{X})$ has property C, it is a boundary set in $\mathscr{E}$ (even its measure is zero). Therefore there exists an arbitrarily small number $\eta > 0$ which is not a value of the function f, i.e. there exists no x such that $|x - x_0| = \eta$. Hence the set $\underset{x}{E}[|x - x_0| < \eta]$ is both closed

(¹) Compare W. Sierpiński, *Sur un ensemble non dénombrable dont toute image continue est de mesure nulle*, Fund. Math. 11 (1928), p. 304. Extensions of this theorem to arbitrary measures are due to Poprougénko and Szpilrajn-Marczewski. See Szpilrajn-Marczewski, *Remarques sur les fonctions complètement additives*, Fund. Math. 22 (1934), p. 311. See also Section IX.

(²) E. Szpilrajn-Marczewski, *La dimension et la mesure*, Fund. Math. 28 (1937), p. 82.

and open, contains x_0 and has diameter $\leqslant 2\eta$. This proves that $\dim_{x_0} \mathscr{X} = 0$.

THEOREM 9. *Every space $\mathscr{X}$ with property* C, *and hence every continuous image of a ν space, is totally imperfect*(¹).

Proof. Suppose that the space $\mathscr{X}$ contains a set P homeomorphic to the Cantor set $\mathscr{C}$. Let f be a continuous mapping of P onto a set of a positive measure, for example, onto the interval $\mathscr{I}$ (compare § 16, II, Corollary 6a). We can therefore assume that

$$\text{measure } [f(P)] = 1. \tag{3}$$

Since the function f is uniformly continuous, there exists for every n a $\lambda_n > 0$ such that the condition $|x - x'| < \lambda_n$ implies condition $|f(x) - f(x')| < 1/2^{n+1}$. Since the set P (as a subset of $\mathscr{X}$) has property C, we have

$$P = A_1 \cup A_2 \cup \ldots, \quad A_n = \bar{A}_n, \quad \delta(A_n) < \lambda_n,$$

hence

$$f(P) = f(A_1) \cup f(A_2) \cup \ldots, \quad \text{measure } [f(A_n)] \leqslant \delta[f(A_n)] \leqslant 1/2^{n+1}.$$

It follows that

$$\text{measure } [f(P)] \leqslant 1/4 + 1/8 + \ldots = 1/2,$$

which contradicts (3).

VIII. Relation to the Baire property in the restricted sense.

THEOREM 1. *A set E with property ν situated in a complete separable space does not have the Baire property in the restricted sense unless it is countable.*

Proof. If E has the Baire property with respect to $\bar{E}$, then E is the union of a G_δ-set with a set P of the first category in $\bar{E}$ (§ 11, IV, 2). By § 10, IV, 2, P is of the first category in E, hence it must be countable. Since E is totally imperfect (VII, 1), the G_δ-set is countable. It follows that E is countable.

THEOREM 2. *There exists a one-to-one continuous function f: $\mathscr{N} \to \mathscr{N}$, which maps every set with property* L *onto a set which is always of the first category*(²).

(¹) E. Szpilrajn-Marczewski, *Sur une hypothèse de M. Borel*, Fund. Math. 15 (1930), p. 126.

(²) See N. Lusin, *Sur les ensembles toujours de première catégorie*, Fund. Math. 21 (1933), p. 114.

Proof. According to § 36, I (i) and (j), let f be a one-to-one continuous function defined on $\mathcal{N}$, with values in $\mathcal{N}$, and such that f maps every open subset of $\mathcal{N}$ onto a set of the first category on itself. Let E be a subset of $\mathcal{N}$ with property L and let P be a perfect subset of $\mathcal{N}$. According to II (ii), we have to prove that the set $P \cap f(E)$ is of the first category on P.

Let $Q = f^{-1}(P)$. Then (compare § 3, III, (13))

$$P \cap f(E) = f[E \cap f^{-1}(P)] = f(E \cap Q).$$

Let $G = \mathrm{Int}(Q)$. Since f is continuous, Q is closed and $Q-G$ must be nowhere dense. Therefore

$$\overline{\overline{(E \cap Q)-G}} \leqslant \aleph_0, \quad \text{and hence} \quad \overline{\overline{f\big((E \cap Q)-G\big)}} \leqslant \aleph_0.$$

Thus $f\big((E \cap Q)-G\big)$ is of the first category on P. Since G is open, $f(G)$ is of the first category on itself, hence on P, because $f(G) \subset f(Q) = P$. It follows that $f(E \cap G)$ is of the first category on P and so is the set

$$P \cap f(E) = f(E \cap Q) = f(E \cap G) \cup f(E \cap Q-G).$$

Theorem 2 has the following interesting consequences (by virtue of the continuum hypothesis which implies the existence of uncountable sets with property L)(1).

THEOREM 3. *The Baire property in the restricted sense is not an invariant of the cartesian multiplication by an axis.*

In fact, *if E is an uncountable set with property* L *and if f is the function considered in Theorem* 2, *then the set $\mathcal{N} \times f(E)$ does not have the Baire property in the restricted sense.*

Proof. By Theorem 1, the set E does not have the Baire property in the restricted sense and neither has it (compare § 35, IV, 2) the set

$$Z = \mathop{E}_{xy}\big[\big(y = f(x)\big) \wedge (x \in E)\big]$$

which is homeomorphic to E (compare § 15, V, Theorem 1). The equivalence

$$\{[y = f(x)] \wedge [y \in f(E)]\} \equiv \{[y = f(x)] \wedge (x \in E)\}$$

(1) W. Sierpiński, *Sur un problème de M. Kuratowski concernant la propriété de Baire des ensembles*, Fund. Math. 22 (1934), p. 54. Compare the remark, § 32, IV.

implies that

$$Z = \underset{xy}{E}[y = f(x)] \wedge [\mathcal{N} \times f(E)].$$

Since the set $\underset{xy}{E}[y = f(x)]$ is closed and Z does not have the Baire property in the restricted sense, it follows that $\mathcal{N} \times f(E)$ cannot have it either.

THEOREM 4. *There exists a function f with the Baire property in the restricted sense defined on $\mathcal{N}$ such that the graph of f does not have this property.*

Proof. In the preceding notation, let $K = f(E)$ and define

$$g(y) = \begin{cases} f^{-1}(y) & \text{if} \quad y \in K, \\ -1 & \text{if} \quad y \in \mathcal{N} - K. \end{cases}$$

Since K is of the first category on every perfect set, the function g is continuous on every perfect set apart from sets of the first category on it. That is, g has the Baire property in the restricted sense.

On the other hand, we have

$$Z = \underset{xy}{E}[x = g(y)] - W, \quad \text{where} \quad W = \underset{xy}{E}[(x = -1) \wedge (y \in \mathcal{N} - K)].$$

Since K has the Baire property in the restricted sense, $\mathcal{N} - K$ has it; since W is homeomorphic to $\mathcal{N} - K$, hence W has this property. But Z does not have it, hence $\underset{xy}{E}[x = g(y)]$ does not have it either.

THEOREM 5. *There exists a function h of one variable, with the Baire property in the restricted sense such that the function g of two variables, defined by $g(x, y) = h(y)$ does not have this property.*

For, if h is the characteristic function of K, then g is the characteristic function of $\mathcal{N} \times K$.

THEOREM 6. *There exists a set which is always of the first category without being rarefied.*

Such is the set K. For suppose that K is rarefied. By IV, Theorem 1, E is therefore rarefied and hence (by III, Theorem 1) it must have the Baire property in the restricted sense. But E has also property ν, which contradicts Theorem 1.

IX. Relation of the ν spaces to the general set theory.

THEOREM 1. *Let $\mathscr{X}$ be a ν space and let m be a function (a "measure") which assigns a number $m(X) \geqslant 0$ to every Borel set $X \subset \mathscr{X}$ so that the following conditions are fulfilled:*

(i) *$m(X) = 0$, if X consists of a single point;*

(ii) *$m(X_1 \cup X_2 \cup \ldots) = m(X_1)+m(X_2)+\ldots$, if $X_i \cap X_j = 0$, for $i \neq j$ (countable additivity).*

Under these assumptions $m(\mathscr{X}) = 0$ (1).

Proof. We shall prove first (without using property ν) that every point has an open neighbourhood of an arbitrarily small measure. Let $x_0 \in \mathscr{X}$ and $\varepsilon > 0$; let B_n be an open ball with centre x_0 and radius $1/n$; and let $B_0 = \mathscr{X}$. Denote $D_n = B_n - B_{n+1}$. Then

$$B_n = (x_0) \cup D_n \cup D_{n+1} \cup \ldots, \qquad D_n \cap D_m = 0 \quad \text{for} \quad n \neq m.$$

It follows that $m(B_n) = m(D_n)+m(D_{n+1})+\ldots$. In particular

$$m(\mathscr{X}) = m(D_0)+m(D_1)+\ldots,$$

and therefore there exists an n such that

$$m(D_n)+m(D_{n+1})+\ldots < \varepsilon, \quad \text{i.e.} \quad m(B_n) < \varepsilon.$$

Now let $p_1, p_2, \ldots$ be a sequence dense in $\mathscr{X}$. Let A_k be an open neighbourhood of p_k such that $m(A_k) < \varepsilon/2^k$ and let

$$A = A_1 \cup A_2 \cup \ldots = A_1 \cup (A_2 - A_1) \cup [A_3 - (A_1 \cup A_2)] \cup \ldots.$$

It follows that

$$m(A) = m(A_1)+m(A_2-A_1)+m[A_3-(A_1 \cup A_2)]+\ldots$$
$$\leqslant m(A_1)+m(A_2)+m(A_3)+\ldots \leqslant \varepsilon.$$

The set $\mathscr{X}-A$ is nowhere dense, and hence countable. It follows that

$$m(\mathscr{X}-A) = 0, \quad \text{whence} \quad m(\mathscr{X}) = m(A) \quad \text{and} \quad m(\mathscr{X}) \leqslant \varepsilon.$$

Since ε is an arbitrary number, it follows that $m(\mathscr{X}) = 0$.

---

(1) See E. Szpilrajn-Marczewski, *loc. cit.* Fund. Math. 22, p. 307. The assumption $m(X) \geqslant 0$ is not essential.

The preceding theorem together with the theorem on the existence of ν spaces of power continuum yields a *negative solution of the generalized measure problem*. Namely:

THEOREM 2. *Except for the function identically equal to zero there exists no non-negative function defined for every subset of the interval* $\mathcal{J}$ *and satisfying conditions* (i) *and* (ii)(¹).

Proof. The last statement is a theorem of the general set theory; it will be proved for every space of power continuum provided that we find a single space for which it holds. But we have shown that there are spaces of power continuum for which it holds; these are ν spaces.

Remark. As seen, the preceding argument allows us to derive a theorem of the general set theory from a topological fact (namely, from the existence of ν spaces of power $\mathfrak{c}$). Conversely, the existence of ν spaces of power $\mathfrak{c}$ can be proved (without using the continuum hypothesis) to be equivalent to a theorem of the general set theory (²).

Analogously, the existence of uncountable ν sets can be used to derive the following two significant theorems (³).

There exists a continuous function f defined on a set of power $\mathfrak{c}$ *such that f is not uniformly continuous on any uncountable set.*

There exists a convergent sequence of functions $f_1, f_2, \ldots$ defined on $\mathcal{E}$ which is not uniformly convergent on any uncountable set.

(¹) Theorem of S. Banach and myself, *Sur une généralisation du problème de la mesure*, Fund. Math. 14 (1930), p. 127. See also S. Ulam, *Zur Masstheorie in der allgemeinen Mengenlehre, ibid.* 16(1930), p. 140. The fact that the existence of uncountable ν spaces provides a solution of the generalized measure problem was observed by E. Szpilrajn-Marczewski in his paper of Fund. Math. 22 (1934), p. 304.

(²) See my paper *Sur le rapport des ensembles de M. Lusin à la Théorie générale des Ensembles*, Fund. Math. 22 (1934), p. 315.

(³) See W. Sierpiński, *Hypothèse du continu*, p. 52. These two theorems have been proved using the continuum hypothesis; actually they are equivalent. Let us mention that the second theorem was suggested by a theorem of Egorov which states that every convergent sequence of measurable functions is uniformly convergent apart from a set of arbitrarily small (exterior) measure.

APPENDIX A

SOME APPLICATIONS OF TOPOLOGY TO MATHEMATICAL LOGIC

by A. MOSTOWSKI

I. Classifications of definable sets. In logical research it is often important to determine the structure of certain sets composed of positive integers or of objects which can be enumerated by means of positive integers. For example, the decision problem for a system (S) of axioms is equivalent to the problem as to whether the set of theorems of (S) is recursive, i.e. whether there exists an algorithm which permits us to verify in an automatic manner whether a given expression is a theorem of (S). Similarly, the problem whether a set E of expressions is axiomatizable is equivalent to the problem whether this set is *recursively enumerable*, i.e. whether there exists a finite number of expressions $E_1, \ldots, E_m$ (axioms) in E and a finite number of finitary operations $O_1, \ldots, O_n$ (rules) such that the set E coincide with the set of expressions obtained by applying the operations $O_1, \ldots, O_n$ in a certain order and a certain number of times to the expressions $E_1, \ldots, E_m$.

As a tool in such investigations, a classification of sets of integers has been constructed; it closely resembles the classification of projective sets(1). The lowest class P_0 of this classification is formed by recursive sets and the higher classes are obtained by the operations of projection and complement just as in the theory of projective sets (compare § 38, I). The sets obtained in this way are called *definable*.

In the above classification, the recursively enumerable sets correspond to the analytic sets. As in the theory of analytic sets we prove that the class P_1 of recursively enumerable sets is different from P_0. The analogy between P_1 and the class of analytic sets is not, however, complete. For example, the first separation theorem (§ 39, III) does not hold for P_1 sets(2).

(1) See S. C. Kleene, *Recursive predicates and quantifiers*, Trans. Amer. Math. Soc. 53 (1943), pp. 41–73 and A. Mostowski, *On definable sets of positive integers*, Fund. Math. 34 (1946), pp. 81–112.

(2) See S. C. Kleene, *A symmetric form of Gődel's theorem*, Proc. Nederl. Akad. van Wetensch. 53 (1950), pp. 800–802.

Let us mention some applications of the classification of definable sets. The Gödel theorem on incompleteness of arithmetic is proved to be equivalent to the statement $P_0 \neq P_1$(¹). The recursively enumerable and recursively non-separable sets are used in construction of numerous counterexamples as well as to obtain sharpened versions of the Gödel theorem(²). Formal systems of arithmetic or of the set theory are compared by determining the place occupied by the set of the propositions true in these systems in the classification of definable sets(³).

The classification of definable sets can be extended to transfinite ordinals(⁴). One also considers sets of integers defined by means of expressions involving variables of higher type(⁵). Clearly those sets do not fall under the classification described above.

Thus it is seen that the theory of projective sets, while not directly applicable to the theory of definable sets, serves as a pattern after which one forms a considerable part of the reasoning. Attention has recently been attracted to analogies between the theory of Borel and projective sets on the one hand, and the theory of sets of integers defined in an elementary and non-elementary way on the other hand(⁶).

II. The space of ideals and the proof of completeness of the logic of predicates. One of the most important theorems of the mathematical logic is the *Gödel completeness theorem.* It states that every

(1) See S. C. Kleene, *Recursive predicates*..., *loc. cit.* and A. Mostowski, *loc. cit.*

(2) See A. Grzegorczyk, *Some proofs of undecidability of arithmetic*, Fund. Math. 43 (1956), pp. 166–177; S. C. Kleene, *Recursive predicates*..., *loc. cit.*; A. Mostowski, *loc. cit.* and *Development and applications of the "projective" classification of sets of integers*, Proc. Int. Congr. of Math. Amsterdam 1954, vol. 1; W. A. Uspenskij, *Teorema Gedela i teorija algorifmov* (in Russian) Doklady Akademii Nauk USSR, 91 (1953), pp. 737–740.

(3) See S. C. Kleene, *Recursive predicates*..., *loc. cit.* and A. Mostowski, *A classification of logical systems*, Studia Philosophica 4 (1951), pp. 237–274.

(4) See M. Davis, *Relatively recursive functions and the extended Kleene hierarchy*, Proc. Int. Congr. of Math., Cambridge 1950, vol. 1 (1952), p. 723; S. C. Kleene, *Arithmetical predicates and function quantifiers*, Trans. Amer. Math. Soc. 79 (1955), pp. 312–340 and A. Mostowski, *A classification*..., *loc. cit.*

(5) See S. C. Kleene, *Arithmetical predicates*..., *loc. cit.*

(6) See J. W. Addison, *Analogies in the Borel, Lusin and Kleene hierarchies I*, Bull. Amer. Math. Soc. 61 (1955), p. 75; II *ibidem* p. 171.

consistent set (U) of axioms of the first order (i.e. all of whose variables have the type of individuals) admits at least one model. The fact that there exists even a countable model (for example a model whose elements are positive integers) is known under the name of *Löwenheim-Skolem theorem*. Every model of a system (U) made up of positive integers determines a partition of expression (which do not involve free variables) into two classes: the true expressions and the false expressions, in this model. If we arrange the variables involved in the considered expressions into a sequence $x_1, x_2, \ldots$ and assume that the free variable x_j is interpreted in the model as the integer j, the partition of expressions into true and false ones can be extended to all expressions, even to those which involve free variables. The set $\boldsymbol{J}$ of expressions which are true in the model constitute then a prime ideal of the Boolean algebra formed by the expressions. That is, the set $\boldsymbol{J}$ satisfies the following conditions:

(i) if A and B belong to $\boldsymbol{J}$ and C is an arbitrary expression, then the expressions $A \vee C$ (i.e. A or C) and $A \wedge B$ (i.e. A and B) belong to $\boldsymbol{J}$;

(ii) if $(X \vee Y) \in \boldsymbol{J}$, then at least one of the expressions X, Y belongs to $\boldsymbol{J}$.

Thus every model determines a prime ideal, but, in general, an ideal may not correspond to a model.

By a known theorem(1) of M. H. Stone, the set of prime ideals of an arbitrary Boolean algebra can be regarded as a compact Hausdorff space of dimension 0. By endowing the set of prime ideals of the Boolean algebra of expressions with this topology, Rasiowa and Sikorski(2) have shown that the set of ideals which, in the sense described above, do not correspond to models is of the first category. By using the Baire theorem, one obtains a topological proof of the Gödel and Löwenheim-Skolem theorems(3).

(1) See M. H. Stone, *Boolean algebras and their application to topology*, Proc. Nat. Acad. of Sci. 20 (1934), pp. 107–202.

(2) See H. Rasiowa and R. Sikorski, *A proof of the completeness theorem of Gödel*, Fund. Math. 37 (1950), pp. 193–200.

(3) Topological proofs which do not refer to the Baire theorem are also known; see W. E. Beth, *A topological proof of the theorem of Löwenheim–Skolem–Gödel*, Proc. Nederl. Akad. van Wetensch., Series A, 54 (1951), pp. 436–444; A. Blake, *Canonical expression in Boolean algebra*, Chicago 1938; and L. Rieger, *O sčetnyh obobščennyh δ-algebrah i novom dokazatelstve teoremy Gedela o polnote*, Czechoslovak Math. Journ. 1 (1951), pp. 33–49.

Although there are proofs (such as that of Gödel) which do not involve topology, the topological proofs (and, first of all, the method of Rasiowa and Sikorski) are interesting not only for their methodological aspect, but also for the possibility of obtaining stronger existence theorems than those which can be derived by other methods. Here again (compare § 34, VIII) "the method of category" appears to be extremely efficient.

For instance, C. Ryll-Nardzewski, by using the method of category, has found an elegant proof of a theorem discovered independently and published by S. Orey(1); it states that every axiomatic system which contains the notations of the arithmetic and is consistent with respect to the infinite induction rule(2) possesses at least one model whose integers are isomorphic to the ordinary integers. The idea of this proof was then used in topological proofs of existence of various pathological models for the axiomatic system of the set theory(3).

III. Non-classical logics. Besides the classical system, we also consider, in the mathematical logic, non-classical (multi-valued) systems of logic. In the ordinary logic, the logical functions (negation, conjunction and so on) are defined on a set $\{0, 1\}$ of two elements and assume the values 0 and 1 only (compare § 1, I). Here 0 denotes the logical value "false" and 1 denotes the value "true". The theorems of the propositional calculus are expressions built up from propositional variables and symbols for logical functions which assume the value 1 for all values of variables.

In order to define a *multi-valued logic*, we replace the set $\{0, 1\}$ by another set which may be finite or infinite. This set, together with the operations which define logical operators, is called the *characteristic matrix* of the multi-valued logic in question.

Every multi-valued logic can be characterized by an axiomatic method: it suffices to specify the axioms and rules of inference.

---

(1) See S. Orey, *On ω-consistency and related properties*, Journ. Symb. Logic 21 (1956), pp. 246–252.

(2) This rule has not a finite character; it states that a general proposition "for every integer n, $F(n)$" may be admitted as a theorem, if the propositions $F(1), F(2), \ldots$ are proved.

(3) See A. Mostowski, *On models of axiomatic set-theory*, Bull. Ac. Pol. des Sci., Cl. III, 4 (1956), p. 663–668.

Application of topology to multi-valued logics has become possible after discovering the fact that most of the known multi-valued logics, which were originally defined in an axiomatic way, admit characteristic matrices defined in topological terms. We shall give two examples.

One of the most interesting multi-valued logic is the *intuitionistic logic* created by Brouwer and then axiomatized by Heyting([1]). A simple characteristic matrix for this logic was found by Tarski([2]). Its elements are open subsets of a dense in itself normal separable space E (thus, for example, the open subsets of the euclidean space of any dimension). The conjunction and disjunction in this matrix are interpreted as usual by means of operations of the set theory. Other logical functions differ from the classical Boolean interpretation: The negation is interpreted by means of the function $\mathrm{Int}(E-X)$ and the implication $X \to Y$ by means of the function $\mathrm{Int}\big((E-X) \cup Y\big)$([3]).

Consider now the *modal logic* $S4$ of Lewis and Langford([4]). Besides ordinary logical functions, this logic admits the functions Pp (= it is possible that p), Np (= it is necessary that p) and some other functions definable by means of P and N. A characteristic matrix of the logic $S4$ consists of arbitrary subsets of a dense in itself normal separable topological space; the ordinary logical functions are interpreted as usual, and the functions P and N as the operations $\overline{X}$ and $\mathrm{Int}(X)$([5]).

The purely mathematical definitions of the intuitionistic logic and of the logic $S4$ described above constitute a source of numerous

---

([1]) See A. Heyting, *Die formalen Regeln der intuitionistischen Logik*, Sgb. Preuss. Akad. Wiss. 1930, pp. 42–56.

([2]) See A. Tarski, *Der Aussagenkalkül und die Topologie*, Fund. Math. 31 (1938), pp. 103–134.

([3]) See J. C. C. McKinsey and A. Tarski, *The algebra of topology*, Annals of Math. 45 (1944), pp. 141–191; M. H. Stone, *Topological representations of distributive lattices and Brouwerian logics*, Časopis pro pestovani mat. a fys. 67 (1937), pp. 1–25 and A. Tarski, *loc. cit.*

([4]) See C. I. Lewis and C. H. Langford, *Symbolic logic*, New York–London 1932.

([5]) See J. C. C. McKinsey and A. Tarski, *Some theorems about the sentential calculi of Lewis and Heyting*, Journ. of Symb. Logic 13 (1948), pp. 1–13; Tang Tsao-Chen, *Algebraic postulates and a geometric interpretation for the Lewis calculus of strict implication*, Bull. Amer. Math. Soc. 44 (1938), pp. 737–744.

meta-mathematical theorems on these logics(1). Most of them were not known before the discovery of the topological interpretation.

We pass now to the first order functional calculus based on a multi--valued logic; for example, one can find topological interpretations of the functional calculi based on the intuitionistic(2), modal(3), and some other(4) logics. In these interpretations, the predicates $f(x_1, \ldots, x_n)$ are regarded as functions whose arguments vary in an abstract set and whose values belong to a matrix of the multi-valued logic in question. In the case of the intuitionistic logic, $f(x_1, \ldots, x_n)$ is then an open subset of a topological space E and, in the case of the modal logic, it is an arbitrary subset of E. The universal quantifier in the intuitionistic logic is interpreted as the interior of the intersection and the existential quantifier is interpreted as the union. In the modal logic, the quantifiers are interpreted as the intersection and the union in the sense of the set theory.

For a suitably chosen space E, the logical theorems coincide with the expressions which, in the above interpretations, are identically equal to E(5). This result corresponds to the Tarski theorem referred to above which dealt with the propositional calculus(6). It is worth noting, however, that, in the case of modal logic, the class of spaces E for which the theorem holds is narrower than in the case of modal propositional calculus(7). The analogous question for the intuitionistic logic is open.

Topological interpretations of non-classical functional calculi was studied mainly by Rasiowa and Sikorski. By using it they

---

(1) See J. C. C. McKinsey and A. Tarski, *The Algebra of Topology, loc. cit.*; *On closed elements in closure algebras*, Ann. of Math. 47 (1946), p. 122–162; *Some theorems ..., loc. cit.*; A. Tarski, *Der Aussagenkalkül ..., loc. cit.*

(2) See A. Mostowski, *Proofs of non-deducibility in intuitionistic functional calculus*, Journ. Symb. Logic 13 (1948), pp. 204–207.

(3) See H. Rasiowa, *Algebraic treatment of the functional calculi of Lewis and Heyting*, Fund. Math. 38 (1951), pp. 99–126.

(4) See H. Rasiowa and R. Sikorski, *A proof of the completeness theorem of Gödel*, Fund. Math. 37 (1950), pp. 193–200.

(5) See H. Rasiowa, *loc. cit.*

(6) See A. Tarski, *loc. cit.*

(7) See H. Rasiowa and R. Sikorski, *Algebraic treatment, loc. cit.* and R. Sikorski, Bull. Acad. Pol. des Sci., Cl. III, 4 (1956), pp. 649–650.

obtained a considerable number of meta-mathematical theorems about these calculi and about propositional calculi with quantifiers[1].

We remark finally that a topological interpretation of a part of intuitionistic mathematics was also given by K. Menger[2]; investigations in this direction have not been continued.

IV. Other applications. The logical literature contains various other applications of topology to mathematical logic. We mention, for example, a proof of a theorem of Tarski given by Kuratowski[3]; the theorem states that the notion of well ordering is not definable in first order logic. As was shown by Kuratowski, this theorem is an immediate consequence of the existence of analytic non-borelian sets (compare § 38,VI).

Another application of the topological method was given by Grzegorczyk[4] in the proof of the "uniformity" theorem for *recursive functions*. It reads as follows: If F is a functional which assigns to every positive integer x and every infinite sequence f of positive integers a positive integer $F(f, x)$ in a recursive manner, then there exists a functional $H(h, x)$ such that the conditions:

(i) $f_t \leqslant h_t$, $g_t \leqslant h_t$, for every t,

(ii) $f_t = g_t$, for $t \leqslant H(h, x)$,

imply $F(f, x) = F(g, x)$, for every two sequences f and g.

A different proof of this theorem was given by Kleene[5]; it did not involve topological methods, but used the Brouwer-König "fan" theorem.

---

(1) See H. Rasiowa, *Algebraic models of axiomatic theories*, Fund. Math. 41 (1955), pp. 291–310; H. Rasiowa and R. Sikorski, *Algebraic treatment...*, *ibidem* 40 (1953), *On existential theorems in non-classical functional calculi*, *ibidem* 41 (1954), pp. 21–28 and *An application of lattices to logic*, *ibidem* 42 (1955), pp. 83–100.

(2) See K. Menger, *Bemerkungen zu Grundlagenfragen*, Jahresb. Deutscher Math. Ver. 37 (1938), pp. 81–112.

(3) See K. Kuratowski, *Les types d'ordre définissables et les ensembles boreliens*, Fund. Math. 28 (1937), pp. 97–100.

(4) See A. Grzegorczyk, *Computable functionals*, Fund. Math. 42 (1955), pp. 168–202.

(5) *A note on computable functionals*, Indag. Math. 18 (1956), pp. 275–280.

For other application of topological concepts to the theory of recursive functions, see W. A. Uspenskij, Dokl. Akad. Nauk USSR 103 (1955), pp. 773–776.

APPENDIX B

APPLICATIONS OF TOPOLOGY TO FUNCTIONAL ANALYSIS

by R. SIKORSKI

The study of linear topological spaces is the object of the functional analysis. A *linear space* is a set in which the operations of addition $x+y$ and of multiplication λx (where λ is a number) are defined; these operations are subjected to simple natural axioms (1). A linear space is called a *linear topological space*, if it is provided with a topology in which $x+y$ and λx, regarded as functions of two variables, are continuous.

In the initial period of development of the functional analysis, linear spaces with metrizable topology (*linear metrizable spaces*) were studied and normed linear spaces (2) were mainly considered; a norm $\|x\|$ is understood to be a function with finite non-negative values such that

$$(\|x\| = 0) \equiv (x \text{ is the zero element of the space}),$$

$$\|x+y\| \leqslant \|x\|+\|y\|, \qquad \|\lambda x\| = |\lambda|\cdot\|x\|.$$

A normed space becomes a metric space by defining

$$|x-y| = \|x-y\|.$$

In recent years, non-metrizable linear topological spaces have acquired a great importance (3). As examples, the space of *distributions* of Sobolev–Schwartz (4), and the space of Miku-

(1) Compare S. Banach, *Théorie des opérations linéaires*, Monogr. Mat., Warszawa-Lwów 1932, p. 26. This book is the first systematic treatment of the functional analysis, one of whose founders was its author.

(2) See, e.g. S. Banach, *Sur les opérations dans les ensembles abstraits et leurs applications aux équations intégrales*, Fund. Math. 3 (1922), pp. 7–33.

(3) See, for example, N. Bourbaki, *Espaces vectoriels topologiques*, Paris 1955.

(4) See S. Sobolev, *Méthode nouvelle à résoudre le problème de Cauchy pour les équations hyperboliques normales*, Recueil (Mat. Sbornik) 1 (1936), pp. 39–71; Laurent Schwartz, *Généralisation de la notion de fonction, de dérivation, de transformation de Fourier, et applications mathématiques et physiques*, Ann. Univ. Grenoble 21 (1945), pp. 57–74; of the same author: *Théorie des distributions*, Vol. I, Paris 1950, Vol. II, Paris 1951.

siński *operators*(1) can be quoted. Other important examples of non-metrizable spaces are *normed spaces* with *weak* (sequential or neighbourhood) *topology*, i.e. with the weakest topology(2) which preserves the continuity of all linear functionals. Also ordered spaces with the order topology(3) and spaces with the two-norm convergence(4) are not, in general, metrizable.

The assumption of continuity of the operations which is admitted in the definition of linear topological spaces appears to be an insufficient link between the linearity of the space and its topological structure. It is completed by the axiom on existence of arbitrarily small convex neighbourhoods. This gives the concept of *locally convex spaces*(5); if they are metrizable, they are called B_0^* spaces(6).

The concept of *completeness* (compare § 33, I) is fundamental in the theory of metrizable linear spaces. This concept can also be applied to non-metrizable spaces, but its use is limited there since, in this case, the *Baire theorem* (compare § 34, IV) is no longer valid. The hypothesis of completeness is usually assumed if metrizable linear spaces are examined. Complete normed spaces are also known under the name of *Banach spaces.*

The hypothesis that the space is complete and, in particular, the *Baire theorem* (which is its consequence) is used in an essential way in the proofs of fundamental theorems of the functional

(1) See J. Mikusiński, *Sur les fondements du calcul opératoire*, Studia Math. 11 (1950), p. 41–70; and *Operational calculus* (translated from Polish, Monogr. Mat., Warszawa–Wrocław 1953) Pergamon Press 1959.

(2) In other words, the topology which admits the minimum of open sets.

(3) See L. Kantorovitch, *Lineare halbgeordnete Räume*, Mat. Sbornik 2 (1937), pp. 121–168; and L. Kantorovitch, B. Vulich and A. Pinsker, *Functional analysis in partially ordered spaces*, (in Russian), Moscou-Leningrad 1950.

(4) Compare A. Alexiewicz, *On the two-norm convergence*, Studia Math. 14 (1953), pp. 49–56.

(5) Principles of the theory of linear locally convex topological spaces were founded independently by S. Mazur: *On convex sets and functionals in linear spaces* (in Polish), Lwów 1936, and J. von Neumann, *On complete topological spaces*, Trans. Amer. Math. Soc. 37 (1935), pp. 1–20.

(6) The theory of B_0^* spaces is due to S. Mazur and W. Orlicz. Compare their communications in Ann. Soc. Pol. Math. 12 (1933), pp. 119–120 and 16 (1938), p. 195, as well as their joint papers, *Sur les espaces métriques linéaires*, Studia Math. 10 (1948), pp. 184–208 and 13 (1953), pp. 137–179.

analysis such as: the *interior mapping principle*(1), the *closed graph* theorem(2), the theorem on continuity of the limit of a convergent sequence of continuous linear transformations(3), the well-known Banach–Steinhaus theorem(4). The Baire theorem is often used in the proofs of existence theorems, usually by means of the Banach–Steinhaus theorem (for instance, in the proof of existence of a continuous function with a divergent Fourier series)(5).

In spite of the fact that linear spaces are never compact and that the locally compact linear spaces may be identified with the euclidean spaces $\mathscr{E}^n$, the notion of compactness is of great importance. Thus, for example, an important role in the theory of linear operations is played by the *completely continuous operations*; these are the linear operations that carry bounded sets(6) into compact sets(7).

The notion of compactness plays an important part in the theory of *normed rings*(8) (*Banach algebras*), i.e. in the theory of Banach spaces with an operation of multiplication, subjected to the usual axioms. The fundamental concept in the theory of normed rings is that of additive and multiplicative functional (or the equivalent notion of maximal ideal). The set of these functionals (or, what amounts to the same, the set of maximal ideals) constitutes a topological compact space.

Closed spheroids in infinitely dimensional Banach space are not compact. Under additional assumptions about the space, however, they become compact in the weak topology (*reflexive spaces*). This fact implies the frequent use of the notion of compactness in the weak topology.

(1) See S. Banach, *Opérations linéaires*, p. 38, Theorem 3.

(2) *Ibid.*, Theorem 7.

(3) *Ibid.*, p. 23, Theorem 4.

(4) S. Banach and H. Steinhaus, *Sur le principe de la condensation de singularités*, Fund. Math. 9 (1927), pp. 50–61. Compare the Banach's monograph quoted above, p. 80, Theorem 5.

(5) See A. Zygmund, *Trigonometrical series*, Monogr. Mat., Warszawa–Lwów 1935, p. 172, 8.31.

(6) A subset of a linear topological space is said to be *bounded*, if every sequence $x_1, x_2, \ldots$ selected from it is bounded, that is, if $\lambda_n x_n$ converges to zero, whenever the sequence λ_n converges to zero.

(7) Compare S. Banach, *Opérations linéaires*, p. 96.

(8) See S. Mazur, *Sur les anneaux linéaires*, C. R. Acad. Sci. Paris 207(1938), pp. 1025–1027. This was actually the first paper on normed rings. Their theory was developed by I. Gelfand, *Normierte Ringe*, Mat. Sbornik 9 (1941), pp. 3–24.

Also the *Baire property* (compare § 11, I) is used in the functional analysis[1]. Thus, for instance, an important theorem on F spaces states that a linear set with the Baire property is either of the first category, or is equal to the entire space[2] (compare also § 13, XII).

The *fixed point theorems* are of a great importance in the nonlinear functional analysis and in its applications to differential and integral equations. We quote firstly the following Banach theorem[3]: *Let f be a continuous mapping of a complete space onto its subset such that $|f(x)-f(x')| \leqslant \lambda|x-x'|$, where $\lambda < 1$. Then there exists a unique point x_0 such that $f(x_0) = x_0$.*

Another fundamental theorem used often in applications is the Schauder theorem[4]: *Every continuous mapping f of a closed convex subset A of a Banach space onto a compact subset of A admits a fixed point.* The proof of this theorem is based on the classical Brouwer fixed-point theorem concerning the n-dimensional simplex (§ 28, III).

The application of topological theorems concerning *homology* and *homotopy* properties leads to very deep results. We mention here first the Leray–Schauder method of the *degree of a mapping* (of a subset of a Banach space into itself) of the form $I+A$, where I is the identity and A is a completely continuous operation[5].

We also mention the *Borsuk theorem on antipodes*[6] and the Lusternik–Schnirelman theorem which states that the sphere $\mathcal{S}_n$ cannot be decomposed into the union of $n+1$ sets of diameter

---

(1) Compare S. Banach, *Opérations linéaires*, p. 22–23.

(2) *Ibid.*, p. 21, Theorem 1.

(3) See S. Banach, Fund. Math. 3, *op. cit.*, p. 160.

(4) See J. Schauder, *Der Fixpunktsatz in Funktionalräumen*, Studia Math. 2 (1930), pp. 171–180. An extension to non-metrizable spaces is due to A. Tychonov, *Ein Fixpunktsatz*, Math. Ann. 111 (1935). For a generalization of the Banach and Schauder theorems see also M. Krasnosielskii, *Two remarks on the method of successive approximations* (in Russian), Uspiechy Mat. Nauk 10, 1, (1955).

(5) See J. Leray and J. Schauder, *Topologie et équations fonctionnelles*, Ann. Ec. Norm. Sup. 13 (1934), pp. 45–78.

(6) See K. Borsuk, *Drei Sätze über die n-dimensionale euklidische Sphäre*, Fund. Math. 20 (1933), pp. 177–190.

$< \delta(\mathscr{S}_n)$[1]. The topological theorems used here concern finite dimensions. They are, however, applicable to infinitely dimensional subsets of Banach spaces by virtue of the fact that a completely continuous mapping can be approximated with mappings onto sets of finite dimensions[2].

In our exposition of applications of the topology to the functional analysis we have confined ourselves to some general methods. There are, however, various applications of a more special character. Let us mention for instance, that the theorem by which every compact metric space is a continuous image of the Cantor set $\mathscr{C}$ is applied to the Banach–Mazur theorem[3] which asserts that a separable Banach space is equivalent with a closed linear subset of the space $\mathscr{E}^{\mathscr{I}}$ (comp. § 22, II, remark 2). We also mention an application of the concept of analytic set (§ 38, I) to a generalization of a theorem of Saks on linear operations[4].

---

(1) See L. Lusternik and L. Schnirelman, *Méthodes topologiques dans les problèmes variationnels*, Issliedov. Inst. Mat. i Mech., Moscow 1930, p. 30 (Lemma 1).

(2) See A. Granas, *On the Schauder theorem on the invariance of domains* (in Russian), Bull. Acad. Pol. Sc. 10 (1962), pp. 233–238; *An extension to functional spaces of Borsuk–Ulam theorem on antipodes*, *ibidem* pp. 81–86; *A note on compact deformations in functional spaces*, *ibidem* pp. 87–90; *Homotopy extension theorem in Banach spaces and some of its applications to the theory of nonlinear equations* (in Russian), *ibidem* 7 (1959), p. 387–394; *On the disconnection of Banach spaces*, *ibidem* 7 (1959), p. 395–399; *On continuous mappings of open sets in Banach space*, *ibidem* 6 (1958), p. 25–29; *Über einen Satz von K. Borsuk*, *ibidem* 5 (1957), p. 959–962. See also K. Gęba, A. Granas, and A. Jankowski, *Some theorems on the sweeping in Banach spaces*, Bull. Acad. Pol. Sc. 7 (1959), p. 539–544; M. Krasnosielskii, *Topological methods in the theory of non-linear integral equations* (in Russian), Moscow 1956.

(3) Compare S. Banach, *Opérations linéaires*, p. 185.

(4) See A. Alexiewicz, *A theorem on the structure of linear operations*, Studia Math. 14 (1953), pp. 1–12; S. Saks, *On some functionals*, Trans. Amer. Math. Soc. 35 (1932), pp. 549–556 and 41 (1937), pp. 160–170.

LIST OF IMPORTANT SYMBOLS

AUTHOR INDEX

SUBJECT INDEX